Computer Arithmetic and Verilog HDL Fundamentals

By the same author:

DIGITAL COMPUTER ARITHMETIC: Design and Implementation

SEQUENTIAL LOGIC: Analysis and Synthesis

VERILOG HDL: Digital Design and Modeling

DIGITAL DESIGN and Verilog HDL Fundamentals

THE COMPUTER CONSPIRACY
A novel

Computer Arithmetic and Verilog HDL Fundamentals

Joseph Cavanagh

Santa Clara University
California, USA

CRC Press is an imprint of the
Taylor & Francis Group, an **informa** business

CRC Press
Taylor & Francis Group
6000 Broken Sound Parkway NW, Suite 300
Boca Raton, FL 33487-2742

International Standard Book Number: 978-1-4398-1124-5 (Hardback)

Library of Congress Cataloging-in-Publication Data

Cavanagh, Joseph J. F.
Computer arithmetic and Verilog HDL fundamentals / Joseph Cavanagh.
p. cm.
Includes bibliographical references and index.
ISBN 978-1-4398-1124-5
1. Digital electronics. 2. Computer arithmetic. 3. Verilog (Computer hardware description language) I. Title.

TK7868.D5C3937 2010
004.01'513--dc22 2009029354

Visit the Taylor & Francis Web site at
http://www.taylorandfrancis.com

and the CRC Press Web site at
http://www.crcpress.com

To Dr. Ivan Pesic
for generously providing the SILOS Simulation Environment software
for all of my books that use Verilog HDL and for his continued support

PREFACE

The purpose of this book is to present computer arithmetic for fixed-point, decimal, and floating-point number representations for the operations of addition, subtraction, multiplication, and division, and to implement those operations using the Verilog Hardware Description Language (HDL). The Verilog HDL language provides a means to model a digital system at many levels of abstraction.

The four basic operations are implemented in the execution unit of a computer which includes the arithmetic and logic unit (ALU). The execution unit is the focal point of the computer and performs all of the arithmetic and logical operations.

Addition and subtraction for all three number representations are relatively simple; however, multiplication and division are comparatively more complex. The arithmetic algorithms for the three number representations are presented in sufficient detail to permit ease of understanding.

The different modeling constructs supported by Verilog are described in detail. Numerous examples are designed in each chapter for specific operations using the appropriate number representation, including both combinational and clocked sequential arithmetic circuits. Emphasis is placed on the detailed theory and design of various arithmetic circuits using applicable algorithms. The Verilog HDL design projects include the design module implemented using built-in primitives, dataflow modeling, behavioral modeling or structural modeling, the test bench module, the outputs obtained from the simulator, and the waveforms obtained from the simulator that illustrate the complete functional operation of the design.

The book is intended to be tutorial, and as such, is comprehensive and self contained. All designs are carried through to completion — nothing is left unfinished or partially designed. Each chapter includes numerous problems of varying complexity to be designed by the reader.

Chapter 1 covers the number systems of different radices, such as binary, octal, binary-coded octal, decimal, binary-coded decimal, hexadecimal, and binary-coded hexadecimal. The chapter also presents the number representations of sign magnitude, diminished-radix complement, and radix complement.

Chapter 2 presents a review of logic design fundamentals, including Boolean algebra and minimization techniques for switching functions. The minimization techniques include algebraic minimization using Boolean algebra, Karnaugh maps, map-entered variables, the Quine-McCluskey algorithm, and the Petrick algorithm. Various combinational logic macro functions are also presented. These include multiplexers of different sizes and types, such as linear-select multiplexers and nonlinear-select multiplexers. The chapter also shows a one-to-one correspondence between the data input numbers d_i of a multiplexer and the minterm locations in a

Karnaugh map. Decoders and encoders are presented together with comparators. Sequential logic includes *SR* latches, *D* flip-flops, and *JK* flip-flops. Counters of different moduli are designed for both count-up and count-down counters. The complete design process for Moore and Mealy synchronous sequential machines is presented.

Chapter 3 introduces Verilog HDL, which will be used throughout the book to design the arithmetic circuits. Verilog HDL is the state-of-the-art method for designing digital and computer systems and is ideally suited to describe both combinational and clocked sequential arithmetic circuits. Verilog provides a clear relationship between the language syntax and the physical hardware. The Verilog simulator used in this book is easy to learn and use, yet powerful enough for any application. It is a logic simulator — called SILOS — developed by Silvaco International for use in the design and verification of digital systems. The SILOS simulation environment is a method to quickly prototype and debug any logic function. It is an intuitive environment that displays every variable and port from a module to a logic gate. SILOS allows single-stepping through the Verilog source code, as well as drag-and-drop ability from the source code to a data analyzer for waveform generation and analysis. This chapter introduces the reader to the different modeling techniques, including built-in primitives for logic primitive gates and user-defined primitives for larger logic functions. The three main modeling methods of dataflow modeling, behavioral modeling, and structural modeling are introduced.

Chapter 4 presents fixed-point addition. The different categories of addition circuits are: ripple-carry addition, carry lookahead addition, carry-save addition, memory-based addition, carry-select addition, and serial addition. A ripple-carry adder is not considered a high-speed adder, but requires less logic than a high-speed adder using the carry lookahead technique. Using the carry lookahead method, a considerable increase in speed can be realized by expressing the carry-out $cout_i$ of any stage$_i$ as a function of the two operand bits a_i and b_i and the carry-in cin_{-1} to the low-order stage$_0$ of the adder. Carry-save adders (CSAs) save the carry from propagating to the next higher-order stage in an n-bit adder. They can be used to add multiple bits of the same weight from multiple operands or to add multiple n-bit operands. With the advent of high-density, high-speed memories, addition can be easily accomplished by applying the augend and addend as address inputs to the memory — the outputs are the sum. A carry-select adder is not as fast as the carry lookahead adder, however it has a unique organization that is interesting. The carry-select principle produces two sums that are generated simultaneously. One sum assumes that the carry-in to that group was a 0; the other sum assumes that the carry-in was a 1. The predicted carry is obtained using the carry lookahead technique which selects one of the two sums. If a minimal amount of hardware is a prerequisite and speed is not essential, then a serial adder may be utilized. A serial adder adds the augend and addend one bit per clock pulse — thus, eight clock pulses are required to add two bytes of data.

Chapter 5 presents fixed-point subtraction, which like addition, is relatively simple. It also shares much of the same circuitry as addition. The two operands for subtraction are the minuend and the subtrahend — the subtrahend is subtracted from the minuend. Computers use an adder for the subtraction operation by adding the radix complement of the subtrahend to the minuend. A 2s complementer is used that produces either the uncomplemented version of the addend for addition or the 2s complement version of the subtrahend for subtraction. Subtraction can be performed in all three number representations: sign magnitude, diminished-radix complement, and radix complement; however, radix complement is the easiest and most widely used method for subtraction in any radix.

Chapter 6 presents several methods for fixed-point multiplication. The multiplicand is multiplied by the multiplier to generate the product. In all methods the product is usually $2n$ bits in length. The sequential add-shift method is a common approach for low-speed multiplication, but usually requires that the multiplier be positive. The Booth algorithm is an effective technique for multiplying operands that are in 2s complement representation, including the case where the multiplier is negative. Unlike the sequential add-shift method, it treats both positive and negative operands uniformly; that is, the multiplicand and multiplier can both be negative or positive. Another method is bit-pair recoding that represents a speedup technique that is derived from the Booth algorithm and assures that an n-bit multiplier will have no more than $n/2$ partial products. It also treats both positive and negative multipliers uniformly; that is, there is no need to 2s complement the multiplier before multiplying, or to 2s complement the product after multiplying. A table lookup multiplier permits a multiply operation for unsigned operands and positive signed operands. For signed operands, the multiplier must be positive — the multiplicand can be positive or negative. Multiplication can be accomplished by using a table that contains different versions of the multiplicand to be added to the partial products. Firmware loads the multiplicand table in memory and the multiplier prior to the multiply operation. With the advent of high-capacity, high-speed random-access memories (RAMs), multiplication using RAM may be a viable option. The multiplicand and multiplier are used as address inputs to the memory — the outputs are the product. The operands can be either unsigned or signed numbers in 2s complement representation. Multiple-operand multiplication is also presented.

Chapter 7 covers several methods of fixed-point division, including the sequential shift-add/subtract restoring division method in which the partial remainder is restored to its previous value if it is negative. The speed of the division algorithm can be increased by modifying the algorithm to avoid restoring the partial remainder in the event that a negative partial remainder occurs. This method allows both a positive partial remainder and a negative partial remainder to be utilized in the division process. In nonrestoring division, a negative partial remainder is not restored to the previous value but is used unchanged in the following cycle. SRT division was developed independently by Sweeney, Robertson, and Tocher as a way to increase the speed of a divide operation. It was intended to improve radix-2 floating-point arithmetic by shifting over strings of 0s or 1s in much the same way as the Booth algorithm shifts over strings of 0s in a multiply operation. SRT division is designed in this chapter using a table lookup approach and using the **case** statement, which is a multiple-way conditional branch. Convergence division is also covered in which a multiplier is used in the division process; thus, the method is referred to as multiplicative division. A combinational array can be used for division in much the same way as an array was used for multiplication. This is an extremely fast division operation, because the array is entirely combinational — the only delay is the propagation delay through the gates.

Chapter 8 presents decimal addition, in which each operand is represented by a 4-bit binary-coded decimal (BCD) digit. The most common code in BCD arithmetic is the 8421 code. Since four binary bits have $2^4 = 16$ combinations (0000 – 1111) and the range for a single decimal digit is 0 – 9, six of the sixteen combinations (1010 – 1111) are invalid for BCD. These invalid BCD digits must be converted to valid digits by adding six to the digit. This is the concept for addition with sum correction. The adder must include correction logic for intermediate sums that are greater than or equal to 1010 in radix 2. An alternative approach to determining whether to add six to correct an invalid decimal number is to use a multiplexer. This approach uses two fixed-point adders — one adder adds the two operands; the other adder always adds six to the intermediate sum that is obtained from the first adder. A multiplexer then selects the correct sum. A memory can be used to correct the intermediate sum for a decimal add operation. This is a low-capacity memory containing only 32 four-bit words. One memory is required for each decade. Another interesting approach to decimal addition is to bias one of the decimal operands prior to the add operation — in this case the augend — and then to remove the bias, if necessary, at the end of the operation depending on certain intermediate results. This is, in effect, preprocessing and postprocessing the decimal digits.

Chapter 9 introduces decimal subtraction. The (BCD) code is not self-complementing as is the radix 2 fixed-point number representation; that is, the $r - 1$ complement cannot be acquired by inverting each bit of the 4-bit BCD digit. Therefore, a 9s complementer must be designed that provides the same function as the diminished-radix complement for the fixed-point number representation. Thus, subtraction in BCD is essentially the same as in fixed-point binary; that is, add the *r*s (10s) complement of the subtrahend to the minuend. As in fixed-point arithmetic, the decimal adder can be designed with minor modifications so that the operations of addition and subtraction can be performed by the same hardware.

Chapter 10 covers decimal multiplication. The multiplication algorithms for decimal operands are similar to those for binary operands. However, the algorithms for decimal multiplication are more complex than those for fixed-point multiplication. This is because decimal digits consist of four bits and have values in the range of 0 to 9, whereas fixed-point digits have values of 0 or 1. This chapter includes a binary-to-decimal converter, which is used to convert a fixed-point multiplication product to decimal. Decimal multiplication can be easily implemented with high-speed, high-capacity memories (including read-only memories), as designed in this chapter. The speed of decimal multiplication can be increased by using a table lookup method. This is similar to the table lookup method presented for fixed-point multiplication, in which the multiplicand table resides in memory.

Chapter 11 presents decimal division. Two methods of restoring division are given. A straightforward method to perform binary-coded decimal division is to implement the design using the fixed-point restoring division algorithm and then convert the resulting quotient and remainder to BCD. This is a simple process and involves only a few lines of Verilog code. A multiplexer can be used to bypass the restoring process. The divisor is not added to the partial remainder to restore the previous partial remainder; instead the previous partial remainder — which is unchanged — is used. This avoids the time required for addition. Decimal division using the table lookup method is also covered. This is equivalent to the binary search technique used in programming.

Chapter 12 covers floating-point addition. The material presented in this chapter is based on the Institute of Electrical and Electronics Engineers (IEEE) Standard for Binary Floating-Point Arithmetic IEEE Std 754-1985 (Reaffirmed 1990). Floating-point numbers consist of the following three fields: a sign bit *s*, an exponent *e*, and a fraction *f*. Unbiased and biased exponents are explained. Examples are given that clarify the technique for adding floating-point numbers. Fraction overflow and fraction underflow are defined. The floating-point addition algorithm is given in a step-by-step procedure complete with flowcharts. A floating-point adder is implemented using behavioral modeling.

Chapter 13 addresses floating-point subtraction and presents several numerical examples and flowcharts that graphically portray the steps required for true addition and true subtraction for floating-point operands. Subtraction can yield a result that is either true addition or true subtraction. True addition produces a result that is the sum of the two operands disregarding the signs; true subtraction produces a result that is the difference of the two operands disregarding the signs. Five behavioral modules are presented that illustrate subtraction operations which yield either true addition or true subtraction.

Chapter 14 presents floating-point multiplication, including double biased exponents. Numerical examples and flowcharts that describe the steps required for floating-point multiplication are also included. Two behavioral modules are designed: one using the Verilog HDL multiplication operator (*) and one using the sequential add-shift method.

Chapter 15 covers floating-point division, including the generation of a zero-biased exponent. Exponent overflow and exponent underflow are presented together with flowcharts and numerical examples that illustrate the division process.

Chapter 16 presents additional floating-point topics, which include rounding methods, such as truncation rounding, adder-based rounding, and von Neumann rounding. Guard bits are also covered, which are required in order to achieve maximum accuracy in the result of a floating-point operation. Guard bits are positioned to the right of the low-order fraction bit and may consist of one or more bits. Verilog HDL is used to design three methods for implementing adder-based rounding. The first method uses a memory to generate the desired rounded fraction in a behavioral module. The second method uses dataflow modeling as the implementation technique for designing combinational logic to generate the desired rounded fraction. The third method uses behavioral modeling without utilizing a memory. This design adds two floating-point operands and adjusts the sum based upon the adder-based rounding algorithm.

Chapter 17 presents additional topics in computer arithmetic that supplement the previous topics in the book. Included is residue checking, which is a method of checking the accuracy of an arithmetic operation and is done in parallel with the operation. Several examples are given that illustrate the concept of residue checking. Residue checking is implemented in both dataflow and structural modeling. Parity checking is discussed as it applies to shift registers. Shift registers that perform the operations of shift left logical (SLL), shift left algebraic (SLA), shift right logical

(SRL), or shift right algebraic (SRA) can be checked using parity bits. Parity prediction is covered, which can be considered a form of checksum error detection. The checksum is the sum derived from the application of an algorithm that is calculated before and after an operation to ensure that the data is free from errors. Condition codes for addition are introduced for high-speed addition. For high-speed addition, it is desirable to have the following condition codes generated in parallel with the add operation: sum < 0; sum $= 0$; sum > 0; and overflow. The generation of condition codes is implemented with a dataflow module. Logical and algebraic shifters are covered and designed using behavioral and structural modeling. Two arithmetic and logic units are designed using mixed-design modeling and behavioral modeling. A count-down counter is implemented using behavioral modeling. Different types of shift registers are designed using behavioral modeling. These include the following types: a parallel-in, serial-out; serial-in, serial-out; parallel-in, serial-in, serial-out; and serial-in, parallel-out.

Appendix A presents a variety of designs for logic gates and logic macro functions to supplement the Verilog HDL designs that were implemented throughout the book. These designs range from simple AND gates and OR gates to code converters and an adder/subtractor.

Appendix B presents a brief discussion on event handling using the event queue. Operations that occur in a Verilog module are typically handled by an event queue. Appendix C presents a procedure to implement a Verilog project. Appendix D contains the solutions to select problems in each chapter.

The material presented in this book represents more than two decades of computer equipment design by the author. The book is not intended as a text on logic design, although this subject is reviewed where applicable. It is assumed that the reader has an adequate background in combinational and sequential logic design. The book presents Verilog HDL with numerous design examples for fixed-point, decimal, and floating-point number representations to help the reader thoroughly understand this popular HDL and how it applies to computer arithmetic.

This book is designed for practicing electrical engineers, computer engineers, and computer scientists; for graduate students in electrical engineering, computer engineering, and computer science; and for senior-level undergraduate students.

A special thanks to Dr. Ivan Pesic, CEO of Silvaco International, for allowing use of the SILOS Simulation Environment software for the examples in this book. SILOS is an intuitive, easy to use, yet powerful Verilog HDL simulator for logic verification.

I would like to express my appreciation and thanks to the following people who gave generously of their time and expertise to review the manuscript and submit comments: Professor Daniel W. Lewis, Department of Computer Engineering, Santa Clara University who supported me in all my endeavors; Dr. Geri Lamble; and Steve Midford. Thanks also to Nora Konopka and the staff at Taylor & Francis for their support.

Non-Verilog HDL figures and Verilog HDL figures can be downloaded at:
http://www.crcpress.com/product/isbn/9781439811245

Joseph Cavanagh

CONTENTS

Preface xv

Chapter 1 Number Systems and Number Representations 1

1.1 Number Systems 1
1.1.1 Binary Number System 4
1.1.2 Octal Number System 6
1.1.3 Decimal Number System 8
1.2 Number Representations 12
1.2.1 Sign Magnitude 13
1.2.2 Diminished-Radix Complement 15
1.2.3 Radix Complement 18
1.3 Problems 22

Chapter 2 Logic Design Fundamentals 25

2.1 Boolean Algebra 25
2.2 Minimization Techniques 32
2.2.1 Algebraic Minimization 32
2.2.2 Karnaugh Maps 33
2.2.3 Quine-McCluskey Algorithm 39
2.3 Combinational Logic 44
2.3.1 Multiplexers 47
2.3.2 Decoders 53
2.3.3 Encoders 56
2.3.4 Comparators 58
2.4 Sequential Logic 60
2.4.1 Counters 62
2.4.2 Moore Machines 71
2.4.3 Mealy Machines 78
2.5 Problems 84

Chapter 3 Introduction to Verilog HDL 93

3.1 Built-In Primitives 94
3.2 User-Defined Primitives 108
3.3 Dataflow Modeling 118
3.3.1 Continuous Assignment 118
3.4 Behavioral Modeling 129
3.4.1 Initial Statement 129

3.4.2 Always Statement ... 129
3.4.3 Intrastatement Delay ... 133
3.4.4 Interstatement Delay ... 133
3.4.5 Blocking Assignments ... 133
3.4.6 Nonblocking Assignments ... 136
3.4.7 Conditional Statements ... 138
3.4.8 Case Statement ... 141
3.4.9 Loop Statements ... 150
3.5 Structural Modeling ... 154
3.5.1 Module Instantiation ... 154
3.5.2 Design Examples ... 155
3.6 Problems ... 179

Chapter 4 Fixed-Point Addition ... 183

4.1 Ripple-Carry Addition ... 184
4.2 Carry Lookahead Addition ... 191
4.3 Carry-Save Addition ... 201
4.3.1 Multiple-Bit Addition ... 201
4.3.2 Multiple-Operand Addition ... 206
4.4 Memory-Based Addition ... 212
4.5 Carry-Select Addition ... 216
4.6 Serial Addition ... 227
4.7 Problems ... 234

Chapter 5 Fixed-Point Subtraction ... 237

5.1 Twos Complement Subtraction ... 238
5.2 Ripple-Carry Subtraction ... 243
5.3 Carry Lookahead Addition/Subtraction ... 250
5.4 Behavioral Addition/Subtraction ... 267
5.5 Problems ... 271

Chapter 6 Fixed-Point Multiplication ... 275

6.1 Sequential Add-Shift Multiplication ... 276
6.1.1 Sequential Add-Shift Multiplication Hardware Algorithm ... 278
6.1.2 Sequential Add-Shift Multiplication — Version 1 ... 282
6.1.3 Sequential Add-Shift Multiplication — Version 2 ... 285
6.2 Booth Algorithm Multiplication ... 289
6.3 Bit-Pair Recoding Multiplication ... 304

6.4 Array Multiplication 318
6.5 Table Lookup Multiplication 329
6.6 Memory-Based Multiplication 339
6.7 Multiple-Operand Multiplication 344
6.8 Problems 353

Chapter 7 Fixed-Point Division 359

7.1 Sequential Shift-Add/Subtract Restoring Division 360
7.1.1 Restoring Division — Version 1 362
7.1.2 Restoring Division — Version 2 368
7.2 Sequential Shift-Add/Subtract Nonrestoring Division 374
7.3 SRT Division 382
7.3.1 SRT Division Using Table Lookup 393
7.3.2 SRT Division Using the Case Statement 397
7.4 Multiplicative Division 402
7.5 Array Division 408
7.6 Problems 423

Chapter 8 Decimal Addition 427

8.1 Addition with Sum Correction 427
8.2 Addition Using Multiplexers 437
8.3 Addition with Memory-Based Correction 444
8.4 Addition with Biased Augend 454
8.5 Problems 460

Chapter 9 Decimal Subtraction 463

9.1 Subtraction Examples 464
9.2 Two-Decade Addition/Subtraction Unit for A+B and A–B ... 467
9.3 Two-Decade Addition/Subtraction Unit for A+B, A–B, and B–A 481
9.4 Problems 491

Chapter 10 Decimal Multiplication 493

10.1 Binary-to-BCD Conversion 493
10.2 Multiplication Using Behavioral Modeling 495
10.3 Multiplication Using Structural Modeling 498
10.4 Multiplication Using Memory 510

10.5 Multiplication Using Table Lookup 524
10.6 Problems 528

Chapter 11 Decimal Division 529

11.1 Restoring Division — Version 1 529
11.2 Restoring Division — Version 2 538
11.3 Division Using Table Lookup 545
11.4 Problems 550

Chapter 12 Floating-Point Addition 551

12.1 Floating-Point Format 552
12.2 Biased Exponents 554
12.3 Floating-Point Addition 557
12.4 Overflow and Underflow 560
12.5 General Floating-Point Organization 561
12.6 Verilog HDL Implementation 564
12.7 Problems 569

Chapter 13 Floating-Point Subtraction 571

13.1 Numerical Examples 573
13.2 Flowcharts 581
13.3 Verilog HDL Implementations 584
13.3.1 True Addition 584
13.3.2 True Subtraction — Version 1 589
13.3.3 True Subtraction — Version 2 593
13.3.4 True Subtraction — Version 3 598
13.3.5 True Subtraction — Version 4 603
13.4 Problems 608

Chapter 14 Floating-Point Multiplication 611

14.1 Double Bias 613
14.2 Flowcharts 614
14.3 Numerical Examples 616
14.4 Verilog HDL Implementations 618
14.4.1 Floating-Point Multiplication — Version 1 618
14.4.2 Floating-Point Multiplication — Version 2 624
14.5 Problems 631

Chapter 15 Floating-Point Division 633

15.1 Zero Bias 635
15.2 Exponent Overflow/Underflow 638
15.3 Flowcharts 641
15.4 Numerical Examples 643
15.5 Problems 646

Chapter 16 Additional Floating-Point Topics 649

16.1 Rounding Methods 649
16.1.1 Truncation Rounding 650
16.1.2 Adder-Based Rounding 651
16.1.3 Von Neumann Rounding 653
16.2 Guard Bits 654
16.3 Verilog HDL Implementations 654
16.3.1 Adder-Based Rounding Using Memory 655
16.3.2 Adder-Based Rounding Using Combinational Logic 660
16.3.3 Adder-Based Rounding Using Behavioral Modeling 668
16.3.4 Combined Truncation, Adder-Based, and von Neumann Rounding 674
16.4 Problems 680

Chapter 17 Additional Topics in Computer Arithmetic 685

17.1 Residue Checking 686
17.1.1 Dataflow Modeling 690
17.1.2 Structural Modeling 693
17.2 Parity-Checked Shift Register 717
17.3 Parity Prediction 723
17.4 Condition Codes for Addition 738
17.5 Logical and Algebraic Shifters 747
17.5.1 Behavioral Modeling 748
17.5.2 Structural Modeling 753
17.6 Arithmetic and Logic Units 760
17.6.1 Four-Function Arithmetic and Logic Unit 760
17.6.2 Sixteen-Function Arithmetic and Logic Unit 764
17.7 Count-Down Counter 771
17.8 Shift Registers 775
17.8.1 Parallel-In, Serial-Out Shift Register 775
17.8.2 Serial-In, Serial-Out Shift Register 778

17.8.3 Parallel-In, Serial-In, Serial-Out Shift Register 782
17.8.4 Serial-In, Parallel-Out Shift Register 787
17.9 Problems 795

Appendix A Verilog HDL Designs for Select Logic Functions 801

A.1 AND Gate 801
A.2 NAND Gate 806
A.3 OR Gate 809
A.4 NOR Gate 811
A.5 Exclusive-OR Function 814
A.6 Exclusive-NOR Function 818
A.7 Multiplexers 822
A.8 Decoders 825
A.9 Encoders 829
A.10 Priority Encoder 833
A.11 Binary-to-Gray Code Converter 836
A.12 Adder/Subtractor 843

Appendix B Event Queue 849

B.1 Event Handling for Dataflow Assignments 849
B.2 Event Handling for Blocking Assignments 854
B.3 Event Handling for Nonblocking Assignments 857
B.4 Event Handling for Mixed Blocking and Nonblocking Assignments 861

Appendix C Verilog HDL Project Procedure 865

Appendix D Answers to Select Problems 867

Chapter 1 Number Systems and Number Representations 867
Chapter 2 Logic Design Fundamentals 869
Chapter 3 Introduction to Verilog HDL 873
Chapter 4 Fixed-Point Addition 883
Chapter 5 Fixed-Point Subtraction 887
Chapter 6 Fixed-Point Multiplication 891
Chapter 7 Fixed-Point Division 897
Chapter 8 Decimal Addition 903
Chapter 9 Decimal Subtraction 907

Chapter 10 Decimal Multiplication 908
Chapter 11 Decimal Division 912
Chapter 12 Floating-Point Addition 913
Chapter 13 Floating-Point Subtraction 915
Chapter 14 Floating-Point Multiplication 918
Chapter 15 Floating-Point Division 924
Chapter 16 Additional Floating-Point Topics 926
Chapter 17 Additional Topics in Computer Arithmetic 932

Index 943

1.1 Number Systems
1.2 Number Representations
1.3 Problems

1

Number Systems and Number Representations

Computer arithmetic is a subset of computer architecture and will be presented for fixed-point, decimal, and floating-point number representations. Each number representation will be presented in relation to the four operations of addition, subtraction, multiplication, and division. Different methods will be presented for each operation in each of the three number representations, together with the detailed theory and design.

Computer arithmetic differs from real arithmetic primarily in precision. This is due to the fixed-word-length registers which allow only finite-precision results. Rounding may also be required which further reduces precision.

1.1 Number Systems

Numerical data are expressed in various positional number systems for each *radix* or *base*. A *positional number system* encodes a vector of n bits in which each bit is weighted according to its position in the vector. The encoded vector is also associated with a radix r, which is an integer greater than or equal to 2. A number system has exactly r digits in which each bit in the radix has a value in the range of 0 to $r - 1$, thus the highest digit value is one less than the radix. For example, the binary radix has two digits which range from 0 to 1; the octal radix has eight digits which range from 0 to 7. An n-bit integer A is represented in a positional number system as follows:

$$A = (a_{n-1} a_{n-2} a_{n-3} \dots a_1 a_0) \tag{1.1}$$

where $0 \le a_i \le r - 1$. The high-order and low-order digits are a_{n-1} and a_0, respectively. The number in Equation 1.1 (also referred to as a vector or operand) can represent positive integer values in the range 0 to $r^n - 1$. Thus, a positive integer A is written as

$$A = a_{n-1} r^{n-1} + a_{n-2} r^{n-2} + a_{n-3} r^{n-3} + \dots + a_1 r^1 + a_0 r^0 \tag{1.2}$$

The value for A can be represented more compactly as

$$A = \sum_{i=0}^{n-1} a_i r^i \tag{1.3}$$

The expression of Equation 1.2 can be extended to include fractions. For example,

$$A = a_{n-1} r^{n-1} + \dots + a_1 r^1 + a_0 r^0 + a_{-1} r^{-1} + a_{-2} r^{-2} + \dots + a_{-m} r^{-m} \tag{1.4}$$

Equation 1.4 can be represented as

$$A = \sum_{i=-m}^{n-1} a_i r^i \tag{1.5}$$

Adding 1 to the highest digit in a radix r number system produces a sum of 0 and a carry of 1 to the next higher-order column. Thus, counting in radix r produces the following sequence of numbers:

$$0, 1, 2, \dots, (r - 1), 10, 11, 12, \dots, 1(r - 1), \dots.$$

Table 1.1 shows the counting sequence for different radices. The low-order digit will always be 0 in the set of r digits for the given radix. The set of r digits for various radices is given in Table 1.2. In order to maintain one character per digit, the numbers 10, 11, 12, 13, 14, and 15 are represented by the letters A, B, C, D, E, and F, respectively.

Table 1.1 Counting Sequence for Different Radices

Decimal	$r = 2$	$r = 4$	$r = 8$
0	0	0	0
1	1	1	1
2	10	2	2
3	11	3	3
4	100	10	4
5	101	11	5
6	110	12	6
7	111	13	7
8	1000	20	10
9	1001	21	11
10	1010	22	12
11	1011	23	13
12	1100	30	14
13	1101	31	15
14	1110	32	16
15	1111	33	17
16	10000	100	20
17	10001	101	21

Table 1.2 Character Sets for Different Radices

Radix (base)	Character Sets for Different Radices
2	{0, 1}
3	{0, 1, 2}
4	{0, 1, 2, 3}
5	{0, 1, 2, 3, 4}
6	{0, 1, 2, 3, 4, 5}
7	{0, 1, 2, 3, 4, 5, 6}
8	{0, 1, 2, 3, 4, 5, 6, 7}
9	{0, 1, 2, 3, 4, 5, 6, 7, 8}
10	{0, 1, 2, 3, 4, 5, 6, 7, 8, 9}
11	{0, 1, 2, 3, 4, 5, 6, 7, 8, 9, A}
12	{0, 1, 2, 3, 4, 5, 6, 7, 8, 9, A, B}
13	{0, 1, 2, 3, 4, 5, 6, 7, 8, 9, A, B, C}
14	{0, 1, 2, 3, 4, 5, 6, 7, 8, 9, A, B, C, D}
15	{0, 1, 2, 3, 4, 5, 6, 7, 8, 9, A, B, C, D, E}
16	{0, 1, 2, 3, 4, 5, 6, 7, 8, 9, A, B, C, D, E, F}

Example 1.1 Count from decimal 0 to 25 in radix 5. Table 1.2 indicates that radix 5 contains the following set of four digits: {0, 1, 2, 3, 4}. The counting sequence in radix 5 is:

$$\begin{aligned}
000, 001, 002, 003, 004 &= (0 \times 5^2) + (0 \times 5^1) + (4 \times 5^0) = 4_{10} \\
010, 011. 012, 013, 014 &= (0 \times 5^2) + (1 \times 5^1) + (4 \times 5^0) = 9_{10} \\
020, 021, 022, 023, 024 &= (0 \times 5^2) + (2 \times 5^1) + (4 \times 5^0) = 14_{10} \\
030, 031, 032, 033, 034 &= (0 \times 5^2) + (3 \times 5^1) + (4 \times 5^0) = 19_{10} \\
040, 041, 042, 043, 044 &= (0 \times 5^2) + (4 \times 5^1) + (4 \times 5^0) = 24_{10} \\
100 &= (1 \times 5^2) + (0 \times 5^1) + (0 \times 5^0) = 25_{10}
\end{aligned}$$

Example 1.2 Count from decimal 0 to 25 in radix 12. Table 1.2 indicates that radix 12 contains the following set of twelve digits: {0, 1, 2, 3, 4, 5, 6, 7, 8, 9, A, B}. The counting sequence in radix 12 is:

$$\begin{aligned}
00, 01, 02, 03, 04, 05, 06, 07, 08, 09, 0A, 0B &= (0 \times 12^1) + (11 \times 12^0) = 11_{10} \\
10, 11, 12, 13, 14, 15, 16, 17, 18, 19, 1A, 1B &= (1 \times 12^1) + (11 \times 12^0) = 23_{10} \\
20, 21 &= (2 \times 12^1) + (1 \times 12^0) = 25_{10}
\end{aligned}$$

1.1.1 Binary Number System

The radix is 2 in the *binary number system*; therefore, only two digits are used: 0 and 1. The low-value digit is 0 and the high-value digit is $(r-1) = 1$. The binary number system is the most conventional and easily implemented system for internal use in a digital computer; therefore, most digital computers use the binary number system. There is a disadvantage when converting to and from the externally used decimal system; however, this is compensated for by the ease of implementation and the speed of execution in binary of the four basic operations: addition, subtraction, multiplication, and division. The radix point is implied within the internal structure of the computer; that is, there is no specific storage element assigned to contain the radix point.

The weight assigned to each position of a binary number is as follows:

$$2^{n-1} 2^{n-2} \ldots 2^3\ 2^2\ 2^1\ 2^0 .\ 2^{-1} 2^{-2} 2^{-3} \ldots 2^{-m}$$

where the integer and fraction are separated by the radix point (binary point). The decimal value of the binary number 1011.101_2 is obtained by using Equation 1.4, where $r = 2$ and $a_i \in \{0,1\}$ for $-m \le i \le n-1$. Therefore,

$$\begin{array}{cccccccc}
2^3 & 2^2 & 2^1 & 2^0 & . & 2^{-1} & 2^{-2} & 2^{-3} \\
1 & 0 & 1 & 1 & . & 1 & 0 & 1_2
\end{array}
\begin{aligned}
&= (1 \times 2^3) + (0 \times 2^2) + (1 \times 2^1) + (1 \times 2^0) + \\
&\quad (1 \times 2^{-1}) + (0 \times 2^{-2}) + (1 \times 2^{-3}) \\
&= 11.625_{10}
\end{aligned}$$

Digital systems are designed using bistable storage devices that are either reset (logic 0) or set (logic 1). Therefore, the binary number system is ideally suited to represent numbers or states in a digital system, since radix 2 consists of the alphabet 0 and 1. These bistable devices can be concatenated to any length n to store binary data. For example, to store one byte (eight bits) of data, eight bistable storage devices are required as shown in Figure 1.1 for the value 0110 1011 (107_{10}). Counting in binary is shown in Table 1.3, which shows the weight associated with each of the four binary positions. Notice the alternating groups of 1s in Table 1.3. A binary number is a group of n bits that can assume 2^n different combinations of the n bits. The range for n bits is 0 to $2^n - 1$.

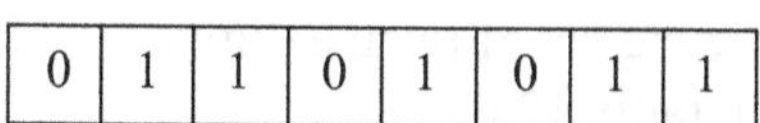

Figure 1.1 Concatenated 8-bit storage elements.

Table 1.3 Counting in Binary

Decimal	Binary			
	8	4	2	1
	2^3	2^2	2^1	2^0
0	0	0	0	0
1	0	0	0	1
2	0	0	1	0
3	0	0	1	1
4	0	1	0	0
5	0	1	0	1
6	0	1	1	0
7	0	1	1	1
8	1	0	0	0
9	1	0	0	1
10	1	0	1	0
11	1	0	1	1
12	1	1	0	0
13	1	1	0	1
14	1	1	1	0
15	1	1	1	1

The binary weights for the bit positions of an 8-bit integer are shown in Table 1.4; the binary weights for a an 8-bit fraction are shown in Table 1.5.

Table 1.4 Binary Weights for an 8-Bit Integer

2^7	2^6	2^5	2^4	2^3	2^2	2^1	2^0
128	64	32	16	8	4	2	1

Table 1.5 Binary Weights for an 8-Bit Fraction

2^{-1}	2^{-2}	2^{-3}	2^{-4}	2^{-5}	2^{-6}	2^{-7}	2^{-8}
1/2	1/4	1/8	1/16	1/32	1/64	1/128	1/256
0.5	0.25	0.125	0.0625	0.03125	0.015625	0.0078125	0.00390625

Each 4-bit binary segment has a weight associated with the segment and is assigned the value represented by the low-order bit of the corresponding segment, as shown in the first row of Table 1.6. The 4-bit binary number in each segment is then multiplied by the value of the segment. Thus, the binary number 0010 1010 0111 1100 0111 is equal to the decimal number $59{,}335_{10}$ as shown below.

$$(2 \times 8192) + (10 \times 4096) + (7 \times 256) + (12 \times 16) + (7 \times 1) = 59{,}335_{10}$$

Table 1.6 Weight Associated with 4-Bit Binary Segments

8192	4096	256	16	1
0001	0001	0001	0001	0001
0010	1010	0111	1100	0111

1.1.2 Octal Number System

The radix is 8 in the *octal number system*; therefore, eight digits are used, 0 through 7. The low-value digit is 0 and the high-value digit is $(r - 1) = 7$. The weight assigned to each position of an octal number is as follows:

$$8^{n-1}\, 8^{n-2} \;\ldots\; 8^3\, 8^2\, 8^1\, 8^0 .\, 8^{-1} 8^{-2} 8^{-3} \;\ldots\; 8^{-m}$$

where the integer and fraction are separated by the radix point (octal point). The decimal value of the octal number 217.6_8 is obtained by using Equation 1.4, where $r = 8$ and $a_i \in \{0,1,2,3,4,5,6,7\}$ for $-m \le i \le n - 1$. Therefore,

$$\begin{array}{cccccl} 8^2 & 8^1 & 8^0 & . & 8^{-1} & \\ 2 & 1 & 7 & . & 6_8 & = (2 \times 8^2) + (1 \times 8^1) + (7 \times 8^0) + (6 \times 8^{-1}) \\ & & & & & = 143.75_{10} \end{array}$$

When a count of 1 is added to 7_8, the sum is zero and a carry of 1 is added to the next higher-order column on the left. Counting in octal is shown in Table 1.7, which shows the weight associated with each of the three octal positions.

Table 1.7 Counting in Octal

Decimal	Octal		
	64	8	1
	8^2	8^1	8^0
0	0	0	0
1	0	0	1
2	0	0	2
3	0	0	3
4	0	0	4
5	0	0	5
6	0	0	6
7	0	0	7
8	0	1	0
9	0	1	1
...		...	
14	0	1	6
15	0	1	7
16	0	2	0
17	0	2	1
...		...	
22	0	2	6
23	0	2	7
24	0	3	0
25	0	3	1
...		...	
30	0	3	6
31	0	3	7
...		...	
84	1	2	4
...		...	
242	3	6	2
...		...	
377	5	7	1

Binary-coded octal Each octal digit can be encoded into a corresponding binary number. The highest-valued octal digit is 7; therefore, three binary digits are required to represent each octal digit. This is shown in Table 1.8, which lists the eight decimal digits (0 through 7) and indicates the corresponding octal and binary-coded octal (BCO) digits. Table 1.8 also shows octal numbers of more than one digit.

Table 1.8 Binary-Coded Octal Numbers

Decimal	Octal	Binary-Coded Octal		
0	0			000
1	1			001
2	2			010
3	3			011
4	4			100
5	5			101
6	6			110
7	7			111
8	10		001	000
9	11		001	001
10	12		001	010
11	13		001	011
...	...		...	
20	24		010	100
21	25		010	101
...	...		...	
100	144	001	100	100
101	145	001	100	101
...	...		...	
267	413	100	001	011
...	...		...	
385	601	110	000	001

1.1.3 Decimal Number System

The radix is 10 in the *decimal number system*; therefore, ten digits are used, 0 through 9. The low-value digit is 0 and the high-value digit is $(r-1) = 9$. The weight assigned to each position of a decimal number is as follows:

$$10^{n-1}\,10^{n-2}\,\ldots\;10^{3}\;10^{2}\;10^{1}\;10^{0}.\;10^{-1}\,10^{-2}\,10^{-3}\;\ldots\;10^{-m}$$

where the integer and fraction are separated by the radix point (decimal point). The value of 6537_{10} is immediately apparent; however, the value is also obtained by using Equation 1.4, where $r = 10$ and $a_i \in \{0,1,2,3,4,5,6,7,8,9\}$ for $-m \leq i \leq n-1$. That is,

$$\begin{array}{llll} 10^3 & 10^2 & 10^1 & 10^0 \\ 6 & 3 & 5 & 7_{10} = (6 \times 10^3) + (3 \times 10^2) + (5 \times 10^1) + (7 \times 10^0) \end{array}$$

When a count of 1 is added to decimal 9, the sum is zero and a carry of 1 is added to the next higher-order column on the left. The following example contains both an integer and a fraction:

$$\begin{array}{llllll} 10^3 & 10^2 & 10^1 & 10^0 & . & 10^{-1} \\ 5 & 4 & 3 & 6 & . & 5 = (5 \times 10^3) + (4 \times 10^2) + (3 \times 10^1) + (6 \times 10^0) + (5 \times 10^{-1}) \end{array}$$

Binary-coded decimal Each decimal digit can be encoded into a corresponding binary number; however, only ten decimal digits are valid The highest-valued decimal digit is 9, which requires four bits in the binary representation. Therefore, four binary digits are required to represent each decimal digit. This is shown in Table 1.9, which lists the ten decimal digits (0 through 9) and indicates the corresponding binary-coded decimal (BCD) digits. Table 1.9 also shows BCD numbers of more than one decimal digit.

Table 1.9 Binary-Coded Decimal Numbers

Decimal	Binary-Coded Decimal
0	0000
1	0001
2	0010
3	0011
4	0100
5	0101
6	0110
7	0111
8	1000
9	1001
10	0001 0000
11	0001 0001
12	0001 0010
...	...
124	0001 0010 0100
...	...
365	0011 0110 0101

1.1.4 Hexadecimal Number System

The radix is 16 in the *hexadecimal number system*; therefore, 16 digits are used, 0 through 9 and A through F, where by convention A, B, C, D, E, and F correspond to decimal 10, 11, 12, 13, 14, and 15, respectively. The low-value digit is 0 and the high-value digit is $(r - 1) = 15$ (F). The weight assigned to each position of a hexadecimal number is as follows:

$$16^{n-1}16^{n-2} \ldots 16^3\ 16^2\ 16^1\ 16^0 .\ 16^{-1}16^{-2}16^{-3} \ldots 16^{-m}$$

where the integer and fraction are separated by the radix point (hexadecimal point). The decimal value of the hexadecimal number $6A8C.D416_{16}$ is obtained by using Equation 1.4, where $r = 16$ and $a_i \in \{0,1,2,3,4,5,6,7,8,9,A,B,C,D,E,F\}$ for $-m \leq i \leq n - 1$. Therefore,

$$\begin{array}{llllllllll} 16^3 & 16^2 & 16^1 & 16^0 & . & 16^{-1} & 16^{-2} & 16^{-3} & 16^{-4} & \\ 6 & A & 8 & C & . & D & 4 & 1 & 6 & = (6 \times 16^3) + (10 \times 16^2) + (8 \times 16^1) \\ &&&&&&&&& \quad + (12 \times 16^0) + (13 \times 16^{-1}) + (4 \times 16^{-2}) \\ &&&&&&&&& \quad + (1 \times 16^{-3}) + (6 \times 16^{-4}) \\ &&&&&&&&& = 27{,}276.828460692_{10} \end{array}$$

When a count of 1 is added to hexadecimal F, the sum is zero and a carry of 1 is added to the next higher-order column on the left.

Binary-coded hexadecimal Each hexadecimal digit corresponds to a 4-bit binary number as shown in Table 1.10. All 2^4 values of the four binary bits are used to represent the 16 hexadecimal digits. Table 1.10 also indicates hexadecimal numbers of more than one digit. Counting in hexadecimal is shown in Table 1.11. Table 1.12 summarizes the characters used in the four number systems: binary, octal, decimal, and hexadecimal.

Table 1.10 Binary-Coded Hexadecimal Numbers

Decimal	Hexadecimal	Binary-Coded Hexadecimal
0	0	0000
1	1	0001
2	2	0010
3	3	0011
4	4	0100
5	5	0101

Continued on next page

Table 1.10 Binary-Coded Hexadecimal Numbers

Decimal	Hexadecimal	Binary-Coded Hexadecimal
6	6	0110
7	7	0111
8	8	1000
9	9	1001
10	A	1010
11	B	1011
12	C	1100
13	D	1101
14	E	1110
15	F	1111
...	...	...
124	7C	0111 1100
...	...	...
365	16D	0001 0110 1101

Table 1.11 Counting in Hexadecimal

Decimal	Hexadecimal		
	256	16	1
	16^2	16^1	16^0
0	0	0	0
1	0	0	1
2	0	0	2
3	0	0	3
4	0	0	4
5	0	0	5
6	0	0	6
7	0	0	7
8	0	0	8
9	0	0	9
10	0	0	A
11	0	0	B
12	0	0	C
13	0	0	D
14	0	0	E
15	0	0	F

Continued on next page

Table 1.11 Counting in Hexadecimal

Decimal	Hexadecimal		
	256	16	1
	16^2	16^1	16^0
16	0	1	0
17	0	1	1
...		...	
26	0	1	A
27	0	1	B
...		...	
30	0	1	E
31	0	1	F
...		...	
256	1	0	0
...		...	
285	1	1	D
...		...	
1214	4	B	E

Table 1.12 Digits Used for Binary, Octal, Decimal, and Hexadecimal Number Systems

0	1	2	3	4	5	6	7	8	9	A	B	C	D	E	F
Binary															
Octal															
Decimal															
Hexadecimal															

1.2 Number Representations

The material presented thus far covered only positive numbers. However, computers use both positive and negative numbers. Since a computer cannot recognize a plus (+) or a minus (–) symbol, an encoding method must be established to represent the sign of a number in which both positive and negative numbers are distributed as evenly as possible.

There must also be a method to differentiate between positive and negative numbers; that is, there must be an easy way to test the sign of a number. Detection of a number with a zero value must be straightforward. The leftmost (high-order) digit is usually reserved for the sign of the number. Consider the following number A with radix r:

$$A = (a_{n-1}\, a_{n-2}\, a_{n-3}\, \cdots\, a_2\, a_1\, a_0)_r$$

where digit a_{n-1} has the following value:

$$A = \begin{cases} 0 & \text{if } A \geq 0 \\ r-1 & \text{if } A < 0 \end{cases} \tag{1.6}$$

The remaining digits of A indicate either the true magnitude or the magnitude in a complemented form. There are three conventional ways to represent positive and negative numbers in a positional number system: sign magnitude, diminished-radix complement, and radix complement. In all three number representations, the high-order digit is the sign of the number, according to Equation 1.6.

$$\begin{aligned} 0 &= \text{ positive} \\ r-1 &= \text{ negative} \end{aligned}$$

1.2.1 Sign Magnitude

In this representation, an integer has the following decimal range:

$$-(r^{n-1} - 1) \text{ to } + (r^{n-1} - 1) \tag{1.7}$$

where the number zero is considered to be positive. Thus, a positive number A is represented as

$$A = (0\, a_{n-2} a_{n-3}\, \cdots\, a_1 a_0)_r \tag{1.8}$$

and a negative number with the same absolute value as

$$A' = [(r-1)\, a_{n-2} a_{n-3}\, \cdots\, a_1 a_0]_r \tag{1.9}$$

In sign-magnitude notation, the positive version $+A$ differs from the negative version $-A$ only in the sign digit position. The magnitude portion $a_{n-2} a_{n-3}\, \cdots\, a_1 a_0$ is identical for both positive and negative numbers of the same absolute value.

There are two problems with sign-magnitude representation. First, there are two representations for the number 0; specifically +0 and –0; ideally there should be a unique representation for the number 0. Second, when adding two numbers of opposite signs, the magnitudes of the numbers must be compared to determine the sign of the result. This is not necessary in the other two methods that are presented in subsequent sections. Sign-magnitude notation is used primarily for representing fractions in floating-point notation.

Examples of sign-magnitude notation are shown below using 8-bit binary numbers and decimal numbers that represent both positive and negative values. Notice that the magnitude parts are identical for both positive and negative numbers for the same radix.

Radix 2

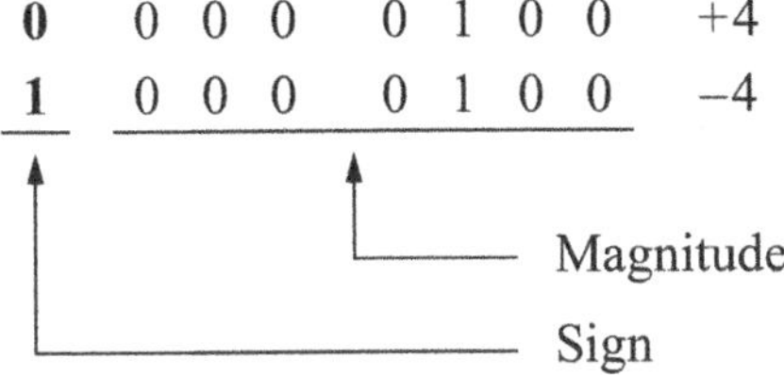

0 0 0 0 1 1 0 1 . 1 0 1 +13.625
1 0 0 0 1 1 0 1 . 1 0 1 –13.625

0 1 0 1 0 1 1 0 . 0 1 1 +86.375
1 1 0 1 0 1 1 0 . 0 1 1 –86.375

Radix 10

0 7 4 3 +743
9 7 4 3 –743

where 0 represents a positive number in radix 10, and 9 (r – 1) represents a negative number in radix 10. Again, the magnitudes of both numbers are identical.

0 6 7 8 4 +6784
9 6 7 8 4 –6784

1.2.2 Diminished-Radix Complement

This is the *(r – 1) complement* in which the radix is diminished by 1 and an integer has the following decimal range:

$$-(r^{n-1} - 1) \text{ to } +(r^{n-1} - 1) \tag{1.10}$$

which is the same as the range for sign-magnitude integers, although the numbers are represented differently, and where the number zero is considered to be positive. Thus, a positive number A is represented as

$$A = (0\ a_{n-2}a_{n-3}\ \cdots\ a_1a_0)_r \tag{1.11}$$

and a negative number as

$$A' = [(r-1)\ a_{n-2}'a_{n-3}'\ \cdots\ a_1'a_0']_r \tag{1.12}$$

where

$$a_i' = (r-1) - a_i \tag{1.13}$$

In binary notation ($r = 2$), the diminished-radix complement ($r - 1 = 2 - 1 = 1$) is the 1s complement. Positive and negative integers have the ranges shown below and are represented as shown in Equation 1.11 and Equation 1.12, respectively.

Positive integers: 0 to $2^{n-1} - 1$
Negative integers: 0 to $-(2^{n-1} - 1)$

To obtain the 1s complement of a binary number, simply complement (invert) all the bits. Thus, $0011\ 1100_2$ ($+60_{10}$) becomes $1100\ 0011_2$ (-60_{10}). To obtain the value of a positive binary number, the 1s are evaluated according to their weights in the positional number system, as shown below.

2^7	2^6	2^5	2^4	2^3	2^2	2^1	2^0	
0	0	1	1	1	1	0	0	$+60_{10}$

To obtain the value of a negative binary number, the 0s are evaluated according to their weights in the positional number system, as shown below.

2^7	2^6	2^5	2^4	2^3	2^2	2^1	2^0	
1	1	0	0	0	0	1	1	-60_{10}

When performing arithmetic operations on two operands, comparing the signs is straightforward, because the leftmost bit is a 0 for positive numbers and a 1 for negative numbers. There is, however, a problem when using the diminished-radix complement. There is a dual representation of the number zero, because a word of all 0s (+0) becomes a word of all 1s (–0) when complemented. This does not allow the requirement of having a unique representation for the number zero to be attained. The examples shown below represent the diminished-radix complement for different radices.

Example 1.3 The binary number 1101_2 will be 1s complemented. The number has a decimal value of –2. To obtain the 1s complement, subtract each digit in turn from 1 (the highest number in the radix), as shown below (Refer to Equation 1.12 and Equation 1.13). Or in the case of binary, simply invert each bit. Therefore, the 1s complement of 1101_2 is 0010_2, which has a decimal value of +2.

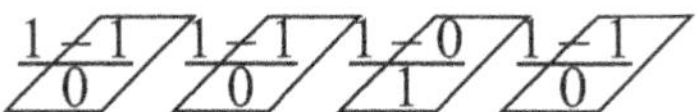

To verify the operation, add the negative and positive numbers to obtain 1111_2, which is zero in 1s complement notation.

$$\begin{array}{r} 1\ 1\ 0\ 1 \\ +)\ 0\ 0\ 1\ 0 \\ \hline 1\ 1\ 1\ 1 \end{array}$$

Example 1.4 Obtain the diminished-radix complement (9s complement) of 08752.43_{10}, where 0 is the sign digit indicating a positive number. The 9s complement is obtained by using Equation 1.12 and Equation 1.13. When a number is complemented in any form, the number is negated. Therefore, the sign of the complemented radix 10 number is $(r - 1) = 9$. The remaining digits of the number are obtained by using Equation 1.13 such that each digit in the complemented number is obtained by subtracting the given digit from 9. Therefore, the 9s complement of 08752.43_{10} is

$$\underset{9}{\underline{9-0}}\quad \underset{1}{\underline{9-8}}\quad \underset{2}{\underline{9-7}}\quad \underset{4}{\underline{9-5}}\quad \underset{7}{\underline{9-2}}\,.\,\underset{5}{\underline{9-4}}\quad \underset{6}{\underline{9-3}}$$

where the sign digit is $(r - 1) = 9$. If the above answer is negated, then the original number will be obtained. Thus, the 9s complement of $91247.56_{10} = 08752.43_{10}$; that is, the 9s complement of -1247.56_{10} is $+8752.43_{10}$, as written in conventional sign magnitude notation for radix 10.

Example 1.5 The diminished-radix complement of the positive decimal number 06784_{10} will be 9s complemented. To obtain the 9s complement, subtract each digit in turn from 9 (the highest number in the radix), as shown below to obtain the negative number with the same absolute value. The sign of the positive number is 0 and the sign of the negative number is 9 (refer to Equation 1.11 and Equation 1.12).

$$\frac{9-0}{9} \quad \frac{9-6}{3} \quad \frac{9-7}{2} \quad \frac{9-8}{1} \quad \frac{9-4}{5}$$

To verify the operation, add the negative and positive numbers to obtain 99999_{10}, which is zero in 9s complement notation.

$$\begin{array}{rccccc} & 0 & 6 & 7 & 8 & 4 \\ +) & 9 & 3 & 2 & 1 & 5 \\ \hline & 9 & 9 & 9 & 9 & 9 \end{array}$$

Example 1.6 The diminished-radix complement of the positive radix 8 number 05734_8 will be 7s complemented. To obtain the 7s complement, subtract each digit in turn from 7 (the highest number in the radix), as shown below to obtain the negative number with the same absolute value. The sign of the positive number is 0 and the sign of the negative number is 7 (refer to Equation 1.11 and Equation 1.12).

$$\frac{7-0}{7} \quad \frac{7-5}{2} \quad \frac{7-7}{0} \quad \frac{7-3}{4} \quad \frac{7-4}{3}$$

To verify the operation, add the negative and positive numbers to obtain 77777_8, which is zero in 7s complement notation.

$$\begin{array}{rccccc} & 0 & 5 & 7 & 3 & 4 \\ +) & 7 & 2 & 0 & 4 & 3 \\ \hline & 7 & 7 & 7 & 7 & 7 \end{array}$$

Example 1.7 The diminished-radix complement of the positive radix 16 number $0A7C4_{16}$ will be 15s complemented. To obtain the 15s complement, subtract each digit in turn from 15 (the highest number in the radix), as shown below to obtain the negative number with the same absolute value. The sign of the positive number is 0 and the sign of the negative number is F (refer to Equation 1.11 and Equation 1.12).

$$\frac{F-0}{F} \quad \frac{F-A}{5} \quad \frac{F-7}{8} \quad \frac{F-C}{3} \quad \frac{F-4}{B}$$

To verify the operation, add the negative and positive numbers to obtain $FFFFF_{16}$, which is zero in 15s complement notation.

$$
\begin{array}{rccccc}
 & 0 & A & 7 & C & 4 \\
+) & F & 5 & 8 & 3 & B \\
\hline
 & F & F & F & F & F
\end{array}
$$

1.2.3 Radix Complement

This is the *r complement*, where an integer has the following decimal range:

$$-(r^{n-1}) \text{ to } +(r^{n-1} - 1) \tag{1.14}$$

where the number zero is positive. A positive number A is represented as

$$A = (0\ a_{n-2}a_{n-3}\ \cdots\ a_1a_0)_r \tag{1.15}$$

and a negative number as

$$(A')_{+1} = \{[(r-1)\ a_{n-2}'a_{n-3}'\ \cdots\ a_1'a_0'] + 1\}_r \tag{1.16}$$

where A' is the diminished-radix complement. Thus, the radix complement is obtained by adding 1 to the diminished-radix complement; that is, $(r-1) + 1 = r$. Note that all three number representations have the same format for positive numbers and differ only in the way that negative numbers are represented, as shown in Table 1.13.

Table 1.13 Number Representations for Positive and Negative Integers of the Same Absolute Value for Radix *r*

Number Representation	Positive Numbers	Negative Numbers
Sign magnitude	$0\ a_{n-2}a_{n-3}\ \cdots\ a_1a_0$	$(r-1)\ a_{n-2}a_{n-3}\ \cdots\ a_1a_0$
Diminished-radix complement	$0\ a_{n-2}a_{n-3}\ \cdots\ a_1a_0$	$(r-1)\ a_{n-2}'a_{n-3}'\ \cdots\ a_1'a_0'$
Radix complement	$0\ a_{n-2}a_{n-3}\ \cdots\ a_1a_0$	$(r-1)\ a_{n-2}'a_{n-3}'\ \cdots\ a_1'a_0' + 1$

Another way to define the radix complement of a number is shown in Equation 1.17, where n is the number of digits in A.

$$(A')_{+1} = r^n - A_r \tag{1.17}$$

For example, assume that $A = 0101\ 0100_2\ (+84_{10})$. Then, using Equation 1.17,

$2^8 = 256_{10} = 10000\ 0000_2$. Thus, $256_{10} - 84_{10} = 172_{10}$

$(A')_{+1} = 2^8 - (0101\ 0100)$

$$\begin{array}{r} 1\ 0\ 0\ 0\ 0\ 0\ 0\ 0\ 0 \\ -)\quad 0\ 1\ 0\ 1\ 0\ 1\ 0\ 0 \\ \hline 1\ 0\ 1\ 0\ 1\ 1\ 0\ 0 \end{array}$$

As can be seen from the above example, to generate the radix complement for a radix 2 number, keep the low-order 0s and the first 1 unchanged and complement (invert) the remaining high-order bits. To obtain the value of a negative number in radix 2, the 0s are evaluated according to their weights in the positional number system, then 1 is added to the value obtained.

2^7	2^6	2^5	2^4	2^3	2^2	2^1	2^0	
1	0	1	0	1	1	0	0	
80				3 + 1 = 4				-84_{10}

Table 1.14 and Table 1.15 show examples of the three number representations for positive and negative numbers in radix 2. Note that the positive numbers are identical for all three number representations; only the negative numbers change.

Table 1.14 Number Representations for Positive and Negative Integers in Radix 2

Number representation	$+127_{10}$	-127_{10}
Sign magnitude	**0** 111 1111	**1** 111 1111
Diminished-radix complement (1s)	**0** 111 1111	**1** 000 0000
Radix complement (2s)	**0** 111 1111	**1** 000 0001

Table 1.15 Number Representations for Positive and Negative Integers in Radix 2

Number representation	$+54_{10}$	-54_{10}
Sign magnitude	**0** 011 0110	**1** 011 0110
Diminished-radix complement (1s)	**0** 011 0110	**1** 100 1001
Radix complement (2s)	**0** 011 0110	**1** 100 1010

There is a unique zero for binary numbers in radix complement, as shown below. When the number zero is 2s complemented, the bit configuration does not change. The 2s complement is formed by adding 1 to the 1s complement.

$$\begin{array}{rcccccccc}
\text{Zero in 2s complement} = & 0 & 0 & 0 & 0 & 0 & 0 & 0 & 0 \\
\text{Form the 1s complement} = & 1 & 1 & 1 & 1 & 1 & 1 & 1 & 1 \\
\text{Add 1} = & & & & & & & & 1 \\
\hline
 & 0 & 0 & 0 & 0 & 0 & 0 & 0 & 0
\end{array}$$

Example 1.8 Convert -20_{10} to binary and obtain the 2s complement. Then obtain the 2s complement of $+20_{10}$ using the fast method of keeping the low-order 0s and the first 1 unchanged as the number is scanned from right to left, then inverting all remaining bits.

$$\underbrace{1110\ 1\underline{100}}_{-20} \xrightarrow{\text{2s}} \underbrace{0001\ 0\underline{100}}_{+20} \xrightarrow{\text{2s}} \underbrace{1110\ 1\underline{100}}_{-20}$$

Example 1.9 Obtain the radix complement (10s complement) of the positive number 08752.43_{10}. Determine the 9s complement as in Example 1.5, then add 1. The 10s complement of 08752.43_{10} is the negative number 91247.57_{10}.

$$\begin{array}{rccccccc}
 & \frac{9-0}{9} & \frac{9-8}{1} & \frac{9-7}{2} & \frac{9-5}{4} & \frac{9-2}{7} . & \frac{9-4}{5} & \frac{9-3}{6} \\
+) & & & & & & & 1 \\
\hline
 & 9 & 1 & 2 & 4 & 7 . & 5 & 7
\end{array}$$

Adding 1 to the 9s complement in this example is the same as adding 10^{-2} ($.01_{10}$). To verify that the radix complement of 08752.43_{10} is 91247.57_{10}, the sum of the two numbers should equal zero for radix 10. This is indeed the case, as shown below.

$$\begin{array}{r} 08752.43 \\ +)\ \underline{91247.57} \\ 00000.00 \end{array}$$

Example 1.10 Obtain the 10s complement of 0.4572_{10} by adding a power of ten to the 9s complement of the number. The 9s complement of 0.4572_{10} is

$$9-0=9 \quad 9-4=5 \quad 9-5=4 \quad 9-7=2 \quad 9-2=7$$

Therefore, the 10s complement of $0.4572_{10} = 9.5427_{10} + 10^{-4} = 9.5428_{10}$. Adding the positive and negative numbers again produces a zero result.

Example 1.11 Obtain the radix complement of $1111\ 1111_2$ (-1_{10}). The answer can be obtained using two methods: add 1 to the 1s complement or keep the low-order 0s and the first 1 unchanged, then invert all remaining bits. Since there are no low-order 0s, only the rightmost 1 is unchanged. Therefore, the 2s complement of $1111\ 1111_2$ is $0000\ 0001_2$ ($+1_{10}$).

Example 1.12 Obtain the 8s complement of 04360_8. First form the 7s complement, then add 1. The 7s complement of 04360_8 is 73417_8. Therefore, the 8s complement of 04360_8 is $73417_8 + 1$, as shown below, using the rules for octal addition. Adding the positive and negative numbers results in a sum of zero.

$$\begin{array}{rccccc} & 7 & 3 & 4 & 1 & 7 \\ +) & & & & & 1 \\ \hline & 7 & 3 & 4 & 2 & 0 \end{array}$$

Example 1.13 Obtain the 16s complement of $F8A5_{16}$. First form the 15s complement, then add 1. The 15s complement of $F8A5_{16}$ is $075A_{16}$. Therefore, the 16s complement of $F8A5_{16} = 075A_{16} + 1 = 075B_{16}$. Adding the positive and negative numbers results in a sum of zero.

Example 1.14 Obtain the 4s complement of 0231_4. The rules for obtaining the radix complement are the same for any radix: generate the diminished-radix complement, then add 1. Therefore, the 4s complement of $0231_4 = 3102_4 + 1 = 3103_4$. To verify the result, add $0231_4 + 3103_4 = 0000_4$, as shown below using the rules for radix 4 addition.

$$\begin{array}{rcccc} & 0 & 2 & 3 & 1 \\ +) & 3 & 1 & 0 & 3 \\ \hline & 0 & 0 & 0 & 0 \end{array}$$

1.3 Problems

1.1 Rewrite the following hexadecimal number in binary and octal notation: 7C64B

1.2 Convert the octal number 5476_8 to radix 10.

1.3 Convert the following octal number to hexadecimal: 63354_8:

1.4 Convert the hexadecimal number $4AF9_{16}$ to radix 10.

1.5 Convert the following radix 2 number to a decimal value, then convert the radix 2 number to hexadecimal: $1011\ 0111\ .\ 1111_2$

1.6 Convert the unsigned binary number 1100.110_2 to radix 10.

1.7 Obtain the decimal value of the following numbers: 011011.110_2, 674.7_8, $AD.2_{16}$

1.8 Convert the octal number 173.25_8 to radix 10.

1.9 Convert the following decimal number to a 16-bit binary number: $+127.5625_{10}$

1.10 Convert the following decimal number to octal and hexadecimal notation: 130.21875_{10}

1.11 Convert 122.13_{-4} to radix 3.

1.12 Convert 375.54_8 to radix 3.

1.13 The numbers shown below are in 2s complement representation. Convert the numbers to sign-magnitude representation for radix 2 with the same numerical value using eight bits.

2s complement	Sign magnitude
0111 1111	
1000 0001	
0000 1111	
1111 0001	
1111 0000	

1.14 Obtain the radix complement of $F8B6_{16}$.

1.15 Obtain the diminished-radix complement of 0778_9.

1.16 Obtain the sign magnitude, diminished-radix complement, and radix complement number representations for radix 2 for the following decimal number: –136.

1.17 Obtain the sign magnitude, diminished-radix complement, and radix complement number representations for radix 10 for the following decimal number: –136.

1.18 Obtain the sign magnitude, diminished-radix complement, and radix complement number representations for radix 2 for the following decimal numbers: –113 and +54

1.19 Determine the range of numbers for the following number representations for radix 2:

Sign magnitude
Diminished-radix complement
Radix complement

1.20 Obtain the *r*s complement of the following numbers: $A736_{16}$ and 5620_8:

1.21 Obtain the diminished-radix complement and the radix complement of the following numbers: 9834_{10}, 1000_{10}, and 0000_{10}.

1.22 Given the hexadecimal number $ABCD_{16}$, perform the following steps:

(a) Obtain the *r*s complement of $ABCD_{16}$.
(b) Convert $ABCD_{16}$ to binary.
(c) Obtain the 2s complement of the binary number.
(d) Convert the 2s complement to hexadecimal.

1.23 Obtain the 1s complement and the 2s complement of the following radix 2 numbers: 1000 1000, 0001 1111, 0000 0000, and 1111 1111.

1.24 Convert the following unsigned radix 2 numbers to radix 10:
$100\ 0001.111_2$
$1111\ 1111.1111_2$

1.25 Obtain the decimal value of the following numbers:
1110.1011_2
431.32_5

1.26 Convert π to radix 2 using 13 binary digits.

1.27 Generate the decimal integers 0 – 15 for the following positional number systems: radix 4 and radix 12.

1.28 Convert the following integers to radix 3 and radix 7: 10_{10} and 111_4.

2.1 *Boolean Algebra*
2.2 *Minimization Techniques*
2.3 *Combinational Logic*
2.4 *Sequential Logic*
2.5 *Problems*

2

Logic Design Fundamentals

It is assumed that the reader has an adequate background in combinational and sequential logic analysis and synthesis; therefore, this chapter presents only a brief review of the basic concepts of the synthesis, or design, of combinational and sequential logic. The chapter begins with a presentation of the axioms and theorems of Boolean algebra — the axioms are also referred to as postulates. The minimization of logic functions can be realized by applying the principles of Boolean algebra. Other minimization techniques are also reviewed. These include Karnaugh maps, the Quine-McCluskey algorithm, and the Petrick algorithm.

Combinational logic design of different macro functions is presented. These include multiplexers, decoders, encoders, and comparators, all of which can be used in the design of arithmetic circuits. Sequential logic design includes different types of storage elements that can be used in the design of registers, counters of various moduli, Moore machines, and Mealy machines. These sequential circuits can be used to store data and to control the sequencing of arithmetic operations.

2.1 Boolean Algebra

In 1854, George Boole introduced a systematic treatment of the logic operations AND, OR, and NOT, which is now called Boolean algebra. The symbols (or operators) used for the algebra and the corresponding function definitions are listed in Table 2.1. The table also includes the exclusive-OR function, which is characterized by the three operations of AND, OR, and NOT. Table 2.2 illustrates the truth tables for the Boolean operations AND, OR, NOT, exclusive-OR, and exclusive-NOR, where z_1 is the result of the operation.

Table 2.1 Boolean Operators for Variables x_1 and x_2

Operator	Function	Definition
•	AND	$x_1 \bullet x_2$ (Also $x_1 x_2$)
+	OR	$x_1 + x_2$
'	NOT (negation)	x_1'
⊕	Exclusive-OR	$(x_1 x_2') + (x_1' x_2)$

Table 2.2 Truth Table for AND, OR, NOT, Exclusive-OR, and Exclusive-NOR Operations

AND		OR		NOT		Exclusive-OR		Exclusive-NOR	
$x_1 x_2$	z_1	$x_1 x_2$	z_1	x_1	z_1	$x_1 x_2$	z_1	$x_1 x_2$	z_1
0 0	0	0 0	0	0	1	0 0	0	0 0	1
0 1	0	0 1	1	1	0	0 1	1	0 1	0
1 0	0	1 0	1			1 0	1	1 0	0
1 1	1	1 1	1			1 1	0	1 1	1

The AND operator, which corresponds to the Boolean product, is also indicated by the symbol "$\wedge$" ($x_1 \wedge x_2$) or by no symbol if the operation is unambiguous. Thus, $x_1 x_2$, $x_1 \bullet x_2$, and $x_1 \wedge x_2$ are all read as "x_1 AND x_2." The OR operator, which corresponds to the Boolean sum, is also specified by the symbol "$\vee$." Thus, $x_1 + x_2$ and $x_1 \vee x_2$ are both read as "x_1 OR x_2." The symbol for the complement (or negation) operation is usually specified by the prime " ' " symbol immediately following the variable (x_1'), by a bar over the variable ($\overline{x}_1$), or by the symbol "$\neg$" ($\neg x_1$). This book will use the symbols defined in Table 2.1.

Boolean algebra is a deductive mathematical system that can be defined by a set of variables, a set of operators, a set of axioms (or postulates), and a set of theorems. An *axiom* is a statement that is universally accepted as true; that is, the statement needs no proof, because its truth is obvious. The axioms of Boolean algebra form the basis from which the theorems and other properties can be derived.

Most axioms and theorems are characterized by two laws. Each law is the dual of the other. The principle of duality specifies that the *dual* of an algebraic expression can be obtained by interchanging the binary operators • and +, and by interchanging the identity elements 0 and 1. Since the primary emphasis of this book is the design of computer arithmetic circuits, the axioms and theorems of combinational logic are presented without proof.

Boolean algebra is an algebraic structure consisting of a set of elements B, together with two binary operators • and + and a unary operator ', such that the following axioms are true, where the notation $x_1 \in X$ is read as "x_1 is an element of the set X":

Axiom 1: Boolean set definition The set B contains at least two elements x_1 and x_2, where $x_1 \neq x_2$.

Axiom 2: Closure laws For every $x_1, x_2 \in B$,

(a) $x_1 + x_2 \in B$
(b) $x_1 \bullet x_2 \in B$

Axiom 3: Identity laws There exists two unique *identity elements* 0 and 1, where 0 is an identity element with respect to the Boolean sum and 1 is an identity element with respect to the Boolean product. Thus, for every $x_1 \in B$,

(a) $x_1 + 0 = 0 + x_1 = x_1$
(b) $x_1 \bullet 1 = 1 \bullet x_1 = x_1$

Axiom 4: Commutative laws The commutative laws specify that the order in which the variables appear in a Boolean expression is irrelevant — the result is the same. Thus, for every $x_1, x_2 \in B$,

(a) $x_1 + x_2 = x_2 + x_1$
(b) $x_1 \bullet x_2 = x_2 \bullet x_1$

Axiom 5: Associative laws The associative laws state that three or more variables can be combined in an expression using Boolean multiplication or addition and that the order of the variables can be altered without changing the result. Thus, for every $x_1, x_2, x_3 \in B$,

(a) $(x_1 + x_2) + x_3 = x_1 + (x_2 + x_3)$
(b) $(x_1 \bullet x_2) \bullet x_3 = x_1 \bullet (x_2 \bullet x_3)$

Axiom 6: Distributive laws The distributive laws for Boolean algebra are similar, in many respects, to those for traditional algebra. The interpretation, however, is different and is a function of the Boolean product and the Boolean sum. This is a very useful axiom in minimizing Boolean functions. For every $x_1, x_2, x_3 \in B$,

(a) The operator + is distributive over the operator • such that,
$x_1 + (x_2 \bullet x_3) = (x_1 + x_2) \bullet (x_1 + x_3)$

(b) The operator • is distributive over the operator + such that,
$x_1 \bullet (x_2 + x_3) = (x_1 \bullet x_2) + (x_1 \bullet x_3)$

Axiom 7: Complementation laws For every $x_1 \in B$, there exists an element x_1' (called the complement of x_1), where $x_1' \in B$, such that,

(a) $x_1 + x_1' = 1$
(b) $x_1 \bullet x_1' = 0$

The theorems presented below are derived from the axioms and are listed in pairs, where applicable, in which each theorem in the pair is the dual of the other.

Theorem 1: 0 and 1 associated with a variable Every variable in Boolean algebra can be characterized by the identity elements 0 and 1. Thus, for every $x_1 \in B$,

(a) $x_1 + 1 = 1$
(b) $x_1 \bullet 0 = 0$

Theorem 2: 0 and 1 complement The 2-valued Boolean algebra has two distinct identity elements 0 and 1, where $0 \neq 1$. The operations using 0 and 1 are as follows:

$0 + 0 = 0$ $\quad\quad$ $0 + 1 = 1$
$1 \bullet 1 = 1$ $\quad\quad$ $1 \bullet 0 = 0$

A corollary to Theorem 2 specifies that element 1 satisfies the requirements of the complement of element 0, and vice versa. Thus, each identity element is the complement of the other.

(a) $0' = 1$
(b) $1' = 0$

Theorem 3: Idempotent laws Idempotency relates to a nonzero mathematical quantity which, when applied to itself for a binary operation, remains unchanged. Thus, if $x_1 = 0$, then $x_1 + x_1 = 0 + 0 = 0$ and if $x_1 = 1$, then $x_1 + x_1 = 1 + 1 = 1$. Therefore, one of the elements is redundant and can be discarded. The dual is true for the operator •. The idempotent laws eliminate redundant variables in a Boolean expression and can be extended to any number of identical variables. This law is also referred to as the *law of tautology*, which precludes the needless repetition of the variable. For every $x_1 \in B$,

(a) $x_1 + x_1 = x_1$
(b) $x_1 \bullet x_1 = x_1$

Theorem 4: Involution law The involution law states that the complement of a complemented variable is equal to the variable. There is no dual for the involution law. The law is also called the law of double complementation. Thus, for every $x_1 \in B$,

$$x_1'' = x_1$$

Theorem 5: Absorption law 1 This version of the absorption law states that some 2-variable Boolean expressions can be reduced to a single variable without altering the result. Thus, for every $x_1, x_2 \in B$,

(a) $x_1 + (x_1 \bullet x_2) = x_1$
(b) $x_1 \bullet (x_1 + x_2) = x_1$

Theorem 6: Absorption law 2 This version of the absorption law is used to eliminate redundant variables from certain Boolean expressions. Absorption law 2 eliminates a variable or its complement and is a very useful law for minimizing Boolean expressions.

(a) $x_1 + (x_1' \bullet x_2) = x_1 + x_2$
(b) $x_1 \bullet (x_1' + x_2) = x_1 \bullet x_2$

Theorem 7: DeMorgan's laws DeMorgan's laws are also useful in minimizing Boolean functions. DeMorgan's laws convert the complement of a sum term or a product term into a corresponding product or sum term, respectively. For every $x_1, x_2 \in B$,

(a) $(x_1 + x_2)' = x_1' \bullet x_2'$
(b) $(x_1 \bullet x_2)' = x_1' + x_2'$

Parts (a) and (b) of DeMorgan's laws represent expressions for NOR and NAND gates, respectively. DeMorgan's laws can be generalized for any number of variables, such that,

(a) $(x_1 + x_2 + \ldots + x_n)' = x_1' \bullet x_2' \bullet \ldots \bullet x_n'$
(b) $(x_1 \bullet x_2 \bullet \ldots \bullet x_n)' = x_1' + x_2' + \ldots + x_n'$

When applying DeMorgan's laws to an expression, the operator • takes precedence over the operator +. For example, using DeMorgan's law to complement the Boolean expression $x_1 + x_2x_3$ yields the following:

$$(x_1 + x_2x_3)' = [x_1 + (x_2x_3)]'$$
$$= x_1'(x_2' + x_3')$$

Note that: $(x_1 + x_2x_3)' \neq x_1' \bullet x_2' + x_3'$

Minterm A minterm is the Boolean product of n variables and contains all n variables of the function exactly once, either true or complemented. For example, for the function $z_1(x_1, x_2, x_3)$, $x_1 x_2' x_3$ is a minterm.

Maxterm A maxterm is the Boolean sum of n variables and contains all n variables of the function exactly once, either true or complemented. For example, for the function $z_1(x_1, x_2, x_3)$, $(x_1 + x_2' + x_3)$ is a maxterm.

Product term A product term is the Boolean product of variables containing a subset of the possible variables or their complements. For example, for the function $z_1(x_1, x_2, x_3)$, $x_1' x_3$ is a product term, because it does not contain all the variables.

Sum term A sum term is the Boolean sum of variables containing a subset of the possible variables or their complements. For example, for the function $z_1(x_1, x_2, x_3)$, $(x_1' + x_3)$ is a sum term, because it does not contain all the variables.

Sum of minterms A sum of minterms is an expression in which each term contains all the variables, either true or complemented. For example,

$$z_1(x_1, x_2, x_3) = x_1' x_2 x_3 + x_1 x_2' x_3' + x_1 x_2 x_3$$

is a Boolean expression in a sum-of-minterms form. This particular form is also referred to as a *minterm expansion*, a *standard sum of products*, a *canonical sum of products*, or a *disjunctive normal form*. Since each term is a minterm, the expression for z_1 can be written in a more compact sum-of-minterms form as $z_1(x_1, x_2, x_3) = \Sigma_m(3,4,7)$, where each term is converted to its minterm value. For example, the first term in the expression is $x_1' x_2 x_3$, which corresponds to binary 011, representing minterm 3.

Sum of products A sum of products is an expression in which at least one term does not contain all the variables; that is, at least one term is a proper subset of the possible variables or their complements. For example,

$$z_1(x_1, x_2, x_3) = x_1' x_2 x_3 + x_2' x_3' + x_1 x_2 x_3$$

is a sum of products for the function z_1, because the second term does not contain the variable x_1.

Product of maxterms A product of maxterms is an expression in which each term contains all the variables, either true or complemented. For example,

$$z_1(x_1, x_2, x_3) = (x_1' + x_2 + x_3)(x_1 + x_2' + x_3')(x_1 + x_2 + x_3)$$

is a Boolean expression in a product-of-maxterms form. This particular form is also referred to as a *maxterm expansion*, a *standard product of sums*, a *canonical product of sums*, or a *conjunctive normal form*. Since each term is a maxterm, the expression for z_1 can be written in a more compact product-of-maxterms form as $z_1(x_1, x_2, x_3) = \Pi_M(0,3,4)$, where each term is converted to its maxterm value.

Product of sums A product of sums is an expression in which at least one term does not contain all the variables; that is, at least one term is a proper subset of the possible variables or their complements. For example,

$$z_1(x_1, x_2, x_3) = (x_1' + x_2 + x_3)\,(x_2' + x_3')\,(x_1 + x_2 + x_3)$$

is a product of sums for the function z_1, because the second term does not contain the variable x_1.

Summary of Boolean algebra axioms and theorems Table 2.3 provides a summary of the axioms and theorems of Boolean algebra. Each of the laws listed in the table is presented in pairs, where applicable, in which each law in the pair is the dual of the other.

Table 2.3 Summary of Boolean Algebra Axioms and Theorems

Axiom or Theorem	Definition
Axiom 1: Boolean set definition	$x_1, x_2 \in B$
Axiom 2: Closure laws	(a) $x_1 + x_2 \in B$ (b) $x_1 \bullet x_2 \in B$
Axiom 3: Identity laws	(a) $x_1 + 0 = 0 + x_1 = x_1$ (b) $x_1 \bullet 1 = 1 \bullet x_1 = x_1$
Axiom 4: Commutative laws	(a) $x_1 + x_2 = x_2 + x_1$ (b) $x_1 \bullet x_2 = x_2 \bullet x_1$
Axiom 5: Associative laws	(a) $(x_1 + x_2) + x_3 = x_1 + (x_2 + x_3)$ (b) $(x_1 \bullet x_2) \bullet x_3 = x_1 \bullet (x_2 \bullet x_3)$
Axiom 6: Distributive laws	(a) $x_1 + (x_2 \bullet x_3) = (x_1 + x_2) \bullet (x_1 + x_3)$ (b) $x_1 \bullet (x_2 + x_3) = (x_1 \bullet x_2) + (x_1 \bullet x_3)$
Axiom 7: Complementation laws	(a) $x_1 + x_1' = 1$ (b) $x_1 \bullet x_1' = 0$
Theorem 1: 0 and 1 associated with a variable	(a) $x_1 + 1 = 1$ (b) $x_1 \bullet 0 = 0$
Theorem 2: 0 and 1 complement	(a) $0' = 1$ (b) $1' = 0$

Continued on next page

Table 2.3 Summary of Boolean Algebra Axioms and Theorems

Axiom or Theorem	Definition
Theorem 3: Idempotent laws	(a) $x_1 + x_1 = x_1$
	(b) $x_1 \bullet x_1 = x_1$
Theorem 4: Involution law	$x_1'' = x_1$
Theorem 5: Absorption law 1	(a) $x_1 + (x_1 \bullet x_2) = x_1$
	(b) $x_1 \bullet (x_1 + x_2) = x_1$
Theorem 6: Absorption law 2	(a) $x_1 + (x_1' \bullet x_2) = x_1 + x_2$
	(b) $x_1 \bullet (x_1' + x_2) = x_1 \bullet x_2$
Theorem 7: DeMorgan's laws	(a) $(x_1 + x_2)' = x_1' \bullet x_2'$
	(b) $(x_1 \bullet x_2)' = x_1' + x_2'$

2.2 Minimization Techniques

This section will present various techniques for minimizing a Boolean function. A Boolean function is an algebraic representation of digital logic. Each term in an expression represents a logic gate and each variable in a term represents an input to a logic gate. It is important, therefore, to have the fewest number of terms in a Boolean equation and the fewest number of variables in each term. A Boolean equation with a minimal number of terms and variables reduces not only the number of logic gates, but also the delay required to generate the function.

2.2.1 Algebraic Minimization

The number of terms and variables that are necessary to generate a Boolean function can be minimized by algebraic manipulation. Since there are no specific rules or algorithms to use for minimizing a Boolean function, the procedure is inherently heuristic in nature. The only method available is an empirical procedure utilizing the axioms and theorems which is based solely on experience and observation without reference to theoretical principles. The examples which follow illustrate the process for minimizing a Boolean function using the axioms and theorems of Boolean algebra.

Example 2.1 The following expression will be converted to a sum-of-products form: $[(x_1 + x_2)' + x_3]'$

$$[(x_1 + x_2)' + x_3]' = [x_1'x_2' + x_3]' \qquad \text{DeMorgan's law}$$

$$= (x_1 + x_2)x_3' \qquad \text{DeMorgan's law}$$

$$= x_1x_3' + x_2x_3' \qquad \text{Distributive law}$$

Example 2.2 DeMorgan's law will be applied to the following expression: $(x_1x_2' + x_3'x_4 + x_5x_6)'$

$$(x_1x_2' + x_3'x_4 + x_5x_6)' = (x_1x_2')' (x_3'x_4)' (x_5x_6)'$$
$$= (x_1' + x_2)(x_3 + x_4')(x_5' + x_6')$$

Example 2.3 The following function will be minimized to a sum-of-products form using Boolean algebra:

$$z_1 = x_1'x_2x_3' + x_2(x_1' + x_3) + x_2x_3'(x_1 + x_2)'$$

$$z_1 = x_1'x_2x_3' + x_1'x_2 + x_2x_3 + x_2x_3'(x_1'x_2')$$ Distributive law and Commutative law and DeMorgan's law

$$= x_1'x_2x_3' + x_1'x_2 + x_2x_3$$ Complementation law

$$= x_1'x_2(x_3' + 1) + x_2x_3$$ Distributive law

$$= x_1'x_2 + x_2x_3$$ Theorem 1

2.2.2 Karnaugh Maps

A Karnaugh map provides a geometrical representation of a Boolean function. The Karnaugh map is arranged as an array of squares (or cells) in which each square represents a binary value of the input variables. The map is a convenient method of obtaining a minimal number of terms with a minimal number of variables per term for a Boolean function. A Karnaugh map presents a clear indication of function minimization without recourse to Boolean algebra and will generate a minimized expression in either a sum-of-products form or a product-of-sums form.

Figure 2.1 shows Karnaugh maps for two, three, four, and five variables. Each square in the maps corresponds to a unique minterm. The maps for three or more variables contain column headings that are represented in the *Gray code* format; the maps for four or more variables contain column and row headings that are represented in Gray code. Using the Gray code to designate column and row headings permits physically adjacent squares to be also logically adjacent; that is, to differ by only one variable. Map entries that are adjacent can be combined into a single term.

For example, the expression $z_1 = x_1x_2'x_3 + x_1x_2x_3$, which corresponds to minterms 5 and 7 in Figure 2.1(b), reduces to $z_1 = x_1x_3(x_2' + x_2) = x_1x_3$ using the distributive and complementation laws. Thus, if 1s are entered in minterm locations 5 and 7, then the two minterms can be combined into the single term x_1x_3.

Similarly, in Figure 2.1(c), if 1s are entered in minterm locations 4, 6, 12, and 14, then the four minterms combine as x_2x_4'. That is, only variables x_2 and x_4' are common to all four squares — variables x_1 and x_3 are discarded by the complementation law. The minimized expression obtained from the Karnaugh map can be verified algebraically by listing the four minterms as a sum-of-minterms expression, then applying the appropriate laws of Boolean algebra as shown in Example 2.4.

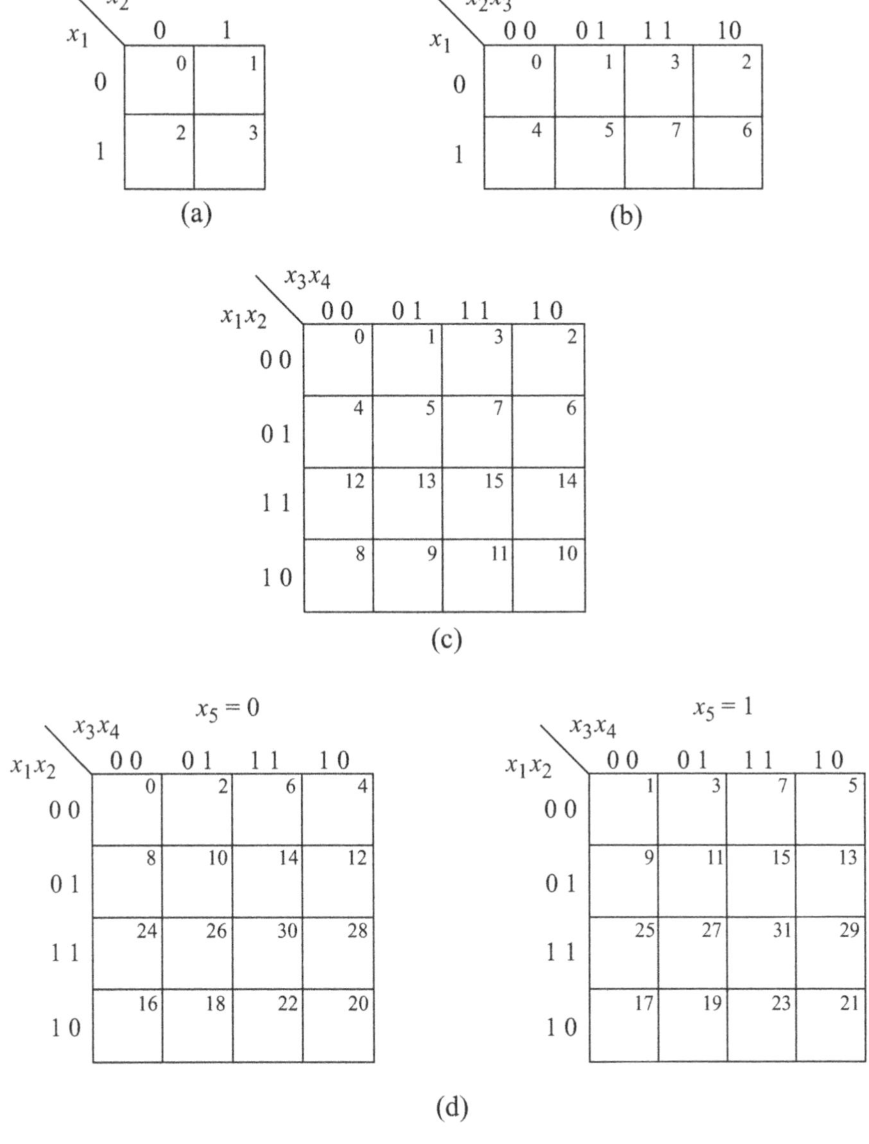

Figure 2.1 Karnaugh maps showing minterm locations: (a) two variables; (b) three variables; (c) four variables; (d) five variables; and (e) alternative map for five variables.

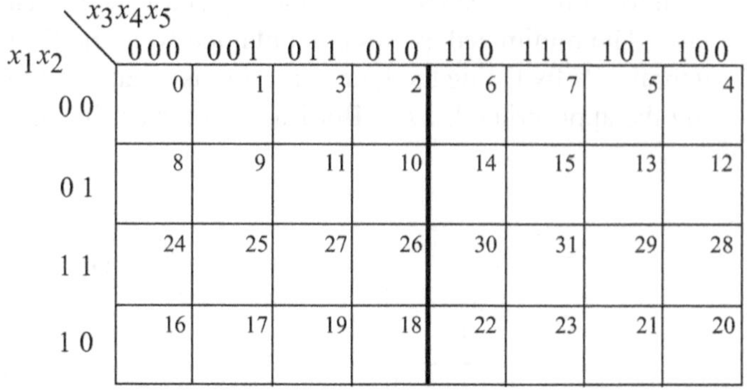

x_1x_2 \ $x_3x_4x_5$	000	001	011	010	110	111	101	100
00	0	1	3	2	6	7	5	4
01	8	9	11	10	14	15	13	12
11	24	25	27	26	30	31	29	28
10	16	17	19	18	22	23	21	20

(e)

Figure 2.1 (Continued)

Example 2.4 The following expression will be minimized using Boolean algebra:

$$x_1'x_2x_3'x_4' + x_1'x_2x_3x_4' + x_1x_2x_3'x_4' + x_1x_2x_3x_4'$$
$$= x_2x_4'(x_1'x_3' + x_1'x_3 + x_1x_3' + x_1x_3)$$
$$= x_2x_4'$$

Example 2.5 The following function will be minimized using a 4-variable Karnaugh map:

$$z_1(x_1, x_2, x_3, x_4) = x_2x_3' + x_2x_3x_4' + x_1x_2'x_3 + x_1x_3x_4'$$

The minimized result will be obtained in both a sum-of-products form and a product-of-sums form. To plot the function on the Karnaugh map, 1s are entered in the minterm locations that represent the product terms. For example, the term x_2x_3' is represented by the 1s in minterm locations 4, 5, 12, and 13. Only variables x_2 and x_3' are common to these four minterm locations. The term $x_2x_3x_4'$ is entered in minterm locations 6 and 14.

When the function has been plotted, a minimal set of prime implicants can be obtained that represents the function. The largest grouping of 1s should always be combined, where the number of 1s in a group is a power of 2. The grouping of 1s is shown in Figure 2.2 and the resulting equation in Equation 2.1 in a sum-of-products notation.

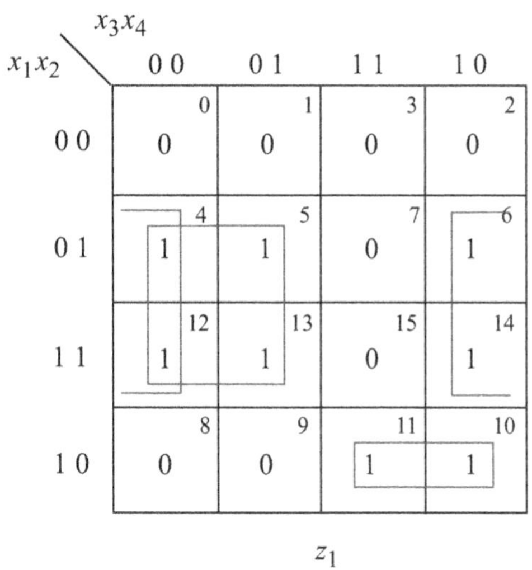

Figure 2.2 Karnaugh map representation for the function $z_1(x_1, x_2, x_3, x_4) = x_2x_3' + x_2x_3x_4' + x_1x_2'x_3 + x_1x_3x_4'$.

$$z_1(x_1,x_2,x_3,x_4) = x_2x_3' + x_2x_4' + x_1x_2'x_3 \tag{2.1}$$

The minimal product-of-sums expression can be obtained by combining the 0s in Figure 2.2 to form sum terms in the same manner as the 1s were combined to form product terms. However, since 0s are being combined, each sum term must equal 0. Thus, the four 0s in row $x_1x_2 = 00$ in Figure 2.2 combine to yield the sum term $(x_1 + x_2)$. In a similar manner, the remaining 0s are combined to yield the product-of-sums expression shown in Equation 2.2. When combining 0s to obtain sum terms, a variable value of 1 is treated as false and a variable value of 0 is treated as true. Thus, minterm locations 7 and 15 have variables $x_2x_3x_4 = 111$, providing a sum term of $(x_2' + x_3' + x_4')$.

$$z_1(x_1,x_2,x_3,x_4) = (x_1 + x_2)(x_2 + x_3)(x_2' + x_3' + x_4') \tag{2.2}$$

Equation 2.1 and Equation 2.2 both specify the conditions where z_1 is equal to 1. For example, consider the first term of Equation 2.1. If $x_2\,x_3 = 10$, then Equation 2.1 yields $z_1 = 1 + \cdots + 0$ which generates a value of 1 for z_1. Applying $x_2\,x_3 = 10$ to Equation 2.2 will cause every term to be equal to 1, such that, $z_1 = (1)\,(1)\,(1) = 1$.

Figure 2.1(d) illustrates a 5-variable Karnaugh map. To determine adjacency, the left map is superimposed on the right map. Any cells that are then physically adjacent are also logically adjacent and can be combined. Since x_5 is the low-order variable, the left map contains only even-numbered minterms; the right map is characterized by odd-numbered minterms. If 1s are entered in minterm locations 28, 29, 30, and 31, the four cells combine to yield the term $x_1 x_2 x_3$.

Figure 2.1(e) illustrates an alternative configuration for a Karnaugh map for five variables. The map hinges along the vertical centerline and folds like a book. Any squares that are then physically adjacent are also logically adjacent. For example, if 1s are entered in minterm locations 24, 25, 28, and 29, then the four squares combine to yield the term $x_1 x_2 x_4'$.

Some minterm locations in a Karnaugh map may contain unspecified entries which can be used as either 1s or 0s when minimizing the function. These "*don't care*" entries are indicated by a dash (–) in the map. A typical situation which includes "don't care" entries is a Karnaugh map used to represent the BCD numbers. This requires a 4-variable map in which minterm locations 10 through 15 contain unspecified entries, since digits 10 through 15 are invalid for BCD.

Map-entered variables Variables may also be entered in a Karnaugh map as map-entered variables, together with 1s and 0s. A map of this type is more compact than a standard Karnaugh map, but contains the same information. A map containing map-entered variables is particularly useful in analyzing and synthesizing synchronous sequential machines. When variables are entered in a Karnaugh map, two or more squares can be combined only if the squares are adjacent and contain the same variable(s).

Example 2.6 The following Boolean equation will be minimized using a 3-variable Karnaugh map with x_4 as a map-entered variable, as shown in Figure 2.3:

$$z_1(x_1, x_2, x_3, x_4) = x_1 x_2' x_3 x_4' + x_1 x_2 + x_1' x_2' x_3' x_4' + x_1' x_2' x_3' x_4$$

Note that instead of $2^4 = 16$ squares, the map of Figure 2.3 contains only $2^3 = 8$ squares, since only three variables are used in constructing the map. To facilitate plotting the equation in the map, the variable that is to be entered is shown in parenthesis as follows:

$$z_1(x_1, x_2, x_3, x_4) = x_1 x_2' x_3 (x_4') + x_1 x_2 + x_1' x_2' x_3' (x_4') + x_1' x_2' x_3' (x_4)$$

The first term in the equation for z_1 is $x_1 x_2' x_3$ (x_4') and indicates that the variable x_4' is entered in minterm location 5 $(x_1 x_2' x_3)$. The second term $x_1 x_2$ is plotted in the usual manner: 1s are entered in minterm locations 6 and 7. The third term specifies that the variable x_4' is entered in minterm location 0 $(x_1' x_2' x_3')$. The fourth term also applies to minterm 0, where x_4 is entered. The expression in minterm location 0, therefore, is $x_4' + x_4$.

$x_1 \backslash x_2x_3$	0 0	0 1	1 1	1 0
0	0 $x_4' + x_4$	1 0	3 0	2 0
1	4 0	5 x_4'	7 1	6 1

z_1

Figure 2.3 Karnaugh map for Example 2.6 using x_4 as a map-entered variable.

To obtain the minimized equation for z_1 in a sum-of-products form, 1s are combined in the usual manner; variables are combined only if the minterm locations containing the variables are adjacent and the variables are identical. Consider the expression $x_4' + x_4$ in minterm location 0. Since $x_4' + x_4 = 1$, minterm 0 equates to $x_1'x_2'x_3'$. The entry of 1 in minterm location 7 can be restated as $1 + x_4'$ without changing the value of the entry (Theorem 1). This allows minterm locations 5 and 7 to be combined as $x_1 x_3 x_4'$. Finally, minterms 6 and 7 combine to yield the term x_1x_2. The minimized equation for z_1 is shown in Equation 2.3.

$$z_1 = x_1'x_2'x_3' + x_1x_3x_4' + x_1x_2 \tag{2.3}$$

Example 2.7 The following Boolean equation will be minimized using x_4 and x_5 as map-entered variables:

$$\begin{aligned} z_1 = {} & x_1'x_2'x_3'(x_4x_5') + x_1'x_2 + x_1'x_2'x_3'(x_4x_5) + x_1x_2'x_3'(x_4x_5) \\ & + x_1x_2'x_3 + x_1x_2'x_3'(x_4') + x_1x_2'x_3'(x_5') \end{aligned}$$

Figure 2.4 shows the map entries for Example 2.7. The expression $x_4x_5' + x_4x_5$ in minterm location 0 reduces to x_4; the 1 entry in minterm location 2 can be expanded to $1 + x_4$ without changing the value in location 2. Therefore, locations 0 and 2 combine as $x_1'x_3'x_4$. The expression $x_4x_5 + x_4' + x_5'$ in minterm location 4 reduces to 1. Thus, the 1 entries in the map combine in the usual manner to yield Equation 2.4.

Karnaugh maps are ideally suited for 2-, 3-, 4-, or 5-variable Boolean functions. For six or more variables, a more systematic minimization technique is recommended — the Quine-McCluskey algorithm, which is presented in the next section.

x_1 \ x_2x_3	0 0	0 1	1 1	1 0
0	(0) $x_4x_5' + x_4x_5$	(1) 0	(3) 1	(2) 1
1	(4) $x_4x_5 + x_4' + x_5'$	(5) 1	(7) 0	(6) 0

z_1

Figure 2.4 Karnaugh map for Example 2.7 using x_4 and x_5 as map-entered variables.

$$z_1 = x_1'x_3'x_4 + x_1'x_2 + x_1x_2' \tag{2.4}$$

2.2.3 Quine-McCluskey Algorithm

The Quine-McCluskey algorithm is a tabular method of obtaining a minimal set of prime implicants that represents the Boolean function. Because the process is inherently algorithmic, the technique is easily implemented with a computer program. The method consists of two steps: first obtain a set of prime implicants for the function; then obtain a minimal set of prime implicants that represents the function.

The rationale for the Quine-McCluskey method relies on the repeated application of the distributive and complementation laws. For example, for a 4-variable function, minterms $x_1x_2x_3'x_4$ and $x_1x_2x_3'x_4'$ are adjacent because they differ by only one variable. The two minterms can be combined, therefore, into a single product term as follows:

$$\begin{aligned} &x_1x_2x_3'x_4 + x_1x_2x_3'x_4' \\ &= x_1x_2x_3'(x_4 + x_4') \\ &= x_1x_2x_3' \end{aligned}$$

The resulting product term is specified as $x_1x_2x_3x_4 = 110-$, where the dash (–) represents the variable that has been removed. The process repeats for all minterms in the function. Two product terms with dashes in the same position can be further combined into a single term if they differ by only one variable. Thus, the terms $x_1x_2x_3x_4 = 110-$ and $x_1x_2x_3x_4 = 100-$ combine to yield the term $x_1x_2x_3x_4 = 1-0-$, which corresponds to x_1x_3'.

The minterms are initially grouped according to the number of 1s in the binary representation of the minterm number. Comparison of minterms then occurs only between adjacent groups of minterms in which the number of 1s in each group differs by one. Minterms in adjacent groups that differ by only one variable can then be combined.

Example 2.8 The following function will be minimized using the Quine-McCluskey method: $z_1(x_1, x_2, x_3, x_4) = \Sigma_m(0,1,3,6,7,8,9,14)$. The first step is to list the minterms according to the number of 1s in the binary representation of the minterm number. Table 2.4 shows the listing of the various groups. Minterms that combine cannot be prime implicants; therefore, a check (✓) symbol is placed beside each minterm that combines with another minterm. When all lists in the table have been processed, the terms that have no check marks are prime implicants.

Table 2.4 Minterms Listed in Groups for Example 2.8

List 1			List 2			List 3		
Group	Minterms	$x_1x_2x_3x_4$	Group	Minterms	$x_1x_2x_3x_4$	Group	Minterms	$x_1x_2x_3x_4$
0	0	0 0 0 0 ✓	0	0,1	0 0 0 – ✓	0	0,1,8,9	– 0 0 –
				0,8	– 0 0 0 ✓			
1	1	0 0 0 1 ✓	1	1,3	0 0 – 1			
	8	1 0 0 0 ✓		1,9	– 0 0 1 ✓			
				8,9	1 0 0 – ✓			
2	3	0 0 1 1 ✓	2	3,7	0 – 1 1			
	6	0 1 1 0 ✓		6,7	0 1 1 –			
	9	1 0 0 1 ✓		6,14	– 1 1 0			
3	7	0 1 1 1 ✓						
	14	1 1 1 0 ✓						

Consider List 1 in Table 2.4. Minterm 0 differs by only one variable with each minterm in Group 1. Therefore, minterms 0 and 1 combine as 000–, as indicated in the first entry in List 2 and minterms 0 and 8 combine to yield –000, as shown in the second row of List 2. Next, compare minterms in List 1, Group 1 with those in List 1, Group 2. It is apparent that the following pairs of minterms combine because they differ by only one variable: (1,3), (1,9), and (8,9) as shown in List 2, Group 1. Minterms 1 and 3 are in adjacent groups and can combine because they differ by only one variable. The resulting term is 00–1. Minterms 1 and 6 cannot combine, because they differ by more than one variable. Minterms 1 and 9 combine as –001 and minterms 8 and 9 combine to yield 100–.

In a similar manner, minterms in the remaining groups are compared for possible adjacency. Note that those minterms that combine differ by a power of 2 in the decimal value of their minterm number. For example, minterms 6 and 14 combine as –110, because they differ by a power of 2 ($2^3 = 8$). Note also that the variable x_1 which is removed is located in column 2^3, where the binary weights of the four variables are $x_1x_2x_3x_4 = 2^3\ 2^2\ 2^1\ 2^0$.

List 3 is derived in a similar manner to that of List 2. However, only those terms that are in adjacent groups and have dashes in the same column can be compared. For example, the terms 0,1 (000–) and 8,9 (100–) both contain dashes in column x_4 and differ by only one variable. Thus, the two terms can combine into a single product term as $x_1x_2x_3x_4 = -00-$ ($x_2'x_3'$). If the dashes are in different columns, then the two terms do not represent product terms of the same variables and thus, cannot combine into a single product term.

When all comparisons have been completed, some terms will not combine with any other term. These terms are indicated by the absence of a check symbol and are designated as prime implicants. For example, the term $x_1x_2x_3x_4 = 00{-}1$ ($x_1'x_2'x_4$) in List 2 cannot combine with any term in either the previous group or the following group. Thus, $x_1'x_2'x_4$ is a prime implicant. The following terms represent prime implicants: $x_1'x_2'x_4$, $x_1'x_3x_4$, $x_1'x_2x_3$, $x_2x_3x_4'$ and $x_2'x_3'$.

Some of the prime implicants may be redundant, since the minterms covered by a prime implicant may also be covered by one or more other prime implicants. Therefore, the second step in the algorithm is to obtain a minimal set of prime implicants that covers the function. This is accomplished by means of a *prime implicant chart* as shown in Figure 2.5(a). Each column of the chart represents a minterm and each row of the chart represents a prime implicant. The first row of Figure 2.5(a) is specified by the minterm grouping of (1,3), which corresponds to the prime implicant $x_1'x_2'x_4$ (00–1). Since prime implicant $x_1'x_2'x_4$ covers minterms 1 and 3, an × is placed in columns 1 and 3 in the corresponding prime implicant row. The remaining rows are completed in a similar manner. Consider the last row which corresponds to prime implicant $x_2'x_3'$. Since prime implicant $x_2'x_3'$ covers minterms 0, 1, 8, and 9, an × is placed in the minterm columns 0, 1, 8, and 9.

A single × appearing in a column indicates that only one prime implicant covers the minterm. The prime implicant, therefore, is an *essential prime implicant*. In Figure 2.5(a), there are two essential prime implicants: $x_2x_3x_4'$ and $x_2'x_3'$. A horizontal line is drawn through all ×s in each essential prime implicant row. Since prime implicant $x_2x_3x_4'$ covers minterm 6, there is no need to have prime implicant $x_1'x_2x_3$ also cover minterm 6. Therefore, a vertical line is drawn through all ×s in column 6, as shown in Figure 2.5(a). For the same reason, a vertical line is drawn through all ×s in column 1 for the second essential prime implicant $x_2'x_3'$.

The only remaining minterms not covered by a prime implicant are minterms 3 and 7. Minterm 3 is covered by prime implicants $x_1'x_2'x_4$ and $x_1'x_3x_4$; minterm 7 is covered by prime implicants $x_1'x_3x_4$ and $x_1'x_2x_3$, as shown in Figure 2.5(b). Therefore, a minimal cover for minterms 3 and 7 consists of the *secondary essential prime implicant* $x_1'x_3x_4$. The complete minimal set of prime implicants for the function z_1 is shown in Equation 2.5. The minimized expression for z_1 can be verified by plotting the function on a Karnaugh map, as shown in Figure 2.6.

$$z_1(x_1,x_2,x_3,x_4) = x_2x_3x_4' + x_2'x_3' + x_1'x_3x_4 \qquad (2.5)$$

Prime implicants		Minterms 0	1	3	6	7	8	9	14
1,3	$(x_1'x_2'x_4)$		×	×					
3,7	$(x_1'x_3x_4)$			×		×			
6,7	$(x_1'x_2x_3)$				×	×			
* 6,14	$(x_2x_3x_4')$				×				⊗
* 0,1,8,9	$(x_2'x_3')$	⊗	×				×	×	

(a)

Prime implicants		Minterms 0	1	3	6	7	8	9	14
1,3	$(x_1'x_2'x_4)$			×					
3,7	$(x_1'x_3x_4)$			×		×			
6,7	$(x_1'x_2x_3)$					×			

(b)

Figure 2.5 Prime implicant chart for Example 2.8: (a) essential and nonessential prime implicants and (b) secondary essential prime implicants with minimal cover provided by prime implicant 3, 7 $(x_1'x_3x_4)$.

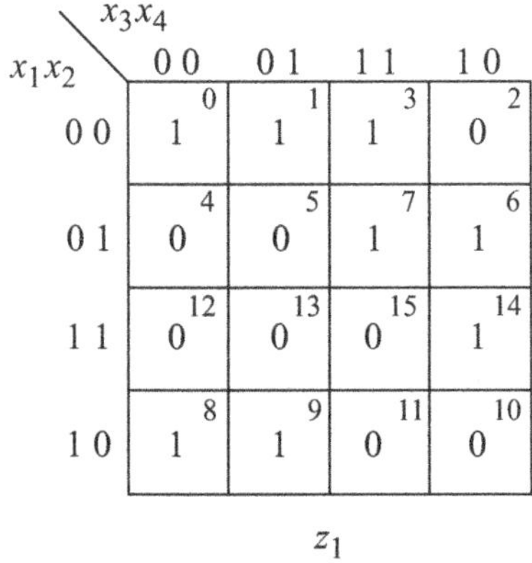

Figure 2.6 Karnaugh map for Example 2.8.

Functions which include unspecified entries ("don't cares") are handled in a similar manner. The tabular representation of step 1 lists all the minterms, including "don't cares." The "don't care" conditions are then utilized when comparing minterms in adjacent groups. In step 2 of the algorithm, only the minterms containing specified entries are listed — the "don't care" minterms are not used. Then the minimal set of prime implicants is found as described in Example 2.8.

Petrick algorithm The function may not always contain an essential prime implicant, or the secondary essential prime implicants may not be intuitively obvious, as they were in Example 2.8. The technique for obtaining a minimal cover of secondary prime implicants is called the *Petrick algorithm* and can best be illustrated by an example.

Example 2.9 Given the prime implicant chart of Figure 2.7 for function z_1, it is obvious that there are no essential prime implicants, since no minterm column contains a single ×. It is observed that minterm m_i is covered by prime implicants pi_1 or pi_2; m_j is covered by pi_1 or pi_3; m_k is covered by pi_2 or pi_4; m_l is covered by pi_2 or pi_3; and m_m is covered by pi_1 or pi_4. Since the function is covered only if all minterms are covered, Equation 2.6 represents this requirement.

Prime implicants	Minterms m_i	m_j	m_k	m_l	m_m
pi_1	×	×			×
pi_2	×		×	×	
pi_3		×		×	
pi_4			×		×

Figure 2.7 Prime implicant chart for Example 2.9.

$$\text{Function is covered} = (pi_1 + pi_2)\,(pi_1 + pi_3)\,(pi_2 + pi_4)\,(pi_2 + pi_3)\,(pi_1 + pi_4) \quad (2.6)$$

Equation 2.6 can be reduced by Boolean algebra or by a Karnaugh map to obtain a minimal set of prime implicants that represents the function. Figure 2.8 illustrates the Karnaugh map in which the sum terms of Equation 2.6 are plotted. The map is then used to obtain a minimized expression that represents the different combinations of prime implicants in which all minterms are covered. Equation 2.7 lists the product terms specified as prime implicants in a sum-of-products notation.

$pi_1 pi_2$ \ $pi_3 pi_4$	0 0	0 1	1 1	1 0
0 0	0	0	0	0
0 1	0	0	1	0
1 1	1	1	1	1
1 0	0	0	1	0

Figure 2.8 Karnaugh map in which the sum terms of Equation 2.6 are entered as 0s.

$$\text{Function is covered} = pi_1\, pi_2 + pi_2\, pi_3\, pi_4 + pi_1\, pi_3\, pi_4 \qquad (2.7)$$

The first term of Equation 2.7 represents the fewest number of prime implicants to cover the function. Thus, function z_1 will be completely specified by the expression $z_1 = pi_1\, pi_2$. From any covering equation, the term with the fewest number of variables is chosen to provide a minimal set of prime implicants. Assume, for example, that prime implicant $pi_1 = x_i x_j' x_k$ and that $pi_2 = x_l' x_m x_n$. Thus, the sum-of-products expression is $z_1 = x_i x_j' x_k + x_l' x_m x_n$.

2.3 Combinational Logic

Synthesis of combinational logic consists of translating a set of network specifications into minimized Boolean equations and then to generate a logic diagram from the equations using the logic primitives of AND, OR, and NOT. The equations are independent of any logic family and portray the functional operation of the network. The logic primitives can be realized by either AND gates, OR gates, and inverters, or by *functionally complete gates* such as, NAND or NOR gates.

Example 2.10 The equation shown below will be synthesized using NAND gates only.

$$z_1 = x_2 x_4' + x_1' x_3 x_4 + x_2 x_3 (x_1 \oplus x_4')$$

The equation can be expanded to a sum-of-products expression, as shown below, then designed using NAND gates, as shown in Figure 2.9.

$$z_1 = x_2x_4' + x_1'x_3x_4 + x_2x_3(x_1 \oplus x_4')$$
$$z_1 = x_2x_4' + x_1'x_3x_4 + x_1x_2x_3x_4 + x_1'x_2x_3x_4'$$

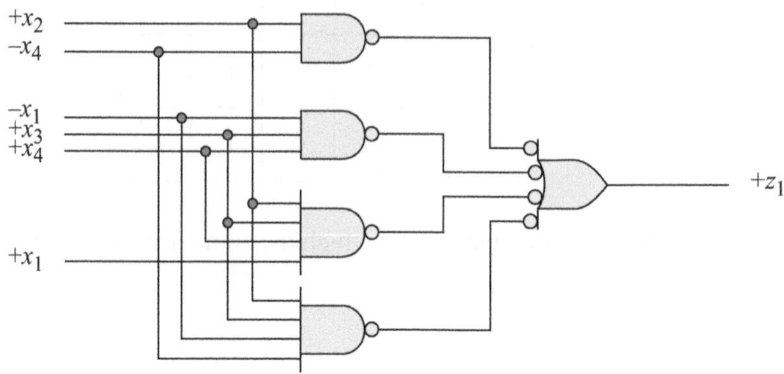

Figure 2.9 Logic diagram for Example 2.10.

There are different ways to indicate the active level (or assertion) of a signal. Table 2.5 lists various methods used by companies and textbooks. This book will use the $+x_1$ and $-x_1$ method. The AND function can be represented three ways, as shown in Figure 2.10, using an AND gate, a NAND gate, and a NOR gate. Although only two inputs are shown, both AND and OR circuits can have three or more inputs. The plus (+) and minus (–) symbols that are placed to the left of the variables indicate a high or low voltage level, respectively. This indicates the asserted (or active) voltage level for the variables; that is, the *logical 1* (or true) state, in contrast to the *logical 0* (or false) state.

Table 2.5 Assertion Levels

Active high assertion	$+x_1$	x_1	x_1(H)	x_1	x_1	x_1
Active low assertion	$-x_1$	$\neg x_1$	x_1(L)	$*x_1$	$\overline{x_1}$	x_1'

Thus, a signal can be asserted either plus or minus, depending upon the active condition of the signal at that point. For example, Figure 2.10(a) specifies that the AND function will be realized when both input x_1 and input x_2 are at their more positive potential, thus generating an output at its more positive potential. The word *positive* as used here does not necessarily mean a positive voltage level, but merely the more positive of two voltage levels. Therefore, the output of the AND gate of Figure 2.10(a) can be written as $+(x_1 \bullet x_2)$.

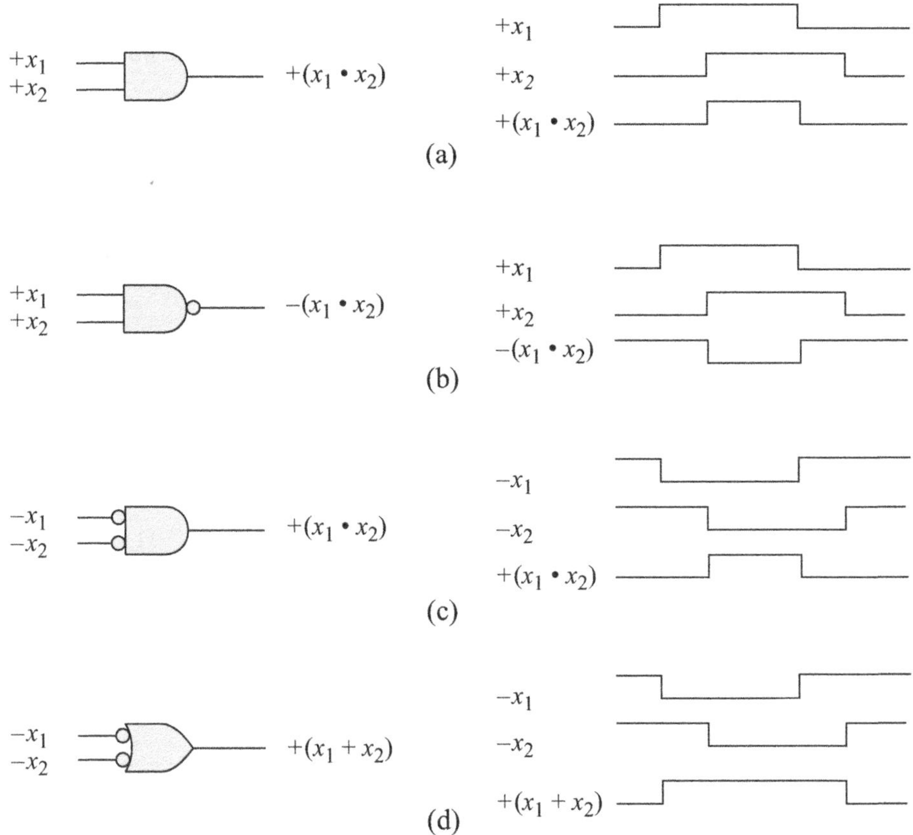

Figure 2.10 Logic symbols and waveforms for AND, NAND, NOR, and NAND (negative-input OR) gates.

To illustrate that a plus level does not necessarily mean a positive voltage level, consider two logic families: transistor-transistor logic (TTL) and emitter-coupled logic (ECL). The TTL family uses a +5 volt power supply. A plus level is approximately +3.5 volts and above; a minus level is approximately +0.2 volts. The ECL family uses a –5.2 volt power supply. A plus level is approximately –0.95 volts; a minus level is approximately –1.7 volts. Although –0.95 volts is a negative voltage, it is the more positive of the two ECL voltages.

The logic symbol of Figure 2.10(b) is a NAND gate in which inputs x_1 and x_2 must both be at their more positive potential for the output to be at its more negative potential. A small circle (or wedge symbol for IEEE Standard 91-1984 logic functions) at the input or output of a logic gate indicates a more negative potential. The output of the NAND gate can be written as $-(x_1 \bullet x_2)$.

Figure 2.10(c) illustrates a NOR gate used for the AND function. In this case, inputs x_1 and x_2 must be active (or asserted) at their more negative potential in order for the output to be at its more positive potential. Thus, the output can be written as $+(x_1 \bullet x_2)$. Figure 2.10(d) shows a NAND gate used for the OR function. Either input x_1 or x_2 (or both) must be at its more negative potential to assert the output at its more positive potential.

Example 2.11 The equation shown below will be synthesized using exclusive-OR gates only. The equation is represented in a sum-of-products form and will be minimized using Boolean algebra to obtain an equivalent equation using only the exclusive-OR function. The resulting logic diagram is shown in Figure 2.11.

$$\begin{aligned} z_1 = {} & x_1'x_2'x_3'x_4 + x_1'x_2'x_3x_4' + x_1'x_2x_3'x_4' \\ & + x_1'x_2x_3x_4 + x_1x_2x_3'x_4 + x_1x_2x_3x_4' \\ & + x_1x_2'x_3'x_4' + x_1x_2'x_3x_4 \end{aligned}$$

$$\begin{aligned} z_1 &= x_1'x_2'(x_3'x_4 + x_3x_4') + x_1'x_2(x_3'x_4' + x_3x_4) + x_1x_2(x_3'x_4 + x_3x_4') \\ &\quad + x_1x_2'(x_3'x_4' + x_3x_4) \\ &= x_1'x_2'(x_3 \oplus x_4) + x_1'x_2(x_3 \oplus x_4)' + x_1x_2(x_3 \oplus x_4) + x_1x_2'(x_3 \oplus x_4)' \\ &= (x_1 \oplus x_2)'(x_3 \oplus x_4) + (x_1 \oplus x_2)(x_3 \oplus x_4)' \\ &= x_1 \oplus x_2 \oplus x_3 \oplus x_4 \end{aligned}$$

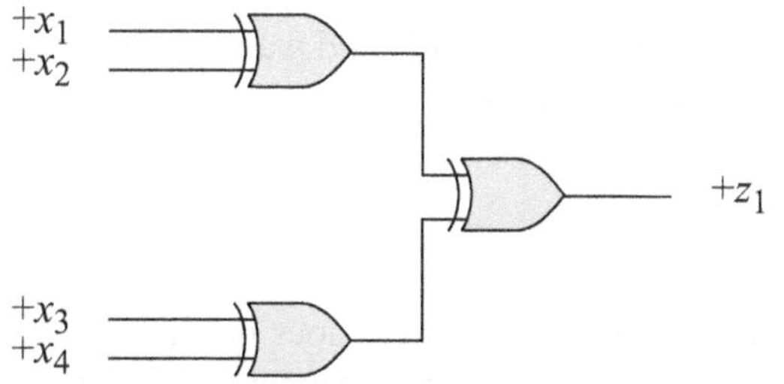

Figure 2.11 Logic diagram for Example 2.11.

2.3.1 Multiplexers

A multiplexer is a logic macro device that allows digital information from two or more data inputs to be directed to a single output. Data input selection is controlled by a set of select inputs that determine which data input is gated to the output. The select inputs are labeled $s_0, s_1, s_2, \cdots, s_i, \cdots, s_{n-1}$, where s_0 is the low-order select input with a binary weight of 2^0 and s_{n-1} is the high-order select input with a binary weight of 2^{n-1}. The data inputs are labeled $d_0, d_1, d_2, \cdots, d_j, \cdots, d_{2^n-1}$. Thus, if a multiplexer has n select inputs, then the number of data inputs will be 2^n and will be labeled d_0 through d_{2^n-1}. For example, if $n = 2$, then the multiplexer has two select inputs s_0 and s_1 and four data inputs d_0, d_1, d_2, and d_3.

Figure 2.12 shows four typical multiplexers drawn in the ANSI/IEEE Std. 91-1984 format. Consider the 4:1 multiplexer in Figure 2.12(b). If $s_1 s_0 = 00$, then data input d_0 is selected and its value is propagated to the multiplexer output z_1. Similarly, if $s_1 s_0 = 01$, then data input d_1 is selected and its value is directed to the multiplexer output. The equation that represents output z_1 in the 4:1 multiplexer of Figure 2.12(b) is shown in Equation 2.8. Output z_1 assumes the value of d_0 if $s_1 s_0 = 00$, as indicated by the term $s_1's_0'd_0$. Likewise, z_1 assumes the value of d_1 when $s_1 s_0 = 01$, as indicated by the term $s_1's_0d_1$.

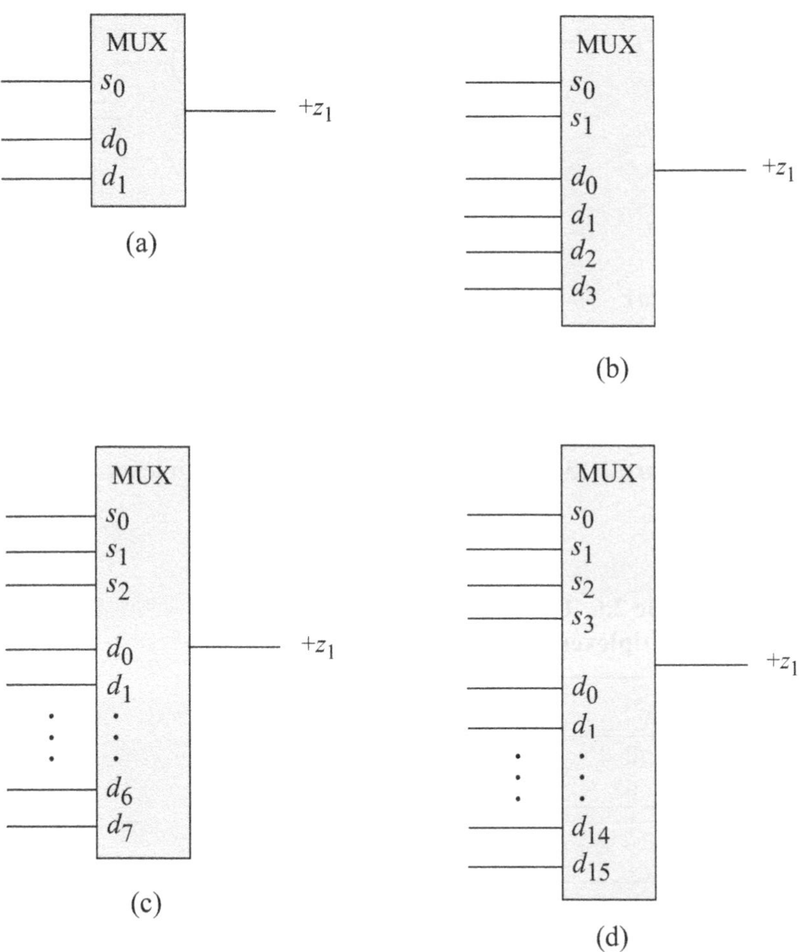

Figure 2.12 ANSI/IEEE Std. 91-1984 symbols for multiplexers: (a) 2:1 multiplexer; (b) 4:1 multiplexer; (c) 8:1 multiplexer; and (d) 16:1 multiplexer.

$$z_1 = s_1's_0'd_0 + s_1's_0d_1 + s_1s_0'd_2 + s_1s_0d_3 \tag{2.8}$$

The logic diagram for a 4:1 multiplexer is shown in Figure 2.13. There can also be an *enable* input which gates the selected data input to the output. Each of the four data inputs d_0, d_1, d_2, and d_3 is connected to a separate 3-input AND gate. The select inputs s_0 and s_1 are decoded to select a particular AND gate. The output of each AND gate is applied to a 4-input OR gate that provides the single output z_1. The truth table for the 4:1 multiplexer is shown in Table 2.6. Input lines that are not selected cannot be transferred to the output and are listed as "don't cares."

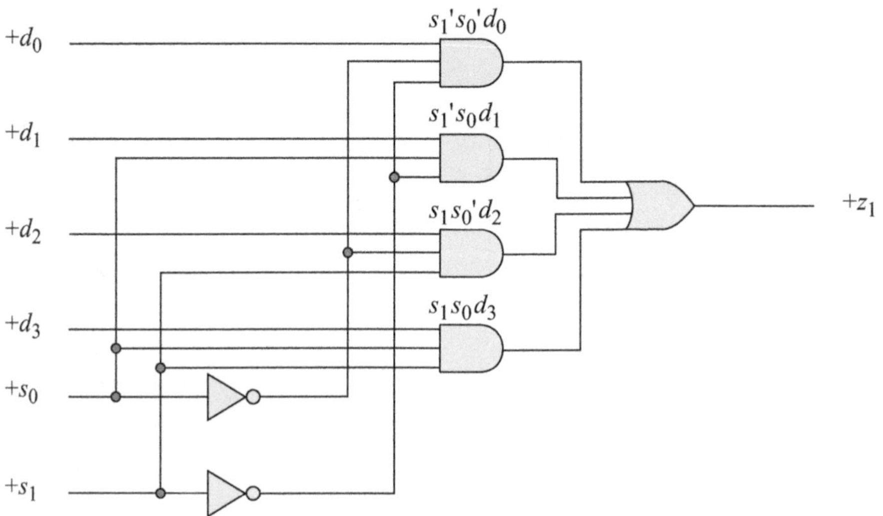

Figure 2.13 Logic diagram for a 4:1 multiplexer.

Table 2.6 Truth Table for the 4:1 Multiplexer of Figure 2.13

s_0	s_1	d_3	d_2	d_1	d_0	z_1
0	0	–	–	–	0	0
0	0	–	–	–	1	1
0	1	–	–	0	–	0
0	1	–	–	1	–	1
1	0	–	0	–	–	0
1	0	–	1	–	–	1
1	1	0	–	–	–	0
1	1	1	–	–	–	1

Example 2.12 There is a one-to-one correspondence between the data input numbers d_i of a multiplexer and the minterm locations in a Karnaugh map. For example, Figure 2.14 shows a Karnaugh map and a 4:1 multiplexer. Minterm location 0 corresponds to data input d_0 of the multiplexer; minterm location 1 corresponds to data input d_1; minterm location 2 corresponds to data input d_2; and minterm location 3 corresponds to data input d_3. The Karnaugh map and the multiplexer implement Equation 2.9, where x_2 is the low-order variable.

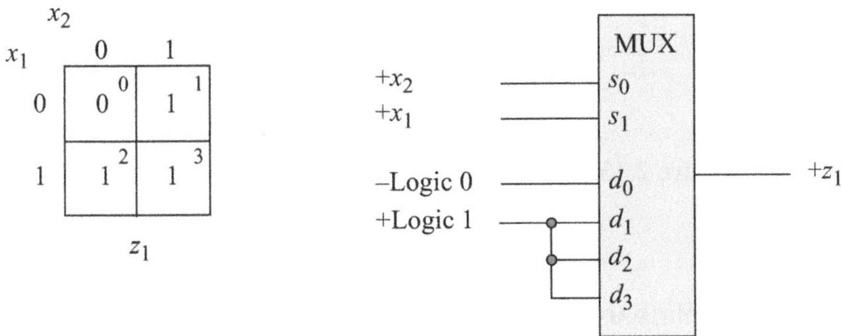

Figure 2.14 One-to-one correspondence between a Karnaugh map and a multiplexer.

$$
\begin{aligned}
z_1 &= x_1'x_2 + x_1x_2' + x_1x_2 \\
&= x_1 + x_2 \qquad (2.9)
\end{aligned}
$$

Linear-select multiplexers The multiplexer examples described thus far have been classified as *linear-select multiplexers*, because all of the variables of the Karnaugh map coordinates have been utilized as the select inputs for the multiplexer. Since there is a one-to-one correspondence between the minterms of a Karnaugh map and the data inputs of a multiplexer, designing the input logic is relatively straightforward: Simply assign the values of the minterms in the Karnaugh map to the corresponding multiplexer data inputs with the same subscript.

Example 2.13 Multiplexers can also be used with Karnaugh maps containing map-entered variables. Equation 2.10 is plotted on the Karnaugh map shown in Figure 2.15(a) using x_3 as a map-entered variable. Figure 2.15(b) shows the implementation using a 4:1 multiplexer.

$$z_1 = x_1x_2(x_3') + x_1x_2'(x_3) + x_1'x_2 \qquad (2.10)$$

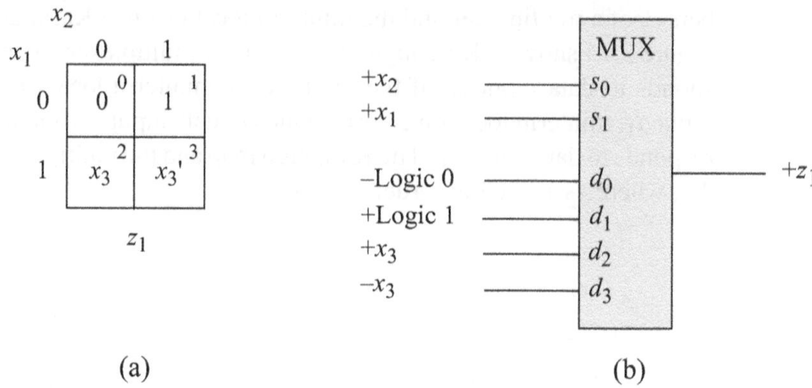

Figure 2.15 Multiplexer using a map-entered variable.

Nonlinear-select multiplexers In the previous subsection, the logic for the function was implemented with linear-select multiplexers. Although the logic functioned correctly according to the equations, the designs illustrated an inefficient use of the 2^p:1 multiplexers. Smaller multiplexers with fewer data inputs could be effectively utilized with a corresponding reduction in machine cost.

If the number of unique entries in a Karnaugh map satisfies the expression of Equation 2.11, where u is the number of unique entries and p is the number of select inputs, then at most a $(2^p \div 2)$:1 multiplexer will satisfy the requirements. This is referred to as a *nonlinear-select multiplexer.*

$$1 < u \geq (2^p \div 2) \tag{2.11}$$

If, however, $u > 2^p \div 2$, then a 2^p:1 multiplexer is necessary. The largest multiplexer with which to economically implement the logic is a 16:1 multiplexer, and then only if the number of distinct entries in the Karnaugh map warrants a multiplexer of this size. Other techniques, such as a programmable logic device (PLD) implementation, would make more efficient use of current technology.

If a multiplexer has unused data inputs — corresponding to unused states in the input map — then these unused inputs can be connected to logically adjacent multiplexer inputs. The resulting linked set of inputs can be addressed by a common select variable. Thus, in a 4:1 multiplexer, if data input $d_2 = 1$ and d_3 = "don't care," then d_2 and d_3 can both be connected to a logic 1. The two inputs can now be selected by $s_1 s_0 = 10$ or 11; that is, $s_1 s_0 = 1–$. Also, multiple multiplexers containing the same number of data inputs should be addressed by the same select input variables, if possible. This permits the utilization of noncustom technology, where multiplexers in the same integrated circuit share common select inputs.

Example 2.14 The Karnaugh map of Figure 2.16 can be implemented with a 4:1 nonlinear-select multiplexer for the function z_1. Variables x_2 and x_3 will connect to select inputs s_1 and s_0, respectively. When select inputs $s_1 s_0 = x_2 x_3 = 00$, data input d_0 is selected; therefore, $d_0 = 0$. When select inputs $s_1 s_0 = x_2 x_3 = 01$, data input d_1 is selected and contains the complement of x_1; therefore, $d_1 = x_1'$. When select inputs $s_1 s_0 = x_2 x_3 = 10$, data input d_2 is selected; therefore, $d_2 = 1$. When $s_1 s_0 = x_2 x_3 = 11$, data input d_3 is selected and contains the same value as x_1; therefore, $d_3 = x_1$. The logic diagram is shown in Figure 2.17.

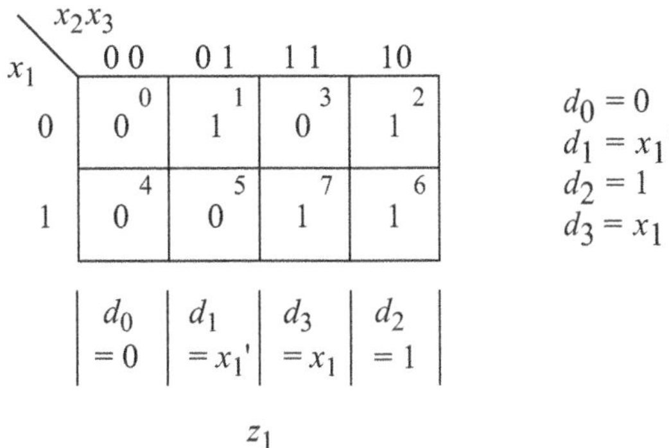

Figure 2.16 Karnaugh map for Example 2.14 which will be implemented by a 4:1 nonlinear-select multiplexer.

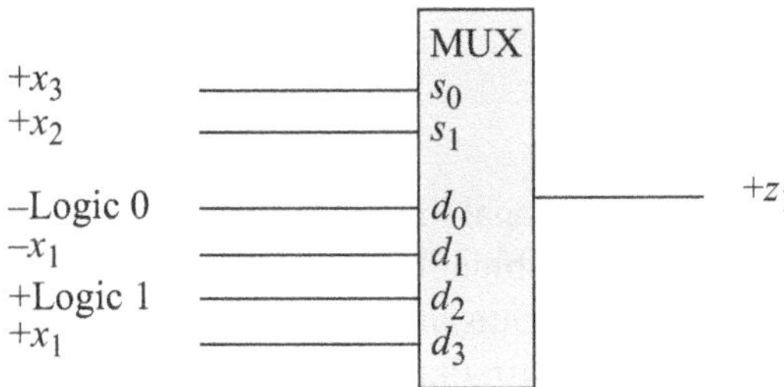

Figure 2.17 A 4:1 nonlinear-select multiplexer to implement the Karnaugh map of Figure 2.16.

Since there are two unique entries in the Karnaugh map of Figure 2.16, any permutation should produce similar results for output z_1; that is, no additional logic. Figure 2.18 shows one permutation in which the minterm locations are physically moved, but remain logically the same. The multiplexer configuration is shown in Figure 2.19.

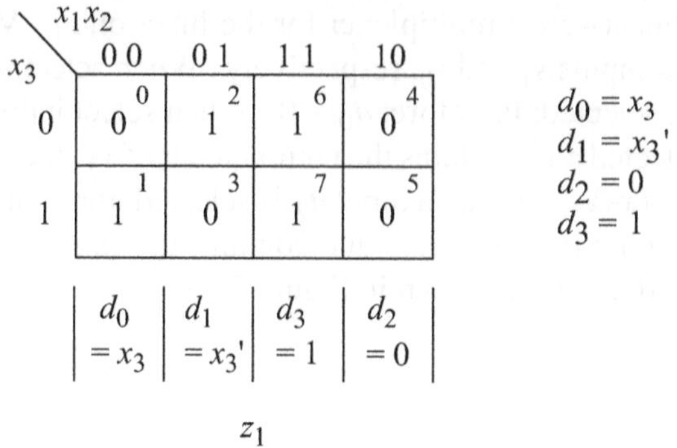

Figure 2.18 A permutation of the Karnaugh map of Figure 2.16.

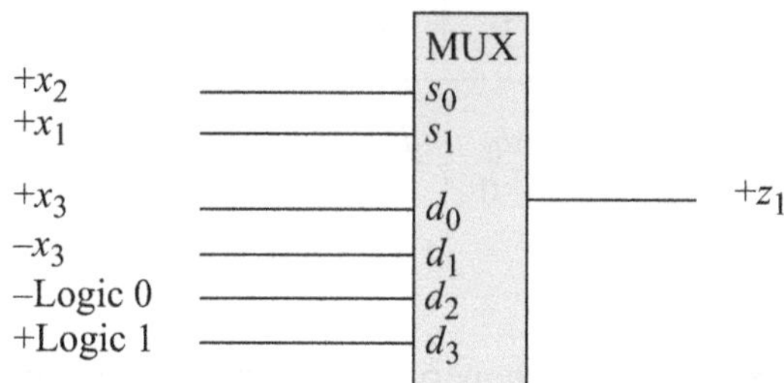

Figure 2.19 A 4:1 nonlinear-select multiplexer to implement the Karnaugh map of Figure 2.18.

2.3.2 Decoders

A decoder is a combinational logic macro that is characterized by the following property: For every valid combination of inputs, a unique output is generated. In general, a decoder has n binary inputs and m mutually exclusive outputs, where $2^n \geq m$. Each output represents a minterm that corresponds to the binary representation of the input vector. Thus, $z_i = m_i$, where m_i is the ith minterm of the n input variables. For example, if $n = 3$ and $x_1x_2x_3 = 101$, then output z_5 is asserted. A decoder with n inputs, therefore, has a maximum of 2^n outputs. Because the outputs are mutually exclusive, only one output is active for each different combination of the inputs. The decoder outputs may be asserted high or low. Decoders have many applications in digital engineering, ranging from instruction decoding to memory addressing to code conversion.

A 3:8 decoder is shown in Figure 2.20 which decodes a binary number into the corresponding octal number. The three inputs are x_1, x_2, and x_3 with binary weights of 2^2, 2^1, and 2^0, respectively. The decoder generates an output that corresponds to the decimal value of the binary inputs. For example, if $x_1x_2x_3 = 110$, then output z_6 is asserted high.

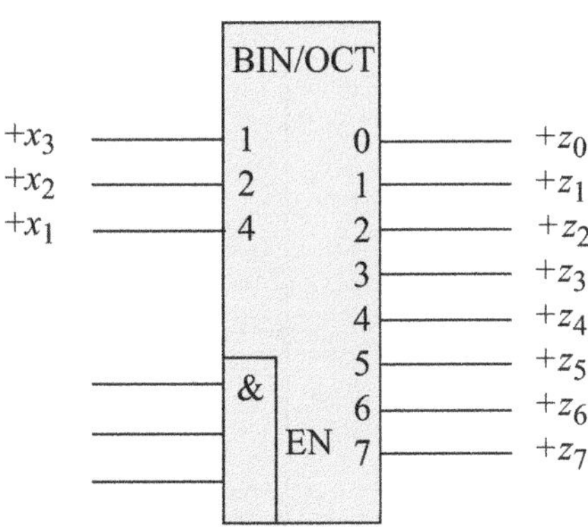

Figure 2.20 A binary-to-octal decoder.

A decoder may also have an enable function which allows the selected output to be asserted. The enable function may be a single input or an AND gate with two or more inputs. Figure 2.20 illustrates an enable input consisting of an AND gate with three inputs. If the enable function is deasserted, then the decoder outputs are deasserted. The 3:8 decoder generates all eight minterms z_0 through z_7 of three binary variables x_1, x_2, and x_3. The truth table for the decoder is shown in Table 2.7 and indicates the asserted output that represents the corresponding minterm.

Table 2.7 Truth Table for the 3:8 Decoder of Figure 2.20

$x_1 x_2 x_3$	z_0	z_1	z_2	z_3	z_4	z_5	z_6	z_7
0 0 0	**1**	0	0	0	0	0	0	0
0 0 1	0	**1**	0	0	0	0	0	0
0 1 0	0	0	**1**	0	0	0	0	0
0 1 1	0	0	0	**1**	0	0	0	0
1 0 0	0	0	0	0	**1**	0	0	0
1 0 1	0	0	0	0	0	**1**	0	0
1 1 0	0	0	0	0	0	0	**1**	0
1 1 1	0	0	0	0	0	0	0	**1**

The internal logic for the binary-to-octal decoder of Figure 2.20 is shown in Figure 2.21. The *enable* gate allows for additional logic functions to control the assertion of the active-high outputs.

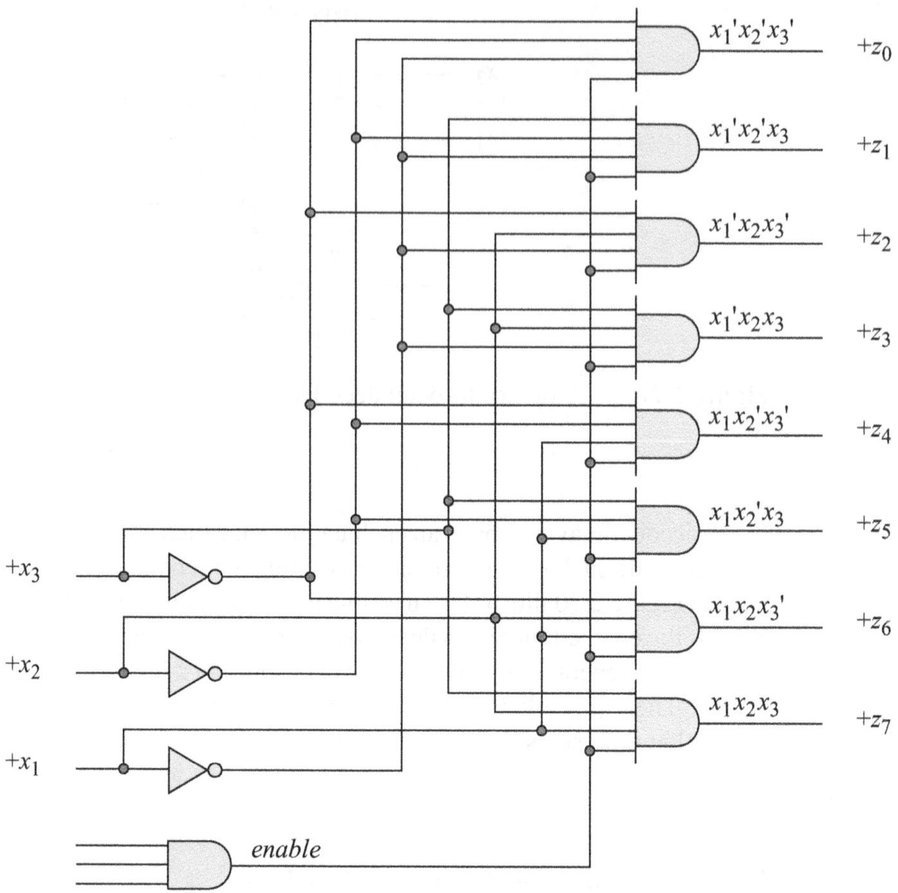

Figure 2.21 Internal logic for the binary-to-octal decoder of Figure 2.20.

Example 2.15 One decoder will be used to implement a Boolean function. The disjunctive normal form equation shown in Equation 2.12 will be synthesized with a 3:8 decoder and one OR gate, as shown in Figure 2.22 with active-low outputs. The terms in Equation 2.12 represent minterms m_4, m_6, m_5, and m_2, respectively. The equation can also be represented as the following sum of minterms expression: $\Sigma_m(2, 4, 5, 6)$. The outputs of the decoder correspond to the eight minterms associated with the three variables $x_1x_2x_3$. Therefore, Equation 2.12 is implemented by ORing decoder outputs 2, 4, 5, and 6.

$$z_1(x_1, x_2, x_3) = x_1x_2'x_3' + x_1x_2x_3' + x_1x_2'x_3 + x_1'x_2x_3' \qquad (2.12)$$

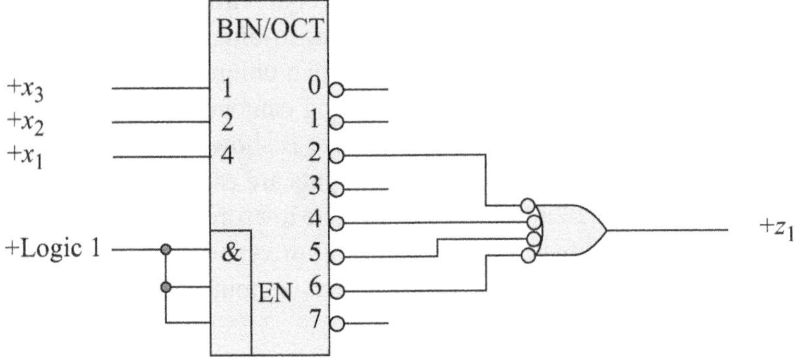

Figure 2.22 Implementation of Equation 2.12 using a 3:8 decoder.

2.3.3 Encoders

An encoder is a macro logic circuit with n mutually exclusive inputs and m binary outputs, where $n \leq 2^m$. The inputs are mutually exclusive to prevent errors from appearing on the outputs. The outputs generate a binary code that corresponds to the active input value. The function of an encoder can be considered to be the inverse of a decoder; that is, the mutually exclusive inputs are encoded into a corresponding binary number.

A general block diagram for an n:m encoder is shown in Figure 2.23. An encoder is also referred to as a code converter. In the label of Figure 2.23, *X* corresponds to the input code and *Y* corresponds to the output code. The general qualifying label *X/Y* is replaced by the input and output codes, respectively such as, *OCT/BIN* for an octal-to-binary code converter. Only one input x_i is asserted at a time. The decimal value of x_i is encoded as a binary number which is specified by the m outputs.

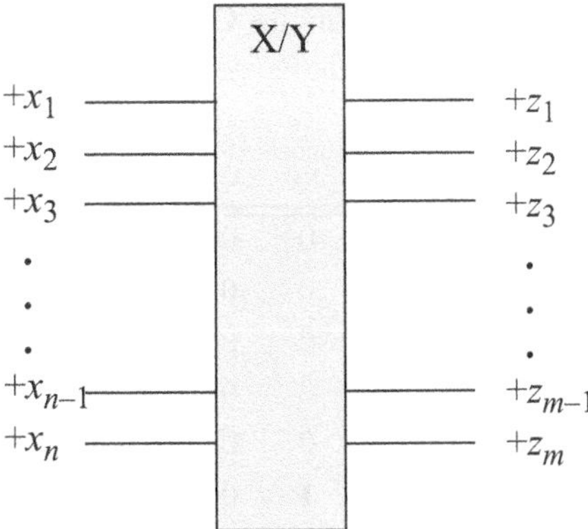

Figure 2.23 An $n : m$ encoder or code converter.

An 8:3 octal-to-binary encoder and a BCD-to-binary encoder are shown in Figure 2.24(a) and Figure 2.24(b), respectively. Although there are 2^8 possible input combinations of eight variables for the octal-to-binary encoder, only eight combinations are valid. The eight inputs each generate a unique octal code word in binary. If the outputs are to be enabled, then the gating can occur at the output gates.

The truth table for an 8:3 encoder is shown in Table 2.8. The encoder can be implemented with OR gates whose inputs are established from the truth table, as shown in Equation 2.13 and Figure 2.25. The low-order output z_3 is asserted when one of the following inputs is active: x_1, x_3, x_5, or x_7; output z_2 is asserted when one of the following inputs is active: x_2, x_3, x_6, or x_7; output z_1 is asserted when one of the following inputs is active: x_4, x_5, x_6, or x_7.

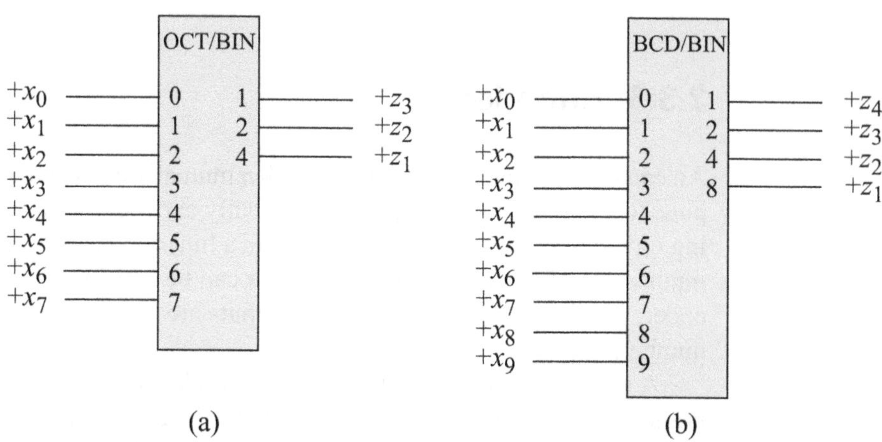

Figure 2.24 Encoders: (a) octal-to-binary and (b) BCD-to-binary.

Table 2.8 Truth Table for an Octal-To-Binary Encoder

Inputs								Outputs		
x_0	x_1	x_2	x_3	x_4	x_5	x_6	x_7	z_1	z_2	z_3
1	0	0	0	0	0	0	0	0	0	0
0	**1**	0	0	0	0	0	0	0	0	1
0	0	**1**	0	0	0	0	0	0	1	0
0	0	0	**1**	0	0	0	0	0	1	1
0	0	0	0	**1**	0	0	0	1	0	0
0	0	0	0	0	**1**	0	0	1	0	1
0	0	0	0	0	0	**1**	0	1	1	0
0	0	0	0	0	0	0	**1**	1	1	1

$$z_3 = x_1 + x_3 + x_5 + x_7$$

$$z_2 = x_2 + x_3 + x_6 + x_7$$

$$z_1 = x_4 + x_5 + x_6 + x_7 \qquad (2.13)$$

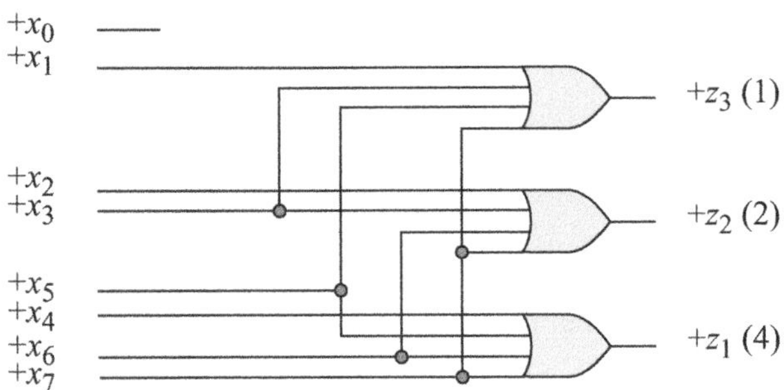

Figure 2.25 Logic diagram for an 8:3 encoder.

It was stated previously that encoder inputs are mutually exclusive. There may be situations, however, where more than one input can be active at a time. Then a priority must be established to select and encode a particular input. This is referred to as a *priority encoder*. Usually the input with the highest valued subscript is selected as highest priority for encoding. Thus, if x_i and x_j are active simultaneously and $i < j$, then x_j has priority over x_i. For example, assume that the octal-to-binary encoder of Figure 2.24(a) is a priority encoder. If inputs x_1, x_5, and x_7 are asserted simultaneously, then the outputs will indicate the binary equivalent of decimal 7 such that, $z_3 z_2 z_1 = 111$.

2.3.4 Comparators

A comparator is a logic macro circuit that compares the magnitude of two n-bit binary numbers X_1 and X_2. Therefore, there are $2n$ inputs and three outputs that indicate the relative magnitude of the two numbers. The outputs are mutually exclusive, specifying $X_1 < X_2$, $X_1 = X_2$, or $X_1 > X_2$. Figure 2.26 shows a general block diagram of a comparator.

The design of a comparator is relatively straightforward. Consider two 3-bit unsigned operands $X_1 = x_{11} x_{12} x_{13}$ and $X_2 = x_{21} x_{22} x_{23}$, where x_{13} and x_{23} are the low-order bits of X_1 and X_2, respectively. Three equations will now be derived to represent the three outputs; one equation each for $X_1 < X_2$, $X_1 = X_2$, and $X_1 > X_2$. Comparison occurs in a left-to-right manner beginning at the high-order bits of the two operands.

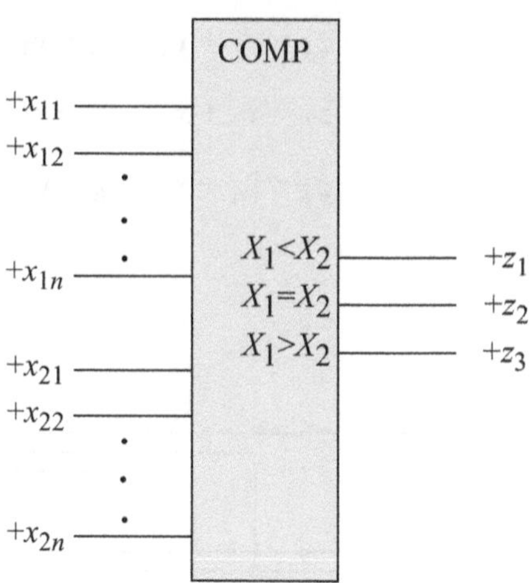

Figure 2.26 General block diagram of a comparator.

Operand X_1 will be less than X_2 if $x_{11}\,x_{21} = 01$. Thus, X_1 cannot be more than 011 while X_2 cannot be less than 100, indicating that $X_1 < X_2$. Therefore, the first term of the equation for $X_1 < X_2$ is $x_{11}'x_{21}$. If, however, $x_{11} = x_{21}$, then the relative magnitude depends on the values of x_{12} and x_{22}. The equality of two bits is represented by the exclusive-NOR function, also called the *equality function*. Thus, the second term in the equation for $X_1 < X_2$ is $(x_{11} \oplus x_{21})'x_{12}'x_{22}$. The analysis continues in a similar manner for the remaining bits of the two operands.

The equation for $X_1 < X_2$ is shown in Equation 2.14. The equality of X_1 and X_2 is true if and only if each bit-pair is equal, where $x_{11} = x_{21}, x_{12} = x_{22}$, and $x_{13} = x_{23}$; that is, $X_1 = X_2$ if and only if $x_{1i} = x_{2i}$ for $i = 1, 2, 3$. This is indicated by the Boolean product of three equality functions as shown in Equation 2.14 for $X_1 = X_2$. The final equation, which specifies $X_1 > X_2$, is obtained in a manner analogous to that for $X_1 < X_2$. If the high-order bits are $x_{11}x_{21} = 10$, then it is immediately apparent that $X_1 > X_2$. Using the equality function with the remaining bits yields the equation for $X_1 > X_2$ as shown in Equation 2.14. The design process is modular and can be extended to accommodate any size operands in a well-defined regularity. Two n-bit operands will contain column subscripts of 11 through $1n$ and 21 through $2n$, where n specifies the low-order bits.

$$(X_1 < X_2) = x_{11}'x_{21} + (x_{11} \oplus x_{21})'x_{12}'x_{22} + (x_{11} \oplus x_{21})'(x_{12} \oplus x_{22})'x_{13}'x_{23}$$

$$(X_1 = X_2) = (x_{11} \oplus x_{21})'(x_{12} \oplus x_{22})'(x_{13} \oplus x_{23})'$$

$$(X_1 > X_2) = x_{11}x_{21}' + (x_{11} \oplus x_{21})'x_{12}x_{22}' + (x_{11} \oplus x_{21})'(x_{12} \oplus x_{22})'x_{13}x_{23}' \qquad (2.14)$$

2.4 Sequential Logic

This section will briefly review the operating characteristics of the *SR* latch, the *D* flip-flop, and the *JK* flip-flop, then proceed to synchronous counters, and then to the definition and synthesis of Moore and Mealy sequential machines. A latch is a level-sensitive storage element in which a change to an input signal affects the output directly without recourse to a clock input. The set (s) and reset (r) inputs may be active high or active low. The *D* flip-flop and the *JK* flip-flop, however, are triggered on the application of a clock signal and are positive- or negative-edge-triggered devices.

SR latch The *SR* latch is usually implemented using either NAND gates or NOR gates, as shown in Figure 2.27(a) and Figure 2.27(b), respectively. When a negative pulse (or level) is applied to the *–set* input of the NAND gate latch, the output $+y_1$ becomes active at a high voltage level. This high level is also connected to the input of NAND gate 2. Since the set and reset inputs cannot both be active simultaneously, the *–reset* input is at a high level, providing a low voltage level on the output of gate 2 which is fed back to the input of gate 1. The negative feedback, therefore, provides a second set input to the latch. The original set pulse can now be removed and the latch will remain set. Concurrent set and reset inputs represent an invalid condition, since both outputs will be at the same voltage level; that is, outputs $+y_1$ and $-y_1$ will both be at the more positive voltage level — an invalid state for a bistable device with complementary outputs.

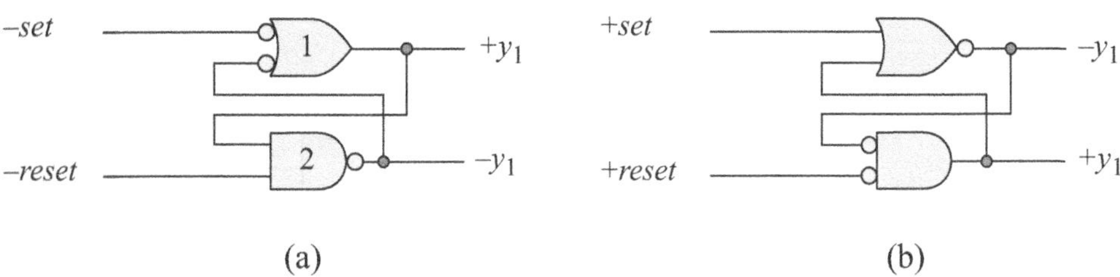

Figure 2.27 *SR* latches: (a) using NAND gates and (b) using NOR gates.

If the NAND gate latch is set, then a low voltage level on the *–reset* input will cause the output of gate 2 to change to a high level which is fed back to gate 1. Since both inputs to gate 1 are now at a high level, the $+y_1$ and $-y_1$ outputs will change to a low and high level, respectively, which is the reset state for the latch. The excitation equation is shown in Equation 2.15 where $Y_{j(t)}$ and $Y_{k(t+1)}$ are the present state and next state of the latch, respectively.

$$Y_{k(t+1)} = S + R' \, Y_{j(t)} \tag{2.15}$$

D flip-flop A D flip-flop is an edge-triggered device with one data input and one clock input. Figure 2.28 illustrates a positive-edge-triggered D flip-flop. The $+y_1$ output will assume the state of the D input at the next positive clock transition. After the occurrence of the clock's positive edge, any change to the D input will not affect the output until the next active clock transition. The excitation equation is shown in Equation 2.16.

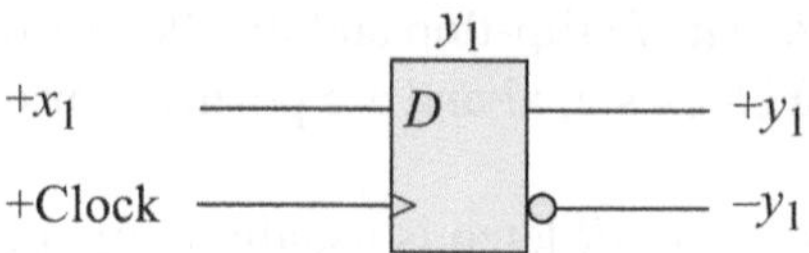

Figure 2.28 A positive-edge-triggered D flip-flop.

$$Y_{k(t+1)} = D \tag{2.16}$$

JK flip-flop The JK flip-flop is also an edge-triggered storage device. The active clock transition can be either the positive or negative edge. Figure 2.29 illustrates a negative-edge-triggered JK flip-flop. The functional characteristics of the JK data inputs are defined in Table 2.9. Table 2.10 shows an excitation table in which a particular state transition predicates a set of values for J and K. This table is especially useful in the synthesis of synchronous sequential machines.

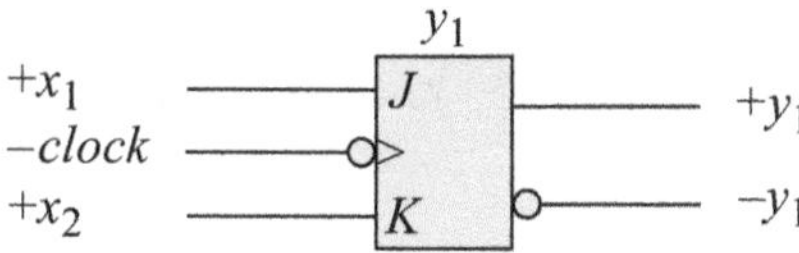

Figure 2.29 A negative-edge-triggered JK flip-flop.

Table 2.9 *JK* Functional Characteristic Table

JK	Function
0 0	No change
0 1	Reset
1 0	Set
1 1	Toggle

Table 2.10 Excitation Table for a *JK* Flip-Flop

Present state $Y_{j(t)}$	Next state $Y_{k(t+1)}$	Data inputs JK
0	0	0 –
0	1	1 –
1	0	– 1
1	1	– 0

The excitation equation for a *JK* flip-flop is derived from Table 2.10 and is shown in Equation 2.17.

$$Y_{k(t+1)} = Y_{j(t)}'J + Y_{j(t)}K' \tag{2.17}$$

2.4.1 Counters

Counters are one of the simplest types of sequential machines, requiring only one input in most cases. The single input is a clock pulse. A *counter* is constructed from one or more flip-flops that change state in a prescribed sequence upon the application of a series of clock pulses. The sequence of states in a counter may generate a binary count, a binary-coded decimal (BCD) count, or any other counting sequence. The counting sequence does not have to be sequential.

Counters are used for counting the number of occurrences of an event and for general timing sequences. A block diagram of a synchronous counter is shown in Figure 2.30. The diagram depicts a typical counter consisting of combinational input logic for the δ next-state function, storage elements, and combinational output logic for the λ output function. Input logic is required when an initial count must be loaded into the counter. The input logic then differentiates between a clock pulse that is used for loading and a clock pulse that is used for counting. Not all counters are implemented with input and output logic, however. Some counters contain only storage elements that are connected in cascade.

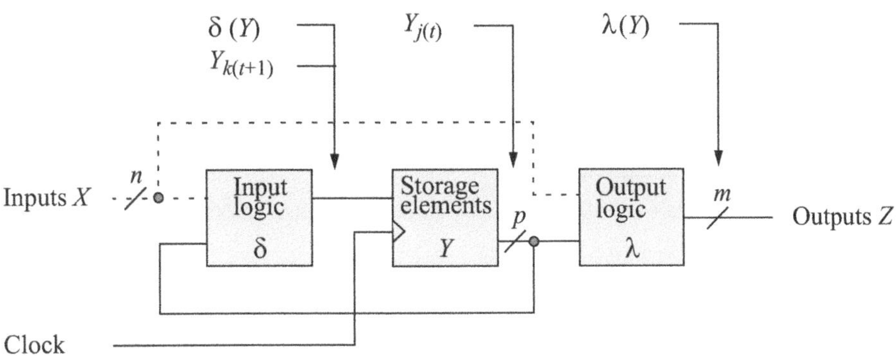

Figure 2.30 Counter block diagram.

Counters can be designed as count-up counters, in which the counting sequence increases numerically, or as count-down counters, in which the counting sequence decreases numerically. A counter may also be designed as both a count-up counter and a count-down counter, the mode of operation being controlled by a separate input.

Any counting sequence is a valid sequence for a counter, depending upon the application. Binary counters are the most common, and increment or decrement in a binary sequence, such as, 0000, 0001, 0010, 0011, . . . , 1110, 1111, 0000, . . . , which represents a modulo-16 binary counter, where the terminal count is followed by the initial count.

The counting sequence for a BCD counter is modulo-10: 0000, 0001, 0010, 0011, . . . , 1000, 1001, 0000, ... There are also counters that are classified as Gray code counters or as Johnson counters. A *Gray code* counter has the unique characteristic where only one stage of the counter changes state with each clock pulse, thus providing a glitch-free counting sequence. A *Johnson counter* is designed by connecting the complement of the output of the final storage element to the input of the first storage element. This feedback connection generates the singular sequence of states shown in Table 2.11 for four bits, where y_4 is the low-order bit. A 4-bit Johnson counter produces a total of eight states; a 5-bit counter yields ten states. In general, an n-stage Johnson counter will generate $2n$ states, where n is the number of storage elements in the counter.

Table 2.11 Four-Bit Johnson Counter

y_1	y_2	y_3	y_4
0	0	0	0
1	0	0	0
1	1	0	0
1	1	1	0
1	1	1	1
0	1	1	1
0	0	1	1
0	0	0	1
0	0	0	0

Modulo-10 count-up counter The simplest counter is the *binary counter* which counts in an increasing binary sequence from $y_1y_2 \ldots y_p = 00 \ldots 0$ to $y_1y_2 \ldots y_p = 11 \ldots 1$, then returns to zero. This section will design a modulo-10 counter using D flip-flops. The counting sequence is: $y_1y_2y_3y_4 = 0000$, 0001, 0010, 0011, 0100, 0101, 0110, 0111, 1000, 1001, 0000, ...

The input maps represent the δ next-state function from which the equations are generated for the data input logic of the flip-flops. Since D flip-flops are used in the design, the derivation of the input maps is relatively straightforward and can be generated directly from the next-state table shown in Table 2.12. The input maps can also be generated from a state diagram, which is a graphical representation of the counting sequence. The input maps for the modulo-10 counter are shown in Figure 2.31.

Table 2.12 Next-State Table for a Modulo-10 Counter

Present State $y_1y_2y_3y_4$	Next State $y_1y_2y_3y_4$	Flip-Flop Inputs Dy_1	Dy_2	Dy_3	Dy_4
0 0 0 0	0 0 0 1	0	0	0	1
0 0 0 1	0 0 1 0	0	0	1	0
0 0 1 0	0 0 1 1	0	0	1	1
0 0 1 1	0 1 0 0	0	1	0	0
0 1 0 0	0 1 0 1	0	1	0	1
0 1 0 1	0 1 1 0	0	1	1	0
0 1 1 0	0 1 1 1	0	1	1	1
0 1 1 1	1 0 0 0	1	0	0	0
1 0 0 0	1 0 0 1	1	0	0	1
1 0 0 1	0 0 0 0	0	0	0	0

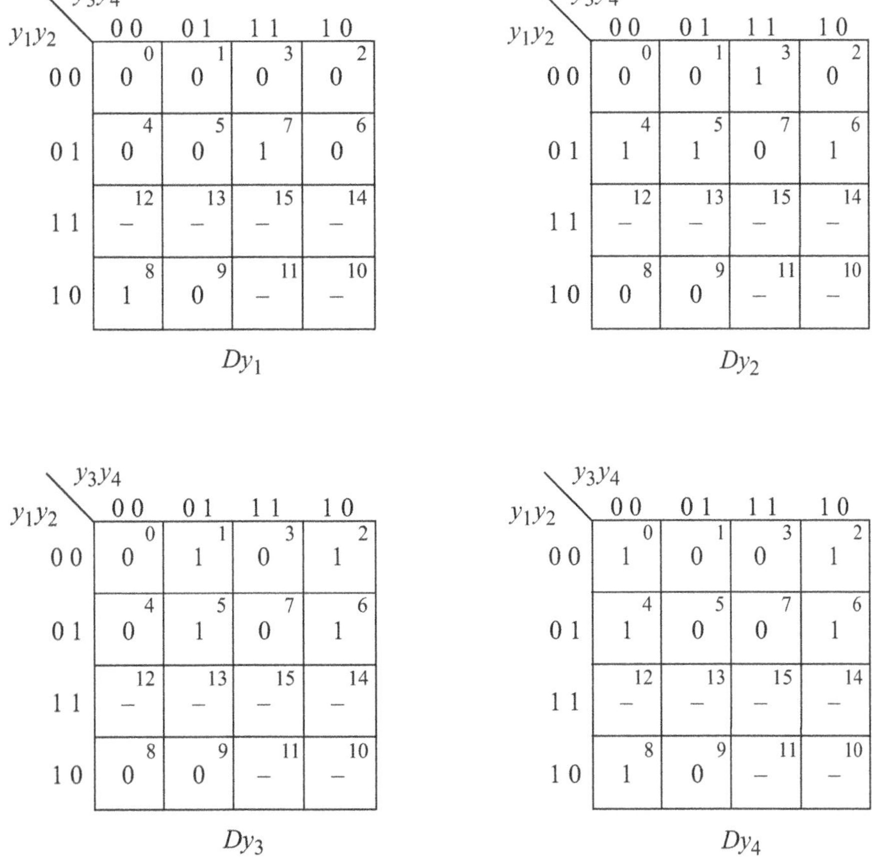

Figure 2.31 Input maps for the modulo-10 counter of Table 2.12.

Since this is a modulo-10 counter, minterm locations 10, 11, 12, 13, 14, and 15 are invalid and treated as "don't cares." The equations for the D inputs of flip-flops y_1, y_2, y_3, and y_4 are obtained directly from the input maps and are shown in Equation 2.18. Note that in Table 2.12, the next state for flip-flop y_4 always toggles; therefore, the input for flip-flop y_4 can be implemented by connecting the complemented output to the D input. The logic diagram is shown in Figure 2.32.

$$
\begin{aligned}
Dy_1 &= y_2 y_3 y_4 + y_1 y_4' \\
Dy_2 &= y_2' y_3 y_4 + y_2 y_3' + y_2 y_4' \\
Dy_3 &= y_3 y_4' + y_1' y_3' y_4 \\
Dy_4 &= y_4'
\end{aligned}
\tag{2.18}
$$

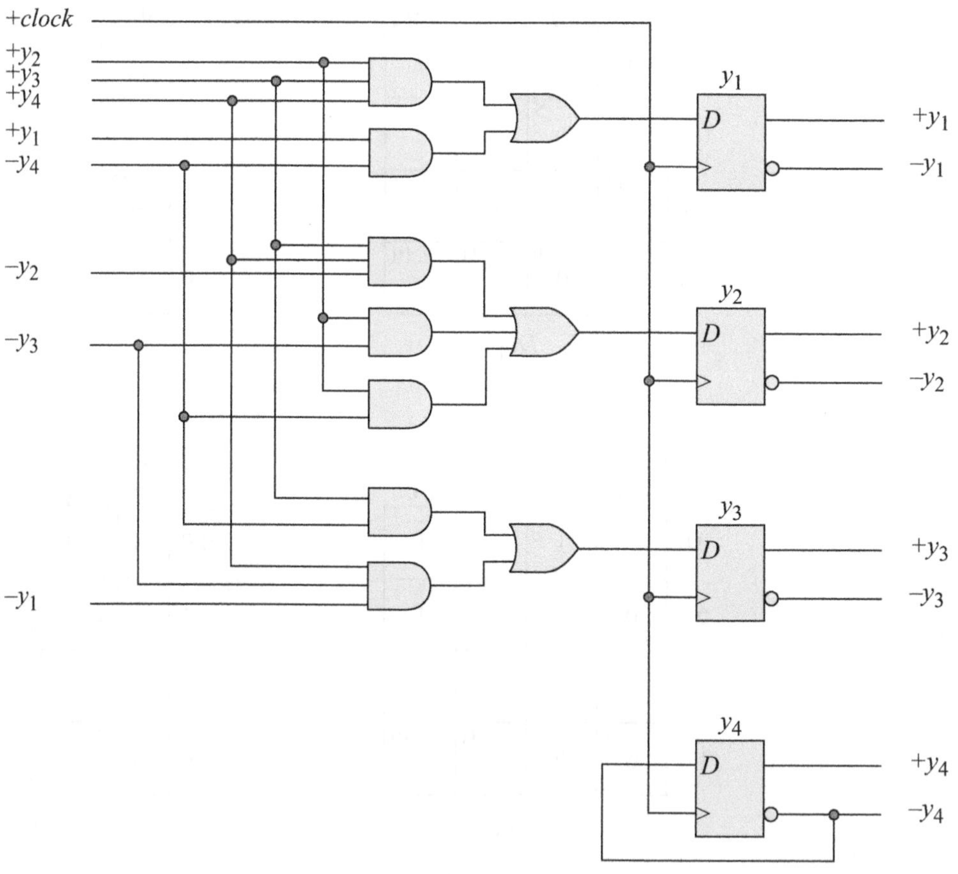

Figure 2.32 Logic diagram for the modulo-10 counter.

Modulo-8 count-down counter It is sometimes desirable to have a counter that counts down to zero from a predetermined initial count. A counter of this type will be used in the section describing a sequential add-shift multiplier. A modulo-8 counter will now be designed using *JK* flip-flops that counts down to zero from an initial count of $y_1y_2y_3 = 111$. The next-state table is shown in Table 2.13 and the *JK* excitation table is shown in Table 2.14.

The values for *J* and *K* in the next-state table are obtained by applying the state transition sequences of Table 2.14 to the present state and next state values for each flip-flop in the next-state table. For example, in the first row of Table 2.13, flip-flop y_1 sequences from $y_1 = 1$ to $y_1 = 1$, providing *JK* values of $Jy_1 Ky_1 = -0$.

The Karnaugh maps are shown in Figure 2.33 and are obtained directly from the next-state table. The logic diagram is shown in Figure 2.34. It is assumed that the counter is initially set to $y_1y_2y_3 = 111$, where y_3 is the low-order flip-flop. The next-state table indicates that flip-flop y_3 toggles with each active clock transition; therefore, $Jy_3\ Ky_3 = 11$, which toggles flip-flop y_3. The equations for the *JK* inputs are shown in Equation 2.19.

Table 2.13 Next-State Table for the Modulo-8 Count-Down Counter

Present State $y_1\ y_2\ y_3$	Next State $y_1\ y_2\ y_3$	Jy_1	Ky_1	Jy_2	Ky_2	Jy_3	Ky_3
1 1 1	1 1 0	–	0	–	0	–	1
1 1 0	1 0 1	–	0	–	1	1	–
1 0 1	1 0 0	–	0	0	–	–	1
1 0 0	0 1 1	–	1	1	–	1	–
0 1 1	0 1 0	0	–	–	0	–	1
0 1 0	0 0 1	0	–	–	1	1	–
0 0 1	0 0 0	0	–	0	–	–	1
0 0 0	1 1 1	1	–	1	–	1	–

Table 2.14 Excitation Table for a *JK* Flip-Flop

Present state $Y_{j(t)}$	Next state $Y_{k(t+1)}$	Data inputs *J K*
0	0	0 –
0	1	1 –
1	0	– 1
1	1	– 0

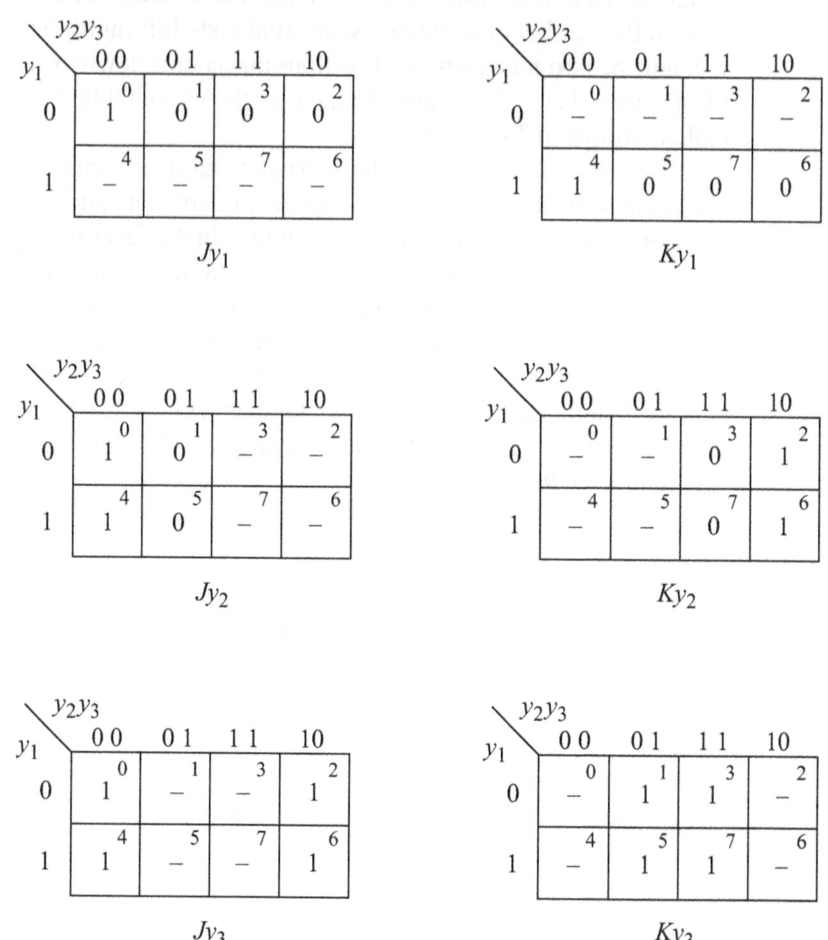

Figure 2.33 Karnaugh maps for the modulo-8 count-down counter.

$$Jy_1 = y_2'y_3'$$

$$Ky_1 = y_2'y_3'$$

$$Jy_2 = y_3'$$

$$Ky_2 = y_3'$$

$$Jy_3 = \text{Logic } 1$$

$$Ky_3 = \text{Logic } 1 \quad (2.19)$$

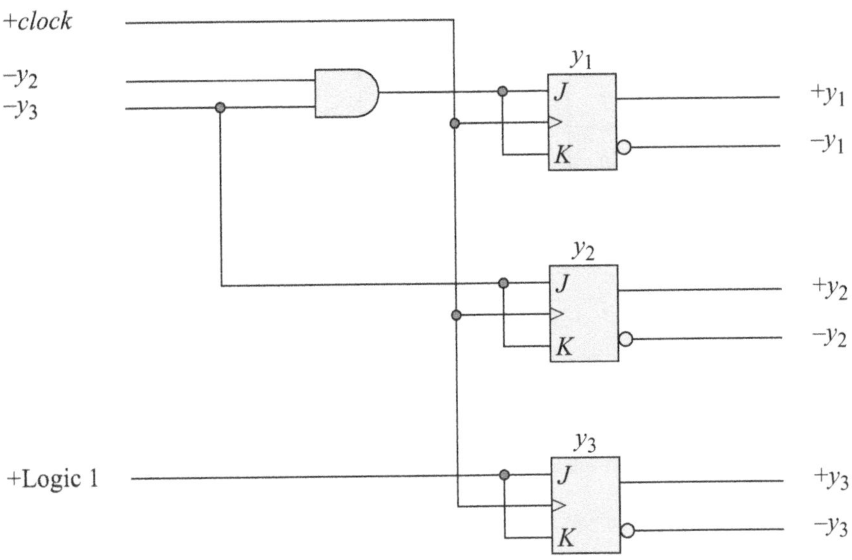

Figure 2.34 Logic diagram for the modulo-8 count-down counter.

Nonsequential counter A series of nonoverlapping, disjoint pulses can be generated by means of a counter that counts in a nonsequential sequence. A counter of this type is useful in a digital system when it is desired to have separate states for certain functions. For example, a peripheral control unit containing an embedded microprocessor may require the following three states:

State 1: Communicate with the input/output channel of a computer to send or receive data.

State 2: Perform internal processing.

State 3: Communicate with the input/output device to send or receive data.

The counting sequence for a counter of this type is as follows: $y_1y_2y_3 = 100, 010, 001$ in which each 1 bit represents a unique state. This sequence can easily be extended to any number of nonoverlapping pulses. The Karnaugh maps for a counter that generates three disjoint pulses are shown in Figure 2.35 and the equations are shown in Equation 2.20. The logic diagram, using *D* flip-flops, is shown in Figure 2.36. It is assumed that the counter is initialized to $y_1y_2y_3 = 100$.

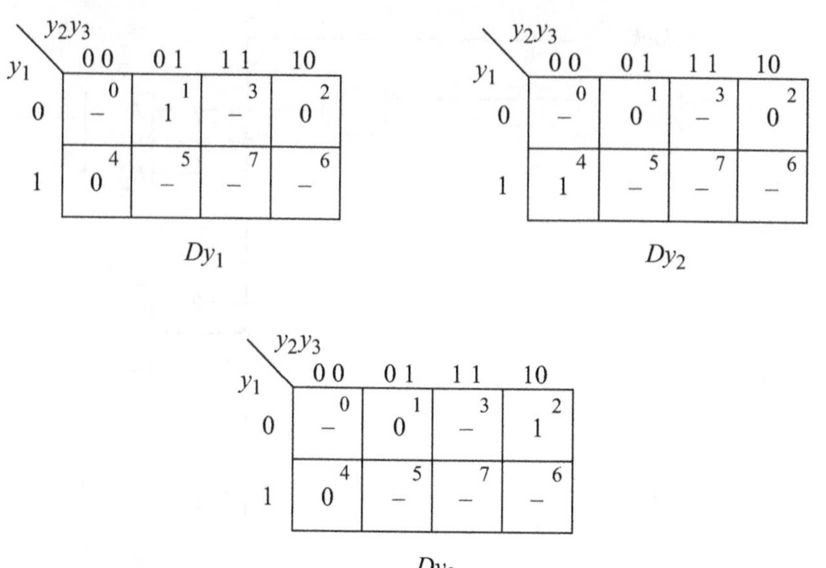

Figure 2.35 Karnaugh maps for the nonsequential counter.

$$
\begin{aligned}
Dy_1 &= y_3 \\
Dy_2 &= y_1 \\
Dy_3 &= y_2
\end{aligned}
\quad (2.20)
$$

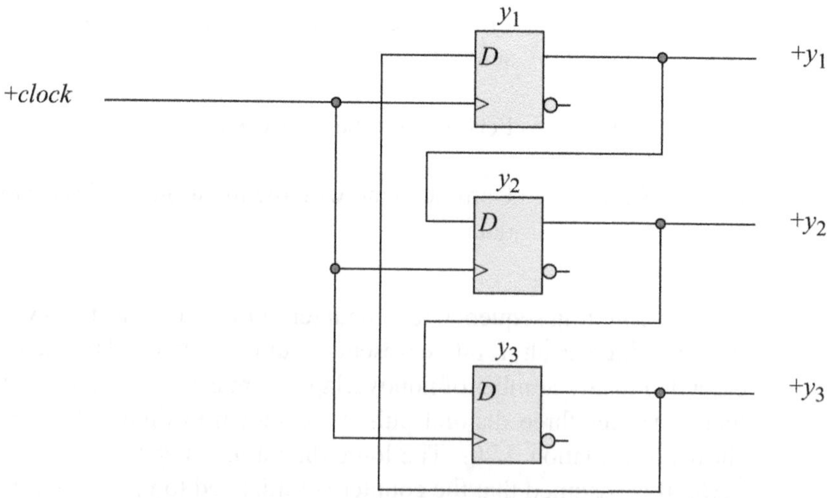

Figure 2.36 Logic diagram for the nonsequential counter.

Pulse generator An alternative method to generate discrete pulses is to use a shift register plus additional logic. This method will be used to create a design that will generate eight nonoverlapping pulses to provide eight unique states. The shift register is an 8-bit serial-in, parallel-out shift register. There is no need for a next-state table or Karnaugh maps for this design — knowledge of how a shift register operates is all that is required. The logic diagram is shown in Figure 2.37 and the resulting outputs are shown in Figure 2.38.

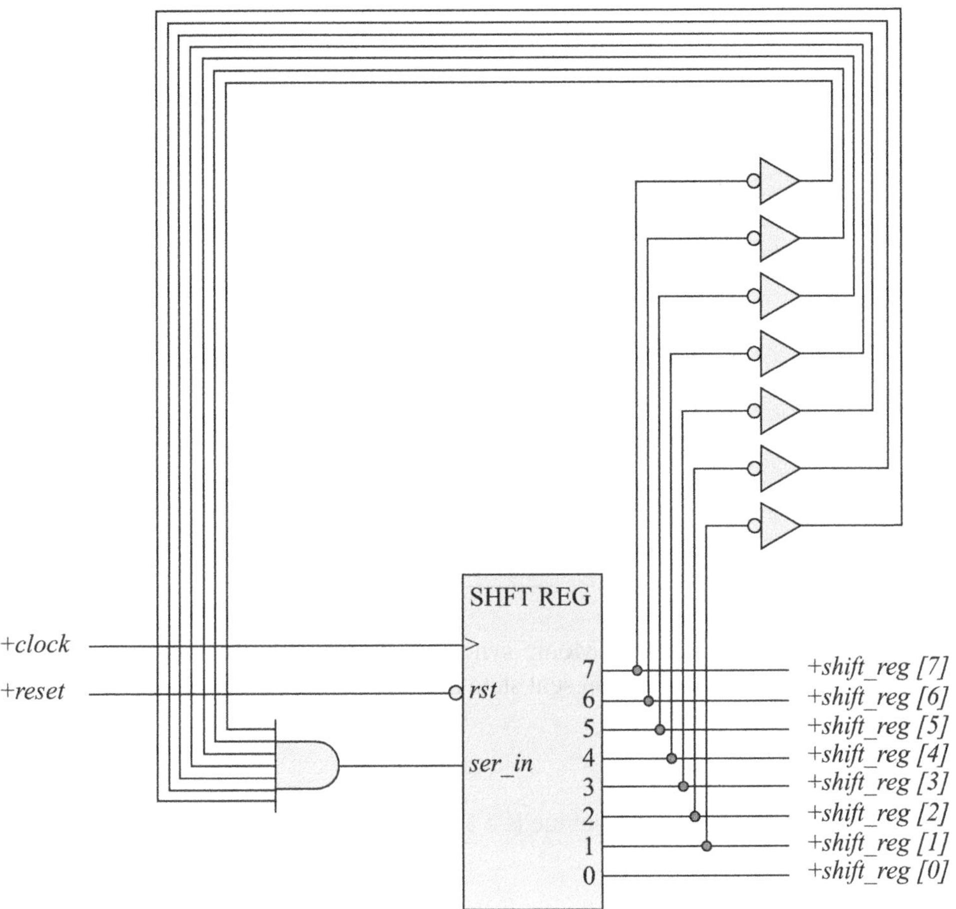

Figure 2.37 Logic diagram to generate eight nonoverlapping, disjoint pulses.

```
shift_reg [7] = 10000000        shift_reg [2] = 00000100
shift_reg [6] = 01000000        shift_reg [1] = 00000010
shift_reg [5] = 00100000        shift_reg [0] = 00000001
shift_reg [4] = 00010000        shift_reg [7] = 10000000
shift_reg [3] = 00001000        shift_reg [6] = 01000000
```

Figure 2.38 Outputs for the shifter pulse generator.

2.4.2 Moore Machines

Moore machines may be synchronous or asynchronous sequential machines in which the output function λ produces an output vector Z_r, which is determined by the present state only, and is not a function of the present inputs. The general configuration of a Moore machine is shown in Figure 2.39. The next-state function δ is an $(n + p)$-input, p-output switching function. The output function λ is a p-input, m-output switching function. If a Moore machine has no data input, then it is referred to as an *autonomous* machine. Autonomous circuits are independent of the inputs. The clock signal is not considered as a data input. An autonomous Moore machine is an important class of synchronous sequential machines, the most common application being a counter, as discussed in the previous section. Although a Moore machine may be synchronous or asynchronous, this section pertains to synchronous organizations only.

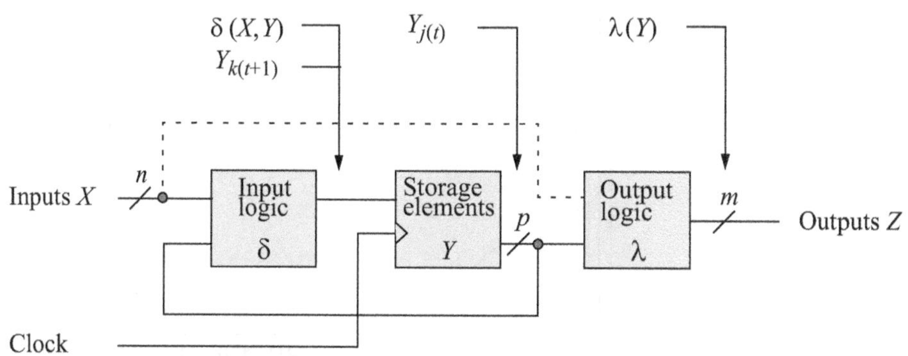

Figure 2.39 Moore synchronous sequential machine in which the outputs are a function of the present state only.

A Moore machine is a 5-tuple and can be formally defined as follows:

$$M = (X, Y, Z, \delta, \lambda)$$

where

1. X is a nonempty finite set of inputs such that,
   $X = \{X_0, X_1, X_2, \cdots, X_{2^n-2}, X_{2^n-1}\}$

2. Y is a nonempty finite set of states such that,
   $Y = \{Y_0, Y_1, Y_2, \cdots, Y_{2^p-2}, Y_{2^p-1}\}$

3. Z is a nonempty finite set of outputs such that,
   $Z = \{Z_0, Z_1, Z_2, \cdots, Z_{2^m-2}, Z_{2^m-1}\}$

4. $\delta(X, Y) : X \times Y \rightarrow Y$

5. $\lambda(Y) : Y \rightarrow Z$

The next state of a Moore machine is determined by both the present inputs and the present state. Thus, the next-state function δ is a function of the input alphabet X and the state alphabet Y, and maps the Cartesian product of X and Y into Y. For a Moore machine, the outputs can be asserted for segments of the clock period rather than for the entire clock period only. This is illustrated in Figure 2.40 where the positive clock transitions define the clock cycles, and hence, the state times. Two clock cycles are shown, one for the present state $Y_{j(t)}$ and one for the next state $Y_{k(t+1)}$.

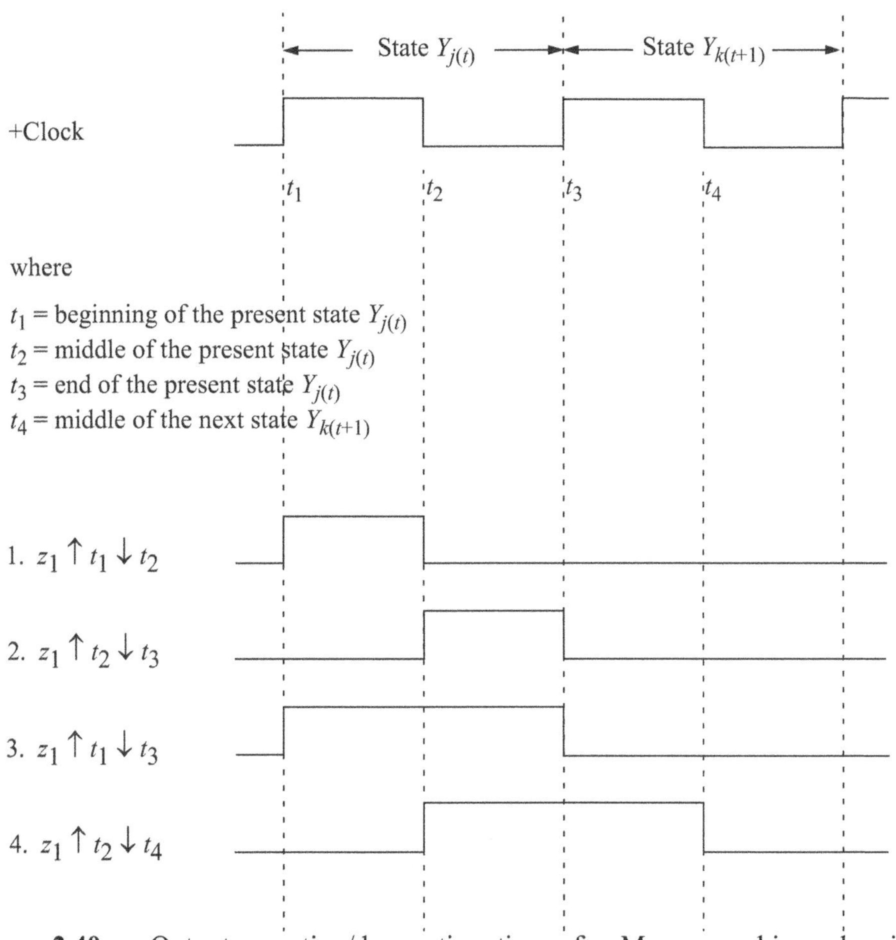

Figure 2.40 Output assertion/deassertion times for Moore machines showing clock pulses; definition of assertion/deassertion times; and assertion/deassertion statements with corresponding asserted outputs.

Example 2.16 A Moore machine which accepts serial data in the form of 3-bit words on an input line x_1 will be synthesized using D flip-flops. There is one bit space between contiguous words, as shown below,

$$x_1 = \cdots \left| b_1 b_2 b_3 \right| \quad \left| b_1 b_2 b_3 \right| \quad \left| b_1 b_2 b_3 \right| \cdots$$

where $b_i = 0$ or 1. Whenever a word contains the bit pattern $b_1 b_2 b_3 = 111$, the machine will assert output z_1 during the bit time between words according to the following assertion/deassertion statement:

$$z_1 \uparrow t_2 \downarrow t_3$$

An example of a valid word in a series of words is shown below. Notice that the output signal is displaced in time with respect to the input sequence and occurs one state time later. The steps that follow depict the procedure that is used to design both Moore and Mealy synchronous sequential machines.

$$x_1 = \cdots \left| 0\,0\,1 \right| \quad \left| 1\,0\,1 \right| \quad \left| 0\,1\,1 \right| \quad \left| 1\,1\,1 \right| \quad \left| 0\,1\,0 \right| \cdots$$

Output $z_1 \uparrow t_2 \downarrow t_3$

State diagram This is an extremely important step, since all remaining steps depend upon a state diagram which correctly represents the machine specifications. Generating an accurate state diagram is thus a pivotal step in the synthesis of synchronous sequential machines. The state diagram for this example is illustrated in Figure 2.41, which graphically describes the machine's behavior. Seven states are required, providing four state levels — one level for each bit in the 3-bit words and one level for the bit space between words.

The flip-flop names are positioned alongside the state symbol. In Figure 2.41, the machine is designed using three flip-flops which are designated as $y_1 y_2 y_3$, where y_3 is the low-order flip-flop. Directly beneath the flip-flop names, the *state code* is specified. The state code represents the state of the individual flip-flops.

The machine is reset to an initial state which is labeled a in Figure 2.41. Since x_1 is a serial input line, the state transition can proceed in only one of two directions from state a, depending upon the value of x_1: to state b if $x_1 = 0$, or to state c if $x_1 = 1$. Both paths represent a test of the first bit (b_1) of a word and both state transitions occur on the rising edge of the clock signal.

Since the state transition from a to b occurs only if $x_1 = 0$, any bit sequence consisting of $b_1 b_2 b_3 = 000$ through 011 will proceed from state a to state b. Since this path will never generate an output (the first bit is invalid), there is no need to test the value of x_1 in states b or d. States b and d, together with state f, are required only to maintain four state levels in the machine, where the first three levels represent one bit each of the 3-bit word as follows: states a, b, and d correspond to bits b_1, b_2, and b_3, respectively. State f, which is the fourth level, corresponds to the bit space between words. This assures that the clocking will remain synchronized for the following word. From state d, the machine proceeds to state f and then returns to state a where the first bit of the next word is checked.

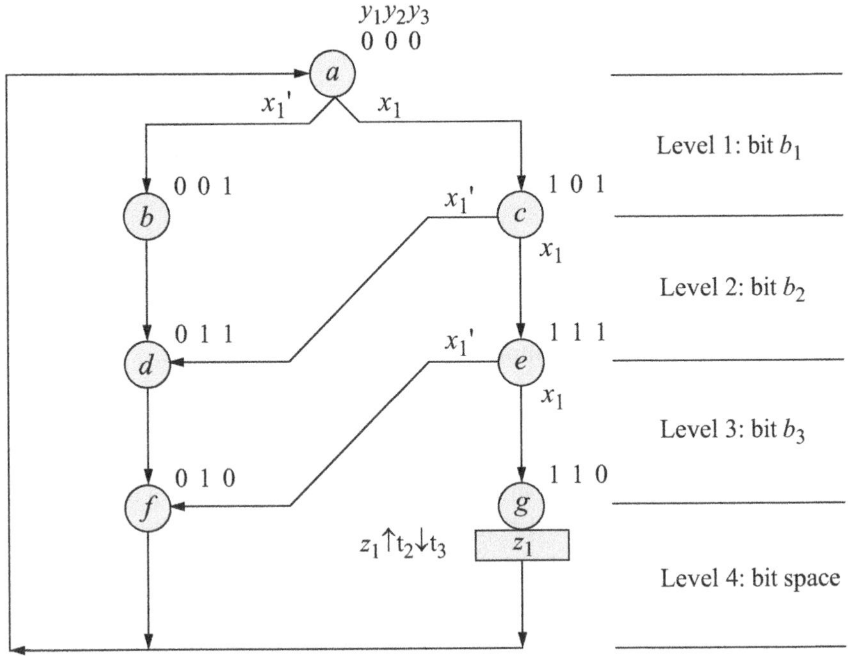

Figure 2.41 State diagram for the Moore machine of Example 2.16, which generates an output z_1 whenever a 3-bit word $x_1 = 111$.

The path a, c, e, g depicts a valid word which culminates in the assertion of output z_1 in state g. All other sequences of three bits will not generate an output, but each path must maintain four state levels. This guarantees that state a will always test the first bit of each 3-bit word. Output z_1 is a function of the state alphabet only, and thus, the state diagram represents a Moore machine.

Whenever possible, state codes should be assigned such that there are a maximal number of adjacent 1s in the flip-flop input maps. This allows more minterm locations to be combined, resulting in minimized input equations in a sum-of-products form. State codes are adjacent if they differ in only one variable. For example, state codes $y_1y_2y_3 = 101$ and 100 are adjacent, because only y_3 changes. Thus, minterm locations 101 and 100 can be combined into one term. However, state codes $y_1y_2y_3 = 101$ and 110 are not adjacent, because two variables change: flip-flops y_2 and y_3.

The rules shown below are useful in assigning state codes such that there will be a maximal number of 1s in adjacent squares of the input maps, thus minimizing the δ next-state logic. It should be noted, however, that these rules do not guarantee a minimum solution with the fewest number of terms and the fewest number of variables per term. There may be several sets of state code assignments that meet the adjacency requirements, but not all will result in a minimally reduced set of input equations.

1. When a state has two possible next states, then the two next states should be adjacent; that is, if an input causes a state transition from state Y_i to either Y_j or Y_k, then Y_j and Y_k should be assigned adjacent state codes.

2. When two states have the same next state, the two states should be adjacent; that is, if Y_i and Y_j both have Y_k as a next state, then Y_i and Y_j should be assigned adjacent state codes.

3. A third rule is useful in minimizing the λ output logic. States which have the same output should have adjacent state code assignments; that is, if states Y_i and Y_j both have z_1 as an output, then Y_i and Y_j should be adjacent. This allows for a larger grouping of 1s in the output map.

Input maps The input maps, also called excitation maps, represent the δ next-state logic for the flip-flop inputs. The input maps are constructed from the state diagram and are shown in Figure 2.42 using input x_1 as a map-entered-variable. Refer to the input map for flip-flop y_1. Since the purpose of an input map is to obtain the flip-flop input equations by combining 1s in the minterm locations, the variable x_1 is entered as the value in minterm location $y_1y_2y_3 = 000$. That is, y_1 has a next value of 1 if and only if x_1 has a value of 1. In a similar manner, the entries for the remaining squares of y_1 are obtained and also for the input maps for y_2 and y_3. The input equations are listed in Equation 2.21.

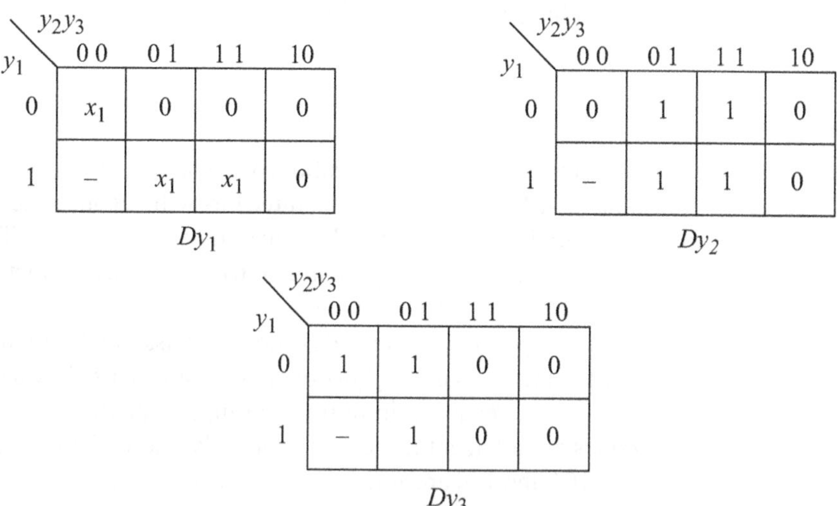

Figure 2.42 Input maps for the Moore machine of Figure 2.41.

$$Dy_1 = y_2'y_3'x_1 + y_1y_3x_1$$

$$Dy_2 = y_3$$

$$Dy_3 = y_2' \tag{2.21}$$

Output map Outputs from a synchronous sequential machine can be asserted at a variety of different times depending on the machine specifications. In some cases, the output assertion and deassertion may not be specified, giving substantial flexibility in the design of the λ output logic. A contributing factor in considering the output design is the possibility of glitches. Glitches are spurious electronic signals caused by varying gate delays or improper design techniques, in which the design was not examined in sufficient detail using "worst case" circuit conditions. Glitches are more predominant in Moore machines where the outputs are a function of the state alphabet only.

The output map for the Moore machine of Figure 2.41, using state code adjacency requirements, is obtained from the state diagram and is shown in Figure 2.43. Note that input x_1 is not used as a map-entered-variable, because the outputs for a Moore machine are a function of the present state only. The equation for output z_1 is shown in Equation 2.22.

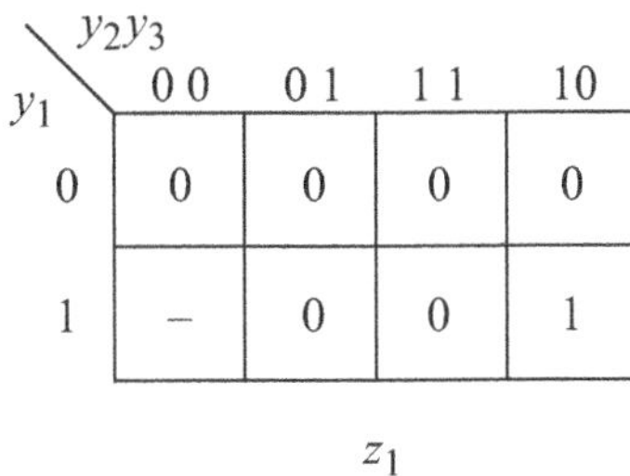

Figure 2.43 Output map for the Moore machine of Figure 2.41.

$$z_1 = y_1y_3' \tag{2.22}$$

The unused state $y_1y_2y_3 = 100$ can be used to minimize the equation for z_1, since the assertion of z_1 occurs at time t_2, long after the machine has stabilized. The state flip-flops are clocked on the positive clock transition at time t_1. A state change causes the machine to move from one stable state to another stable state. If more than one flip-flop changes state during this transition, then it is possible for the machine to momentarily pass through a transient state before reaching the destination stable state. This period of instability occurs immediately after the rising edge of the clock and has a duration of only a small percentage of the clock cycle. The machine has certainly stabilized by time t_2.

Since the assertion/deassertion statement for z_1 is $\uparrow t_2 \downarrow t_3$, the machine has stabilized in its destination state before the specified assertion of z_1. Thus, the "don't care" state $y_1y_2y_3 = 100$ can be used to minimize the equation for z_1 without regard for any momentary transition through state 100, which would otherwise produce a glitch on z_1.

Logic diagram The logic diagram is implemented from the input equation of Equation 2.21 using positive-edge-triggered D flip-flops, and from the output equation of Equation 2.22, as shown in Figure 2.44. In Figure 2.44, output z_1 is asserted at time t_2 and deasserted at time t_3 by the application of the $-clock$ signal to the active-high AND gate inputs. If output z_1 were implemented using NOR logic for the AND function, then the $+clock$ signal would be used to provide the requisite assertion and deassertion times for z_1, because the AND gate function has active-low inputs.

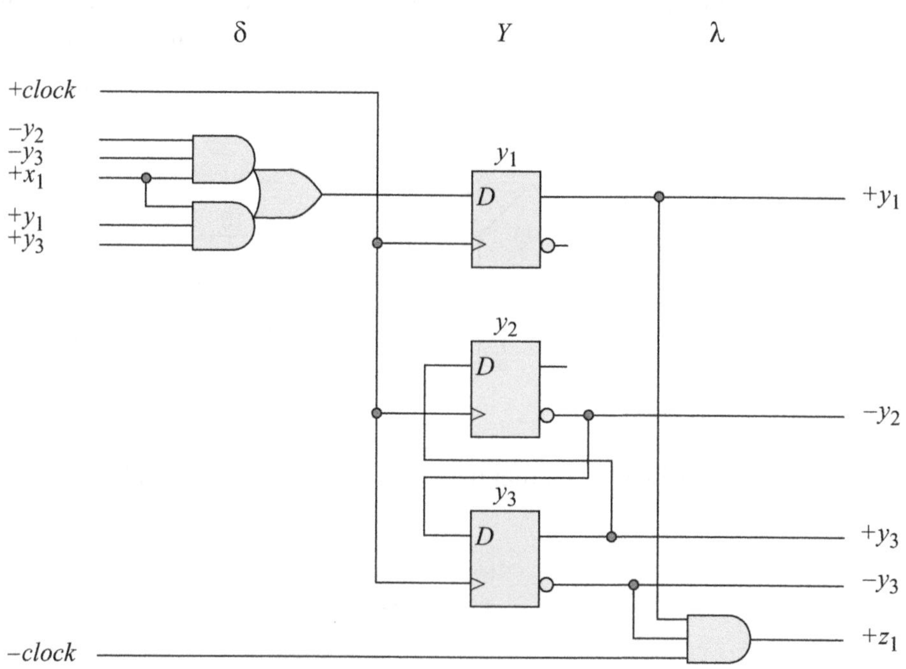

Figure 2.44 Logic diagram for the Moore machine of Figure 2.41 using D flip-flops. Output z_1 is asserted when an input sequence of 111 has been detected in a 3-bit word on a serial data line x_1. Output z_1 is asserted at time t_2 and deasserted at time t_3.

2.4.3 Mealy Machines

Mealy machines may be synchronous or asynchronous sequential machines in which the output function λ produces an output vector $Z_{r(t)}$ which is determined by both the present input vector $X_{i(t)}$ and the present state of the machine $Y_{j(t)}$. The general configuration of a Mealy machine is shown in Figure 2.45. The next-state function δ is an $(n + p)$-input, p-output switching function. The output function λ is an $(n + p)$-input, m-output switching function. A Mealy machine is not an autonomous machine, because the outputs are a function of the input signals. Although a Mealy machine may be synchronous or asynchronous, this section pertains to synchronous organizations only.

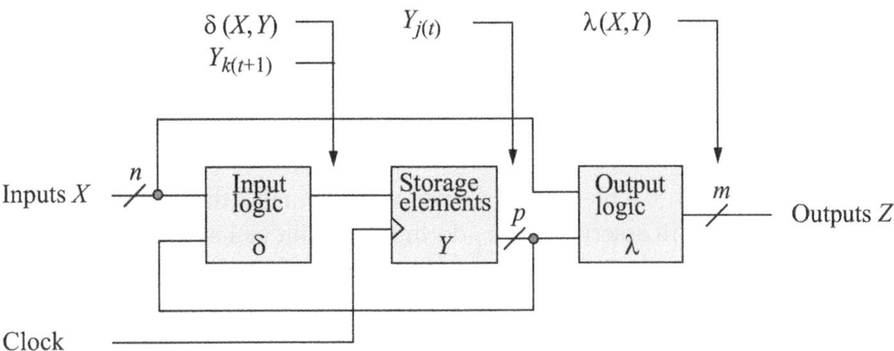

Figure 2.45 Mealy machine in which the outputs are a function of both the present state and the present inputs.

A Mealy machine is a 5-tuple and can be formally defined as follows:

$$M = (X, Y, Z, \delta, \lambda)$$

where

1. X is a nonempty finite set of inputs such that,
   $$X = \{X_0, X_1, X_2, \cdots, X_{2^n-2}, X_{2^n-1}\}$$

2. Y is a nonempty finite set of states such that,
   $$Y = \{Y_0, Y_1, Y_2, \cdots, Y_{2^p-2}, Y_{2^p-1}\}$$

3. Z is a nonempty finite set of outputs such that,
   $$Z = \{Z_0, Z_1, Z_2, \cdots, Z_{2^m-2}, Z_{2^m-1}\}$$

4. $\delta(X, Y) : X \times Y \rightarrow Y$

5. $\lambda(X, Y) : Y \rightarrow Z$

The definitions for Mealy and Moore machines are the same, except for part 5, which shows that the outputs of a Mealy machine are a function of both the inputs and the present state, whereas, the outputs of a Moore machine are a function of the present state only. A Moore machine, therefore, is a special case of a Mealy machine.

Example 2.17 This example is similar to the Moore machine of Example 2.16, but will be synthesized as a Mealy machine using negative-edge-triggered *JK* flip-flops. This allows for a correlation between the design methodologies of the two state machines.

A synchronous sequential machine will be designed that accepts serial data on an input line x_1 which consists of 3-bit words. The words are contiguous with no space between adjacent words. The machine is controlled by a periodic clock, where one clock period is equal to one bit cell. The format for the 3-bit words is

$$x_1 = \cdots \left| b_1b_2b_3 \right| b_1b_2b_3 \left| b_1b_2b_3 \right| \cdots$$

where $b_i = 0$ or 1. Whenever a word contains the bit pattern $b_1b_2b_3 = 111$, the machine will assert output z_1 during the b_3 bit cell according to the following assertion/deassertion statement: $z_1 \uparrow t_2 \downarrow t_3$. Thus, z_1 is active for the last half of bit cell b_3. An example of a valid word in a series of words is as follows:

$$x_1 = \cdots \left| 001 \right| 101 \left| 011 \right| 101 \left| 111 \right| 010 \left| \cdots \right.$$

$$z_1 \uparrow t_2 \downarrow t_3$$

Notice that the output signal — compared with the Moore machine — occurs during the third bit period of a valid word. The Moore output was displaced in time by one clock period. The excitation table for a *JK* flip-flop is reproduced in Table 2.15 for convenience. The state diagram is shown in Figure 2.46.

Table 2.15 Excitation Table for a *JK* Flip-Flop

Present state $Y_{j(t)}$	Next state $Y_{k(t+1)}$	Data inputs JK
0	0	0 –
0	1	1 –
1	0	– 1
1	1	– 0

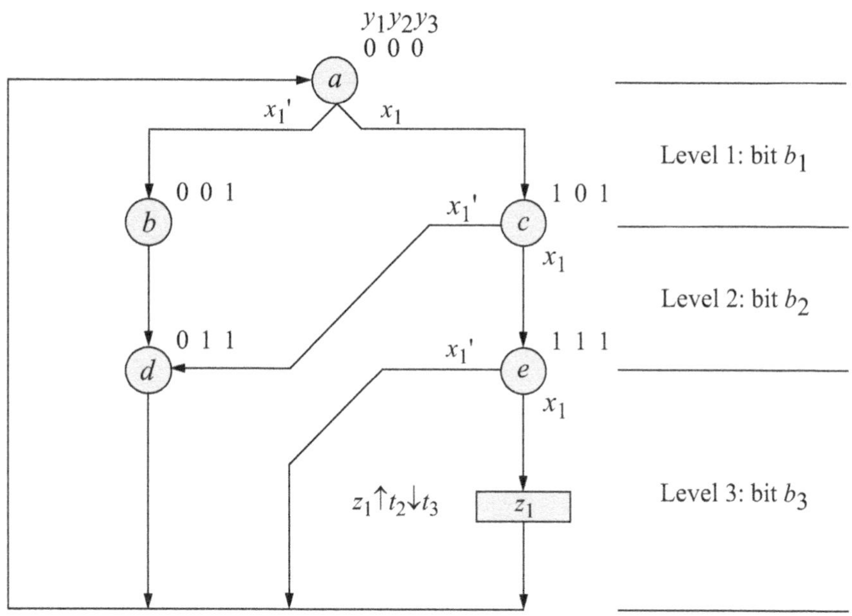

Figure 2.46 State diagram for the Mealy machine of Example 2.17, which generates an output z_1 whenever a 3-bit word $x_1 = 111$.

In many cases — especially where design time is to be kept to a minimum — the next-state table can be bypassed, as is the case here. The input maps are then derived directly from the state diagram using the *JK* flip-flop excitation table.

To construct the input maps from the state diagram, first the "don't care" symbol is entered in all unused states by placing dashes in locations 010, 100, and 110. Refer to Figure 2.46 for the procedure which follows. In state *a* ($y_1y_2y_3 = 000$) with $x_1 = 0$, the state transition for y_1 is from 0 to 0 (state *b*) and from 0 to 1 (state *c*) if $x_1 = 1$. Therefore, y_1 assumes the value of x_1 for any transition from state *a*.

Since the values for Jy_1 and Ky_1 are not readily apparent, a partial next-state table is required for state *a*, as shown below in Table 2.16, in which the only next state under consideration is for y_1. A similar situation occurs for y_1 in state *c* ($y_1 = 101$), as shown in Table 2.17.

Table 2.16 Partial Next-State Table to Determine the Values for Jy_1Ky_1 in State *a* ($y_1y_2y_3 = 000$)

Present State $y_1y_2y_3$	Input x_1	Next state y_1	Flip-flop inputs Jy_1	Ky_1
<u>0</u> 0 0	0	<u>0</u>	0	–
<u>0</u> 0 0	1	<u>1</u>	1	–
			x_1	–

Table 2.17 Partial Next-State Table to Determine the Values for Jy_1Ky_1 in State c ($y_1y_2y_3$ = 101)

Present State $y_1y_2y_3$	Input x_1	Next state y_1	Flip-flop inputs Jy_1 Ky_1
1 0 1	0	0	– 1
1 0 1	1	1	– 0
			– x_1'

The input maps are shown in Figure 2.47 using x_1 as a map-entered variable. The input equations are shown in Equation 2.23.

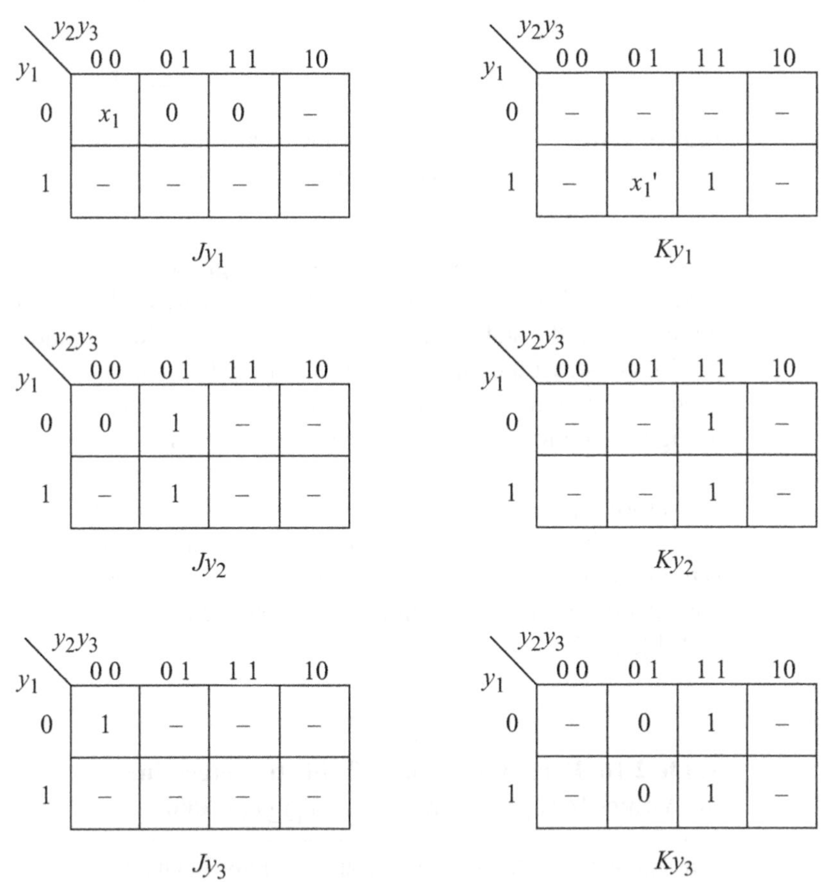

Figure 2.47 Input maps for the Mealy machine of Figure 2.46 using *JK* flip-flops and adjacent state codes for state pairs (b, c) and (d, e). Input x_1 is a map-entered variable.

$$Jy_1 = y_3'x_1$$
$$Ky_1 = y_2 + x_1'$$

$$Jy_2 = y_3$$
$$Ky_2 = \text{Logic 1}$$

$$Jy_3 = \text{Logic 1}$$
$$Ky_3 = y_2 \tag{2.23}$$

The output map is shown in Figure 2.48 using x_1 as a map-entered variable and the output equation is shown in Equation 2.24. The logic diagram is shown in Figure 2.49 and is implemented from the input equations and output equation of Equation 2.23 and Equation 2.24, respectively, using three negative-edge-triggered *JK* flip-flops. Output z_1 is asserted at time t_2 and deasserted at time t_3 by the application of the $-$*clock* signal to the positive-input AND gate which generates z_1. When the $-$*clock* signal attains a high voltage level at time t_2 in state *d* ($y_1y_2y_3 = 111$), z_1 is asserted, assuming that x_1 is already asserted.

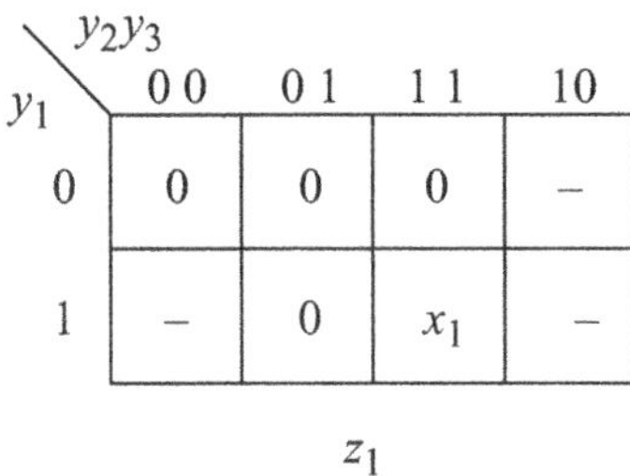

Figure 2.48 Output map for the Mealy machine of Figure 2.46 using x_1 as a map-entered variable.

$$z_1 = y_1y_2x_1 \tag{2.24}$$

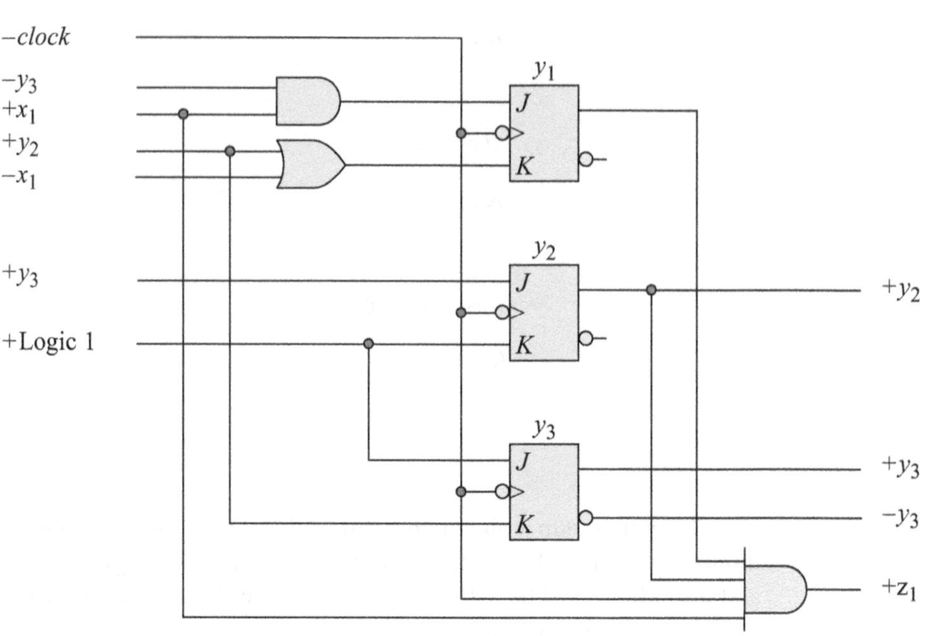

Figure 2.49 Logic diagram for the Mealy machine of Figure 2.46 using *JK* flip-flops. Output z_1 is asserted when an input sequence of 111 is detected on a serial input line x_1. Output z_1 is asserted at time t_2 and deasserted at time t_3.

This section on sequential logic has briefly reviewed the operational characteristics of the *SR* latch, the *D* flip-flop, and the *JK* flip-flop, which are used in synchronous and asynchronous sequential machines. Section 2.4 also provided an overview of the design methodologies used in the implementation of counters of different moduli. To complete the presentation of sequential logic, the design of Moore and Mealy finite-state machines was examined

The synthesis procedure utilizes a hierarchical method — also referred to as a *top-down* approach — for machine design. This is a systematic and orderly procedure that commences with the machine specifications and advances down through increasing levels of detail to arrive at a final logic diagram. Thus, the machine is decomposed into modules which are independent of previous and following modules, yet operate together as a cohesive system to satisfy the machine's requirements.

2.5 Problems

2.1 Minimize the following equation using Boolean algebra:

$$z_1 = x_1'x_3'x_4' + x_1'x_3x_4' + x_1x_3'x_4' + x_2x_3x_4 + x_1x_3x_4'$$

2.2 Indicate which of the following statements will always generate a logic 1:

(a) $x_1 + x_1'x_2x_3' + x_3 + x_1'x_4 + x_1'x_3'$
(b) $x_2'x_3x_4 + x_1'x_3' + x_2x_4' + x_1x_2x_3$
(c) $x_1x_2x_3'x_4 + x_1'x_2'x_3x_4' + x_1x_2'x_3x_4'$
(d) $x_2x_3'x_4 + x_1x_3x_4' + x_2x_3' + x_4'$

2.3 Indicate whether the following statement is true or false:

$$x_1'x_2'x_3' + x_1x_2x_3' = x_3'$$

2.4 Indicate which answer below represents the product-of-maxterms for the following function: $z_1(x_1, x_2, x_3, x_4) = x_3$

(a) $\Pi_M(0, 1, 4, 5, 8, 9, 12, 13)$
(b) $\Pi_M(1, 2, 3, 4, 5, 6, 7, 8)$
(c) $\Pi_M(2, 3, 6, 7, 10, 11, 14, 15)$
(d) $\Pi_M(0, 2, 4, 5, 6, 8, 12, 15)$
(e) None of the above

2.5 Minimize each of the following expressions:

(a) $(x_1'x_2 + x_3)(x_1'x_2 + x_3)'$
(b) $(x_1x_2' + x_3' + x_4x_5')(x_1x_2' + x_3')$
(c) $(x_1'x_3 + x_2 + x_4 + x_5)(x_1'x_3 + x_4)'$

2.6 Minimize the following expression as a sum of products:

$$(x_2 + x_3' + x_4)(x_1 + x_2 + x_4)(x_3' + x_4)$$

2.7 Prove algebraically that $x_1 + 1 = 1$ using only the axioms.

2.8 Determine if $A = B$ by expanding A and B into sum-of-minterms expressions.

$$A = x_3(x_1 + x_2) + x_1x_2$$
$$B = x_1x_3 + x_2x_3 + x_1x_2$$

2.9 Use Boolean algebra to convert the following function into a product of maxterms:

$$z_1(x_1, x_2, x_3) = x_1x_2 + x_2'x_3'$$

2.10 Obtain the minimized product-of-sums expression for the function z_1 represented by the Karnaugh map shown below.

$x_5 = 0$

x_1x_2 \ x_3x_4	0 0	0 1	1 1	1 0
0 0	0 (0)	1 (2)	1 (6)	0 (4)
0 1	1 (8)	1 (10)	1 (14)	1 (12)
1 1	0 (24)	1 (26)	0 (30)	0 (28)
1 0	0 (16)	1 (18)	1 (22)	1 (20)

$x_5 = 1$

x_1x_2 \ x_3x_4	0 0	0 1	1 1	1 0
0 0	0 (1)	1 (3)	1 (7)	0 (5)
0 1	1 (9)	1 (11)	1 (15)	1 (13)
1 1	1 (25)	1 (27)	0 (31)	0 (29)
1 0	1 (17)	1 (19)	1 (23)	0 (21)

z_1

2.11 Plot the following expression on a Karnaugh map and obtain the minimum sum-of-products expression:

$$z_1 = x_1'x_2(x_3'x_4' + x_3'x_4) + x_1x_2(x_3'x_4' + x_3'x_4) + x_1x_2'x_3'x_4$$

2.12 Obtain the minimized expression for the function z_1 in a sum-of-products form from the Karnaugh map shown below.

x_1x_2 \ x_3x_4	0 0	0 1	1 1	1 0
0 0	b	0	–	1
0 1	a	–	1	0
1 1	1	–	1	0
1 0	1	–	0	b

z_1

2.13 Write the equation for a logic circuit that generates a logic 1 output whenever a 4-bit unsigned binary number is greater than six. The equation is to be in a minimum sum-of-products notation and a minimum product-of-sums notation.

2.14 Given the Karnaugh map shown below for the function z_1, obtain the minimized expression for z_1 in a sum-of-products form.

x_1x_2 \ x_3x_4	0 0	0 1	1 1	1 0
0 0	1	0	0	$a + a'$
0 1	0	a	0	0
1 1	0	$a + a'$	a'	0
1 0	b	b'	0	b

z_1

2.15 Obtain the minimal product-of-sums notation for the expression shown below. Use any minimization technique.

$$z_1(x_1, x_2, x_3, x_4) = x_1'x_3x_4' + x_1'x_2'x_4' + x_1'x_2x_3' + x_2x_3'x_4 + x_1x_2'x_4$$

2.16 Obtain the disjunctive normal form for the expression shown below.

$$z_1 = x_3 + (x_1' + x_2)(x_1 + x_2')$$

2.17 Convert the following expression to a product-of-sums form using any method:

$$z_1 = x_1x_2' + x_2x_3x_4' + x_1'x_2x_4$$

2.18 Obtain the disjunctive normal form for the function shown below using any method. Then use Boolean algebra to minimize the function.

$$z_1(x_1, x_2, x_3, x_4) = x_1x_2x_4 + x_1x_2'x_3$$

2.19 Determine if the following equation is true or false:

$$(x_1x_2) \oplus (x_1x_3) = x_1(x_2 \oplus x_3)'$$

2.20 Plot the following expression on a Karnaugh map, then obtain the minimized expression in a sum-of-products form and a product-of-sums form:

$$z_1(x_1, x_2, x_3, x_4, x_5) = x_1 x_3 x_4 (x_2 + x_4 x_5') + (x_2' + x_4)(x_1 x_3' + x_5)$$

2.21 Obtain the minimum sum-of-products expression for the Quine-McCluskey prime implicant table shown below, where $f(x_1, x_2, x_3, x_4, x_5)$.

Prime implicants		Minterms									
		0	1	3	7	15	16	18	19	23	31
0 0 0 0 –	$(x_1' x_2' x_3' x_4')$	×	×								
0 0 0 – 1	$(x_1' x_2' x_3' x_5)$		×	×							
– 0 – 1 1	$(x_2' x_4 x_5)$			×	×				×	×	
– – 1 1 1	$(x_3 x_4 x_5)$				×	×				×	×
1 0 0 1 –	$(x_1 x_2' x_3' x_4)$							×	×		
1 0 0 – 0	$(x_1 x_2' x_3' x_5')$						×	×			
– 0 0 0 0	$(x_2' x_3' x_4' x_5')$	×					×				

2.22 Use the Quine-McCluskey algorithm to obtain the minimized expression for the following function:

$$z_1(x_1, x_2, x_3, x_4) = \Sigma_{\mathrm{m}}(0, 2, 5, 6, 7, 8, 10, 13, 15)$$

2.23 Given the chart shown below, use the Petrick algorithm to find a minimal set of prime implicants for the function.

Prime implicants	Minterms						
	a	b	c	d	e	f	g
A	×	×	×			×	×
B	×		×	×	×		
C		×				×	
D				×	×		×

2.24 Minimize the equation shown below then implement the equation: (a) using NAND gates and (b) NOR gates. Output z_1 is to be asserted high in both cases.

$$z_1 = x_1' x_2 (x_3' x_4' + x_3' x_4) + x_1 x_2 (x_3' x_4' + x_3' x_4) + x_1 x_2' x_3' x_4$$

2.25 Obtain the minimized equation for the logic diagram shown below.

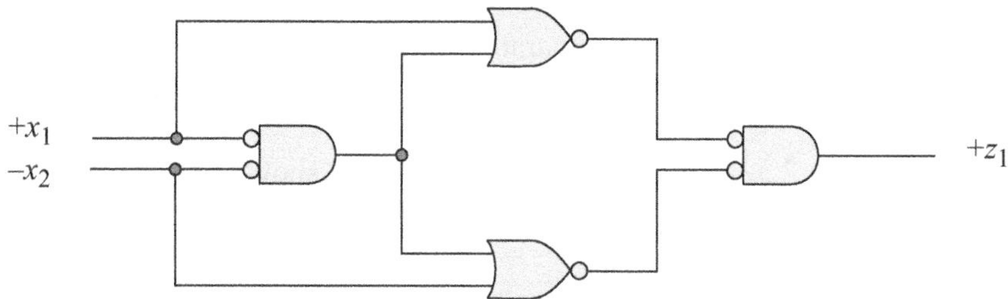

2.26 Design a logic circuit that generates an output if a 4-bit number $z_1 = x_1 x_2 x_3 x_4$ satisfies the criteria shown below. Output z_1 is to be asserted at a high voltage level. Use only NOR gates.

$$4 \leq z_1 < 9$$
$$10 < z_1 < 14$$

2.27 Given the following equation for the function z_1, obtain the Karnaugh map using a and b as map-entered variables, the minimized equation for z_1, and the logic diagram using a linear-select multiplexer:

$$z_1 = x_1'x_2'x_3 + x_1x_2'x_3'a'b' + x_1x_2'x_3'ab + x_1x_2'x_3a + x_1x_2'x_3b + x_1x_2x_3'ab$$

2.28 Given the Karnaugh map shown below for the function z_1, implement the function using a nonlinear-select multiplexer and additional logic, if necessary.

x_1 \ x_2x_3	0 0	0 1	1 1	1 0
0	1 (0)	1 (1)	ab (3)	– (2)
1	– (4)	1 (5)	$a' + b'$ (7)	$(ab)'$ (6)

z_1

2.29 Implement the equation shown below using a decoder and a minimal amount of additional logic. The decoder has active-low outputs.

$$z_1(x_1, x_2, x_3) = x_1'x_2'x_3 = x_1x_3' + x_2x_3 + x_1x_2'x_3' + x_1x_2$$

2.30 Implement an octal-to-binary encoder using only NOR gates. The outputs are active low.

2.31 Use comparators and additional logic to determine if a 4-bit number N is within the range $3 \leq N \leq 12$. Use the least amount of logic. Show two solutions.

2.32 Given the input equations shown below for a counter that uses three D flip-flops $y_1y_2y_3$, obtain the counting sequence. The counter is reset initially such that $y_1y_2y_3 = 000$.

$$Dy_1 = y_1'y_2 + y_2'y_3$$
$$Dy_2 = y_2y_3' + y_1y_2 + y_1'y_2'y_3$$
$$Dy_3 = y_1'y_2' + y_1'y_3 + y_1y_2y_3'$$

2.33 Design a counter using D flip-flops that counts in the following sequence:

$y_1y_2y_3 = 000, 010, 100, 110, 101, 101, \ldots$

2.34 Design a 3-bit Gray code counter using JK flip-flops. The Gray code counts in the following sequence:

$y_1y_2y_3 = 000, 001, 011, 010, 110, 111, 101, 100, 000, \ldots$

2.35 Obtain the state diagram for the Moore synchronous sequential machine shown below. The flip-flops are reset initially; that is, $y_1y_2 = 00$.

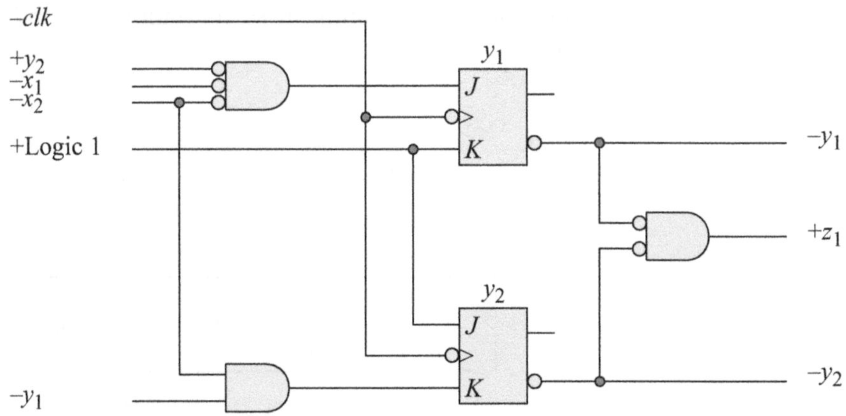

2.36 A Moore synchronous sequential machine receives 3-bit words on a parallel input bus as shown below, where x_3 is the low-order bit. There are five outputs labeled z_1, z_2, z_3, z_4, and z_5, where the subscripts indicate the decimal value of the corresponding unsigned binary input word. Design the machine using AND gates, OR gates, and positive-edge-triggered D flip-flops. The outputs are asserted at time t_2 and deasserted at time t_3.

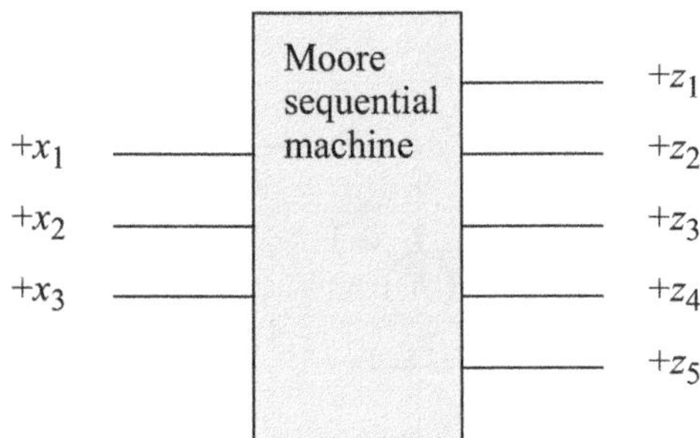

2.37 Draw the state diagram for the Mealy synchronous sequential machine shown below, where z_1 is the output. Assume that the flip-flops are set initially; that is, $y_1y_2 = 11$.

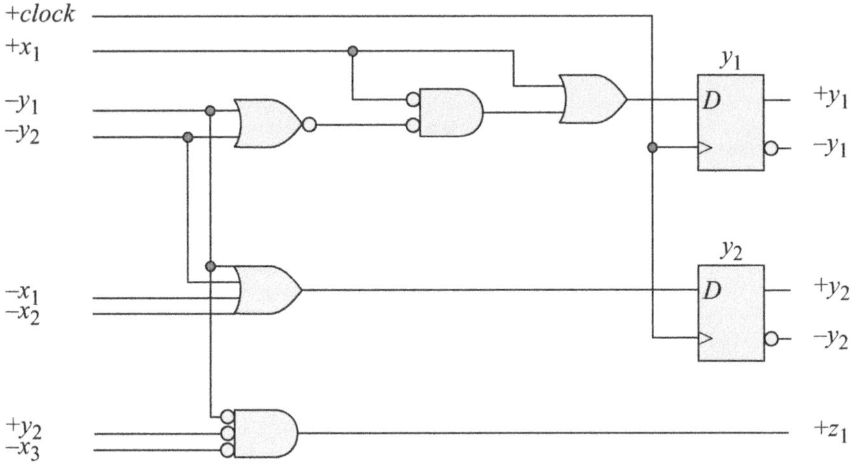

2.38 Given the state diagram shown below for a synchronous sequential machine containing Moore- and Mealy-type outputs, design the machine using linear-select multiplexers for the δ next-state logic and D flip-flops for the storage elements. Use x_1 and x_2 as map-entered variables. Output z_1 is asserted at time t_2 and deasserted at time t_3.

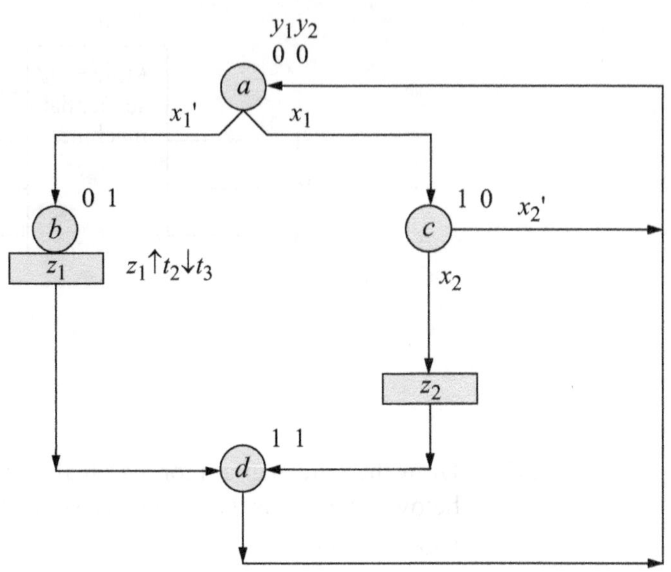
y_1y_2
0 0
a
x_1'
x_1
0 1
b
1 0
c
x_2'
z_1
$z_1\uparrow t_2\downarrow t_3$
x_2
z_2
1 1
d

3.1 *Built-In Primitives*
3.2 *User-Defined Primitives*
3.3 *Dataflow Modeling*
3.4 *Behavioral Modeling*
3.5 *Structural Modeling*
3.6 *Problems*

3

Introduction to Verilog HDL

This chapter covers the design of combinational and sequential logic using the Verilog hardware description language (HDL). An HDL is a method of designing digital hardware by means of software. A considerable saving of time can be realized when designing systems using an HDL. This offers a competitive advantage by reducing the time-to-market for a system. Another advantage is that the design can be simulated and tested for correct functional operation before implementing the system in hardware. Any errors that are found during simulation can be corrected before committing the design to expensive hardware implementation.

This chapter provides a brief introduction to the design methodologies and modeling constructs of the Verilog hardware description language (HDL). Modules and ports will be presented. Modules are the basic units that describe the design of the Verilog hardware. Ports allow the modules to communicate with the external environment; that is, other modules and input/output signals. Different methods will be presented for designing test benches. Test benches are used to apply input vectors to the module in order to test the functional operation of the module in a simulation environment.

Five different types of module constructs will be described together with applications of each type of module. Modules will be designed using built-in logic primitives, used-defined logic primitives, dataflow modeling, behavioral modeling, and structural modeling. Examples will be shown for each type of modeling.

A *module* is the basic unit of design in Verilog. It describes the functional operation of some logical entity and can be a stand-alone module or a collection of modules that are instantiated into a structural module. *Instantiation* means to use one or more lower-level modules in the construction of a higher-level structural module. A module can be a logic gate, an adder, a multiplexer, a counter, or some other logical function.

A module consists of declarative text which specifies the function of the module using Verilog constructs; that is, a Verilog module is a software representation of the physical hardware structure and behavior. The declaration of a module is indicated by the keyword **module** and is always terminated by the keyword **endmodule**.

A Verilog module defines the information that describes the relationship between the inputs and outputs of a logic circuit. The inputs and outputs are defined by the keywords **input** and **output**. A structural module will have one or more instantiations of other modules or logic primitives.

Verilog has predefined logical elements called *primitives*. These built-in logic primitives are structural elements that can be instantiated into a larger design to form a more complex structure. Examples of built-in logic primitives are the logical operations of AND, OR, XOR, and NOT. Built-in primitives are discussed in more detail in Section 3.1.

3.1 Built-In Primitives

Verilog has a profuse set of built-in primitive gates that are used to model nets. The single output of each gate is declared as type **wire**. The inputs are declared as type **wire** or as type **reg** depending on whether they were generated by a structural or behavioral module. This section presents a design methodology that is characterized by a low level of abstraction, where the logic hardware is described in terms of gates. Designing logic at this level is similar to designing logic by drawing gate symbols — there is a close correlation between the logic gate symbols and the Verilog built-in primitive gates. Each predefined primitive is declared by a keyword. Section 3.1 through Section 3.5 continue modeling logic units at progressively higher levels of abstraction.

The multiple-input gates are **and**, **nand**, **or**, **nor**, **xor**, and **xnor,** which are built-in primitive gates used to describe a net and which have one or more scalar inputs, but only one scalar output. The output signal is listed first, followed by the inputs in any order. The outputs are declared as **wire;** the inputs can be declared as either **wire** or **reg**. The gates represent combinational logic functions and can be instantiated into a module, as follows, where the instance name is optional:

gate_type inst1 (output, input_1, input_2, . . . , input_*n*);

Two or more instances of the same type of gate can be specified in the same construct, as shown below. Note that only the last instantiation has a semicolon terminating the line. All previous lines are terminated by a comma.

gate_type inst1 (output_1, input_11, input_12, . . . , input_1*n*),
inst2 (output_2, input_21, input_22, . . . , input_2*n*),

.

.

.

inst*m* (output_*m*, input_*m*1, input_*m*2, . . . , input_*mn*);

Several examples will now be presented to exemplify the design methodologies using built-in logic primitives. The examples will first be designed using traditional methods and then implemented using built-in primitives. The examples will include: a product-of-sums circuit, detecting if a number is within a certain range, a 3:8 decoder, and a 3-bit comparator.

Example 3.1 The following product-of-sums equation will be implemented using NOR gates only:

$$z_1 = (x_1 + x_2')(x_2 + x_3 + x_4')(x_1' + x_3' + x_4)$$

The Karnaugh map that represents the equation for z_1 is shown in Figure 3.1 in which zeros are entered for the sum terms in the equation. The logic diagram is shown in Figure 3.2 and is designed directly from the equation for z_1. The design module is shown in Figure 3.3 using NOR built-in primitives for the OR gates and the AND gates.

The test bench is shown in Figure 3.4. Since there are four inputs to the product-of-sums circuit, all 16 combinations of four variables must be applied to the circuit. This is accomplished by a **for** loop statement, which is similar in construction to a **for** loop in the C programming language.

Following the keyword **begin** is the name of the block: *apply_stimulus*. In this block, a 5-bit **reg** variable is declared called *invect*. This guarantees that all 16 combinations of the four inputs will be tested by the **for** loop, which applies input vectors of $x_1x_2x_3x_4 = 0000$ through 1111 to the circuit. The **for** loop stops when the pattern 10000 is detected by the test segment (*invect* < 16). If only a 4-bit vector were applied, then the expression (*invect* < 16) would always be true and the loop would never terminate. The increment segment of the **for** loop does not support an increment designated as *invect*++; therefore, the long notation must be used: *invect* = *invect* + *1*.

Communication between the test bench module and the design module is accomplished by instantiating the module into the test bench. The name of the instantiation must be the same as the module under test, in this case, *log_eqn_pos_nor2*. This is followed by an instance name (*inst1*) followed by a left parenthesis. The $.x_1$ variable refers to a port in the design module that corresponds to a port (x_1) in the test bench.

The outputs obtained from the test bench are shown in Figure 3.5. Note that the outputs correspond directly with the zeros in the maxterm locations and with the ones in the minterm locations in the Karnaugh map. Thus, the outputs correspond to an entry of 1 in minterm locations 0, 2, 3, 8, 11, 12, 13, 15.

The waveforms are shown in Figure 3.6 and agree with the Karnaugh map entries and the outputs.

x_1x_2 \ x_3x_4	0 0	0 1	1 1	1 0
0 0	1 (0)	0 (1)	1 (3)	1 (2)
0 1	0 (4)	0 (5)	0 (7)	0 (6)
1 1	1 (12)	1 (13)	1 (15)	0 (14)
1 0	1 (8)	0 (9)	1 (11)	0 (10)

z_1

Figure 3.1 Karnaugh map for Example 3.1.

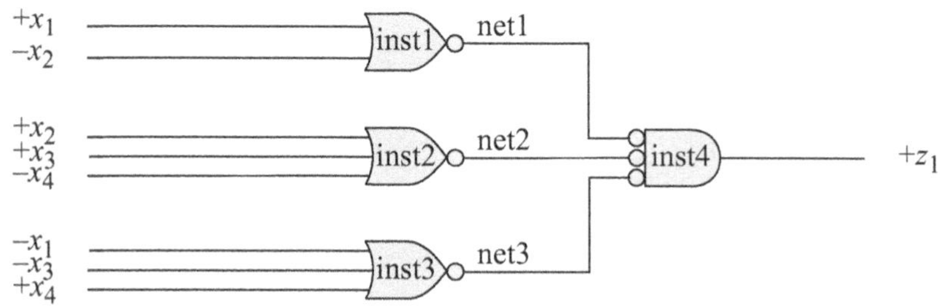

Figure 3.2 Logic diagram for Example 3.1.

```
//built-in primitives for a pos equation
module log_eqn_pos_nor2 (x1, x2, x3, x4, z1);

input x1, x2, x3, x4;
output z1;

nor    inst1 (net1, x1, ~x2),
       inst2 (net2, x2, x3, ~x4),
       inst3 (net3, ~x1, ~x3, x4);

nor    inst4 (z1, net1, net2, net3);

endmodule
```

Figure 3.3 Design module using built-in primitives for Example 3.1.

```
//test bench for pos equation
module log_eqn_pos_nor2_tb;

//inputs are reg for test bench
reg x1, x2, x3, x4;

//outputs are wire for test bench
wire z1;

//apply input vectors
initial
begin: apply_stimulus
   reg [4:0] invect;
   for (invect=0; invect<16; invect=invect+1)
      begin
         {x1, x2, x3, x4} = invect [4:0];
         #10 $display ("x1 x2 x3 x4 = %b, z1 = %b",
                              {x1, x2, x3, x4}, z1);
      end
end

//instantiate the module into the test bench
log_eqn_pos_nor2 inst1 (
   .x1(x1),
   .x2(x2),
   .x3(x3),
   .x4(x4),
   .z1(z1)
   );

endmodule
```

Figure 3.4 Test bench for Figure 3.3 of Example 3.1.

```
x1 x2 x3 x4 = 0000, z1 = 1  |  x1 x2 x3 x4 = 1000, z1 = 1
x1 x2 x3 x4 = 0001, z1 = 0  |  x1 x2 x3 x4 = 1001, z1 = 0
x1 x2 x3 x4 = 0010, z1 = 1  |  x1 x2 x3 x4 = 1010, z1 = 0
x1 x2 x3 x4 = 0011, z1 = 1  |  x1 x2 x3 x4 = 1011, z1 = 1
x1 x2 x3 x4 = 0100, z1 = 0  |  x1 x2 x3 x4 = 1100, z1 = 1
x1 x2 x3 x4 = 0101, z1 = 0  |  x1 x2 x3 x4 = 1101, z1 = 1
x1 x2 x3 x4 = 0110, z1 = 0  |  x1 x2 x3 x4 = 1110, z1 = 0
x1 x2 x3 x4 = 0111, z1 = 0  |  x1 x2 x3 x4 = 1111, z1 = 1
```

Figure 3.5 Outputs for Figure 3.3 of Example 3.1.

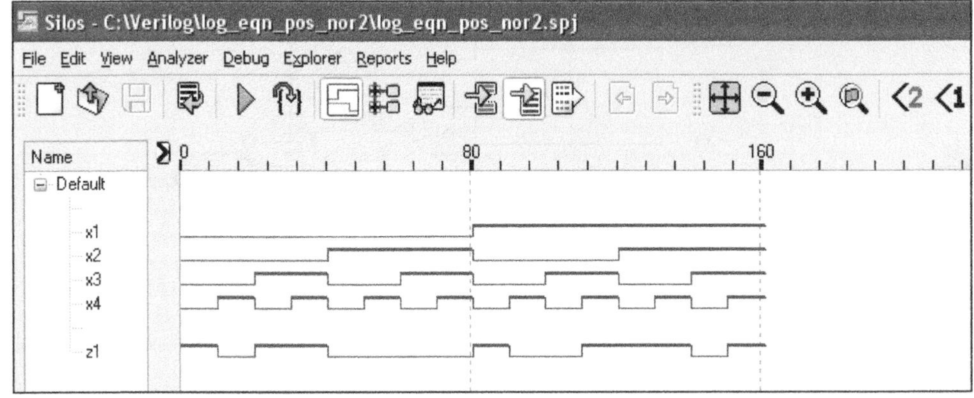

Figure 3.6 Waveforms for Figure 3.3 of Example 3.1.

Example 3.2 A circuit will be designed using built-in primitives that generates an output whenever a 4-bit number N is within the following range:

$$3 < N \leq 8$$
$$10 \leq N < 15$$

The Karnaugh map is shown in Figure 3.7 that represents the above number range. The equation is shown in Equation 3.1. The circuit will be designed using NAND logic gates in a sum-of-products form as shown in Figure 3.8.

x_1x_2 \ x_3x_4	0 0	0 1	1 1	1 0
0 0	[0] 0	[1] 0	[3] 0	[2] 0
0 1	[4] 1	[5] 1	[7] 1	[6] 1
1 1	[12] 1	[13] 1	[15] 0	[14] 1
1 0	[8] 1	[9] 0	[11] 1	[10] 1

z_1

Figure 3.7 Karnaugh map for Example 3.2.

$$z_1 = x_1'x_2 + x_2x_3' + x_1x_4' + x_1x_2'x_3 \tag{3.1}$$

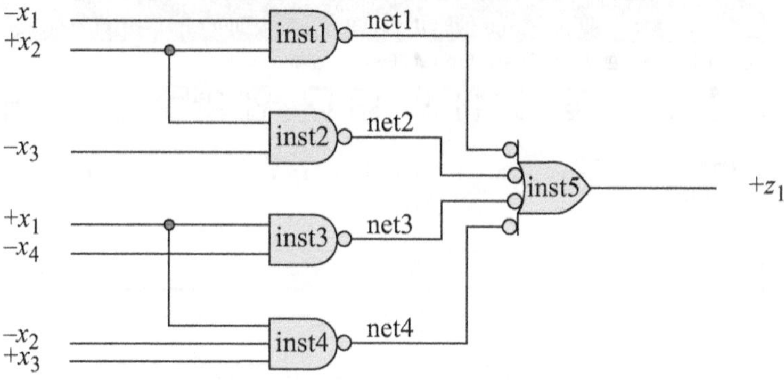

Figure 3.8 Logic diagram for Example 3.2.

The design module is shown in Figure 3.9 using NAND built-in primitives. The test bench is shown in Figure 3.10 using all 16 combinations of four variables. The outputs and waveforms are shown in Figure 3.11 and Figure 3.12, respectively.

```
//built-in primitive number range
module num_range3 (x1, x2, x3, x4, z1);

//list inputs and output
input x1, x2, x3, x4;
output z1;

//design the logic using built-in primitives
nand  inst1 (net1, ~x1, x2),
      inst2 (net2, x2, ~x3),
      inst3 (net3, x1, ~x4),
      inst4 (net4, x1, ~x2, x3);
nand  inst5 (z1, net1, net2, net3, net4);
endmodule
```

Figure 3.9 Design module for Example 3.2 using built-in primitives.

```
//test bench for number range module
module num_range3_tb;

//inputs are reg outputs are wire for test bench
reg x1, x2, x3, x4;
wire z1;
//continued on next page
```

Figure 3.10 Test bench for Figure 3.9 of Example 3.2.

```
//apply input vectors
initial
begin: apply_stimulus
   reg [4:0] invect;
   for (invect=0; invect<16; invect=invect+1)
      begin
         {x1, x2, x3, x4} = invect [4:0];
         #10 $display ("x1 x2 x3 x4 = %b, z1 = %b",
                        {x1, x2, x3, x4}, z1);
      end
end

//instantiate the module into the test bench
num_range3 inst1 (
   .x1(x1),
   .x2(x2),
   .x3(x3),
   .x4(x4),
   .z1(z1)
   );

endmodule
```

Figure 3.10 (Continued)

```
x1 x2 x3 x4 = 0000, z1 = 0
x1 x2 x3 x4 = 0001, z1 = 0
x1 x2 x3 x4 = 0010, z1 = 0
x1 x2 x3 x4 = 0011, z1 = 0
x1 x2 x3 x4 = 0100, z1 = 1
x1 x2 x3 x4 = 0101, z1 = 1
x1 x2 x3 x4 = 0110, z1 = 1
x1 x2 x3 x4 = 0111, z1 = 1
x1 x2 x3 x4 = 1000, z1 = 1
x1 x2 x3 x4 = 1001, z1 = 0
x1 x2 x3 x4 = 1010, z1 = 1
x1 x2 x3 x4 = 1011, z1 = 1
x1 x2 x3 x4 = 1100, z1 = 1
x1 x2 x3 x4 = 1101, z1 = 1
x1 x2 x3 x4 = 1110, z1 = 1
x1 x2 x3 x4 = 1111, z1 = 0
```

Figure 3.11 Outputs for Figure 3.9 of Example 3.2.

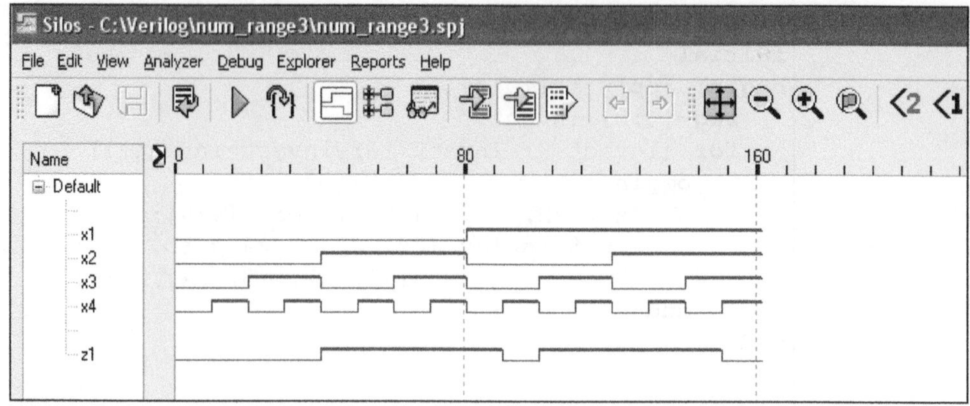

Figure 3.12 Waveforms for Figure 3.9 of Example 3.2.

Example 3.3 A binary-to-octal 3:8 decoder will be designed using built-in primitives. The 3:8 decoder that was designed in Chapter 2 will be used in the Verilog implementation and is shown in Figure 3.13.

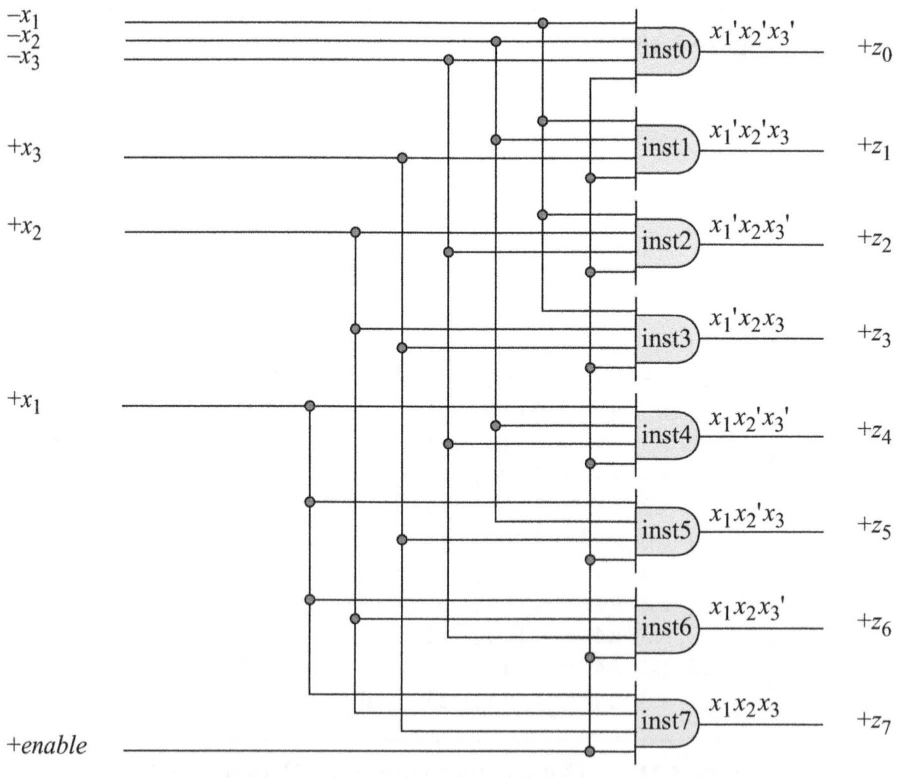

Figure 3.13 Logic diagram for a binary-to-octal 3:8 decoder.

The decoder will be designed using AND gate built-in primitives. The inputs $x_1x_2x_3$ are available in both active-high and active-low voltage levels. There is also an active-high *enable*. The design module is shown in Figure 3.14 and the test bench is shown in Figure 3.15. Note that the outputs are a vector *z[7:0]*, where *z[0]* is the low-order bit. In this test bench, the input vectors are applied individually every 10 time units (#10). The outputs and waveforms are shown in Figure 3.16 and Figure 3.17, respectively.

```
//built-in primitive 3:8 decoder
module decoder_3to8_bip (x1, x2, x3, enable, z);

input x1, x2, x3, enable;
output [7:0] z;

and   inst0 (z[0], ~x1, ~x2, ~x3, enable),
      inst1 (z[1], ~x1, ~x2, x3, enable),
      inst2 (z[2], ~x1, x2, ~x3, enable),
      inst3 (z[3], ~x1, x2, x3, enable),
      inst4 (z[4], x1, ~x2, ~x3, enable),
      inst5 (z[5], x1, ~x2, x3, enable),
      inst6 (z[6], x1, x2, ~x3, enable),
      inst7 (z[7], x1, x2, x3, enable);

endmodule
```

Figure 3.14 Design module for a 3:8 decoder using AND gate built-in primitives.

```
//test bench for the 3:8 decoder
module decoder_3to8_bip_tb;

reg x1, x2, x3, enable;
wire [7:0] z;

//display variables
initial
$monitor ("x1 x2 x3 = %b, z = %b", {x1, x2, x3}, z);

//apply input vectors
initial
begin
   #0    enable = 1'b1;
         x1=1'b0;    x2=1'b0;    x3=1'b0;

//continued on next page
```

Figure 3.15 Test bench for the 3:8 decoder.

```
    #10    x1=1'b0;    x2=1'b0;    x3=1'b1;
    #10    x1=1'b0;    x2=1'b1;    x3=1'b0;
    #10    x1=1'b0;    x2=1'b1;    x3=1'b1;
    #10    x1=1'b1;    x2=1'b0;    x3=1'b0;
    #10    x1=1'b1;    x2=1'b0;    x3=1'b1;
    #10    x1=1'b1;    x2=1'b1;    x3=1'b0;
    #10    x1=1'b1;    x2=1'b1;    x3=1'b1;
    #10    $stop;
end

//instantiate the module into the test bench
decoder_3to8_bip inst1 (
    .x1(x1),
    .x2(x2),
    .x3(x3),
    .enable(enable),
    .z(z)
    );
endmodule
```

Figure 3.15 (Continued)

```
x1 x2 x3 = 000, z = 00000001  | x1 x2 x3 = 100, z = 00010000
x1 x2 x3 = 001, z = 00000010  | x1 x2 x3 = 101, z = 00100000
x1 x2 x3 = 010, z = 00000100  | x1 x2 x3 = 110, z = 01000000
x1 x2 x3 = 011, z = 00001000  | x1 x2 x3 = 111, z = 10000000
```

Figure 3.16 Outputs for the 3:8 decoder.

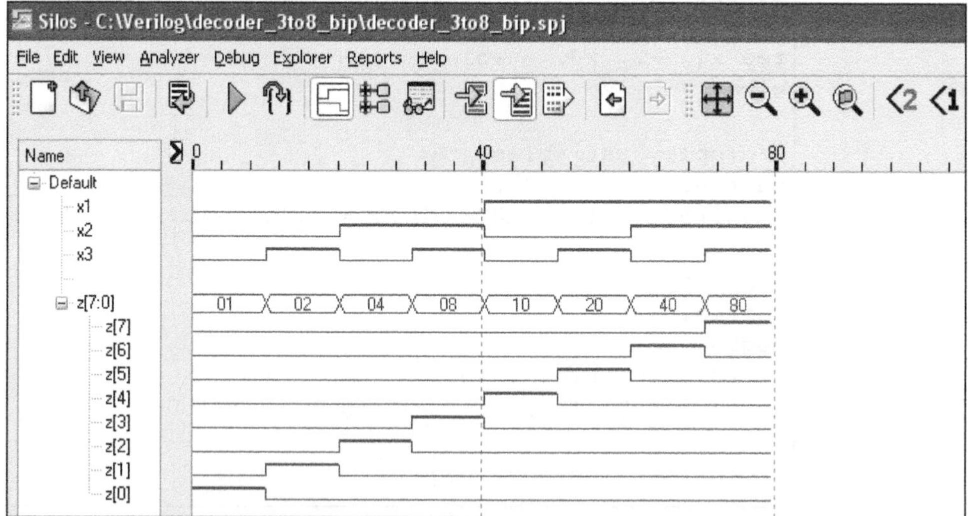

Figure 3.17 Waveforms for the 3:8 decoder.

Example 3.4 A 3-bit binary comparator will be designed using built-in primitives. The operands are 3-bit vectors $A = a_2a_1a_0$ and $B = b_2b_1b_0$, where a_0 and b_0 are the low-order bits of A and B, respectively. There are three outputs indicating the relative magnitude of the two operands: $(A < B)$, $(A = B)$, and $(A > B)$.

The equations to obtain the outputs are shown in Equation 3.2. The logic diagram is shown in Figure 3.18 as obtained directly from Equation 3.2. The design module, test bench module, outputs, and waveforms are shown in Figure 3.19, Figure 3.20, Figure 3.21, and Figure 3.22, respectively. To thoroughly test a module, all combinations of the inputs should be applied to the module in the test bench.

$$(A < B) = a_2'b_2 + (a_2 \oplus b_2)'a_1'b_1 + (a_2 \oplus b_2)'(a_1 \oplus b_1)'a_0'b_0$$

$$(A = B) = (a_2 \oplus b_2)'(a_1 \oplus b_1)'(a_0 \oplus b_0)'$$

$$(A > B) = a_2b_2' + (a_2 \oplus b_2)'a_1b_1' + (a_2 \oplus b_2)'(a_1 \oplus b_1)'a_0b_0' \tag{3.2}$$

Referring to Equation 3.2 for $(A < B)$, the term $a_2'b_2$ indicates that if the high-order bits of A and B are 0 and 1, respectively, then A must be less than B. If the high-order bits of A and B are equal, then the relative magnitude of A and B depends upon the next lower-order bits a_1 and b_1. This is indicated by the second term of the equation for $(A < B)$.

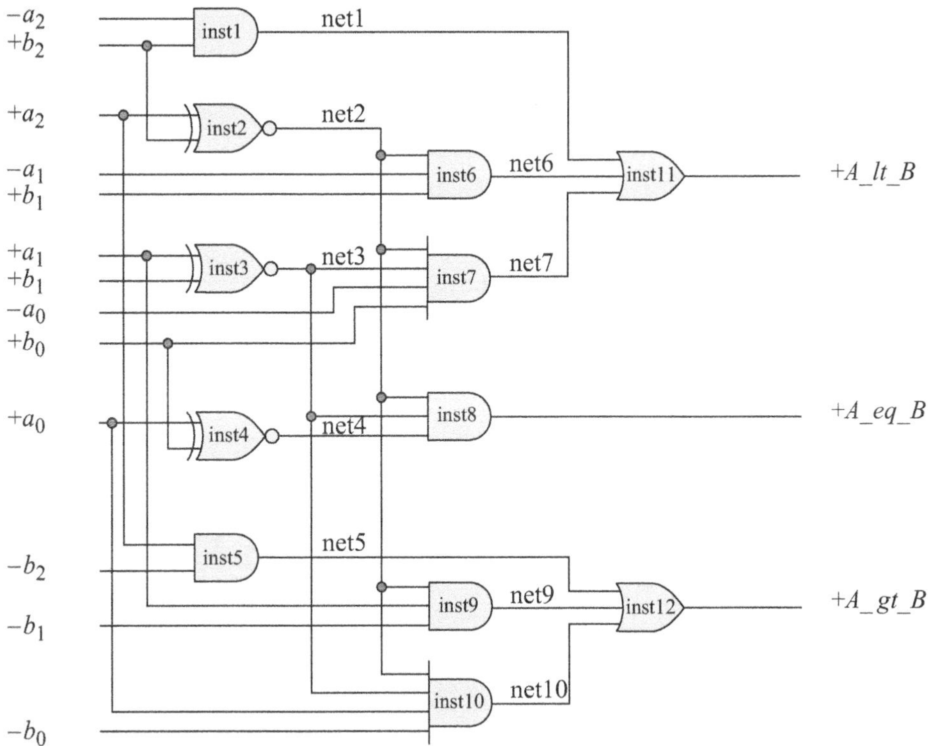

Figure 3.18 Logic diagram for a 3-bit comparator.

```
//built-in primitives for 3-bit comparator
module comparator3_bip (a2, a1, a0, b2, b1, b0,
                         a_lt_b, a_eq_b, a_gt_b);

input a2, a1, a0, b2, b1, b0;
output a_lt_b, a_eq_b, a_gt_b;

and   inst1(net1, ~a2, b2);
xnor  inst2(net2, a2, b2);
xnor  inst3(net3, a1, b1);
xnor  inst4(net4, a0, b0);
and   inst5(net5, a2, ~b2);
and   inst6(net6, net2, ~a1, b1);
and   inst7(net7, net2, net3, ~a0, b0);
and   inst9(net9, net2, a1, ~b1);
and   inst10(net10, net2, net3, a0, ~b0);

and   inst8(a_eq_b, net2, net3, net4);
or    inst11(a_lt_b, net1, net6, net7);
or    inst12(a_gt_b, net5, net9, net10);

endmodule
```

Figure 3.19 Design module for the 3-bit comparator using built-in primitives.

```
//test bench for 3-bit comparator
module comparator3_bip_tb;

reg a2, a1, a0, b2, b1, b0;
wire a_lt_b, a_eq_b, a_gt_b;

//apply input vectors
initial
begin: apply_stimulus
   reg [6:0] invect;
   for (invect=0; invect<64; invect=invect+1)
      begin
         {a2, a1, a0, b2, b1, b0} = invect [6:0];
         #10 $display ("a2 a1 a0 = %b, b2 b1 b0 = %b,
                  a_lt_b = %b, a_eq_b = %b, a_gt_b = %b",
                  {a2, a1, a0}, {b2, b1, b0},
                  a_lt_b, a_eq_b, a_gt_b);
      end
end
//continued on next page
```

Figure 3.20 Test bench for the 3-bit comparator.

```
//instantiate the module into the test bench
comparator3_bip inst1 (
   .a2(a2),
   .a1(a1),
   .a0(a0),
   .b2(b2),
   .b1(b1),
   .b0(b0),
   .a_lt_b(a_lt_b),
   .a_eq_b(a_eq_b),
   .a_gt_b(a_gt_b)
   );

endmodule
```

Figure 3.20 (Continued)

```
a2 a1 a0 = 000, b2 b1 b0 = 000, a_lt_b=0, a_eq_b=1, a_gt_b=0
a2 a1 a0 = 000, b2 b1 b0 = 001, a_lt_b=1, a_eq_b=0, a_gt_b=0
a2 a1 a0 = 000, b2 b1 b0 = 010, a_lt_b=1, a_eq_b=0, a_gt_b=0
a2 a1 a0 = 000, b2 b1 b0 = 011, a_lt_b=1, a_eq_b=0, a_gt_b=0
a2 a1 a0 = 000, b2 b1 b0 = 100, a_lt_b=1, a_eq_b=0, a_gt_b=0
a2 a1 a0 = 000, b2 b1 b0 = 101, a_lt_b=1, a_eq_b=0, a_gt_b=0
a2 a1 a0 = 000, b2 b1 b0 = 110, a_lt_b=1, a_eq_b=0, a_gt_b=0
a2 a1 a0 = 000, b2 b1 b0 = 111, a_lt_b=1, a_eq_b=0, a_gt_b=0

a2 a1 a0 = 001, b2 b1 b0 = 000, a_lt_b=0, a_eq_b=0, a_gt_b=1
a2 a1 a0 = 001, b2 b1 b0 = 001, a_lt_b=0, a_eq_b=1, a_gt_b=0
a2 a1 a0 = 001, b2 b1 b0 = 010, a_lt_b=1, a_eq_b=0, a_gt_b=0
a2 a1 a0 = 001, b2 b1 b0 = 011, a_lt_b=1, a_eq_b=0, a_gt_b=0
a2 a1 a0 = 001, b2 b1 b0 = 100, a_lt_b=1, a_eq_b=0, a_gt_b=0
a2 a1 a0 = 001, b2 b1 b0 = 101, a_lt_b=1, a_eq_b=0, a_gt_b=0
a2 a1 a0 = 001, b2 b1 b0 = 110, a_lt_b=1, a_eq_b=0, a_gt_b=0
a2 a1 a0 = 001, b2 b1 b0 = 111, a_lt_b=1, a_eq_b=0, a_gt_b=0

a2 a1 a0 = 010, b2 b1 b0 = 000, a_lt_b=0, a_eq_b=0, a_gt_b=1
a2 a1 a0 = 010, b2 b1 b0 = 001, a_lt_b=0, a_eq_b=0, a_gt_b=1
a2 a1 a0 = 010, b2 b1 b0 = 010, a_lt_b=0, a_eq_b=1, a_gt_b=0
a2 a1 a0 = 010, b2 b1 b0 = 011, a_lt_b=1, a_eq_b=0, a_gt_b=0
a2 a1 a0 = 010, b2 b1 b0 = 100, a_lt_b=1, a_eq_b=0, a_gt_b=0
a2 a1 a0 = 010, b2 b1 b0 = 101, a_lt_b=1, a_eq_b=0, a_gt_b=0
a2 a1 a0 = 010, b2 b1 b0 = 110, a_lt_b=1, a_eq_b=0, a_gt_b=0
a2 a1 a0 = 010, b2 b1 b0 = 111, a_lt_b=1, a_eq_b=0, a_gt_b=0

//continued on next page
```

Figure 3.21 Outputs for the 3-bit comparator.

```
a2 a1 a0 = 011, b2 b1 b0 = 000, a_lt_b=0, a_eq_b=0, a_gt_b=1
a2 a1 a0 = 011, b2 b1 b0 = 001, a_lt_b=0, a_eq_b=0, a_gt_b=1
a2 a1 a0 = 011, b2 b1 b0 = 010, a_lt_b=0, a_eq_b=0, a_gt_b=1
a2 a1 a0 = 011, b2 b1 b0 = 011, a_lt_b=0, a_eq_b=1, a_gt_b=0
a2 a1 a0 = 011, b2 b1 b0 = 100, a_lt_b=1, a_eq_b=0, a_gt_b=0
a2 a1 a0 = 011, b2 b1 b0 = 101, a_lt_b=1, a_eq_b=0, a_gt_b=0
a2 a1 a0 = 011, b2 b1 b0 = 110, a_lt_b=1, a_eq_b=0, a_gt_b=0
a2 a1 a0 = 011, b2 b1 b0 = 111, a_lt_b=1, a_eq_b=0, a_gt_b=0

a2 a1 a0 = 100, b2 b1 b0 = 000, a_lt_b=0, a_eq_b=0, a_gt_b=1
a2 a1 a0 = 100, b2 b1 b0 = 001, a_lt_b=0, a_eq_b=0, a_gt_b=1
a2 a1 a0 = 100, b2 b1 b0 = 010, a_lt_b=0, a_eq_b=0, a_gt_b=1
a2 a1 a0 = 100, b2 b1 b0 = 011, a_lt_b=0, a_eq_b=0, a_gt_b=1
a2 a1 a0 = 100, b2 b1 b0 = 100, a_lt_b=0, a_eq_b=1, a_gt_b=0
a2 a1 a0 = 100, b2 b1 b0 = 101, a_lt_b=1, a_eq_b=0, a_gt_b=0
a2 a1 a0 = 100, b2 b1 b0 = 110, a_lt_b=1, a_eq_b=0, a_gt_b=0
a2 a1 a0 = 100, b2 b1 b0 = 111, a_lt_b=1, a_eq_b=0, a_gt_b=0

a2 a1 a0 = 101, b2 b1 b0 = 000, a_lt_b=0, a_eq_b=0, a_gt_b=1
a2 a1 a0 = 101, b2 b1 b0 = 001, a_lt_b=0, a_eq_b=0, a_gt_b=1
a2 a1 a0 = 101, b2 b1 b0 = 010, a_lt_b=0, a_eq_b=0, a_gt_b=1
a2 a1 a0 = 101, b2 b1 b0 = 011, a_lt_b=0, a_eq_b=0, a_gt_b=1
a2 a1 a0 = 101, b2 b1 b0 = 100, a_lt_b=0, a_eq_b=0, a_gt_b=1
a2 a1 a0 = 101, b2 b1 b0 = 101, a_lt_b=0, a_eq_b=1, a_gt_b=0
a2 a1 a0 = 101, b2 b1 b0 = 110, a_lt_b=1, a_eq_b=0, a_gt_b=0
a2 a1 a0 = 101, b2 b1 b0 = 111, a_lt_b=1, a_eq_b=0, a_gt_b=0

a2 a1 a0 = 110, b2 b1 b0 = 000, a_lt_b=0, a_eq_b=0, a_gt_b=1
a2 a1 a0 = 110, b2 b1 b0 = 001, a_lt_b=0, a_eq_b=0, a_gt_b=1
a2 a1 a0 = 110, b2 b1 b0 = 010, a_lt_b=0, a_eq_b=0, a_gt_b=1
a2 a1 a0 = 110, b2 b1 b0 = 011, a_lt_b=0, a_eq_b=0, a_gt_b=1
a2 a1 a0 = 110, b2 b1 b0 = 100, a_lt_b=0, a_eq_b=0, a_gt_b=1
a2 a1 a0 = 110, b2 b1 b0 = 101, a_lt_b=0, a_eq_b=0, a_gt_b=1
a2 a1 a0 = 110, b2 b1 b0 = 110, a_lt_b=0, a_eq_b=1, a_gt_b=0
a2 a1 a0 = 110, b2 b1 b0 = 111, a_lt_b=1, a_eq_b=0, a_gt_b=0

a2 a1 a0 = 111, b2 b1 b0 = 000, a_lt_b=0, a_eq_b=0, a_gt_b=1
a2 a1 a0 = 111, b2 b1 b0 = 001, a_lt_b=0, a_eq_b=0, a_gt_b=1
a2 a1 a0 = 111, b2 b1 b0 = 010, a_lt_b=0, a_eq_b=0, a_gt_b=1
a2 a1 a0 = 111, b2 b1 b0 = 011, a_lt_b=0, a_eq_b=0, a_gt_b=1
a2 a1 a0 = 111, b2 b1 b0 = 100, a_lt_b=0, a_eq_b=0, a_gt_b=1
a2 a1 a0 = 111, b2 b1 b0 = 101, a_lt_b=0, a_eq_b=0, a_gt_b=1
a2 a1 a0 = 111, b2 b1 b0 = 110, a_lt_b=0, a_eq_b=0, a_gt_b=1
a2 a1 a0 = 111, b2 b1 b0 = 111, a_lt_b=0, a_eq_b=1, a_gt_b=0
```

Figure 3.21 (Continued)

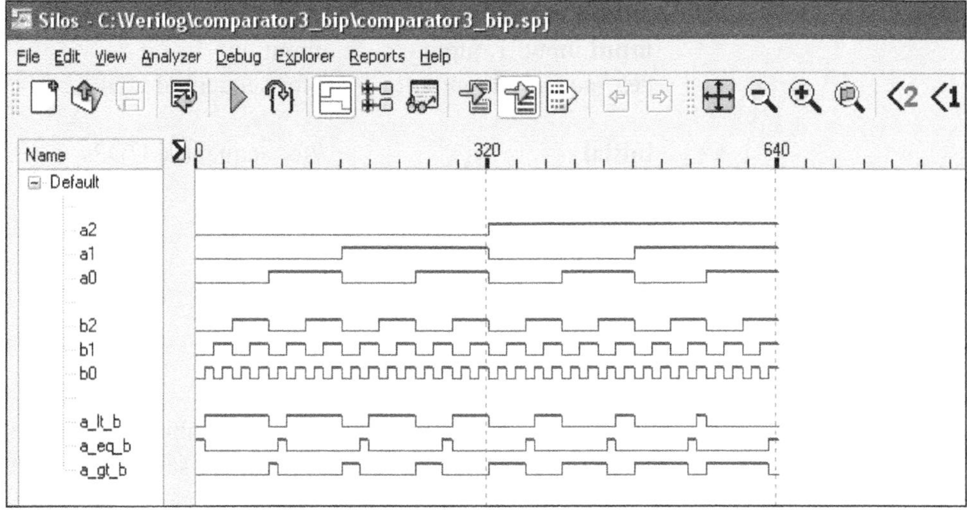

Figure 3.22 Waveforms for the 3-bit comparator.

3.2 User-Defined Primitives

In addition to built-in primitives, Verilog provides the ability to design primitives according to user specifications. These are called *user-defined primitives* (UDPs) and are usually a higher-level logic function than built-in primitives. They are independent primitives and do not instantiate other primitives or modules. UDPs are instantiated into a module the same way as built-in primitives; that is, the syntax for a UDP instantiation is the same as that for a built-in primitive instantiation. A UDP is defined outside the module into which it is instantiated. There are two types of UDPs: combinational and sequential. Sequential primitives include level-sensitive and edge-sensitive circuits

The syntax for a UDP is similar to that for declaring a module. The definition begins with the keyword **primitive** and ends with the keyword **endprimitive**. The UDP contains a name and a list of ports, which are declared as **input** or **output**. For a sequential UDP, the output port is declared as **reg**. UDPs can have one or more scalar inputs, but only one scalar output. The output port is listed first in the terminal list followed by the input ports, in the same way that the terminal list appears in built-in primitives. UDPs do not support **inout** ports.

The UDP table is an essential part of the internal structure and defines the functionality of the circuit. It is a lookup table similar in concept to a truth table. The table begins with the keyword **table** and ends with the keyword **endtable**. The contents of the table define the value of the output with respect to the inputs. The syntax for a UDP is shown below.

```
primitive udp_name (output, input_1, input_2, . . . , input_n);
    output output;
    input input_1, input_2, . . . , input_n;
    reg sequential_output;          //for sequential UDPs

    initial                         //for sequential UDPs

    table
        state table entries
    endtable
endprimitive
```

Examples will now be presented that include combinational UDPs and sequential UDPs. UDPs are not compiled separately. They are saved in the same project as the module with a .v extension; for example, *udp_and.v*.

Example 3.5 A majority circuit generates logic 1 output whenever the majority of the inputs is at a logic 1 level; otherwise, the output is a logic 0. Therefore, a majority circuit must have an odd number of inputs in order to have a majority of the inputs be at the same logic level.

A 5-input majority circuit will be designed using the Karnaugh map of Figure 3.23, where a 1 entry indicates that the majority of the inputs is a logic 1. An equation is not necessary to design a majority circuit — the output is a logic 1 if the majority of the inputs is a logic 1. Whenever three inputs are at the same logic level, the state of the remaining two inputs is irrelevant — they may be 0s or 1s. The module is shown in Figure 3.24 where the symbol (?) indicates a "don't care" condition. The test bench is shown in Figure 3.25, and the outputs are shown in Figure 3.26.

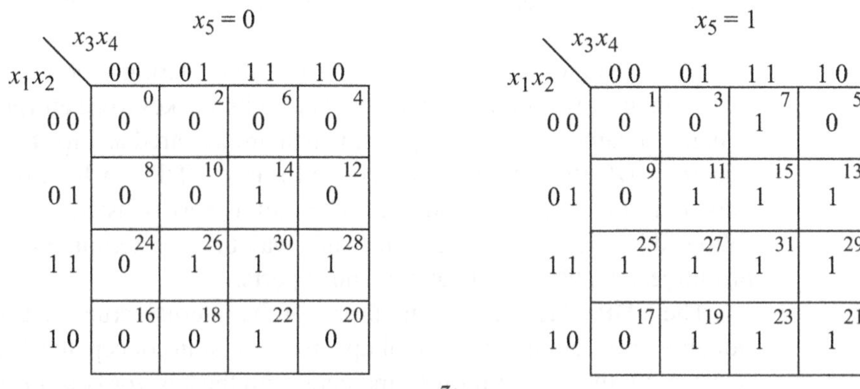

Figure 3.23 Karnaugh map for the majority circuit of Example 3.5.

```
//five-input majority circuit as a udp.  Save as a .v file
primitive udp_maj5 (z1, x1, x2, x3, x4, x5);

output z1;
input x1, x2, x3, x4, x5;

table
//inputs are in same order as input list
// x1 x2 x3 x4 x5  :  z1;
   0  0  0  ?  ?   :  0;
   0  0  ?  0  ?   :  0;
   0  0  ?  ?  0   :  0;
   0  ?  0  0  ?   :  0;
   0  ?  ?  0  0   :  0;
   0  ?  0  ?  0   :  0;
   ?  0  0  0  ?   :  0;
   ?  ?  0  0  0   :  0;
   ?  0  0  ?  0   :  0;
   ?  0  ?  0  0   :  0;
   ?  ?  0  0  0   :  0;

   1  1  1  ?  ?   :  1;
   1  1  ?  1  ?   :  1;
   1  1  ?  ?  1   :  1;
   1  ?  1  1  ?   :  1;
   1  ?  ?  1  1   :  1;
   1  ?  1  ?  1   :  1;
   ?  1  1  1  ?   :  1;
   ?  ?  1  1  1   :  1;
   ?  1  1  ?  1   :  1;
   ?  1  ?  1  1   :  1;
   ?  ?  1  1  1   :  1;
endtable

endprimitive
```

Figure 3.24 A UDP module for a 5-input majority circuit.

```
//udp_maj5 test bench
module udp_maj5_tb;
reg x1, x2, x3, x4, x5;      //inputs are reg for tb
wire z1;                      //outputs are wire

//continued on next page
```

Figure 3.25 Test bench for the 5-input majority circuit.

```
initial
begin: name //a name is required for this method
   reg [5:0] invect;
   for (invect = 0; invect < 32; invect = invect + 1)
      begin
         {x1, x2, x3, x4, x5} = invect [4:0];
         #10 $display ("x1x2x3x4x5 = %b%b%b%b%b, z1=%b",
                              x1, x2, x3, x4, x5, z1);

      end
end

//instantiation must be done by position, not by name.
udp_maj5 inst1 (z1, x1, x2, x3, x4, x5);

endmodule
```

Figure 3.25 (Continued)

```
x1x2x3x4x5 = 00000, z1=0        x1x2x3x4x5 = 10000, z1=0
x1x2x3x4x5 = 00001, z1=0        x1x2x3x4x5 = 10001, z1=0
x1x2x3x4x5 = 00010, z1=0        x1x2x3x4x5 = 10010, z1=0
x1x2x3x4x5 = 00011, z1=0        x1x2x3x4x5 = 10011, z1=1
x1x2x3x4x5 = 00100, z1=0        x1x2x3x4x5 = 10100, z1=0
x1x2x3x4x5 = 00101, z1=0        x1x2x3x4x5 = 10101, z1=1
x1x2x3x4x5 = 00110, z1=0        x1x2x3x4x5 = 10110, z1=1
x1x2x3x4x5 = 00111, z1=1        x1x2x3x4x5 = 10111, z1=1
x1x2x3x4x5 = 01000, z1=0        x1x2x3x4x5 = 11000, z1=0
x1x2x3x4x5 = 01001, z1=0        x1x2x3x4x5 = 11001, z1=1
x1x2x3x4x5 = 01010, z1=0        x1x2x3x4x5 = 11010, z1=1
x1x2x3x4x5 = 01011, z1=1        x1x2x3x4x5 = 11011, z1=1
x1x2x3x4x5 = 01100, z1=0        x1x2x3x4x5 = 11100, z1=1
x1x2x3x4x5 = 01101, z1=1        x1x2x3x4x5 = 11101, z1=1
x1x2x3x4x5 = 01110, z1=1        x1x2x3x4x5 = 11110, z1=1
x1x2x3x4x5 = 01111, z1=1        x1x2x3x4x5 = 11111, z1=1
```

Figure 3.26 Outputs for the 5-input majority circuit.

Example 3.6 Most counters count in either a count-up or count-down sequence. Still other counters can be designed for a unique application in which the counting sequence is neither entirely up nor entirely down. These have a nonsequential counting sequence that is prescribed by external requirements. Such a counter has a counting sequence as follows: $y_1y_2y_3y_4$ = 0000, 1000, 1100, 1110, 1111, 0111, 0011, 0001, 0000, and is classified as a *Johnson counter*. The counter is reset initially to

$y_1y_2y_3y_4 = 0000$. The unspecified states can be regarded as "don't care" states in order to minimize the δ next-state logic. The inverted output of the last flip-flop is fed back to the *D* input of the first flip-flop.

The logic diagram for a 4-bit Johnson counter is shown in Figure 3.27 using positive-edge-triggered *D* flip-flops. The *D* flip-flop will be designed as a user-defined primitive, then instantiated four times into the design module of the Johnson counter. The *D* flip-flop is shown in Figure 3.28 as a UDP.

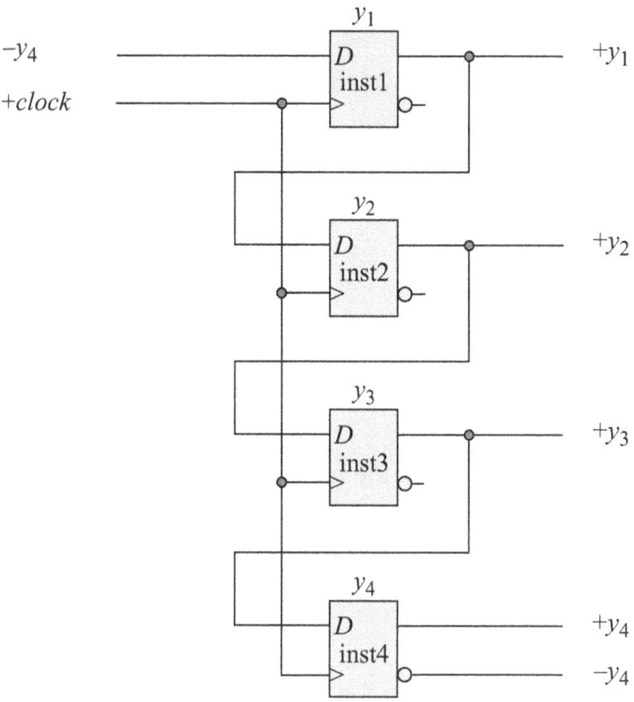

Figure 3.27 Logic diagram for a 4-bit Johnson counter.

```
//a positive-edge-sensitive D flip-flop
primitive udp_dff_edge1 (q, d, clk, rst_n);

input d, clk, rst_n;
output q;

reg q;        //q is internal storage

//initialize q to 0
initial
   q = 0;
//continued on next page
```

Figure 3.28 A user-defined primitive for a *D* flip-flop.

```
//define state table
table
//inputs are in the same order as the input list
// d     clk   rst_n : q  : q+;    q+ is the next state
   0     (01)  1     : ?  : 0;     //(01) is rising edge
   1     (01)  1     : ?  : 1;     //rst_n = 1 means no rst
   1     (0x)  1     : 1  : 1;     //(0x) is no change
   0     (0x)  1     : 0  : 0;
   ?     (?0)  1     : ?  : -;     //ignore negative edge
//reset case when rst_n is 0 and clk has any transition
   ?     (??)  0     : ?  : 0;     //rst_n = 0 means reset
//reset case when rst_n is 0.  d & clk can be anything, q+=0
   ?     ?     0     : ?  : 0;
//reset case when 0 --> 1 transition on rst_n.  Hold q+ state
   ?     ?     (01)  : ?  : -;
//non-reset case when d has any trans, but clk has no trans
   (??)  ?     1     : ?  : -;     //clk = ?, means no edge
endtable
endprimitive
```

Figure 3.28 (Continued)

The design module for the Johnson counter is shown in Figure 3.29 which instantiates the user-defined primitive *udp_dff_edge1* four times to implement the Johnson counter. The test bench is shown in Figure 3.30. The outputs and waveforms are shown in Figure 3.31 and Figure 3.32, respectively.

```
//udp for a 4-bit johnson counter
module ctr_johnson4 (rst_n, clk, y1, y2, y3, y4);
input rst_n, clk;
output y1, y2, y3, y4;

//instantiate D flip-flop for y1
udp_dff_edge1 inst1 (y1, ~y4, clk, rst_n);

//instantiate D flip-flop for y2
udp_dff_edge1 inst2 (y2, y1, clk, rst_n);

//instantiate D flip-flop for y3
udp_dff_edge1 inst3 (y3, y2, clk, rst_n);

//instantiate D flip-flop for y4
udp_dff_edge1 inst4 (y4, y3, clk, rst_n);
endmodule
```

Figure 3.29 A Johnson counter designed using a UDP for a *D* flip-flop.

```
//test bench for the 4-bit johnson counter
module ctr_johnson4_tb;

reg clk, rst_n;           //inputs are reg for tb
wire y1, y2, y3, y4;      //outputs are wire for tb

initial
$monitor ("count = %b", {y1, y2, y3, y4});

initial                   //define clk
begin
   clk = 1'b0;
   forever
      #10clk = ~clk;
end

initial                   //define reset
begin
   #0 rst_n = 1'b0;
   #5 rst_n = 1'b1;
   #200  $stop;
end

ctr_johnson4 inst1 (      //instantiate the module
   .rst_n(rst_n),
   .clk(clk),
   .y1(y1),
   .y2(y2),
   .y3(y3),
   .y4(y4)
   );
endmodule
```

Figure 3.30 Test bench for the 4-bit UDP Johnson counter.

```
count = 0000
count = 1000
count = 1100
count = 1110
count = 1111
count = 0111
count = 0011
count = 0001
count = 0000
count = 1000
```

Figure 3.31 Outputs for the 4-bit UDP Johnson counter.

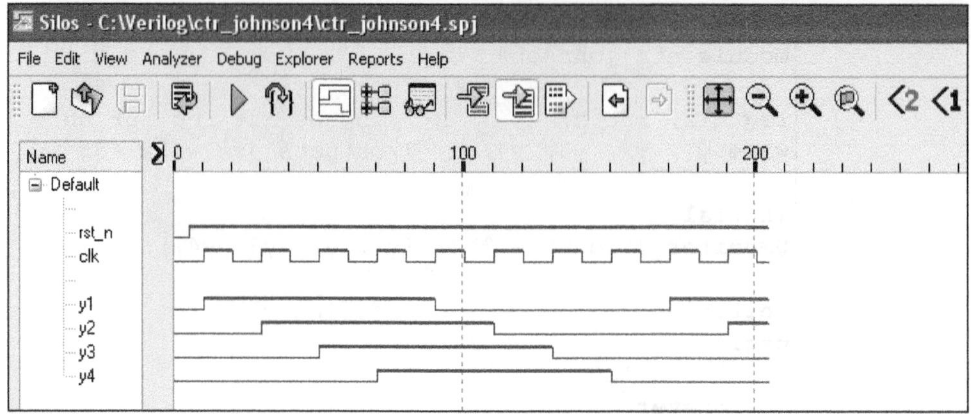

Figure 3.32 Waveforms for the 4-bit UDP Johnson counter.

Example 3.7 The logic circuit shown in Figure 3.33 will be designed using NOR gates that were designed as user-defined primitives. The equation that represents output z_1 is shown in Equation 3.3. The UDP design module for the NOR gate is shown in Figure 3.34 and the design module for the circuit is shown in Figure 3.35. The test bench module, the outputs, and the waveforms are shown in Figure 3.36, Figure 3.37, and Figure 3.38, respectively.

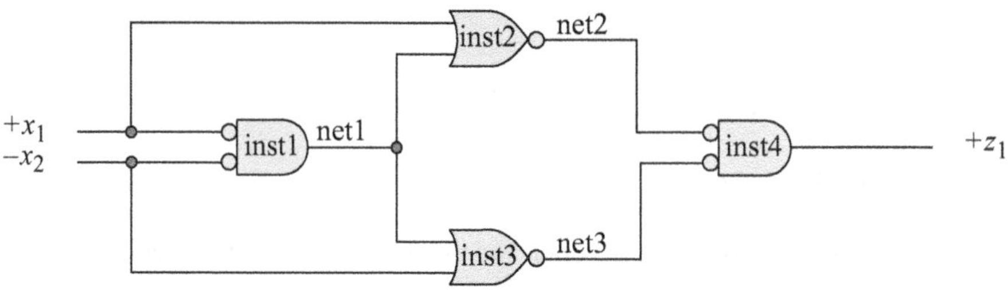

Figure 3.33 Logic diagram for Example 3.7.

$$\begin{aligned} z_1 &= [x_1 + (x_1'x_2)]\,[x_2' + (x_1'x_2)] \\ &= (x_1 + x_2)\,(x_2' + x_1') \\ &= (x_1 + x_2)x_2' + (x_1 + x_2)x_1' \\ &= x_1x_2' + x_1'x_2 \\ &= x_1 \oplus x_2 \end{aligned} \tag{3.3}$$

```
//2-input NOR gate as a user-defined primitive
primitive udp_nor2 (z1, x1, x2);

input x1, x2;
output z1;

//define state table
table
//inputs are in the same order as the input list
// x1 x2 :  z1;   comment is for readability
   0  0  :  1;
   0  1  :  0;
   1  0  :  0;
   1  1  :  0;
endtable
endprimitive
```

Figure 3.34 A NOR gate designed as a user-defined primitive.

```
//module for logic diagram using NOR logic
//user-defined primitives
module log_diag_eqn5 (x1, x2, z1);

input x1, x2;
output z1;

//instantiate the udps
udp_nor2 inst1 (net1, x1, ~x2);
udp_nor2 inst2 (net2, x1, net1);
udp_nor2 inst3 (net3, net1, ~x2);
udp_nor2 inst4 (z1, net2, net3);
endmodule
```

Figure 3.35 Design module for the logic diagram of Figure 3.33.

```
//test bench for logic diagram equation 5
module log_diag_eqn5_tb;

reg x1, x2;
wire z1;
//continued on next page
```

Figure 3.36 Test bench for design module of Figure 3.35.

```
//display variables
initial
$monitor ("x1 x2 = %b %b, z1 = %b", x1, x2, z1);

//apply input vectors
initial
begin
   #0    x1 = 1'b0;x2 = 1'b0;
   #10   x1 = 1'b0;x2 = 1'b1;
   #10   x1 = 1'b1;x2 = 1'b0;
   #10   x1 = 1'b1;x2 = 1'b1;

   #10   $stop;
end

//instantiate the module into the test bench
log_diag_eqn5 inst1 (
   .x1(x1),
   .x2(x2),
   .z1(z1)
   );
endmodule
```

Figure 3.36 (Continued)

```
x1 x2 = 0 0, z1 = 0
x1 x2 = 0 1, z1 = 1
x1 x2 = 1 0, z1 = 1
x1 x2 = 1 1, z1 = 0
```

Figure 3.37 Outputs for design module of Figure 3.35.

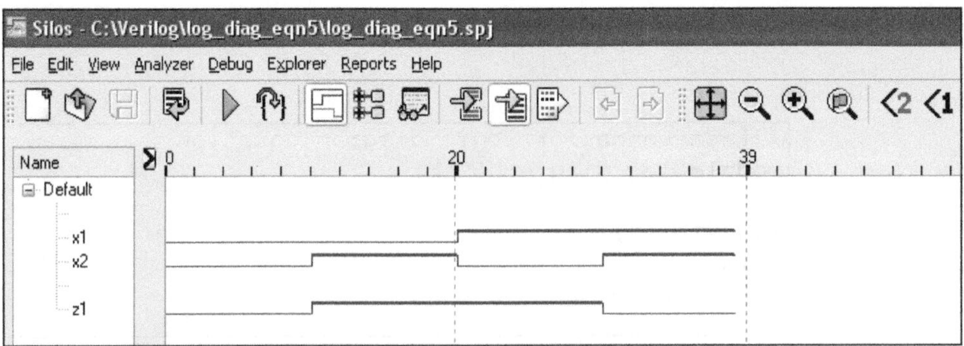

Figure 3.38 Waveforms for design module of Figure 3.35.

3.3 Dataflow Modeling

Gate-level modeling is an intuitive approach to digital design because it corresponds one-to-one with conventional digital logic design at the gate level. Dataflow modeling is similar to designing with built-in primitives, but at a slightly higher level of abstraction. Design automation tools are used to create gate-level logic from dataflow modeling by a process called *logic synthesis*. Register transfer level (RTL) is a combination of dataflow modeling and behavioral modeling and characterizes the flow of data through logic circuits.

3.3.1 Continuous Assignment

The *continuous assignment* statement models dataflow behavior and is used to design combinational logic without using gates and interconnecting nets. Continuous assignment statements provide a Boolean correspondence between the right-hand side expression and the left-hand side target. The continuous assignment statement uses the keyword **assign** and has the following syntax with optional drive strength and delay:

assign [drive_strength] [delay] left-hand side target = right-hand side expression

The continuous assignment statement assigns a value to a net (**wire**) that has been previously declared — it cannot be used to assign a value to a register. Therefore, the left-hand target must be a scalar or vector net or a concatenation of scalar and vector nets. The operands on the right-hand side can be registers, nets, or function calls. The registers and nets can be declared as either scalars or vectors.

Shown below are examples of continuous assignment statements for scalar nets, vector nets, and a concatenation of scalar and vector nets.

assign $z_1 = x_1 \ \& \ x_2 \ \& \ x_3$;

where the symbol (&) is the AND function.

assign $z_1 = x_1 \ \wedge \ x_2$;

where the symbol (^) is the exclusive-OR function.

assign $z_1 = (x_1 \ \& \ x_2) \mid x_3$;

where the symbol (|) is the OR function.

assign sum = a + b + cin

where the symbol (+) is addition, *sum* is a 9-bit vector to accommodate the *sum* and carry-out, *a* and *b* are 8-bit vectors, and *cin* is a scalar:

assign {cout, sum} = a + b + cin;

where the symbols ({ }) indicate concatenation; therefore, {cout, sum} is a concatenation of a scalar net and a vector net, respectively and *a* and *b* are 4-bit vectors, and *cin* and *cout* are scalars

The **assign** statement continuously monitors the right-hand side expression. If a variable changes value, then the expression is evaluated and the result is assigned to the target after any specified delay. If no delay is specified, then the default delay is zero. The continuous assignment statement can be considered to be a form of behavioral modeling, because the behavior of the circuit is specified, not the implementation. Several examples will now be presented which illustrate the concepts of dataflow modeling.

Example 3.8 From the Karnaugh map shown in Figure 3.39, the equation for the function z_1 is shown in Equation 3.3 in a sum-of-products notation. The logic diagram is shown in Figure 3.40. The circuit will be implemented using the continuous assignment statement of dataflow modeling.

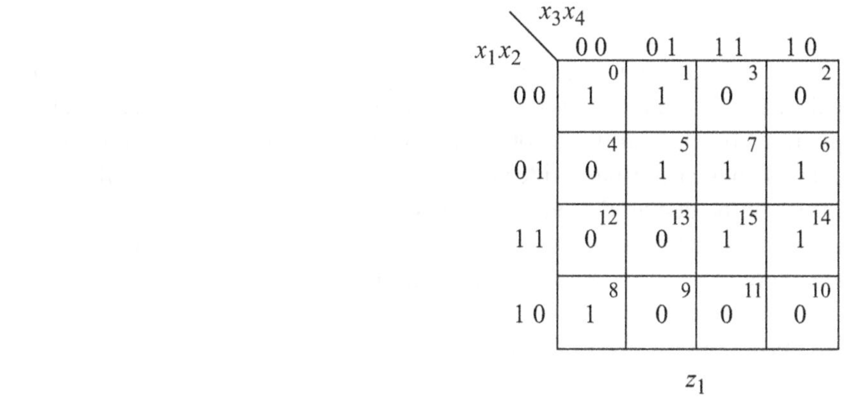

x_1x_2 \ x_3x_4	0 0	0 1	1 1	1 0
0 0	1 (0)	1 (1)	0 (3)	0 (2)
0 1	0 (4)	1 (5)	1 (7)	1 (6)
1 1	0 (12)	0 (13)	1 (15)	1 (14)
1 0	1 (8)	0 (9)	0 (11)	0 (10)

z_1

Figure 3.39 Karnaugh map for Example 3.8.

$$z_1 = x_2'x_3'x_4' + x_1'x_3'x_4 + x_2x_3 \quad (3.3)$$

The design module is shown in Figure 3.41. The same continuous assignment statement can be used for multiple assignments by placing a comma at the end of all statements except the last statement, which is terminated by a semicolon. The test bench is shown in Figure 3.42. The outputs and waveforms are shown in Figure 3.43 and Figure 3.44, respectively.

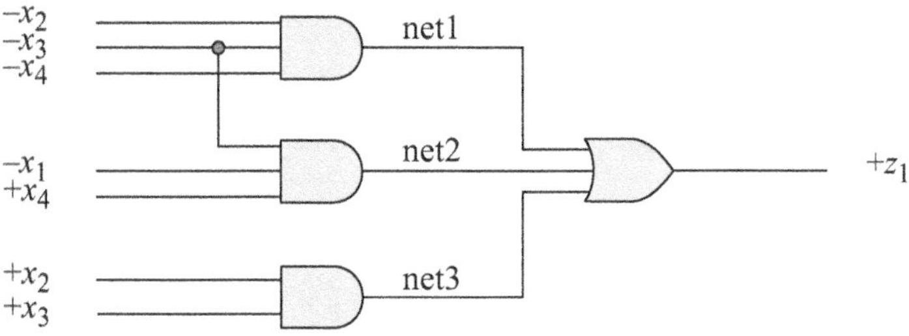

Figure 3.40 Logic diagram for Example 3.8.

```
//dataflow for a sum-of-products equation
module sop_eqn_df2 (x1, x2, x3, x4, z1);

//define inputs and output
input x1, x2, x3, x4;
output z1;

//define internal nets
wire net1, net2, net3;

//design logic
assign   net1 = ~x2 & ~x3 & ~x4,
         net2 = ~x1 & ~x3 & x4,
         net3 = x2 & x3;

assign   z1 = net1 | net2 | net3;

endmodule
```

Figure 3.41 Dataflow module for Figure 3.40 using continuous assignment.

```
//test bench for the dataflow sop
module sop_eqn_df2_tb;

reg x1, x2, x3, x4;
wire z1;

//continued on next page
```

Figure 3.42 Test bench for the dataflow module of Figure 3.41.

```
initial  //apply input vectors and display variables
begin: apply_stimulus
   reg [4:0] invect;
   for (invect=0; invect<16; invect=invect+1)
      begin
         {x1, x2, x3, x4} = invect [4:0];
         #10 $display ("x1 x2 x3 x4 = %b, z1 = %b",
                           {x1, x2, x3, x4}, z1);
      end
end

sop_eqn_df2 inst1 (     //instantiate the module
   .x1(x1),
   .x2(x2),
   .x3(x3),
   .x4(x4),
   .z1(z1)
   );
endmodule
```

Figure 3.42 (Continued)

```
x1 x2 x3 x4 = 0000, z1 = 1
x1 x2 x3 x4 = 0001, z1 = 1
x1 x2 x3 x4 = 0010, z1 = 0
x1 x2 x3 x4 = 0011, z1 = 0
x1 x2 x3 x4 = 0100, z1 = 0
x1 x2 x3 x4 = 0101, z1 = 1
x1 x2 x3 x4 = 0110, z1 = 1
x1 x2 x3 x4 = 0111, z1 = 1

x1 x2 x3 x4 = 1000, z1 = 1
x1 x2 x3 x4 = 1001, z1 = 0
x1 x2 x3 x4 = 1010, z1 = 0
x1 x2 x3 x4 = 1011, z1 = 0
x1 x2 x3 x4 = 1100, z1 = 0
x1 x2 x3 x4 = 1101, z1 = 0
x1 x2 x3 x4 = 1110, z1 = 1
x1 x2 x3 x4 = 1111, z1 = 1
```

Figure 3.43 Outputs for the dataflow module of Figure 3.41.

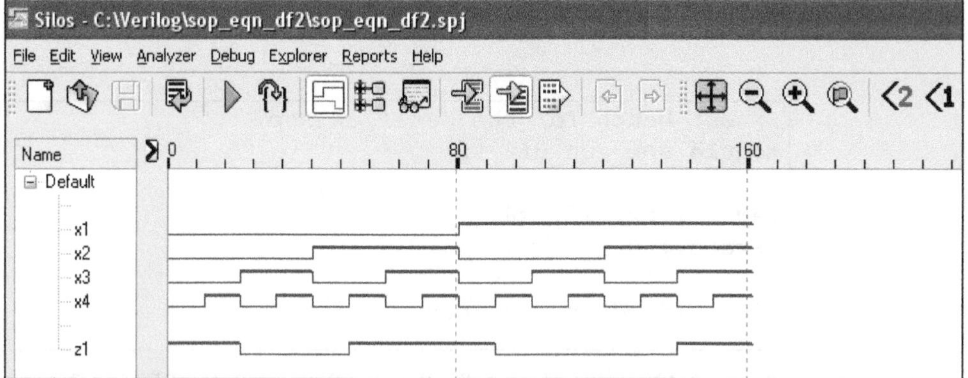

Figure 3.44 Waveforms for the dataflow module of Figure 3.41.

Example 3.9 Example 3.8 will be repeated as a product of sums using NAND gates. The zeros are combined in the Karnaugh map of Figure 3.39 to provide Equation 3.10. The logic diagram is shown in Figure 3.45. The design module, test bench module, outputs, and waveforms are shown in Figure 3.46, Figure 3.47, Figure 3.48, and Figure 3.49, respectively. The outputs and waveforms are identical for both examples.

$$z_1 = (x_2 + x_3')(x_2' + x_3 + x_4)(x_1' + x_3 + x_4') \tag{3.10}$$

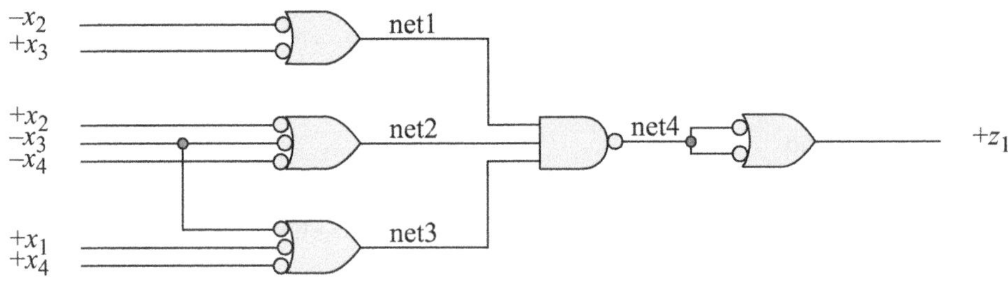

Figure 3.45 Logic diagram for Example 3.9.

```
//dataflow for a product-of-sums equation
module pos_eqn_df (x1, x2, x3, x4, z1);

//define inputs and output
input x1, x2, x3, x4;
output z1;

//define internal nets
wire net1, net2, net3, net4;

//design the logic using continuous assignment
assign   net1 = (x2 | ~x3),
         net2 = (~x2 | x3 | x4),
         net3 = (~x1 | x3 | ~x4),
         net4 = ~(net1 & net2 & net3);

assign   z1 = (~net4 | ~net4);

endmodule
```

Figure 3.46 Dataflow module for Example 3.9.

```
//test bench for product-of-sums equation
module pos_eqn_df_tb;

reg x1, x2, x3, x4;
wire z1;

//apply input vectors and display variables
initial
begin: apply_stimulus
   reg [4:0] invect;
   for (invect=0; invect<16; invect=invect+1)
      begin
         {x1, x2, x3, x4} = invect [4:0];
         #10 $display ("x1 x2 x3 x4 = %b, z1 = %b",
                        {x1, x2, x3, x4}, z1);
      end
end

//instantiate the module into the test bench
pos_eqn_df inst1 (
   .x1(x1),
   .x2(x2),
   .x3(x3),
   .x4(x4),
   .z1(z1)
   );

endmodule
```

Figure 3.47 Test bench for the dataflow module of Figure 3.46.

```
x1 x2 x3 x4 = 0000, z1 = 1    x1 x2 x3 x4 = 1000, z1 = 1
x1 x2 x3 x4 = 0001, z1 = 1    x1 x2 x3 x4 = 1001, z1 = 0
x1 x2 x3 x4 = 0010, z1 = 0    x1 x2 x3 x4 = 1010, z1 = 0
x1 x2 x3 x4 = 0011, z1 = 0    x1 x2 x3 x4 = 1011, z1 = 0
x1 x2 x3 x4 = 0100, z1 = 0    x1 x2 x3 x4 = 1100, z1 = 0
x1 x2 x3 x4 = 0101, z1 = 1    x1 x2 x3 x4 = 1101, z1 = 0
x1 x2 x3 x4 = 0110, z1 = 1    x1 x2 x3 x4 = 1110, z1 = 1
x1 x2 x3 x4 = 0111, z1 = 1    x1 x2 x3 x4 = 1111, z1 = 1
```

Figure 3.48 Outputs for the dataflow module of Figure 3.46.

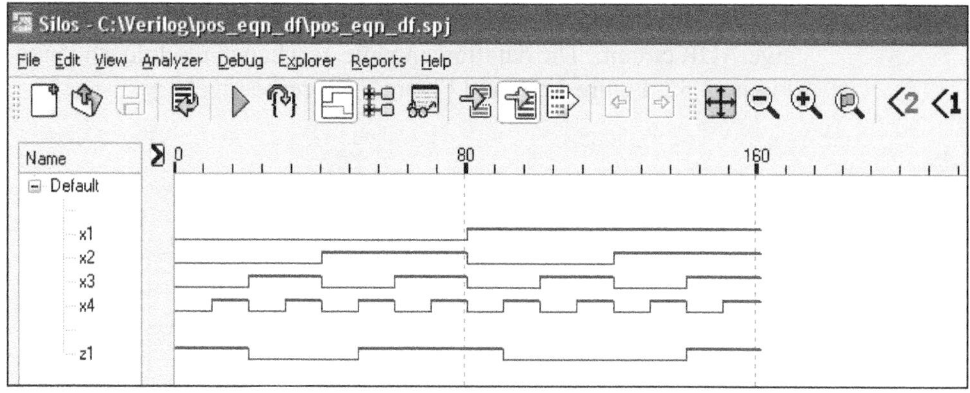

Figure 3.49 Waveforms for the dataflow module of Figure 3.46.

Example 3.10 A logic circuit will be designed to activate segment *a* only, for the 7-segment LED shown below. The inputs to the circuit are x_1, x_2, x_3, and x_4, where x_4 is the low-order input. The circuit will be a dataflow module implemented using the continuous assignment construct. Table 3.1 shows the digits that require the activation of segment *a*.

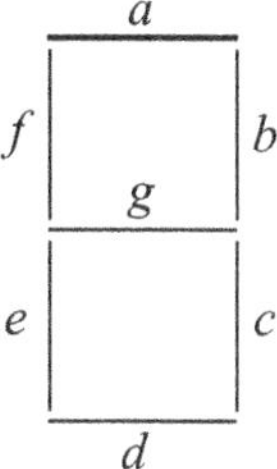

Table 3.1 Digits That Require the Activation of Segment *a*

Decimal Digit	Segment *a*	Decimal Digit	Segment *a*
0	1	8	1
1	0	9	1
2	1	10	Invalid
3	1	11	Invalid
4	0	12	Invalid
5	1	13	Invalid
6	1	14	Invalid
7	1	15	Invalid

The Karnaugh map is shown in Figure 3.50 and the equation is shown in Equation 3.4. The logic diagram is shown in Figure 3.51 using one NAND gate and one exclusive-NOR circuit. The dataflow module, test bench module, and outputs are shown in Figure 3.52, Figure 3.53, and Figure 3.54, respectively.

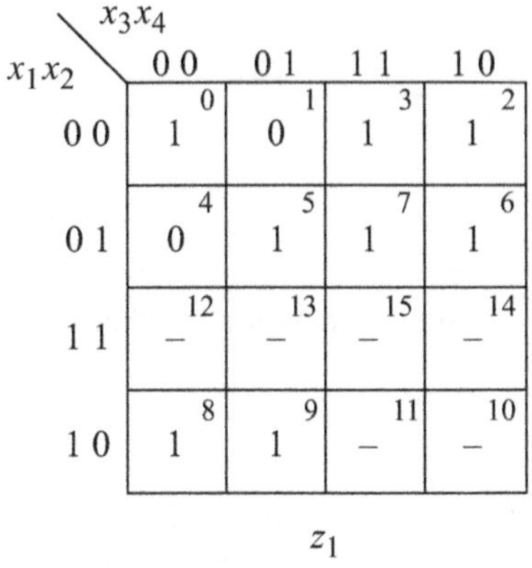

x_1x_2 \ x_3x_4	0 0	0 1	1 1	1 0
0 0	1 (0)	0 (1)	1 (3)	1 (2)
0 1	0 (4)	1 (5)	1 (7)	1 (6)
1 1	– (12)	– (13)	– (15)	– (14)
1 0	1 (8)	1 (9)	– (11)	– (10)

z_1

Figure 3.50 Karnaugh map for Example 3.10.

$$\begin{aligned} z_1 &= x_1 + x_3 + x_2x_4 + x_2'x_4' \\ &= x_1 + x_3 + (x_2 \oplus x_4)' \end{aligned} \tag{3.4}$$

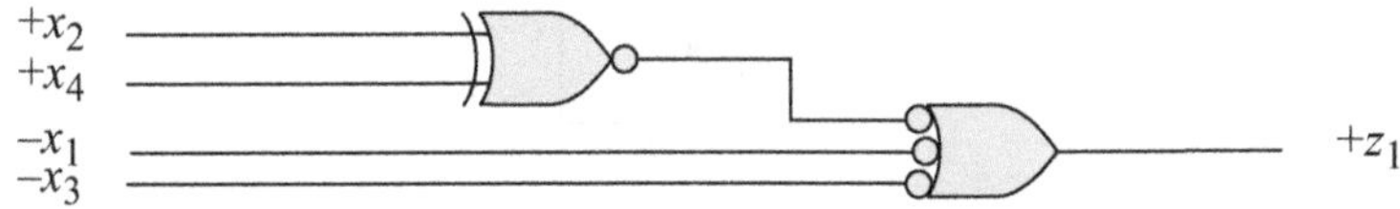

Figure 3.51 Logic diagram to activate segment *a* of a 7-segment LED.

```
//dataflow to activate segment a
module seg_a_df (x1, x2, x3, x4, z1);

input x1, x2, x3, x4;
output z1;
wire net1;

assign   net1 = ~(x2 ^ x4);        //exclusive-NOR
assign   z1 = (net1 | x1 | x3);
endmodule
```

Figure 3.52 Dataflow module to activate segment *a* of a 7-segment LED.

```
//test bench for segment a dataflow module
module seg_a_df_tb;

reg x1, x2, x3, x4;
wire z1;

initial     //apply input vectors and display variables
begin: apply_stimulus
   reg [4:0] invect;
   for (invect=0; invect<16; invect=invect+1)
      begin
         {x1, x2, x3, x4} = invect [4:0];
         #10 $display ("x1 x2 x3 x4 = %b, z1 = %b",
                              {x1, x2, x3, x4}, z1);
      end
end

seg_a_df inst1 (  //instantiate the module into the test bench
   .x1(x1),
   .x2(x2),
   .x3(x3),
   .x4(x4),
   .z1(z1)
   );
endmodule
```

Figure 3.53 Test bench for the dataflow module of Figure 3.52.

```
x1 x2 x3 x4 = 0000, z1 = 1
x1 x2 x3 x4 = 0001, z1 = 0
x1 x2 x3 x4 = 0010, z1 = 1
x1 x2 x3 x4 = 0011, z1 = 1
x1 x2 x3 x4 = 0100, z1 = 0
x1 x2 x3 x4 = 0101, z1 = 1
x1 x2 x3 x4 = 0110, z1 = 1
x1 x2 x3 x4 = 0111, z1 = 1
x1 x2 x3 x4 = 1000, z1 = 1
x1 x2 x3 x4 = 1001, z1 = 1
--------------------------
x1 x2 x3 x4 = 1010, z1 = 1
x1 x2 x3 x4 = 1011, z1 = 1
x1 x2 x3 x4 = 1100, z1 = 1      Invalid
x1 x2 x3 x4 = 1101, z1 = 1
x1 x2 x3 x4 = 1110, z1 = 1
x1 x2 x3 x4 = 1111, z1 = 1
```

Figure 3.54 Outputs for the dataflow module of Figure 3.52.

Example 3.11 Example 3.10 will be repeated using NOR logic in a product-of-sums notation for dataflow modeling. Equation 3.5 shows the equation as obtained from the Karnaugh map of Figure 3.50. The logic diagram is shown in Figure 3.55 and the dataflow module, test bench, and outputs are shown in Figure 3.56, Figure 3.57, and Figure 3.58, respectively.

$$z_1 = (x_2' + x_3 + x_4)(x_1 + x_2 + x_3 + x_4') \quad (3.5)$$

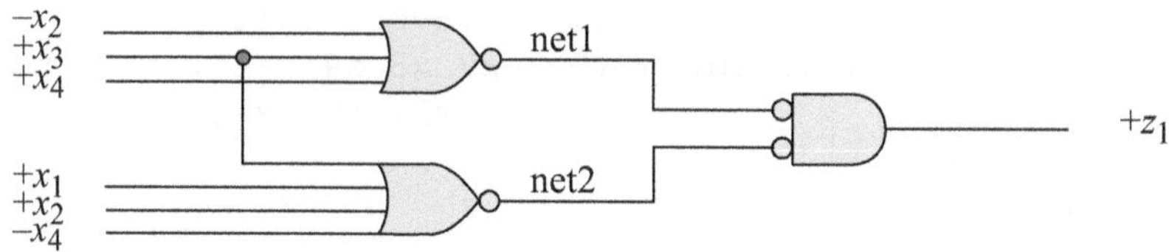

Figure 3.55 Logic diagram for Example 3.11.

```
//dataflow pos for segment a
module pos_eqn_df2 (x1, x2, x3, x4, z1);

input x1, x2, x3, x4;
output z1;

wire net1, net2;

assign   net1 = ~(~x2 | x3 | x4),
         net2 = ~(x1 | x2 | x3 | ~x4);

assign   z1 = (~net1 & ~net2);

endmodule
```

Figure 3.56 Dataflow module for Example 3.11.

```
//test bench for pos for segment a
module pos_eqn_df2_tb;

reg x1, x2, x3, x4;
wire z1;
//continued on next page
```

Figure 3.57 Test bench for Figure 3.56.

```
//apply input vectors and display variables
initial
begin: apply_stimulus
   reg [4:0] invect;
   for (invect=0; invect<16; invect=invect+1)
      begin
         {x1, x2, x3, x4} = invect [4:0];
         #10 $display ("x1 x2 x3 x4 = %b, z1 = %b",
                           {x1, x2, x3, x4}, z1);
      end
end

//instantiate the module into the test bench
pos_eqn_df2 inst1 (
   .x1(x1),
   .x2(x2),
   .x3(x3),
   .x4(x4),
   .z1(z1)
   );

endmodule
```

Figure 3.57 (Continued)

```
x1 x2 x3 x4 = 0000, z1 = 1
x1 x2 x3 x4 = 0001, z1 = 0
x1 x2 x3 x4 = 0010, z1 = 1
x1 x2 x3 x4 = 0011, z1 = 1
x1 x2 x3 x4 = 0100, z1 = 0
x1 x2 x3 x4 = 0101, z1 = 1
x1 x2 x3 x4 = 0110, z1 = 1
x1 x2 x3 x4 = 0111, z1 = 1
x1 x2 x3 x4 = 1000, z1 = 1
x1 x2 x3 x4 = 1001, z1 = 1
--------------------------
x1 x2 x3 x4 = 1010, z1 = 1
x1 x2 x3 x4 = 1011, z1 = 1
x1 x2 x3 x4 = 1100, z1 = 0      Invalid
x1 x2 x3 x4 = 1101, z1 = 1
x1 x2 x3 x4 = 1110, z1 = 1
x1 x2 x3 x4 = 1111, z1 = 1
```

Figure 3.58 Outputs for the dataflow module of Figure 3.56.

Notice that the outputs are identical except for minterm location $x_1x_2x_3x_4 = 1100$ (12). In the outputs of Figure 3.54 for Example 3.10, minterm 12 contains a 1 in order to minimize the equation for output z_1. In the outputs of Figure 3.58, however, minterm 12 contains a 0 in order to minimize the product-of-sums equation shown in Equation 3.5. Minterm 12 is treated as a "don't care" condition in both cases due to the invalid minterms 10 through 15; therefore, the outputs of both examples are identical for the valid minterms and the circuits operate in an identical manner.

3.4 Behavioral Modeling

This section describes the *behavior* of a digital system and is not concerned with the direct implementation of logic gates but more with the architecture of the system. This is an algorithmic approach to hardware implementation and represents a higher level of abstraction than previous modeling methods. The constructs in behavioral modeling closely resemble those used in the C programming language.

Verilog contains two structured procedure statements or behaviors: **initial** and **always**. A behavior may consist of a single statement or a block of statements delimited by the keywords **begin** . . . **end**. A module may contain multiple **initial** and **always** statements. These statements are the basic statements used in behavioral modeling and execute concurrently starting at time zero in which the order of execution is not important. All other behavioral statements are contained inside these structured procedure statements. The keywords **initial** and **always** specify a behavior and the statements within a behavior are classified as *behavioral* or *procedural.*

3.4.1 Initial Statement

All statements within an **initial** statement comprise an **initial** block. An **initial** statement executes only once beginning at time zero, then suspends execution. An **initial** statement provides a method to initialize and monitor variables before the variables are used in a module; it is also used to generate waveforms. For a given time unit, all statements within the **initial** block execute sequentially. Execution or assignment is controlled by the # symbol. The syntax for an **initial** statement is shown below.

initial [optional timing control] procedural statement or
block of procedural statements

3.4.2 Always Statement

The **always** statement executes the behavioral statements within the **always** block repeatedly in a looping manner and begins execution at time zero. Execution of the statements continues indefinitely until the simulation is terminated. The syntax for the **always** statement is shown below.

> **always** [optional timing control] procedural statement or
> block of procedural statements

An **always** statement is often used with an *event control list* — or *sensitivity list* — to execute a sequential block. When a change occurs to a variable in the sensitivity list, the statement or block of statements in the **always** block is executed. The keyword **or** is used to indicate multiple events. When one or more inputs change state, the statement in the **always** block is executed. The **begin . . . end** keywords are necessary only when there is more than one behavioral statement. Target variables used in an **always** statement are declared as type **reg**.

Example 3.12 A 5-input majority circuit will be designed using behavioral modeling with the **always** statement. The inputs are labeled $x_1x_2x_3x_4x_5$; the output is z_1. Figure 3.59 illustrates a Karnaugh map in which the minterm entries of 1 specify that a majority of the inputs are at a logic 1 level. The equation for output z_1 is shown in Equation 3.6.

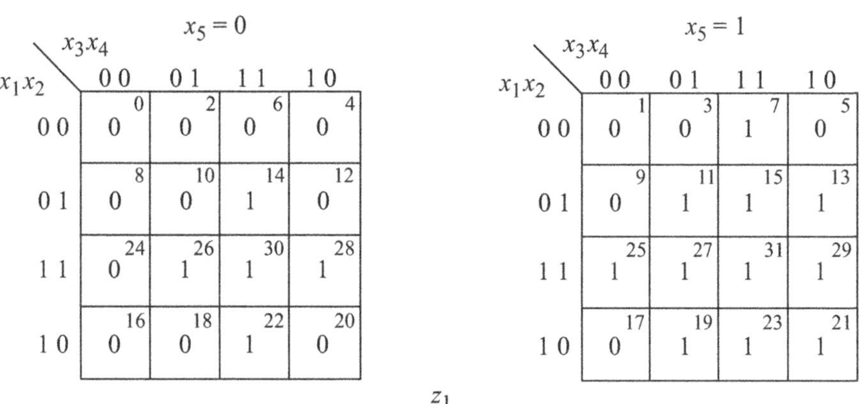

Figure 3.59 Karnaugh map for the majority circuit of Example 3.12.

$$\begin{aligned} z_1 = {} & x_3x_4x_5 + x_2x_3x_5 + x_2x_4x_5 + x_1x_3x_5 + \\ & x_1x_4x_5 + x_1x_2x_3 + x_1x_2x_4 + x_2x_3x_4 + \\ & x_1x_3x_4 + x_1x_2x_5 \end{aligned} \tag{3.6}$$

The behavioral design module is shown in Figure 3.60. The entry of #5 to the immediate right of the equal sign specifies that the output is delayed by five time units to allow for the propagation delay — inertial delay — of the AND gate. The sensitivity list in the **always** statement lists the five inputs; whenever one or more of the inputs changes value, the equation for z_1 is executed.

The test bench is shown in Figure 3.61. The system function **$time** obtains the current simulation time and is displayed in the outputs of Figure 3.62 every seven time units. The waveforms are shown in Figure 3.63 and clearly show the propagation delay of seven time units that occurs when an input changes; that is, output z_1 is asserted seven time units after an input changes value.

```
//behavioral 5-input majority circuit
module maj5_bh (x1, x2, x3, x4, x5, z1);

input x1, x2, x3, x4, x5;
output z1;

wire x1, x2, x3, x4, x5;
reg z1;

always @ (x1 or x2 or x3 or x4 or x5)
   z1 = #5 (x3 & x4 & x5) | (x2 & x3 & x5) | (x2 & x4 & x5) |
           (x1 & x3 & x5) | (x1 & x4 & x5) | (x1 & x2 & x3) |
           (x1 & x2 & x4) | (x2 & x3 & x4) | (x1 & x3 & x4) |
           (x1 & x2 & x5);
endmodule
```

Figure 3.60 Behavioral module for a 5-input majority circuit.

```
//test bench for the 5-input majority circuit
module maj5_bh_tb;

reg x1, x2, x3, x4, x5;
wire z1;

initial       //apply vectors and display variables
begin: apply_stimulus
   reg [5:0] invect;
   for (invect=0; invect<32; invect=invect+1)
      begin
         {x1, x2, x3, x4, x5} = invect [5:0];
         #7 $display ($time, "input = %b, z1 = %b",
                        {x1, x2, x3, x4, x5}, z1);
      end
end           //continued on next page
```

Figure 3.61 Test bench for the 5-input majority circuit.

```
//instantiate the module into the test bench
maj5_bh inst1 (
   .x1(x1),
   .x2(x2),
   .x3(x3),
   .x4(x4),
   .x5(x5),
   .z1(z1)
   );
endmodule
```

Figure 3.61 (Continued)

```
7      input = 00000, z1 = 0   |  119   input = 10000, z1 = 0
14     input = 00001, z1 = 0   |  126   input = 10001, z1 = 0
21     input = 00010, z1 = 0   |  133   input = 10010, z1 = 0
28     input = 00011, z1 = 0   |  140   input = 10011, z1 = 1
35     input = 00100, z1 = 0   |  147   input = 10100, z1 = 0
42     input = 00101, z1 = 0   |  154   input = 10101, z1 = 1
49     input = 00110, z1 = 0   |  161   input = 10110, z1 = 1
56     input = 00111, z1 = 1   |  168   input = 10111, z1 = 1
63     input = 01000, z1 = 0   |  175   input = 11000, z1 = 0
70     input = 01001, z1 = 0   |  182   input = 11001, z1 = 1
77     input = 01010, z1 = 0   |  189   input = 11010, z1 = 1
84     input = 01011, z1 = 1   |  196   input = 11011, z1 = 1
91     input = 01100, z1 = 0   |  203   input = 11100, z1 = 1
98     input = 01101, z1 = 1   |  210   input = 11101, z1 = 1
105    input = 01110, z1 = 1   |  217   input = 11110, z1 = 1
112    input = 01111, z1 = 1   |  224   input = 11111, z1 = 1
```

Figure 3.62 Outputs for the 5-input majority circuit.

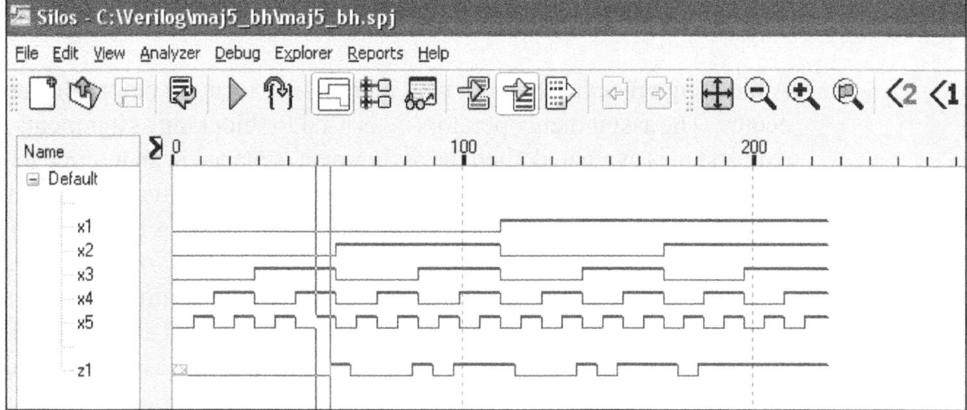

Figure 3.63 Waveforms for the 5-input majority circuit.

3.4.3 Intrastatement Delay

A procedural assignment may have an optional delay. A delay appearing to the right of an assignment operator is called an intrastatement delay. It is the delay by which the right-hand result is delayed before assigning it to the left-hand target. In the example below, the expression (x_1 & x_2) is evaluated, a delay of five time units is taken, then the result is assigned to z_1.

$$z_1 = \#5\ (x_1\ \&\ x_2);$$

One purpose for an intrastatement delay is to simulate the delay through a logic gate. In the above example, the propagation delay through the AND gate is five time units.

3.4.4 Interstatement Delay

Interstatement delay is the delay taken before a statement is executed. In the code segment shown below, the delay given in the second statement specifies that when the first statement has finished executing, wait five time units before executing the second statement.

$$z_1 = (x_1 + x_2)\ x_3;$$
$$\#5\ z_2 = x_4 \wedge x_5;$$

If no delays are specified in a procedural assignment, then there is zero delay in the assignment.

3.4.5 Blocking Assignments

A blocking procedural assignment completes execution before the next statement executes. The assignment operator (=) is used for blocking assignments. The right-hand expression is evaluated, then the assignment is placed in an internal temporary register called the *event queue* and scheduled for assignment. If no time units are specified, the scheduling takes place immediately. The event queue is covered in Appendix A.

In the code segment below, an interstatement delay of one time unit is specified. Execution of the statement is delayed by the timing control of one time unit. At time units $t + 1$, the right-hand expression for z_1 is evaluated and assigned to z_1. The execution of any following statements is blocked until the assignment occurs.

```
initial
    begin
        #1  z1 = x1 & x2;
            z2 = x2 | x3;
    end
```

Example 3.13 This example uses behavioral modeling to illustrates the use of blocking assignments with intrastatement and interstatement delays. The design module is shown in Figure 3.64, in which the statements for z_1 and z_2 have intrastatement delays of two time units; the statements for z_3 and z_4 have interstatement delays of two time units. The test bench, shown in Figure 3.65, assigns all possible combinations of the inputs. The outputs and waveforms are shown in Figure 3.66 and Figure 3.67, respectively.

In Figure 3.64, when either x_1, x_2, or x_3 changes value, z_1 executes immediately, but the assignment to z_1 is delayed by two time units. The statement for z_2 is blocked until the assignment to z_1 has completed. The execution of the statement for z_3 is delayed by two time units, at which time the statement is executed and the result is assigned to z_3 immediately. This is evident in the waveforms of Figure 3.67.

```
//behavioral blocking assignment using
//intrastatement and interstatement delays
module blocking6 (x1, x2, x3, z1, z2, z3, z4);

input x1, x2, x3;
output z1, z2, z3, z4;

reg z1, z2, z3, z4;

always @ (x1 or x2 or x3)
begin
   z1 = #2 (x1 ^ x2) & x3;
   z2 = #2 ~(x1 ^ x2) | x3;
end

always @ (x1 or x2 or x3)
begin
   #2 z3 = x1 & x2 & x3;
   #2 z4 = x1 ^ x2 ^ x3;
end

endmodule
```

Figure 3.64 Behavioral module illustrating blocking assignments with intrastatement and interstatement delays.

```
//test bench for blocking assignment using
//intrastatement and interstatement delays
module blocking6_tb;

reg x1, x2, x3;
wire z1, z2, z3, z4;

//apply input vectors and display variables
initial
begin: apply_stimulus
   reg [3:0] invect;
   for (invect=0; invect<8; invect=invect+1)
      begin
         {x1, x2, x3} = invect [3:0];
         #10 $display ("x1 x2 x3 = %b, z1 z2 z3 z4 = %b",
                            {x1, x2, x3}, {z1, z2, z3, z4});
      end
end

//instantiate the module into the test bench
blocking6 inst1 (
   .x1(x1),
   .x2(x2),
   .x3(x3),
   .z1(z1),
   .z2(z2),
   .z3(z3),
   .z4(z4)
   );
endmodule
```

Figure 3.65 Test bench for the behavioral module of Figure 3.64.

```
                z1 = #2 (x1 ^ x2) & x3
                z2 = #2 ~(x1 ^ x2) | x3
             #2 z3 = x1 & x2 & x3
             #2 z4 = x1 ^ x2 ^ x3
x1 x2 x3 = 000, z1 z2 z3 z4 = 0100
x1 x2 x3 = 001, z1 z2 z3 z4 = 0101
x1 x2 x3 = 010, z1 z2 z3 z4 = 0001
x1 x2 x3 = 011, z1 z2 z3 z4 = 1100
x1 x2 x3 = 100, z1 z2 z3 z4 = 0001
x1 x2 x3 = 101, z1 z2 z3 z4 = 1100
x1 x2 x3 = 110, z1 z2 z3 z4 = 0100
x1 x2 x3 = 111, z1 z2 z3 z4 = 0111
```

Figure 3.66 Outputs for the behavioral module of Figure 3.64.

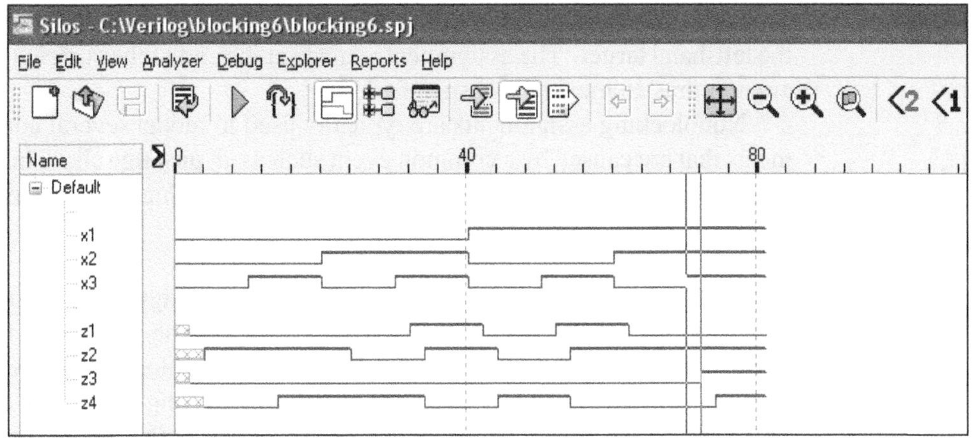

Figure 3.67 Waveforms for the behavioral module of Figure 3.64.

3.4.6 Nonblocking Assignments

The assignment symbol (<=) is used to represent a nonblocking procedural assignment. Nonblocking assignments allow the scheduling of assignments without blocking execution of the following statements in a sequential procedural block. A nonblocking assignment is used to synchronize assignment statements so that they appear to execute at the same time. In the code segment shown below using blocking assignments, the result is indeterminate because both **always** blocks execute concurrently resulting in a race condition. Depending on the simulator implementation, either $x_1 = x_2$ would be executed before $x_2 = x_3$ or vice versa.

```
always @ (posedge clk)
    x1 = x2;

always @ (posedge clk)
    x2 = x3;
```

The race condition is solved by using nonblocking assignments as shown below.

```
always @ (posedge clk)
    x1 <= x2;

always @ (posedge clk)
    x2 <= x3;
```

The Verilog simulator schedules a nonblocking assignment statement to execute, then proceeds to the next statement in the block without waiting for the previous nonblocking statement to complete execution. That is, the right-hand expression is evaluated and the value is stored in the event queue and is *scheduled* to be assigned to the left-hand target. The assignment is made at the end of the current time step if there are no intrastatement delays specified.

Nonblocking assignments are typically used to model several concurrent assignments that are caused by a common event such as @ **posedge** clk. The order of the assignments is irrelevant because the right-hand side evaluations are stored in the event queue before any assignments are made.

Example 3.14 This example will model register assignments using blocking and nonblocking constructs with intrastatement delays. The first three statements in the **initial** block of the module shown in Figure 3.68 use blocking assignments and execute sequentially at time 0. Because of the nonblocking behavioral construct in the **initial** block, the next three statements are processed at the same simulation time, but are scheduled to execute at different times due to the intrastatement delays.

```
//behavioral to illustrate blocking
//and nonblocking assignments
module block_nonblock (data_reg_a, data_reg_b, data_reg_c);

output [7:0] data_reg_a, data_reg_b, data_reg_c;
reg [7:0] data_reg_a, data_reg_b, data_reg_c;

initial
begin
   data_reg_a = 8'b0111_1100;
   data_reg_b = 8'b1111_0000;
   data_reg_c = 8'b1111_1111;

   data_reg_a [2:0] <= #5 3'b111;
   data_reg_b [7:0] <= #10 {data_reg_b [7:4], 4'b1111};
   data_reg_c [7:0] <= #15 {2'b11, data_reg_a [5:0]};
end

endmodule
```

Figure 3.68 Module to illustrate the operation of blocking and nonblocking assignments.

The waveforms of Figure 3.69 show the assignments to the registers based on their scheduling in the event queue. Register *data_reg_a* is set to a value of 7C hexadecimal (7CH) by the first blocking assignment. The statement *data_reg_a[2:0] <= #5 3'b111* is scheduled to execute at time unit 5, which changes *data_reg_a* to 7FH by replacing the low-order three bits with *3'b111*. Register *data_reg_b* is initially set to a value of F0H by a blocking assignment, but is changed to FFH by the second nonblocking assignment, which is scheduled to execute at time unit 10. This statement concatenates *data_reg_b [7:4]* with *4'b1111*. Register *data_reg_c* is initially set to a value of FFH by a blocking assignment, but is changed to FCH by concatenating *2'b11* with the low-order six bits of *data_reg_a* using a nonblocking assignment.

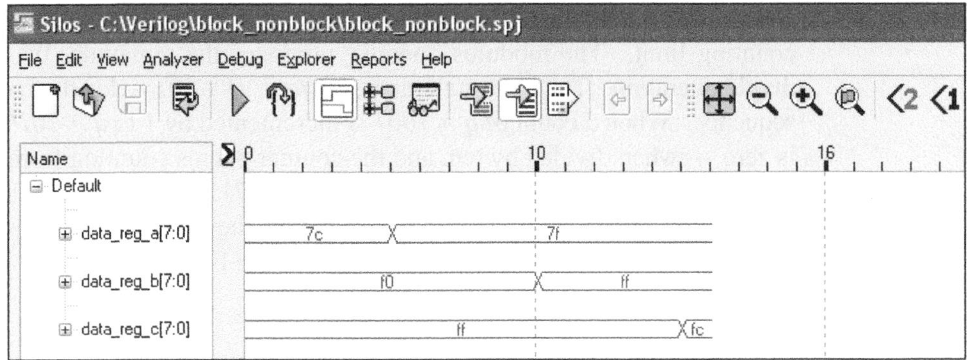

Figure 3.69 Waveform for the behavioral module of Figure 3.68.

3.4.7 Conditional Statements

Conditional statements alter the flow within a behavior based upon certain conditions. The choice among alternative statements depends on the Boolean value of an expression. The alternative statements can be a single statement or a block of statements delimited by the keywords **begin** . . . **end**. The keywords **if** and **else** are used in conditional statements. There are three categories of the conditional statement as shown below. A true value is 1 or any nonzero value; a false value is 0, **x** (unknown), or **z** (high impedance). If the evaluation is false, then the next expression in the activity flow is evaluated.

```
//no else statement
if (expression) statement1;          //if expression is true, then statement1 is executed.


//one else statement                 //choice of two statements. Only one is executed.
if (expression) statement1;          //if expression is true, then statement1 is executed.
else statement2;                     //if expression is false, then statement2 is executed.


//nested if-else if-else             //choice of multiple statements. Only one is executed.
if (expression1) statement1;         //if expression1 is true, then statement1 is executed.
else if (expression2) statement2;    //if expression2 is true, then statement2 is executed.
else if (expression3) statement3;    //if expression3 is true, then statement3 is executed.
else default statement;
```

Example 3.15 Behavioral modeling with the continuous assignment will be used to design a modulo-10 counter (*q[3:0]*) that counts sequentially from 0000 to 1001, then begins again at 0000. An AND gate will be connected to the counter outputs *q[2]* and *q[1]*; an exclusive-OR circuit will be connected to the counter outputs *q[2]*, *q[1]*, and *q[0]*; an OR gate will be connected to the counter outputs *q[3]* and *q[0]*.

The design module is shown in Figure 3.70 using the keywords **if** and **else** in the conditional statements. The arithmetic operator modulus (%) is used to define the counting limit. The modulus operator produces the remainder that results from a divide operation. Thus, the statement $q = (q + 1) \% 10$ yields the required counting sequence. When a count of $q = 1001$ is incremented by 1 to $q = 1010$, the remainder is zero — when divided by ten, and the counter begins counting from zero again.

The test bench module is shown in Figure 3.71. The outputs are shown in Figure 3.72 in which the outputs z_1, z_2, and z_3 generate the appropriate values as described in the design module. The waveforms are shown in Figure 3.73.

```
//behavioral modulo-10 counter with logic gating
module ctr_mod10_logic (rst_n, clk, q, z1, z2, z3);


input rst_n, clk;
output [3:0] q;
output z1, z2, z3;

wire rst_n, clk;
reg [3:0] q;
wire z1, z2, z3;


//define counting sequence
always @ (posedge clk or negedge rst_n)
begin
   if (rst_n == 0)
      q = 4'b0000;
   else
      q = (q + 1) % 10;
end


//define outputs
assign    z1 = q[2] & q[1],
          z2 = q[2] ^ q[1] ^ q[0],
          z3 = q[3] | q[0];


endmodule
```

Figure 3.70 Behavioral module for a modulo-10 counter with output gating.

```
//test bench for modulo-10 counter
module ctr_mod10_logic_tb;

reg rst_n, clk;
wire [3:0] q;
wire z1, z2, z3;

//display outputs
initial
$monitor ("count = %b, z1 z2 z3 = %b", q, {z1, z2, z3});

//define reset
initial
begin
   #0 rst_n = 1'b0;
   #5 rst_n = 1'b1;
end

//define clock
initial
begin
   clk = 1'b0;
   forever
      #10clk = ~clk;
end

//define length of simulation
initial
begin
   #200  $finish;
end

//instantiate the module into the test bench
ctr_mod10_logic inst1 (
   .rst_n(rst_n),
   .clk(clk),
   .q(q),
   .z1(z1),
   .z2(z2),
   .z3(z3)
   );

endmodule
```

Figure 3.71 Test bench for the modulo-10 counter of Figure 3.70.

```
z1 = q[2] & q[1]
z2 = q[2] ^ q[1] ^ q[0]
z3 = q[3] | q[0]

count = 0000, z1 z2 z3 = 000
count = 0001, z1 z2 z3 = 011
count = 0010, z1 z2 z3 = 010
count = 0011, z1 z2 z3 = 001
count = 0100, z1 z2 z3 = 010
count = 0101, z1 z2 z3 = 001
count = 0110, z1 z2 z3 = 100
count = 0111, z1 z2 z3 = 111
count = 1000, z1 z2 z3 = 001
count = 1001, z1 z2 z3 = 011
count = 0000, z1 z2 z3 = 000
```

Figure 3.72 Outputs for the modulo-10 counter of Figure 3.70.

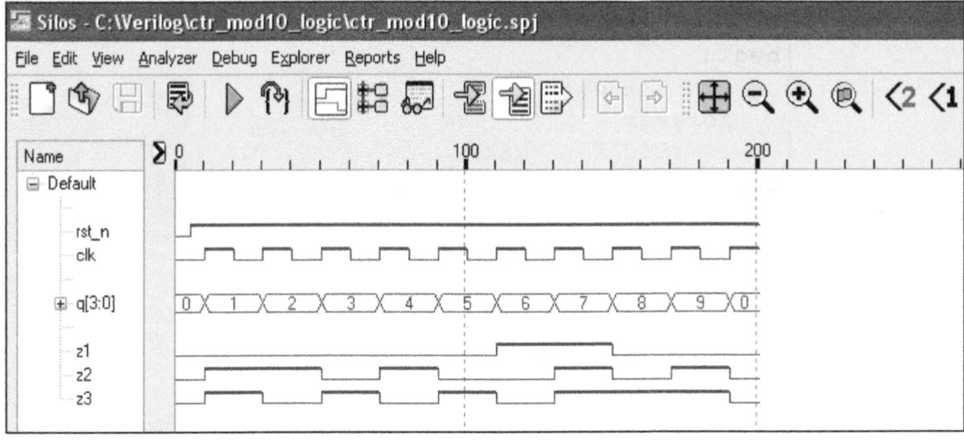

Figure 3.73 Waveforms for the modulo-10 counter of Figure 3.70.

3.4.8 Case Statement

The **case** statement is an alternative to the **if** . . . **else if** construct and may simplify the readability of the Verilog code. The **case** statement is a multiple-way conditional branch. It executes one of several different procedural statements depending on the comparison of an expression with a case item. The expression and the case item are compared bit-by-bit and must match exactly. The statement that is associated with a case item may be a single procedural statement or a block of statements delimited by the keywords **begin** . . . **end.** The **case** statement has the following syntax:

```
case (expression)
    case_item1 : procedural_statement1;
    case_item2 : procedural_statement2;
    case_item3 : procedural_statement3;
                 .
                 .
                 .
    case_itemn : procedural_statementn;
    default : default_statement;
endcase
```

The case expression may be an expression or a constant. The case items are evaluated in the order in which they are listed. If a match occurs between the case expression and the case item, then the corresponding procedural statement, or block of statements, is executed. If no match occurs, then the optional default statement is executed.

Example 3.16 A 16-bit counter will be designed using behavioral modeling with the **case** statement that counts in the following hexadecimal sequence: 0000, 8000, C000, E000, F000, F800, FC00, FE00, FF00, FF80, FFC0, FFE0, FFF0, FFF8, FFFC, FFFE, FFFF, 0000. The behavioral module is shown in Figure 3.74. Whenever the variable *count* changes, the **case** statement determines the *next_count* from the present count.

The test bench is shown in Figure 3.75. The length of simulation is 320 time units and is terminated by the system task **$finish**, which causes the simulator to exit the module and return control to the operating system. The outputs are shown in Figure 3.76 and the waveforms are shown in Figure 3.77.

```
//behavioral counter using the case statement
module ctr_triangle (rst_n, clk, count);

input rst_n, clk;
output [15:0] count;

wire rst_n, clk;
reg [15:0] count, next_count;

always @ (posedge clk or negedge rst_n)
begin
   if (~rst_n)     //if the reset = 0
      count = 16'h0000;
   else
      count = next_count;
end                  //continued on next page
```

Figure 3.74 Behavioral module for the counter of Example 3.16.

```
//define the counting sequence
always @ (count)
begin
   case (count)
      16'h0000 : next_count = 16'h8000;
      16'h8000 : next_count = 16'hc000;
      16'hc000 : next_count = 16'he000;
      16'he000 : next_count = 16'hf000;
      16'hf000 : next_count = 16'hf800;
      16'hf800 : next_count = 16'hfc00;
      16'hfc00 : next_count = 16'hfe00;
      16'hfe00 : next_count = 16'hff00;
      16'hff00 : next_count = 16'hff80;
      16'hff80 : next_count = 16'hffc0;
      16'hffc0 : next_count = 16'hffe0;
      16'hffe0 : next_count = 16'hfff0;
      16'hfff0 : next_count = 16'hfff8;
      16'hfff8 : next_count = 16'hfffc;
      16'hfffc : next_count = 16'hfffe;
      16'hfffe : next_count = 16'hffff;
      default  : next_count = 16'h0000;
   endcase
end

endmodule
```

Figure 3.74 (Continued)

```
//test bench for counter triangle
module ctr_triangle_tb;

reg rst_n, clk;
wire [15:0] count;

//display count
initial
$monitor ("count = %b", count);

//define clock
initial
begin
   clk = 1'b0;
   forever
      #10 clk = ~clk;
end                    //continued on next page
```

Figure 3.75 Test bench for the counter of Example 3.16.

```
//define reset
initial
begin
   #0 rst_n = 1'b0;
   #5 rst_n = 1'b1;
end

//define length of simulation
initial
begin
   #320 $finish;
end

//instantiate the module into the test bench
ctr_triangle inst1 (
   .rst_n(rst_n),
   .clk(clk),
   .count(count)
   );

endmodule
```

Figure 3.75 (Continued)

```
count = 0000000000000000
count = 1000000000000000
count = 1100000000000000
count = 1110000000000000
count = 1111000000000000
count = 1111100000000000
count = 1111110000000000
count = 1111111000000000
count = 1111111100000000
count = 1111111110000000
count = 1111111111000000
count = 1111111111100000
count = 1111111111110000
count = 1111111111111000
count = 1111111111111100
count = 1111111111111110
count = 1111111111111111
```

Figure 3.76 Outputs for the counter of Example 3.16.

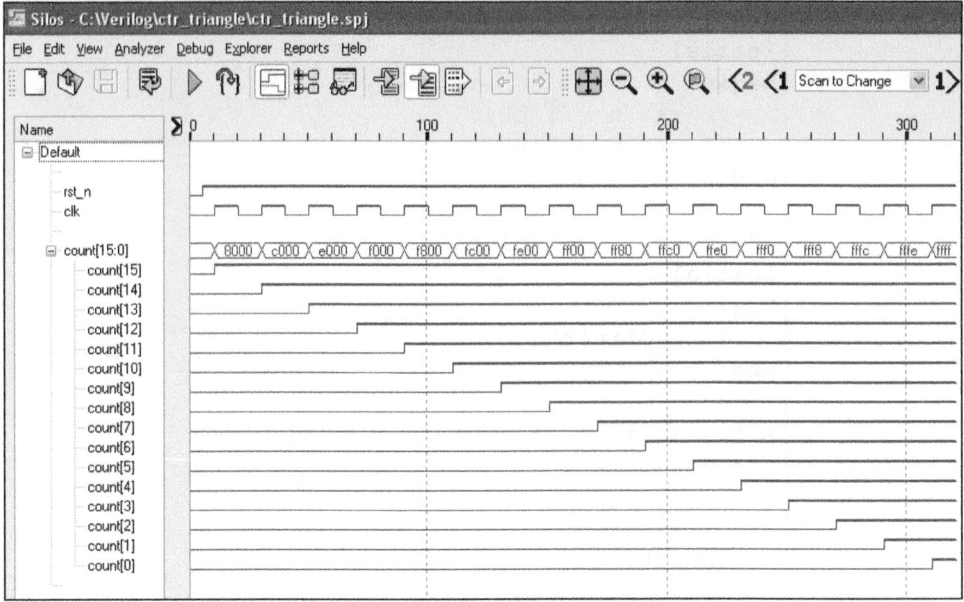

Figure 3.77 Waveforms for the counter of Example 3.16.

Example 3.17 A Moore synchronous sequential machine will be designed using the state diagram of Figure 3.78. Since the state code assignment precludes the possibility of glitches on any of the outputs, the assertion/deassertion of all outputs is $\uparrow t_1 \downarrow t_3$. A logic diagram is not required, since behavioral modeling will be used to implement the Moore machine. As stated previously, behavioral modeling implements a design based on the architecture — or behavior — of the machine and leaves the logic implementation details to the synthesis tool.

The behavioral module is shown in Figure 3.79 and uses the **parameter** keyword to define constants, in this case, the state codes. The **assign** continuous assignment is used to define the outputs and the **case** statement determines the next state based on the current state and the present inputs.

The test bench module is shown in Figure 3.80 and defines the input sequence to allow the machine to proceed through all states and generate the appropriate outputs. State transitions occur on the positive edge of the machine clock. The system task **$random** is used in the test bench to randomly select a value for x_1 or x_2 from the values of 0 and 1, because some state transitions are independent of the values for x_1 or x_2. For example, the transition from state *a* to state *b* does not depend on the value of x_2; the transition from states *c* or *d* to state *e* are independent of assertion or deassertion of x_1 or x_2.

The outputs are shown in Figure 3.81 and the waveforms are shown in Figure 3.82. The waveforms show the various states through which the machine sequences and shows the asserted outputs in their respective states. The outputs are asserted for for the entire state time; that is, $\uparrow t_1 \downarrow t_3$.

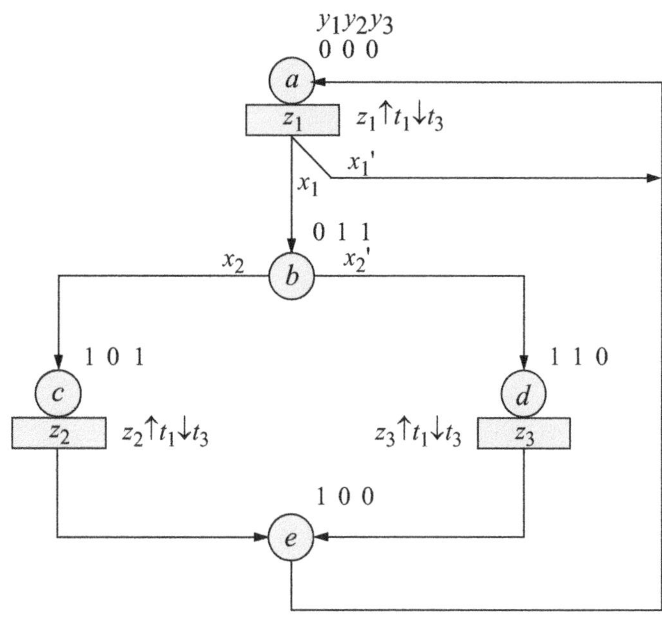

Figure 3.78 State diagram for the Moore machine of Example 3.17.

```
//behavioral moore ssm
module moore_ssm23 (rst_n, clk, x1, x2, y, z1, z2, z3);

input rst_n, clk, x1, x2;
output [1:3] y;
output z1, z2, z3;

reg [1:3] y, next_state;
wire z1, z2, z3;

//assign state codes
parameter    state_a = 3'b000,
             state_b = 3'b011,
             state_c = 3'b101,
             state_d = 3'b110,
             state_e = 3'b100;

//continued on next page
```

Figure 3.79 Design module for the Moore machine of Example 3.17.

```
//set next state
always @ (posedge clk)
begin
   if (~rst_n)
      y <= state_a;
   else
      y <= next_state;
end

assign   z1 = (~y[1] & ~y[2] & ~y[3]), //define outputs
         z2 = (y[1] & ~y[2] & y[3]),
         z3 = (y[1] & y[2] & ~y[3]);

//determine the next state
always @ (x1 or x2 or y)
begin
   case (y)
      state_a:
         if (x1==0)
            next_state = state_a;
         else
            next_state = state_b;

      state_b:
         if (x2==0)
            next_state = state_d;
         else
            next_state = state_c;

      state_c: next_state = state_e;
      state_d: next_state = state_e;
      state_e: next_state = state_a;
      default: next_state = state_a;
   endcase
end
endmodule
```

Figure 3.79 (Continued)

```
//test bench for moore ssm
module moore_ssm23_tb;

reg rst_n, clk, x1, x2;
wire [1:3] y;
wire z1, z2, z3;         //continued on next page
```

Figure 3.80 Test bench for the Moore machine of Example 3.17.

```
//display variables
initial
$monitor ("x1 x2 = %b, state = %b, z1 z2 z3 = %b",
          {x1, x2}, y, {z1, z2, z3});

//define clock
initial
begin
   clk = 1'b0;
   forever
      #10 clk = ~clk;
end

//define input sequence
initial
begin
   #0    rst_n = 1'b0;
         x1 = 1'b0;
         x2 = 1'b0;

   #5    rst_n = 1'b1;

         x1 = 1'b0;      x2 = $random;
         @ (posedge clk)    //go to state_a (000); assert z1

         x1 = 1'b1;      x2 = $random;
         @ (posedge clk)    //go to state_b (011)

         x1 = $random;   x2 = 1'b0;
         @ (posedge clk)    //go to state_d (110); assert z3

         x1 = $random;   x2 = $random;
         @ (posedge clk)    //go to state_e (100)

         x1 = $random;   x2 = $random;
         @ (posedge clk)    //go to state_a (000); assert z1

         x1 = 1'b1;      x2 = $random;
         @ (posedge clk)    //go to state_b (011)

         x1 = $random;   x2 = 1'b1;
         @ (posedge clk)    //go to state_c (101); assert z2

         x1 = $random;   x2 = $random;
         @ (posedge clk)    //go to state_e (100)

//continued on next page
```

Figure 3.80 (Continued)

```
          x1 = $random;   x2 = $random;
          @ (posedge clk)    //go to state_a (000); assert z1

          x1 = 1'b0;        x2 = $random;
          @ (posedge clk)    //go to state_a (000); assert z1


    #10   $stop;


end


//instantiate the module into the test bench
moore_ssm23 inst1 (
   .rst_n(rst_n),
   .clk(clk),
   .x1(x1),
   .x2(x2),
   .y(y),
   .z1(z1),
   .z2(z2),
   .z3(z3)
   );


endmodule
```

Figure 3.80 (Continued)

```
x1 x2 = 00, state = xxx, z1 z2 z3 = xxx
x1 x2 = 11, state = 000, z1 z2 z3 = 100
x1 x2 = 10, state = 011, z1 z2 z3 = 000
x1 x2 = 11, state = 110, z1 z2 z3 = 001
x1 x2 = 11, state = 100, z1 z2 z3 = 000
x1 x2 = 10, state = 000, z1 z2 z3 = 100
x1 x2 = 11, state = 011, z1 z2 z3 = 000
x1 x2 = 10, state = 101, z1 z2 z3 = 010
x1 x2 = 11, state = 100, z1 z2 z3 = 000
x1 x2 = 00, state = 000, z1 z2 z3 = 100
```

Figure 3.81 Outputs for the Moore machine of Example 3.17.

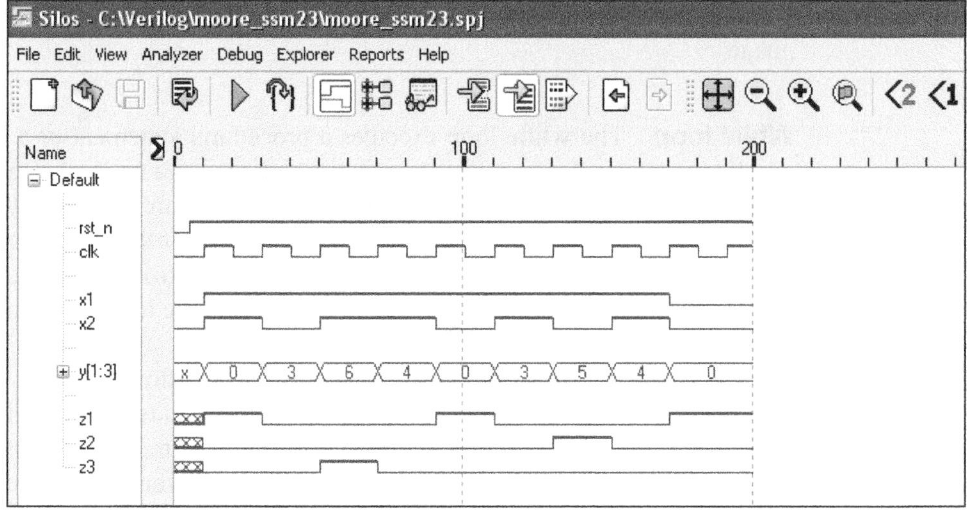

Figure 3.82 Waveforms for the Moore machine of Example 3.17.

3.4.9 Loop Statements

There are four types of loop statements in Verilog: **for**, **while**, **repeat**, and **forever**. Loop statements must be placed within an **initial** or an **always** block and may contain delay controls. The loop constructs allow for repeated execution of procedural statements within an **initial** or an **always** block.

For loop The **for** loop was presented in the test benches of previous examples; therefore, no Verilog code will be shown here. A brief review, however, will be given. The **for** loop contains three parts:

1. An *initial* condition to assign a value to a register control variable. This is executed once at the beginning of the loop to initialize a register variable that controls the loop.

2. A *test* condition to determine when the loop terminates. This is an expression that is executed before the procedural statements of the loop to determine if the loop should execute. The loop is repeated as long as the expression is true. If the expression is false, the loop terminates and the activity flow proceeds to the next statement in the module.

3. An *assignment* to modify the control variable, usually an increment or a decrement. This assignment is executed after each execution of the loop and before the next test to terminate the loop.

The **for** loop is generally used when there is a known beginning and an end to a loop. The **for** loop is similar in function to the **for** loop in the C programming language.

While loop The **while** loop executes a procedural statement or a block of procedural statements as long as a Boolean expression returns a value of true. When the procedural statements are executed, the Boolean expression is reevaluated. The loop is executed until the expression returns a value of false. If the evaluation of the expression is false, then the **while** loop is terminated and control is passed to the next statement in the module. If the expression is false before the loop is initially entered, then the **while** loop is not executed.

The Boolean expression may contain any of the following types: arithmetic, logical, relational, equality, bitwise, reduction, shift, concatenation, replication, or conditional. If the **while** loop contains multiple procedural statements, then they are delimited by the keywords **begin** . . . **end**. The syntax for a **while** statement is as follows:

> **while** (expression)
>   procedural statement or block of procedural statements

Example 3.18 This example demonstrates the use of the **while** construct to determine the numerical value of an 8-bit register *reg_a*. The module is shown in Figure 3.83. The variable *value* is declared as type **integer** and is used to obtain the cumulative value. The first **begin** keyword must have a name associated with the keyword because this declaration is allowed only with named blocks.

The register is initialized to a value of 27 (*8'b0001_1011*). Alternatively, the register can be loaded from any other register. If *reg_a* contains a 1 bit in any bit position, then the **while** loop is executed, because the register has a nonzero value. If *reg_a* contains all zeroes, then the **while** loop is terminated. As long as *reg_a > 0*, the **while** loop executes. If a value of 1 (true) is returned — indicating that register *a* is nonzero — *value* is incremented by one and the value of the register is decreased by one.

The **$display** system task then displays the value of the register based upon the position of the 1s in the register. The value of *reg_a* is shown in Figure 3.84.

Repeat loop The **repeat** loop executes a procedural statement or a block of procedural statements a specified number of times. The **repeat** construct can contain a constant, an expression, a variable, or a signed value. The syntax for the **repeat** loop is as follows:

> **repeat** (loop count expression)
>   procedural statement or block of procedural statements

If the loop count is **x** (unknown value) or **z** (high impedance), then the loop count is treated as zero. The value of the loop count expression is evaluated once at the beginning of the loop.

```
//example of a while loop
//determine the value of a register
module reg_value;

integer value;

initial
begin: determine_value
   reg [7:0] reg_a;
   value = 0;
   reg_a = 8'b0001_1011;

   while (reg_a > 0)
      begin
         value = value + 1;
         reg_a = reg_a - 1;
         $display ("value = %d", value);
      end
end

endmodule
```

Figure 3.83 Module to illustrate the use of the **while** construct.

```
value = 1          value = 10          value = 19
value = 2          value = 11          value = 20
value = 3          value = 12          value = 21
value = 4          value = 13          value = 22
value = 5          value = 14          value = 23
value = 6          value = 15          value = 24
value = 7          value = 16          value = 25
value = 8          value = 17          value = 26
value = 9          value = 18          value = 27
```

Figure 3.84 Outputs for the module of Figure 3.83 that determines the value of a register using the **while** loop.

Example 3.19 An example of the **repeat** loop is shown Figure 3.85, in which two 8-bit registers are added to yield a sum of eight bits. Register *reg_a* is initialized to a value of eight; *reg_b* is initialized to a value of two. The add operation is repeated eight times and register *b* is incremented by one after each add operation. The outputs are shown in Figure 3.86.

```
//example of the repeat keyword
module add_regs_repeat;

reg [7:0] reg_a, reg_b, sum;

initial
begin
   reg_a = 8'b0000_1000;
   reg_b = 8'b0000_0010;

   repeat (8)
   begin
      sum = reg_a + reg_b;
      $display ("reg_a=%b, reg_b=%b, sum=%b",
                  reg_a, reg_b, sum);
      reg_b = reg_b + 1;
   end
end

endmodule
```

Figure 3.85 Module to illustrate the use of the **repeat** construct.

```
reg_a=00001000, reg_b=00000010, sum=00001010
reg_a=00001000, reg_b=00000011, sum=00001011
reg_a=00001000, reg_b=00000100, sum=00001100
reg_a=00001000, reg_b=00000101, sum=00001101
reg_a=00001000, reg_b=00000110, sum=00001110
reg_a=00001000, reg_b=00000111, sum=00001111
reg_a=00001000, reg_b=00001000, sum=00010000
reg_a=00001000, reg_b=00001001, sum=00010001
```

Figure 3.86 Outputs for the module of Figure 3.85 that adds two 8-bit registers using the **repeat** loop.

Forever loop The **forever** loop was presented in the test benches of previous examples to generate a series of clock pulses; therefore, no Verilog code will be shown here. A brief review, however, will be given. The **forever** loop executes the procedural statement continuously until the system tasks **$finish** or **$stop** are encountered. It can also be terminated by the **disable** statement.

The **disable** statement is a procedural statement; therefore, it must be used within an **initial** or an **always** block. It is used to prematurely terminate a block of procedural statements or a system task. When a **disable** statement is executed, control is transferred to the statement immediately following the procedural block or task.

The **forever** loop is similar to a **while** loop in which the expression always evaluates to true (1). A timing control must be used with the **forever** loop; otherwise, the simulator would execute the procedural statement continuously without advancing the simulation time. The syntax of the **forever** loop is as follows:

```
forever
    procedural statement
```

The **forever** statement is typically used for clock generation together with the system task **$finish**. The variable *clk* will toggle every *n* time units for a period of 2*n* time units. The length of simulation is determined by the timing control system task **$finish**.

3.5 Structural Modeling

Structural modeling consists of instantiation of one or more of the following design objects:

- Built-in primitives
- User-defined primitives (UDPs)
- Design modules

Instantiation means to use one or more lower-level modules — including logic primitives — that are interconnected in the construction of a higher-level structural module. A module can be a logic gate, an adder, a multiplexer, a counter, or some other logical function. The objects that are instantiated are called *instances*. Structural modeling is described by the interconnection of these lower-level logic primitives or modules. The interconnections are made by wires that connect primitive terminals or module ports.

3.5.1 Module Instantiation

Design modules were instantiated into every test bench module in previous examples. The ports of the design module were instantiated by name and connected to the corresponding net names of the test bench. Each named instantiation was of the form

.design_module_port_name (test_bench_module_net_name)

Design module ports can be instantiated by name explicitly or by position. Instantiation by position is not recommended when a large number of ports are involved.

Instantiation by name precludes the possibility of making errors in the instantiation process. Modules cannot be nested, but they can be instantiated into other modules.

Structural modeling is analogous to placing the instances on a logic diagram and then connecting them by wires. When instantiating built-in primitives, an instance name is optional; however, when instantiating a module, an instance name must be used. Instances that are instantiated into a structural module are connected by nets of type **wire**.

A structural module may contain behavioral statements (**always**), continuous assignment statements (**assign**), built-in primitives (**and**, **or**, **nand**, **nor**, etc.), UDPs (*mux4*, *half_adder*, *adder4*, etc.), design modules, or any combination of these objects. Design modules can be instantiated into a higher-level structural module in order to achieve a hierarchical design.

Each module in Verilog is either a top-level (higher-level) module or an instantiated module. There is only one top-level module and it is not instantiated anywhere else in the design project. Instantiated primitives or modules, however, can be instantiated many times into a top-level module and each instance of a module is unique and has a unique instance name.

3.5.2 Design Examples

Examples will now be presented that illustrate the structural modeling technique. These examples include a logic circuit to determine if a binary number is within a prescribed range, a modulo-10 counter, and a Moore synchronous sequential machine. Each example will be completely designed in detail and will include appropriate theory where applicable.

Example 3.20 A circuit will be designed using structural modeling to determine if a 4-bit number z_1 satisfies the following criteria:

$$2 < z_1 \leq 6$$
$$10 \leq z_1 < 15$$

The Karnaugh map is shown in Figure 3.87 and the equations for z_1 are shown in Equation 3.7 in both a sum-of-products form and a product-of-sums form. The circuit will be implemented in both forms for comparison. The logic diagram for the sum-of-products design is shown in Figure 3.88. The following logic gates will be designed using dataflow modeling, then instantiated into the structural module: *and2_df*, *and3_df*, and *or4_df*. The dataflow modules for the three gates are shown in Figure 3.89, Figure 3.90, and Figure 3.91. The structural design module for the sum-of-products form is shown in Figure 3.92 and the test bench is shown in Figure 3.93. The outputs are shown in Figure 3.94.

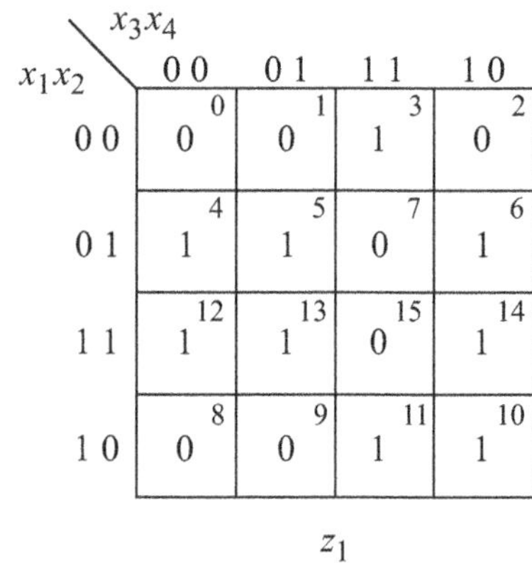

Figure 3.87 Karnaugh map for Example 3.20.

$$z_1 = x_2x_3' + x_2x_4' + x_1x_2'x_3 + x_2'x_3x_4$$

$$z_1 = (x_2 + x_3)(x_2' + x_3' + x_4')(x_1 + x_2 + x_4) \tag{3.7}$$

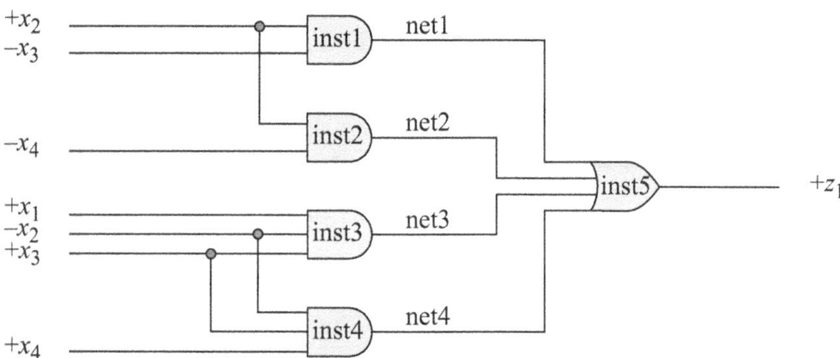

Figure 3.88 Logic diagram for the sum-of-products form for Example 3.20.

```
//dataflow 2-input and gate
module and2_df (x1, x2, z1);

input x1, x2;            //list inputs and output
output z1;

wire x1, x2;             //define signals as wire for dataflow
wire z1;

assign z1 = x1 & x2;     //continuous assign for dataflow

endmodule
```

Figure 3.89 Dataflow module for a 2-input AND gate.

```
//dataflow 3-input and gate
module and3_df (x1, x2, x3, z1);

input x1, x2, x3;        //list inputs and output
output z1;

wire x1, x2, x3;         //define signals as wire for dataflow
wire z1;

assign z1 = x1 & x2 & x3;//continuous assign for dataflow

endmodule
```

Figure 3.90 Dataflow module for a 3-input AND gate.

```
//or4 dataflow
module or4_df (x1, x2, x3, x4, z1);

input x1, x2, x3, x4;
output z1;

wire x1, x2, x3, x4;
wire z1;

assign z1 = x1 | x2 | x3 | x4;

endmodule
```

Figure 3.91 Dataflow module for a 4-input OR gate.

```
//structural sum of products 2 < z1 <= 6; 10 <= z1 < 15
module sop_struc2 (x1, x2, x3, x4, z1);

input x1, x2, x3, x4;
output z1;

wire x1, x2, x3, x4;
wire z1;

//define internal wires
wire net1, net2, net3, net4;

//instantiate the logic
and2_df inst1 (
   .x1(x2),
   .x2(~x3),
   .z1(net1)
   );

and2_df inst2 (
   .x1(x2),
   .x2(~x4),
   .z1(net2)
   );

and3_df inst3 (
   .x1(x1),
   .x2(~x2),
   .x3(x3),
   .z1(net3)
   );

and3_df inst4 (
   .x1(~x2),
   .x2(x3),
   .x3(x4),
   .z1(net4)
   );

or4_df inst5 (
   .x1(net1),
   .x2(net2),
   .x3(net3),
   .x4(net4),
   .z1(z1)
   );
endmodule
```

Figure 3.92 Structural sum-of-products module for Example 3.20.

In the structural module of Figure 3.92, instance *inst1* instantiates a 2-input AND gate labeled *and2_df*. Input x_1 of the AND gate, indicated by $(.x_1)$ connects to variable $+x_2$ indicated by (x_2) in the structural module. Input x_2 of the AND gate indicated by $(.x_2)$ connects to variable $-x_3$ indicated by $(\sim x_3)$ of the structural module. There is a one-to-one correspondence between the AND gate represented by instantiation *inst1* of the logic diagram and the AND gate represented by instantiation *inst1* in the structural module.

```
//test bench for sop_struc2
module sop_struc2_tb;

//define inputs and output
reg x1, x2, x3, x4;
wire z1;

//apply input vectors and display variables
initial
begin: apply_stimulus
   reg [4:0] invect;
   for (invect = 0; invect < 16; invect = invect + 1)
      begin
         {x1, x2, x3, x4} = invect [4:0];
         #10 $display ("x1 x2 x3 x4 = %b, z1 = %b",
                           {x1, x2, x3, x4}, z1);
      end
end

//instantiate the module into the test bench
sop_struc2 inst1 (
   .x1(x1),
   .x2(x2),
   .x3(x3),
   .x4(x4),
   .z1(z1)
   );

endmodule
```

Figure 3.93 Test bench for the structural module of Figure 3.92 for Example 3.20.

```
x1 x2 x3 x4 = 0000, z1 = 0        x1 x2 x3 x4 = 1000, z1 = 0
x1 x2 x3 x4 = 0001, z1 = 0        x1 x2 x3 x4 = 1001, z1 = 0
x1 x2 x3 x4 = 0010, z1 = 0        x1 x2 x3 x4 = 1010, z1 = 1
x1 x2 x3 x4 = 0011, z1 = 1        x1 x2 x3 x4 = 1011, z1 = 1
x1 x2 x3 x4 = 0100, z1 = 1        x1 x2 x3 x4 = 1100, z1 = 1
x1 x2 x3 x4 = 0101, z1 = 1        x1 x2 x3 x4 = 1101, z1 = 1
x1 x2 x3 x4 = 0110, z1 = 1        x1 x2 x3 x4 = 1110, z1 = 1
x1 x2 x3 x4 = 0111, z1 = 0        x1 x2 x3 x4 = 1111, z1 = 0
```

Figure 3.94 Outputs for the structural module of Figure 3.92 for Example 3.20.

The logic diagram for the product-of-sums design is shown in Figure 3.95. The following logic gates will be designed using dataflow modeling, then instantiated into the structural module: *nor2_df* and *nor3_df*. The dataflow modules for the two gates are shown in Figure 3.96 and Figure 3.97. The structural design module for the product-of-sums form is shown in Figure 3.98 and the test bench is shown in Figure 3.99. The outputs are shown in Figure 3.100.

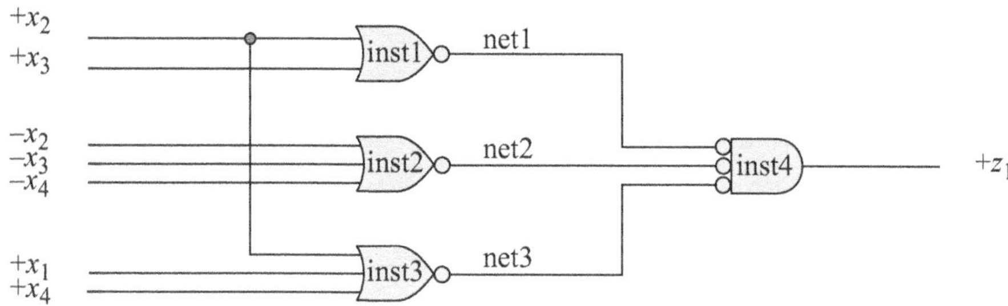

Figure 3.95 Logic diagram for the product-of-sums form for Example 3.20.

```
//dataflow 2-input nor gate
module nor2_df (x1, x2, z1);

input x1, x2;
output z1;

wire x1, x2;
wire z1;

assign z1 = ~(x1 | x2);
endmodule
```

Figure 3.96 Dataflow module for a 2-input NOR gate.

```
//dataflow 3-input nor gate
module nor3_df (x1, x2, x3, z1);

input x1, x2, x3;
output z1;

wire x1, x2;
wire z1;

assign z1 = ~(x1 | x2 | x3);
endmodule
```

Figure 3.97 Dataflow module for a 3-input NOR gate.

```
//structural product of sums 2 < z1 <= 6; 10 <= z1 < 15
module pos_struc2 (x1, x2, x3, x4, z1);

input x1, x2, x3, x4;
output z1;

//define internal nets
wire net1, net2, net3;

//instantiate the logic gates
nor2_df inst1 (
   .x1(x2),
   .x2(x3),
   .z1(net1)
   );

nor3_df inst2 (
   .x1(~x2),
   .x2(~x3),
   .x3(~x4),
   .z1(net2)
   );

nor3_df inst3 (
   .x1(x2),
   .x2(x1),
   .x3(x4),
   .z1(net3)
   );

//continued on next page
```

Figure 3.98 Structural product-of-sums module for Example 3.20.

```
nor3_df inst4 (
   .x1(net1),
   .x2(net2),
   .x3(net3),
   .z1(z1)
   );

endmodule
```

Figure 3.98 (Continued)

```
//test bench for sop_struc2
module pos_struc2_tb;

//define inputs and output
reg x1, x2, x3, x4;
wire z1;

//apply input vectors and display variables
initial
begin: apply_stimulus
   reg [4:0] invect;
   for (invect = 0; invect < 16; invect = invect + 1)
      begin
         {x1, x2, x3, x4} = invect [4:0];
         #10 $display ("x1 x2 x3 x4 = %b, z1 = %b",
                        {x1, x2, x3, x4}, z1);
      end
end

//instantiate the module into the test bench
pos_struc2 inst1 (
   .x1(x1),
   .x2(x2),
   .x3(x3),
   .x4(x4),
   .z1(z1)
   );

endmodule
```

Figure 3.99 Test bench for the structural module of Figure 3.98 for Example 3.20.

```
x1 x2 x3 x4 = 0000, z1 = 0        x1 x2 x3 x4 = 1000, z1 = 0
x1 x2 x3 x4 = 0001, z1 = 0        x1 x2 x3 x4 = 1001, z1 = 0
x1 x2 x3 x4 = 0010, z1 = 0        x1 x2 x3 x4 = 1010, z1 = 1
x1 x2 x3 x4 = 0011, z1 = 1        x1 x2 x3 x4 = 1011, z1 = 1
x1 x2 x3 x4 = 0100, z1 = 1        x1 x2 x3 x4 = 1100, z1 = 1
x1 x2 x3 x4 = 0101, z1 = 1        x1 x2 x3 x4 = 1101, z1 = 1
x1 x2 x3 x4 = 0110, z1 = 1        x1 x2 x3 x4 = 1110, z1 = 1
x1 x2 x3 x4 = 0111, z1 = 0        x1 x2 x3 x4 = 1111, z1 = 0
```

Figure 3.100 Outputs for the structural module of Figure 3.98 for Example 3.20.

Example 3.21 A counter will be designed with *D* flip-flops using structural modeling that counts in the sequence shown in Table 3.2, where y_4 is the low-order bit. The following modules will be designed for instantiation into the structural module: a dataflow 4-input OR gate *or4_df*, a dataflow 2-input exclusive-OR circuit *xor2_df*; and a behavioral positive-edge-triggered *D* flip-flop *d_ff_bh*.

Table 3.2 Counting Sequence for the Counter of Example 3.21

y_1	y_2	y_3	y_4
0	0	0	0
0	0	0	1
0	0	1	1
0	1	0	1
0	1	1	1
1	0	0	1
1	0	1	1
1	1	0	1
1	1	1	1
0	0	1	0
0	1	0	0
0	1	1	0
1	0	0	0
1	0	1	0
1	1	0	0
1	1	1	0
0	0	0	0

The Karnaugh maps are shown in Figure 3.101 and the equations for the D flip-flops are shown in Equation 3.8. The logic diagram is shown in Figure 3.102 indicating the instantiation names for the logic functions and the net names for the interconnecting wires. The dataflow modules for the 4-input OR gate and the 2-input exclusive-OR circuit are shown in Figure 3.103 and Figure 3.104, respectively. The behavioral module for the positive-edge-triggered D flip-flop is shown in Figure 3.105.

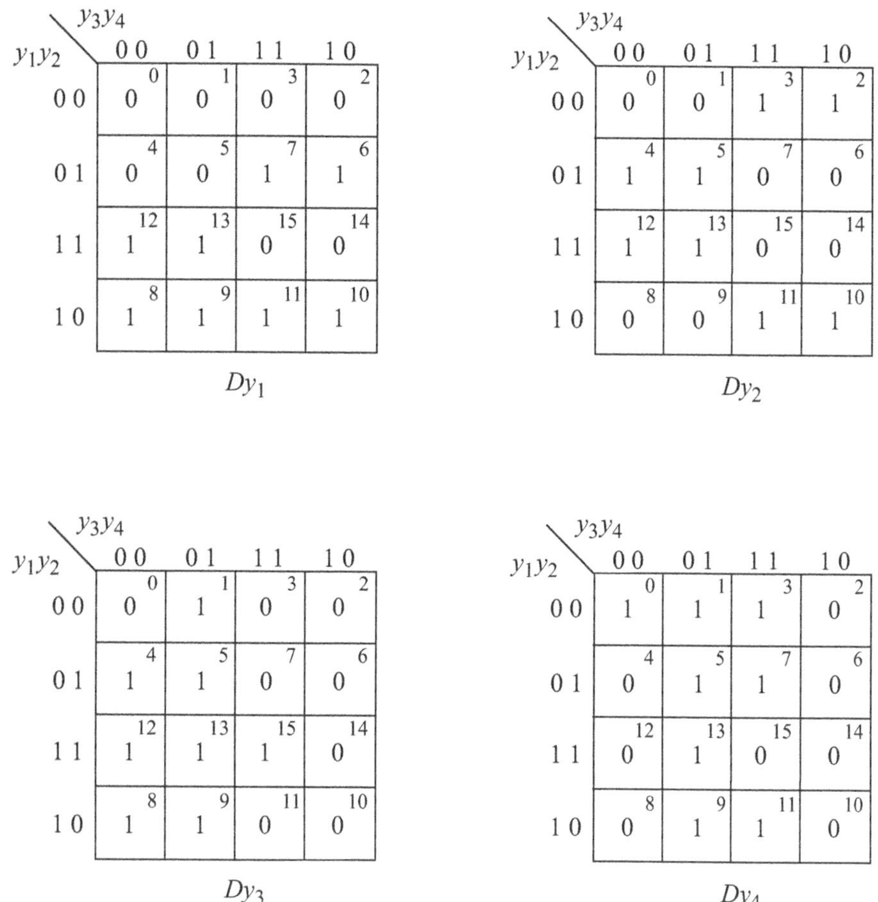

Figure 3.101 Karnaugh maps for the counter of Example 3.21.

The structural module for the counter is shown in Figure 3.106, in which the instantiation names and the net names correspond directly with those in the logic diagram. The test bench is shown in Figure 3.107. The counting sequence is displayed by the **$monitor** system task which continuously monitors the values of the variables specified in the parameter list — in this case, the values of the counter flip-flops. The **forever** statement generates clock pulses with a period of 24 time units. The length of simulation is determined by the system task **$stop**, which occurs after 370 time units.

The outputs are shown in Figure 3.108 which display the counting sequence in binary. The waveforms are shown in Figure 3.109 which shows the counting sequence in hexadecimal notation.

$$
\begin{aligned}
Dy_1 &= y_1y_3' + y_1y_2' + y_1'y_2y_3 \\
Dy_2 &= y_2y_3' + y_2'y_3 = y_2 \oplus y_3 \\
Dy_3 &= y_2y_3' + y_1y_3' + y_3'y_4 + y_1y_2y_4 \\
Dy_4 &= y_3'y_4 + y_1'y_4 + y_2'y_4 + y_1'y_2'y_3'
\end{aligned}
\quad (3.8)
$$

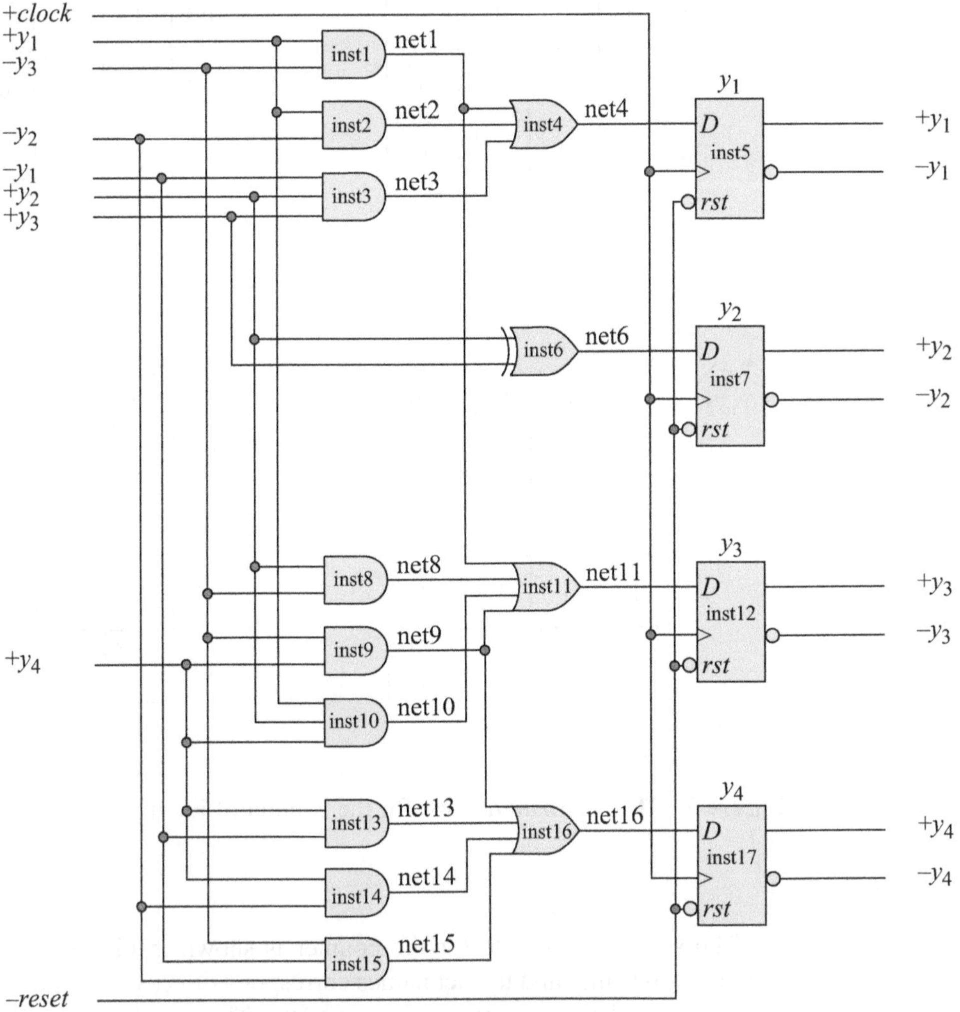

Figure 3.102 Logic diagram for the counter of Example 3.21.

```
//dataflow or4
module or4_df (x1, x2, x3, x4, z1);

input x1, x2, x3, x4;
output z1;

wire x1, x2, x3, x4;
wire z1;

assign z1 = x1 | x2 | x3 | x4;
endmodule
```

Figure 3.103 Dataflow module for a 4-input OR gate.

```
//dataflow xor2_df
module xor2_df (x1, x2, z1);

input x1, x2;
output z1;

wire x1, x2;
wire z1;

assign z1 = x1 ^ x2;
endmodule
```

Figure 3.104 Dataflow module for a 2-input exclusive-OR circuit.

```
//behavioral D flip-flop
module d_ff_bh (rst_n, clk, d, q, q_n);

input rst_n, clk, d;
output q, q_n;

wire rst_n, clk, d;
reg q;

assign q_n = ~q;

always @ (rst_n or posedge clk)
begin
   if (rst_n == 0)
         q <= 1'b0;
   else q <= d;
end
endmodule
```

Figure 3.105 Behavioral module for a positive-edge-triggered *D* flip-flop.

```
//structural odd even counter
module ctr_odd_evn2 (rst_n, clk, y);

input rst_n, clk;
output [1:4] y;

wire rst_n, clk;
wire [1:4] y;

//define internal wires
wire  net1, net2, net3, net4, net6,
      net8, net9, net10, net11,
      net13, net14, net15, net16;

//instantiate the logic for flip-flop y[1]
and2_df inst1 (
   .x1(y[1]),
   .x2(~y[3]),
   .z1(net1)
   );

and2_df inst2 (
   .x1(y[1]),
   .x2(~y[2]),
   .z1(net2)
   );

and3_df inst3 (
   .x1(~y[1]),
   .x2(y[2]),
   .x3(y[3]),
   .z1(net3)
   );

or3_df inst4 (
   .x1(net1),
   .x2(net2),
   .x3(net3),
   .z1(net4)
   );

d_ff_bh inst5 (
   .rst_n(rst_n),
   .clk(clk),
   .d(net4),
   .q(y[1])
   );             //continued on next page
```

Figure 3.106 Structural module for the counter of Example 3.21.

```
//instantiate the logic for flip-flop y[2]
xor2_df inst6 (
   .x1(y[2]),
   .x2(y[3]),
   .z1(net6)
   );

d_ff_bh inst7 (
   .rst_n(rst_n),
   .clk(clk),
   .d(net6),
   .q(y[2])
   );

//instantiate the logic for flip-flop y[3]
and2_df inst8 (
   .x1(y[2]),
   .x2(~y[3]),
   .z1(net8)
   );

and2_df inst9 (
   .x1(~y[3]),
   .x2(y[4]),
   .z1(net9)
   );

and3_df inst10 (
   .x1(y[1]),
   .x2(y[2]),
   .x3(y[4]),
   .z1(net10)
   );

or4_df inst11 (
   .x1(net1),
   .x2(net8),
   .x3(net9),
   .x4(net10),
   .z1(net11)
   );

d_ff_bh inst12 (
   .rst_n(rst_n),
   .clk(clk),
   .d(net11),
   .q(y[3])
   );                    //continued on next page
```

Figure 3.106 (Continued)

```
//instantiate the logic for flip-flop y[4]
and2_df inst13 (
   .x1(~y[1]),
   .x2(y[4]),
   .z1(net13)
   );

and2_df inst14 (
   .x1(~y[2]),
   .x2(y[4]),
   .z1(net14)
   );

and3_df inst15 (
   .x1(~y[1]),
   .x2(~y[2]),
   .x3(~y[3]),
   .z1(net15)
   );

or4_df inst16 (
   .x1(net9),
   .x2(net13),
   .x3(net14),
   .x4(net15),
   .z1(net16)
   );

d_ff_bh inst17 (
   .rst_n(rst_n),
   .clk(clk),
   .d(net16),
   .q(y[4])
   );
endmodule
```

Figure 3.106 (Continued)

```
//test bench for odd even counter
module ctr_odd_evn2_tb;

reg rst_n, clk;
wire [1:4] y;
//continued on next page
```

Figure 3.107 Test bench for the counter of Figure 3.106.

```
//display count
initial
$monitor ("count = %b", y);

//generate reset
initial
begin
   #0    rst_n = 1'b0;
   #2    rst_n = 1'b1;
end

//generate clock
initial
begin
   clk = 1'b0;
   forever
      #12   clk = ~clk;
end

//determine length of simulation
initial
   #370  $stop;

//instantiate the module into the test bench
ctr_odd_evn2 inst1 (
   .rst_n(rst_n),
   .clk(clk),
   .y(y)
   );

endmodule
```

Figure 3.107 (Continued)

```
count = 0000          count = 0010
count = 0001          count = 0100
count = 0011          count = 0110
count = 0101          count = 1000
count = 0111          count = 1010
count = 1001          count = 1100
count = 1011          count = 1110
count = 1101          count = 0000
count = 1111
```

Figure 3.108 Outputs for the structural module counter of Figure 3.106.

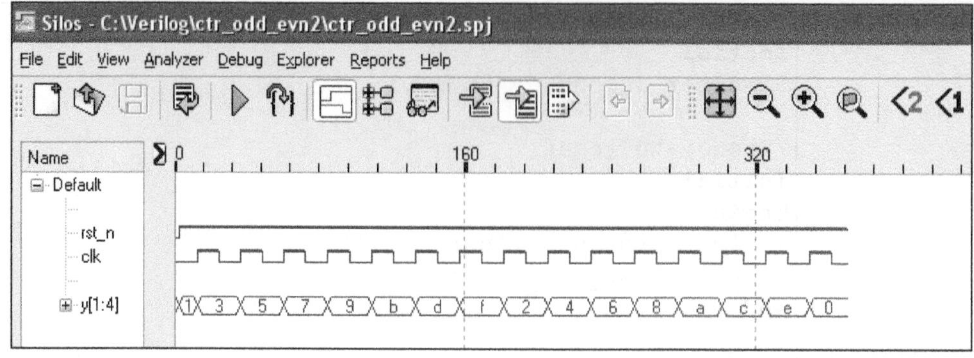

Figure 3.109 Waveforms for the structural module counter of Figure 3.106.

Example 3.22 Using structural modeling, a Moore synchronous sequential machine will be designed that receives 3-bit words x_1, x_2, and x_3 on a parallel input bus, where input x_3 is the low-order bit. There are five outputs z_1, z_2, z_3, z_4, and z_5, where the subscripts indicate the decimal value of the corresponding unsigned binary input word. The outputs are asserted at time t_2 and deasserted at time t_3. The machine will be designed using positive-edge-triggered D flip-flops and additional AND gates and OR gates. The state diagram is shown in Figure 3.110.

The Karnaugh maps for the D flip-flops are shown in Figure 3.111 and the equations for the D inputs are shown in Equation 3.9. The logic diagram is shown in Figure 3.112 indicating the instantiation names and the net names which correspond to the instantiation names and net names in the structural module.

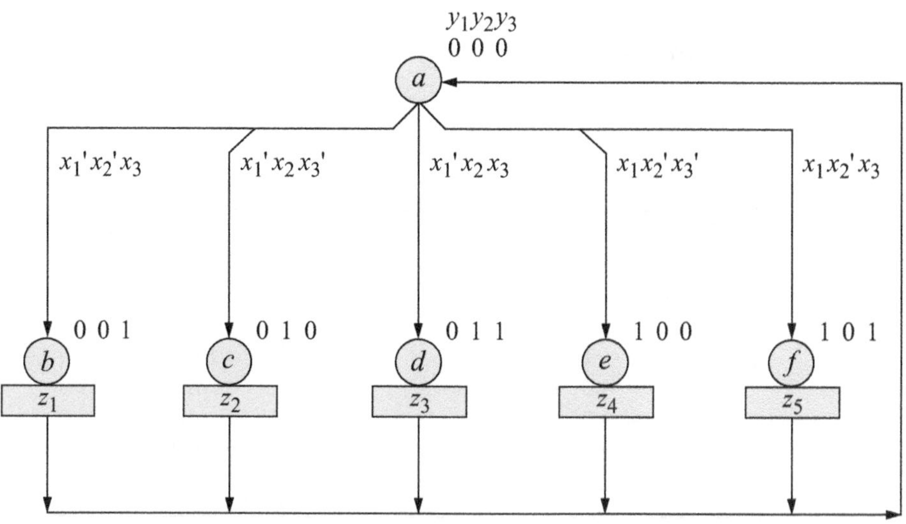

Figure 3.110 State diagram for the Moore machine of Example 3.22.

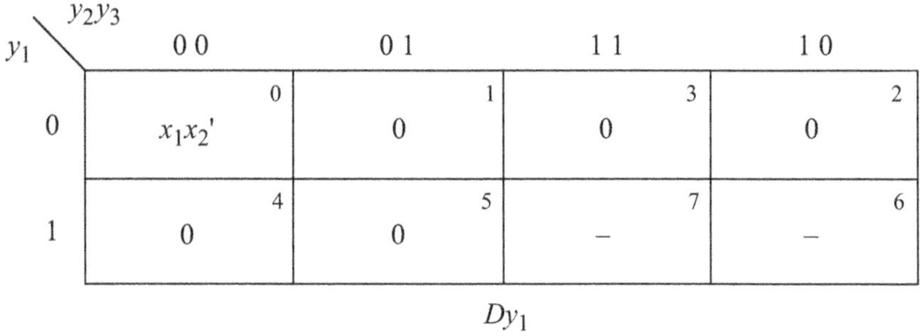

y_1 \ y_2y_3	00	01	11	10
0	$x_1'x_2$ [0]	0 [1]	0 [3]	0 [2]
1	0 [4]	0 [5]	– [7]	– [6]

Dy_2

y_1 \ y_2y_3	00	01	11	10
0	$x_3(x_1' + x_2')$ [0]	0 [1]	0 [3]	0 [2]
1	0 [4]	0 [5]	– [7]	– [6]

Dy_3

Figure 3.111 Karnaugh maps for the Moore machine of Example 3.22.

$$
\begin{aligned}
Dy_1 &= y_1'y_2'y_3'x_1x_2' \\
Dy_2 &= y_1'y_2'y_3'x_1'x_2 \\
Dy_3 &= y_1'y_2'y_3'x_1'x_3 + y_1'y_2'y_3'x_2'x_3
\end{aligned}
\tag{3.9}
$$

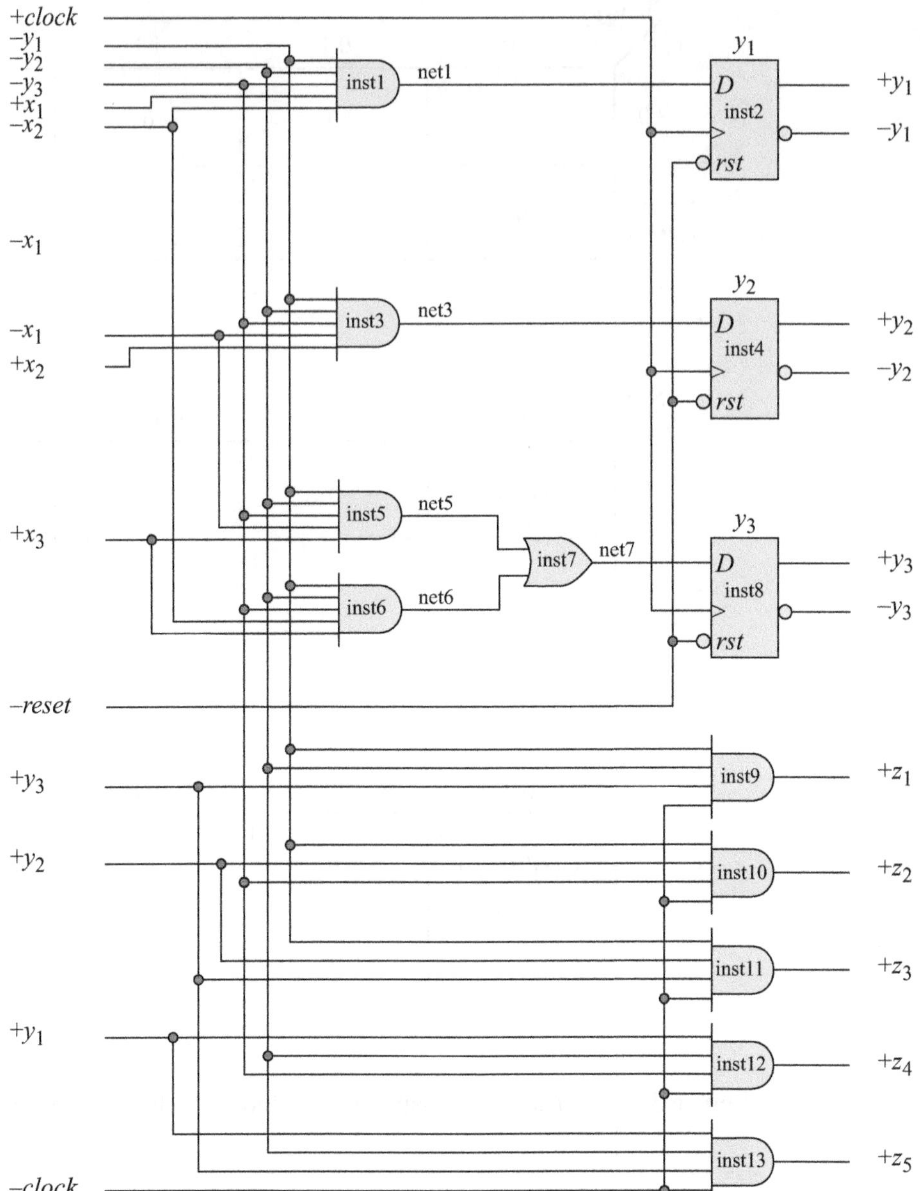

Figure 3.112 Logic diagram for the Moore machine of Example 3.22.

The structural module for this Moore machine is shown in Figure 3.113. The test bench, shown in Figure 3.114, sequences the machine through all of the states in the state diagram. The system task **$random** generates a random value for the inputs where appropriate, because the value is irrelevant. This occurs for a transition from states *b*, *c*, *d*, *e*, or *f* to state *a*.

The outputs are shown in Figure 3.115 and the waveforms are shown in Figure 3.116, which gives a clear indication of the state transition sequences and the associated outputs. As can be seen, the outputs are asserted during the last half of the clock period.

```
//structural moore ssm
module moore_ssm22 (rst_n, clk, x1, x2, x3, y,
                    z1, z2, z3, z4, z5);

//define inputs and outputs
input rst_n, clk, x1, x2, x3;
output [1:3] y;
output z1, z2, z3, z4, z5;

//define internal wires
wire net1, net3, net5, net6, net7;

//instantiate the logic for flip-flop y[1]
and5_df inst1 (
   .x1(~y[1]),
   .x2(~y[2]),
   .x3(~y[3]),
   .x4(x1),
   .x5(~x2),
   .z1(net1)
   );

d_ff_bh inst2 (
   .rst_n(rst_n),
   .clk(clk),
   .d(net1),
   .q(y[1])
   );

//instantiate the logic for flip-flop y[2]
and5_df inst3 (
   .x1(~y[1]),
   .x2(~y[2]),
   .x3(~y[3]),
   .x4(~x1),
   .x5(x2),
   .z1(net3)
   );


//continued on next page
```

Figure 3.113 Structural module for the Moore machine of Example 3.22.

```
d_ff_bh inst4 (
   .rst_n(rst_n),
   .clk(clk),
   .d(net3),
   .q(y[2])
   );

//instantiate the logic for flip-flop y[3]
and5_df inst5 (
   .x1(~y[1]),
   .x2(~y[2]),
   .x3(~y[3]),
   .x4(~x1),
   .x5(x3),
   .z1(net5)
   );

and5_df inst6 (
   .x1(~y[1]),
   .x2(~y[2]),
   .x3(~y[3]),
   .x4(~x2),
   .x5(x3),
   .z1(net6)
   );

or2_df inst7 (
   .x1(net5),
   .x2(net6),
   .z1(net7)
   );

d_ff_bh inst8 (
   .rst_n(rst_n),
   .clk(clk),
   .d(net7),
   .q(y[3])
   );

//instantiate the logic outputs z1, z2, z3, z4, and z5
and4_df inst9 (
   .x1(~y[1]),
   .x2(~y[2]),
   .x3(y[3]),
   .x4(~clk),
   .z1(z1)
   );              //continued on next page
```

Figure 3.113 (Continued)

```
and4_df inst10 (
   .x1(~y[1]),
   .x2(y[2]),
   .x3(~y[3]),
   .x4(~clk),
   .z1(z2)
   );

and4_df inst11 (
   .x1(~y[1]),
   .x2(y[2]),
   .x3(y[3]),
   .x4(~clk),
   .z1(z3)
   );

and4_df inst12 (
   .x1(y[1]),
   .x2(~y[2]),
   .x3(~y[3]),
   .x4(~clk),
   .z1(z4)
   );

and4_df inst13 (
   .x1(y[1]),
   .x2(~y[2]),
   .x3(y[3]),
   .x4(~clk),
   .z1(z5)
   );

endmodule
```

Figure 3.113 (Continued)

```
//test bench for moore ssm22
module moore_ssm22_tb;

reg rst_n, clk, x1, x2, x3;

wire [1:3] y;
wire z1, z2, z3, z4, z5;

//continued on next page
```

Figure 3.114 Test bench for the Moore machine of Figure 3.113.

```
//display inputs and outputs
initial
$monitor ("x1 x2 x3 = %b, state = %b, z1 z2 z3 z4 z5 = %b",
          {x1, x2, x3}, y, {z1, z2, z3, z4, z5});

//define clock
initial
begin
   clk = 1'b0;
   forever
      #10   clk = ~clk;
end

//define input sequence
initial
begin
   #0    rst_n = 1'b0;  //reset to state_a (000)
         x1=1'b0; x2=1'b0; x3=1'b0;
   #5    rst_n = 1'b1;

   x1=1'b0;    x2=1'b0;    x3=1'b1;
   @ (posedge clk)      //go to state_b (001); assert z1

   x1=$random; x2=$random; x3=$random;
   @ (posedge clk)      //go to state_a (000)

   x1=1'b0;    x2=1'b1;    x3=1'b0;
   @ (posedge clk)      //go to state_c (010); assert z2

   x1=$random; x2=$random; x3=$random;
   @ (posedge clk)      //go to state_a (000)

   x1=1'b0;    x2=1'b1;    x3=1'b1;
   @ (posedge clk)      //go to state_d (011); assert z3

   x1=$random; x2=$random; x3=$random;
   @ (posedge clk)      //go to state_a (000)

   x1=1'b1;    x2=1'b0;    x3=1'b0;
   @ (posedge clk)      //go to state_e (100); assert z4

   x1=$random; x2=$random; x3=$random;
   @ (posedge clk)      //go to state_a (000)

   x1=1'b1;    x2=1'b0;    x3=1'b1;
   @ (posedge clk)      //go to state_f (101); assert z5
//continued on next page
```

Figure 3.114 (Continued)

```
   x1=$random; x2=$random; x3=$random;
   @ (posedge clk)      //go to state_a (000)

   #10   $stop;
end

//instantiate the module into the test bench
moore_ssm22 inst1 (
   .rst_n(rst_n),
   .clk(clk),
   .x1(x1),
   .x2(x2),
   .x3(x3),
   .y(y),
   .z1(z1),
   .z2(z2),
   .z3(z3),
   .z4(z4),
   .z5(z5)
   );

endmodule
```

Figure 3.114 (Continued)

```
x1 x2 x3 = 000, state = 000, z1 z2 z3 z4 z5 = 00000
x1 x2 x3 = 001, state = 000, z1 z2 z3 z4 z5 = 00000
x1 x2 x3 = 011, state = 001, z1 z2 z3 z4 z5 = 00000
x1 x2 x3 = 011, state = 001, z1 z2 z3 z4 z5 = 10000
x1 x2 x3 = 010, state = 000, z1 z2 z3 z4 z5 = 00000
x1 x2 x3 = 111, state = 010, z1 z2 z3 z4 z5 = 00000
x1 x2 x3 = 111, state = 010, z1 z2 z3 z4 z5 = 01000
x1 x2 x3 = 011, state = 000, z1 z2 z3 z4 z5 = 00000
x1 x2 x3 = 101, state = 011, z1 z2 z3 z4 z5 = 00000
x1 x2 x3 = 101, state = 011, z1 z2 z3 z4 z5 = 00100
x1 x2 x3 = 100, state = 000, z1 z2 z3 z4 z5 = 00000
x1 x2 x3 = 101, state = 100, z1 z2 z3 z4 z5 = 00000
x1 x2 x3 = 101, state = 100, z1 z2 z3 z4 z5 = 00010
x1 x2 x3 = 101, state = 000, z1 z2 z3 z4 z5 = 00000
x1 x2 x3 = 101, state = 101, z1 z2 z3 z4 z5 = 00000
x1 x2 x3 = 101, state = 101, z1 z2 z3 z4 z5 = 00001
x1 x2 x3 = 101, state = 000, z1 z2 z3 z4 z5 = 00000
```

Figure 3.115 Outputs for the Moore machine of Figure 3.113.

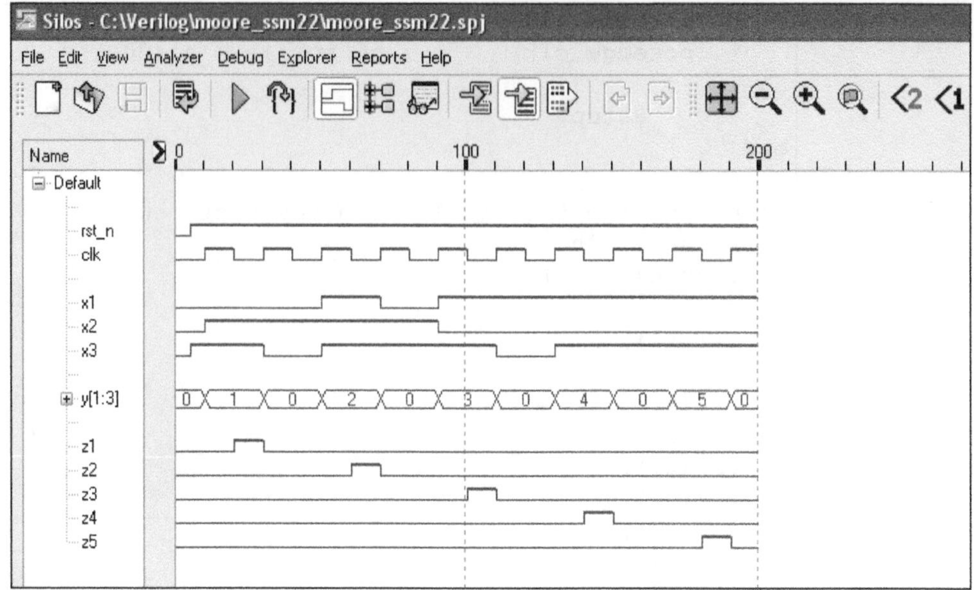

Figure 3.116 Waveforms for the Moore machine of Figure 3.113.

3.6 Problems

3.1 Given the equation shown below, obtain the minimized equation for z_1 in a sum-of-products notation and implement the equation using NAND gate built-in primitives. Then obtain the minimized equation for z_1 in a product-of-sums notation and implement the equation using NAND gate built-in primitives. Obtain the design module, the test bench module, and the outputs. Output z_1 is asserted high in both cases.

$$z_1(x_1, x_2, x_3, x_4) = \Sigma_{\text{m}}(1, 4, 7, 9, 11, 13) + \Sigma_{\text{d}}(5, 14, 15)$$

3.2 Design the circuit for the equation shown below using built-in primitives. Obtain the design module, the test bench module, and outputs.

$$z_1 = [x_1 x_2 + (x_1 \oplus x_2)]\, x_3$$

3.3 Minimize the following function using any minimization method, then use only NAND gate and NOR gate built-in primitives to implement the function:

$$z_1 = x_1{}' x_2 + x_3 x_4 + (x_1 + x_2)'\, [x_1 x_3 x_4 + (x_2 x_5)']$$

The inputs and output are asserted high. Obtain the design module, the test bench module using all five variables, and the outputs.

3.4 Obtain the minimized equation for a circuit that generates an output z_1 whenever a 4-bit unsigned binary number N meets the following requirements:

N is an even number or N is evenly divisible by three.

The format for N is: $N = n_3\,n_2\,n_1\,n_0$, where n_0 is the low-order bit. Design the necessary user-defined primitives that are required for the design. Obtain the design module, the test bench module, and the outputs.

3.5 Obtain the equation for a logic circuit that will generate a logic 1 whenever a 4-bit unsigned binary number N satisfies the following criteria:

$N = x_1x_2x_3x_4$ (low order)
$2 < N \leq 6$
$11 \leq N < 14$

Using NOR user-defined primitives, obtain the design module, test bench, and the outputs. Output z_1 will be active high if the above conditions are met.

3.6 Use only user-defined primitives to design the Moore synchronous sequential machine that operates according to the state diagram shown below. Use positive-edge-triggered D flip-flops and any additional logic functions. Obtain the design module, the test bench module, the outputs, and the waveforms.

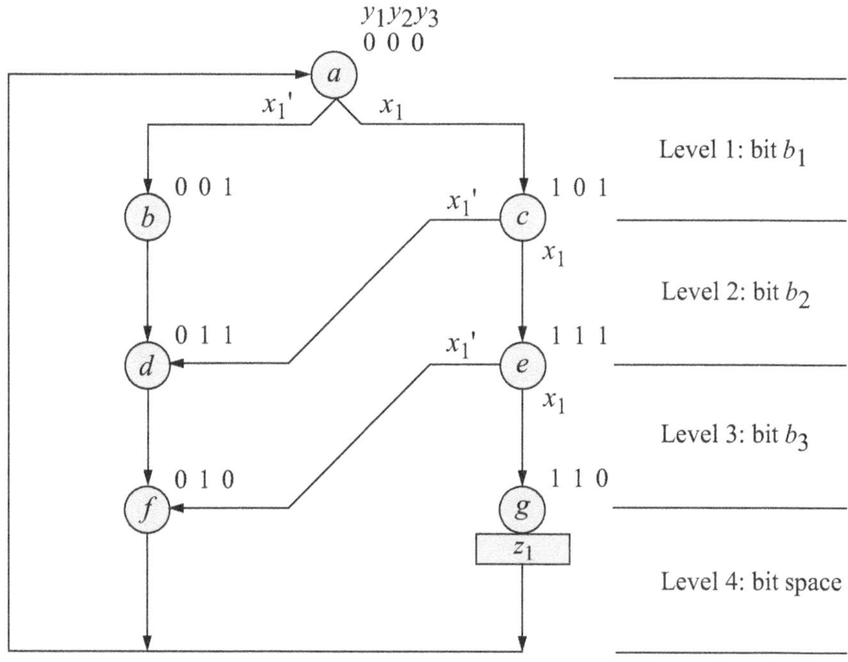

3.7 Design a circuit using dataflow modeling that generates an output whenever a 4-bit number $z_1 = x_1x_2x_3x_4$ has a value that is a power of two or a value that is evenly divisible by three. Input x_4 is the low-order variable. Obtain the design module using AND gates and OR gates, the test bench module, and the outputs.

3.8 Design a dataflow module that will generate a logic 1 when two 4-bit unsigned binary operands are unequal. The operands are: a = *[3:0]* and b = *[3:0]*; the output is z_1. Obtain the design module, the test bench module for 16 combinations of the two operands, and the outputs.

3.9 Use dataflow modeling to design an 8-bit odd parity generator. Obtain the design module, the test bench module for 16 combinations of the input vector, and the outputs.

3.10 Use behavioral modeling to design a circuit that has three 8-bit registers and three 8-bit outputs z_1, z_2, and z_3. The outputs are defined as follows:

z_1 = register *a* added to register *b*
z_2 = register *a* ORed with register *b*
z_3 = register *a* exclusive-ORed with register *b*

3.11 Use behavioral modeling to design a 4-bit counter whose counting sequence is determined by a control input x_1. If $x_1 = 0$, the counter counts up from 0000 to 1110 in increments of two; if $x_1 = 1$, the counter counts down from 1111 to 0000 in increments of one. Obtain the behavioral module, the test bench module, the outputs, and the waveforms.

3.12 Use behavioral modeling to design a Mealy synchronous sequential machine that operates according to the state diagram shown on the following page. The outputs are asserted during the last half of the clock cycle. Obtain the behavioral module, the test bench module, the outputs, and the waveforms.

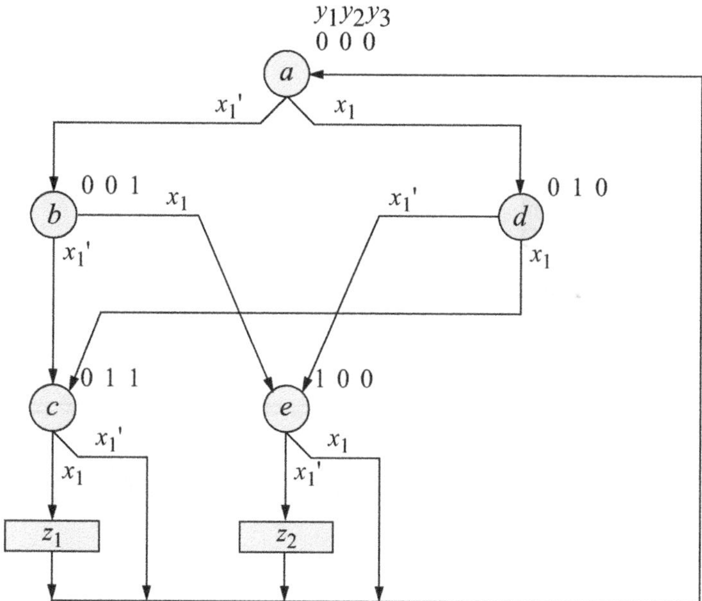

3.13 Using structural modeling, design a count-up modulo-10 counter with positive-edge-triggered *D* flip-flops. The logic gates are to be designed using dataflow modeling; the *D* flip-flops are to be designed using behavioral modeling.

Obtain the Karnaugh maps, the input equations, and the logic diagram. Then obtain the design module, the test bench module, the outputs, and the waveforms.

3.14 Use structural modeling to design a modulo-8 count-down counter with positive-edge-triggered *JK* flip-flops that were designed using behavioral modeling. All logic gates are to be designed using dataflow modeling. The counter is to be initialized to a count of $y_1y_2y_3 = 111$, where y_3 is the low-order flip-flop.

Obtain the next-state table, the Karnaugh maps, the *JK* input equations, and the logic diagram. Then design the structural design module, the test bench module, the outputs, and the waveforms.

3.15 Use structural modeling to design a Mealy synchronous sequential machine that generates an output z_1 whenever a serial input line x_1 contains a sequence of three consecutive 1s. Overlapping sequences are allowed.

Use *JK* flip-flops for the storage elements that were designed using behavioral modeling. Use dataflow modeling for the logic gates. Generate the state diagram, the *JK* input maps and equations, and the logic diagram. Obtain the structural design module, the test bench module, the outputs, and the waveforms.

4.1 Ripple-Carry Addition
4.2 Carry Lookahead Addition
4.3 Carry-Save Addition
4.4 Memory-Based Addition
4.5 Carry-Select Addition
4.6 Serial Adder
4.7 Problems

Fixed-Point Addition

Fixed-point addition is undoubtedly the easiest arithmetic operation to implement in any number representation. In fixed-point operations, the radix point is in a fixed location in the operand. The radix point (or binary point for radix 2) is to the immediate right of the low-order bit for integers, or to the immediate left of the high-order bit for fractions. The operands in a computer can be expressed by any of the following number representations: unsigned, sign-magnitude, diminished-radix complement, or radix complement.

Addition of two binary operands treats both signed and unsigned operands the same — there is no distinction between the two types of numbers during the add operation. If the numbers are signed, then the sign bit can be extended to the left indefinitely without changing the value of the number. An n-bit signed number A is shown in Equation 4.1, where the leftmost digit a_{n-1} is the sign bit. The sign bit for any radix is 0 for positive numbers and $r-1$ for negative numbers, as shown in Equation 4.2.

$$A = a_{n-1}\ a_{n-2}\ \ldots\ a_1\ a_2 \tag{4.1}$$

$$A = \begin{cases} 0 & \text{for } A \geq 0 \\ r-1 & \text{for } A < 0 \end{cases} \tag{4.2}$$

The operands for addition are the *augend* and the *addend*, where the addend is added to the augend and the sum replaces the augend in most computers — the addend is unchanged. The rules for radix 2 addition are shown in Table 4.1. An example of binary addition is shown in Figure 4.1.

The sum of column 1 is 2_{10} (10_2); therefore, the sum is 0 with a carry of 1 to column 2. The sum of column 2 is 4_{10} (100_2); therefore, the sum is 0 with a carry of 0 to column 3 and a carry of 1 to column 4. The sum of column 3 is 4_{10} (100_2); therefore, the sum is 0 with a carry of 0 to column 4 and a carry of 1 to column 5. The sum of column 4 is 2_{10} (10_2); therefore, the sum is 0 with a carry of 1 to column 5. The sum of column 5 is 2 (10_2); therefore, the sum is 0 with a carry to column 6. The unsigned values of the binary operands are shown in the rightmost column together with the resulting sum.

Table 4.1 Rules for Binary Addition

+	0	1
0	0	1
1	1	0^1

(1) 1 + 1 = 0 with a carry to the next higher-order column.

Column	6	5	4	3	2	1	Radix 10 values
			0	1	1	0	6
			0	1	0	1	5
			1	1	1	1	15
+)		$_{11}$	0_{10}	1_0	1_1	0	6
	1	0	0	0	0	0	32

Figure 4.1 Example of binary addition.

4.1 Ripple-Carry Addition

A ripple-carry adder is not considered a high-speed adder, but requires less logic than a high-speed adder using the carry lookahead technique. An n-stage ripple adder requires n full adders. A full adder can be implemented using two half adders. A *half adder* is a combinational circuit that adds two operand bits and produces two outputs: sum and carry-out. A *full adder* is a combinational circuit that adds two operand bits plus a carry-in bit. The carry-in bit represents the carry-out of the previous lower-order stage. A full adder produces two outputs: sum and carry-out.

The truth tables for a half adder and full adder are shown in Table 4.2 and Table 4.3, respectively. The corresponding equations for the sum and carry-out are listed in Equation 4.3 and Equation 4.4 and are obtained directly from the truth tables. The logic diagram for a half adder is shown in Figure 4.2 and for a full adder in Figure 4.3 using two half adders. A higher speed full adder can be realized if the sum-of-products expression is utilized, providing only two gate delays.

Table 4.2 Truth Table for a Half Adder

a_i	b_i	$cout_i$	sum_i
0	0	0	0
0	1	0	1
1	0	0	1
1	1	1	0

Table 4.3 Truth Table for a Full Adder

a_i	b_i	cin_{i-1}	$cout_i$	sum_i
0	0	0	0	0
0	0	1	0	1
0	1	0	0	1
0	1	1	1	0
1	0	0	0	1
1	0	1	1	0
1	1	0	1	0
1	1	1	1	1

$$\begin{aligned} sum_i &= a_i'b_i + a_ib_i' \\ &= a_i \oplus b_i \\ \\ cout_i &= a_ib_i \end{aligned} \tag{4.3}$$

$$\begin{aligned} sum_i &= a_i'b_i'cin_{i-1} + a_i'b_icin_{i-1}' + a_ib_i'cin_{i-1}' + a_ib_icin_{i-1} \\ &= a_i \oplus b_i \oplus cin_{i-1} \\ \\ cout_i &= a_i'b_icin_{i-1} + a_ib_i'cin_{i-1} + a_ib_icin_{i-1}' + a_ib_icin_{i-1} \\ &= a_ib_i + (a_i \oplus b_i)cin_{i-1} \end{aligned} \tag{4.4}$$

From Table 4.3, the carry-out for the full adder, $cout_i$, can also be written as a sum of products

$$a_ib_i + a_icin_{i-1} + b_icin_{i-1}$$

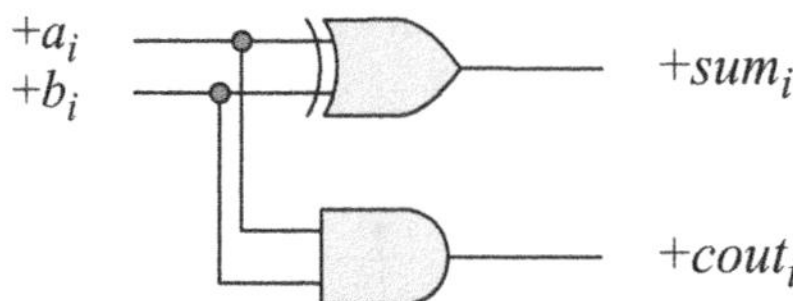

Figure 4.2 Logic diagram for a half adder.

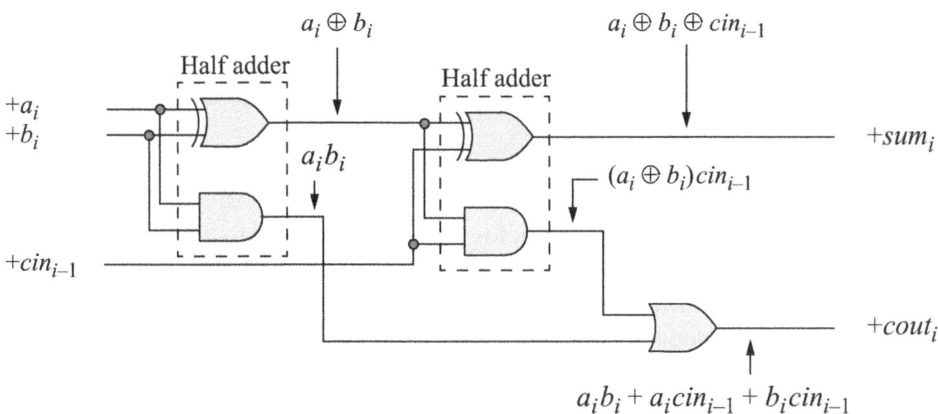

Figure 4.3 Logic diagram for a full adder using two half adders.

The logic symbol for a full adder for any stage $_i$ is shown in Figure 4.4, where the inputs are the augend a_i, the addend b_i, and the carry-in from the previous lower-order stage cin_{i-1}. The outputs are the sum, sum_i, and the carry-out, $cout_i$. The logic diagram for a 4-bit ripple adder is shown in Figure 4.5 in which the carries propagate (or ripple) through the adder.

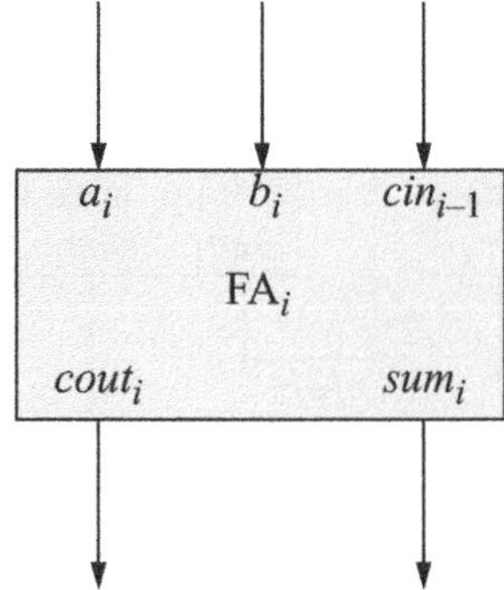

Figure 4.4 Logic symbol for a full adder.

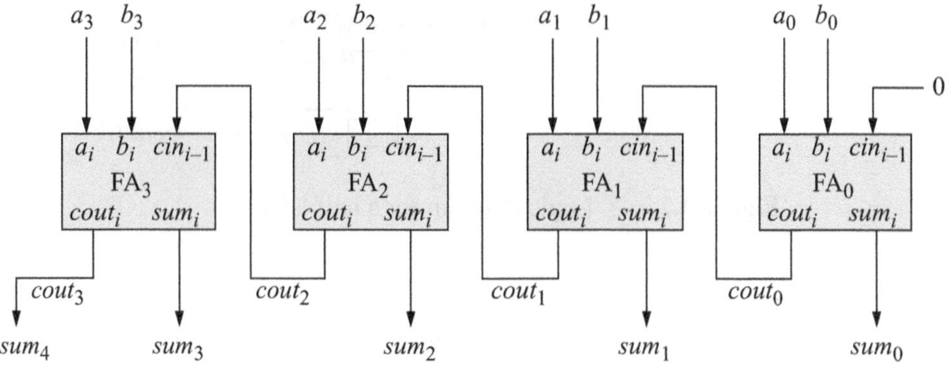

Figure 4.5 Logic diagram for a 4-bit ripple-carry adder.

The organization of Figure 4.5 will be used to implement a 4-bit ripple-carry adder using structural modeling. Figure 4.6 replicates the logic diagram of Figure 4.5 and shows the instantiation names and net names. A full adder will be designed using dataflow modeling, then instantiated four times into the structural module. The full adder dataflow module is shown in Figure 4.7 with inputs *a*, *b*, and *cin*, and outputs *sum* and *cout*.

The structural module for the 4-bit ripple-carry adder is shown in Figure 4.8. The inputs are two 4-bit vectors, *a[3:0]* and *b[3:0]*, where *a[0]* and *b[0]* are the low-order bits of *A* and *B*, respectively, and a scalar input *cin*. The outputs are a 4-bit vector *sum[3:0]* and a scalar output *cout*. The internal nets are represented by a 4-bit vector *c[3:0]*, which connects the carries between the adder stages. The test bench, outputs, and waveforms are shown in Figure 4.9, Figure 4.10, and Figure 4.11, respectively.

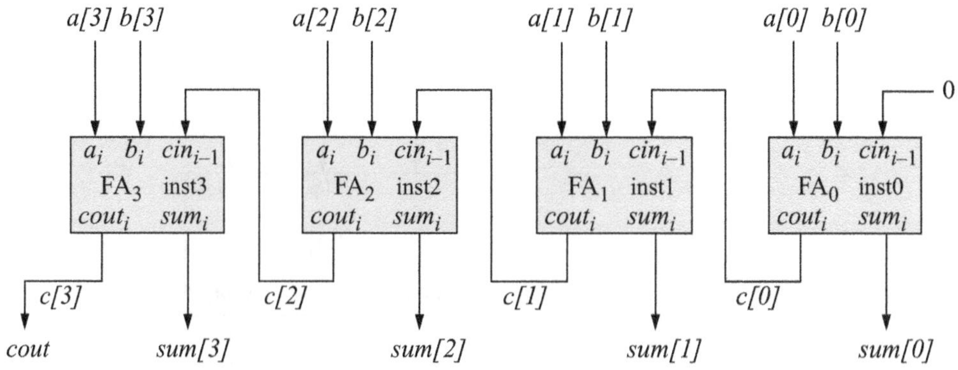

Figure 4.6 Logic diagram for the 4-bit ripple adder showing instantiation names and net names.

```
//dataflow full adder
module full_adder (a, b, cin, sum, cout);

//list all inputs and outputs
input a, b, cin;
output sum, cout;

//define wires
wire a, b, cin;
wire sum, cout;

//continuous assign
assign sum = (a ^ b) ^ cin;
assign cout = cin & (a ^ b) | (a & b);

endmodule
```

Figure 4.7 Dataflow module for a full adder.

```
//structural 4_bit ripple-carry counter
module adder_ripple4_struc2 (a, b, cin, sum, cout);

input [3:0] a, b;
input cin;
output [3:0] sum;
output cout;

wire [3:0] a, b;
wire cin;
wire [3:0] sum;
wire cout;

wire [3:0] c;          //define internal nets for carries

assign cout = c[3];

full_adder inst0 (
   .a(a[0]),
   .b(b[0]),
   .cin(cin),
   .sum(sum[0]),
   .cout(c[0])
   );
//continued on next page
```

Figure 4.8 Structural module for the 4-bit ripple-carry adder.

```
full_adder inst1 (
   .a(a[1]),
   .b(b[1]),
   .cin(c[0]),
   .sum(sum[1]),
   .cout(c[1])
   );

full_adder inst2 (
   .a(a[2]),
   .b(b[2]),
   .cin(c[1]),
   .sum(sum[2]),
   .cout(c[2])
   );

full_adder inst3 (
   .a(a[3]),
   .b(b[3]),
   .cin(c[2]),
   .sum(sum[3]),
   .cout(c[3])
   );

endmodule
```

Figure 4.8 (Continued)

```
//test bench for 4-bit ripple-carry adder
module adder_ripple4_struc2_tb;

//define inputs
reg [3:0] a, b;
reg cin;

//define outputs
wire [3:0] sum;
wire cout;

initial
$monitor ("a=%b, b=%b, cin=%b, cout=%b, sum=%b",
            a, b, cin, cout, sum);

//continued on next  page
```

Figure 4.9 Test bench for the 4-bit ripple-carry adder.

```
initial
begin
   #0 a = 4'b0000;    b = 4'b0001;    cin = 1'b0;
   #10a = 4'b0011;    b = 4'b0100;    cin = 1'b1;
   #10a = 4'b0111;    b = 4'b0101;    cin = 1'b0;
   #10a = 4'b1011;    b = 4'b1100;    cin = 1'b1;
   #10a = 4'b0110;    b = 4'b0100;    cin = 1'b0;
   #10a = 4'b0101;    b = 4'b0100;    cin = 1'b1;
   #10a = 4'b1111;    b = 4'b1111;    cin = 1'b1;
   #10a = 4'b1000;    b = 4'b1000;    cin = 1'b1;
   #10a = 4'b1100;    b = 4'b1100;    cin = 1'b0;
   #10a = 4'b1001;    b = 4'b0101;    cin = 1'b1;
   #10a = 4'b0111;    b = 4'b0111;    cin = 1'b0;

   #10   $stop;
end

//instantiate the module into the test bench
adder_ripple4_struc2 inst1 (
   .a(a),
   .b(b),
   .cin(cin),
   .sum(sum),
   .cout(cout)
   );

endmodule
```

Figure 4.9 (Continued)

```
a=0000, b=0001, cin=0, cout=0, sum=0001
a=0011, b=0100, cin=1, cout=0, sum=1000
a=0111, b=0101, cin=0, cout=0, sum=1100
a=1011, b=1100, cin=1, cout=1, sum=1000
a=0110, b=0100, cin=0, cout=0, sum=1010
a=0101, b=0100, cin=1, cout=0, sum=1010
a=1111, b=1111, cin=1, cout=1, sum=1111
a=1000, b=1000, cin=1, cout=1, sum=0001
a=1100, b=1100, cin=0, cout=1, sum=1000
a=1001, b=0101, cin=1, cout=0, sum=1111
a=0111, b=0111, cin=0, cout=0, sum=1110
```

Figure 4.10 Outputs for the 4-bit ripple-carry adder.

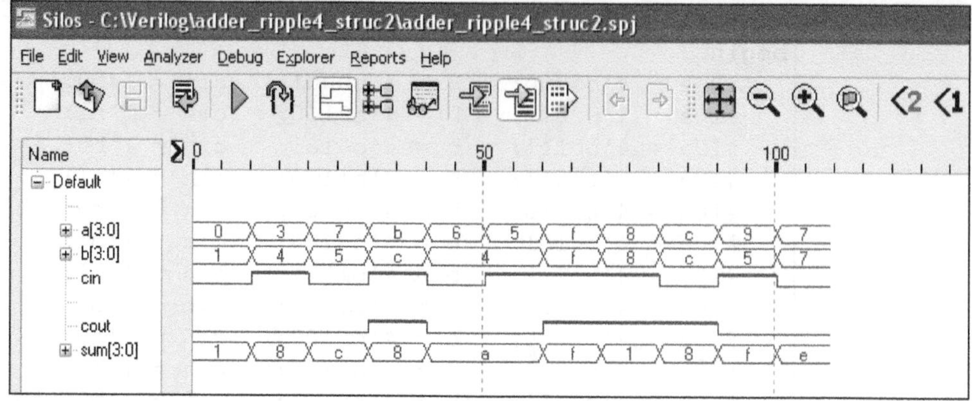

Figure 4.11 Waveforms for the 4-bit ripple-carry adder.

4.2 Carry Lookahead Addition

The speed limitation in the ripple adder arises from specifying $cout_i$ as a function of the carry-out from the previous lower-order stage $cout_{i-1}$. A considerable increase in speed can be realized by expressing the carry-out $cout_i$ of any stage$_i$ as a function of the two operand bits, a_i and b_i, and the carry-in cin_{-1} to the low-order stage$_0$ of the adder, where the adder is an n-bit adder $n_{-1}\ n_{-2} \ldots n_1\ n_0$. The Karnaugh map that represents the carry-out from stage$_i$ is shown in Figure 4.12, which yields Equation 4.5.

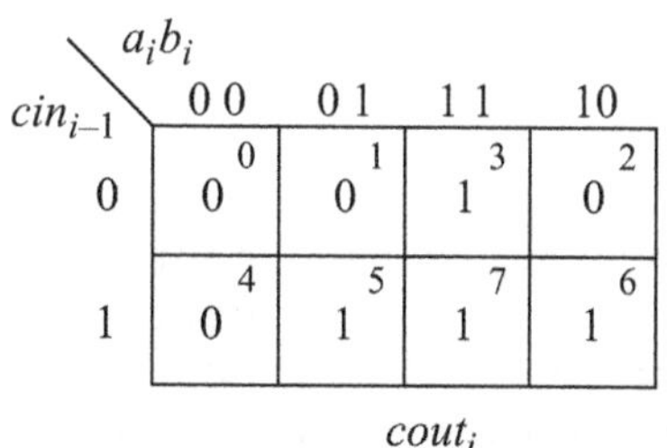

Figure 4.12 Karnaugh map for the carry-out of stage$_i$ of an n-bit adder.

$$\begin{aligned} cout_i &= a_i'b_icin_{i-1} + a_ib_i'cin_{i-1} + a_ib_icin_{i-1}' + a_ib_icin_{i-1} \\ &= a_ib_i + (a_i \oplus b_i)cin_{i-1} \end{aligned} \quad (4.5)$$

Equation 4.5 states that a carry will be generated whenever $a_i = b_i = 1$, or when either $a_i = 1$ or $b_i = 1$ — but not both — with $c_{i-1} = 1$. Note that if $a_i = b_i = 1$, then this represents a generate function, not a propagate function. Verilog requires a propagate function to be the exclusive-OR of a_i and b_i. A technique will now be presented that increases the speed of the carry propagation in a parallel adder. The carries entering all the bit positions of the adder can be generated simultaneously by a *carry lookahead* generator. This results in a constant addition time that is independent of the length of the adder. Two auxiliary functions are defined as follows:

Generate $G_i = a_i b_i$
Propagate $P_i = a_i \oplus b_i$

The carry *generate* function, G_i, reflects the condition where a carry is generated at the *i*th stage. The carry *propagate* function, P_i, is true when the *i*th stage will pass through (or propagate) the incoming carry cin_{i-1} to the next higher $stage_{i+1}$. Equation 4.5 can now be restated as Equation 4.6.

$$\begin{aligned} cout_i &= a_i b_i + (a_i \oplus b_i)\, cin_{i-1} \\ &= G_i + P_i\, cin_{i-1} \end{aligned} \tag{4.6}$$

Equation 4.6 indicates that the generate G_i and propagate P_i functions for any carry out $cout_i$ can be obtained independently and in parallel when the operand inputs are applied to the *n*-bit adder. The equation can be applied recursively to obtain a set of carry-out equations in terms of the variables G_i, P_i, and $cin_{\_1}$ for a 4-bit adder, where $cin_{\_1}$ is the carry-in to the low-order $stage_0$ of the adder. The equations are shown in Equation 4.7.

$$cout_0 = G_0 + P_0\, cin_{\_1}$$

$$\begin{aligned} cout_1 &= G_1 + P_1\, cout_0 \\ &= G_1 + P_1\,(G_0 + P_0\, cin_{\_1}) \\ &= G_1 + P_1 G_0 + P_1 P_0\, cin_{\_1} \end{aligned}$$

$$\begin{aligned} cout_2 &= G_2 + P_2\, cout_1 \\ &= G_2 + P_2\,(G_1 + P_1 G_0 + P_1 P_0\, cin_{\_1}) \\ &= G_2 + P_2 G_1 + P_2 P_1 G_0 + P_2 P_1 P_0\, cin_{\_1} \end{aligned} \tag{4.7}$$

Continued on next page

$$\begin{aligned} cout_3 = {} & G_3 + P_3\, cout_2 \\ = {} & G_3 + P_3(G_2 + P_2G_1 + P_2P_1G_0 + P_2P_1P_0\, cin_{-1}) \\ = {} & G_3 + P_3\, G_2 + P_3P_2G_1 + P_3P_2P_1G_0 + P_3P_2P_1P_0\, cin_{-1} \end{aligned} \tag{4.7}$$

Examples of the generate and propagate functions are shown in Figure 4.13 for two 8-bit operands in 2s complement notation.

$A =$	1	1	0	0	1	1	0	1	−51
+) $B =$	0	1	0	1	1	1	0	0	+92
	G'	G	G'	G'	G	G	G'	G'	
	P	P'	P'	P	P'	P'	P'	P	
								0	$cin = 0$
$cout = 1$	0	0	1	0	1	0	0	1	+41

$A =$	0	0	1	0	0	1	0	1	+37
+) $B =$	0	1	0	0	1	1	1	1	+79
	G'	G'	G'	G'	G'	G	G'	G	
	P'	P	P	P'	P	P'	P	P'	
								1	$cin = 1$
$cout = 0$	0	1	1	1	0	1	0	1	+117

Figure 4.13 Examples of generate and propagate functions.

Consider the expression for $cout_2$ in Equation 4.7 to further explain the generate and propagate functions to produce a carry-out. Each of the product terms shown below will produce a carry-out of 1 for $cout_2$.

$cout_2 =$	G_2	+	P_2G_1	+	$P_2P_1G_0$	+	$P_2P_1P_0$
	1		0 1		0 1 1		1 0 0
	1		1 1		1 0 1		0 1 1
							1 ← cin_{-1}
1 ←	0	1 ←	0 0	1 ←	0 0 0	1 ←	0 0 0

It can be seen from Equation 4.7 that each carry is now an expression consisting of only three gate delays: one delay each for the generate and propagate functions, one delay to AND the generate and propagate functions, and one delay to OR all of the product terms. If a high-speed full adder is used in the implementation, then the sum bits can be generated with only two gate delays, providing a maximum of only five delays for an add operation. This technique provides an extremely fast addition of two n-bit operands. Equation 4.7 can be restated more compactly as shown in Equation 4.8.

$$cout_i = G_i + \sum_{j=0}^{i-1} \left(\prod_{k=j+1}^{i} P_k \right) G_j + \prod_{k=0}^{i} P_k \, cin_{-1} \tag{4.8}$$

Group generate and propagate As n becomes large, the number of inputs to the high-order gates also becomes large, which may be a problem for some technologies. The problem can be alleviated to some degree by partitioning the adder stages into 4-bit groups. Additional auxiliary functions can then be defined for *group generate* and *group propagate*, as shown in Equation 4.9 for group$_j$, which consists of individual adder stages $i + 3$ through i. In this method, each group of four adders is considered as a unit with its individual group carry sent to the next higher-order group.

$$\begin{aligned} \text{Group generate:} \quad GG_j &= G_{i+3} + P_{i+3}G_{i+2} + P_{i+3}P_{i+2}G_{i+1} + \\ &\quad P_{i+3}P_{i+2}P_{i+1}G_i \\ \text{Group propagate:} \quad GP_j &= P_{i+3}P_{i+2}P_{i+1}P_i \end{aligned} \tag{4.9}$$

The group generate GG_j signifies a carry that is generated out of the high-order $(i + 3)$ bit position that originated from within the group. The group propagate GP_j indicates that a carry was propagated through the group. The group carry can now be written in terms of the group generate and group propagate functions, as shown in Equation 4.10. The term GC_{j-1} is the carry-in to the group from the previous lower-order group. If group$_j$ is the low-order group, then $GC_{j-1} = cin_{-1}$.

$$\text{Group carry:} \quad GC_j = GG_j + GP_jGC_{j-1} \tag{4.10}$$

Section generate and propagate If the fan-in limitation is still a problem for very large operands, then the group generate and group propagate concept can be extended to partition four groups into one section. For a 64-bit adder, there would be four sections with four groups per section, with four full adders per group. Two additional auxiliary functions can now be defined as *section generate* and *section propagate* for section$_k$, as shown in Equation 4.11.

$$\text{Section generate:} \quad SG_k = GG_{j+3} + GP_{j+3}GG_{j+2} + GP_{j+3}GP_{j+2}GG_{j+1} + GP_{j+3}GP_{j+2}GP_{j+1}GG_j$$

$$\text{Section propagate:} \quad SP_k = GP_{j+3}GP_{j+2}GP_{j+1}GP_j \qquad (4.11)$$

The section generate SG_k signifies a carry that is generated out of the high-order $(j + 3)$ position that originated from within the section. The section propagate SP_k indicates that a carry was propagated through the section. The section carry can now be written in terms of the section generate and section propagate functions, as shown in Equation 4.12. The term SC_{k-1} is the carry-in to the section from the previous lower-order section. If section$_k$ is the low-order section, then $SC_{k-1} = cin_{-1}$.

$$\text{Section carry:} \quad SC_k = SG_k + SP_kSC_{k-1} \qquad (4.12)$$

The carry-out of the high-order section SC_{k+3} is also the carry-out of the adder and can be written as $cout_{n-1}$. This section has presented a method to increase the speed of addition by partitioning the adder into sections and groups and developing carry lookahead logic within individual groups and within individual sections. Figure 4.14 shows a block diagram of a 64-bit adder consisting of four sections, with four groups per section, and four full adders per group.

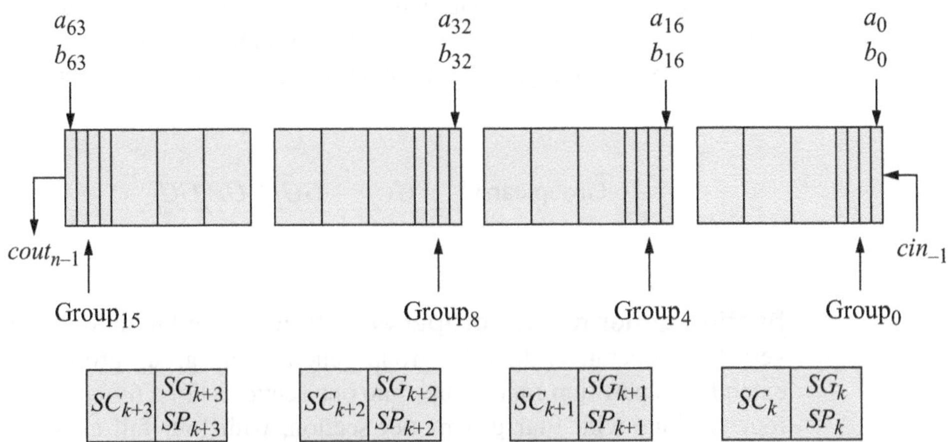

Figure 4.14 Block diagram of a 64-bit carry lookahead adder.

An 8-bit carry lookahead adder will be designed using dataflow modeling. The block diagram of the adder is shown in Figure 4.15. The adder has two groups of four bits per group. The high-order group has operands *a[7:4]* and *b[7:4]* that produce a sum of *sum[7:4]*; the low-order group has operands *a[3:0]* and *b[3:0]* that produce a sum of *sum[3:0]*. The carry-in is *cin*; the carry-out is *cout*, which is the carry-out of bit 7.

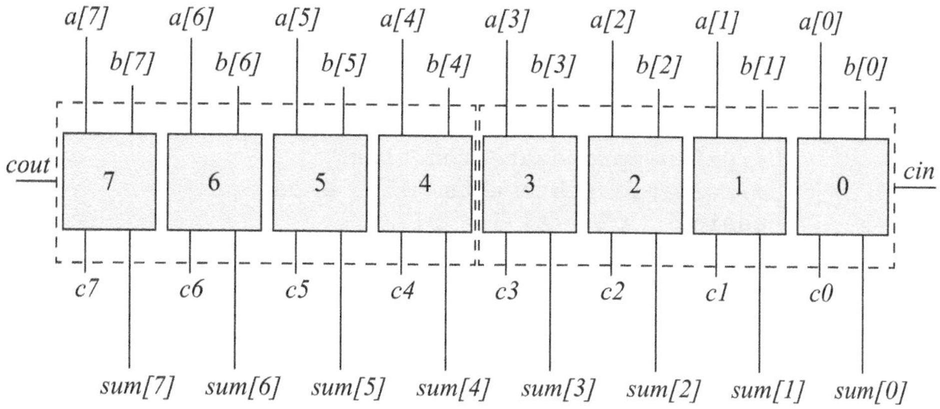

Figure 4.15 An 8-bit carry lookahead adder.

The dataflow module is shown in Figure 4.16 using the **assign** statement for the generate function, the propagate function, the internal carries, and the sum. Note that the propagate functions are the exclusive-OR of the respective augend and addend bits. The carry equations c_i correspond with the carry equations of Equation 4.7. The test bench module is shown in Figure 4.17 for 12 values of *A*, *B*, and *cin*. The outputs are shown in Figure 4.18 in decimal notation and the waveforms are shown in Figure 4.19 in hexadecimal notation.

```
//dataflow 8-bit carry lookahead adder
module adder_cla8 (a, b, cin, sum, cout);

input [7:0] a, b;        //input/output port declaration
input cin;
output [7:0] sum;
output cout;

wire g7, g6, g5, g4, g3, g2, g1, g0;//define internal wires
wire p7, p6, p5, p4, p3, p2, p1, p0;
wire c7, c6, c5, c4, c3, c2, c1, c0;
//continued on next page
```

Figure 4.16 Dataflow module for the 8-bit carry lookahead adder.

```
//define generate functions
//multiple statements using 1 assign
assign    g7 = a[7] & b[7],
          g6 = a[6] & b[6],
          g5 = a[5] & b[5],
          g4 = a[4] & b[4],
          g3 = a[3] & b[3],
          g2 = a[2] & b[2],
          g1 = a[1] & b[1],
          g0 = a[0] & b[0];


//define propagate functions
//multiple statements using 1 assign
assign    p7 = a[7] ^ b[7],
          p6 = a[6] ^ b[6],
          p5 = a[5] ^ b[5],
          p4 = a[4] ^ b[4],
          p3 = a[3] ^ b[3],
          p2 = a[2] ^ b[2],
          p1 = a[1] ^ b[1],
          p0 = a[0] ^ b[0];


//obtain the carry equations for low order
assign    c0 = g0 | (p0 & cin),
          c1 = g1 | (p1 & g0) | (p1 & p0 & cin),
          c2 = g2 | (p2 & g1) | (p2 & p1 & g0) |
                (p2 & p1 & p0 & cin),
          c3 = g3 | (p3 & g2) | (p3 & p2 & g1) |
                (p3 & p2 & p1 & g0) |
                (p3 & p2 & p1 & p0 & cin);


//obtain the carry equations for high order
assign    c4 = g4 | (p4 & c3),
          c5 = g5 | (p5 & g4) | (p5 & p4 & c3),
          c6 = g6 | (p6 & g5) | (p6 & p5 & g4) |
                (p6 & p5 & p4 & c3),
          c7 = g7 | (p7 & g6) | (p7 & p6 & g5) |
                (p7 & p6 & p5 & g4) |
                (p7 & p6 & p5 & p4 & c3);

//continued on next page
```

Figure 4.16 (Continued)

```
//obtain the sum equations
assign   sum[0] = p0 ^ cin,
         sum[1] = p1 ^ c0,
         sum[2] = p2 ^ c1,
         sum[3] = p3 ^ c2,
         sum[4] = p4 ^ c3,
         sum[5] = p5 ^ c4,
         sum[6] = p6 ^ c5,
         sum[7] = p7 ^ c6;

//obtain cout
assign   cout = c7;

endmodule
```

Figure 4.16 (Continued)

```
//test bench for dataflow 8-bit carry lookahead adder
module adder_cla8_tb;

reg [7:0] a, b;
reg cin;

wire [7:0] sum;
wire cout;

//display signals
initial
$monitor ("a = %d, b = %d, cin = %b, cout = %b, sum = %d",
            a, b, cin, cout, sum);

//apply stimulus
initial
begin
   #0    a = 8'b0000_0000;
         b = 8'b0000_0000;
         cin = 1'b0;    //cout = 0, sum = 0000_0000

   #10   a = 8'b0000_0001;
         b = 8'b0000_0010;
         cin = 1'b0;    //cout = 0, sum = 0000_0011

//continued on next page
```

Figure 4.17 Test bench for the 8-bit carry lookahead adder.

```
	#10	a = 8'b0000_0010;
		b = 8'b0000_0110;
		cin = 1'b0;	//cout = 0, sum = 0000_1000

	#10	a = 8'b0000_0111;
		b = 8'b0000_0111;
		cin = 1'b0;	//cout = 0, sum = 0000_1110

	#10	a = 8'b0000_1001;
		b = 8'b0000_0110;
		cin = 1'b0;	//cout = 0, sum = 0000_1111

	#10	a = 8'b0000_1100;
		b = 8'b0000_1100;
		cin = 1'b0;	//cout = 0, sum = 0001_1000

	#10	a = 8'b0000_1111;
		b = 8'b0000_1110;
		cin = 1'b0;	//cout = 0, sum = 0001_1101

	#10	a = 8'b0000_1110;
		b = 8'b0000_1110;
		cin = 1'b1;	//cout = 0, sum = 0001_1101

	#10	a = 8'b0000_1111;
		b = 8'b0000_1111;
		cin = 1'b1;	//cout = 0, sum = 0001_1111

	#10	a = 8'b1111_0000;
		b = 8'b0000_1111;
		cin = 1'b1;	//cout = 1, sum = 0000_0000

	#10	a = 8'b0111_0000;
		b = 8'b0000_1111;
		cin = 1'b1;	//cout = 0, sum = 1000_0000

	#10	a = 8'b1011_1000;
		b = 8'b0100_1111;
		cin = 1'b1;	//cout = 1, sum = 0000_1000

	#10	$stop;
end

//continued on next page
```

Figure 4.17 (Continued)

```
adder_cla8 inst1 (         //instantiate the module
    .a(a),
    .b(b),
    .cin(cin),
    .sum(sum),
    .cout(cout)
    );

endmodule
```

Figure 4.17 (Continued)

```
a = 0,    b = 0,    cin = 0, cout = 0, sum = 0
a = 1,    b = 2,    cin = 0, cout = 0, sum = 3
a = 2,    b = 6,    cin = 0, cout = 0, sum = 8
a = 7,    b = 7,    cin = 0, cout = 0, sum = 14
a = 9,    b = 6,    cin = 0, cout = 0, sum = 15
a = 12,   b = 12,   cin = 0, cout = 0, sum = 24
a = 15,   b = 14,   cin = 0, cout = 0, sum = 29
a = 14,   b = 14,   cin = 1, cout = 0, sum = 29
a = 15,   b = 15,   cin = 1, cout = 0, sum = 31
a = 240,  b = 15,   cin = 1, cout = 1, sum = 0
a = 112,  b = 15,   cin = 1, cout = 0, sum = 128
a = 184,  b = 79,   cin = 1, cout = 1, sum = 8
```

Figure 4.18 Outputs for the 8-bit carry lookahead adder.

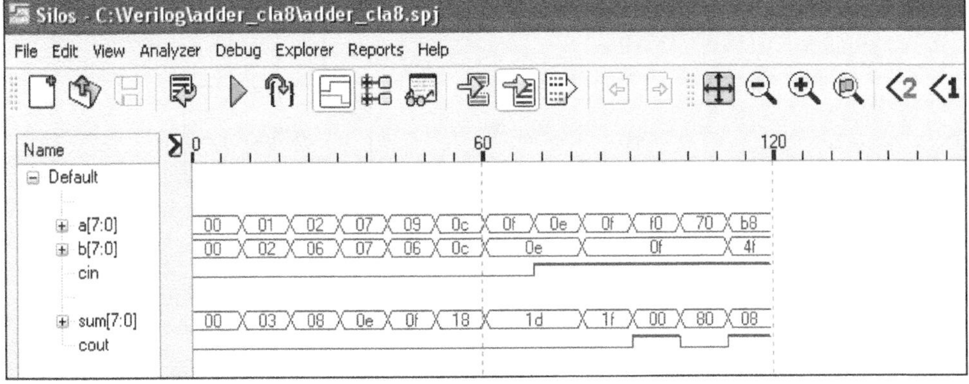

Figure 4.19 Outputs for the 8-bit carry lookahead adder.

This section has presented a technique for implementing a high-speed adder and is used primarily for large adders. If less hardware is required and lower speed is not a prohibitive factor, then one or more levels of carry lookahead logic can be removed. This would allow partial carry lookahead to be applied — the remaining carries would use the ripple-carry technique.

4.3 Carry-Save Addition

Carry-save adders (CSAs) save the carry from propagating to the next higher-order stage in an n-bit adder. They can be used to add multiple bits of the same weight from multiple operands or to add multiple n-bit operands. An adder using the carry-save technique results in a high-speed add operation. A carry-save adder is simply a full adder and is also referred to as a *Wallace tree*.

4.3.1 Multiple-Bit Addition

Figure 4.20 illustrates a carry-save adder with three inputs and two outputs that can be used to implement a Wallace tree for use in high-speed fixed-point array multipliers. The inputs are order-independent; the outputs indicate the sum and carry-out. The carry-out has a value of 1 whenever the majority of the inputs has a value of 1. The dataflow module for the carry-save full adder is shown in Figure 4.21. The test bench and outputs are shown in Figure 4.22 and Figure 4.23, respectively.

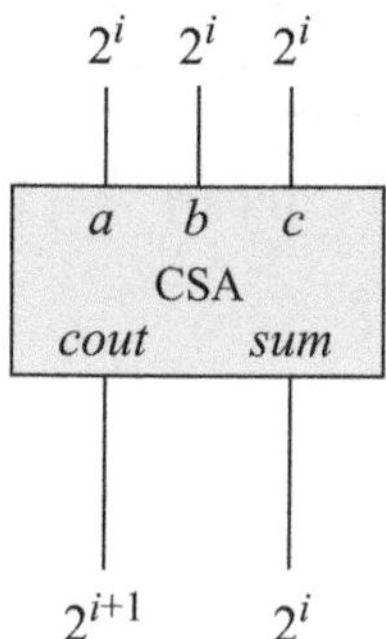

Figure 4.20 A carry-save full adder.

```
//dataflow carry-save full adder
module csa_full_adder (a, b, c, sum, cout);

input a, b, c;            //list all inputs and outputs
output sum, cout;

wire a, b, c;             //define wires
wire sum, cout;

assign sum = (a ^ b) ^ c;  //continuous assign
assign cout = c & (a ^ b) | (a & b);

endmodule
```

Figure 4.21 Dataflow module for a carry-save full adder.

```
//test bench for carry-save full adder
module csa_full_adder_tb;

reg a, b, c;
wire sum, cout;

//apply input vectors
initial
begin: apply_stimulus
   reg [3:0] invect;
   for (invect = 0; invect < 8; invect = invect + 1)
      begin
         {a, b, c} = invect [3:0];
         #10 $display ("a b c = %b, cout = %b, sum = %b",
                        {a, b, c}, cout, sum);
      end
end

//instantiate the module into the test bench
csa_full_adder inst1 (
   .a(a),
   .b(b),
   .c(c),
   .sum(sum),
   .cout(cout)
   );

endmodule
```

Figure 4.22 Test bench for the carry-save full adder.

```
a b c = 000, cout = 0, sum = 0     a b c = 100, cout = 0, sum = 1
a b c = 001, cout = 0, sum = 1     a b c = 101, cout = 1, sum = 0
a b c = 010, cout = 0, sum = 1     a b c = 110, cout = 1, sum = 0
a b c = 011, cout = 1, sum = 0     a b c = 111, cout = 1, sum = 1
```

Figure 4.23 Outputs for the carry-save full adder.

Wallace tree to add five bits A Wallace tree containing three CSAs that adds five bits from column 2^i is shown in Figure 4.24, where the inputs are represented by a 5-bit vector *in[4:0]*. The sum of the inputs is represented by the output vector *sum_w[2:0]* in columns 2^{i+2}, 2^{i+1}, and 2^i, respectively, where *sum_w[0]* is the low-order bit. Figure 4.24 also shows the instantiation names and the net names.

The carry-out and the sum that are generated from the same CSA cannot then be added subsequently in the same CSA — this would result in an incorrect sum. The carry-save full adder dataflow module of Figure 4.21 will be instantiated three times into a structural module as shown in Figure 4.25 for the 5-input, 3-output Wallace tree. The test bench and outputs for the 5-input Wallace tree are shown in Figure 4.26 and Figure 4.27, respectively.

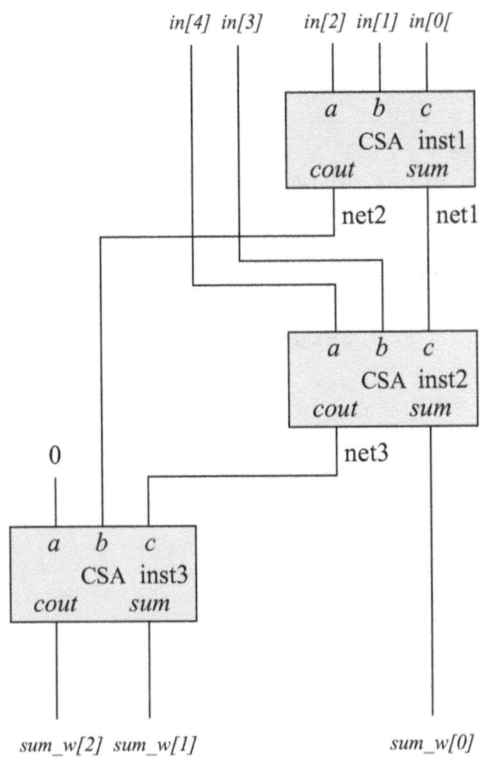

Figure 4.24 Five-input Wallace tree containing three carry-save full adders that adds five inputs in the same column from five operands.

```
//structural 5-input wallace tree
module wallace_tree_5_inputs (in, sum_w);

input [4:0] in;
output [2:0] sum_w;

wire [4:0] in;
wire [2:0] sum_w;


//define internal nets
wire net1, net2, net3;

//instantiate the carry-save full adders
csa_full_adder inst1 (
   .a(in[2]),
   .b(in[1]),
   .c(in[0]),
   .sum(net1),
   .cout(net2)
   );

csa_full_adder inst2 (
   .a(in[4]),
   .b(in[3]),
   .c(net1),
   .sum(sum_w[0]),
   .cout(net3)
   );

csa_full_adder inst3 (
   .a(1'b0),
   .b(net2),
   .c(net3),
   .sum(sum_w[1]),
   .cout(sum_w[2])
   );

endmodule
```

Figure 4.25 Structural module for the 5-input Wallace tree.

```
//test bench for 5-input wallace tree
module wallace_tree_5_inputs_tb;

reg [4:0] in;                //define inputs and outputs
wire [2:0] sum_w;

//apply input vectors and display variables
initial
begin: apply_stimulus
   reg [5:0] invect;
   for (invect = 0; invect < 32; invect = invect + 1)
      begin
         in = invect [5:0];
         #10 $display ("inputs = %b, sum = %b",
                           in, sum_w);
      end
end

wallace_tree_5_inputs nst1 (  //instantiate the module
   .in(in),
   .sum_w(sum_w)
   );

endmodule
```

Figure 4.26 Test bench for the 5-input Wallace tree.

```
inputs = 00000, sum = 000      inputs = 10000, sum = 001
inputs = 00001, sum = 001      inputs = 10001, sum = 010
inputs = 00010, sum = 001      inputs = 10010, sum = 010
inputs = 00011, sum = 010      inputs = 10011, sum = 011
inputs = 00100, sum = 001      inputs = 10100, sum = 010
inputs = 00101, sum = 010      inputs = 10101, sum = 011
inputs = 00110, sum = 010      inputs = 10110, sum = 011
inputs = 00111, sum = 011      inputs = 10111, sum = 100
inputs = 01000, sum = 001      inputs = 11000, sum = 010
inputs = 01001, sum = 010      inputs = 11001, sum = 011
inputs = 01010, sum = 010      inputs = 11010, sum = 011
inputs = 01011, sum = 011      inputs = 11011, sum = 100
inputs = 01100, sum = 010      inputs = 11100, sum = 011
inputs = 01101, sum = 011      inputs = 11101, sum = 100
inputs = 01110, sum = 011      inputs = 11110, sum = 100
inputs = 01111, sum = 100      inputs = 11111, sum = 101
```

Figure 4.27 Outputs for the 5-input Wallace tree.

4.3.2 Multiple-Operand Addition

The delay of adding multiple operands can be reduced by using carry-save adders in a tree configuration. This negates the slower method of adding two operands, obtaining the sum, then adding the sum to a third operand. Using CSAs, the sums and carries from each column of each operand are obtained independently of the other columns. An example is shown in Figure 4.28 by adding five operands *a[3:0]*, *b[3:0]*, *c[3:0]*, *d[3:0]*, and *e[3:0]*. The 5-input CSA of Figure 4.24 will be used together with the 3-input CSA of Figure 4.20 to add the five operands *A*, *B*, *C*, *D*, and *E*, as shown in Figure 4.29.

			2^3	2^2	2^1	2^0	Decimal value
a[3:0] =			0	1	1	1	7
b[3:0] =			1	0	1	0	10
c[3:0] =			0	1	0	1	5
d[3:0] =			1	0	0	1	9
e[3:0] =			0	0	1	1	3
Sum =	1	0	0	0	1	0	34

Figure 4.28 Example of adding five operands.

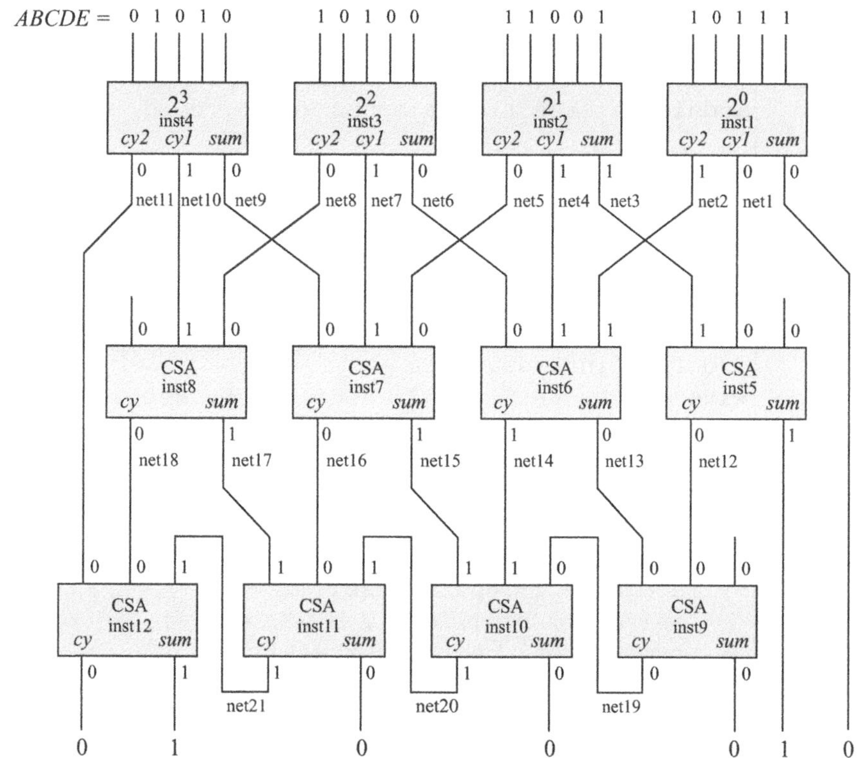

Figure 4.29 Carry-save adder Wallace tree to add five operands.

The 5-operand Wallace tree of Figure 4.29 will be designed using structural modeling. The *wallace_tree_5_inputs* module will be instantiated four times into the structural module; the *csa_full_adder* module will be instantiated eight times into the structural module. The structural module to add five operands will be designed from the logic diagram of Figure 4.29, which shows the instantiation names and the net names that will be used in the structural module.

The structural module is shown in Figure 4.30. Instantiation *inst1* instantiates the 5-input Wallace tree, whose inputs are the vector *in[4:0]*. Therefore, the expression *.in({a[0], b[0], c[0], d[0], e[0]})* in the structural module specifies that the input port vector *.in[4:0]* of the 5-input Wallace tree corresponds to the low-order bits of the five operands *a[0]*, *b[0]*, *c[0]*, *d[0]*, and *e[0]* in the structural module.

In a similar manner, the expression *.sum_w({net2, net1, sum_w[0]})* specifies that the 3-bit output port vector *sum_w[2:0]* of the 5-input Wallace tree corresponds to the 3-bit output vector *{net2, net1, sum_w[0]}* of the Wallace tree specified by instantiation *inst1* in the structural module.

Instantiation *inst5* instantiates the CSA full adder, whose inputs are the scalar variables *a*, *b*, and *c*. Thus, input port *.a* of the CSA full adder corresponds to a logic 0 value — the inputs are order-independent. Input port *.b* corresponds to *net1* of the structural module and input port *.c* corresponds to *net3* of the structural module.

The *.sum* output port of the CSA full adder corresponds to *sum_w[1]* of the structural module, and the *.cout* output port of the CSA full adder corresponds to *net12* of the structural module.

```
//structural wallace tree for 5 operands
module wallace_tree_5_opnds (a, b, c, d, e, sum_w);

input [3:0] a, b, c, d, e;
output [6:0] sum_w;

wire [3:0] a, b, c, d, e;
wire [6:0] sum_w;

//define internal nets
wire    net1, net2, net3, net4, net5, net6,
        net7, net8, net9, net10, net11,
        net12, net13, net14, net15, net16,
        net17, net18, net19, net20, net21;

//instantiate the 5-input wallace tree
wallace_tree_5_inputs inst1 (
   .in({a[0], b[0], c[0], d[0], e[0]}),
   .sum_w({net2, net1, sum_w[0]})
   );
//continued on next page
```

Figure 4.30 Structural module for the Wallace tree that adds five 4-bit operands.

```
wallace_tree_5_inputs inst2 (
   .in({a[1], b[1], c[1], d[1], e[1]}),
   .sum_w({net5, net4, net3})
   );

wallace_tree_5_inputs inst3 (
   .in({a[2], b[2], c[2], d[2], e[2]}),
   .sum_w({net8, net7, net6})
   );

wallace_tree_5_inputs inst4 (
   .in({a[3], b[3], c[3], d[3], e[3]}),
   .sum_w({net11, net10, net9})
   );

//instantiate the carry-save full adder
csa_full_adder inst5 (
   .a(1'b0),
   .b(net1),
   .c(net3),
   .sum(sum_w[1]),
   .cout(net12)
   );

csa_full_adder inst6 (
   .a(net2),
   .b(net4),
   .c(net6),
   .sum(net13),
   .cout(net14)
   );

csa_full_adder inst7 (
   .a(net5),
   .b(net7),
   .c(net9),
   .sum(net15),
   .cout(net16)
   );

csa_full_adder inst8 (
   .a(net8),
   .b(net10),
   .c(1'b0),
   .sum(net17),
   .cout(net18)
   );                //continued on next page
```

Figure 4.30 (Continued)

```
csa_full_adder inst9 (
   .a(net12),
   .b(1'b0),
   .c(net13),
   .sum(sum_w[2]),
   .cout(net19)
   );

csa_full_adder inst10 (
   .a(net19),
   .b(net14),
   .c(net15),
   .sum(sum_w[3]),
   .cout(net20)
   );

csa_full_adder inst11 (
   .a(net20),
   .b(net16),
   .c(net17),
   .sum(sum_w[4]),
   .cout(net21)
   );

csa_full_adder inst12 (
   .a(net21),
   .b(net18),
   .c(net11),
   .sum(sum_w[5]),
   .cout(sum_w[6])
   );

endmodule
```

Figure 4.30 (Continued)

The test bench module is shown in Figure 4.31, which applies seven sets of input vectors to the five 4-bit operands. The outputs are shown in Figure 4.32, which displays the values of the five operands and their resulting sums. The concept of Wallace trees presented in this section can be easily expanded to any size operands and any number of operands. Wallace trees are used extensively in the calculation of partial products for array multipliers in order to minimize the time required for a multiply operation. Array multipliers are presented in Chapter 6 which covers fixed-point multiplication.

```
//test bench for 5-operand wallace tree
module wallace_tree_5_opnds_tb;

//define inputs and outputs
reg [3:0] a, b, c, d, e;
wire [6:0] sum_w;

//display outputs
initial
$monitor ("a=%b, b=%b, c=%b, d=%b, e=%b, sum=%b",
            a, b, c, d, e, sum_w);

//apply input vectors
initial
begin
   #0     a = 4'b0001;
          b = 4'b0001;
          c = 4'b0001;
          d = 4'b0001;
          e = 4'b0001;   //sum = 000 0101

   #10    a = 4'b0111;
          b = 4'b1010;
          c = 4'b0101;
          d = 4'b1001;
          e = 4'b0011;   //sum = 010 0010

   #10    a = 4'b1111;
          b = 4'b1010;
          c = 4'b1101;
          d = 4'b1001;
          e = 4'b1011;   //sum = 011 1010

   #10    a = 4'b0110;
          b = 4'b1110;
          c = 4'b1111;
          d = 4'b1000;
          e = 4'b1001;   //sum = 011 0100

   #10    a = 4'b1111;
          b = 4'b1111;
          c = 4'b1111;
          d = 4'b1111;
          e = 4'b1111;   //sum = 100 1011

//continued on next page
```

Figure 4.31 Test bench for the Wallace tree module of Figure 4.30.

```
   #10    a = 4'b1000;
          b = 4'b1000;
          c = 4'b1000;
          d = 4'b1000;
          e = 4'b1000;    //sum = 010 1000

   #10    a = 4'b1100;
          b = 4'b1100;
          c = 4'b1100;
          d = 4'b1100;
          e = 4'b1100;    //sum = 011 1100

   #10    $stop;
end

//instantiate the module into the test bench
wallace_tree_5_opnds inst1 (
   .a(a),
   .b(b),
   .c(c),
   .d(d),
   .e(e),
   .sum_w(sum_w)
   );

endmodule
```

Figure 4.31 (Continued)

```
a=0001, b=0001, c=0001, d=0001, e=0001, sum=0000101

a=0111, b=1010, c=0101, d=1001, e=0011, sum=0100010

a=1111, b=1010, c=1101, d=1001, e=1011, sum=0111010

a=0110, b=1110, c=1111, d=1000, e=1001, sum=0110100

a=1111, b=1111, c=1111, d=1111, e=1111, sum=1001011

a=1000, b=1000, c=1000, d=1000, e=1000, sum=0101000

a=1100, b=1100, c=1100, d=1100, e=1100, sum=0111100
```

Figure 4.32 Outputs for the Wallace tree module of Figure 4.30.

4.4 Memory-Based Addition

With the advent of high-density, high-speed memories, addition can be easily accomplished by applying the augend and addend as address inputs to the memory — the outputs are the sum. Memories can be represented in Verilog by an array of registers and are declared using a **reg** data type as follows:

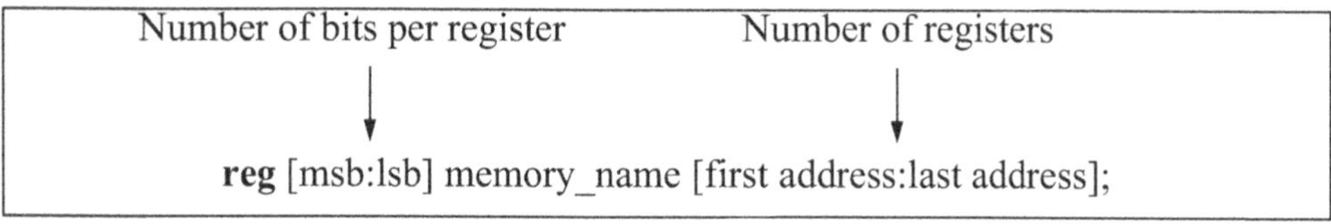

A 32-word memory with one byte per word would be declared as follows:

reg [7:0] memory_name [0:31];

An array can have only two dimensions. Memories must be declared as **reg** data types, not as **wire** data types. A register can be assigned a value using one statement, as shown below. Register *reg_name* is assigned the value $0111\ 1010\ 1011\ 0101_2$.

reg [15:0] reg_name;
reg_name = 16'h7ab5;

Values can be stored in memories by assigning a value to each word individually, as shown below for a data cache of eight registers with 8 bits per register.

reg [7:0] data_cache [0:7];

Alternatively, memories can be initialized by means of one of the following system tasks:

$readmemb	for binary data
$readmemh	for hexadecimal data

A behavioral module will be designed to add two 3-bit operands using a memory as a table lookup device. A text file is prepared for the specified memory in either binary or hexadecimal format. The file is created and saved as a separate file in the project folder without the **.v** extension. The system task reads the file and loads the contents into memory. The contents of *opnds.add* are loaded into the memory *mem_add* beginning at location 0 by the following two statements:

reg [3:0] mem_add [0:63];

$readmemb ("opnds.add", mem_add);

In order to keep the address space to a reasonable size, only 3-bit operands are used. The contents of the *opnds.add* file are shown in Figure 4.33, where the 6-bit address is shown for reference. The leftmost three digits of the address represent the augend; the rightmost three digits represent the addend — the space is shown only for clarity. The address bits are not part of the memory contents — only the four bits for the sum are entered into the *opnds.add* file. A block diagram of the memory is shown in Figure 4.34 with a 6-bit address *opnds[5:0]* consisting of a 3-bit augend *aug[2:0]* concatenated with 3-bit addend *add[2:0]*. There is a 4-bit output vector *sum[3:0]* containing the sum.

address aug add	sum 3210	address aug add	sum 3210
000 000	**0000**	**100 000**	**0100**
000 001	0001	100 001	0101
000 010	0010	100 010	0110
000 011	0011	100 011	0111
000 100	0100	100 100	1000
000 101	0101	100 101	1001
000 110	0110	100 110	1010
000 111	0111	100 111	1011
001 000	**0001**	**101 000**	**0101**
001 001	0010	101 001	0110
001 010	0011	101 010	0111
001 011	0100	101 011	1000
001 100	0101	101 100	1001
001 101	0110	101 101	1010
001 110	0111	101 110	1011
001 111	1000	101 111	1100
010 000	**0010**	**110 000**	**0110**
010 001	0011	110 001	0111
010 010	0100	110 010	1000
010 011	0101	110 011	1001
010 100	0110	110 100	1010
010 101	0111	110 101	1011
010 110	1000	110 110	1100
010 111	1001	110 111	1101
011 000	**0011**	**111 000**	**0111**
011 001	0100	111 001	1000
011 010	0101	111 010	1001
011 011	0110	111 011	1010
011 100	0111	111 100	1011
011 101	1000	111 101	1100
011 110	1001	111 110	1101
011 111	1010	111 111	1110

Figure 4.33 Contents of memory *opnds.add* that are loaded into memory *mem_add*.

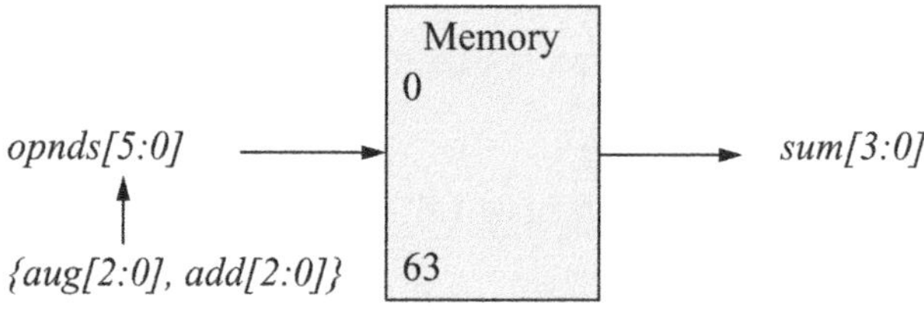

Figure 4.34 Block diagram of memory that adds two 3-bit operands.

The behavioral module to add the augend and addend using memory is shown in Figure 4.35 using binary data for the file called *opnds.add*. The test bench is shown in Figure 4.36 using ten different vectors that represent the augend and addend. The outputs are shown in Figure 4.37.

There are two ports in Figure 4.35: input port *opnds[5:0]* and output port *sum[3:0]*, which is declared as type **reg** because it operates as a storage element. The memory is defined as an array of 64 four-bit registers by the following statement:

reg [3:0] mem_add [0:63];

An **initial** procedural construct is used to load data from the *opnds.add* file into the memory *mem_add* by means of the system task **$readmemb**. The **initial** statement executes only once to initialize the memory. An **always** procedural construct is then used to read the contents of the memory based on the value of the concatenated operands; that is, *sum[3:0]* receives the contents of the memory at the address specified by the augend and addend. The variable *opnds* is an event control used in the **always** statement — when the operands change, the statement in the **begin** . . . **end** block is executed.

```
//behavioral to add two operands
module mem_sum (opnds, sum);

input [5:0] opnds;    //list inputs and outputs
output [3:0] sum;

wire [5:0] opnds;     //augend and addend to address 64 words
reg [3:0] sum;

//define memory size
//mem_add is an array of 64 four-bit registers
reg [3:0] mem_add [0:63];
//continued on next page
```

Figure 4.35 Behavioral module using memory for addition.

```
//define memory contents
//load mem_add from file opnds.add
initial
begin
   $readmemb ("opnds.add", mem_add);
end

//use the operands to access the memory
always @ (opnds)
begin
   sum = mem_add [opnds];
end
endmodule
```

Figure 4.35 (Continued)

```
//test bench for mem_sum module
module mem_sum_tb;

reg [5:0] opnds;
wire [3:0] sum;

//display variables
initial
$monitor ("augend_addend = %b, sum = %b", opnds, sum);

//apply input vectors for augend and addend
initial
begin
   #0    opnds = 6'b000_011;  //augend_addend
   #10   opnds = 6'b001_100;
   #10   opnds = 6'b010_100;
   #10   opnds = 6'b011_110;
   #10   opnds = 6'b100_100;
   #10   opnds = 6'b101_101;
   #10   opnds = 6'b110_101;
   #10   opnds = 6'b100_110;
   #10   opnds = 6'b111_010;
   #10   opnds = 6'b111_100;
   #10   opnds = 6'b111_110;
   #10   opnds = 6'b111_111;
   #10   $stop;
end                            //continued on next page
```

Figure 4.36 Test bench for the behavioral module to add two operands using memory.

```
//instantiate the module into the test bench
mem_sum inst1 (
   .opnds(opnds),
   .sum(sum)
   );

endmodule
```

Figure 4.36 (Continued)

```
augend_addend = 000011, sum = 0011
augend_addend = 001100, sum = 0101
augend_addend = 010100, sum = 0110
augend_addend = 011110, sum = 1001
augend_addend = 100100, sum = 1000
augend_addend = 101101, sum = 1010
augend_addend = 110101, sum = 1011
augend_addend = 100110, sum = 1010
augend_addend = 111010, sum = 1001
augend_addend = 111100, sum = 1011
augend_addend = 111110, sum = 1101
augend_addend = 111111, sum = 1110
```

Figure 4.37 Outputs for the behavioral module to add two operands using memory.

4.5 Carry-Select Addition

A carry-select adder is not as fast as the carry lookahead adder, however it has a unique organization that is interesting. Each section of the adder is partitioned into two 4-bit groups consisting of identical 4-bit adders to which the same 4-bit operands are applied. Since the low-order stage of an adder does not usually have a carry-in, the low-order section consists of only one 4-bit adder with a carry-in of 0.

The carry-select principle produces two sums that are generated simultaneously. One sum assumes that the carry-in to that group was a 0; the other sum assumes that the carry-in was a 1. The predicted carry is obtained using the carry lookahead technique which selects one of the two sums.

The logic diagram to add two 12-bit operands is shown in Figure 4.38. The carry-out *cout1* of section 0 determines which sum is selected in section 1. If *cout1* is a 0, then the sum obtained by the adder in section 1 whose carry-in is 0 is selected and sent to a 2:1 multiplexer. If *cout1* is 1, then the sum obtained by the adder whose carry-in is 1 is selected and sent to a 2:1 multiplexer.

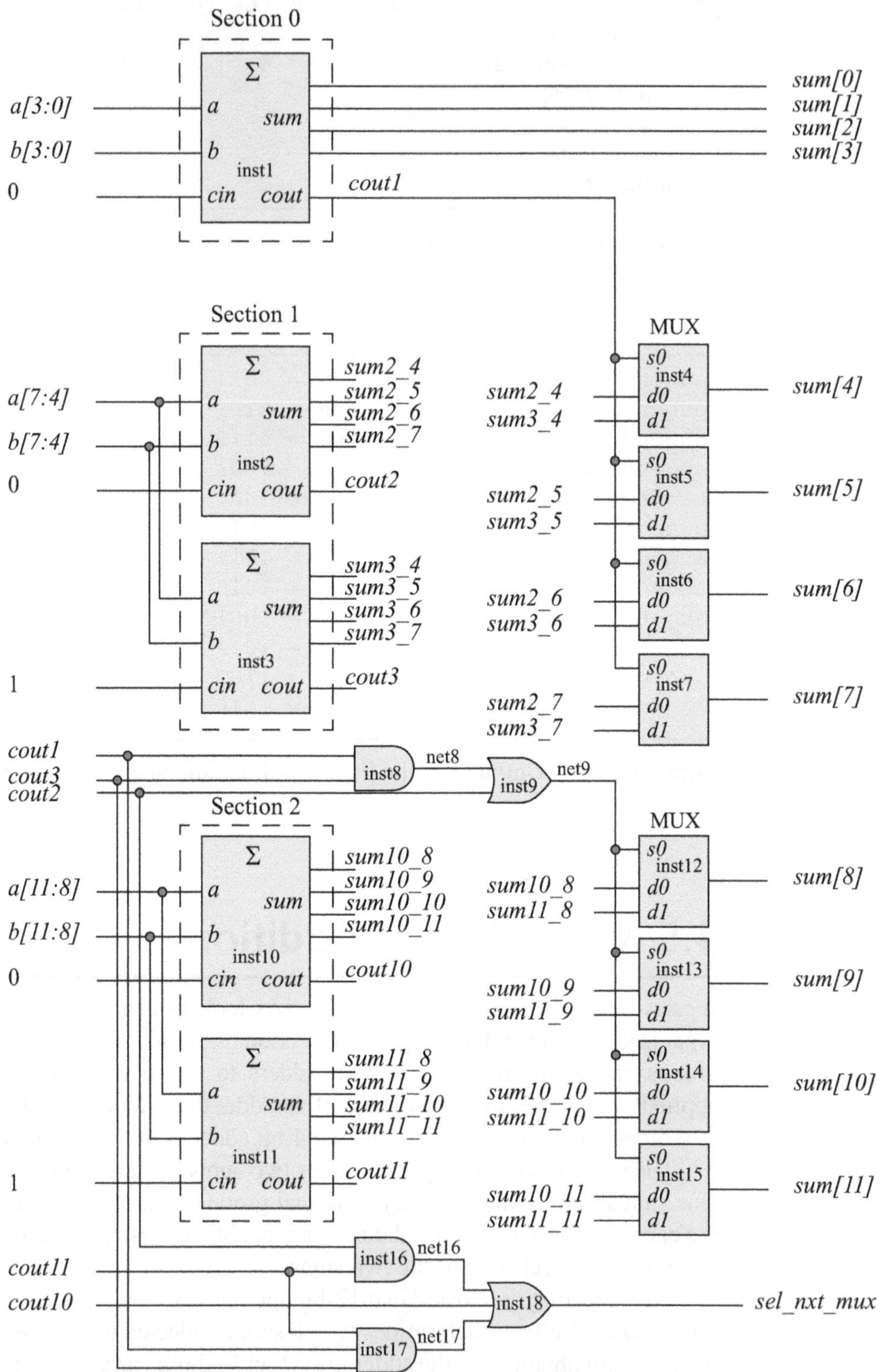

Figure 4.38 Carry-select adder to add two 12-bit operands.

If the carry-out of any adder is 1, whose carry-in is assumed to be 0, then this constitutes a generate function. If the carry-out of any adder is 1, whose carry-in is assumed to be 1, then this constitutes a propagate function. Consider section 0. If the carry-out *cout1* is a 1, then section 0 comprises a generate function; otherwise, section 0 is a propagate function. The carry equations for section 0, section 1, and section 2 are shown in Equation 4.13, where c_{-1} is the carry-in to the adder and assumed to be 0. The carry-select adder shown in Figure 4.38 can be easily expanded to accommodate operands of any size.

$$
\begin{aligned}
SC_0 &= G_0 + P_0 c_{-1} \\
&= cout1
\end{aligned}
$$

$$
\begin{aligned}
SC_1 &= G_1 + P_1 G_0 + P_1 P_0 c_{-1} \\
&= cout2 + (cout3)(cout1)
\end{aligned}
$$

$$
\begin{aligned}
SC_2 &= G_2 + P_2 G_1 + P_2 P_1 G_0 + P_2 P_1 P_0 c_{-1} \\
&= cout10 + (cout2)(cout11) + (cout1)(cout3)(cout11) \qquad (4.13)
\end{aligned}
$$

The following modules will be designed, then instantiated into a structural module to implement the carry-select adder: a behavioral 4-bit adder, *adder4*; a dataflow 2:1 multiplexer, *mux2_df*; a dataflow 2-input AND gate, *and2_df*; a dataflow 3-input AND gate, *and3_df*; a dataflow 2-input OR gate, *or2_df*; and a dataflow 3-input OR gate, *or3_df*.

The 4-bit adder module is shown in Figure 4.39, the 2:1 multiplexer module is shown in Figure 4.40, the 2-input AND gate module is shown in Figure 4.41, the 3-input AND gate module is shown in Figure 4.42, the 2-input OR gate module is shown in Figure 4.43, and the 3-input OR gate module is shown in Figure 4.44.

```
//behavioral model for a 4-bit adder
module adder4 (a, b, cin, sum, cout);

//define inputs and outputs
input [3:0] a, b;
input cin;
output [3:0] sum;
output cout;
//continued on next page
```

Figure 4.39 Behavioral module for a 4-bit adder.

```
wire [3:0] a, b;
wire cin;
reg [3:0] sum;
reg cout;

always @ (a or b or cin)
begin
   sum  = a + b + cin;
   cout = (a[3] & b[3]) |
          ((a[3] | b[3]) & (a[2] & b[2])) |
          ((a[3] | b[3]) & (a[2] | b[2]) & (a[1] & b[1])) |
          ((a[3] | b[3]) & (a[2] | b[2]) & (a[1] |
          b[1]) & (a[0] & b[0])) |
          ((a[3] | b[3]) & (a[2] | b[2]) & (a[1] |
          b[1]) & (a[0] | b[0]) & cin);
end
endmodule
```

Figure 4.39 (Continued)

```
//dataflow 2:1 multiplexer
module mux2_df (sel, data, z1);

//define inputs and output
input sel;
input [1:0] data;
output z1;

assign z1 = (~sel & data[0]) | (sel & data[1]);
endmodule
```

Figure 4.40 Dataflow module for a 2:1 multiplexer.

```
//dataflow 2-input and gate
module and2_df (x1, x2, z1);

input x1, x2;
output z1;

wire x1, x2;
wire z1;

//continuous assign for dataflow
assign z1 = x1 & x2;
endmodule
```

Figure 4.41 Dataflow module for a 2-input AND gate.

```
//and3 dataflow
module and3_df (x1, x2, x3, z1);

input x1, x2, x3;        //list inputs and output
output z1;

wire x1, x2, x3;         //define signals as wire for dataflow
wire z1;

assign z1 = x1 & x2 & x3;//continuous assign for dataflow

endmodule
```

Figure 4.42 Dataflow module for a 3-input AND gate.

```
//dataflow or2
module or2_df (x1, x2, z1);

input x1, x2;          //list inputs and output
output z1;

wire x1, x2;           //define signals as wire for dataflow
wire z1;

assign z1 = x1 | x2; //continuous assign for dataflow

endmodule
```

Figure 4.43 Dataflow module for a 2-input OR gate.

```
//or3 dataflow
module or3_df (x1, x2, x3, z1);

input x1, x2, x3;
output z1;

wire x1, x2, x3;
wire z1;

assign z1 = x1 | x2 | x3;

endmodule
```

Figure 4.44 Dataflow module for a 3-input OR gate.

The structural module for the carry-select adder is shown in Figure 4.45. The 4-bit adder is instantiated five times and the multiplexer is instantiated eight times. The test bench is shown in Figure 4.46 listing twelve input vectors for augend *a[11:0]* and addend *b[11:0]*. The outputs are shown in Figure 4.47 and the waveforms are shown in Figure 4.48.

```
//structural carry-select adder
module adder_carry_sel (a, b, sum, sel_nxt_mux);

input [11:0] a, b;
output [11:0] sum;
output sel_nxt_mux;

wire [11:0] a, b;
wire [11:0] sum;
wire sel_nxt_mux;

//define internal nets
wire   sum2_7, sum2_6, sum2_5, sum2_4, cout2, cout1,
       sum3_7, sum3_6, sum3_5, sum3_4, cout3, net8,
       net9, sum10_11, sum10_10, sum10_9, sum10_8, cout10,
       sum11_11, sum11_10, sum11_9, sum11_8, cout11,
       net16, net17;

//instantiate the 4-bit adder for a[3:0] and b[3:0]
adder4 inst1 (
   .a(a[3:0]),
   .b(b[3:0]),
   .cin(1'b0),
   .sum(sum[3:0]),
   .cout(cout1)
   );

//instantiate the 4-bit adder for a[7:4] and b[7:4]
adder4 inst2 (
   .a(a[7:4]),
   .b(b[7:4]),
   .cin(1'b0),
   .sum({sum2_7, sum2_6, sum2_5, sum2_4}),
   .cout(cout2)
   );

//continued on next page
```

Figure 4.45 Structural module for the 12-bit carry-select adder.

```
adder4 inst3 (
   .a(a[7:4]),
   .b(b[7:4]),
   .cin(1'b1),
   .sum({sum3_7, sum3_6, sum3_5, sum3_4}),
   .cout(cout3)
   );

//instantiate the 2:1 muxs for sum[7:4]
mux2_df inst4 (
   .sel(cout1),
   .data({sum3_4, sum2_4}),
   .z1(sum[4])
   );

mux2_df inst5 (
   .sel(cout1),
   .data({sum3_5, sum2_5}),
   .z1(sum[5])
   );

mux2_df inst6 (
   .sel(cout1),
   .data({sum3_6, sum2_6}),
   .z1(sum[6])
   );

mux2_df inst7 (
   .sel(cout1),
   .data({sum3_7, sum2_7}),
   .z1(sum[7])
   );

//instantiate the logic to select muxs for sum[11] thru sum[8]
and2_df inst8 (
   .x1(cout1),
   .x2(cout3),
   .z1(net8)
   );

or2_df inst9 (
   .x1(cout2),
   .x2(net8),
   .z1(net9)
   );

//continued on next page
```

Figure 4.45 (Continued)

```
//instantiate the 4-bit adder for a[11:8] and b[11:8]
adder4 inst10 (
   .a(a[11:8]),
   .b(b[11:8]),
   .cin(1'b0),
   .sum({sum10_11, sum10_10, sum10_9, sum10_8}),
   .cout(cout10)
   );


adder4 inst11 (
   .a(a[11:8]),
   .b(b[11:8]),
   .cin(1'b1),
   .sum({sum11_11, sum11_10, sum11_9, sum11_8}),
   .cout(cout11)
   );


//instantiate the 2:1 muxs for sum[11:8]
mux2_df inst12 (
   .sel(net9),
   .data({sum11_8, sum10_8}),
   .z1(sum[8])
   );

mux2_df inst13 (
   .sel(net9),
   .data({sum11_9, sum10_9}),
   .z1(sum[9])
   );

mux2_df inst14 (
   .sel(net9),
   .data({sum11_10, sum10_10}),
   .z1(sum[10])
   );

mux2_df inst15 (
   .sel(net9),
   .data({sum11_11, sum10_11}),
   .z1(sum[11])
   );

//continued on next page
```

Figure 4.45 (Continued)

```
//instantiate the logic for sel_nxt_mux
and2_df inst16 (
    .x1(cout2),
    .x2(cout11),
    .z1(net16)
    );

and3_df inst17 (
    .x1(cout1),
    .x2(cout3),
    .x3(cout11),
    .z1(net17)
    );

or3_df inst18 (
    .x1(cout10),
    .x2(net16),
    .x3(net17),
    .z1(sel_nxt_mux)
    );

endmodule
```

Figure 4.45 (Continued)

```
//test bench for carry-select adder
module adder_carry_sel_tb;

//define inputs and outputs
reg [11:0] a, b;
wire [11:0] sum;
wire sel_nxt_mux;

//display outputs
initial
$monitor ("a = %b, b = %b, sum = %b, sel_nxt_mux = %b",
            a, b, sum, sel_nxt_mux);

//apply input vectors
initial
begin
    #0    a = 12'b0000_0000_0001;
          b = 12'b0000_0000_0010;

//continued on next page
```

Figure 4.46 Test bench for the 12-bit carry-select adder.

```
   #10   a = 12'b0000_0001_0010;
         b = 12'b0000_0000_0011;

   #10   a = 12'b0001_0001_0010;
         b = 12'b0000_0000_0011;

   #10   a = 12'b0011_0011_0010;
         b = 12'b0001_0100_0011;

   #10   a = 12'b1111_0000_0111;
         b = 12'b1000_1101_1000;

   #10   a = 12'b1111_1111_1111;
         b = 12'b1111_1111_1111;

   #10   a = 12'b1010_1010_1010;
         b = 12'b0101_0101_0101;

   #10   a = 12'b0111_1011_1010;
         b = 12'b0011_0110_1101;

   #10   a = 12'b1111_0111_0110;
         b = 12'b0011_1100_0111;

   #10   a = 12'b1100_1010_0111;
         b = 12'b0101_0011_0101;

   #10   a = 12'b0011_0111_1011;
         b = 12'b1011_0011_0111;

   #10   a = 12'b1011_0111_1100;
         b = 12'b0100_1000_0110;

   #10   $stop;
end

//instantiate the module into the test bench
adder_carry_sel inst1 (
   .a(a),
   .b(b),
   .sum(sum),
   .sel_nxt_mux(sel_nxt_mux)
   );

endmodule
```

Figure 4.46 (Continued)

```
a   = 000000000001,
b   = 000000000010,
sum = 000000000011, sel_nxt_mux = 0

a   = 000000010010,
b   = 000000000011,
sum = 000000010101, sel_nxt_mux = 0

a   = 000100010010,
b   = 000000000011,
sum = 000100010101, sel_nxt_mux = 0

a   = 001100110010,
b   = 000101000011,
sum = 010001110101, sel_nxt_mux = 0

a   = 111100000111,
b   = 100011011000,
sum = 011111011111, sel_nxt_mux = 1

a   = 111111111111,
b   = 111111111111,
sum = 111111111110, sel_nxt_mux = 1

a   = 101010101010,
b   = 010101010101,
sum = 111111111111, sel_nxt_mux = 0

a   = 011110111010,
b   = 001101101101,
sum = 101100100111, sel_nxt_mux = 0

a   = 111101110110,
b   = 001111000111,
sum = 001100111101, sel_nxt_mux = 1

a   = 110010100111,
b   = 010100110101,
sum = 000111011100, sel_nxt_mux = 1

a   = 001101111011,
b   = 101100110111,
sum = 111010110010, sel_nxt_mux = 0

a   = 101101111100,
b   = 010010000110,
sum = 000000000010, sel_nxt_mux = 1
```

Figure 4.47 Outputs for the 12-bit carry-select adder.

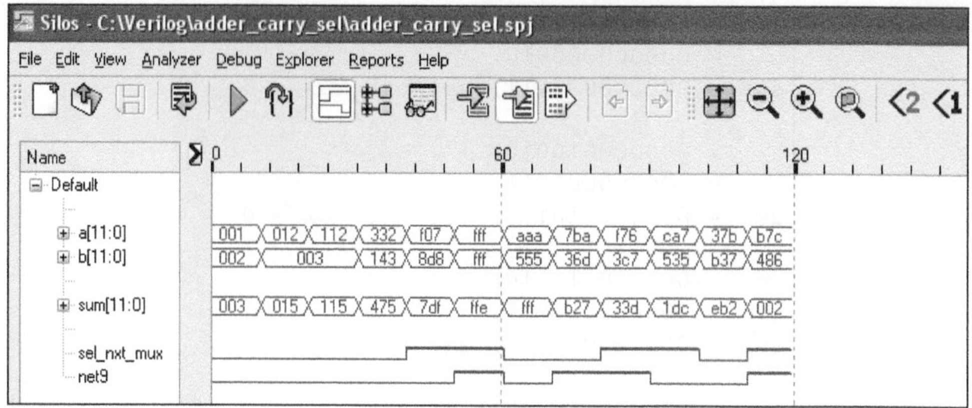

Figure 4.48 Waveforms for the 12-bit carry-select adder.

4.6 Serial Addition

If a minimal amount of hardware is a prerequisite and speed is not essential, then a serial adder may be utilized. A serial adder adds the augend and addend one bit per clock pulse — thus, eight clock pulses are required to add two bytes of data. Two parallel-in, serial-out shift registers are loaded with the augend and addend prior to the shift operation. The logic diagram for a serial adder is shown in Figure 4.49.

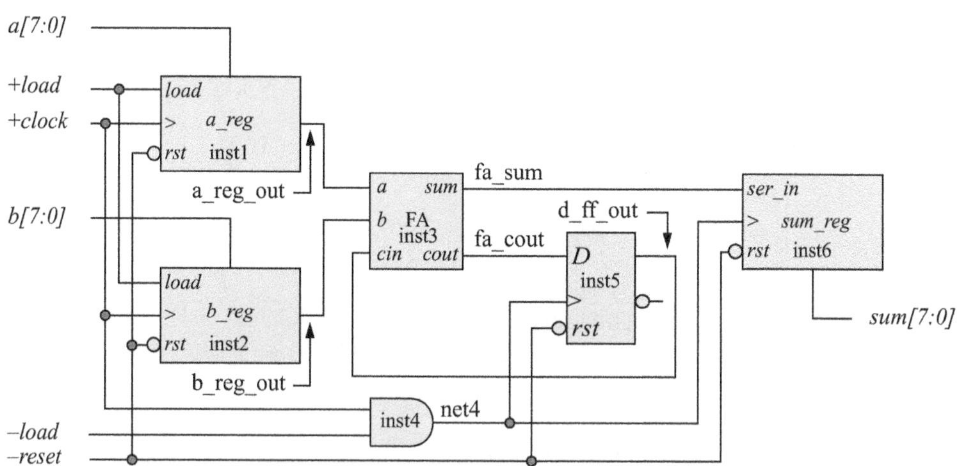

Figure 4.49 Logic diagram for a serial adder.

The serial outputs of the shift registers connect to the *a* and *b* inputs of a full adder. The *sum* output of the full adder connects to the serial input of a serial-in, parallel-out shift register. Alternatively, the *sum* output can be connected to the serial input of the *a_reg*, which must now be a more universal shift register with parallel-in, serial-out and serial-in, parallel-out capabilities. The *carry-out* of the full adder connects to the data input of a *D* flip-flop whose output represents the *carry-in* to the full adder.

This method can be used to add any number of bits simply by increasing the size of all shift registers. If the augend and addend are both n bits $[(n-1): 0]$, then the result can be fully contained in n bits except for overflow, which can be detected in the usual way, as shown in Equation 4.14. The high-order bits of *a_reg*$(n-1)$ and *b_reg*$(n-1)$ can be saved in two *D* flip-flops, then compared with *sum_reg*$(n-1)$ at the completion of the shift operation, as shown in the equation. The behavioral modules for the parallel-in, serial-out shift register and the serial-in, parallel-out shift register are shown in Figure 4.50 and Figure 4.51, respectively. The dataflow module for the full adder is shown in Figure 4.52.

$$\begin{aligned}\text{Overflow} = {} & \{(a\_reg[n-1])(b\_reg[n-1])(sum\_reg[n-1])\}' + \\ & \{(a\_reg[n-1])'(b\_reg[n-1])'(sum\_reg[n-1])\}\end{aligned} \tag{4.14}$$

```
//behavioral parallel-in, serial-out shift register
module shift_reg_piso8 (rst_n, clk, load, x, y, z1);

input rst_n, clk, load;
input [7:0] x;
output [7:0] y;
output z1;

reg [7:0] y;
assign z1 = y[0];

always @ (rst_n)
begin
   if (rst_n == 0)
      y <= 8'b0000_0000;

end
always @ (posedge clk)
begin
   y[7] <= ((load && x[7]) || (~load && 1'b0));
   y[6] <= ((load && x[6]) || (~load && y[7]));
   y[5] <= ((load && x[5]) || (~load && y[6]));
   y[4] <= ((load && x[4]) || (~load && y[5]));
   y[3] <= ((load && x[3]) || (~load && y[4]));
   y[2] <= ((load && x[2]) || (~load && y[3]));
   y[1] <= ((load && x[1]) || (~load && y[2]));
   y[0] <= ((load && x[0]) || (~load && y[1]));
end
endmodule
```

Figure 4.50 Behavioral module for a parallel-in, serial-out shift register.

```
//behavioral serial-in parallel-out shift register
module shift_reg_sipo (rst_n, clk, ser_in, shift_reg);

input rst_n, clk, ser_in;
output [7:0] shift_reg;

reg [7:0] shift_reg;

always @ (rst_n)
begin
   if (rst_n == 0)
      shift_reg <= 8'b0000_0000;
end

always @ (posedge clk)
begin
   shift_reg [7] <= ser_in;
   shift_reg [6] <= shift_reg [7];
   shift_reg [5] <= shift_reg [6];
   shift_reg [4] <= shift_reg [5];
   shift_reg [3] <= shift_reg [4];
   shift_reg [2] <= shift_reg [3];
   shift_reg [1] <= shift_reg [2];
   shift_reg [0] <= shift_reg [1];
end
endmodule
```

Figure 4.51 Behavioral module for a serial-in, parallel-out shift register.

```
//dataflow full adder
module full_adder (a, b, cin, sum, cout);

//list all inputs and outputs
input a, b, cin;
output sum, cout;

//define wires
wire a, b, cin;
wire sum, cout;

//continuous assign
assign sum = (a ^ b) ^ cin;
assign cout = cin & (a ^ b) | (a & b);

endmodule
```

Figure 4.52 Dataflow module for a full adder.

The behavioral module for the *D* flip-flop is shown in Figure 4.53. The structural module for the serial adder is shown in Figure 4.54, in which all of the previously designed modules are instantiated, including a dataflow module for a 2-input AND gate. The test bench, shown in Figure 4.55, applies two sets of operands for the augend and addend. In the first set, the augend is 3_{10} (0000 0011) and the addend is 8_{10} (0000 1000), resulting in a sum of 11_{10} (0000 1011). In the second set, the augend is 25_{10} (0001 1001) and the addend is 34_{10} (0010 0010), resulting in a sum of 59_{10} (0011 1011). The outputs and waveforms are shown in Figure 4.56 and Figure 4.57, respectively.

```
//behavioral D flip-flop
module d_ff_bh (rst_n, clk, d, q, q_n);

input rst_n, clk, d;
output q, q_n;

wire rst_n, clk, d;
reg q;

assign q_n = ~q;

always @ (rst_n or posedge clk)
begin
   if (rst_n == 0)
      q <= 1'b0;
   else q <= d;
end
endmodule
```

Figure 4.53 Behavioral module for a *D* flip-flop.

```
//structural for serial adder
module adder_serial (rst_n, clk, load, a, b, sum);

input rst_n, clk, load;
input [7:0] a, b;
output [7:0] sum;

//define internal nets
wire  a_reg_out, b_reg_out, fa_sum, fa_cout,
      net4, d_ff_out;
//continued on next page
```

Figure 4.54 Structural module for the serial adder.

```
//instantiate the parallel-in, serial-out shift registers
shift_reg_piso8 inst1 (
   .rst_n(rst_n),
   .clk(clk),
   .load(load),
   .x(a),
   .z1(a_reg_out)
   );

shift_reg_piso8 inst2 (
   .rst_n(rst_n),
   .clk(clk),
   .load(load),
   .x(b),
   .z1(b_reg_out)
   );

//instantiate the full adder
full_adder inst3 (
   .a(a_reg_out),
   .b(b_reg_out),
   .cin(d_ff_out),
   .sum(fa_sum),
   .cout(fa_out)
   );

//instantiate the d flip-flop and logic
and2_df inst4 (
   .x1(clk),
   .x2(~load),
   .z1(net4)
   );

d_ff_bh inst5 (
   .rst_n(rst_n),
   .clk(clk),
   .d(fa_out),
   .q(d_ff_out)
   );

//instantiate the serial-in, parallel-out shift register
shift_reg_sipo inst6 (
   .rst_n(rst_n),
   .clk(clk),
   .ser_in(fa_sum),
   .shift_reg(sum)
   );

endmodule
```

Figure 4.54 (Continued)

```
//test bench for the serial adder
module adder_serial_tb;

reg rst_n, clk, load;
reg [7:0] a, b;
wire [7:0] sum;

//define clock
initial
begin
   clk = 1'b0;
   forever
      #10 clk = ~clk;
end

//display variables
initial
$monitor ("a = %b, b = %b, sum = %b", a, b, sum);

//apply input vectors
initial
begin
   #0    rst_n = 1'b0;
   #2    rst_n = 1'b1;

   #3    a = 8'b0000_0011;
         b = 8'b0000_1000;

         load = 1'b1;
   #15   load = 1'b0;
//------------------------------
   #165  rst_n = 1'b0;
   #2    rst_n = 1'b1;

   #2    a = 8'b0001_1001;
         b = 8'b0010_0010;

         load = 1'b1;
   #15   load = 1'b0;
end

//determine length of simulation
initial
begin
   #370  $stop;
end
//continued on next page
```

Figure 4.55 Test bench for the serial adder.

```
//instantiate the module into the test bench
adder_serial inst1 (
   .rst_n(rst_n),
   .clk(clk),
   .load(load),
   .a(a),
   .b(b),
   .sum(sum)
   );

endmodule
```

Figure 4.55 (Continued)

```
a = xxxxxxxx, b = xxxxxxxx, sum = 00000000
a = 00000011, b = 00001000, sum = 00000000
a = 00000011, b = 00001000, sum = 10000000
a = 00000011, b = 00001000, sum = 11000000
a = 00000011, b = 00001000, sum = 01100000
a = 00000011, b = 00001000, sum = 10110000
a = 00000011, b = 00001000, sum = 01011000
a = 00000011, b = 00001000, sum = 00101100
a = 00000011, b = 00001000, sum = 00010110
a = 00000011, b = 00001000, sum = 00001011 = 11
a = 00000011, b = 00001000, sum = 00000000
       3                8

------------------------------------------

a = 00011001, b = 00100010, sum = 00000000
a = 00011001, b = 00100010, sum = 10000000
a = 00011001, b = 00100010, sum = 11000000
a = 00011001, b = 00100010, sum = 01100000
a = 00011001, b = 00100010, sum = 10110000
a = 00011001, b = 00100010, sum = 11011000
a = 00011001, b = 00100010, sum = 11101100
a = 00011001, b = 00100010, sum = 01110110
a = 00011001, b = 00100010, sum = 00111011 = 59
       25               34
```

Figure 4.56 Outputs for the serial adder.

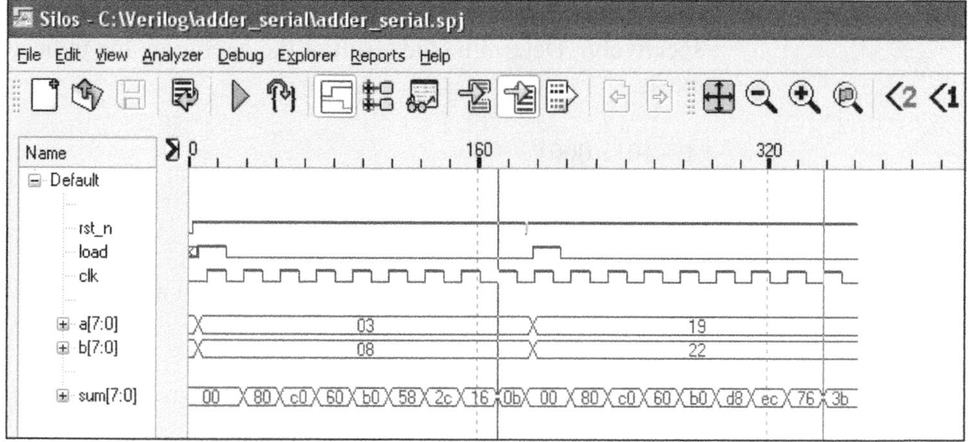

Figure 4.57 Waveforms for the serial adder.

4.7 Problems

4.1 For $n = 8$, let A and B be two fixed-point binary numbers in 2s complement representation, where

$A = 0110\ 1101$
$B = 1100\ 1111$

Find $A + B' + 1 + B$, where B' is the 1s complement of B.

4.2 Perform the arithmetic operations shown below with fixed-point binary numbers in 2s complement representation. In each case, indicate if there is an overflow.

(a)
$$\begin{array}{r} 0100\ 0000 \\ +)\ \underline{0100\ 0000} \end{array}$$

(b)
$$\begin{array}{r} 0011\ 0110 \\ +)\ \underline{1110\ 0011} \end{array}$$

(c)
$$\begin{array}{r} +64_{10} \\ +63_{10} \end{array}$$

4.3 Let A and B be two binary integers in 2s complement representation as shown below, where A' and B' are the diminished radix complement of A and B, respectively. Determine the result of the operation and indicate if an overflow exists.

$A = 1011\ 0001$
$B = 1110\ 0100$

4.4 Add the following numbers and show the sum as an integer in radix 8: $925_{10} + 634_8$

4.5 Add the following numbers and obtain the sum as an integer in radix 4: $201_{10} + 321_4$

4.6 Obtain the group generate and group propagate functions for stage $i + 4$ and stage $i + 5$ of a 6-bit carry lookahead adder.

4.7 Write the equations for two ways to detect overflow for two 8-bit (7:0) operands A and B in radix 2.

4.8 Use only carry-save full adders in a Wallace tree configuration to design a circuit that will add the 2^ith bit of four different operands. Use the fewest number of carry-save adders. Then obtain the structural module, the test bench module, the outputs, and the waveforms.

4.9 Use only carry-save full adders in a Wallace tree configuration to design a circuit that will add the 2^ith bit of seven different operands. Use the fewest number of carry-save adders. Then obtain the structural module, the test bench module, and the outputs for the first 16 inputs and the last 16 inputs.

4.10 Design an 8-bit serial adder that also detects overflow. Draw the logic diagram and implement the adder as a structural module. Obtain the test bench module and apply two sets of input vectors for the augend and addend. The first vector will have no overflow; the second vector will generate an overflow. Obtain the outputs, and the waveforms. The overflow bit will fluctuate as the variables shift serially, but will become valid at the completion of the shift operation.

5.1 Twos Complement Subtraction
5.2 Ripple-Carry Subtraction
5.3 Carry Lookahead Addition/Subtraction
5.4 Behavioral Addition/Subtraction
5.5 Problems

5

Fixed-Point Subtraction

Like fixed-point addition, fixed-point subtraction is relatively simple. It also shares much of the same circuitry as addition. The two operands for subtraction are the minuend and the subtrahend — the subtrahend is subtracted from the minuend according to the rules shown in Table 5.1 for radix 2. Subtraction can be performed in all three number representations: sign magnitude, diminished-radix complement, and radix complement; however, radix complement is the easiest and most widely used method for subtraction in any radix. Example 5.1 elaborates on the rules for binary subtraction.

Table 5.1 Truth Table for Subtraction

0 – 0	=	0	
0 – 1	=	1	with a borrow from the next higher-order minuend
1 – 0	=	1	
1 – 1	=	0	

Example 5.1 Two 8-bit operands are shown below to illustrate the rules for radix 2 subtraction in which all four combinations of two bits are provided. The borrow from the minuend in column 2^1 to the minuend in column 2^0 changes column 2^1 from 1 0 to 0 0, as shown below; that is, the operation of column 2^1 the becomes 0 – 0 = 0.

1 0

to

0 0

	2^7	2^6	2^5	2^4	2^3	2^2	2^1	2^0
A (Minuend) = +54	0	0	1	1	0	1	1	0
–) *B* (Subtrahend) = +37	0	0	1	0	0	1	0	1
D (Difference) = +17	0	0	0	1	0	0	0	1

5.1 Twos Complement Subtraction

The above method of direct subtraction is useful only to demonstrate the paper-and-pencil method. Computers use an adder for the subtraction operation by adding the radix complement of the subtrahend to the minuend. Thus, let *A* and *B* be two *n*-bit operands, where *A* is the minuend and *B* is the subtrahend as follows:

$$A = a_{n-1}\ a_{n-2} \ldots a_1\ a_0$$

$$B = b_{n-1}\ b_{n-2} \ldots b_1\ b_0$$

Therefore, $A - B = A + (B' + 1)$, where B' is the 1s complement of *B*. Example 5.1 will be repeated in Example 5.2 using this method.

Example 5.2 Subtrahend *B* (+37) will be subtracted from minuend *A* (+54) to obtain a difference of +17 by adding the 2s complement (or negation) of the subtrahend, as shown below. The carry-out can be ignored because overflow is not possible when adding a negative number to a positive number.

	2^7	2^6	2^5	2^4	2^3	2^2	2^1	2^0
A (Minuend) = +54	0	0	1	1	0	1	1	0
+) *B* (Subtrahend) = –37	1_1	1_1	0_1	1_1	1_1	0_1	1	1
D (Difference) = +17	0	0	0	1	0	0	0	1

The diminished-radix $(r - 1)$ complement of an n-digit number A is defined as

$$A' = (r^n - 1) - A$$

Thus, the $(r - 1)$ complement of the radix 2 number 0110 is

$$\begin{aligned}(r^n - 1) - A &= (2^4 - 1) - 6 \\ &= 15 - 6 \\ &= 9 \\ &= 1001\end{aligned}$$

The rs complement is the $(r - 1)$ complement plus 1. Therefore, the rs complement of an n-digit number A is defined as

$$\begin{aligned}A'_{+1} &= r^n - A \\ &= A' + 1\end{aligned}$$

Thus, the rs complement of the radix 2 number 0110 is

$$\begin{aligned}A'_{+1} &= r^n - A \\ &= 16 - 6 \\ &= 10 \\ &= 1010\end{aligned}$$

Subtraction is accomplished by adding the rs complement of the subtrahend to the minuend. Let $A = 0111$ and $B = 0100$. Therefore,

$$\begin{aligned}A - B &= A - r^n + r^n - B \\ &= A - r^n + (r^n - B) \\ &= A - r^n + B'_{+1} \\ &= (7 - 2^4) + 12 \\ &= (7 - 16) + 12 \\ &= -9 + 12 \\ &= 3\end{aligned}$$

Using the $(r - 1)$ complement for subtraction is accomplished as follows:

$$A - B = A + B' \text{ (}A\text{ plus the 1s complement of }B\text{)}$$

The $(r - 1)$ complement, however, is not used for subtraction because it may result in an incorrect result requiring one more addition cycle to obtain a correct result. For example, the radix 2 operation $[A = 1111\ 1001\ (-6)] - [B = 1110\ 1101\ (-18)] = +12$ yields an incorrect result initially, as shown below. If a subtraction produces a carry out of the high-order bit position, then the carry must be added to the intermediate result. This is called an *end-around carry.* Another reason why the $(r-1)$ complement is not used is because there is a double representation for the number zero. For example, for $r = 2$ and $n = 8$ the number zero = 0000 0000 and the 1s complement is 1111 1111. This is not true for the rs complement. The number zero is 0000 0000 and the 2s complement is also 0000 0000.

```
          A =  1111   1001                      –6
    +)   B' =  0001   0010                     +18
  ┌── 1 ◄───── 0000   1011                     +11
  └──────────►           1   ◄──── End-around carry
               0000   1100                     +12
```

If the carry-out $c_{n-1} = 0$, then the result is correct; if the carry-out $c_{n-1} = 1$, then the result is incorrect and the carry is added to the intermediate result.

There are four cases to consider when subtracting two numbers of unequal magnitude in rs complement: $(+A) - (-B)$, $(-A) - (+B)$, $(+A) - (+B)$, and $(-A) - (-B)$. In case 1 where $(+A) - (-B)$, after B has been rs complemented, both operands are positive. If $A + B < r^{n-1}$, there will be no carry into the high-order digit position and thus, no overflow. If $A + B \geq r^{n-1}$, an overflow will occur. For example, let A = 0011 0001 (+49) and B = 1000 0100 (–124), where $r = 2$ and $n = 8$. Then

$$A + B = 49 + 124 = 173 \geq r^{n-1} = 2^{8-1} = 2^7 = 128$$

```
        A =  0011  0001        +49
   –)   B =  1000  0100       –124
                  ↓
        A =  0011  0001        +49
   +)   B =  0111  1100       +124
           0 1010  1101       +173
```

In case 2 where $(-A) - (+B)$, after B has been rs complemented, both operands are negative. Since the signs of both operands are negative, there will always be a carry out of the high-order digit position. Since $A'_{+1} = r^n - A$ and $B'_{+1} = r^n - B$,

$$\text{Difference} = (r^n - A) + (r^n - B) = 2r^n - (A + B)$$

If $A + B \leq r^{n-1}$, then the result will be negative with no overflow. If $A + B > r^{n-1}$, then an overflow will occur. For example, let A = 1011 0111 (–73) and B = 0101 1100 (+92), where $r = 2$ and $n = 8$. Then

$$A + B = 73 + 92 = 165 > r^{n-1} = 2^{8-1} = 2^7 = 128$$

$$
\begin{array}{rrr}
 & A = 1011\ \ 0111 & -73 \\
-) & B = \underline{0101\ \ 1100} & +92
\end{array}
$$

$$\downarrow$$

$$
\begin{array}{rrr}
 & A = 1011\ \ 0111 & -73 \\
+) & B = \underline{1010\ \ 0100} & -92 \\
 & 1\ \ 0101\ \ 1011 & -165
\end{array}
$$

In case 3 and case 4 where $(+A) - (+B)$, and $(-A) - (-B)$, respectively, the resulting addition of a positive and a negative number produces a difference that is within the range of the two numbers. Therefore, no overflow can occur. Several examples will now be presented using the rs complement for radix 2 that illustrate case 1 $(+A) - (-B)$ and case 2 $(-A) - (+B)$. Some examples produce no overflow and some examples produce an overflow.

Example 5.3 For case 1 where $(+A) - (-B)$, let $A = +4$ and $B = -4$ for $n = 4$. Therefore, $A + B = 8 \geq r^{n-1}$ (8) will produce an overflow, as shown below.

$$
\begin{array}{rr}
 & 0100 \\
-) & \underline{1100}
\end{array}
\longrightarrow
\begin{array}{rr}
 & 0100 \\
+) & \underline{0100} \\
 & 1000
\end{array}
$$

Example 5.4 For case 1 where $(+A) - (-B)$, let $A = +5$ and $B = -4$ for $n = 4$. Therefore, $A + B = 9 \geq r^{n-1}$ (8) will produce an overflow, as shown below.

$$
\begin{array}{rr}
 & 0101 \\
-) & \underline{1100}
\end{array}
\longrightarrow
\begin{array}{rr}
 & 0101 \\
+) & \underline{0100} \\
 & 1001
\end{array}
$$

Example 5.5 For case 1 where $(+A) - (-B)$, let $A = +6$ and $B = -1$ for $n = 4$. Therefore, $A + B = 7 < r^{n-1}$ (8) will produce no overflow, as shown below.

$$
\begin{array}{rr}
 & 0110 \\
-) & \underline{1111}
\end{array}
\longrightarrow
\begin{array}{rr}
 & 0110 \\
+) & \underline{0001} \\
 & 0111
\end{array}
$$

Example 5.6 For case 2 where $(-A) - (+B)$, let $A = -4$ and $B = +3$ for $n = 4$. Therefore, $A + B = 7 \leq r^{n-1}$ (8) will produce no overflow, as shown below.

```
     1100                         1100
-)   0011     ------->       +)   1101
                              1   1001
```

Example 5.7 For case 2 where $(-A) - (+B)$, let $A = -4$ and $B = +4$ for $n = 4$. Therefore, $A + B = 8 \leq r^{n-1}$ (8) will produce no overflow, as shown below.

```
     1100                         1100
-)   0100     ------->       +)   1100
                              1   1000
```

Example 5.8 For case 2 where $(-A) - (+B)$, let $A = -6$ and $B = +3$ for $n = 4$. Therefore, $A + B = 9 > r^{n-1}$ (8) will produce an overflow, as shown below.

```
     1010                         1010
-)   0011     ------->       +)   1101
                              1   0111
```

The rules for subtracting in radix 2 also apply to any radix. In the following two examples, subtraction is performed in radix 16 using the *r*s complement:

Example 5.9 Let the minuend be A = 9C (156_{10}) and the subtrahend B = A4 (164_{10}), yielding a difference of F8 (-8_{10}). The *r*s complement of A4 is

$$15 - 10 = 5; \qquad 15 - 4 = 11 + 1 = \text{C}$$

```
     9C                         9C
-)   A4     ------->       +)   5C
                                F8
```

Example 5.10 Let the minuend be A = 8B7D ($35{,}709_{10}$) and the subtrahend B = A3CF ($41{,}935_{10}$), yielding a difference of E7AE ($-6{,}226_{10}$). The *r*s complement of A3CF is

$$15 - 10 = 5; \qquad 15 - 3 = \text{C}; \qquad 15 - 12 = 3; \qquad 15 - 15 = 0 + 1 = 1$$

	8B7D	→		8B7D
–)	A3CF		+)	5C31
				E7AE

5.2 Ripple-Carry Subtraction

Like fixed-point addition, subtraction in radix 2 is relatively easy compared to multiplication and division. The adder used in addition can be modified slightly to accommodate subtraction. It was stated previously that subtraction can be accomplished by adding the radix complement of the subtrahend to the minuend, where the radix complement is formed by adding 1 to the diminished-radix complement. The diminished-radix complement simply inverts all the bits of the subtrahend. This can be achieved by using the exclusive-OR to invert the bits of the subtrahend.

An inverter could be used to invert each subtrahend bit, but this would not allow the noninverted addend bits to be used for addition. The logic that inverts the subtrahend bits should also allow for addition. The exclusive-OR operation for two variables is defined as

$$A \oplus B = AB' + A'B$$

Thus, $A \oplus B = 1$ only if $A \neq B$. The truth table for the exclusive-OR function is shown in Table 5.2. Note that when $m = 1$, B is inverted; when $m = 0$, B is noninverted. Therefore, the variable m can be used as a *mode control* input to determine whether the operation is addition or subtraction. If the mode control line is zero, then the operation is addition; if the mode control line is one, then the operation is subtraction.

Table 5.2 Rules for the Exclusive-OR Operation

		b	
	$\oplus$	0	1
m	0	0	1
	1	1	0

This section presents the design of an 8-bit fixed-point ripple adder/subtractor. The design of a carry lookahead adder/subtractor is similar except that the carry logic uses the carry lookahead technique. It is desirable to have the adder unit perform both addition and subtraction since there is no advantage to having a separate adder and subtractor. A ripple-carry adder will be modified so that it can perform subtraction while still maintaining the ability to add. In order to form the 2s complement from the 1s complement, the carry-in to the low-order stage of the adder will be a 1 if subtraction is to be performed. The logic diagram is shown in Figure 5.1. Overflow is detected if the carries out of bit 6 and bit 7 are different.

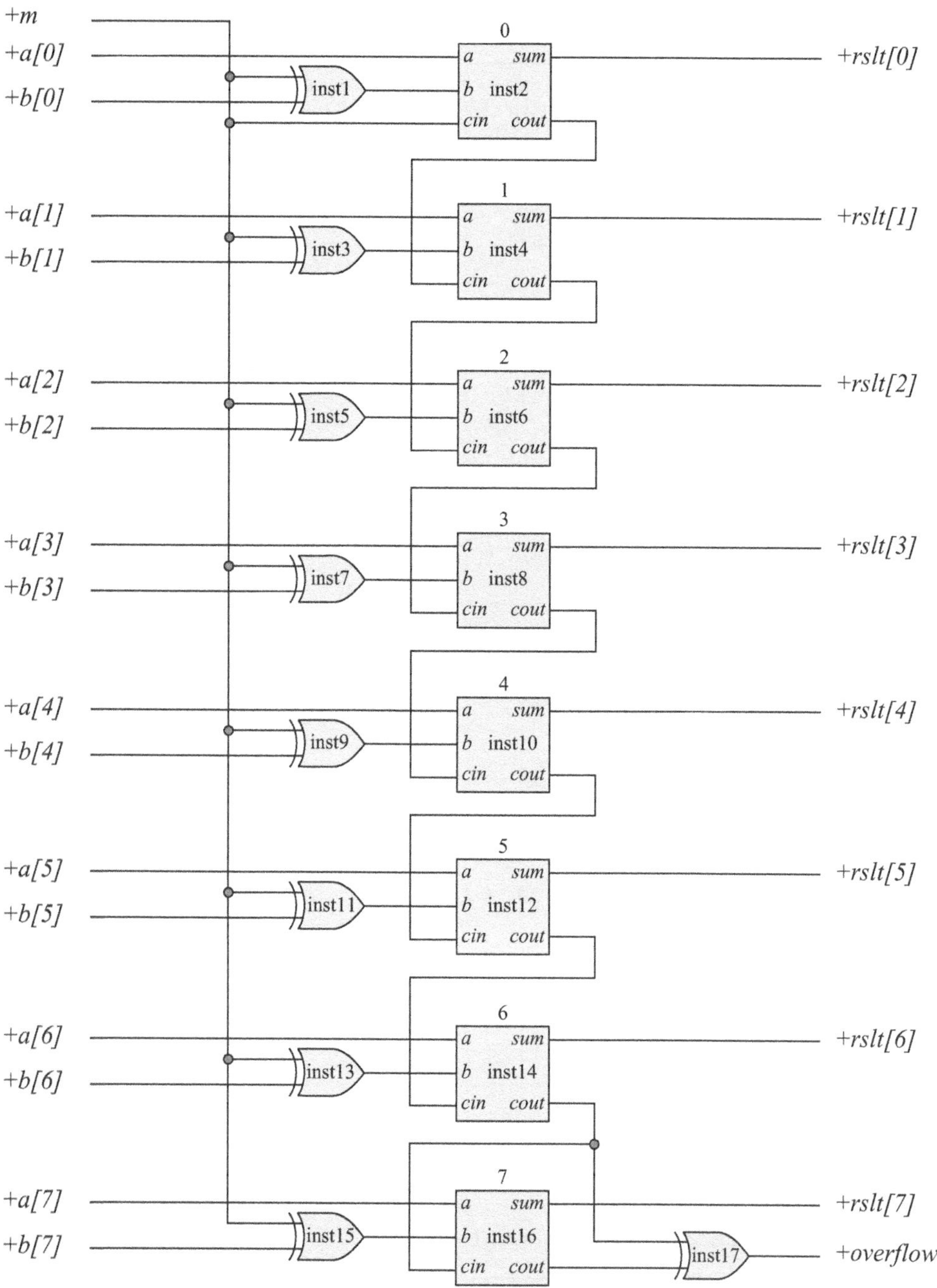

Figure 5.1 Logic diagram for an 8-bit ripple-carry adder/subtractor.

The structural module is shown in Figure 5.2. The test bench module is shown in Figure 5.3, which applies eight input vectors for addition, eight input vectors for subtraction, two input vectors to cause an overflow for addition, and two input vectors to cause an overflow for subtraction. The outputs are shown in Figure 5.4 and the waveforms are shown in Figure 5.5.

```
//structural 8-bit ripple-carry adder/subtractor
module adder_subt_ripple8 (a, b, m, rslt, cout, ovfl);

input [7:0] a, b;
input m;
output [7:0] rslt, cout;
output ovfl;

//define internal nets
wire  net1, net3, net5, net7,
      net9, net11, net13, net15;

//instantiate the xor and the full adder for FA0
xor inst1 (net1, b[0], m);

full_adder inst2 (
   .a(a[0]),
   .b(net1),
   .cin(m),
   .sum(rslt[0]),
   .cout(cout[0])
   );

//instantiate the xor and the full adder for FA1
xor inst3 (net3, b[1], m);

full_adder inst4 (
   .a(a[1]),
   .b(net3),
   .cin(cout[0]),
   .sum(rslt[1]),
   .cout(cout[1])
   );

//instantiate the xor and the full adder for FA2
xor inst5 (net5, b[2], m);

//continued on next page
```

Figure 5.2 Structural module for the 8-bit ripple-carry adder/subtractor.

```
full_adder inst6 (
   .a(a[2]),
   .b(net5),
   .cin(cout[1]),
   .sum(rslt[2]),
   .cout(cout[2])
   );

//instantiate the xor and the full adder for FA3
xor inst7 (net7, b[3], m);

full_adder inst8 (
   .a(a[3]),
   .b(net7),
   .cin(cout[2]),
   .sum(rslt[3]),
   .cout(cout[3])
   );

//instantiate the xor and the full adder for FA4
xor inst9 (net9, b[4], m);

full_adder inst10 (
   .a(a[4]),
   .b(net9),
   .cin(cout[3]),
   .sum(rslt[4]),
   .cout(cout[4])
   );

//instantiate the xor and the full adder for FA5
xor inst11 (net11, b[5], m);

full_adder inst12 (
   .a(a[5]),
   .b(net11),
   .cin(cout[4]),
   .sum(rslt[5]),
   .cout(cout[5])
   );

//instantiate the xor and the full adder for FA6
xor inst13 (net13, b[6], m);

//continued on next page
```

Figure 5.2 (Continued)

```
full_adder inst14 (
   .a(a[6]),
   .b(net13),
   .cin(cout[5]),
   .sum(rslt[6]),
   .cout(cout[6])
   );

//instantiate the xor and the full adder for FA7
xor inst15 (net15, b[7], m);

full_adder inst16 (
   .a(a[7]),
   .b(net15),
   .cin(cout[6]),
   .sum(rslt[7]),
   .cout(cout[7])
   );

//instantiate the xor gate to detect overflow
xor inst17 (ovfl, cout[6], cout[7]);

endmodule
```

Figure 5.2 (Continued)

```
//test bench for structural adder-subtractor
module adder_subt_ripple8_tb;

reg [7:0] a, b;
reg m;
wire [7:0] rslt, cout;
wire  ovfl;

initial     //display variables
$monitor ("a=%b, b=%b, m=%b, rslt=%b, ovfl=%b",
            a, b, m, rslt, ovfl);

initial     //apply input vectors
begin
//addition
   #0    a = 8'b0000_0000; b = 8'b0000_0001; m = 1'b0;
   #10   a = 8'b0010_1101; b = 8'b1100_0101; m = 1'b0;
   #10   a = 8'b0000_0110; b = 8'b0000_0001; m = 1'b0;
   #10   a = 8'b0000_0101; b = 8'b0011_0001; m = 1'b0;
//continued on next page
```

Figure 5.3 Test bench for the 8-bit ripple-carry adder/subtractor.

```
    #10    a = 8'b1000_0000; b = 8'b0101_1100; m = 1'b0;
    #10    a = 8'b1110_1101; b = 8'b0101_0101; m = 1'b0;
    #10    a = 8'b1111_1111; b = 8'b1111_1111; m = 1'b0;
    #10    a = 8'b1111_1111; b = 8'b1111_0001; m = 1'b0;

//subtraction
    #10    a = 8'b0000_0000; b = 8'b0000_0001; m = 1'b1;
    #10    a = 8'b0010_1101; b = 8'b0000_0101; m = 1'b1;
    #10    a = 8'b0000_0110; b = 8'b0000_0001; m = 1'b1;
    #10    a = 8'b0001_0101; b = 8'b0011_0001; m = 1'b1;
    #10    a = 8'b1000_0000; b = 8'b1001_1100; m = 1'b1;
    #10    a = 8'b1110_1101; b = 8'b0101_0101; m = 1'b1;
    #10    a = 8'b1111_1111; b = 8'b1111_1111; m = 1'b1;
    #10    a = 8'b1110_1111; b = 8'b1111_0001; m = 1'b1;

//overflow
    #10    a = 8'b0111_1111; b = 8'b0101_0101; m = 1'b0;
    #10    a = 8'b1010_1101; b = 8'b1011_0101; m = 1'b0;
    #10    a = 8'b0110_0110; b = 8'b1100_0001; m = 1'b1;
    #10    a = 8'b1000_0101; b = 8'b0010_0001; m = 1'b1;

    #10    $stop;
end

//instantiate the module into the test bench
adder_subt_ripple8 inst1 (
    .a(a),
    .b(b),
    .m(m),
    .rslt(rslt),
    .cout(cout),
    .ovfl(ovfl)
    );
endmodule
```

Figure 5.3 (Continued)

```
Addition
a=00000000, b=00000001, m=0, rslt=00000001, ovfl=0
a=00101101, b=11000101, m=0, rslt=11110010, ovfl=0
a=00000110, b=00000001, m=0, rslt=00000111, ovfl=0
a=00000101, b=00110001, m=0, rslt=00110110, ovfl=0
a=10000000, b=01011100, m=0, rslt=11011100, ovfl=0
a=11101101, b=01010101, m=0, rslt=01000010, ovfl=0
a=11111111, b=11111111, m=0, rslt=11111110, ovfl=0
a=11111111, b=11110001, m=0, rslt=11110000, ovfl=0//next page
```

Figure 5.4 Outputs for the 8-bit ripple-carry adder/subtractor.

```
Subtraction
a=00000000, b=00000001, m=1, rslt=11111111, ovfl=0
a=00101101, b=00000101, m=1, rslt=00101000, ovfl=0
a=00000110, b=00000001, m=1, rslt=00000101, ovfl=0
a=00010101, b=00110001, m=1, rslt=11100100, ovfl=0
a=10000000, b=10011100, m=1, rslt=11100100, ovfl=0
a=11101101, b=01010101, m=1, rslt=10011000, ovfl=0
a=11111111, b=11111111, m=1, rslt=00000000, ovfl=0
a=11101111, b=11110001, m=1, rslt=11111110, ovfl=0

Overflow for addition
a=01111111, b=01010101, m=0, rslt=11010100, ovfl=1
a=10101101, b=10110101, m=0, rslt=01100010, ovfl=1

Overflow for subtraction
a=01100110, b=11000001, m=1, rslt=10100101, ovfl=1
a=10000101, b=00100001, m=1, rslt=01100100, ovfl=1
```

Figure 5.4 (Continued)

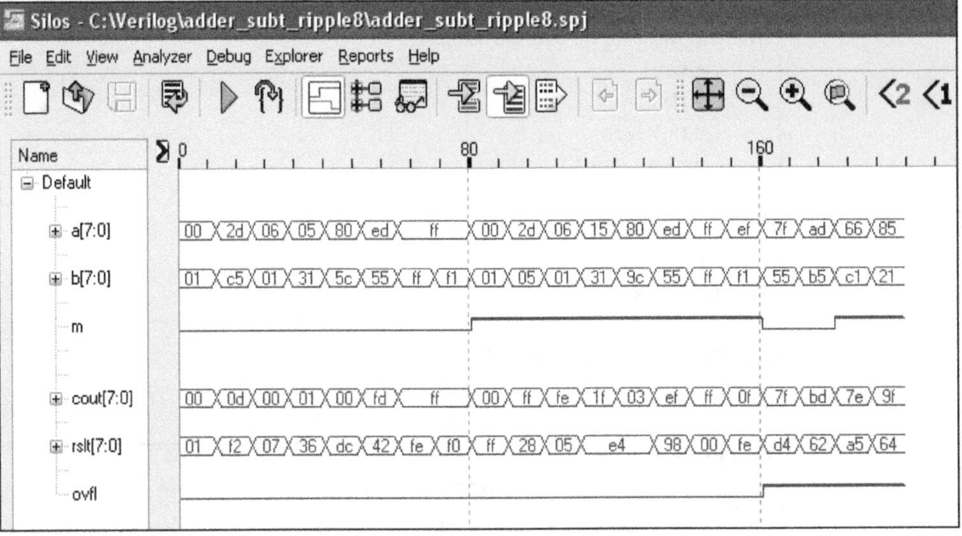

Figure 5.5 Waveforms for the 8-bit ripple-carry adder/subtractor.

5.3 Carry Lookahead Addition/Subtraction

The hardware required for fixed-point addition can be easily expanded to accommodate fixed-point subtraction. Subtraction of two operands, minuend A and subtrahend B, can be accomplished by adding the 2s complement of the subtrahend to the minuend; that is, $A - B = A + (B' + 1)$, where $(B' + 1)$ is the 2s complement of the subtrahend B. This section will present the structural design of an 8-bit carry lookahead adder/subtractor comprised of two groups with four adder stages per group.

A 2s complementer will be initially designed that produces either the uncomplemented version of the addend for addition or the 2s complement version of the subtrahend for subtraction. The 2s complementer will then be instantiated into the structural module of the carry lookahead adder/subtractor. Once the 2s complementer has been instantiated into the structural module, high-speed addition can take place to produce the sum or difference — the 2s complementer simply provides different operands to the adder.

The block diagram of the 2s complementer is shown in Figure 5.6 together with the block diagram of the adder/subtractor. The inputs are the minuend *a[7:0]*, the subtrahend *b[7:0]*, and a mode control which indicates whether the operation is to be addition (*mode* = 0) or subtraction (*mode* = 1). The outputs of the 2s complementer are the result of the operation and are labeled *rslt[7:0]*. The outputs of the adder/subtractor are the sum or difference, *sum_diff[7:0]*; the carry-out, *cout*; and an overflow indication, *ovfl*.

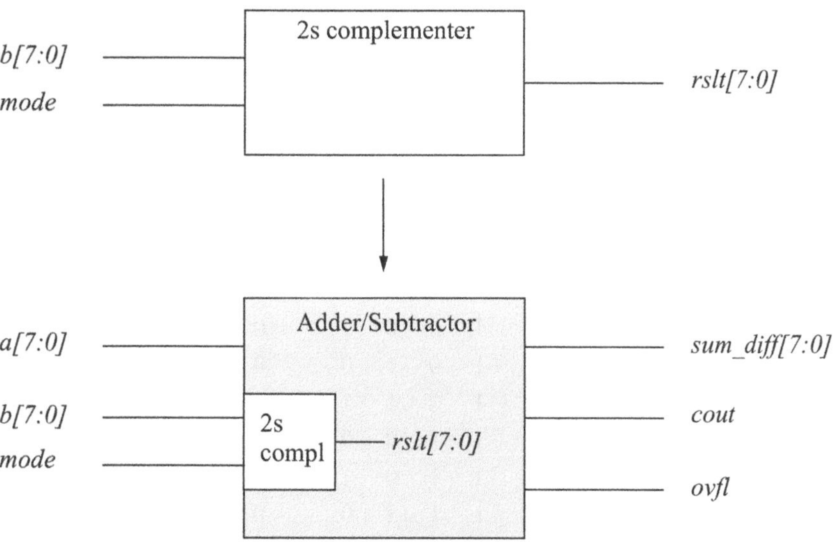

Figure 5.6 Block diagram of an 8-bit carry lookahead adder/subtractor.

The truth table for the 2s complementer is shown in Table 5.3, which lists the first eight entries and the last eight entries of the 256 possible values of eight variables. The logic diagram for the 2s complementer is shown in Figure 5.7 together with the instantiation names and the net names. The logic diagram illustrates an iterative network of identical cells, except for the low-order bit *rslt[0]* and the high-order bit *rslt[7]*.

Alternatively, a parallel network may be implemented; however, the fan-in requirements of the high-order OR gates increases proportionately. For example, the equations for net 15 and net 18, resulting from the OR gate instantiations *inst15* and *inst18*, would be as follows:

$$net15 = b[5] + b[4] + b[3] + b[2] + b[1] + b[0]$$

$$net18 = b[6] + b[5] + b[4] + b[3] + b[2] + b[1] + b[0]$$

Table 5.3 Truth Table for the 2s Complementer

Inputs *b[7:0]*								Outputs *rslt[7:0]*							
7	6	5	4	3	2	1	0	7	6	5	4	3	2	1	0
0	0	0	0	0	0	0	0	0	0	0	0	0	0	0	0
0	0	0	0	0	0	0	1	1	1	1	1	1	1	1	1
0	0	0	0	0	0	1	0	1	1	1	1	1	1	1	0
0	0	0	0	0	0	1	1	1	1	1	1	1	1	0	1
0	0	0	0	0	1	0	0	1	1	1	1	1	1	0	0
0	0	0	0	0	1	0	1	1	1	1	1	1	0	1	1
0	0	0	0	0	1	1	0	1	1	1	1	1	0	1	0
0	0	0	0	0	1	1	1	1	1	1	1	1	0	0	1
							. . .								
1	1	1	1	1	0	0	0	0	0	0	0	1	0	0	0
1	1	1	1	1	0	0	1	0	0	0	0	0	1	1	1
1	1	1	1	1	0	1	0	0	0	0	0	0	1	1	0
1	1	1	1	1	0	1	1	0	0	0	0	0	1	0	1
1	1	1	1	1	1	0	0	0	0	0	0	0	1	0	0
1	1	1	1	1	1	0	1	0	0	0	0	0	0	1	1
1	1	1	1	1	1	1	0	0	0	0	0	0	0	1	0
1	1	1	1	1	1	1	1	0	0	0	0	0	0	0	1

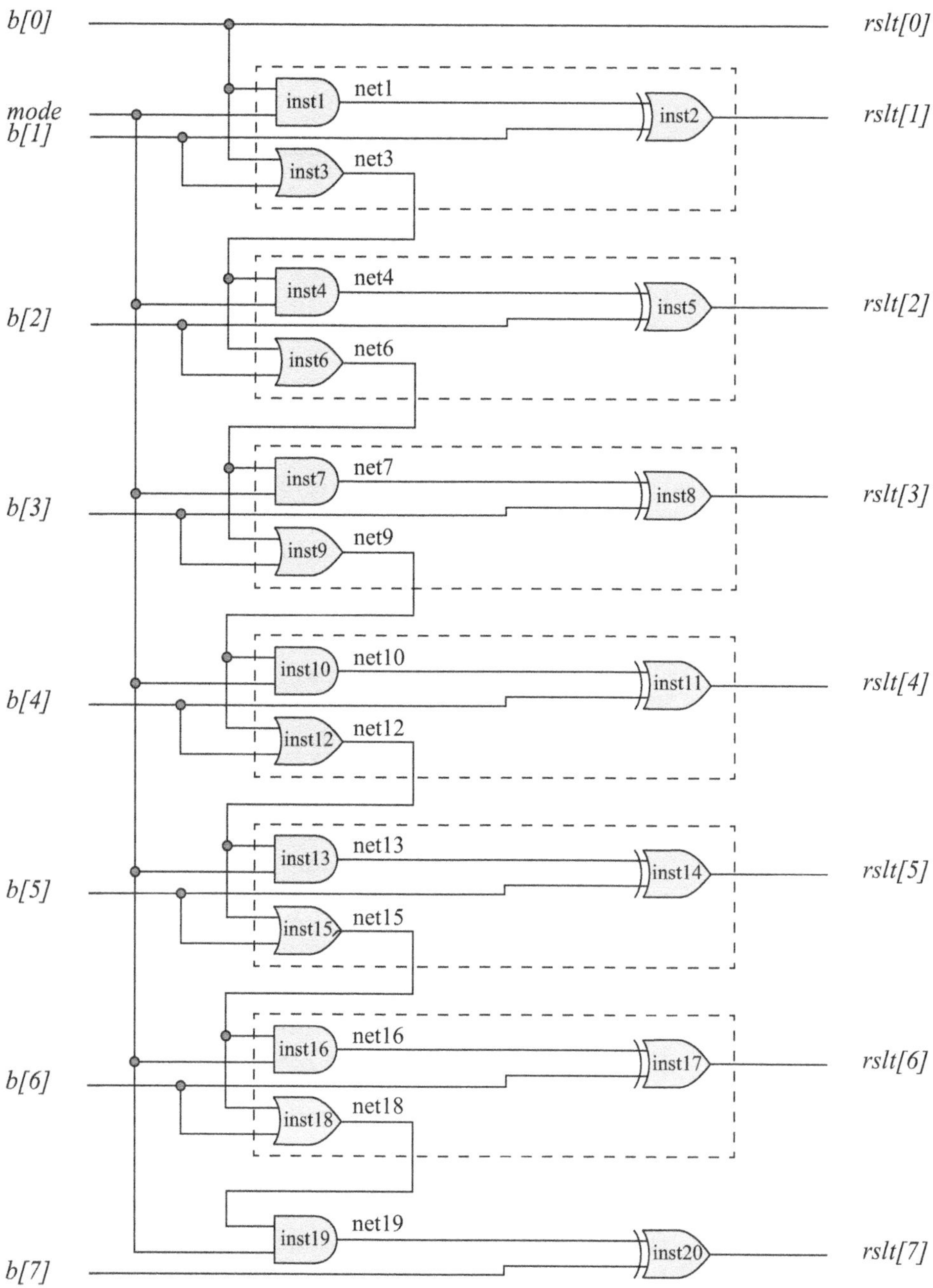

Figure 5.7 Logic diagram for the 2s complementer.

The design module for the 2s complementer is shown in Figure 5.8 using dataflow modeling and built-in primitives. The test bench, shown in Figure 5.9, applies 256 input vectors to the design module. The outputs are shown in Figure 5.10 for 40 input vectors that indicate both uncomplemented and 2s complemented versions of the input vector *b[7:0]*.

```
//dataflow and built-in primitives for an
//8-bit 2s complementer for add and subtract
module twos_compl_add_sub_8 (b, mode, rslt);

input [7:0] b;
input mode;
output [7:0] rslt;

assign rslt[0] = b[0];

and    inst1   (net1, b[0], mode);
xor    inst2   (rslt[1], b[1], net1);
or     inst3   (net3, b[1], b[0]);

and    inst4   (net4, net3, mode);
xor    inst5   (rslt[2], b[2], net4);
or     inst6   (net6, b[2], net3);

and    inst7   (net7, net6, mode);
xor    inst8   (rslt[3], b[3], net7);
or     inst9   (net9, b[3], net6);

and    inst10  (net10, net9, mode);
xor    inst11  (rslt[4], b[4], net10);
or     inst12  (net12, b[4], net9);

and    inst13  (net13, net12, mode);
xor    inst14  (rslt[5], b[5], net13);
or     inst15  (net15, b[5], net12);

and    inst16  (net16, net15, mode);
xor    inst17  (rslt[6], b[6], net16);
or     inst18  (net18, b[6], net15);

and    inst19  (net19, net18, mode);
xor    inst20  (rslt[7], b[7], net19);

endmodule
```

Figure 5.8 Design module for the 8-bit 2s complementer using dataflow and built-in primitives.

```
//test bench for 8-bit 2s complementer for add and sub
module twos_compl_add_sub_8_tb;

reg [7:0] b;
reg mode;
wire [7:0] rslt;

//apply input vectors and display variables
initial
begin: apply_stimulus
   reg [9:0] invect;
   for (invect = 0; invect < 512; invect = invect + 1)
      begin
         {b, mode} = invect [9:0];
         #10 $display ("b = %b, mode = %b, rslt = %b",
                              b, mode, rslt);
      end
end

//instantiate the module into the test bench
twos_compl_add_sub_8 inst1 (
   .b(b),
   .mode(mode),
   .rslt(rslt)
   );
endmodule
```

Figure 5.9 Test bench for the 8-bit 2s complementer.

```
b = 00000000, mode = 0, rslt = 00000000
b = 00000000, mode = 1, rslt = 00000000
b = 00000001, mode = 0, rslt = 00000001
b = 00000001, mode = 1, rslt = 11111111
b = 00000010, mode = 0, rslt = 00000010
b = 00000010, mode = 1, rslt = 11111110
b = 00000011, mode = 0, rslt = 00000011
b = 00000011, mode = 1, rslt = 11111101

b = 00000100, mode = 0, rslt = 00000100
b = 00000100, mode = 1, rslt = 11111100
b = 00000101, mode = 0, rslt = 00000101
b = 00000101, mode = 1, rslt = 11111011
b = 00000110, mode = 0, rslt = 00000110
b = 00000110, mode = 1, rslt = 11111010
b = 00000111, mode = 0, rslt = 00000111
b = 00000111, mode = 1, rslt = 11111001//continued next page
```

Figure 5.10 Outputs for the 8-bit 2s complementer.

```
b = 00011100, mode = 0, rslt = 00011100
b = 00011100, mode = 1, rslt = 11100100
b = 00011101, mode = 0, rslt = 00011101
b = 00011101, mode = 1, rslt = 11100011
b = 00011110, mode = 0, rslt = 00011110
b = 00011110, mode = 1, rslt = 11100010
b = 00011111, mode = 0, rslt = 00011111
b = 00011111, mode = 1, rslt = 11100001

b = 11110001, mode = 0, rslt = 11110001
b = 11110001, mode = 1, rslt = 00001111
b = 11110010, mode = 0, rslt = 11110010
b = 11110010, mode = 1, rslt = 00001110
b = 11110011, mode = 0, rslt = 11110011
b = 11110011, mode = 1, rslt = 00001101
b = 11110100, mode = 0, rslt = 11110100
b = 11110100, mode = 1, rslt = 00001100

b = 11111100, mode = 0, rslt = 11111100
b = 11111100, mode = 1, rslt = 00000100
b = 11111101, mode = 0, rslt = 11111101
b = 11111101, mode = 1, rslt = 00000011
b = 11111110, mode = 0, rslt = 11111110
b = 11111110, mode = 1, rslt = 00000010
b = 11111111, mode = 0, rslt = 11111111
b = 11111111, mode = 1, rslt = 00000001
```

Figure 5.10 (Continued)

A more detailed block diagram of the carry lookahead adder/subtractor is shown in Figure 5.11 depicting the two groups of four full adders per group. Group 0 consists of full adders 0 through 3; group 1 consists of full adders 4 through 7. The inputs are the augend/minuend *a[7:0]* and the addend/subtrahend *rslt[7:0]*, which are the outputs of the 2s complementer. The carry-out of group 0 is *grp0_cy*; the carry-out of group 1 is *cout*, the carry-out of the adder.

Since this is a structural design, the logic diagram of the carry lookahead adder/subtractor will be designed first using gates and full adders, then implemented with built-in primitives and module instantiations of the full adder. The logic diagram is illustrated in Figure 5.12 showing the instantiation names and the net names.

The structural module is shown in Figure 5.13, and the test bench is shown in Figure 5.14, in which several vectors are applied for *a[7:0]* and *b[7:0]* for addition (*mode* = 0) and subtraction (*mode* = 1). The outputs are shown in Figure 5.15 displaying the two operands, the state of the *mode* input, the sum or difference, and the state of the overflow output.

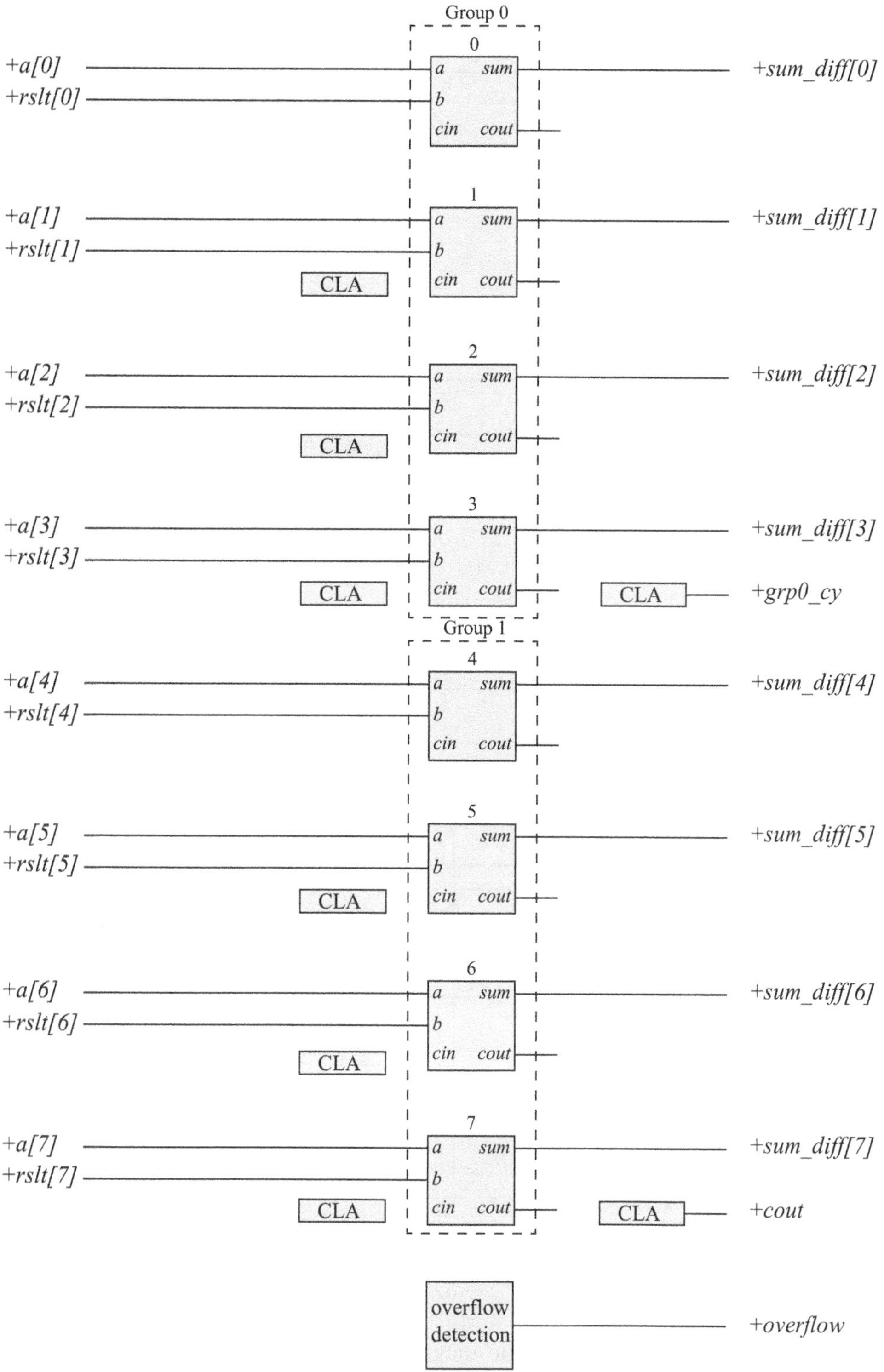

Figure 5.11 Detailed block diagram of an 8-bit carry lookahead adder/subtractor.

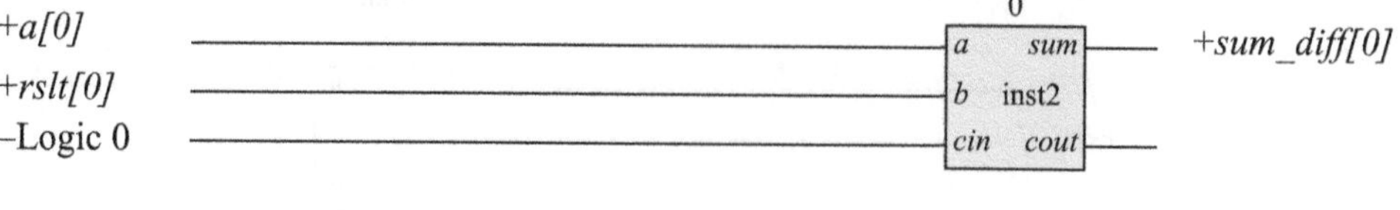

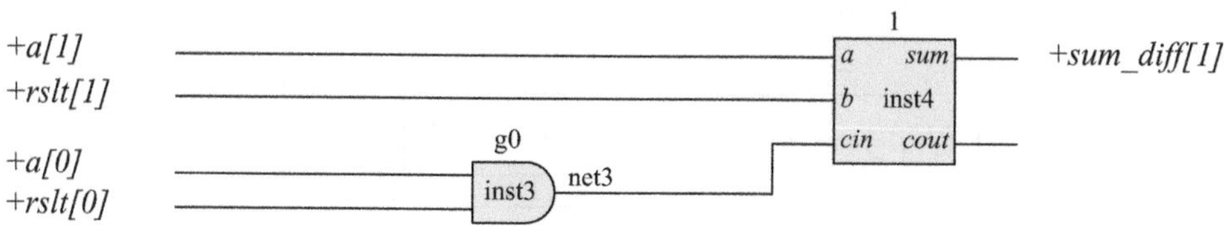

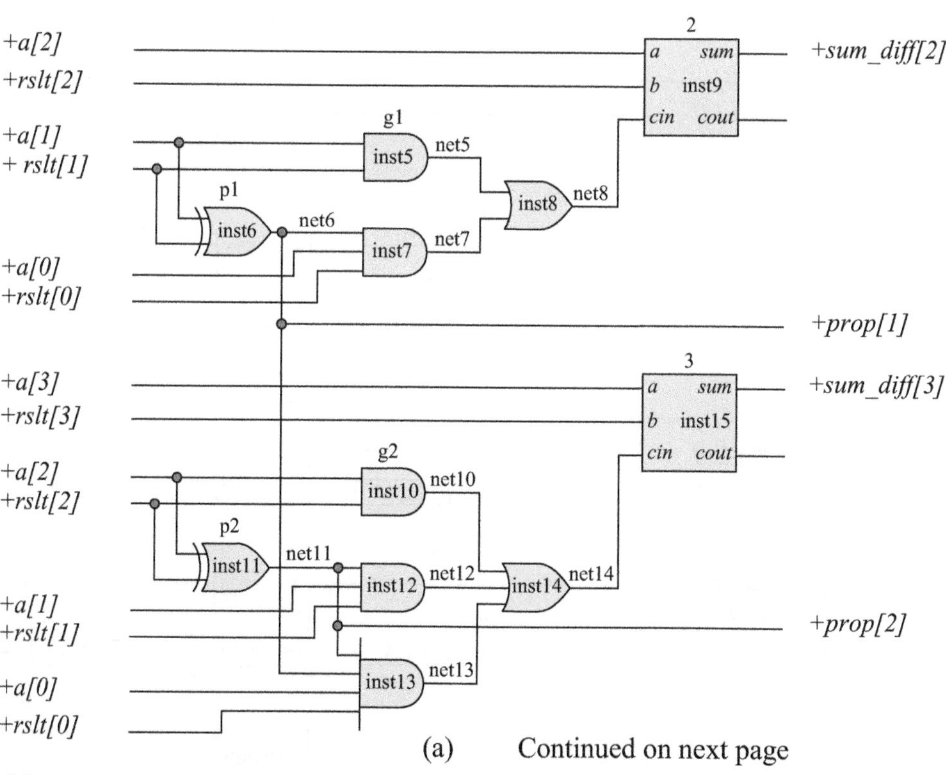

(a) Continued on next page

Figure 5.12 Logic diagram for the 8-bit carry lookahead adder/subtractor: (a) group 0 logic and full adders; (b) group 0 carry generation; (c) group 1 logic and full adders; and (d) group 1 carry generation and overflow detection.

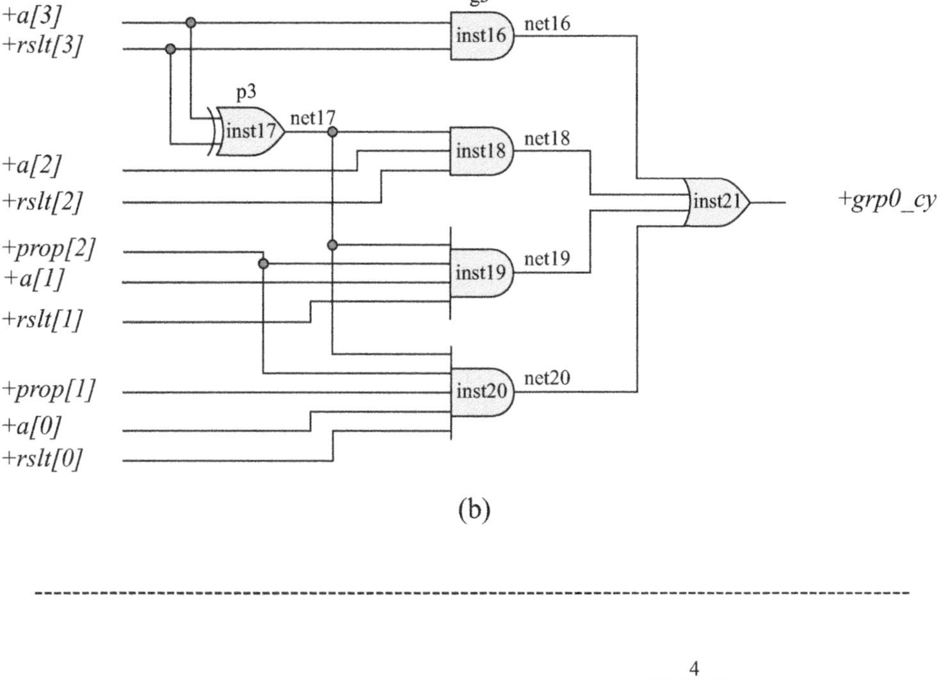

(b)

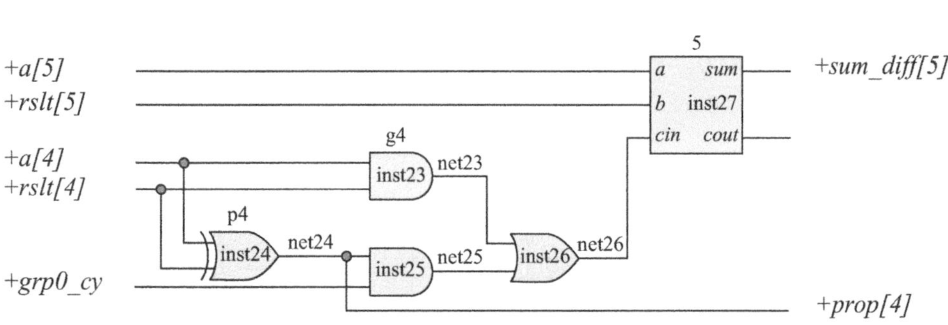

(c) Continued on next page

Figure 5.12 (Continued)

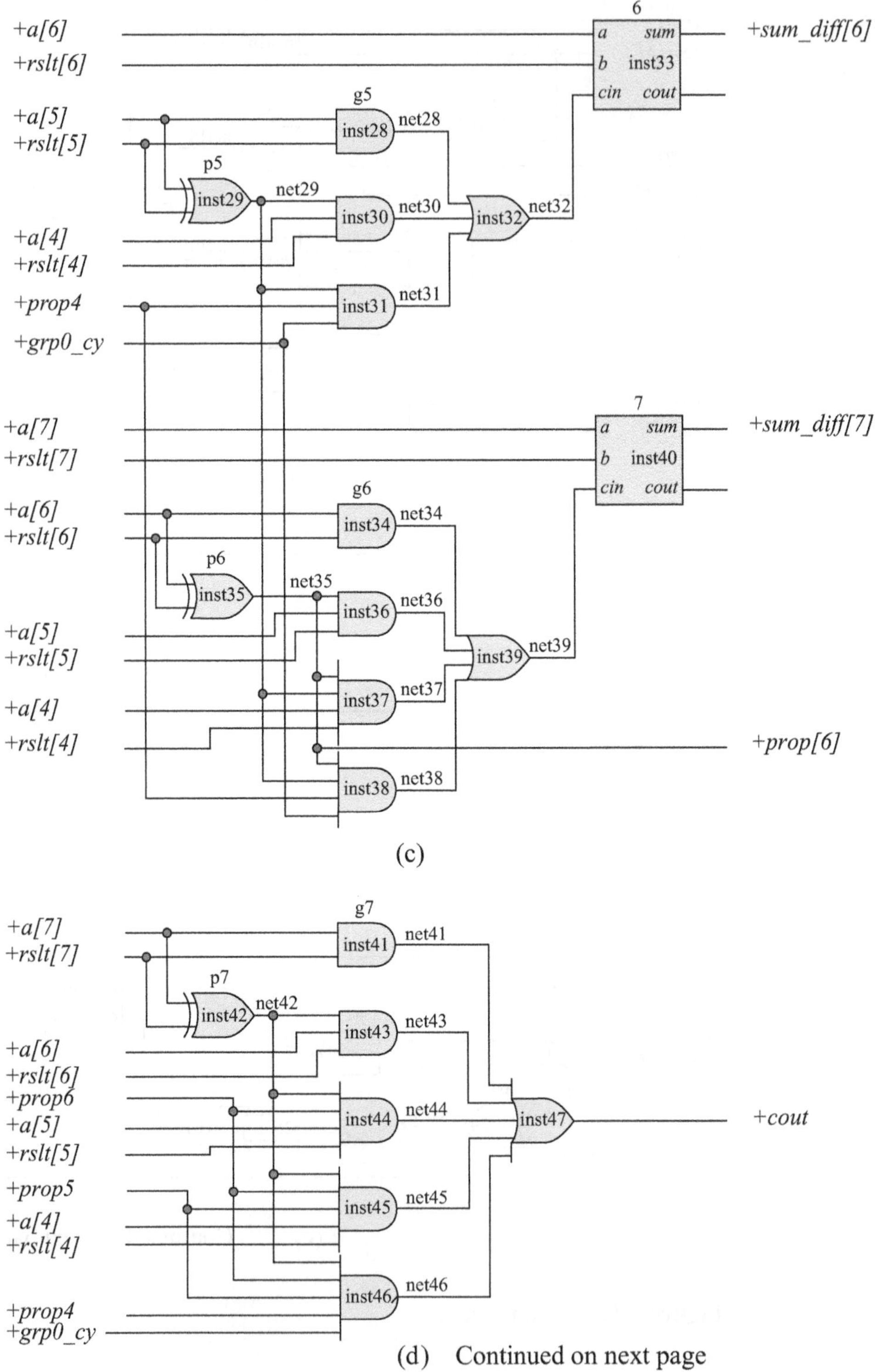

(c)

(d) Continued on next page

Figure 5.12 (Continued)

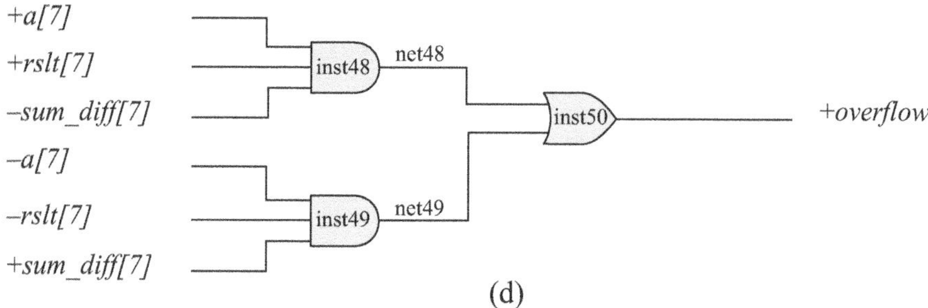

Figure 5.12 (Continued)

Once the outputs of the 2s complementer are available, all carries generated by the carry lookahead logic are available after 5 gate delays. The carry lookahead technique requires a significant increase in the amount of logic compared to the ripple-carry method, but offers a significant increase in speed.

```
//structural and built-in primitives for an
//8-bit carry lookahead adder/subtractor
module add_sub_cla8 (a, b, mode, sum_diff, cout, ovfl);

input [7:0] a, b;
input mode;

output [7:0] sum_diff;
output cout, ovfl;

//define internal nets
wire [7:0] rslt;
wire  net3, net5, net6, net7, net8,
      net10, net11, net12, net13, net14,
      net16, net17, net18, net19, net20,
      net23, net24, net25, net26,
      net28, net29, net30, net31, net32,
      net34, net35, net36, net37, net38, net39,
      net41, net42, net43, net44, net45, net46,
      net48, net49;

//instantiate the 2s complementer
twos_compl_add_sub_8 inst1 (
   .b(b),
   .mode(mode),
   .rslt(rslt)
   );          //continued on next page
```

Figure 5.13 Structural module for 8-bit the carry lookahead adder/subtractor.

```
//-------------------------------------------------
//instantiate the logic for group 0
//-------------------------------------------------
//instantiate the logic for FA0 of group 0
full_adder inst2 (
   .a(a[0]),
   .b(rslt[0]),
   .cin(1'b0),
   .sum(sum_diff[0])
   );

//instantiate the logic for FA1 of group 0
and   inst3 (net3, a[0], rslt[0]); //g0

full_adder inst4 (
   .a(a[1]),
   .b(rslt[1]),
   .cin(net3),
   .sum(sum_diff[1])
   );

//instantiate the logic for FA2 of group 0
and   inst5 (net5, a[1], rslt[1]); //g1
xor   inst6 (net6, a[1], rslt[1]); //p1
and   inst7 (net7, net6, a[0], rslt[0]);
or    inst8 (net8, net5, net7); //g1,p1,g0

full_adder inst9 (
   .a(a[2]),
   .b(rslt[2]),
   .cin(net8),
   .sum(sum_diff[2])
   );

//instantiate the logic for FA3 of group 0
and   inst10 (net10, a[2], rslt[2]); //g2
xor   inst11 (net11, a[2], rslt[2]); //p2
and   inst12 (net12, net11, a[1], rslt[1]); //p2,g1
and   inst13 (net13, net11, net6, a[0], rslt[0]); //p2,p1,g0
or    inst14 (net14, net10, net12, net13);//g2,p2,g1,p2,p1,g0

full_adder inst15 (
   .a(a[3]),
   .b(rslt[3]),
   .cin(net14),
   .sum(sum_diff[3])
   );                    //continued on next page
```

Figure 5.13 (Continued)

```
//instantiate the logic for the carry from group 0 grp0_cy
and   inst16 (net16, a[3], rslt[3]); //g3
xor   inst17 (net17, a[3], rslt[3]); //p3
and   inst18 (net18, net17, a[2], rslt[2]); //p3,g2
and   inst19 (net19, net17, net11, a[1], rslt[1]); //p3,p2,g1
and   inst20 (net20, net17, net11, net6, a[0], rslt[0]);
                                    //p3,p2,p1,g0
or    inst21 (grp0_cy, net16, net18, net19, net20);
                              //g3,p3,g2,p3,p2,g1,p3,p2,p1,g0

//--------------------------------------------------
//instantiate the logic for group 1
//--------------------------------------------------
//instantiate the logic for FA4 of group 1
full_adder inst22 (
   .a(a[4]),
   .b(rslt[4]),
   .cin(grp0_cy),
   .sum(sum_diff[4])
   );

//instantiate the logic for FA5 of group 1
and   inst23 (net23, a[4], rslt[4]); //g4
xor   inst24 (net24, a[4], rslt[4]); //p4
and   inst25 (net25, net24, grp0_cy); //p4,grp0_cy
or    inst26 (net26, net23, net25); //g4,p4,grp0_cy

full_adder inst27 (
   .a(a[5]),
   .b(rslt[5]),
   .cin(net26),
   .sum(sum_diff[5])
   );

//instantiate the logic for FA6 of group 1
and   inst28 (net28, a[5], rslt[5]); //g5
xor   inst29 (net29, a[5], rslt[5]); //p5
and   inst30 (net30, net29, a[4], rslt[4]); //p5,g4
and   inst31 (net31, net29, net24, grp0_cy); //p5,p4,grp0_cy
or    inst32 (net32, net28, net30, net31);
                           //g5,p5,g4,p5,p4,grp0_cy
full_adder inst33 (
   .a(a[6]),
   .b(rslt[6]),
   .cin(net32),
   .sum(sum_diff[6])
   );                      //continued on next page
```

Figure 5.13 (Continued)

```
//instantiate the logic for FA7 of group 1
and    inst34 (net34, a[6], rslt[6]); //g6
xor    inst35 (net35, a[6], rslt[6]); //p6
and    inst36 (net36, net35, a[5], rslt[5]); //p6,g5
and    inst37 (net37, net35, net29, a[4], rslt[4]);
                    //p6,p5,g4
and    inst38 (net38, net35, net29, net24, grp0_cy);
                    //p6,p5,p4,grp0_cy
or     inst39 (net39, net34, net36, net37, net38);
                    //g6,p6,g5,p6,p5,g4,p6,p5,p4,grp0_cy


full_adder inst40 (
   .a(a[7]),
   .b(rslt[7]),
   .cin(net39),
   .sum(sum_diff[7])
   );


//instantiate the logic for the carry from group 1 cout
and    inst41 (net41, a[7], rslt[7]); //g7
xor    inst42 (net42, a[7], rslt[7]); //p7
and    inst43 (net43, net42, a[6], rslt[6]); //p7,g6
and    inst44 (net44, net42, net35, a[5], rslt[5]);
                      //p7,p6,g5
and    inst45 (net45, net42, net35, net29, a[4], rslt[4]);
                      //p7,p6,p5,g4
and    inst46 (net46, net42, net35, net29, net24, grp0_cy);
                      //p7,p6,p5,p4,grp0_cy
or     inst47 (cout, net41, net43, net44, net45, net46);
                 //g7,p7,g6,p7,p6,g5,p7,p6,p5,g4,p7,p6,p5,p4,grp0_cy


//---------------------------------------------------------
//instantiate the logic to detect overflow
//---------------------------------------------------------
and    inst48 (net48, a[7], rslt[7], ~sum_diff[7]);
and    inst49 (net49, ~a[7], ~rslt[7], sum_diff[7]);
or     inst50 (ovfl, net48, net49);

endmodule
```

Figure 5.13 (Continued)

```
//test bench for the 8-bit
//carry lookahead adder/subtractor
module add_sub_cla8_tb;

reg [7:0] a, b;
reg mode;
wire [7:0] sum_diff;
wire cout, ovfl;

//display variables
initial
$monitor ("a=%b, b=%b, mode=%b, sum_diff=%b, ovfl=%b",
            a, b, mode, sum_diff, ovfl);

//apply input vectors
initial
begin
   #0    a = 8'b0000_0000; b = 8'b0000_0001; mode = 1'b0;
   #10   a = 8'b0000_0000; b = 8'b0000_0001; mode = 1'b1;

   #10   a = 8'b0000_0110; b = 8'b0000_0011; mode = 1'b0;
   #10   a = 8'b0000_0110; b = 8'b0000_0011; mode = 1'b1;

   #10   a = 8'b0000_1100; b = 8'b0000_0101; mode = 1'b0;
   #10   a = 8'b0000_1100; b = 8'b0000_0101; mode = 1'b1;

   #10   a = 8'b0000_0001; b = 8'b1111_1001; mode = 1'b0;
   #10   a = 8'b0000_0001; b = 8'b1111_1001; mode = 1'b1;

   #10   a = 8'b0000_0001; b = 8'b1000_0001; mode = 1'b0;
   #10   a = 8'b0000_0001; b = 8'b1000_0001; mode = 1'b1;

   #10   a = 8'b1111_0000; b = 8'b0000_0001; mode = 1'b0;
   #10   a = 8'b1111_0000; b = 8'b0000_0001; mode = 1'b1;

   #10   a = 8'b0110_1101; b = 8'b0100_0101; mode = 1'b0;
   #10   a = 8'b0010_1101; b = 8'b0000_0101; mode = 1'b1;

   #10   a = 8'b0000_0110; b = 8'b0000_0001; mode = 1'b0;
   #10   a = 8'b0000_0110; b = 8'b0000_0001; mode = 1'b1;

   #10   a = 8'b0001_0101; b = 8'b0011_0001; mode = 1'b0;
   #10   a = 8'b0001_0101; b = 8'b0011_0001; mode = 1'b1;

   #10   a = 8'b1000_0000; b = 8'b1001_1100; mode = 1'b0;
   #10   a = 8'b1000_0000; b = 8'b1001_1100; mode = 1'b1;
//continued on next page
```

Figure 5.14 Test bench for the 8-bit carry lookahead adder/subtractor.

```
   #10   a = 8'b1110_1101; b = 8'b0101_0101; mode = 1'b0;
   #10   a = 8'b1110_1101; b = 8'b0101_0101; mode = 1'b1;

   #10   a = 8'b1111_1111; b = 8'b1111_1111; mode = 1'b0;
   #10   a = 8'b1111_1111; b = 8'b1111_1111; mode = 1'b1;

   #10   a = 8'b1110_1111; b = 8'b1111_0001; mode = 1'b0;
   #10   a = 8'b1110_1111; b = 8'b1111_0001; mode = 1'b1;

   #10   a = 8'b0111_1111; b = 8'b0101_0101; mode = 1'b0;
   #10   a = 8'b1010_1101; b = 8'b1011_0101; mode = 1'b1;

   #10   a = 8'b0110_0110; b = 8'b1100_0001; mode = 1'b0;
   #10   a = 8'b0110_0110; b = 8'b1100_0001; mode = 1'b1;

   #10   a = 8'b1000_0101; b = 8'b0010_0001; mode = 1'b0;
   #10   a = 8'b1000_0101; b = 8'b0010_0001; mode = 1'b1;

   #10   a = 8'b0111_0110; b = 8'b1101_0101; mode = 1'b0;
   #10   a = 8'b0111_0110; b = 8'b1101_0101; mode = 1'b1;

   #10   a = 8'b0110_0111; b = 8'b1110_0111; mode = 1'b0;
   #10   a = 8'b0110_0111; b = 8'b1110_0111; mode = 1'b1;

   #10   a = 8'b1111_1111; b = 8'b1111_1111; mode = 1'b0;
   #10   a = 8'b1111_1111; b = 8'b1111_1111; mode = 1'b1;

   #10   $stop;

end


//instantiate the module into the test bench
add_sub_cla8 inst1 (
   .a(a),
   .b(b),
   .mode(mode),
   .sum_diff(sum_diff),
   .cout(cout),
   .ovfl(ovfl)
   );

endmodule
```

Figure 5.14 (Continued)

```
a = augend/minuend
b = addend/subtrahend

mode = 0 add
mode = 1 subtract

sum_diff is in 2s complement

a=00000000, b=00000001, mode=0, sum_diff=00000001, ovfl=0
a=00000000, b=00000001, mode=1, sum_diff=11111111, ovfl=0
a=00000110, b=00000011, mode=0, sum_diff=00001001, ovfl=0
a=00000110, b=00000011, mode=1, sum_diff=00000011, ovfl=0
a=00001100, b=00000101, mode=0, sum_diff=00010001, ovfl=0
a=00001100, b=00000101, mode=1, sum_diff=00000111, ovfl=0
a=00000001, b=11111001, mode=0, sum_diff=11111010, ovfl=0
a=00000001, b=11111001, mode=1, sum_diff=00001000, ovfl=0
a=00000001, b=10000001, mode=0, sum_diff=10000010, ovfl=0
a=00000001, b=10000001, mode=1, sum_diff=10000000, ovfl=1
a=11110000, b=00000001, mode=0, sum_diff=11110001, ovfl=0
a=11110000, b=00000001, mode=1, sum_diff=11101111, ovfl=0
a=01101101, b=01000101, mode=0, sum_diff=10110010, ovfl=1
a=00101101, b=00000101, mode=1, sum_diff=00101000, ovfl=0
a=00000110, b=00000001, mode=0, sum_diff=00000111, ovfl=0
a=00000110, b=00000001, mode=1, sum_diff=00000101, ovfl=0
a=00010101, b=00110001, mode=0, sum_diff=01000110, ovfl=0
a=00010101, b=00110001, mode=1, sum_diff=11100100, ovfl=0
a=10000000, b=10011100, mode=0, sum_diff=00011100, ovfl=1
a=10000000, b=10011100, mode=1, sum_diff=11100100, ovfl=0
a=11101101, b=01010101, mode=0, sum_diff=01000010, ovfl=0
a=11101101, b=01010101, mode=1, sum_diff=10011000, ovfl=0
a=11111111, b=11111111, mode=0, sum_diff=11111110, ovfl=0
a=11111111, b=11111111, mode=1, sum_diff=00000000, ovfl=0
a=11101111, b=11110001, mode=0, sum_diff=11100000, ovfl=0
a=11101111, b=11110001, mode=1, sum_diff=11111110, ovfl=0
a=01111111, b=01010101, mode=0, sum_diff=11010100, ovfl=1
a=10101101, b=10110101, mode=1, sum_diff=11111000, ovfl=0
a=01100110, b=11000001, mode=0, sum_diff=00100111, ovfl=0
a=01100110, b=11000001, mode=1, sum_diff=10100101, ovfl=1
a=10000101, b=00100001, mode=0, sum_diff=10100110, ovfl=0
a=10000101, b=00100001, mode=1, sum_diff=01100100, ovfl=1
a=01110110, b=11010101, mode=0, sum_diff=01001011, ovfl=0
a=01110110, b=11010101, mode=1, sum_diff=10100001, ovfl=1
a=01100111, b=11100111, mode=0, sum_diff=01001110, ovfl=0
a=01100111, b=11100111, mode=1, sum_diff=10000000, ovfl=1
a=11111111, b=11111111, mode=0, sum_diff=11111110, ovfl=0
a=11111111, b=11111111, mode=1, sum_diff=00000000, ovfl=0
```

Figure 5.15 Outputs for the 8-bit carry lookahead adder/subtractor.

5.4 Behavioral Addition/Subtraction

If high speed is not a requirement for an adder/subtractor, then a behavioral design module may suffice. Behavioral modeling describes the *behavior* of the system and specifies the architectural implementation of the system; however, it does not describe the implementation of the design at the gate level. This represents a higher level of abstraction than other modeling methods and relinquishes the logic design details to the synthesis tool.

This section will design an 8-bit adder/subtractor using behavioral modeling. The inputs are the augend/minuend *a[7:0]*, the addend/subtrahend *b[7:0]*, and a *mode* control input to determine the operation to be performed, where addition is defined as *mode* = 0 and subtraction is defined as *mode* = 1. There are two outputs: the result of an operation, *rslt[7:0]*, and an overflow indication, *ovfl*.

If the operation is addition, then overflow is defined as shown in Equation 5.1. If the operation is subtraction, then overflow is defined as shown in Equation 5.2, where the variable *neg_b[7]* is the sign bit of the 2s complement of operand *B*.

$$\text{Overflow} = a[7]\ b[7]\ rslt[7]' + a[7]'\ b[7]'\ rslt[7] \tag{5.1}$$

$$\text{Overflow} = a[7]\ neg\_b[7]\ rslt[7]' + a[7]'\ neg\_b[7]'\ rslt[7] \tag{5.2}$$

Since behavioral modeling uses the **always** procedural construct statement, the target variables are declared as type **reg**. The behavioral module is shown in Figure 5.16, which specifies an internal register *neg_b[7:0]* = *B*' + 1 to be used in the overflow equation. Blocking assignments are used in the **begin . . . end** blocks because the statements for *rslt* and *ovfl* are combinational circuits. This also blocks execution of the overflow statement until the result is obtained. The test bench is shown in Figure 5.17, providing several input vectors for the operands *A* and *B*. The outputs are shown in Figure 5.18.

```
//behavioral 8-bit adder/subtractor
module add_sub_bh (a, b, mode, rslt, ovfl);

input [7:0] a, b;
input  mode;
output [7:0] rslt;
output ovfl;            //continued on next page
```

Figure 5.16 Behavioral module for the 8-bit adder/subtractor.

```
wire [7:0] a, b;
wire mode;
reg [7:0] rslt;
reg ovfl;

reg [7:0] neg_b = ~b + 1;

always @ (a or b or mode)
begin
   if (mode == 0)      //add
      begin
         rslt = a + b;
         ovfl =(a[7] & b[7] & ~rslt[7]) |
               (~a[7] & ~b[7] & rslt[7]);
      end
   else                //subtract
      begin
         rslt = a + neg_b;
         ovfl =(a[7] & neg_b[7] & ~rslt[7]) |
               (~a[7] & ~neg_b[7] & rslt[7]);
      end
end

endmodule
```

Figure 5.16 (Continued)

```
//test bench for the 8-bit
//carry lookahead adder/subtractor
module add_sub_bh_tb;

reg [7:0] a, b;
reg mode;

wire [7:0] rslt;
wire ovfl;

//display variables
initial
$monitor ("a=%b, b=%b, mode=%b, result=%b, ovfl=%b",
            a, b, mode, rslt, ovfl);

//continued on next page
```

Figure 5.17 Test bench for the 8-bit adder/subtractor.

```
//apply input vectors
initial
begin
   #0    a = 8'b0000_0000; b = 8'b0000_0001; mode = 1'b0;
   #10   a = 8'b0000_0000; b = 8'b0000_0001; mode = 1'b1;

   #10   a = 8'b0000_0110; b = 8'b0000_0011; mode = 1'b0;
   #10   a = 8'b0000_0110; b = 8'b0000_0011; mode = 1'b1;

   #10   a = 8'b0000_1100; b = 8'b0000_0101; mode = 1'b0;
   #10   a = 8'b0000_1100; b = 8'b0000_0101; mode = 1'b1;

   #10   a = 8'b0000_0001; b = 8'b1111_1001; mode = 1'b0;
   #10   a = 8'b0000_0001; b = 8'b1111_1001; mode = 1'b1;

   #10   a = 8'b0000_0001; b = 8'b1000_0001; mode = 1'b0;
   #10   a = 8'b0000_0001; b = 8'b1000_0001; mode = 1'b1;

   #10   a = 8'b1111_0000; b = 8'b0000_0001; mode = 1'b0;
   #10   a = 8'b1111_0000; b = 8'b0000_0001; mode = 1'b1;

   #10   a = 8'b0110_1101; b = 8'b0100_0101; mode = 1'b0;
   #10   a = 8'b0010_1101; b = 8'b0000_0101; mode = 1'b1;

   #10   a = 8'b0000_0110; b = 8'b0000_0001; mode = 1'b0;
   #10   a = 8'b0000_0110; b = 8'b0000_0001; mode = 1'b1;

   #10   a = 8'b0001_0101; b = 8'b0011_0001; mode = 1'b0;
   #10   a = 8'b0001_0101; b = 8'b0011_0001; mode = 1'b1;

   #10   a = 8'b1000_0000; b = 8'b1001_1100; mode = 1'b0;
   #10   a = 8'b1000_0000; b = 8'b1001_1100; mode = 1'b1;

   #10   a = 8'b1110_1101; b = 8'b0101_0101; mode = 1'b0;
   #10   a = 8'b1110_1101; b = 8'b0101_0101; mode = 1'b1;

   #10   a = 8'b1111_1111; b = 8'b1111_1111; mode = 1'b0;
   #10   a = 8'b1111_1111; b = 8'b1111_1111; mode = 1'b1;

   #10   a = 8'b1110_1111; b = 8'b1111_0001; mode = 1'b0;
   #10   a = 8'b1110_1111; b = 8'b1111_0001; mode = 1'b1;

   #10   a = 8'b0111_1111; b = 8'b0101_0101; mode = 1'b0;
   #10   a = 8'b1010_1101; b = 8'b1011_0101; mode = 1'b1;

//continued on next page
```

Figure 5.17 (Continued)

```
    #10   a = 8'b0110_0110; b = 8'b1100_0001; mode = 1'b0;
    #10   a = 8'b0110_0110; b = 8'b1100_0001; mode = 1'b1;

    #10   a = 8'b1000_0101; b = 8'b0010_0001; mode = 1'b0;
    #10   a = 8'b1000_0101; b = 8'b0010_0001; mode = 1'b1;

    #10   a = 8'b0111_0110; b = 8'b1101_0101; mode = 1'b0;
    #10   a = 8'b0111_0110; b = 8'b1101_0101; mode = 1'b1;

    #10   a = 8'b0110_0111; b = 8'b1110_0111; mode = 1'b0;
    #10   a = 8'b0110_0111; b = 8'b1110_0111; mode = 1'b1;

    #10   a = 8'b1111_1111; b = 8'b1111_1111; mode = 1'b0;
    #10   a = 8'b1111_1111; b = 8'b1111_1111; mode = 1'b1;

    #10   $stop;
end

//instantiate the module into the test bench
add_sub_bh inst1 (
    .a(a),
    .b(b),
    .mode(mode),
    .rslt(rslt),
    .ovfl(ovfl)
    );

endmodule
```

Figure 5.17 (Continued)

```
a=00000000, b=00000001, mode=0, result=00000001, ovfl=0
a=00000000, b=00000001, mode=1, result=11111111, ovfl=0
a=00000110, b=00000011, mode=0, result=00001001, ovfl=0
a=00000110, b=00000011, mode=1, result=00000011, ovfl=0
a=00001100, b=00000101, mode=0, result=00010001, ovfl=0
a=00001100, b=00000101, mode=1, result=00000111, ovfl=0
a=00000001, b=11111001, mode=0, result=11111010, ovfl=0
a=00000001, b=11111001, mode=1, result=00001000, ovfl=0
a=00000001, b=10000001, mode=0, result=10000010, ovfl=0
a=00000001, b=10000001, mode=1, result=10000000, ovfl=1
a=11110000, b=00000001, mode=0, result=11110001, ovfl=0
a=11110000, b=00000001, mode=1, result=11101111, ovfl=0
//continued on next page
```

Figure 5.18 Outputs for the 8-bit adder/subtractor.

```
a=01101101, b=01000101, mode=0, result=10110010, ovfl=1
a=00101101, b=00000101, mode=1, result=00101000, ovfl=0
a=00000110, b=00000001, mode=0, result=00000111, ovfl=0
a=00000110, b=00000001, mode=1, result=00000101, ovfl=0
a=00010101, b=00110001, mode=0, result=01000110, ovfl=0
a=00010101, b=00110001, mode=1, result=11100100, ovfl=0
a=10000000, b=10011100, mode=0, result=00011100, ovfl=1
a=10000000, b=10011100, mode=1, result=11100100, ovfl=0
a=11101101, b=01010101, mode=0, result=01000010, ovfl=0
a=11101101, b=01010101, mode=1, result=10011000, ovfl=0
a=11111111, b=11111111, mode=0, result=11111110, ovfl=0
a=11111111, b=11111111, mode=1, result=00000000, ovfl=0
a=11101111, b=11110001, mode=0, result=11100000, ovfl=0
a=11101111, b=11110001, mode=1, result=11111110, ovfl=0
a=01111111, b=01010101, mode=0, result=11010100, ovfl=1
a=10101101, b=10110101, mode=1, result=11111000, ovfl=0
a=01100110, b=11000001, mode=0, result=00100111, ovfl=0
a=01100110, b=11000001, mode=1, result=10100101, ovfl=1
a=10000101, b=00100001, mode=0, result=10100110, ovfl=0
a=10000101, b=00100001, mode=1, result=01100100, ovfl=1
a=01110110, b=11010101, mode=0, result=01001011, ovfl=0
a=01110110, b=11010101, mode=1, result=10100001, ovfl=1
a=01100111, b=11100111, mode=0, result=01001110, ovfl=0
a=01100111, b=11100111, mode=1, result=10000000, ovfl=1
a=11111111, b=11111111, mode=0, result=11111110, ovfl=0
a=11111111, b=11111111, mode=1, result=00000000, ovfl=0
```

Figure 5.18 (Continued)

5.5 Problems

5.1 Indicate whether the operations shown below produce an overflow. The operands are in 2s complement representation.

(a)
```
    0111 1101
–)  0011 0111
```

(b)
```
    1101 0101
–)  0100 1111
```

5.2 Perform the operation of subtraction on the operands shown below, which are in radix complementation for radix 3.

$$\begin{array}{rl} & 0\,2\,0\,2\,1_3 \\ -) & \underline{2\,2\,1\,0\,0_3} \end{array}$$

5.3 Let A and B be two fixed-point binary integers in 2s complement representation as shown below, where A' and B' are the 1s complement of A and B, respectively. Perform the indicated operations and show the result in eight bits.

$A = 1011\ 0001$ $\qquad$ $B = 1110\ 0100$

$A' + 1 - B' + 1$

5.4 Perform the following binary subtraction using the 1s complement method:

$$\begin{array}{rl} & 1100\ 1100 \\ -) & \underline{0011\ 0011} \end{array}$$

5.5 Perform the operation shown below on the two operands, which are in radix complementation for radix 8. Verify the answer by converting to radix 10.

$$\begin{array}{rl} & 0\,4\,3\,5\,7_8 \\ -) & \underline{0\,2\,6\,1\,2_8} \end{array}$$

5.6 Perform the arithmetic operations shown below with fixed-point binary numbers in 2s complement representation. In each case, indicate if there is an overflow.

(a)
$$\begin{array}{rl} & 1001\ 1000 \\ -) & \underline{0010\ 0010} \end{array}$$

(b)
$$\begin{array}{rl} & 0011\ 0110 \\ -) & \underline{1110\ 0011} \end{array}$$

5.7 Indicate whether the operands shown below generate an overflow for the indicated operations. The numbers are in 2s complement representation.

(a)
$$\begin{array}{rl} & 1011\ 0110 \\ -) & \underline{0101\ 1101} \end{array}$$

(b)
$$\begin{array}{rl} & 0111\ 0011 \\ -) & \underline{1000\ 1100} \end{array}$$

5.8 Let A and B be two fixed-point binary numbers in 2s complement representation, where $A = 1011\ 0001$ and $B = 1110\ 0100$. Determine the result of the following operations: (a) $A - B$ and (b) $A - (B' + 1)$.

5.9 Perform the operations shown below with fixed-point numbers in 2s complement representation, where $n = 7$. In each case, indicate if there is an overflow.

(a) (–63) – (+63)

(b) (–31) – (+33)

5.10 Perform subtraction on the following 2s complement numbers:

(a)
```
     01011 . 0101
 –)  00110 . 1100
```

(b)
```
     000111 . 0110
 –)  110011 . 0011
```

5.11 Perform the subtraction shown below using 8-bit operands in 2s complement representation and indicate if an overflow occurs.

(–62) – (+67)

5.12 Design a 2s complementer for an 8-bit operand *b[7:0]* as an iterative array. Obtain the dataflow design module, the test bench module for at least 20 different input vectors, and the outputs.

5.13 Design an 8-bit serial adder/subtractor. The augend/minuend is *a[7:0]* and the addend/subtrahend is *b[7:0]*. There is a *load* input to load the operands into the parallel-in, serial-out shift registers, and a *mode* input to indicate an add operation (*mode* = 0) or a subtract operation (*mode* = 1). Then implement the adder/subtractor as a structural module. Obtain the test bench and apply input vectors for addition and subtraction. Obtain the outputs and the waveforms.

6.1 *Sequential Add-Shift Multiplication*
6.2 *Booth Algorithm Multiplication*
6.3 *Bit-Pair Recoding Multiplication*
6.4 *Array Multiplication*
6.5 *Table Lookup Multiplication*
6.6 *Memory-Based Multiplication*
6.7 *Multiple-Operand Multiplication*
6.8 *Problems*

6

Fixed-Point Multiplication

Fixed-point multiplication is more complex than either addition or subtraction. Several methods will be presented in this chapter to multiply signed and unsigned operands, including the sequential add-shift method, the Booth algorithm, bit-pair recoding, an array multiplier, table lookup, read-only memory (ROM)-based multiplication, and multiple-operand multiplication. The n-bit *multiplicand A* is multiplied by the n-bit *multiplier B* to produce a $2n$-bit product, as shown below.

Multiplicand: $A = a_{n-1}\, a_{n-2}\, a_{n-3} \ldots a_1\, a_0$

Multiplier: $B = b_{n-1}\, b_{n-2}\, b_{n-3} \ldots b_1\, b_0$

Product: $P = p_{2n-1}\, p_{2n-2}\, p_{2n-3} \ldots p_1\, p_0$

Multiplication is a process of multiplying the multiplicand by the multiplier to produce a product. The general procedure consists of scanning the multiplier from the low-order bit to the high-order bit. If the multiplier bit is a 1, the multiplicand becomes the partial product; if the multiplier bit is a 0, then 0s are entered as the partial product. Each partial product is then shifted left one bit position relative to the previous partial product.

When all of the partial products are obtained, the partial products are then added to produce the $2n$-bit product. The sign of the product is a function of the signs of the operands. If both operands have the same sign, then the sign of the product is positive; if the signs of the operands are different, then the sign of the product is negative.

6.1 Sequential Add-Shift Multiplication

Examples will now be presented that illustrate multiplication using the paper-and-pencil method for the add-shift multiplication technique using two 4-bit operands in 2s complement notation. If the operands are in 2s complement notation, then the sign bit is treated in a manner identical to the other bits; however, the sign bit of the multiplicand is extended left in the partial product to accommodate the $2n$-bits of the product. The only requirement is that the multiplier must be positive — the multiplicand can be either positive or negative.

Example 6.1 Let the multiplicand and multiplier be two positive 4-bit operands as shown below, where *a[3:0]* = 0111 (+7) and *b[3:0]* = 0101 (+5) to produce a product *p[7:0]* = 0010 0011 (+35). A multiplier bit of 1 copies the multiplicand to the partial product; a multiplier bit of 0 enters 0s in the partial product.

Multiplicand *A*					0	1	1	1	+7
Multiplier *B*				×)	0	1	0	1	+5
Partial products	0	0	0	0	0	1	1	1	
	0	0	0	0	0	0	0		
	0	0	0	1	1	1			
	0	0	0	0	0				
Product *P*	0	0	1	0	0	0	1	1	+35

Example 6.2 This example multiplies a positive multiplicand by a negative multiplier to demonstrate that the multiplier must be positive. The multiplicand is *a[3:0]* = 0101 (+5); the multiplier is *b[3:0]* = 1101 (–3). The product should be –15; however, since the multiplier is treated as an unsigned number (1101 = 13), the result is 0100 0001 (65).

Multiplicand *A*					0	1	0	1	+5
Multiplier *B*				×)	1	1	0	1	(–3) 13
Partial products	0	0	0	0	0	1	0	1	
	0	0	0	0	0	0	0		
	0	0	0	1	0	1			
	0	0	1	0	1				
Product *P*	0	1	0	0	0	0	0	1	65

The problem can be resolved by either 2s complementing both operands or by 2s complementing the multiplier, performing the multiplication, then 2s complementing the result. Both methods are shown below.

Multiplicand *A*					1	0	1	1	–5
Multiplier *B*				×)	0	0	1	1	+3
Partial products	1	1	1	1	1	0	1	1	
	1	1	1	1	0	1	1		
	0	0	0	0	0	0			
	0	0	0	0	0				
Product *P*	1	1	1	1	0	0	0	1	–15

Multiplicand *A*					0	1	0	1	+5
Multiplier *B*				×)	0	0	1	1	+3
Partial products	0	0	0	0	0	1	0	1	
	0	0	0	0	1	0	1		
	0	0	0	0	0	0			
	0	0	0	0	0				
	0	0	0	0	1	1	1	1	15
Product *P*	1	1	1	1	0	0	0	1	–15

When both operands are negative, the correct result can be obtained by 2s complementing both operands before the operation begins, since a negative multiplicand multiplied by a negative multiplier yields a positive product. The example below shows the result of not negating both operands, where the multiplicand is –7 and the multiplier is –4; that is, multiplying –7 by 12 to yield –84. Then both operands are 2s complemented to obtain a correct product of +28.

Example 6.3 Let the multiplicand *a[3:0]* = 1001 (–7) and the multiplier *b[3:0]* = 1100 (–4) that will initially yield a product of $-7 \times 12 = -84$. Then both operands will be 2s complemented to produce a multiplication of $+7 \times +4$ to yield a correct product of +28.

Multiplicand A						1	0	0	1	–7
Multiplier B					×)	1	1	0	0	–4
Partial products	0	0	0	0		0	0	0	0	
	0	0	0	0		0	0	0		
	1	1	1	0		0	1			
	1	1	0	0		1				
Product P	1	0	1	0		1	1	0	0	–84

Multiplicand A						0	1	1	1	+7
Multiplier B					×)	0	1	0	0	+4
Partial products	0	0	0	0		0	0	0	0	
	0	0	0	0		0	0	0		
	0	0	0	1		1	1			
	0	0	0	0		0				
Product P	0	0	0	1		1	1	0	0	+28

6.1.1 Sequential Add-Shift Multiplication Hardware Algorithm

In this method, the multiplier must be positive. If the multiplier is negative, then the purpose of the multiplier bits is not always the same during the generation of the partial products. Any low-order 0s and the first 1 bit are treated the same as for a positive multiplier; however, the remaining higher-order bits are complemented and have an inverse effect. If the multiplier is negative, then it can be 2s complemented as shown previously, leaving the multiplicand either positive or negative. Negative multipliers are presented in a later section.

As mentioned previously, an alternative approach is to 2s complement both the multiplicand and multiplier if the multiplier is negative. This is equivalent to multiplying both operands by –1, but does not change the sign of the product.

The organization of a sequential add-shift multiplier is shown in Figure 6.1. Register A contains the n-bit multiplicand; register C contains the carry-out of the adder; register D contains the high-order half of the $2n$-bit product; and register B contains the n-bit multiplier. Registers C, D, and B shift right one bit position in concatenation per cycle. After the sum and carry have been loaded into their respective registers, the following right-shift occurs, where the symbol (•) indicates concatenation:

$$D \bullet B = cout \bullet sum_{n-1} \ldots sum_1\, sum_0 \bullet b_{n-1} \ldots b_2\, b_1$$

Shifting each partial product right one bit position is equivalent to shifting the partial product left one bit position in the paper-and-pencil method. Operands of n-bits require n cycles. The adder is designed using the carry lookahead technique. There is also a sequence counter that determines when the multiplication is finished.

During each cycle, the low-order bit b_0 of the multiplier determines whether the multiplicand is added to the partial product. If $b_0 = 1$, then the multiplicand is added to the partial product in register D; if $b_0 = 0$, then 0s are added to the partial product. In both cases, the sum from the adder is loaded into the D register and the carry-out is loaded into the 1-bit C register. At the completion of the operation, the high-order half of the product resides in register $D = d_{n-1} \ldots d_0$ and the low-order half of the product resides in register $B = b_{n-1} \ldots b_0$.

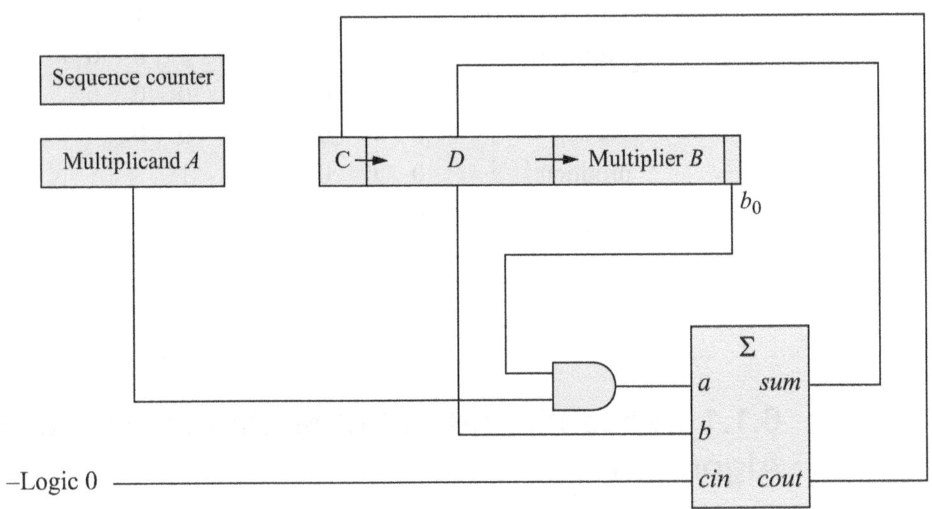

Figure 6.1 Hardware organization of a sequential add-shift multiplier.

Examples will now be presented to illustrate the hardware add-shift technique for both positive and negative multiplicands using a parallel-in, parallel-out register for the multiplicand and a parallel-in, parallel-out shift register for the multiplier.

The n-bit multiplicand is stored in an n-bit register. The multiplier is placed in the low-order n bits of a $2n$-bit shift register that will ultimately contain the $2n$-bit product. The high-order n bits of this shared register are set to zeroes. The low-order multiplier bit determines the operand to be added to the previous partial product. In this version, if the low-order multiplier bit is 0, then no addition takes place and the partial product is shifted right 1 bit position with the high-order bit propagated; otherwise, the multiplicand is added to the partial product and then shifted right 1 bit position together with the carry-out.

Example 6.4 The add-shift multiply hardware algorithm is shown in Figure 6.2 for a multiplicand of +7 and a multiplier of +7. There are n cycles for n-bit operands. During the first cycle, if the low-order multiplier bit is 0, then the shift register is shifted right 1 bit position and the high-order bit of the shift register is propagated to the right. If the low-order multiplier bit is 1, then the multiplicand is added to the high-order n bits of the shift register and the register is shifted right 1 bit position with the sign of the multiplicand placed in the high-order bit position of the shift register.

The sign of the partial product is not determined until the first add-shift cycle, at which time the sign of the multiplicand becomes the sign of the partial product. Alternatively, the sign is determined according to the following rule: Sign = $a_{n-1} \oplus b_{n-1}$. Any carry-out during the add operation is shifted into the resulting partial product during the subsequent shift operation of an add-shift cycle.

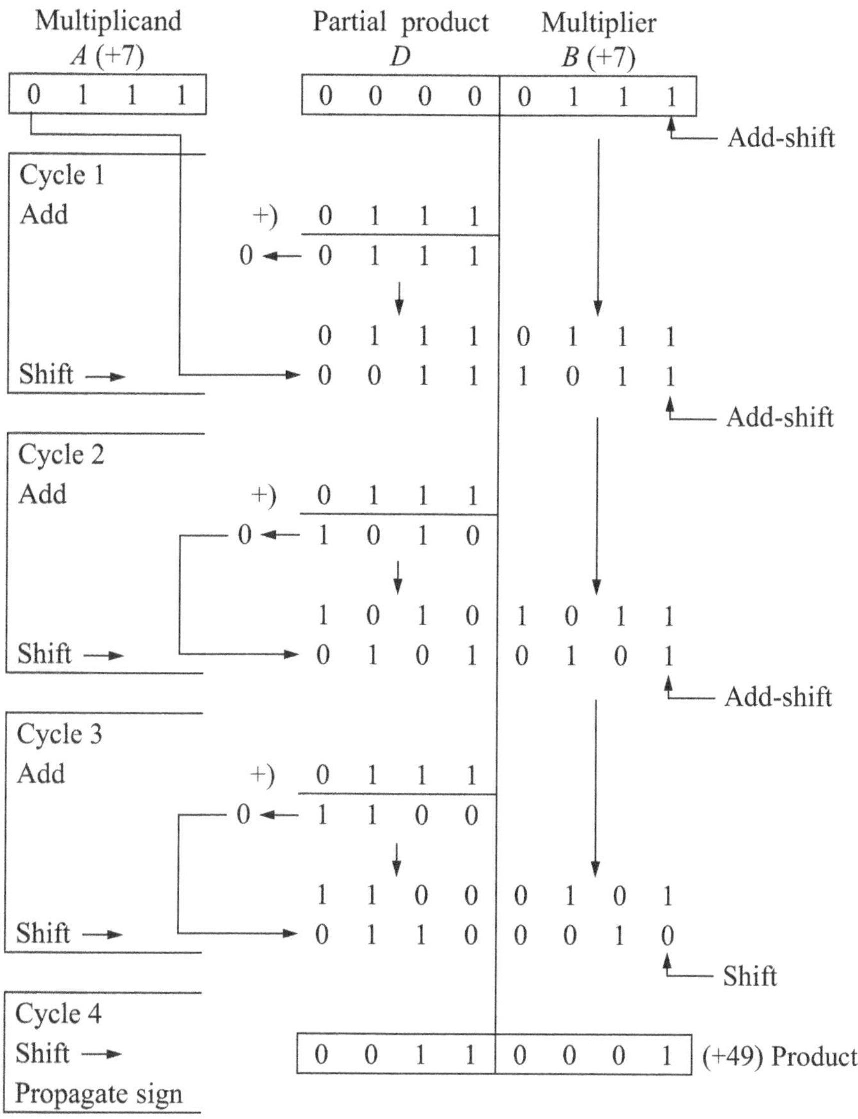

Figure 6.2 Multiplication algorithm for positive operands.

Example 6.5 In Figure 6.3, the multiplicand is negative (–7) and the multiplier is positive (+7) resulting in a product of –49. The sign of the product is again determined during the first add-shift cycle in which the sign of the multiplicand becomes the sign of the product. Alternatively, as shown in Figure 6.3, the sign can be determined by the exclusive-OR of the sign of the multiplicand and the sign of the multiplier.

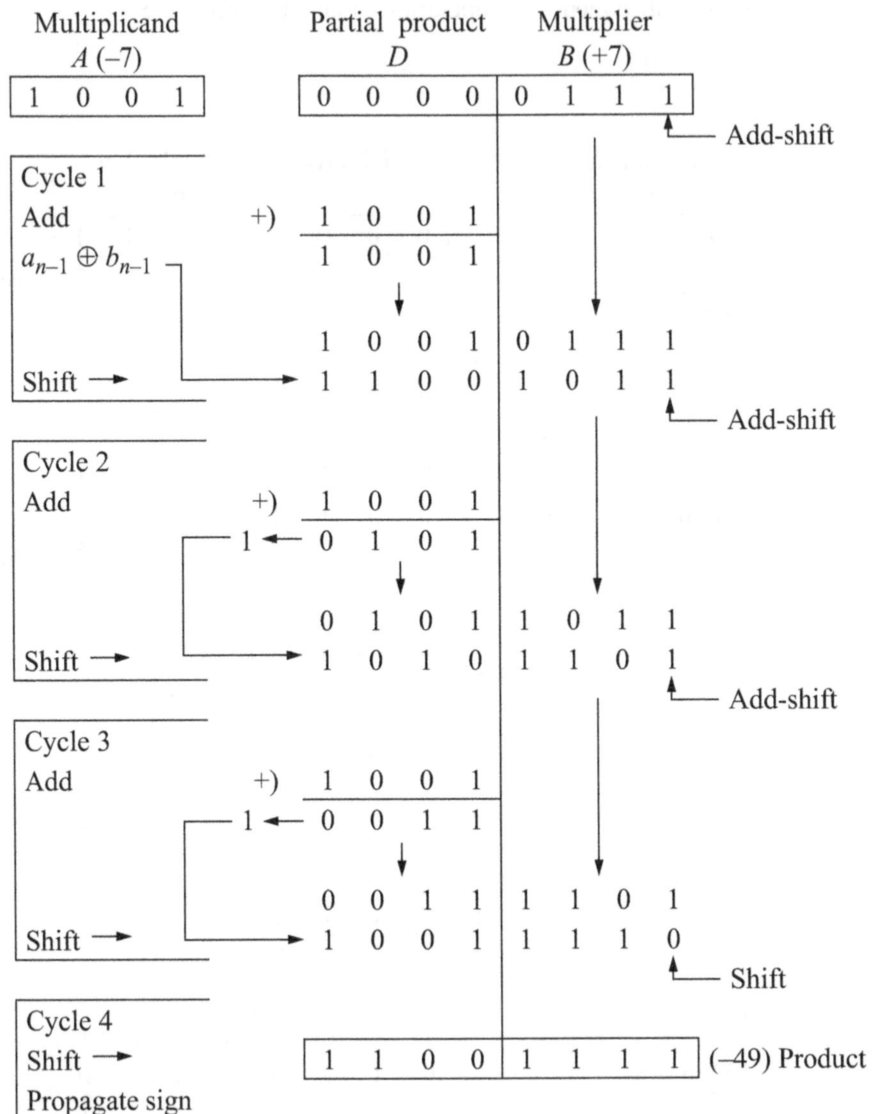

Figure 6.3 Multiplication algorithm for a negative multiplicand and a positive multiplier.

Two versions of a sequential add-shift multiplier will now be designed using Verilog HDL — both versions will be implemented using behavioral modeling. Version 1 will use the organization shown in Figure 6.1 in which zeroes are added to the partial product if the low-order multiplier bit is 0 and then the partial product is shifted right one bit position. Version 2 will simply shift the partial product if the low-order multiplier bit is 0.

6.1.2 Sequential Add-Shift Multiplication — Version 1

The multiplicand is *a[3:0]*, the multiplier is *b[3:0]*, and the product is *prod[7:0]*, which replaces registers *D* and *B* in Figure 6.1. A scalar *start* signal is used to initiate the multiply operation. A count-down sequence counter *count* is initialized to a value of 4 (0100) before the operation begins, because there are four bits in both operands. When the counter reaches a value of 0000, the multiply operation is finished and the 8-bit product is in register *prod[7:0]*.

A comparison is made initially to make certain that both operands are nonzero — if either operand has a value of zero, then the operation is terminated. If the low-order multiplier bit *b[0]* = 0, then a value of 0000 is ANDed with the multiplicand, *a[3:0]*, and added to the high-order half of the partial product, *prod[7:4]*. The result of that operation is concatenated with *prod[3:1]*. Then the product is shifted right one bit position with a 0 assigned to the high-order bit. This sequence repeats until the counter reaches a value of 0000.

The behavioral module is shown in Figure 6.4 and begins execution at the positive edge of the *start* input. The low-order multiplier bit *b[0]* specified by *b_reg[0]* is replicated four times then ANDed to the multiplicand *a*. This result is then added to the high-order four bits of the product *prod[7:4]* and concatenated with *prod[3:1]*, which essentially shifts the partial product right one bit position. The test bench, shown in Figure 6.5, applies several input vectors for the multiplicand, *a[3:0]*, and the multiplier, *b[3:0]*. The outputs and waveforms are shown in Figure 6.6 and Figure 6.7, respectively.

```
//behavioral add-shift multiply
module mul_add_shift3 (a, b, prod, start);

input [3:0] a, b;
input start;
output [7:0] prod;

reg [7:0] prod;
reg [3:0] b_reg;
reg [3:0] count;          //continued on next page
```

Figure 6.4 Behavioral module for version 1 of the sequential add-shift multiplier.

```
always @ (posedge start)
begin
   b_reg = b;
   prod = 0;
   count = 4'b0100;

   if ((a!=0) && (b!=0))
      while (count)
         begin
            prod = {(({4{b_reg[0]}} & a)
                     + prod[7:4]), prod[3:1]};
         b_reg = b_reg >> 1;
         count = count - 1;
         end
end
endmodule
```

Figure 6.4 (Continued)

```
//test bench for add-shift multiplier
module mul_add_shift3_tb;

reg [3:0] a, b;
reg start;
wire [7:0] prod;

initial          //display variables
$monitor ("a = %b, b = %b, prod = %b",
            a, b, prod);

initial          //apply input vectors
begin
   #0    start = 1'b0;
         a = 4'b0110;    b = 4'b0110;
   #10   start = 1'b1;
   #10   start = 1'b0;

   #10   a = 4'b0010;    b = 4'b0110;
   #10   start = 1'b1;
   #10   start = 1'b0;

   #10   a = 4'b0111;    b = 4'b0101;
   #10   start = 1'b1;
   #10   start = 1'b0;         //continued on next page
```

Figure 6.5 Test bench for version 1 of the sequential add-shift multiplier.

```
    #10    a = 4'b0111;    b = 4'b0111;
    #10    start = 1'b1;
    #10    start = 1'b0;

    #10    a = 4'b0101;    b = 4'b0101;
    #10    start = 1'b1;
    #10    start = 1'b0;

    #10    a = 4'b0111;    b = 4'b0011;
    #10    start = 1'b1;
    #10    start = 1'b0;

    #10    a = 4'b0100;    b = 4'b0110;
    #10    start = 1'b1;
    #10    start = 1'b0;

    #10    $stop;
end

//instantiate the module into the test bench
mul_add_shift3 inst1 (
    .a(a),
    .b(b),
    .prod(prod),
    .start(start)
    );

endmodule
```

Figure 6.5 (Continued)

```
a = 0110, b = 0110, prod = xxxxxxxx
a = 0110, b = 0110, prod = 00100100
a = 0010, b = 0110, prod = 00100100
a = 0010, b = 0110, prod = 00001100
a = 0111, b = 0101, prod = 00001100
a = 0111, b = 0101, prod = 00100011
a = 0111, b = 0111, prod = 00100011
a = 0111, b = 0111, prod = 00110001
a = 0101, b = 0101, prod = 00110001
a = 0101, b = 0101, prod = 00011001
a = 0111, b = 0011, prod = 00011001
a = 0111, b = 0011, prod = 00010101
a = 0100, b = 0110, prod = 00010101
a = 0100, b = 0110, prod = 00011000
```

Figure 6.6 Outputs for version 1 of the sequential add-shift multiplier.

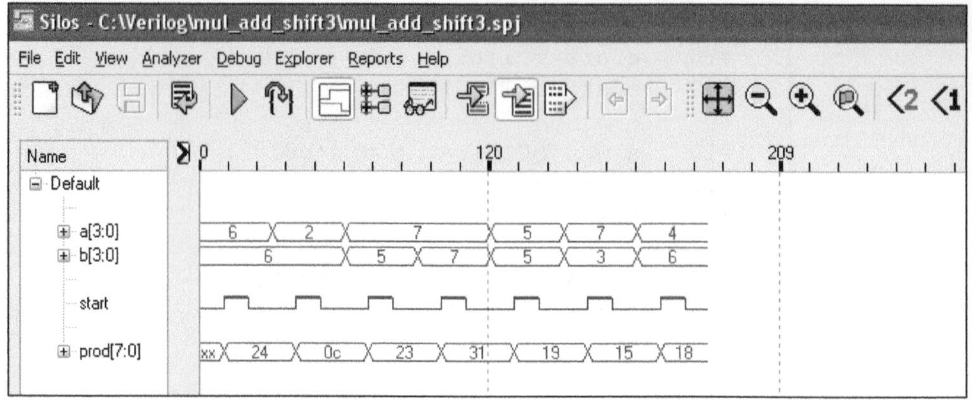

Figure 6.7 Waveforms for version 1 of the sequential add-shift multiplier.

6.1.3 Sequential Add-Shift Multiplication — Version 2

The multiplicand is *a[3:0]*, the multiplier is *b[3:0]*, and the product is *prod[7:0]*. A scalar *start* signal is used to initiate the multiply operation. A count-down sequence counter *count* is initialized to a value of 4 (0100) before the operation begins, since the multiplicand and multiplier are both four bits. When the counter reaches a count of 0000, the multiply operation is finished and the 8-bit product is in register *prod[7:0]*.

The product is initially set to *prod[7:4]* = 0000 and *prod[3:0]* = *b[3:0]*. This method varies slightly from the previous method. If the low-order bit of the product *prod[0]* = 0, then the entire product is shifted right one bit position and the high-order bit extends right one bit position — keeping the sign of the product intact, as shown below, where the symbols ({}) indicate concatenation.

$$Prod = \{prod[7], prod[7:1]\};$$

If the low-order bit of the product *Prod[0]* = 1, then the sign of the multiplicand is concatenated with the 4-bit sum of the multiplicand plus *Prod[7:4]*, which in turn is concatenated with *Prod[3:1]* as shown below. This provides the right shift of one bit position and allows for both positive and negative multiplicands.

$$Prod = \{a[3], (a + Prod[7:4]), Prod[3:1]\};$$

The behavioral module is shown in Figure 6.8 using the **while** loop, as in the previous version. The **while** statement executes a statement or a block of statements while an expression is true, in this case *count* $\neq 0$. The expression is evaluated and a

Boolean value, either true (a logical 1) or false (a logical 0) is returned. If the expression is true, then the procedural statement or block of statements is executed. The **while** loop executes until the expression becomes false, at which time the loop is exited and the next sequential statement is executed. If the expression is false when the loop is entered, then the procedural statement is not executed.

The test bench, shown in Figure 6.9, applies several input vectors for the multiplicand and multiplier, including both positive and negative values for the multiplicand. The outputs are shown Figure 6.10, and the waveforms are shown in Figure 6.11.

```
//behavioral add-shift multiply
module mul_add_shift5 (a, b, prod, start);


input [3:0] a, b;
input start;
output [7:0] prod;

//define internal registers
reg [7:0] prod;
reg [3:0] count;

always @ (posedge start)
begin
   prod [7:4] = 4'b0000;
   prod [3:0] = b;
   count = 4'b0100;

   if ((a!=0) && (b!=0))
      while (count)
         begin
            if (prod[0] == 1'b0)
               begin
                  prod = {prod[7], prod[7:1]};
                  count = count - 1;
               end

            else
               begin
                  prod = {a[3], (a + prod[7:4]), prod[3:1]};
                  count = count - 1;
               end
         end
end
endmodule
```

Figure 6.8 Behavioral module for version 2 of the sequential add-shift multiplier.

```
//test bench for add-shift multiplier
module mul_add_shift5_tb;

reg [3:0] a, b;
reg start;
wire [7:0] prod;

//display variables
initial
$monitor ("a = %b, b = %b, prod = %b",
            a, b, prod);

//apply input vectors
initial
begin
   #0     start = 1'b0;
          a = 4'b0110;    b = 4'b0110;
   #10    start = 1'b1;
   #10    start = 1'b0;

   #10    a = 4'b0010;    b = 4'b0110;
   #10    start = 1'b1;
   #10    start = 1'b0;

   #10    a = 4'b0111;    b = 4'b0101;
   #10    start = 1'b1;
   #10    start = 1'b0;

   #10    a = 4'b0111;    b = 4'b0111;
   #10    start = 1'b1;
   #10    start = 1'b0;

   #10    a = 4'b1111;    b = 4'b0111;
   #10    start = 1'b1;
   #10    start = 1'b0;

   #10    a = 4'b1011;    b = 4'b0111;
   #10    start = 1'b1;
   #10    start = 1'b0;

   #10    a = 4'b1000;    b = 4'b0001;
   #10    start = 1'b1;
   #10    start = 1'b0;

//continued on next page
```

Figure 6.9 Test bench for version 2 of the sequential add-shift multiplier.

```
    #10   a = 4'b1010;    b = 4'b0111;
    #10   start = 1'b1;
    #10   start = 1'b0;

    #10   a = 4'b1101;    b = 4'b0101;
    #10   start = 1'b1;
    #10   start = 1'b0;

    #10   a = 4'b0100;    b = 4'b0111;
    #10   start = 1'b1;
    #10   start = 1'b0;
    #10   $stop;
end

//instantiate the module into the test bench
mul_add_shift5 inst1 (
    .a(a),
    .b(b),
    .prod(prod),
    .start(start)
    );
endmodule
```

Figure 6.9 (Continued)

```
a = 0110, b = 0110, prod = xxxxxxxx
a = 0110, b = 0110, prod = 00100100
a = 0010, b = 0110, prod = 00100100
a = 0010, b = 0110, prod = 00001100
a = 0111, b = 0101, prod = 00001100
a = 0111, b = 0101, prod = 00100011
a = 0111, b = 0111, prod = 00100011
a = 0111, b = 0111, prod = 00110001
a = 1111, b = 0111, prod = 00110001
a = 1111, b = 0111, prod = 11111001
a = 1011, b = 0111, prod = 11111001
a = 1011, b = 0111, prod = 11011101
a = 1000, b = 0001, prod = 11011101
a = 1000, b = 0001, prod = 11111000
a = 1010, b = 0111, prod = 11111000
a = 1010, b = 0111, prod = 11010110
a = 1101, b = 0101, prod = 11010110
a = 1101, b = 0101, prod = 11110001
a = 0100, b = 0111, prod = 11110001
a = 0100, b = 0111, prod = 00011100
```

Figure 6.10 Outputs for version 2 of the sequential add-shift multiplier.

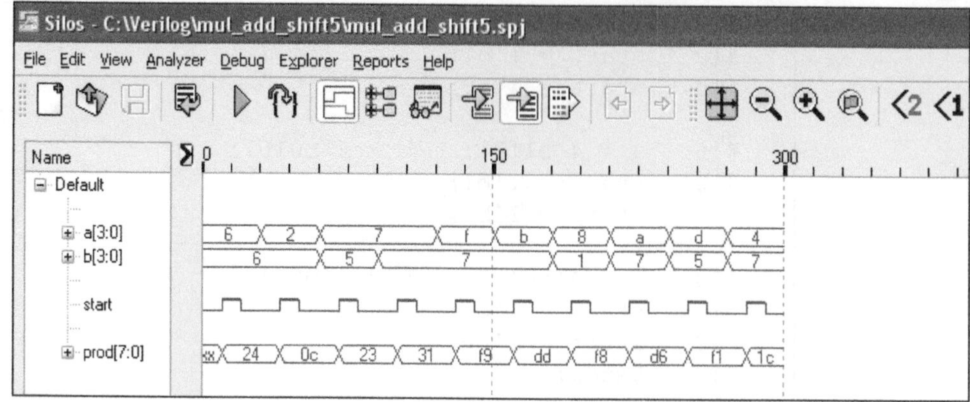

Figure 6.11 Waveforms for version 2 of the sequential add-shift multiplier.

6.2 Booth Algorithm Multiplication

The Booth algorithm is an effective technique for multiplying operands that are in 2s complement representation, including the case where the multiplier is negative. Unlike the sequential add-shift method, it treats both positive and negative operands uniformly; that is, the multiplicand and multiplier can both be negative or positive or of opposite sign as shown below.

Operands	Signs of Operands			
Multiplicand	+	–	+	–
Multiplier	+	+	–	–

In the sequential add-shift method, each multiplier bit generates a version of the multiplicand that is added to the partial product. For large operands, the delay to obtain the product can be significant. The Booth algorithm reduces the number of partial products by shifting over strings of zeros in a recoded version of the multiplier. This method is referred to as *skipping over zeroes*. The increase in speed is proportional to the number of zeroes in the recoded version of the multiplier.

Consider a string of k consecutive 1s in the multiplier as shown below. The multiplier will be recoded so that the k consecutive 1s will be transformed into $k - 1$ consecutive 0s. The multiplier hardware will then shift over the $k - 1$ consecutive 0s without having to generate partial products.

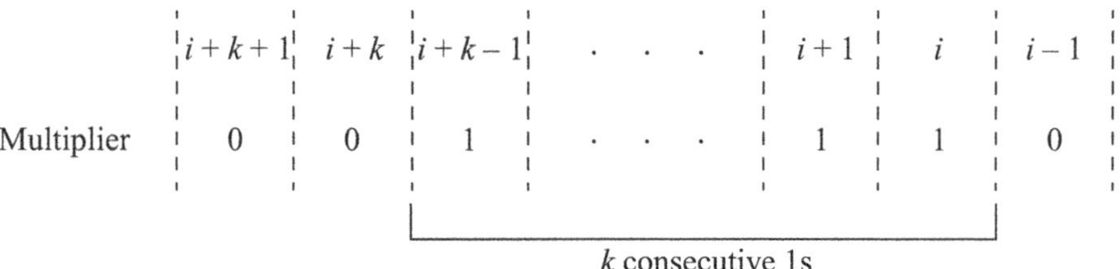

In the sequential add-shift method, the multiplicand would be added k times to the shifted partial product. The number of additions can be reduced by the property of binary strings shown in Equation 6.1.

$$2^{i+k} - 2^i = 2^{i+k-1} + 2^{i+k-2} + \ldots 2^{i+1} + 2^i \tag{6.1}$$

The right-hand side of the equation is a binary string that can be replaced by the difference of two numbers on the left-hand side of the equation. Thus, the k consecutive 1s can be replaced by the following string:

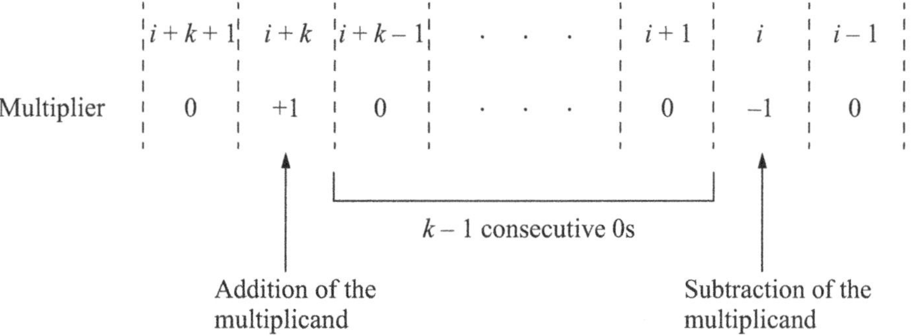

Note that –1 times the left-shifted multiplicand occurs at $0 \rightarrow 1$ boundaries, and +1 times the left-shifted multiplicand occurs at $1 \rightarrow 0$ boundaries as the multiplier is scanned from right to left. The Booth algorithm can be applied to any number of groups of 1s in a multiplier, including the case where the group consists of a single 1 bit. Table 6.1 shows the multiplier recoding table for two consecutive bits that specify which version of the multiplicand will be added to the shifted partial product.

Table 6.1 Booth Multiplier Recoding Table

Multiplier Bit i	Multiplier Bit $i-1$	Version of Multiplicand
0	0	$0 \times$ multiplicand
0	1	$+1 \times$ multiplicand
1	0	$-1 \times$ multiplicand
1	1	$0 \times$ multiplicand

An example will help to clarify the procedure. Let the multiplier be +60 as shown below.

		2^{i+k}							
		2^{i+4}			2^i				
	2^7	2^6	2^5	2^4	2^3	2^2	2^1	2^0	
Multiplier	0	0	1	1	1	1	0	0	+60

$k = 4$ (bits 2^5 through 2^2)

The standard multiplier shown above can be recoded as shown below using the Booth algorithm technique.

	2^7	2^6	2^5	2^4	2^3	2^2	2^1	2^0
Standard multiplier	0	0	+1	+1	+1	+1	0	0
			←	$k = 4$		→		
Recoded multiplier	0	+1	0	0	0	–1	0	0
			←$k - 1$=3→					

The Booth algorithm converts both positive and negative multipliers into a form that generates versions of the multiplicand to be added to the shifted partial products. Since the increase in speed is a function of the bit configuration of the multiplier, the efficiency of the Booth algorithm is obviously data dependent. Examples will now be given to illustrate the Booth algorithm for all combinations of the signs of the multiplicand and multiplier. Then a Booth multiplier will be implemented in Verilog HDL using a combination of dataflow and behavioral modeling.

Example 6.6 An 8-bit positive multiplicand (+27) will be multiplied by an 8-bit positive multiplier (+120) — first using the standard sequential add-shift technique, then using the Booth algorithm to show the reduced number of partial products. In the sequential add-shift technique, there are a total of seven additions; using the Booth algorithm, there are only two additions. The second partial product in the Booth algorithm is the 2s complement of the multiplicand shifted left two bit positions. The third partial product is the result of shifting the multiplicand left over three 0s.

Standard sequential add-shift method																
Multiplicand								0	0	0	1	1	0	1	1	+27
Multiplier							×)	0	1	1	1	1	0	0	0	+120
0	0	0	0	0	0	0	0	0	0	0	0	0	0	0	0	
0	0	0	0	0	0	0	0	0	0	0	0	0	0	0		
0	0	0	0	0	0	0	0	0	0	0	0	0	0			
0	0	0	0	0	0	0	0	1	1	0	1	1				
0	0	0	0	0	0	0	1	1	0	1	1					
0	0	0	0	0	0	1	1	0	1	1						
0	0	0	0	0	1	1	0	1	1							
0	0	0	0	0	0	0	0	0								
0	0	0	0	1	1	0	0	1	0	1	0	1	0	0	0	+3240

Booth algorithm																
Multiplicand								0	0	0	1	1	0	1	1	+27
Recoded multiplier							×)	+1	0	0	0	–1	0	0	0	
0	0	0	0	0	0	0	0	0	0	0	0	0	0	0	0	
1	1	1	1	1	1	1	1	0	0	1	0	1				
0	0	0	0	1	1	0	1	1								
0	0	0	0	1	1	0	0	1	0	1	0	1	0	0	0	+3240

Example 6.7 The Booth algorithm will be used to multiply an 8-bit negative multiplicand 1010 1101 (–83) by an 8-bit positive multiplier 0011 1110 (+62) to yield a product of 1110 1011 1110 0110 (–5146).

Multiplicand		1	0	1	0	1	1	0	1	–83
Multiplier	×)	0	0	1	1	1	1	1	0	+62
										–5146

Booth algorithm																
Multiplicand								1	0	1	0	1	1	0	1	–83
Recoded multiplier							×)	0	+1	0	0	0	0	–1	0	
0	0	0	0	0	0	0	0	0	0	0	0	0	0	0	0	
0	0	0	0	0	0	0	0	1	0	1	0	0	1	1		
1	1	1	0	1	0	1	1	0	1							
1	1	1	0	1	0	1	1	1	1	1	0	0	1	1	0	–5146

Example 6.8 The Booth algorithm will be used to multiply an 8-bit positive multiplicand 0100 1010 (+74) by an 8-bit negative multiplier 1101 1100 (–36) to yield a negative product of 1111 0101 1001 1000 (–2664).

Multiplicand		0	1	0	0	1	0	1	0	+74
Multiplier	×)	1	1	0	1	1	1	0	0	–36
										–2664

Booth algorithm																
Multiplicand								0	1	0	0	1	0	1	0	+74
Recoded multiplier							×)	0	–1	+1	0	0	–1	0	0	
0	0	0	0	0	0	0	0	0	0	0	0	0	0	0	0	
1	1	1	1	1	1	1	0	1	1	0	1	1	0			
0	0	0	0	1	0	0	1	0	1	0						
1	1	1	0	1	1	0	1	1	0							
1	1	1	1	0	1	0	1	1	0	0	1	1	0	0	0	–2664

Example 6.9 The Booth algorithm will be used to multiply an 8-bit negative multiplicand 1010 0010 (–94) by an 8-bit negative multiplier 1101 1100 (–36) to yield a positive product of 0000 1101 0011 1000 (+3384).

Multiplicand		1	0	1	0	0	0	1	0	–94
Multiplier	×)	1	1	0	1	1	1	0	0	–36
										+3384

Booth algorithm																
Multiplicand								1	0	1	0	0	0	1	0	–94
Recoded multiplier							×)	0	–1	+1	0	0	–1	0	0	
0	0	0	0	0	0	0	0	0	0	0	0	0	0	0	0	
0	0	0	0	0	0	0	1	0	1	1	1	1	0			
1	1	1	1	0	1	0	0	0	1	0						
0	0	0	1	0	1	1	1	1	0							
0	0	0	0	1	1	0	1	0	0	1	1	1	0	0	0	+3384

Example 6.10 If the multiplier has a low-order bit of 1, then an implied 0 is placed to the right of the low-order multiplier bit. This provides a boundary of 0 → 1 (–1 times the multiplicand) for the low-order bit as the multiplier is scanned from right to left. This is shown in the example below, which multiplies an 8-bit positive multiplicand 0100 1110 (+78) by an 8-bit positive multiplier 0010 1011 (+43). This example also illustrates a multiplier that has very little advantage over the sequential add-shift method because of alternating 1s and 0s.

Multiplicand		0	1	0	0	1	1	1	0		+78
Multiplier	×)	0	0	1	0	1	0	1	1	[0]	+43
											+3354

Booth algorithm																
Multiplicand								0	1	0	0	1	1	1	0	+78
Recoded multiplier							×)	0	+1	–1	+1	–1	+1	0	–1	
1	1	1	1	1	1	1	1	1	0	1	1	0	0	1	0	
0	0	0	0	0	0	0	1	0	0	1	1	1	0			
1	1	1	1	1	1	0	1	1	0	0	1	0				
0	0	0	0	0	1	0	0	1	1	1	0					
1	1	1	1	0	1	1	0	0	1	0						
0	0	0	1	0	0	1	1	1	0							
0	0	0	0	1	1	0	1	0	0	0	1	1	0	1	0	+3354

The mixed-design module for the Booth algorithm is shown in Figure 6.12 using dataflow modeling and behavioral modeling. The operands are 8-bit vectors *a[7:0]* and *b[7:0]*; the product is a 16-bit result *prod[15:0]*. The following internal wires are defined: *a_ext_pos[15:0]*, which is operand *A* with sign extended, and *a_ext_neg[15:0]*, which is the negation (2s complement) of operand *A* with sign extended.

The following internal registers are defined: *a_neg[7:0]*, which is the negation of operand *A*, and *pp1[15:0]*, *pp2[15:0]*, *pp3[15:0]*, *pp4[15:0]*, *pp5[15:0]*, *pp6[15:0]*, *pp7[15:0]*, and *pp8[15:0]*, which are the partial products to be added together to obtain the product *prod[15:0]*.

The example below illustrates the use of the internal registers using only 4-bit operands for convenience, where the multiplicand is 0111 (+7) and the multiplier is 0101 (+5). The right-most four bits of partial product 1 (*pp1*) are the negation [*a_neg* (2s complement)] of operand *A*, which is generated as a result of the –1 times operand *A* operation. The entire row of partial product 1 (*pp1*) corresponds to *a_neg* with the sign bit extended; that is, *a_ext_neg*.

Multiplicand		0	1	1	1		+7
Multiplier	×)	0	1	0	1	0	+5
							+35

Booth algorithm										
Multiplicand					0	1	1	1	+7	
Recoded multiplier				×)	+1	−1	+1	−1		
a_ext_neg → *pp1*	1	1	1	1	1	0	0	1		← *a_neg*
a_ext_pos → *pp2*	0	0	0	0	1	1	1	0		
pp3	1	1	1	0	0	1	0	0		
pp4	0	0	1	1	1	0	0	0		
	0	0	1	0	0	0	1	1	+35	

The 1s complement of operand *A* is formed by the following statement:

```
assign a_bar = ~a
```

Then the 2s complement is obtained using the **always** statement as follows:

```
always @ (a_bar)
    a_neg = a_bar + 1;
```

These statements specify that whenever *a_bar* changes value, the negation (2s complement) of the multiplicand is obtained by adding 1 to the 1s complement. The 16-bit positive value of operand *A* is obtained by the replication operator in the following statement:

```
assign a_ext_pos = {{8{a[7]}}, a};
```

Replication is a means of performing repetitive concatenation. Replication specifies the number of times to duplicate the expression within the innermost braces, in this case the sign bit of the multiplicand which is duplicated eight times. This is then concatenated with the eight bits of the multiplicand. A similar procedure is used to extend the negation of the multiplicand to 16 bits, as shown below.

```
assign a_ext_neg = {{8{a_neg[7]}}, a_neg};
```

Consider the first **case** statement in which bits 1 and 0 of the multiplier (*b[1:0]*) are evaluated. If *b[1:0]* = 00, then partial product 1 = 0000 0000 0000 0000. Since a $0 \rightarrow 0$ transition specifies $0 \times$ multiplicand, partial product 2 = 0000 0000 0000 0000 also.

If *b[1:0]* = 01, then this implies that a 0 bit be placed to the immediate right of the low-order multiplier bit, as follows: 01**0**. The multiplier is then scanned from right to left. A $0 \rightarrow 1$ transition specifies $-1 \times$ multiplicand; therefore, partial product 1 is the 2s complement of the multiplicand with the sign extended to 16 bits. A $1 \rightarrow 0$ transition specifies $+1 \times$ multiplicand; therefore, partial product 2 is the multiplicand expanded to 16 bits with sign extension, as shown below.

pp2 = {{7{a[7]}}, a[7:0], 1'b0};

If *b[1:0]* = 10, then partial product 1 = 0000 0000 0000 0000. Since a $0 \rightarrow 1$ transition specifies $-1 \times$ multiplicand; therefore, partial product 2 is the 2s complement of the multiplicand with sign extension, as shown below.

pp2 = {a_ext_neg[14:0], 1'b0};

If *b[1:0]* = 11, then this implies that a 0 bit be placed to the immediate right of the low-order multiplier bit, as follows: 11**0**. The multiplier is then scanned from right to left. A $0 \rightarrow 1$ transition specifies $-1 \times$ multiplicand; therefore, partial product 1 is the 2s complement of the multiplicand with sign extension. A $1 \rightarrow 1$ transition specifies $0 \times$ multiplicand; therefore, partial product 2 = 0000 0000 0000 0000.

The remaining pairs of multiplier bits are processed in a similar manner. The product is obtained by adding all of the partial products.

```
//mixed-design for the Booth multiply algorithm
module booth4 (a, b, prod);

input [7:0] a, b;
output [15:0] prod;

wire [7:0] a, b;
wire [15:0] prod;

wire [7:0] a_bar;

//define internal wires
wire [15:0] a_ext_pos;
wire [15:0] a_ext_neg;

//continued on next page
```

Figure 6.12 Mixed-design module for the Booth multiplier.

```
//define internal registers
reg [7:0] a_neg;
reg [15:0] pp1, pp2, pp3, pp4, pp5, pp6, pp7, pp8;

//test b[1:0] ---------------------------------------
assign a_bar = ~a;


always @ (a_bar)
    a_neg = a_bar + 1;


//define 16-bit multiplicand and 2s comp of multiplicand
assign a_ext_pos = {{8{a[7]}}, a};
assign a_ext_neg = {{8{a_neg[7]}}, a_neg};


always @ (b or a_ext_pos or a_ext_neg)
begin
   case (b[1:0])
      2'b00 :
         begin
            pp1 = 16'h00;
            pp2 = 16'h00;
         end

      2'b01 :
         begin
            pp1 = a_ext_neg;
            pp2 = {{7{a[7]}}, a[7:0], 1'b0};
         end

      2'b10 :
         begin
            pp1 = 16'h00;
            pp2 = {a_ext_neg[14:0], 1'b0};
         end

      2'b11 :
         begin
            pp1 = a_ext_neg;
            pp2 = 16'h00;
         end
   endcase

end

//continued on next page
```

Figure 6.12 (Continued)

```
//test b[2:1] ---------------------------------------
always @ (b or a_ext_pos or a_ext_neg)
begin
   case (b[2:1])
      2'b00: pp3 = 16'h00;

      2'b01: pp3 = {a_ext_pos[13:0], 2'b00};

      2'b10: pp3 = {a_ext_neg[13:0], 2'b00};

      2'b11: pp3 = 16'h00;
   endcase
end


//test b[3:2] ---------------------------------------
always @ (b or a_ext_pos or a_ext_neg)
begin
   case (b[3:2])
      2'b00: pp4 = 16'h00;

      2'b01: pp4 = {a_ext_pos[12:0], 3'b000};

      2'b10: pp4 = {a_ext_neg[12:0], 3'b000};

      2'b11: pp4 = 16'h00;
   endcase
end


//test b[4:3] ---------------------------------------
always @ (b or a_ext_pos or a_ext_neg)
begin
   case (b[4:3])
      2'b00: pp5 = 16'h00;

      2'b01: pp5 = {a_ext_pos[11:0], 4'b0000};

      2'b10: pp5 = {a_ext_neg[11:0], 4'b0000};

      2'b11: pp5 = 16'h00;
   endcase
end

//continued on next page
```

Figure 6.12 (Continued)

```
//test b[5:4] ----------------------------------------
always @ (b or a_ext_pos or a_ext_neg)
begin
   case (b[5:4])
      2'b00: pp6 = 16'h00;

      2'b01: pp6 = {a_ext_pos[10:0], 5'b00000};

      2'b10: pp6 = {a_ext_neg[10:0], 5'b00000};

      2'b11: pp6 = 16'h00;
   endcase
end

//test b[6:5] ----------------------------------------
always @ (b or a_ext_pos or a_ext_neg)
begin
   case (b[6:5])
      2'b00: pp7 = 16'h00;

      2'b01: pp7 = {a_ext_pos[9:0], 6'b000000};

      2'b10: pp7 = {a_ext_neg[9:0], 6'b000000};

      2'b11: pp7 = 16'h00;
   endcase
end

//test b[7:6] ----------------------------------------
always @ (b or a_ext_pos or a_ext_neg)
begin
   case (b[7:6])
      2'b00: pp8 = 16'h00;

      2'b01: pp8 = {a_ext_pos[8:0], 7'b0000000};

      2'b10: pp8 = {a_ext_neg[8:0], 7'b0000000};

      2'b11: pp8 = 16'h00;
   endcase
end

assign prod = pp1 + pp2 + pp3 + pp4 + pp5 + pp6 + pp7 + pp8;

endmodule
```

Figure 6.12 (Continued)

The test bench is shown in Figure 6.13, in which the first five sets of input vectors correspond to those of Example 6.6, Example 6.7, Example 6.8, Example 6.9, and Example 6.10. In order to completely test each set of bit pairs for the remaining multipliers, all combinations of the bit pairs are applied as part of the input vectors. That is, all combinations of the following bit pairs are entered as part of the input vectors for the multipliers: *b[1:0]*, *b[2:1]*, *b[3:2]*, *b[4:3]*, *b[5:4]*, *b[6:5]*, and *b[7:6]*.

The outputs are shown in Figure 6.14. The first five sets of outputs correspond to the outputs obtained from the previously mentioned examples. The remaining outputs are the result of applying all combinations of the bit pairs to their respective multiplier input vectors. The outputs also indicate the positive and negative decimal values of the multiplicand and multiplier together with the decimal product. The waveforms are shown in Figure 6.15.

```
//test bench for booth algorithm
module booth4_tb;

//registers for inputs because they hold values
reg [7:0] a, b;
wire [15:0] prod;

//display operands a, b, and product
initial
$monitor ("a = %b, b = %b, prod = %h", a, b, prod);

//apply input vectors
initial
begin
//for examples 6.6 through 6.10
   #0      a = 8'b0001_1011;
           b = 8'b0111_1000;

   #10     a = 8'b1010_1101;
           b = 8'b0011_1110;

   #10     a = 8'b0100_1010;
           b = 8'b1101_1100;

   #10     a = 8'b1010_0010;
           b = 8'b1101_1100;

   #10     a = 8'b0100_1110;
           b = 8'b0010_1011;
//continued on next page
```

Figure 6.13 Test bench module for the Booth multiplier.

```
//test b[1:0] ----------------------------------------
   #10    a = 8'b1100_1100;
          b = 8'b1100_1100;

   #10    a = 8'b1100_1100;
          b = 8'b1100_1101;

   #10    a = 8'b1100_1100;
          b = 8'b1100_1110;

   #10    a = 8'b1100_1100;
          b = 8'b1100_1111;


//test b[2:1] ----------------------------------------

   #10    a = 8'b0111_1111;
          b = 8'b1011_1000;

   #10    a = 8'b1011_0011;
          b = 8'b1100_1011;

   #10    a = 8'b0111_0000;
          b = 8'b0111_0100;

   #10    a = 8'b0111_0000;
          b = 8'b0111_0110;


//test b[3:2] ----------------------------------------
   #10    a = 8'b0111_1111;
          b = 8'b1011_0000;

   #10    a = 8'b1011_0011;
          b = 8'b1100_0111;

   #10    a = 8'b0111_0000;
          b = 8'b0111_1000;

   #10    a = 8'b0111_0000;
          b = 8'b0111_1110;

//continued on next page
```

Figure 6.13 (Continued)

```
//test b[4:3] ----------------------------------------
   #10    a = 8'b0111_1111;
          b = 8'b1010_0000;

   #10    a = 8'b1011_0011;
          b = 8'b1100_1111;

   #10    a = 8'b0111_0000;
          b = 8'b0111_0000;

   #10    a = 8'b0111_0000;
          b = 8'b0111_1110;


//test b[5:4] ----------------------------------------
   #10    a = 8'b0111_1111;
          b = 8'b1000_0000;

   #10    a = 8'b1011_0011;
          b = 8'b1101_1111;

   #10    a = 8'b0111_0000;
          b = 8'b0110_0000;

   #10    a = 8'b0111_0000;
          b = 8'b0111_1110;


//test b[6:5] ----------------------------------------
   #10    a = 8'b0111_1111;
          b = 8'b1000_0000;

   #10    a = 8'b1011_0011;
          b = 8'b1010_1111;

   #10    a = 8'b0111_0000;
          b = 8'b0101_0000;

   #10    a = 8'b0111_0000;
          b = 8'b0111_1110;

//continued on next page
```

Figure 6.13 (Continued)

```
//test b[7:6] ----------------------------------------
   #10   a = 8'b0111_1111;
         b = 8'b0010_0000;

   #10   a = 8'b1011_0011;
         b = 8'b0100_1111;

   #10   a = 8'b0111_0000;
         b = 8'b1011_0000;

   #10   a = 8'b0111_0000;
         b = 8'b1111_1110;

   #10   $stop;
end

//instantiate the module into the test bench
booth4 inst1 (
   .a(a),
   .b(b),
   .prod(prod)
   );

endmodule
```

Figure 6.13 (Continued)

```
a=00011011, b=01111000, prod=0ca8   (+27  × +120 = +3240)
a=10101101, b=00111110, prod=ebe6   (-83  × +62  = -5146)
a=01001010, b=11011100, prod=f598   (+74  × -36  = -2664)
a=10100010, b=11011100, prod=0d38   (-94  × -36  = +3384)
a=01001110, b=00101011, prod=0d1a   (+78  × +43  = +3354)

a=11001100, b=11001100, prod=0a90   (-52  × -52  = +2704)
a=11001100, b=11001101, prod=0a5c   (-52  × -51  = +2652)
a=11001100, b=11001110, prod=0a28   (-52  × -50  = +2600)
a=11001100, b=11001111, prod=09f4   (-52  × -49  = +2458)

a=01111111, b=10111000, prod=dc48   (+127 × -72  = -9144)
a=10110011, b=11001011, prod=0ff1   (-77  × -53  = +4081)
a=01110000, b=01110100, prod=32c0   (+112 × +116 = +12992)
a=01110000, b=01110110, prod=33a0   (+112 × +118 = +13216)

//continued on next page
```

Figure 6.14 Outputs for the Booth multiplier.

```
a=01111111, b=10110000, prod=d850    (+127 × -80  = -10160)
a=10110011, b=11000111, prod=1125    (-77  × -57  = +4389)
a=01110000, b=01111000, prod=3480    (+112 × +120 = +13440)
a=01110000, b=01111110, prod=3720    (+112 × +126 = +14112)

a=01111111, b=10100000, prod=d060    (+127 × -96  = -12192)
a=10110011, b=11001111, prod=0ebd    (-77  × -49  = +3773)
a=01110000, b=01110000, prod=3100    (+112 × +112 = +12544)
a=01110000, b=01111110, prod=3720    (+112 × +126 = +14112)

a=01111111, b=10000000, prod=c080    (+127 × -128 = -16256)
a=10110011, b=11011111, prod=09ed    (-77  × -33  = +2541)
a=01110000, b=01100000, prod=2a00    (+112 × +96  = +10752)
a=01110000, b=01111110, prod=3720    (+112 × +126 = +14112)

a=01111111, b=10000000, prod=c080    (+127 × -128 = -16256)
a=10110011, b=10101111, prod=185d    (-77  × -81  = +6237)
a=01110000, b=01010000, prod=2300    (+112 × +80  = +8960)
a=01110000, b=01111110, prod=3720    (+112 × +126 = +14112)

a=01111111, b=00100000, prod=0fe0    (+127 × +32  = +4064)
a=10110011, b=01001111, prod=e83d    (-77  × +79  = -6083)
a=01110000, b=10110000, prod=dd00    (+112 × -80  = -8960)
a=01110000, b=11111110, prod=ff20    (+112 × -2   = -224)
```

Figure 6.14 (Continued)

6.3 Bit-Pair Recoding Multiplication

The speed increase of the Booth algorithm depended on the bit configuration of the multiplier and is, therefore, data dependent. This section presents a speedup technique that is derived from the Booth algorithm and assures that an n-bit multiplier will have no more than $n/2$ partial products. It also treats both positive and negative multipliers uniformly; that is, there is no need to 2s complement the multiplier before multiplying, or to 2s complement the product after multiplying.

As in the Booth algorithm, bit-pair recoding regards a string of 1s as the difference of two numbers. This property of binary strings is restated in Equation 6.2, where k is the number of consecutive 1s and i is the position of the rightmost 1 in the string of 1s.

$$2^{i+k} - 2^{i} = 2^{i+k-1} + 2^{i+k-2} + \cdots 2^{i+1} + 2^{i} \tag{6.2}$$

Consider the multiplier example introduced in the Booth algorithm in which there were four consecutive 1s for a value of 60, as shown in Figure 6.16. This indicates that the number 0011 1100 (60) has the same value as $2^6 - 2^2 = 60$.

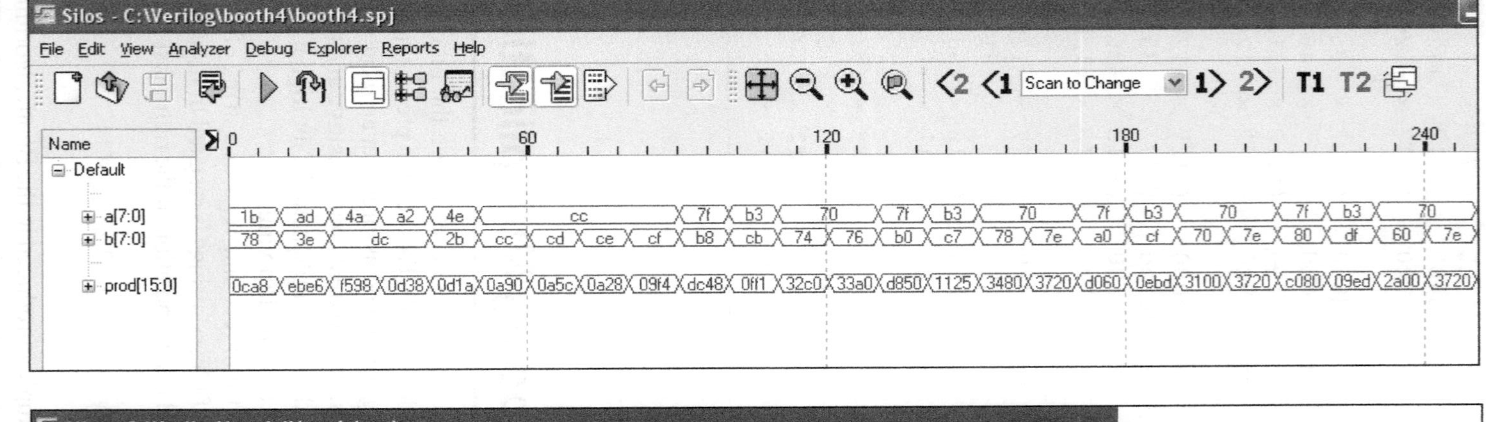

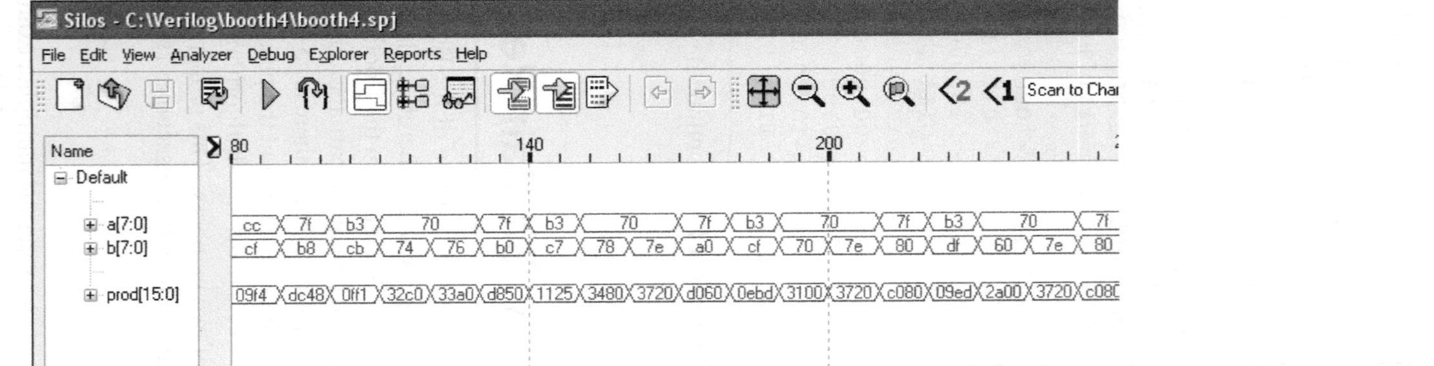

Figure 6.15 Waveforms for the Booth multiplier.

	2^7	2^6	2^5	2^4	2^3	2^2	2^1	2^0	
		2^{i+k}, 2^{i+4}				2^i			
Multiplier	0	0	1	1	1	1	0	0	+60

$k = 4$

Figure 6.16 Multiplier to be used to illustrate the multiplier bit-pair recoding technique.

The same result can be obtained by examining pairs of bits in the multiplier in conjunction with the bit to the immediate right of the bit-pair under consideration, as shown below. Only the binary weights of 2^1 and 2^0 are required because each pair of bits is examined independently of the other pairs. The low-order bit has an implied 0 to the immediate right. This is required so that there will be three bits to consider: the low-order two bits and the implied 0. If the number of bits in the multiplier is not an even number, then the high-order bit is extended to accommodate the leftmost bit pair.

$$\begin{array}{cccccccc l} 2^7 & 2^6 & 2^5 & 2^4 & 2^3 & 2^2 & 2^1 & 2^0 & \\ 0 & 0 & 1 & 1 & 1 & 1 & 0 & 0 & (60) \end{array}$$

$$\downarrow$$

$$\begin{array}{cccccccccc} 2^1 & 2^0 & 2^1 & 2^0 & 2^1 & 2^0 & 2^1 & 2^0 & & \\ 0 & 0 & 1 & 1 & 1 & 1 & 0 & 0 & \boxed{0} & (60) \end{array}$$

Bit-pair 2^1 2^0 is examined in conjunction with the implied 0 to the right of the low-order 0; bit-pair 2^3 2^2 is examined in conjunction with bit 2^1; bit-pair 2^5 2^4 is examined in conjunction with bit 2^3; and bit-pair 2^7 2^6 is examined in conjunction with bit 2^5, where bit 2^7 is the sign extension of the multiplier in order to obtain a bit-pair. Each pair of bits is scanned from right to left with the rightmost bit of each pair used as the column reference for the partial product placement, because it is the center of the three bits under consideration. Using the Booth algorithm technique, the additions and subtractions take place on the multiplier as shown in Figure 6.17.

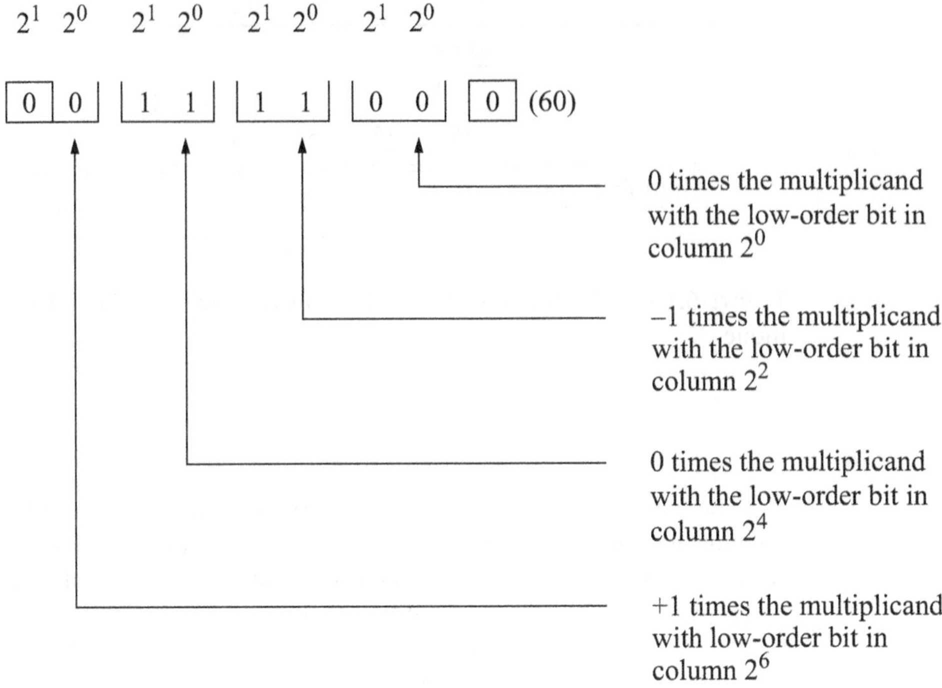

Figure 6.17 Bit-pair recoding for a positive multiplier.

Consider other examples with strings consisting of 010 and 101, as shown in Figure 6.18 and Figure 6.19, respectively.

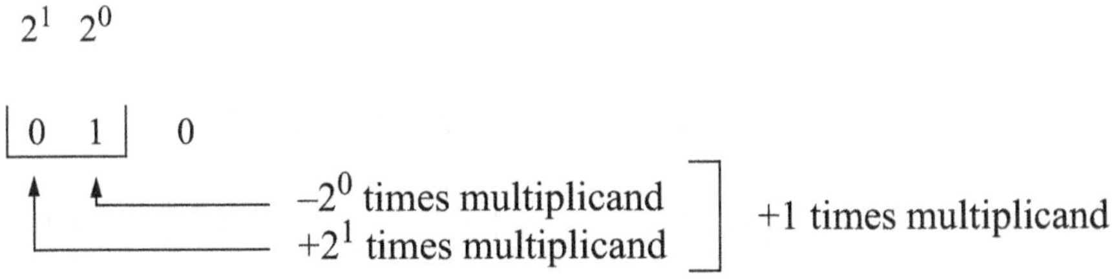

2^1 2^0

0 1 0

-2^0 times multiplicand
$+2^1$ times multiplicand
+1 times multiplicand

Figure 6.18 A string with a single 1 for the beginning and end of a string.

2^1 2^0

1 0 1

$+2^0$ times multiplicand
-2^1 times multiplicand
–1 times multiplicand

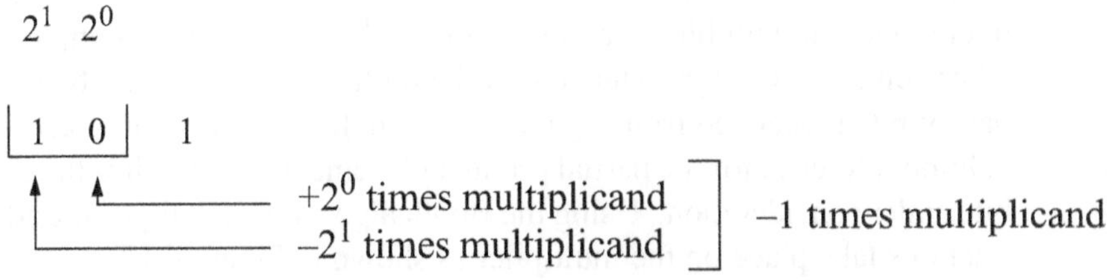

Figure 6.19 A string with a 1 at the beginning and end of a string.

For Figure 6.18 the beginning of the string specifies that -2^0 times the multiplicand is added to the partial product; the end of the string specifies that $+2^1$ times the multiplicand is added to the partial product. The net result of this string is that +1 times the multiplicand is added to the partial product. The net result of Figure 6.19 is that –1 times the multiplicand is added to the partial product.

Since the bit-pair is examined in conjunction with the bit to the immediate right of the bit-pair under consideration, there are a total of eight versions of the multiplicand to be added to or subtracted from the partial product, as shown in Table 6.2. Note that –2 times the multiplicand is the 2s complement of the multiplicand shifted left 1 bit position, and +2 times the multiplicand is the multiplicand shifted left 1 bit position.

Table 6.2 Multiplicand Versions for Multiplier Bit-Pair Recoding

Multiplier Bit-Pair 2^1	Multiplier Bit-Pair 2^0	Multiplier Bit on the Right		
$i+1$	i	$i-1$	Multiplicand Versions	Explanation
0	0	0	0 × multiplicand	No string
0	0	1	+1 × multiplicand	End of string
0	1	0	+1 × multiplicand	Single 1 (+2 –1)
0	1	1	+2 × multiplicand	End of string
1	0	0	–2 × multiplicand	Beginning of string
1	0	1	–1 × multiplicand	End/beginning of string (+1 –2)
1	1	0	–1 × multiplicand	Beginning of string
1	1	1	0 × multiplicand	String of 1s

Figure 6.20 shows a positive multiplicand and a negative multiplier; Figure 6.21 shows a negative multiplicand and a positive multiplier; Figure 6.22 shows a positive multiplicand and a positive multiplier; and Figure 6.23 shows a negative multiplicand and a negative multiplier, all using bit-pair recoding.

The organization of a sequential add-shift multiplier unit using the bit-pair recoding technique is shown in Figure 6.24. The low-order three bits of the multiplier select one of five inputs of the multiplexer-equivalent circuit. The five inputs are: 0 × the multiplicand, +1 × the multiplicand, –1 × the multiplicand, +2 × the multiplicand, and –2 × the multiplicand.

The adder uses the carry lookahead method with a carry-in of 0 and a carry-out that is not used. There is a count-down sequence counter that counts the number of bit pairs (the number of partial products); when the counter reaches a count of zero, the operation is terminated.

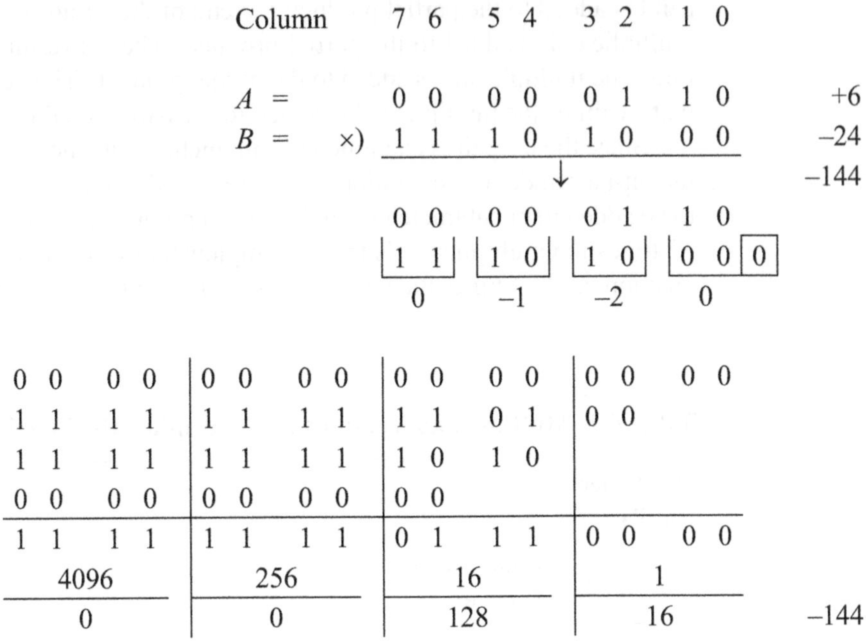

Column		7 6	5 4	3 2	1 0	
A =		0 0	0 0	0 1	1 0	+6
B =	×)	1 1	1 0	1 0	0 0	−24
			↓			−144
		0 0	0 0	0 1	1 0	
		[1 1]	[1 0]	[1 0]	[0 0 [0]]	
		0	−1	−2	0	

0 0 0 0	0 0 0 0	0 0 0 0	0 0 0 0	
1 1 1 1	1 1 1 1	1 1 0 1	0 0	
1 1 1 1	1 1 1 1	1 0 1 0		
0 0 0 0	0 0 0 0	0 0		
1 1 1 1	1 1 1 1	0 1 1 1	0 0 0 0	
4096	256	16	1	
0	0	128	16	−144

Figure 6.20 Bit-pair recoding using a positive multiplicand and a negative multiplier.

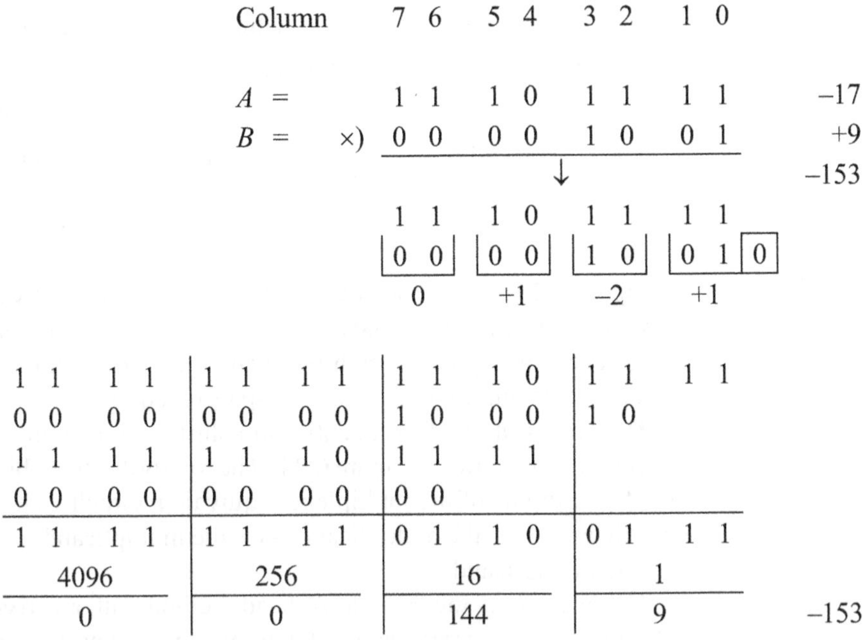

Column		7 6	5 4	3 2	1 0	
A =		1 1	1 0	1 1	1 1	−17
B =	×)	0 0	0 0	1 0	0 1	+9
			↓			−153
		1 1	1 0	1 1	1 1	
		[0 0]	[0 0]	[1 0]	[0 1 [0]]	
		0	+1	−2	+1	

1 1 1 1	1 1 1 1	1 1 1 0	1 1 1 1	
0 0 0 0	0 0 0 0	1 0 0 0	1 0	
1 1 1 1	1 1 1 0	1 1 1 1		
0 0 0 0	0 0 0 0	0 0		
1 1 1 1	1 1 1 1	0 1 1 0	0 1 1 1	
4096	256	16	1	
0	0	144	9	−153

Figure 6.21 Bit-pair recoding using a negative multiplicand and a positive multiplier.

Column		7 6	5 4	3 2	1 0	
A =		0 0	0 1	0 0	1 1	+19
B =	×)	0 0	1 1	0 1	1 0	+54
			↓			+1026
		0 0	0 1	0 0	1 1	
		[0 0]	[1 1]	[0 1]	[1 0] [0]	
		+1	−1	+2	−2	

1	1	1	1	1	1	1	1	1	1	0	1	1	0	1	0	
0	0	0	0	0	0	0	0	1	0	0	1	1	0			
1	1	1	1	1	1	1	0	1	1	0	1					
0	0	0	0	0	1	0	0	1	1							
0	0	0	0	0	1	0	0	0	0	0	0	0	0	1	0	
4096				256				16				1				
0				1024				0				2				+1026

Figure 6.22 Bit-pair recoding using a positive multiplicand and a positive multiplier.

Column		7 6	5 4	3 2	1 0	
A =		1 1	0 0	0 1	1 1	−57
B =	×)	1 0	1 0	1 1	0 1	−83
			↓			+4731
		1 1	0 0	0 1	1 1	
		[1 0]	[1 0]	[1 1]	[0 1] [0]	
		−1	−1	−1	+1	

1	1	1	1	1	1	1	1	1	1	0	0	0	1	1	1	
0	0	0	0	0	0	0	0	1	1	1	0	0	1			
0	0	0	0	0	0	1	1	1	0	0	1					
0	0	0	0	1	1	1	0	0	1							
0	0	0	1	0	0	1	0	0	1	1	1	1	0	1	1	
4096				256				16				1				
4096				512				112				11				+4731

Figure 6.23 Bit-pair recoding using a negative multiplicand and a negative multiplier.

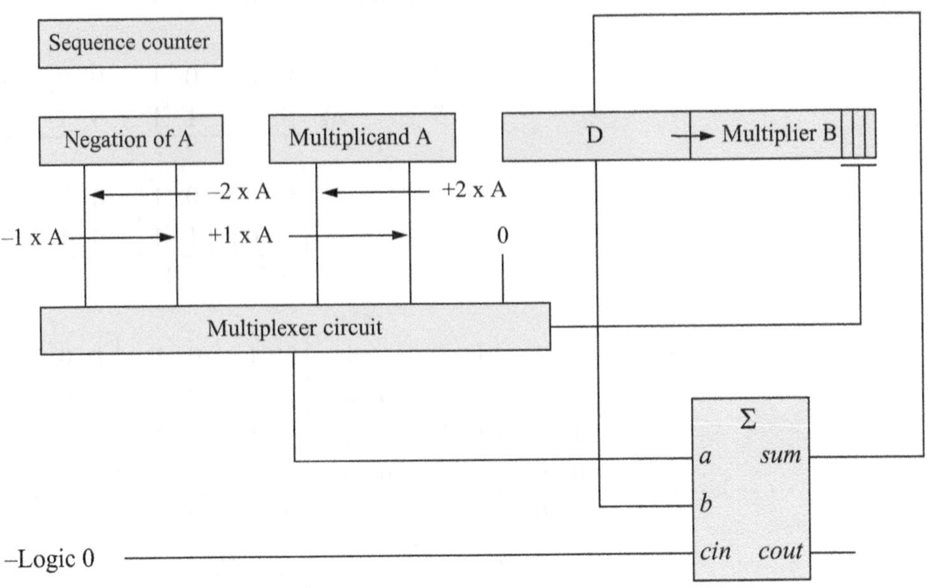

Figure 6.24 Organization of a sequential add-shift multiplier using the bit-pair recoding technique.

The mixed-design module, using behavioral and dataflow modeling, is shown in Figure 6.25. The inputs are the multiplicand, *a[7:0]*, and the multiplier, *b[7:0]*; the output is the product *prod[15:0]*. There are three internal wires defined as: *a_bar[7:0]*, which will be assigned the 1s complement of the multiplicand using the continuous assignment statement; *a_ext_pos[15:0]*, which is the uncomplemented version of the multiplicand with sign extension to 16 bits; and *a_ext_neg[15:0]*, which is the 2s complement (negation) of the multiplicand with sign extension to 16 bits. The internal register is *a_neg[7:0]* which will be assigned the 2s complement of operand *A* by means of the **always** statement, as shown below.

```
always @ (a_bar)
    a_neg = a_bar + 1;
```

The uncomplemented version of *A* is extended to 16 bits by the following statement:

```
a_ext_pos = {{8{a[7]}}, a};
```

where *8{a[7]}* replicates the sign of the multiplicand eight times, which is then concatenated with the 8-bit multiplicand. The 2s complemented version of the multiplicand is extended to 16 bits by the following statement:

a_ext_neg = {{8{a_neg[7]}}, a_neg};

where *8{a_neg[7]}* replicates the sign of the 2s complemented multiplicand eight times, which is then concatenated with the 2s complemented 8-bit multiplicand.

Multiplier bits *b[1:0]* are tested with an implied 0 to the right of bit *b[0]*. Therefore, only four combinations are required to test bits *b[1:0]*. The *sensitivity list* in the **always** statement will cause the statements in the **begin . . . end** block to execute if any variable changes value. The **case** statement is a multiple-way conditional branch. It executes one of several different procedural statements depending on the comparison of an expression with a case item. Refer to Table 6.2 for the description that follows.

If *b[1:0]* = 00(0), then partial product 1 is 0 times the multiplicand which is 16'h0000. If *b[1:0]* = 01(0), then partial product 1 is +1 times the multiplicand, which is *a_ext_pos*. If *b[1:0]* = 10(0), then partial product 1 is –2 times the multiplicand, which is the 2s complement of the multiplicand shifted left one bit position, as shown below.

pp1 = {a_ext_neg[14:8], a_neg, 1'b0};

The above expression concatenates the following bits: bits 14 through 8 of the 2s complement of operand *A*; the eight bits of the 2s complement of *A*; and a zero bit for a total of 16 bits. If *b[1:0]* = 11(0), then partial product 1 is –1 times the multiplicand, which is *a_ext_neg* — the 2s complement of the multiplicand.

Bit 3 and bit 2 of the multiplier are tested next using the case item *b[3:1]*; that is, bits 3 and 2 in conjunction with bit 1. Therefore, all eight combinations of the three bits *b[3:1]* are required for this test to generate partial product 2. Partial product 2 is shifted left two bit positions with respect to partial product 1 with zeroes filling in the vacated rightmost bit positions. Refer to Table 6.2 for the description that follows.

If *b[3:1]* = 000, then partial product 2 is 0 times the multiplicand which is 16'h0000. If *b[3:1]* = 001, then the 1 → 0 boundary specifies partial product 1 as +1 times the multiplicand, which is *{a_ext_pos[13:0], 2'b00}*; that is, bits 13 through 0 of uncomplemented operand *A* concatenated with 00 for a total of sixteen bits. If *b[3:1]* = 010, then the 0 → 1 boundary indicates –1 times the multiplicand, and the 1 → 0 boundary indicates +2 times the multiplicand, resulting in partial product 2 having a value of +1 times the multiplicand.

If *b[3:1]* = 011, then the 1 → 1 boundary specifies 0 times the multiplicand, and the 1 → 0 boundary specifies +2 times the multiplicand, which is the multiplicand shifted left one bit position, as indicated by *{a_ext_pos[12:0], 3'b000}*. If *b[3:1]* = 100, then the 0 → 0 boundary specifies 0 times the multiplicand, and the 0 → 1 boundary specifies –2 times the multiplicand, which is the 2s complemented multiplicand shifted left one bit position, as indicated by *{a_ext_neg[12:0], 3'b000}*.

If *b[3:1]* = 101, then the 1 → 0 boundary specifies +1 times the multiplicand, and the 0 → 1 boundary specifies –2 times the multiplicand, which is the 2s complemented multiplicand shifted left one bit position, resulting in partial product 2 having a value of –1 times the multiplicand, as indicated by *{a_ext_pos[13:0], 2'b00}*.

If *b[3:1]* = 110, then the 0 → 1 boundary specifies –1 times the multiplicand, and the 1 → 1 boundary specifies 0 times the multiplicand, resulting in partial product 2 having a value of –1 times the multiplicand. If *b[3:1]* = 111, then the result is 0 times the multiplicand. The remaining bit pairs are tested in a similar manner.

```
//behavioral and dataflow for bit-pair multiplier
module bit_pair (a, b, prod);

input [7:0] a, b;
output [15:0] prod;

wire [7:0] a, b;
wire [15:0] prod;

//define internal wires
wire [7:0] a_bar;
wire [15:0] a_ext_pos;        //positive A with sign extension
wire [15:0] a_ext_neg;        //A negated with sign extension

//define internal registers
reg [7:0] a_neg;
reg [15:0] pp1, pp2, pp3, pp4;//partial products

//1s complement of A
assign a_bar = ~a;

//2s complement of A
always @ (a_bar)
   a_neg = a_bar + 1;

//define a_ext_pos and a_ext_neg
//extend sign of positive A to 16 bits
assign a_ext_pos = {{8{a[7]}}, a};

//extend sign of negative A to 16 bits
assign a_ext_neg = {{8{a_neg[7]}}, a_neg};

//test b[1:0] ----------------------------------------
always @ (b or a_ext_pos or a_ext_neg)
begin
   case (b[1:0])        //test bits 1 and 0 with 0
         2'b00:            //(000) = 0 times multiplicand
                     pp1 = 16'h0000;

//continued on next page
```

Figure 6.25 Mixed-design module for a multiplier using the bit-pair recoding technique.

```
        2'b01:          //(010) = +1 times multiplicand
                pp1 = a_ext_pos;

        2'b10:          //(100) = -2 times multiplicand
                pp1 = {a_ext_neg[14:8], a_neg, 1'b0};

        2'b11:          //(110) = -1 times multiplicand
                pp1 = a_ext_neg;
   endcase
end

//test b[3:1] ----------------------------------------------
always @ (b or a_ext_pos or a_ext_neg)
begin
   case (b[3:1])     //test bits 3 and 2 with 1
      3'b000:           //0 times multiplicand
                pp2 = 16'h0000;

      3'b001:           //+1 times multiplicand
                pp2 = {a_ext_pos[13:0], 2'b00};

      3'b010:           //+1 times multiplicand
                pp2 = {a_ext_pos[13:0], 2'b00};

      3'b011:           //+2 times multiplicand
                pp2 = {a_ext_pos[12:0], 3'b000};

      3'b100:           //-2 times multiplicand
                pp2 = {a_ext_neg[12:0], 3'b000};

      3'b101:           //-1 times multiplicand
                pp2 = {a_ext_neg[13:0], 2'b00};

      3'b110:           //-1 times multiplicand
                pp2 = {a_ext_neg[13:0], 2'b00};

      3'b111:           //0 times multiplicand
                pp2 = 16'h0000;
   endcase
end

//continued on next page
```

Figure 6.25 (Continued)

```
//test b[5:3] ------------------------------------------------
always @ (b or a_ext_pos or a_ext_neg)
begin
   case (b[5:3])      //test bits 5 and 4 with 3
      3'b000:             //0 times multiplicand
                  pp3 = 16'h0000;

      3'b001:             //+1 times multiplicand
                  pp3 = {a_ext_pos[11:0], 4'b0000};

      3'b010:             //+1 times multiplicand
                  pp3 = {a_ext_pos[11:0], 4'b0000};

      3'b011:             //+2 times multiplicand
                  pp3 = {a_ext_pos[10:0], 5'b00000};

      3'b100:             //-2 times multiplicand
                  pp3 = {a_ext_neg[10:0], 5'b00000};

      3'b101:             //-1 times multiplicand
                  pp3 = {a_ext_neg[11:0], 4'b0000};

      3'b110:             //-1 times multiplicand
                  pp3 = {a_ext_neg[11:0], 4'b0000};

      3'b111:             //0 times multiplicand
                  pp3 = 16'h0000;
   endcase
end

//test b[7:5] ------------------------------------------------
always @ (b or a_ext_pos or a_ext_neg)
begin
   case (b[7:5])      //test bits 7 and 6 with 5
      3'b000:             //0 times multiplicand
                  pp4 = 16'h0000;

      3'b001:             //+1 times multiplicand
                  pp4 = {a_ext_pos[9:0], 6'b000000};

      3'b010:             //+1 times multiplicand
                  pp4 = {a_ext_pos[9:0], 6'b000000};

      3'b011:             //+2 times multiplicand
                  pp4 = {a_ext_pos[8:0], 7'b0000000};

//continued on next page
```

Figure 6.25 (Continued)

```
        3'b100:                 //-2 times multiplicand
                      pp4 = {a_ext_neg[8:0], 7'b0000000};

        3'b101:                 //-1 times multiplicand
                      pp4 = {a_ext_neg[9:0], 6'b000000};

        3'b110:                 //-1 times multiplicand
                      pp4 = {a_ext_neg[9:0], 6'b000000};

        3'b111:                 //0 times multiplicand
                      pp4 = 16'h0000;
   endcase
end

//define product
assign prod = pp1 + pp2 + pp3 + pp4;

endmodule
```

Figure 6.25 (Continued)

The test bench is shown in Figure 6.26 which applies several input vectors for the multiplicand and multiplier, including all combinations of positive and negative operands. The outputs are shown in Figure 6.27 with binary inputs for the operands and hexadecimal outputs for the product. The waveforms are shown in Figure 6.28.

```
//test bench for the bit-pair multiplier
module bit_pair_tb;

reg [7:0] a, b;
wire [15:0] prod;

//display operands a, b, and prod
initial
$monitor ("a = %b, b = %b, prod = %h", a, b, prod);

//apply input vectors
initial
begin
   #0    a = 8'b1100_0111;    //a = -57
         b = 8'b1010_1101;    //b = -83   product = ffc7h
//continued on next page
```

Figure 6.26 Test bench for the bit-pair recoding multiplier of Figure 6.25.

```
   #10   a = 8'b0000_0110;     //a = +6
         b = 8'b1110_1000;     //b = -24    product = ff70h

   #10   a = 8'b1110_1111;     //a = -17
         b = 8'b0000_1001;     //b = +9     product = ff67h

   #10   a = 8'b0001_0011;     //a = +19
         b = 8'b0011_0110;     //b = +54    product = 0402h

   #10   a = 8'b0001_1001;     //a = +25
         b = 8'b0101_1000;     //b = +88    product = 0898h

   #10   a = 8'b0001_1001;     //a = +25
         b = 8'b0110_1110;     //b = +110   product = 0abeh

   #10   a = 8'b0011_1101;     //a = +61
         b = 8'b1110_0011;     //b = -29    product = 1917h

   #10   a = 8'b1100_1111;     //a = -49
         b = 8'b1110_0001;     //b = -31    product = 05efh

   #10   a = 8'b1111_1111;     //a = -1
         b = 8'b1111_1111;     //b = -1     product = 0001h

   #10   a = 8'b1111_1111;     //a = -1
         b = 8'b0000_0001;     //b = +1     product = ffffh

   #10   a = 8'b1100_0111;     //a = -57
         b = 8'b1000_1100;     //b = -116   product = 19d4h

   #10   a = 8'b0011_1110;     //a = +62
         b = 8'b0100_1000;     //b = +72    product = 1170h

   #10   $stop;

end

//instantiate the module into the test bench
bit_pair inst1 (
   .a(a),
   .b(b),
   .prod(prod)
   );

endmodule
```

Figure 6.26 (Continued)

```
a = 11000111, b = 10101101, prod = 127b
a = 00000110, b = 11101000, prod = ff70
a = 11101111, b = 00001001, prod = ff67
a = 00010011, b = 00110110, prod = 0402
a = 00011001, b = 01011000, prod = 0898
a = 00011001, b = 01101110, prod = 0abe
a = 00111101, b = 11100011, prod = f917
a = 11001111, b = 11100001, prod = 05ef
a = 11111111, b = 11111111, prod = 0001
a = 11111111, b = 00000001, prod = ffff
a = 11000111, b = 10001100, prod = 19d4
a = 00111110, b = 01001000, prod = 1170
```

Figure 6.27 Outputs for the bit-pair recoding multiplier of Figure 6.25.

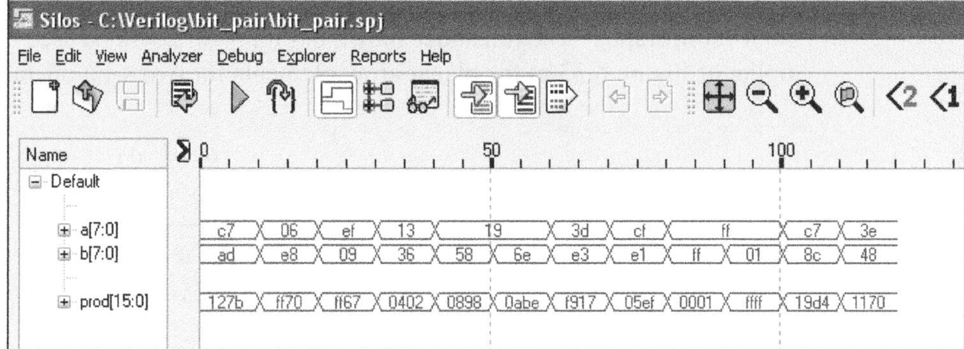

Figure 6.28 Waveforms for the bit-pair recoding multiplier of Figure 6.25.

6.4 Array Multiplication

A hardware array multiplier that permits a very high speed multiply operation for unsigned operands and positive signed operands is presented in this section. For signed operands, the multiplier must be positive — the multiplicand can be positive or negative. The multiplication of two n-bit operands generates a product of $2n$ bits.

The sequential add-shift technique requires less hardware, but is relatively slow when compared to the array multiplier method. In the sequential add-shift method, multiplication of the multiplicand by a 1 bit in the multiplier simply copies the multiplicand. If the multiplier bit is a 1, then the multiplicand is entered in the appropriately shifted position as a partial product to be added to other partial products to form the product. If the multiplier bit is 0, then 0s are entered as a partial product.

Although the array multiplier method is applicable to any size operands, an example will be presented that uses two 4-bit operands as shown in Figure 6.29. The multiplicand is $a\,[3{:}0] = a_3a_2a_1a_0$ and the multiplier is $b[3{:}0] = b_3b_2b_1b_0$, where a_0 and b_0 are the low-order bits of A and B, respectively.

Each bit in the multiplicand is multiplied by the low-order bit, b_0, of the multiplier. This is equivalent to the AND function and generates the first of three partial products. Each bit in the multiplicand is then multiplied by bit b_1 of the multiplier. The resulting partial product is shifted left one bit position. The process is repeated for bit b_2 and bit b_3 of the multiplier. The partial products are then added together to form the product. A carry-out of any column is added to the appropriate next higher-order column.

The array multiplier described in this section assumes that the multiplier is positive, although multipliers can be designed that utilize both positive and negative operands based on a method proposed by Baugh and Wooley. An example is shown in Figure 6.30 with a positive multiplicand and a positive multiplier using the paper-and-pencil method. Since the multiplicand is positive, the partial products are zero-extended to the left to accommodate the 2*n* bits. Figure 6.31 uses a negative multiplicand and a positive multiplier. In this example, the partial products extend the negative sign left to accommodate 2*n* bits. Both examples use the paper-and-pencil method for 4-bit operands. The full adder shown in Figure 6.32 will be used in the array multiplier of Figure 6.33 as the planar array elements for two 4-bit operands.

					a_3	a_2	a_1	a_0
				×)	b_3	b_2	b_1	b_0
Partial product 1					a_3b_0	a_2b_0	a_1b_0	a_0b_0
Partial product 2				a_3b_1	a_2b_1	a_1b_1	a_0b_1	
Partial product 3			a_3b_2	a_2b_2	a_1b_2	a_0b_2		
Partial product 4		a_3b_3	a_2b_3	a_1b_3	a_0b_3			
Product	2^7	2^6	2^5	2^4	2^3	2^2	2^1	2^0

Figure 6.29 General multiply algorithm for two 4-bit operands.

				0	1	1	1	+7
			×)	0	1	1	1	+7
0	0	0	0	0	1	1	1	
0	0	0	0	1	1	1		
0	0	0	1	1	1			
0	0	0	0	0				
0	0	1	1	0	0	0	1	+49

Figure 6.30 Paper-and-pencil method to multiply a positive multiplicand by a positive multiplier.

				1	1	0	0	−4
			×)	0	1	0	1	+5
1	1	1	1	1	1	0	0	
0	0	0	0	0	0	0		
1	1	1	1	0	0			
0	0	0	0	0				
1	1	1	0	1	1	0	0	−20

Figure 6.31 Paper-and-pencil method to multiply a negative multiplicand by a positive multiplier.

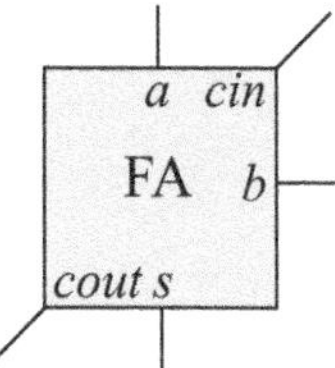

Figure 6.32 Full adder to be used in the array multiplier.

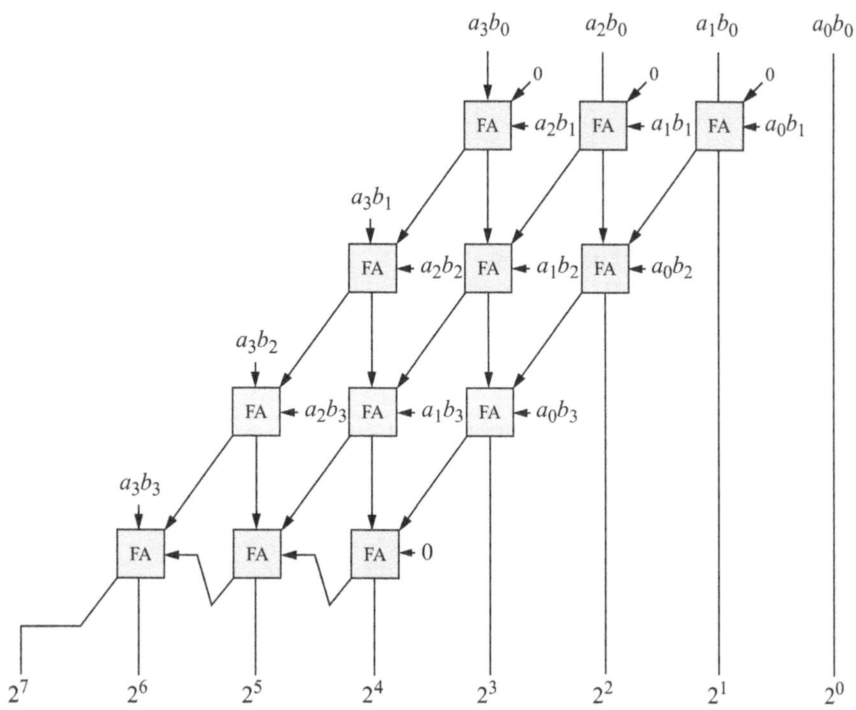

Figure 6.33 Array multiplier for 4-bit operands.

The array multiplier of Figure 6.33 is reproduced in Figure 6.34 showing the instantiation names and net names for the array multiplier structural module. The multiplicand is *a[3:0]*, the multiplier is *b[3:0]*, and the product is *prod[7:0]*. Dataflow modeling will be used to design the logic primitives and the full adder, all of which will be instantiated into the multiplier structural module. The dataflow module for a 2-input AND gate is shown in Figure 6.35 and the dataflow module for a full adder is shown in Figure 6.36.

The structural module for the array multiplier is shown in Figure 6.37 and the test bench module is shown in Figure 6.38 for all 256 combinations of the multiplicand and multiplier. Several outputs are shown in Figure 6.39 in decimal notation.

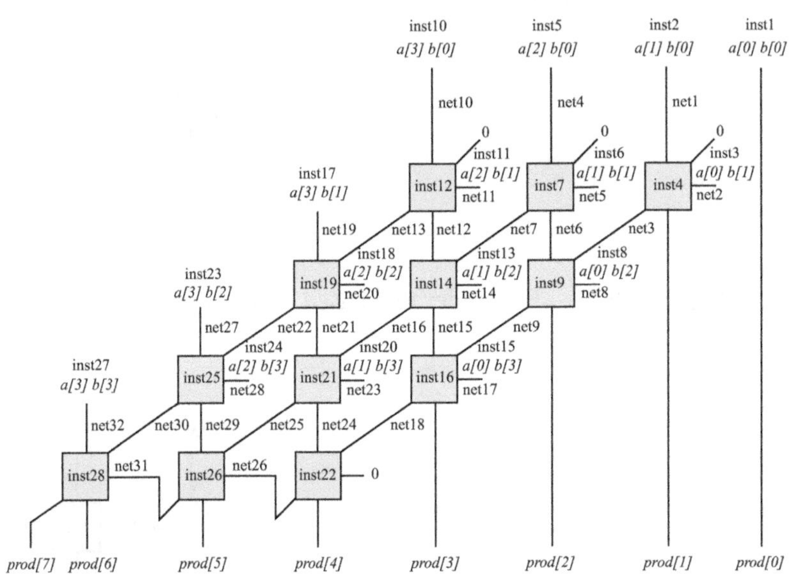

Figure 6.34 Array multiplier for two 4-bit operands showing the instantiation names and net names.

```
//dataflow 2-input and gate
module and2_df (x1, x2, z1);

input x1, x2;
output z1;
wire x1, x2;
wire z1;

assign z1 = x1 & x2;
endmodule
```

Figure 6.35 Dataflow module for a 2-input AND gate.

```
//dataflow full adder
module full_adder (a, b, cin, sum, cout);

//list all inputs and outputs
input a, b, cin;
output sum, cout;

//define wires
wire a, b, cin;
wire sum, cout;

//continuous assign for dataflow
assign sum = (a ^ b) ^ cin;
assign cout = cin & (a ^ b) | (a & b);

endmodule
```

Figure 6.36 Dataflow module for a full adder.

```
//structural array multiplier 4 bits
module array_mul4 (a, b, prod);

//define inputs and outputs
input [3:0] a, b;
output [7:0] prod;

wire [3:0] a, b;
wire [7:0] prod;

//define internal wires
wire  net1, net2, net3, net4, net5, net6, net7, net8,
      net9, net10, net11, net12, net13, net14, net15, net16,
      net17, net18, net19, net20, net21, net22, net23, net24,
      net25, net26, net27, net28, net29, net30, net31, net32;

//instantiate the logic for product[0]
and2_df inst1 (
   .x1(a[0]),
   .x2(b[0]),
   .z1(prod[0])       //product[0]
   );

//continued on next page
```

Figure 6.37 Structural module for a 4-bit array multiplier.

```
//instantiate the logic for product[1]
and2_df inst2 (
   .x1(a[1]),
   .x2(b[0]),
   .z1(net1)
   );

and2_df inst3 (
   .x1(a[0]),
   .x2(b[1]),
   .z1(net2)
   );

full_adder inst4 (
   .a(net1),
   .b(net2),
   .cin(1'b0),
   .sum(prod[1]),      //product[1]
   .cout(net3)
   );

//instantiate the logic for product[2]
and2_df inst5 (
   .x1(a[2]),
   .x2(b[0]),
   .z1(net4)
   );

and2_df inst6 (
   .x1(a[1]),
   .x2(b[1]),
   .z1(net5)
   );

full_adder inst7 (
   .a(net4),
   .b(net5),
   .cin(1'b0),
   .sum(net6),
   .cout(net7)
   );

and2_df inst8 (
   .x1(a[0]),
   .x2(b[2]),
   .z1(net8)
   );                  //continued on next page
```

Figure 6.37 (Continued)

```
full_adder inst9 (
   .a(net6),
   .b(net8),
   .cin(net3),
   .sum(prod[2]),     //product[2]
   .cout(net9)
   );


//instantiate the logic for product[3]
and2_df inst10 (
   .x1(a[3]),
   .x2(b[0]),
   .z1(net10)
   );

and2_df inst11 (
   .x1(a[2]),
   .x2(b[1]),
   .z1(net11)
   );

full_adder inst12 (
   .a(net10),
   .b(net11),
   .cin(1'b0),
   .sum(net12),
   .cout(net13)
   );

and2_df inst13 (
   .x1(a[1]),
   .x2(b[2]),
   .z1(net14)
   );

full_adder inst14 (
   .a(net12),
   .b(net14),
   .cin(net7),
   .sum(net15),
   .cout(net16)
   );

//continued on next page
```

Figure 6.37 (Continued)

```
and2_df inst15 (
   .x1(a[0]),
   .x2(b[3]),
   .z1(net17)
   );

full_adder inst16 (
   .a(net15),
   .b(net17),
   .cin(net9),
   .sum(prod[3]),      //product[3]
   .cout(net18)
   );


//instantiate the logic for product[4]
and2_df inst17 (
   .x1(a[3]),
   .x2(b[1]),
   .z1(net19)
   );

and2_df inst18 (
   .x1(a[2]),
   .x2(b[2]),
   .z1(net20)
   );

full_adder inst19 (
   .a(net19),
   .b(net20),
   .cin(net13),
   .sum(net21),
   .cout(net22)
   );

and2_df inst20 (
   .x1(a[1]),
   .x2(b[3]),
   .z1(net23)
   );

//continued on next page
```

Figure 6.37 (Continued)

```
full_adder inst21 (
   .a(net21),
   .b(net23),
   .cin(net16),
   .sum(net24),
   .cout(net25)
   );

full_adder inst22 (
   .a(net24),
   .b(1'b0),
   .cin(net18),
   .sum(prod[4]),      //product[4]
   .cout(net26)
   );

//instantiate the logic for product[5]
and2_df inst23 (
   .x1(a[3]),
   .x2(b[2]),
   .z1(net27)
   );

and2_df inst24 (
   .x1(a[2]),
   .x2(b[3]),
   .z1(net28)
   );

full_adder inst25 (
   .a(net27),
   .b(net28),
   .cin(net22),
   .sum(net29),
   .cout(net30)
   );

full_adder inst26 (
   .a(net29),
   .b(net26),
   .cin(net25),
   .sum(prod[5]),      //product[5]
   .cout(net31)
   );

//continued on next page
```

Figure 6.37 (Continued)

```
//instantiate the logic for product[6] and product[7]
and2_df inst27 (
   .x1(a[3]),
   .x2(b[3]),
   .z1(net32)
   );

full_adder inst28 (
   .a(net32),
   .b(net31),
   .cin(net30),
   .sum(prod[6]),      //product[6]
   .cout(prod[7])      //product[7]
   );

endmodule
```

Figure 6.37 (Continued)

```
//test bench for array multiplier 4 bits
module array_mul4_tb;

reg [3:0] a, b;
wire [7:0] prod;

//apply input vectors
initial
begin : apply_stimulus
   reg [8:0] invect;
   for (invect=0; invect<256; invect=invect+1)
   begin
      {a, b} = invect [8:0];
      #10 $display ("a = %d, b = %d, product = %d", a, b, prod);
   end
end

//instantiate the module into the test bench
array_mul4 inst1 (
   .a(a),
   .b(b),
   .prod(prod)
   );

endmodule
```

Figure 6.38 Test bench for the array multiplier.

```
a = 1, b = 0,  product = 0
a = 1, b = 1,  product = 1
a = 1, b = 2,  product = 2
a = 1, b = 3,  product = 3

a = 2, b = 4,  product = 8
a = 2, b = 5,  product = 10
a = 2, b = 6,  product = 12
a = 2, b = 7,  product = 14

a = 3, b = 8,  product = 24
a = 3, b = 9,  product = 27
a = 3, b = 10, product = 30
a = 3, b = 11, product = 33

a = 4, b = 12, product = 48
a = 4, b = 13, product = 52
a = 4, b = 14, product = 56
a = 4, b = 15, product = 60

a = 5, b = 0,  product = 0
a = 5, b = 1,  product = 5
a = 5, b = 2,  product = 10
a = 5, b = 3,  product = 15

a = 6, b = 0,  product = 0
a = 6, b = 1,  product = 6
a = 6, b = 2,  product = 12
a = 6, b = 3,  product = 18

a = 7, b = 4,  product = 28
a = 7, b = 5,  product = 35
a = 7, b = 6,  product = 42
a = 7, b = 7,  product = 49

a = 8, b = 8,  product = 64
a = 8, b = 9,  product = 72
a = 8, b = 10, product = 80
a = 8, b = 11, product = 88

a = 9, b = 12, product = 108
a = 9, b = 13, product = 117
a = 9, b = 14, product = 126
a = 9, b = 15, product = 135

a = 10, b = 0,  product = 0
a = 10, b = 1,  product = 10
a = 10, b = 2,  product = 20
a = 10, b = 3,  product = 30

a = 11, b = 4,  product = 44
a = 11, b = 5,  product = 55
a = 11, b = 6,  product = 66
a = 11, b = 7,  product = 77

a = 12, b = 8,  product = 96
a = 12, b = 9,  product = 108
a = 12, b = 10, product = 120
a = 12, b = 11, product = 132

a = 12, b = 12, product = 144
a = 12, b = 13, product = 156
a = 12, b = 14, product = 168
a = 12, b = 15, product = 180

a = 13, b = 8,  product = 104
a = 13, b = 9,  product = 117
a = 13, b = 10, product = 130
a = 13, b = 11, product = 143

a = 14, b = 12, product = 168
a = 14, b = 13, product = 182
a = 14, b = 14, product = 196
a = 14, b = 15, product = 210

a = 14, b = 12, product = 168
a = 14, b = 13, product = 182
a = 14, b = 14, product = 196
a = 14, b = 15, product = 210

a = 15, b = 8,  product = 120
a = 15, b = 9,  product = 135
a = 15, b = 10, product = 150
a = 15, b = 11, product = 165

a = 15, b = 12, product = 180
a = 15, b = 13, product = 195
a = 15, b = 14, product = 210
a = 15, b = 15, product = 225
```

Figure 6.39 Outputs for the array multiplier.

6.5 Table Lookup Multiplication

Multiplication can be accomplished by using a memory that contains different versions of the multiplicand to be added to the partial products. This method is faster than the sequential add-shift technique, because it shifts the partial products three bit positions after the add operation rather than one bit position as in the sequential add-shift method.

The organization of the sequential add-three-bit-shift multiplier is shown in Figure 6.40 for 8-bit multiplicands and multipliers. Firmware loads the multiplicand table in memory and the multiplier prior to the multiply operation. The multiplicand table is addressed by the low-order three bits of the multiplier; therefore, the table contains eight entries as follows: multiplicand times 0 through multiplicand times seven. Each multiplicand version is 11 bits in length to accommodate the highest entry of multiplicand times seven.

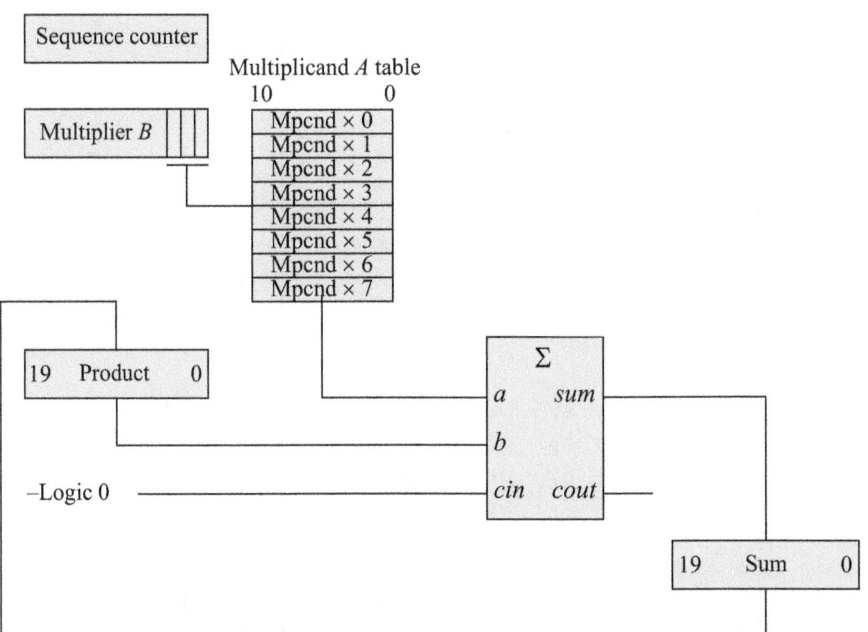

Figure 6.40 Organization for the sequential add-three-bit-shift multiplier.

The multiplier register *B* is a parallel-in, parallel-out right-shift register with a capacity of nine bits to accommodate a multiplier that is divisible by three — the high-order bit is the sign extension. The *sum* register must be wide enough to contain the multiplicand table entries of eleven bits plus an additional nine bits to accommodate the three right shift operations of three bits per shift. The *product* register must be the same width as the *sum* register.

The multiplicand table outputs 10 – 0 connect to the adder *A* inputs 19 – 9 with adder inputs 8 – 0 connected to a logic 0. The product register bits 19 – 0 connect to the adder *B* inputs 19 – 0. The multiplicand version addressed by *b[2:0]* is added to the sum — initially all zeroes. Then the sum is shifted right three bit positions with the sign bit propagated three bit positions and loaded into the *product* register. The *multiplier* register is then shifted right three bit positions to address the multiplicand table and obtain the next version of the multiplicand for the next sequence of add-three-bit-shift operations. A count-down counter determines the aggregate number of add-shift sequences that occur depending on the size of the operands.

Table 6.3 illustrates an example of the table lookup method for a multiplicand of +6 (0000 0110) and a multiplier of +11 (0 0000 1011). The *product* register is initialized to all zeroes. Since the low-order three bits of the multiplier are *b[2:0]* = 011, the multiplicand version that is selected is multiplicand times three. This is added to the product to obtain the sum. The sum is then shifted right three bit positions with the high-order bit propagating right three bits. The process repeats for multiplicand times 1 and multiplicand times 0 to obtain a product of +66.

Table 6.3 Example of Add-Three-Bit-Shift Multiplication

	19 ... 9 Multiplicand Table	8 ... 0	
Product =	00000000000	000000000	
+) Multiplicand × 3	00000010010	000000000	
Sum =	00000010010	000000000	
Shift right 3. Product =	**000**00000010	010000000	
+) Multiplicand × 1	00000000110	000000000	
Sum =	00000001000	010000000	
Shift right 3. Product =	**000**00000001	000010000	
+) Multiplicand × 0	00000000000	000000000	
Sum =	00000000001	000010000	
Shift right 3. Product =	**00000000000**	**001000010**	(+66)

The table lookup multiplication method will be designed using behavioral modeling. The multiplicand table entries are shown in Table 6.4 for a multiplicand of +6 (000 0000 0110). The values of the multiplicand versions are listed for convenience, but only the binary values of multiplicand versions are stored. This table will be stored as a separate file in the project folder as *mem.mpcnd* with no **.v** extension. There is a separate multiplicand table for each different multiplicand. The behavioral module is shown in Figure 6.41.

Table 6.4 Multiplicand Table for a Multiplicand of +6

Multiplicand Versions											
Multiplicand × 0 =	0	0	0	0	0	0	0	0	0	0	0
Multiplicand × 1 =	0	0	0	0	0	0	0	0	1	1	0
Multiplicand × 2 =	0	0	0	0	0	0	0	1	1	0	0
Multiplicand × 3 =	0	0	0	0	0	0	1	0	0	1	0
Multiplicand × 4 =	0	0	0	0	0	0	1	1	0	0	0
Multiplicand × 5 =	0	0	0	0	0	0	1	1	1	1	0
Multiplicand × 6 =	0	0	0	0	0	1	0	0	1	0	0
Multiplicand × 7 =	0	0	0	0	0	1	0	1	0	1	0

```
//behavioral multiplication table lookup
module mul_table_lookup (a, b, start, prod);

input [7:0] a;
input [8:0] b;
input start;
output [19:0] prod;

//define internal registers
reg [19:0] prod;
reg [1:0] count;
reg [19:0] sum;
reg [8:0] b_reg;

//define memory size
//memory is an array of eleven 8-bit registers
reg [10:0] mpcnd_table [0:7];

//define memory contents
//load mpcnd_table from file mem.mpcnd
initial
begin
   $readmemb ("mem.mpcnd", mpcnd_table);
end

//continued on next page
```

Figure 6.41 Behavioral module for the table lookup multiplier using a positive multiplicand.

```
always @ (posedge start)
begin
   prod [19:0] = 20'h00000;
   count = 2'b11;
   b_reg = b;
   if ((a!=0) && (b!=0))
      while (count)
         begin
            sum = {{mpcnd_table[b_reg[2:0]], 9'b0000_0000_0}
                  + prod[19:0]};
            prod = {{3{sum[19]}}, sum[19:3]};
            b_reg = b_reg >> 3;
            count = count - 1;
         end
end
endmodule
```

Figure 6.41 (Continued)

The multiplicand is eleven bits and the multiplier is nine bits producing a product of 20 bits. The multiply operation begins at the positive edge of a scalar input called *start*. The following internal registers are defined: a product, *prod[19:0]*; a 2-bit sequence control counter, *count[1:0]*; a sum, *sum[19:0]*; and a register to be used in place of the multiplier register called *b_reg[8:0]*.

The multiplicand table is defined as an array of eight 11-bit registers called *mpcnd_table*. The multiplicand table *mpcnd_table* is loaded from the file *mem.mpcnd* in memory by the system task **$readmemb** as shown below.

$readmemb ("mem.mpcnd", mpcnd_table);

The multiply operation begins on the positive edge of the *start* signal in the sensitivity list of the **always** statement, at which time three of the internal registers are initialized. Since the multiplicand is nine bits, the sequence counter is initialized to a count of three, because there are three 3-bit sections in the multiplier. The operands are then checked to determine if either the multiplicand or the multiplier is zero. If both operands are nonzero, then the multiply operation begins.

The *sum* register is assigned the sum of the multiplicand addressed by *b_reg[2:0]* concatenated with nine zeroes plus the *product* register *prod[19:0]*, as shown below.

sum = {{mpcnd_table[b_reg[2:0]], 9'b0000_0000_0} + prod[19:0]};

The sum is then assigned to the *product* register *prod[19:0]*, where the sign (positive or negative) is propagated three bit positions to the right by the replication function 3{sum[19]}, then concatenated with the high-order 17 bits of the sum. The multiplier is then shifted right three bit positions to address the multiplicand table and obtain the next version of the multiplicand. The counter decrements by one and the process repeats until the counter reaches a count of zero.

The test bench is shown in Figure 6.42 for a multiplicand of +6 (0000 0110) and seven different multipliers. The outputs are shown in Figure 6.43 and the waveforms are shown in Figure 6.44.

```
//test bench for multiplication using table lookup
module mul_table_lookup_tb;

reg [7:0] a;
reg [8:0] b;
reg start;
wire [19:0] prod;

//display variables
initial
$monitor ("a=%b, b=%b, prod=%b", a, b, prod);

//apply input vectors
initial
begin
   #0     start = 1'b0;

          //6 x 1 = 6 = 00006h
          a = 8'b0000_0110; b = 9'b0_0000_0001;
   #10    start = 1'b1;
   #10    start = 1'b0;

          //6 x 11 = 66 = 00042h
   #10    a = 8'b0000_0110; b = 9'b0_0000_1011;
   #10    start = 1'b1;
   #10    start = 1'b0;

          //6 x 6 = 36 = 00024h
   #10    a = 8'b0000_0110; b = 9'b0_0000_0110;
   #10    start = 1'b1;
   #10    start = 1'b0;

          //6 x 46 = 276 = 00114h
   #10    a = 8'b0000_0110; b = 9'b0_0010_1110;
   #10    start = 1'b1;
   #10    start = 1'b0;

//continued on next page
```

Figure 6.42 Test bench for the table lookup multiplier using a positive multiplicand.

```
            //6 x 159 = 954 = 003bah
    #10     a = 8'b0000_0110; b = 9'b0_1001_1111;
    #10     start = 1'b1;
    #10     start = 1'b0;

            //6 x 170 = 1020 = 003fch
    #10     a = 8'b0000_0110; b = 9'b0_1010_1010;
    #10     start = 1'b1;
    #10     start = 1'b0;

            //6 x 255 = 1530 = 005fah
    #10     a = 8'b0000_0110; b = 9'b0_1111_1111;
    #10     start = 1'b1;
    #10     start = 1'b0;

    #10     $stop;
end

//instantiate the module into the test bench
mul_table_lookup inst1 (
    .a(a),
    .b(b),
    .start(start),
    .prod(prod)
    );

endmodule
```

Figure 6.42 (Continued)

```
a=00000110, b=000000001, prod=00000000000000000110
a=00000110, b=000001011, prod=00000000000000000110
a=00000110, b=000001011, prod=00000000000001000010
a=00000110, b=000000110, prod=00000000000001000010
a=00000110, b=000000110, prod=00000000000000100100
a=00000110, b=000101110, prod=00000000000000100100
a=00000110, b=000101110, prod=00000000000100010100
a=00000110, b=010011111, prod=00000000000100010100
a=00000110, b=010011111, prod=00000000001110111010
a=00000110, b=010101010, prod=00000000001110111010
a=00000110, b=010101010, prod=00000000001111111100
a=00000110, b=011111111, prod=00000000001111111100
a=00000110, b=011111111, prod=00000000010111111010
```

Figure 6.43 Outputs for the table lookup multiplier.

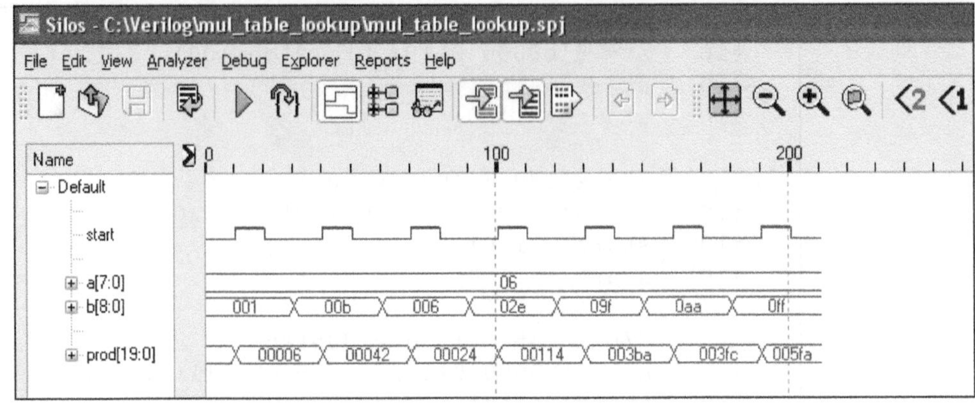

Figure 6.44 Waveforms for the table lookup multiplier.

The behavioral module of Figure 6.41 operates equally well for negative multiplicands as for positive multiplicands. Table 6.5 illustrates an example of the table lookup method for a negative multiplicand of –30 (1110 0010) and a multiplier of +1 (0 0000 0001) to produce a product of –30 (1111 1111 1111 1110 0010). The multiplicand table entries to be used in the behavioral module are shown in Table 6.6 for a multiplicand value of –30 (1110 0010). Figure 6.45 replicates the behavioral module and is used to perform multiplication on negative multiplicands with a variety of multipliers. The test bench is shown in Figure 6.46 for seven different multipliers. The outputs are shown in Figure 6.47, and the waveforms are shown in Figure 6.48.

Table 6.5 Example of Add-Three-Bit-Shift Multiplication

	19 9 Multiplicand Table	8 0	
Product =	00000000000	000000000	
+) Multiplicand × 1	11111100010	000000000	
Sum =	11111100010	000000000	
Shift right 3. Product =	**111**11111100	010000000	
+) Multiplicand × 0	00000000000	000000000	
Sum =	11111111100	010000000	
Shift right 3. Product =	**111**11111111	100010000	
+) Multiplicand × 0	00000000000	000000000	
Sum =	11111111111	100010000	
Shift right 3. Product =	**111**11111111	1**11100010**	(–30)

Table 6.6 Multiplicand Table for a Multiplicand of –30

Multiplicand Versions										
Multiplicand × 0 =	0	0	0	0	0	0	0	0	0	0	0
Multiplicand × 1 =	1	1	1	1	1	1	0	0	0	1	0
Multiplicand × 2 =	1	1	1	1	1	0	0	0	1	0	0
Multiplicand × 3 =	1	1	1	1	0	1	0	0	1	1	0
Multiplicand × 4 =	1	1	1	1	0	0	0	1	0	0	0
Multiplicand × 5 =	1	1	1	0	1	1	0	1	0	1	0
Multiplicand × 6 =	1	1	1	0	1	0	0	1	1	0	0
Multiplicand × 7 =	1	1	1	0	0	1	0	1	1	1	0

```
//behavioral multiplication table lookup
module mul_table_lookup_neg (a, b, start, prod);

input [7:0] a;
input [8:0] b;
input start;
output [19:0] prod;

//define internal registers
reg [19:0] prod;
reg [1:0] count;
reg [19:0] sum;
reg [8:0] b_reg;

//define memory size
//memory is an array of eight 8-bit registers
reg [10:0] mpcnd_table [0:7];

//define memory contents
//load mpcnd_table from file mem.mpcnd
initial
begin
   $readmemb ("mem.mpcnd", mpcnd_table);
end

//continued on next page
```

Figure 6.45 Behavioral module for the table lookup multiplier using a negative multiplicand.

```
always @ (posedge start)
begin
   prod [19:0] = 20'h00000;
   count = 2'b11;
   b_reg = b;
   if ((a!=0) && (b!=0))
      while (count)
         begin
            sum = {{mpcnd_table[b_reg[2:0]], 9'b0000_0000_0}
                  + prod[19:0]};
            prod = {{3{sum[19]}}, sum[19:3]};
            b_reg = b_reg >> 3;
            count = count - 1;
         end
end
endmodule
```

Figure 6.45 (Continued)

```
//test bench for multiplication using table lookup
module mul_table_lookup_neg_tb;

reg [7:0] a;
reg [8:0] b;
reg start;
wire [19:0] prod;

//display variables
initial
$monitor ("a=%b, b=%b, prod=%b", a, b, prod);

//apply input vectors
initial
begin
//multiplicand = -30
   #0    start = 1'b0;

         //-30 x 1 = -30 = fffe2h
         a = 8'b1110_0010; b = 9'b0_0000_0001;
   #10   start = 1'b1;
   #10   start = 1'b0;

//continued on next page
```

Figure 6.46 Test bench module for the table lookup multiplier using a negative multiplicand.

```
        //-30 x 11 = -330 = ffeb6h
    #10     a = 8'b1110_0010; b = 9'b0_0000_1011;
    #10     start = 1'b1;
    #10     start = 1'b0;

        //-30 x 6 = -180 = fff4ch
    #10     a = 8'b1110_0010; b = 9'b0_0000_0110;
    #10     start = 1'b1;
    #10     start = 1'b0;

        //-30 x 46 = -1380 = ffa9ch
    #10     a = 8'b1110_0010; b = 9'b0_0010_1110;
    #10     start = 1'b1;
    #10     start = 1'b0;

        //-30 x 159 = -4770 = fed5eh
    #10     a = 8'b1110_0010; b = 9'b0_1001_1111;
    #10     start = 1'b1;
    #10     start = 1'b0;

        //-30 x 170 = -5100 = fec14h
    #10     a = 8'b1110_0010; b = 9'b0_1010_1010;
    #10     start = 1'b1;
    #10     start = 1'b0;

        //-30 x 255 = -7650 = fe21eh
    #10     a = 8'b1110_0010; b = 9'b0_1111_1111;
    #10     start = 1'b1;
    #10     start = 1'b0;

    #10     $stop;

end


//instantiate the module into the test bench
mul_table_lookup_neg inst1 (
    .a(a),
    .b(b),
    .start(start),
    .prod(prod)
    );

endmodule
```

Figure 6.46 (Continued)

```
a=11100010, b=000000001, prod=11111111111111100010
a=11100010, b=000001011, prod=11111111111111100010
a=11100010, b=000001011, prod=11111111111010110110
a=11100010, b=000000110, prod=11111111111010110110
a=11100010, b=000000110, prod=11111111111101001100
a=11100010, b=000101110, prod=11111111111101001100
a=11100010, b=000101110, prod=11111111101010011100
a=11100010, b=010011111, prod=11111111101010011100
a=11100010, b=010011111, prod=11111110110101011110
a=11100010, b=010101010, prod=11111110110101011110
a=11100010, b=010101010, prod=11111110110000010100
a=11100010, b=011111111, prod=11111110110000010100
a=11100010, b=011111111, prod=11111110001000011110
```

Figure 6.47 Outputs for the table lookup multiplier using a negative multiplicand.

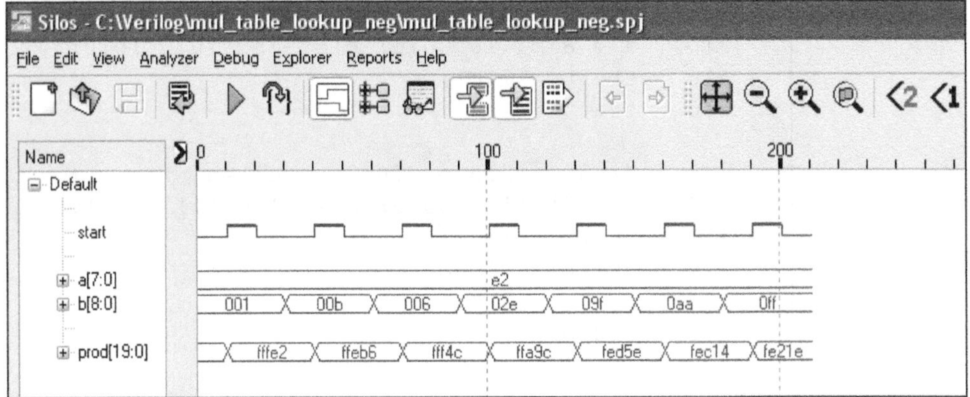

Figure 6.48 Waveforms for the table lookup multiplier using a negative multiplicand.

6.6 Memory-Based Multiplication

With the advent of high-capacity, high-speed random-access memories (RAMs), multiplication using a RAM may be a viable option. The multiplicand and multiplier are used as address inputs to the memory — the outputs are the product. As mentioned previously, memories can be represented in Verilog HDL by an array of registers and are declared using a **reg** data type as follows for the memory *mem_mul*, which is an array of 64 six-bit registers:

```
reg [5:0] mem_mul [0:63];
```

The operands can be either unsigned or signed numbers in 2s complement representation. This design will use signed operands for both the multiplicand and the multiplier. In order to keep the address space to a reasonable size, only 3-bit operands are used. A behavioral module will be designed to multiply the two 3-bit operands using a memory as a table lookup device.

A text file will be prepared for the specified memory in binary format. The file is created and saved as a separate file in the project folder without the **.v** extension. The system task **$readmemb** reads the file and loads the contents into memory. The contents of *opnds.mul* are loaded into the memory *mem_mul* beginning at location 0.

The contents of the *opnds.mul* file are shown in Figure 6.49, where the 6-bit address is shown for reference. The leftmost three digits of the address represent the multiplicand *A*; the rightmost three digits represent the multiplier *B* — the space is shown only for clarity. The address bits are not part of the memory contents — only the six bits for the product *prod* are entered into the *opnds.mul* file. A block diagram of the memory is shown in Figure 6.50 with a 6-bit address consisting of a 3-bit multiplicand, *mpcnd[2:0]*, concatenated with a 3-bit multiplier, *mplyr[2:0]*. There is a 6-bit output vector, *prod[5:0]*, containing the product.

A B	*prod*				
000 000	**000000**	**011 000**	**000000**	**110 000**	**000000**
000 001	000000	011 001	000011	110 001	111110
000 010	000000	011 010	001100	110 010	111100
000 011	000000	011 011	001001	110 011	111010
000 100	000000	011 100	110100	110 100	001000
000 101	000000	011 101	110111	110 101	000110
000 110	000000	011 110	111010	110 110	000100
000 111	000000	011 111	111101	110 111	000010
001 000	**000000**	**100 000**	**000000**	**111 000**	**000000**
001 001	000001	100 001	111100	111 001	111111
001 010	000010	100 010	111000	111 010	111110
001 011	000011	100 011	110100	111 011	111101
001 100	111100	100 100	010000	111 100	000100
001 101	111101	100 101	001100	111 101	000011
001 110	111110	100 110	001000	111 110	000010
001 111	111111	100 111	000100	111 111	000001
010 000	**000000**	**101 000**	**000000**		
010 001	000010	101 001	111101		
010 010	000100	101 010	111010		
010 011	001100	101 011	110111		
010 100	111000	101 100	001100		
010 101	111010	101 101	001001		
010 110	111100	101 110	001100		
010 111	111110	101 111	000011		

Figure 6.49 Contents of memory *opnds.mul* that are loaded into memory *mem_mul*.

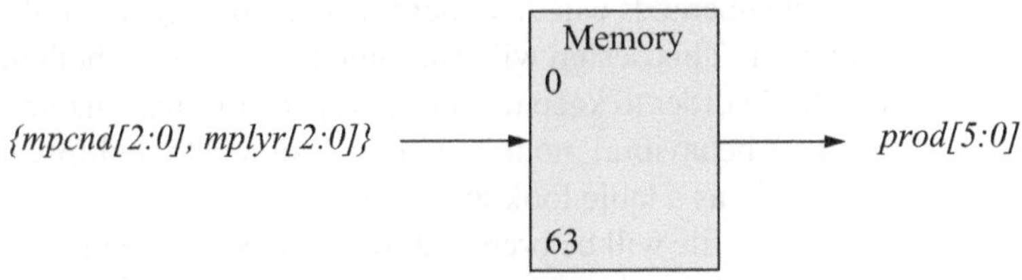

Figure 6.50 Block diagram of a memory that multiplies two 3-bit operands.

The behavioral module to multiply the multiplicand and multiplier using memory is shown in Figure 6.51 using binary data for the file called *opnds.mul*. The test bench is shown in Figure 6.52 using several different vectors that represent the multiplicand and the multiplier. The outputs are shown in Figure 6.53 and the waveforms are shown in Figure 6.54.

There are three ports in Figure 6.51: input port *mpcnd[2:0]* containing the 3-bit multiplicand, input port *mplyr[2:0]* containing the 3-bit multiplier, and output port *prod[5:0]* containing the product, which is declared as type **reg** because it operates as a storage element. The memory is defined as an array of 64 six-bit registers.

An **initial** procedural construct is used to load data from the *opnds.mul* file into the memory *mem_mul* by means of the system task **$readmemb**. The **initial** statement executes only once to initialize the memory. An **always** procedural construct is then used to read the contents of the memory based on the value of the concatenated operands; that is, *prod[5:0]* receives the contents of the memory at the address specified by the multiplicand and the multiplier, as shown below. The variables *mpcnd* and *mplyr* are event control variables used in the **always** statement — when the operands change, the statement in the **begin . . . end** block is executed.

prod = mem_mul [{mpcnd, mplyr}];

```
//behavioral to add two operands
module mem_mul (mpcnd, mplyr, prod);

//list inputs and outputs
input [2:0] mpcnd, mplyr;
output [5:0] prod;

//list wire and reg
wire [2:0] mpcnd, mplyr;//mpcnd and mplyr to address 64 words
reg [5:0] prod;
/continued on next page
```

Figure 6.51 Behavioral module for memory-based multiplication.

```
//define memory size
//mem_mul is an array of 64 six-bit registers
reg [5:0] mem_mul [0:63];

//define memory contents
//load mem_mul from file opnds.mul
initial
begin
   $readmemb ("opnds.mul", mem_mul);
end

//use the operands to access the memory
always @ (mpcnd or mplyr)
begin
   prod = mem_mul [{mpcnd, mplyr}];
end
endmodule
```

Figure 6.51 (Continued)

```
//test bench for mem_mul module
module mem_mul_tb;

reg [2:0] mpcnd, mplyr;
wire [5:0] prod;

//display variables
initial
$monitor ("mpcnd = %b, mplyr = %b, prod = %b",
            mpcnd, mplyr, prod);

//apply input vectors for multiplicand and multiplier
initial
begin
   #0    mpcnd = 3'b000;   mplyr = 3'b011;
   #10   mpcnd = 3'b001;   mplyr = 3'b001;
   #10   mpcnd = 3'b001;   mplyr = 3'b011;
   #10   mpcnd = 3'b001;   mplyr = 3'b100;
   #10   mpcnd = 3'b001;   mplyr = 3'b111;
   #10   mpcnd = 3'b010;   mplyr = 3'b101;
   #10   mpcnd = 3'b010;   mplyr = 3'b111;
   #10   mpcnd = 3'b011;   mplyr = 3'b011;
   #10   mpcnd = 3'b110;   mplyr = 3'b111;
   #10   mpcnd = 3'b100;   mplyr = 3'b010;   //next page
```

Figure 6.52 Test bench for memory-based multiplication.

```
   #10   mpcnd = 3'b100;    mplyr = 3'b101;
   #10   mpcnd = 3'b100;    mplyr = 3'b100;
   #10   mpcnd = 3'b101;    mplyr = 3'b001;
   #10   mpcnd = 3'b101;    mplyr = 3'b011;
   #10   mpcnd = 3'b101;    mplyr = 3'b111;
   #10   mpcnd = 3'b110;    mplyr = 3'b011;
   #10   mpcnd = 3'b110;    mplyr = 3'b111;
   #10   mpcnd = 3'b111;    mplyr = 3'b001;
   #10   mpcnd = 3'b111;    mplyr = 3'b100;
   #10   mpcnd = 3'b111;    mplyr = 3'b101;
   #10   mpcnd = 3'b111;    mplyr = 3'b110;
   #10   mpcnd = 3'b111;    mplyr = 3'b111;
   #10   $stop;
end

mem_mul inst1 (       //instantiate the module
   .mpcnd(mpcnd),
   .mplyr(mplyr),
   .prod(prod)
   );
endmodule
```

Figure 6.52 (Continued)

```
mpcnd = 000, mplyr = 011, prod = 000000
mpcnd = 001, mplyr = 001, prod = 000001
mpcnd = 001, mplyr = 011, prod = 000011
mpcnd = 001, mplyr = 100, prod = 111100
mpcnd = 001, mplyr = 111, prod = 111111
mpcnd = 010, mplyr = 101, prod = 111010
mpcnd = 010, mplyr = 111, prod = 111110
mpcnd = 011, mplyr = 011, prod = 001001
mpcnd = 110, mplyr = 111, prod = 000010
mpcnd = 100, mplyr = 010, prod = 111000
mpcnd = 100, mplyr = 101, prod = 001100
mpcnd = 100, mplyr = 100, prod = 010000
mpcnd = 101, mplyr = 001, prod = 111101
mpcnd = 101, mplyr = 011, prod = 110111
mpcnd = 101, mplyr = 111, prod = 000011
mpcnd = 110, mplyr = 011, prod = 111010
mpcnd = 110, mplyr = 111, prod = 000010
mpcnd = 111, mplyr = 001, prod = 111111
mpcnd = 111, mplyr = 100, prod = 000100
mpcnd = 111, mplyr = 101, prod = 000011
mpcnd = 111, mplyr = 110, prod = 000010
mpcnd = 111, mplyr = 111, prod = 000001
```

Figure 6.53 Outputs for memory-based multiplication.

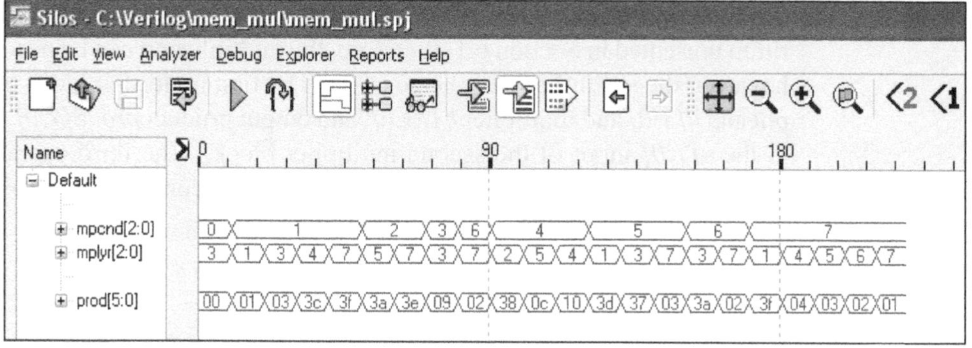

Figure 6.54 Waveform for memory-based multiplication.

6.7 Multiple-Operand Multiplication

This section describes one of various techniques to perform multiplication on three n-bit operands. One method uses two array multipliers in tandem implemented with structural modeling, as shown in the block diagram of Figure 6.55. Array multipliers operate at a high speed; however, for large operands, the amount of hardware required is considerable with a corresponding increase in propagation delay from input to output.

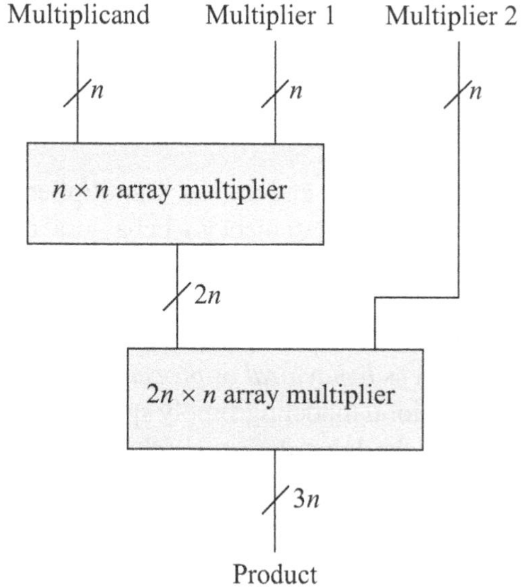

Figure 6.55 Block diagram of a 3-operand array multiplier.

The design of the 3-operand multiplier in this section will be implemented using behavioral modeling. This is an extension of the sequential add-shift multiply algorithm presented in Section 6.1; therefore, the multiplicand can be positive or negative, however, the multiplier must be positive. The first multiplier block has inputs multiplicand *a[3:0]* and multiplier *b1[3:0]*, and output product *prod8[7:0]*, which connects to the *a[7:0]* input of the second multiplier block. The third operand is multiplier *b2[3:0]*, which connects to the *b[3:0]* input of the second multiplier block with output *prod12[11:0]*, as shown in Figure 6.56.

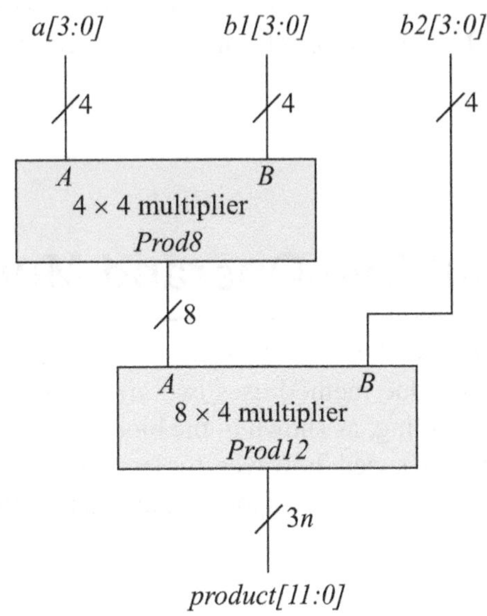

Figure 6.56 Three-operand multiplier to be implemented with behavioral modeling.

The 4×4 multiplier and the 8×4 multiplier are both designed using the **always** statement. The **always** statement is a behavioral construct that executes the behavioral statements within the **always** block in a looping manner and begins execution at time zero. Execution of the statements continues indefinitely until simulation is terminated. The **always** construct specifies a behavior and the statements within the behavior are classified as *behavioral* or *procedural*.

Since behavioral modeling simply specifies the behavior of the system and not the detailed design, the logic design is relegated to the synthesis tool. The behavioral module is shown in Figure 6.57. There are five inputs: operand *a[3:0]*, operand *b1[3:0]*, operand *b2[3:0]*, and two start signals, *start1* and *start2*, that initiate the multiply operations for the 4×4 multiplier block and the 8×4 multiplier block, respectively. The two start signals are offset in time so that the product from the 4×4 multiplier is available as an input to the 8×4 multiplier before the *start2* signal occurs.

There are two outputs: the product from the 4 × 4 block, *prod8[7:0]*, and the product from the 8 × 4 block, *prod12[11:0]*, which is the final product for the multiply operation.

```
//behavioral add-shift multiply for multiple operands
module mul_add_shift_multi (a, b1, b2, start1, start2,
                           prod8, prod12);

input [3:0] a, b1, b2;
input start1, start2;
output [7:0] prod8;
output [11:0] prod12;


//define internal registers
reg [7:0] prod8;
reg [11:0] prod12;
reg [3:0] count;


//design the behavior for the 4 x 4 multiplier
always @ (posedge start1)
begin
   prod8 [7:4] = 4'b0000;
   prod8 [3:0] = b1;
   count = 4'b0100;
      if ((a!=0) && (b1!=0))
         while (count)
            begin
               if (prod8[0] == 1'b0)
                  begin
                     prod8 = {prod8[7], prod8[7:1]};
                     count = count - 1;
                  end


               else
                  begin
                     prod8 = {a[3], (a + prod8[7:4]),
                              prod8[3:1]};
                     count = count - 1;
                  end
            end
end

//continued on next page
```

Figure 6.57 Behavioral module for the 3-operand sequential add-shift multiplier.

```
//design the behavior for the 8 x 4 multiplier
always @ (posedge start2)
begin
   prod12 [11:4] = 8'b0000_0000;
   prod12 [3:0] = b2;
   count = 4'b0100;
      if ((prod8!=0) && (b2!=0))
         while (count)
            begin
               if (prod12[0] == 1'b0)
                  begin
                     prod12 = {prod12[11], prod12[11:1]};
                     count = count - 1;
                  end

               else
                  begin
                     prod12 = {prod8[7],
                               (prod8 + prod12[11:4]),
                              prod12[3:1]};
                     count = count - 1;
                  end
            end
end

endmodule
```

Figure 6.57 (Continued)

The Verilog code for the sequential add-shift multiple-operand multiply algorithm follows the same sequence as the sequential add-shift multiply algorithm described in Section 6.1. If the low-order bit, *b1[0]* (represented by *prod8[0]*) = 0, then the operation is simply a shift right of one bit position, as specified by the statement shown below. The counter is then decremented by one.

prod8 = {prod8[7], prod8[7:1]};

If the low-order bit of the multiplier, *b1[0]* (represented by *prod8[0]*) = 1, then the operation is the concatenation of the following variables: the sign of the multiplicand *a[3]*; the sum of the multiplicand *a* plus the high-order half of multiplier *b1[7:4]* (represented by *prod8[7:4]*; and *b[3:1]* (represented by *prod8[3:1]*). This also represents a shift right of one bit position, as specified by the statement shown below. The counter is then decremented by one.

prod8 = {a[3], (a + prod8[7:4]), prod8[3:1]};

The same sequence applies to the Verilog code for the 8 × 4 multiplier, except that the multiplicand is now *prod8[7:0]* and the multiplier is *b2[3:0]* (represented by *prod12[3:0]*). The test bench is shown in Figure 6.58, which applies several input vectors for the three operands: multiplicand *a[3:0]*, and multipliers *b1[3:0]* and *b2[3:0]*. Both positive and negative multiplicands are utilized.

The outputs are shown in Figure 6.59 displaying the three operands, the intermediate product, *prod8[7:0]*, and the resulting product of the multiplication, *prod12[11:0]*. The waveforms are shown in Figure 6.60.

```
//test bench for add-shift multiplier
module mul_add_shift_multi_tb;

reg [3:0] a, b1, b2;
reg start1, start2;
wire [7:0] prod8;
wire [11:0] prod12;

//display variables
initial
$monitor ("a=%b, b1=%b, b2=%b, start1=%b, start2=%b,
            prod8=%b, prod12=%b",
            a, b1, b2, start1, start2, prod8, prod12);

//apply input vectors
initial
begin
   #0    start1 = 1'b0;
         start2 = 1'b0;

         //7 x 7 x 2 = 98 (062h)
         a=4'b0111;      b1=4'b0111;     b2=4'b0010;
   #10   start1 = 1'b1;
   #10   start1 = 1'b0;
   #10   start2 = 1'b1;
   #10   start2 = 1'b0;

         //3 x 7 x 6 = 126 (07eh)
   #10   a=4'b0011;      b1=4'b0111;     b2= 4'b0110;
   #10   start1 = 1'b1;
   #10   start1 = 1'b0;
   #10   start2 = 1'b1;
   #10   start2 = 1'b0;

//continued on next page
```

Figure 6.58 Test bench module for the 3-operand sequential add-shift multiplier.

```
        //7 x 5 x 6 = 210 (0d2h)
#10     a=4'b0111;       b1=4'b0101;     b2=4'b0110;
#10     start1 = 1'b1;
#10     start1 = 1'b0;
#10     start2 = 1'b1;
#10     start2 = 1'b0;

        //7 x 4 x 5 = 140 (08ch)
#10     a=4'b0111;       b1=4'b0100;     b2=4'b0101;
#10     start1 = 1'b1;
#10     start1 = 1'b0;
#10     start2 = 1'b1;
#10     start2 = 1'b0;

        //-1 x 7 x 1 = -7 (ff9h)
#10     a=4'b1111;       b1=4'b0111;     b2=4'b0001;
#10     start1 = 1'b1;
#10     start1 = 1'b0;
#10     start2 = 1'b1;
#10     start2 = 1'b0;

        //-5 x 6 x 1 = -30 (fe2h)
#10     a=4'b1011;       b1=4'b0110;     b2=4'b0001;
#10     start1 = 1'b1;
#10     start1 = 1'b0;
#10     start2 = 1'b1;
#10     start2 = 1'b0;

        //-6 x 7 x 5 = -210 (f2eh)
#10     a=4'b1010;       b1=4'b0111;     b2=4'b0101;
#10     start1 = 1'b1;
#10     start1 = 1'b0;
#10     start2 = 1'b1;
#10     start2 = 1'b0;

        //-4 x 7 x 3 = -84 (fach)
#10     a=4'b1100;       b1=4'b0111;     b2=4'b0011;
#10     start1 = 1'b1;
#10     start1 = 1'b0;
#10     start2 = 1'b1;
#10     start2 = 1'b0;

//continued on next page
```

Figure 6.58 (Continued)

```
            //-7 x 7 x 7 = -343 (ea9h)
    #10     a=4'b1001;      b1=4'b0111;     b2=4'b0111;
    #10     start1 = 1'b1;
    #10     start1 = 1'b0;
    #10     start2 = 1'b1;
    #10     start2 = 1'b0;

    #20     $stop;
end

//instantiate the module into the test bench
mul_add_shift_multi inst1 (
    .a(a),
    .b1(b1),
    .b2(b2),
    .start1(start1),
    .start2(start2),
    .prod8(prod8),
    .prod12(prod12)
    );
endmodule
```

Figure 6.58 (Continued)

```
a=0111, b1=0111, b2=0010, start1=0, start2=1,
      prod8=00110001, prod12=000001100010

a=0111, b1=0111, b2=0010, start1=0, start2=0,
      prod8=00110001, prod12=000001100010

a=0011, b1=0111, b2=0110, start1=0, start2=0,
      prod8=00110001, prod12=000001100010

a=0011, b1=0111, b2=0110, start1=1, start2=0,
      prod8=00010101, prod12=000001100010

a=0011, b1=0111, b2=0110, start1=0, start2=0,
      prod8=00010101, prod12=000001100010

a=0011, b1=0111, b2=0110, start1=0, start2=1,
      prod8=00010101, prod12=000001111110

a=0011, b1=0111, b2=0110, start1=0, start2=0,
      prod8=00010101, prod12=000001111110
//continued on next page
```

Figure 6.59 Outputs for the 3-operand sequential add-shift multiplier.

```
a=0111, b1=0101, b2=0110, start1=0, start2=0,
      prod8=00010101, prod12=000001111110

a=0111, b1=0101, b2=0110, start1=1, start2=0,
      prod8=00100011, prod12=000001111110

a=0111, b1=0101, b2=0110, start1=0, start2=0,
      prod8=00100011, prod12=000001111110

a=0111, b1=0101, b2=0110, start1=0, start2=1,
      prod8=00100011, prod12=000011010010

a=0111, b1=0101, b2=0110, start1=0, start2=0,
      prod8=00100011, prod12=000011010010

a=0111, b1=0100, b2=0101, start1=0, start2=0,
      prod8=00100011, prod12=000011010010

a=0111, b1=0100, b2=0101, start1=1, start2=0,
      prod8=00011100, prod12=000011010010

a=0111, b1=0100, b2=0101, start1=0, start2=0,
      prod8=00011100, prod12=000011010010

a=0111, b1=0100, b2=0101, start1=0, start2=1,
      prod8=00011100, prod12=000010001100

a=0111, b1=0100, b2=0101, start1=0, start2=0,
      prod8=00011100, prod12=000010001100

a=1111, b1=0111, b2=0001, start1=0, start2=0,
      prod8=00011100, prod12=000010001100

a=1111, b1=0111, b2=0001, start1=1, start2=0,
      prod8=11111001, prod12=000010001100

a=1111, b1=0111, b2=0001, start1=0, start2=0,
      prod8=11111001, prod12=000010001100

a=1111, b1=0111, b2=0001, start1=0, start2=1,
      prod8=11111001, prod12=111111111001

a=1111, b1=0111, b2=0001, start1=0, start2=0,
      prod8=11111001, prod12=111111111001

a=1011, b1=0110, b2=0001, start1=0, start2=0,
      prod8=11111001, prod12=111111111001     //next page
```

Figure 6.59 (Continued)

```
a=1011, b1=0110, b2=0001, start1=1, start2=0,
      prod8=11100010, prod12=111111111001

a=1011, b1=0110, b2=0001, start1=0, start2=0,
      prod8=11100010, prod12=111111111001

a=1011, b1=0110, b2=0001, start1=0, start2=1,
      prod8=11100010, prod12=111111100010

a=1011, b1=0110, b2=0001, start1=0, start2=0,
      prod8=11100010, prod12=111111100010

a=1010, b1=0111, b2=0101, start1=0, start2=0,
      prod8=11100010, prod12=111111100010

a=1010, b1=0111, b2=0101, start1=1, start2=0,
      prod8=11010110, prod12=111111100010

a=1010, b1=0111, b2=0101, start1=0, start2=0,
      prod8=11010110, prod12=111111100010

a=1010, b1=0111, b2=0101, start1=0, start2=1,
      prod8=11010110, prod12=111100101110

a=1010, b1=0111, b2=0101, start1=0, start2=0,
      prod8=11010110, prod12=111100101110

a=1100, b1=0111, b2=0011, start1=0, start2=0,
      prod8=11010110, prod12=111100101110

a=1100, b1=0111, b2=0011, start1=1, start2=0,
      prod8=11100100, prod12=111100101110

a=1100, b1=0111, b2=0011, start1=0, start2=0,
      prod8=11100100, prod12=111100101110

a=1100, b1=0111, b2=0011, start1=0, start2=1,
      prod8=11100100, prod12=111110101100

a=1100, b1=0111, b2=0011, start1=0, start2=0,
      prod8=11100100, prod12=111110101100

a=1001, b1=0111, b2=0111, start1=0, start2=0,
      prod8=11100100, prod12=111110101100

a=1001, b1=0111, b2=0111, start1=1, start2=0,
      prod8=11001111, prod12=111110101100    //next page
```

Figure 6.59 (Continued)

```
a=1001, b1=0111, b2=0111, start1=0, start2=0,
      prod8=11001111, prod12=111110101100

a=1001, b1=0111, b2=0111, start1=0, start2=1,
       prod8=11001111, prod12=111010101001

a=1001, b1=0111, b2=0111, start1=0, start2=0,
      prod8=11001111, prod12=111010101001
```

Figure 6.59 (Continued)

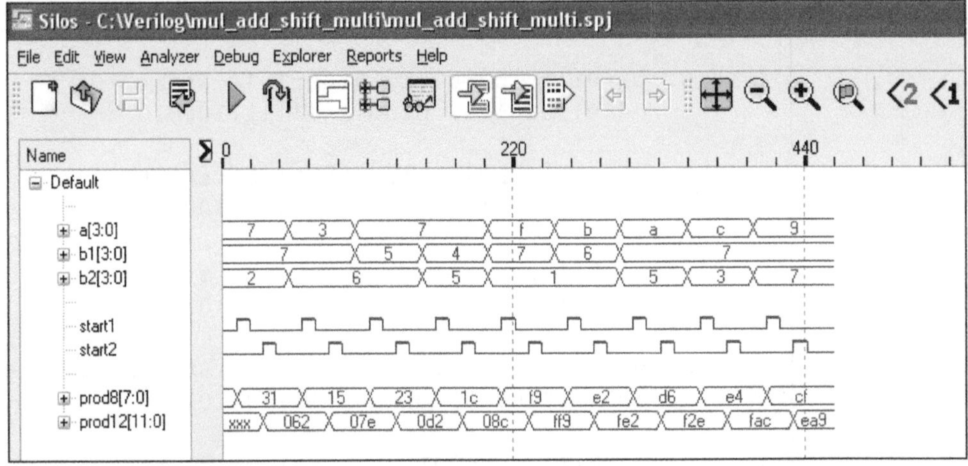

Figure 6.60 Waveforms for the 3-operand sequential add-shift multiplier.

6.8 Problems

6.1 Use the paper-and-pencil method to multiply the operands shown below which are in 2s complement representation.

Multiplicand = 0111
Multiplier = 0101

6.2 Use the paper-and-pencil method to multiply the operands shown below which are in 2s complement representation.

Multiplicand = 0111
Multiplier = 1110

6.3 Use the paper-and-pencil method to multiply the operands shown below which are in 2s complement representation.

Multiplicand = 1010
Multiplier = 0011

6.4 Use the paper-and-pencil method to multiply the operands shown below which are in 2s complement representation.

Multiplicand = 1100
Multiplier = 1010

6.5 Use the paper-and-pencil method to multiply the operands shown below which are in 2s complement representation.

Multiplicand = 01110
Multiplier = 00111

6.6 Use the paper-and-pencil method to multiply the operands shown below which are in 2s complement representation.

Multiplicand = 11111
Multiplier = 01011

6.7 Multiply the two operands shown below using the sequential add-shift technique shown in Figure 6.2.

Multiplicand = 1111
Multiplier = 0101

6.8 Multiply the two operands shown below using the sequential add-shift technique shown in Figure 6.2.

Multiplicand = 0101
Multiplier = 0111

6.9 Use the Booth algorithm to multiply the operands shown below, which are signed operands in 2s complement representation.

Multiplicand	0 1 1 0 1
Multiplier	1 0 1 0 0

6.10 Use the Booth algorithm to multiply the operands shown below, which are signed operands in 2s complement representation.

Multiplicand	0 0 1 1 1
Multiplier	0 1 1 0 1

6.11 Use the Booth algorithm to multiply the operands shown below, which are signed operands in 2s complement representation.

Multiplicand	0 1 0 1 1
Multiplier	1 1 0 1 0

6.12 Use the Booth algorithm to multiply the operands shown below, which are signed operands in 2s complement representation.

Multiplicand	0 0 1 1 1
Multiplier	1 0 1 1 0

6.13 Use the Booth algorithm to multiply the operands shown below, which are signed operands in 2s complement representation.

Multiplicand	1 1 0 0 1
Multiplier	1 0 1 0 1

6.14 Use the Booth algorithm to multiply the operands shown below, which are signed operands in 2s complement representation.

Multiplicand	1 1 1 0 0
Multiplier	1 1 0 1 1

6.15 Use bit-pair recoding to determine the multiplicand multiples to be added for the following multipliers which are in 2s complement representation:

Multiplier	1 0 0 1 1 1
Multiplier	1 0 0 1 1
Multiplier	0 0 1 1 0 1

6.16 Use bit-pair recoding to multiply the following operands which are in 2s complement representation:

Multiplicand	0 0 0 1 1
Multiplier	1 1 1 0 1

6.17 Use bit-pair recoding to multiply the following operands which are in 2s complement representation:

Multiplicand	1 1 0 0 1 0
Multiplier	0 1 1 0 0 1

6.18 Use bit-pair recoding to multiply the following operands which are in 2s complement representation:

Multiplicand	0 1 0 0 1 1 0
Multiplier	1 0 0 1 1 0 1

6.19 Use bit-pair recoding to multiply the following operands which are in 2s complement representation:

Multiplicand	1 1 1 0 0
Multiplier	1 1 0 1 1

6.20 Design an array multiplier that multiplies two 3-bit unsigned fixed-point operands: multiplicand *a[2:0]* and multiplier *b[2:0]*. The product is *p[5:0]*. Obtain the structural design module, the test bench for all combinations of the multiplicand and the multiplier, and the outputs. The inputs and outputs are to be shown in decimal notation.

6.21 Multiply the following two unsigned fixed-point operands using an array multiplier: A (multiplicand) = 0111 and B (multiplier) = 1100. All AND functions are to be truncated as $a_i\, b_j$, for all $i, j = 0, 1, 2, 3$; that is, the actual gates are not drawn. Show the multiplicand and multiplier values for each $a_i\, b_j$ as 1s and 0s. Show all partial products and carries in the array as 1s and 0s. Show the product obtained as a binary number.

6.22 Design an array multiplier that multiplies two 2-bit unsigned fixed-point operands, multiplicand *a[1:0]* and multiplier *b[1:0]*, using only the full adder macro cell shown below. The product is *p[3:0]*. Obtain the structural design module, the test bench for all combinations of the multiplicand and multiplier, the outputs, and the waveforms. The multiplicand, multiplier, and product are to be shown in binary notation.

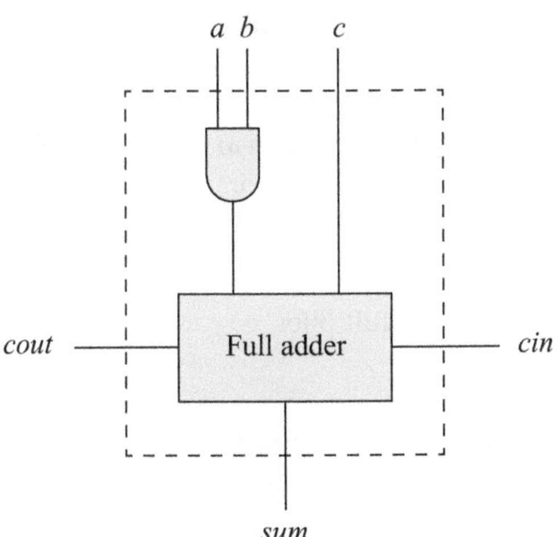

6.23 Design an array multiplier that multiplies two 2-bit unsigned fixed-point operands: multiplicand *a[1:0]* and multiplier *b[1:0]*. The product is *p[3:0]*. Obtain the structural design module using only AND gates and half adders. Obtain the test bench for all combinations of the multiplicand and multiplier, the outputs, and waveforms. The inputs and outputs are to be shown in binary notation.

6.24 Design an array multiplier that multiplies two 3-bit unsigned fixed-point operands: multiplicand *a[2:0]* and multiplier *b[2:0]*. The product is *p[5:0]*. Obtain the structural design module using only AND gates, half adders, and full adders. Obtain the test bench for all combinations of the multiplicand and multiplier and obtain the outputs. The inputs and outputs are to be shown in binary notation.

6.25 Design a structural module to implement a 3-operand array multiplier that multiplies three 2-bit operands. Instantiate into the module a structural module for an array multiplier that multiplies two 2-bit operands. Also instantiate into the module an array multiplier that multiplies a 4-bit operand by a 2-bit operand. These two instantiations will then produce the required structural module that multiplies three 2-bit operands.

Obtain the design module for the array multiplier that multiplies two 2-bit operands, the test bench, and the outputs for all combinations. Obtain the design module for the array multiplier that multiplies a 4-bit operand by a 2-bit operand, the test bench, and the outputs for all combinations.

Obtain the structural design module for the array multiplier that multiplies three 2-bit operands, the test bench, and the outputs for all combinations.

7.1 *Sequential Shift-Add/Subtract Restoring Division*
7.2 *Sequential Shift-Add/Subtract Nonrestoring Division*
7.3 *SRT Division*
7.4 *Multiplicative Division*
7.5 *Array Division*
7.6 *Problems*

7

Fixed-Point Division

In most cases division is slower than multiplication and occurs less frequently. Division is essentially the inverse of multiplication, where the $2n$-bit dividend corresponds to the $2n$-bit product; the n-bit divisor corresponds to the n-bit multiplicand; and the n-bit quotient corresponds to the n-bit multiplier. The equation that represents this concept is shown below and includes the n-bit remainder as one of the variables.

$$2n\text{-bit dividend} = (n\text{-bit divisor} \times n\text{-bit quotient}) + n\text{-bit remainder}$$

Unlike multiplication, division is not commutative; that is, $A/B \neq B/A$, except when $A = B$, where A and B are the dividend and divisor, respectively. This chapter will present sequential shift-subtract/add restoring and sequential shift-subtract/add nonrestoring division techniques. It was shown in the previous chapter that one method for multiplication was the sequential add-shift technique. The algorithm for sequential division is slightly different and uses the shift-add/subtract technique; that is, the dividend is shifted left one bit position, then the divisor is subtracted from the shifted dividend by adding the 2s complement of the divisor.

Also included in this chapter is a high-speed method of division that is similar to the Booth algorithm presented in the previous chapter and is generally referred to as the SRT method, because it was discovered separately by Sweeney, Robertson, and Tocher. This method shifts over strings of 0s and strings of 1s, where a string can consist of one or more bits. Multiplicative division — a form of convergence division — using a high-speed multiplier, and array division using iterative cells are also presented.

7.1 Sequential Shift-Add/Subtract Restoring Division

In general, the operands are as shown below, where A is the $2n$-bit dividend and B is the n-bit divisor. The quotient is Q and the remainder is R, both of which are n bits.

$$A = a_{2n-1}\, a_{2n-2} \ldots a_n\, a_{n-1} \ldots a_1\, a_0$$

$$B = b_{n-1}\, b_{n-2} \ldots b_1\, b_0$$

$$Q = q_{n-1}\, q_{n-2} \ldots q_1\, q_0$$

$$R = r_{n-1}\, r_{n-2} \ldots r_1\, r_0$$

The sign of the quotient is determined by the following equation:

$$q_{n-1} = a_{2n-1} \oplus b_{n-1}$$

The remainder has the same sign as the dividend. The process of division is one of subtract, shift, and compare operations.

Division of two fixed-point operands can be accomplished using the paper-and-pencil method, as shown in Figure 7.1. In this example, the divisor is four bits 0101 (+5) and the dividend is eight bits 0000 1101 (+13), resulting in a quotient of 0010 (+2) and a remainder of 0011 (+3). The division procedure uses a sequential shift-subtract-restore technique.

In the first cycle for this example, the divisor is subtracted from the high-order four bits of the dividend. The result is a partial remainder that is negative — the leftmost bit is 1 — indicating that the divisor is greater than the four high-order bits of the dividend. Therefore, a 0 is placed in the high-order bit position of the quotient. The dividend bits are then restored to their previous values with the next lower-order bit (1) of the dividend being appended to the right of the partial remainder. The divisor is shifted right one bit position and again subtracted from the dividend bits.

This restore-shift-subtract cycle repeats for a total of three cycles until the partial remainder is positive — the leftmost bit is 0, indicating that the divisor is less than the corresponding dividend bits. This results in a no-restore cycle in which the previous partial remainder (0001) is not restored. A 1 bit is placed in the next lower-order quotient bit and the next lower-order dividend bit is appended to the right of the partial remainder. The divisor is again subtracted, resulting in a negative partial remainder, which is again restored by adding the divisor. The 4-bit quotient is 0010 and the 4-bit remainder is 0011.

The results can be verified by multiplying the quotient (0010) by the divisor (0101) and adding the remainder (0011) to obtain the dividend (0000 1101). Thus, $0010 \times 0101 = 1010 + 0011 = 0000\ 1101$.

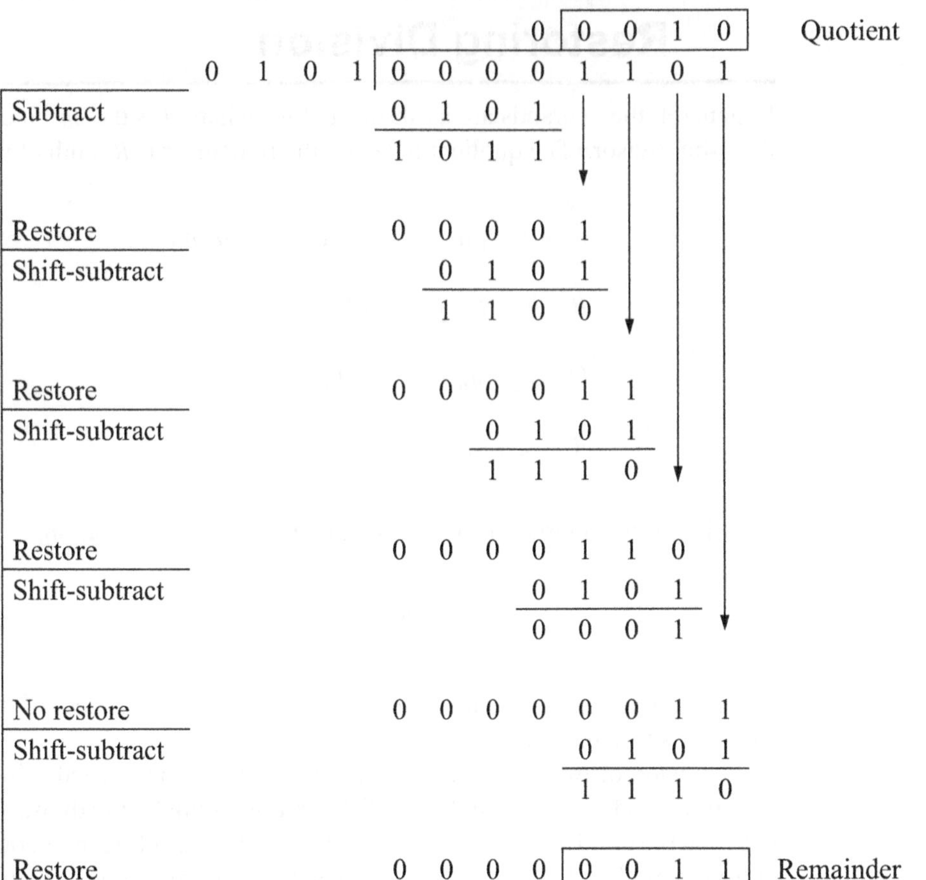

Figure 7.1 Paper-and-pencil example of binary division, where the dividend is +13 and the divisor is +5.

Overflow will occur if the high-order half of the dividend is greater than or equal to the divisor. For example, assume that the high-order half of the dividend is equal to the divisor, as shown below for a dividend of +112 and a divisor of +7, yielding a quotient of 16. The resulting quotient value of 16 cannot be contained in the machine's word size of four bits; therefore, an overflow had occurred. If the high-order half of the dividend is greater than the divisor, then the value of the quotient will be even greater.

B	*A*	*Q*
0111	0111 0000	1 0000

Overflow can be detected by subtracting the divisor from the high-order half of the dividend before the division operation commences. If the result is positive, then an overflow will occur.

7.1.1 Restoring Division — Version 1

The hardware organization for sequential shift-add/subtract restoring division is shown in Figure 7.2. The state of the carry flip-flop C determines the relative magnitudes of the divisor and partial remainder. If $C = 1$, then $A \geq B$ and $q_0 = 1$; if $C = 0$, then $A < B$ and $q_0 = 0$. The partial remainder is then shifted and the process repeats for each bit in the divisor.

If $C = 1$, then the 2s complement of B (subtraction) is added to the partial remainder; if $C = 0$, then the partial remainder is restored to its previous value by adding the divisor to the partial remainder. At the end of each restore cycle the carry flip-flop is set to 1 so that the following cycle will subtract the divisor from the partial remainder. The sequence counter is set to the number of bits in the divisor and counts down to zero, at which time the operation in finished.

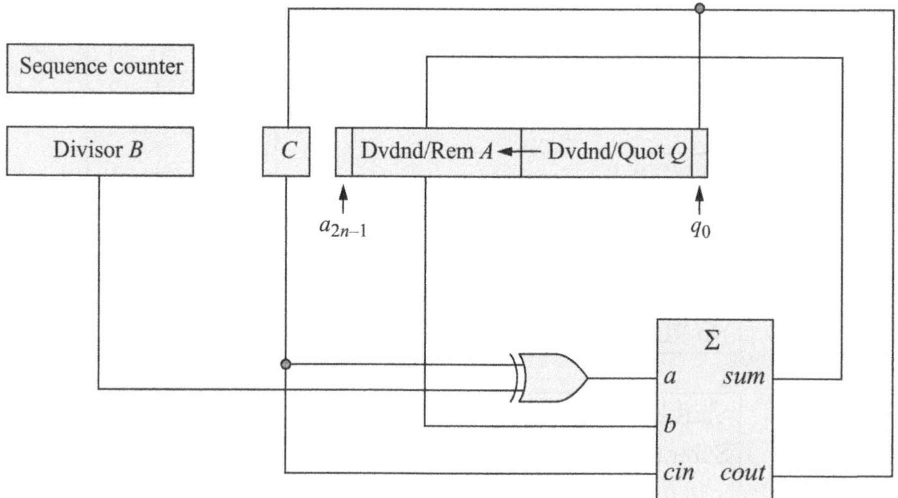

Figure 7.2 Hardware organization for sequential shift-add/subtract restoring division.

An example of binary division using the organization of Figure 7.2 is shown in Figure 7.3 for a dividend of +13 and a divisor of +5. Since there are four bits in the divisor, there are four left-shift operations each followed by a subtract operation. The carry-out of the subtraction is placed in the low-order bit position vacated by the left-shifted partial remainder. A carry-out of 0 indicates that the difference was negative and the partial remainder must be restored by adding the divisor to the partial remainder; a carry-out of 1 indicates that the difference was positive and no restoration is needed. At the completion of the final cycle, the remainder is contained in the high-order half of the dividend and the quotient is contained in the low-order half of the dividend.

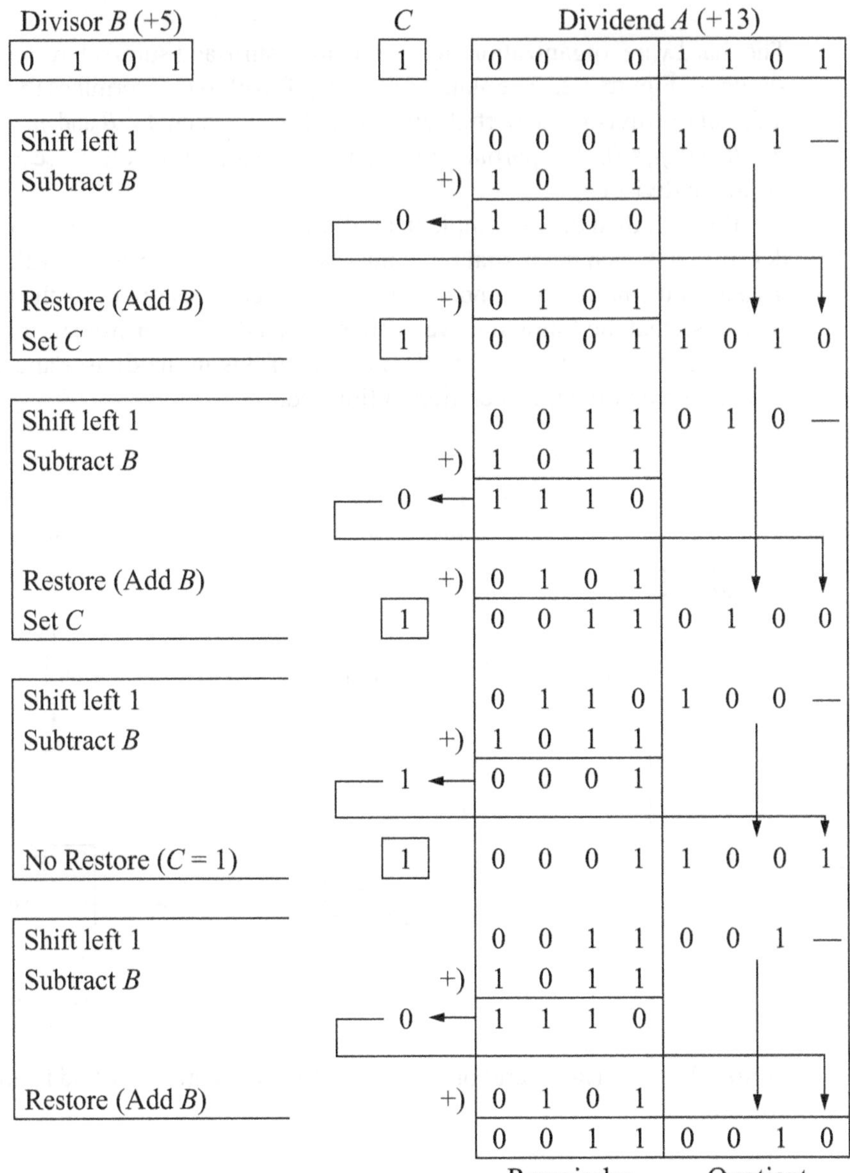

Figure 7.3 Example of sequential shift-add/subtract restoring binary division using the organization shown in Figure 7.2.

The sequential shift-add/subtract restoring division algorithm will now be implemented using Verilog HDL in a mixed-design module incorporating both dataflow and behavioral modeling. The numerical example shown in Figure 7.3 will be modified slightly to illustrate the algorithm as it applies to the Verilog HDL implementation and is shown in Figure 7.4. Examination of the carry flip-flop C will be replaced by testing the state of the sign bit that results from the subtraction of the divisor from the partial remainder. The 8-bit dividend register will be replaced by an 8-bit result register, *rslt[7:0]*, which will ultimately contain the remainder in *rslt[7:4]* and the quotient in *rslt[3:0]*.

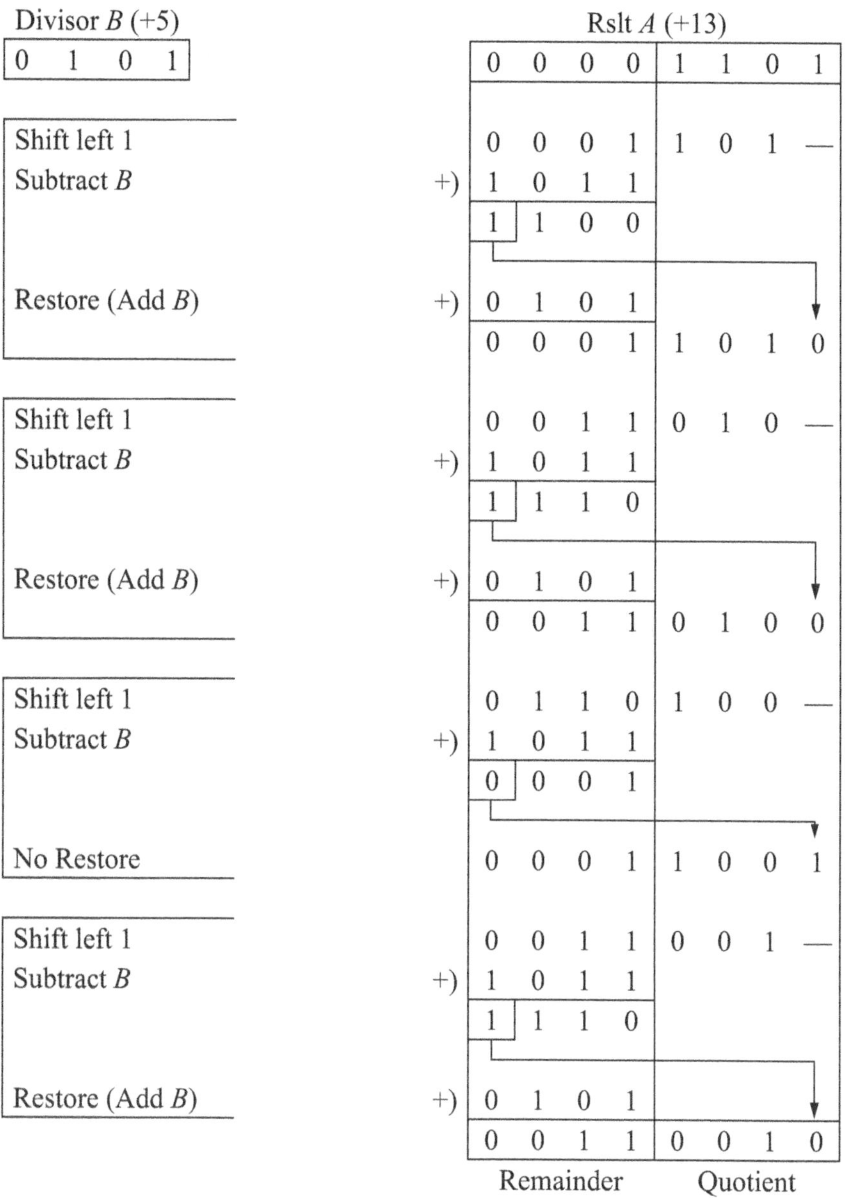

Figure 7.4 Example of sequential shift-add/subtract restoring binary division to be used in the Verilog implementation.

The Verilog HDL design, shown in Figure 7.5, is relatively straightforward and slightly easier to implement than the hardware organization shown in Figure 7.2. The inputs are an 8-bit dividend, *a[7:0]*; a 4-bit divisor, *b[3:0]*; and a scalar signal, *start*, which initiates the divide operation. The output is an 8-bit register *rslt[7:0]* containing the remainder and quotient. Shifting is accomplished by the left-shift operator (<<), as shown below.

rslt = rslt << 1;

Subtracting the divisor from the partial remainder is realized by the following statement, which adds the negation of the divisor to the partial product and concatenates the sum with the low-order four bits from the previous partial remainder:

rslt = {(rslt[7:4] + b_neg), rslt[3:0]};

Then the sign bit (*rslt[7]*) of the sum is tested for a value of 1 or 0. If the sign is 1 (negative), then this indicates that the divisor was greater than the high-order half of the previous partial remainder. Thus, a 0 is placed in the low-order quotient bit. This sequence is executed by the following statements, after which the sequence counter is then decremented by 1:

```
if (rslt[7] == 1)
begin
    rst = {(rslt[7:4] + b), rslt[3:1], 1'b0};
```

If the sign bit (*rslt[7]*) is 0 (positive), then this indicates that the divisor was less than the high-order half of the previous partial remainder. Therefore, no restoration of the partial remainder is required and a 1 is placed in the low-order quotient bit, as shown in the following statement, after which the sequence counter is then decremented by 1:

rslt = {rslt[7:1], 1'b1};

When the sequence counter counts down to zero, the division operation is finished, with the remainder in *rslt[7:4]*, and the quotient in *rslt[3:0]*. The test bench is shown in Figure 7.6 depicting six input vectors for the dividend *A* and divisor *B*. The division operation is initiated on the positive edge of the scalar *start* input signal.

The outputs shown in Figure 7.7, display the dividend, divisor, remainder, and quotient. The waveforms are shown in Figure 7.8.

```
//mixed-design for restoring division
module div_restoring (a, b, start, rslt);
input [7:0] a;
input [3:0] b;
input start;
output [7:0] rslt;        //continued on next page
```

Figure 7.5 Mixed-design module to implement sequential shift-add/subtract restoring division.

```
wire [3:0] b_bar;

//define internal registers
reg [3:0] b_neg;
reg [7:0] rslt;
reg [3:0] count;

assign b_bar = ~b;

always @ (b_bar)
   b_neg = b_bar + 1;

always @ (posedge start)
begin
   rslt = a;
   count = 4'b0100;

   if ((a!=0) && (b!=0))    //if a or b = 0, exit
      while (count)         //else do while loop
      begin
         rslt = rslt << 1;
         rslt = {(rslt[7:4] + b_neg), rslt[3:0]};
            if (rslt[7] == 1)
               begin
                  rslt = {(rslt[7:4] + b), rslt[3:1], 1'b0};
                  count = count - 1;
               end

            else
               begin
                  rslt = {rslt[7:1], 1'b1};
                  count = count - 1;
               end
      end
end

endmodule
```

Figure 7.5 (Continued)

```
//test bench for restoring division
module div_restoring_tb;
reg [7:0] a;
reg [3:0] b;
reg start;
wire [7:0] rslt;

initial      //display variables
$monitor ("a = %b, b = %b, quot = %b, rem = %b",
            a, b, rslt[3:0], rslt[7:4]);

initial      //apply input vectors
begin
   #0    start = 1'b0;
         a = 8'b0000_1101;     b = 4'b0101;
   #10   start = 1'b1;
   #10   start = 1'b0;

   #10   a = 8'b0011_1100;     b = 4'b0111;
   #10   start = 1'b1;
   #10   start = 1'b0;

   #10   a = 8'b0101_0010;     b = 4'b0110;
   #10   start = 1'b1;
   #10   start = 1'b0;

   #10   a = 8'b0011_1000;     b = 4'b0111;
   #10   start = 1'b1;
   #10   start = 1'b0;

   #10   a = 8'b0110_0100;     b = 4'b0111;
   #10   start = 1'b1;
   #10   start = 1'b0;

   #10   a = 8'b0110_1110;     b = 4'b0111;
   #10   start = 1'b1;
   #10   start = 1'b0;
   #10   $stop;
end

div_restoring inst1 (        //instantiate the module
   .a(a),
   .b(b),
   .start(start),
   .rslt(rslt)
   );
endmodule
```

Figure 7.6 Test bench for sequential shift-add/subtract restoring division.

```
a = 00001101, b = 0101, quot = xxxx, rem = xxxx
a = 00001101, b = 0101, quot = 0010, rem = 0011
a = 00111100, b = 0111, quot = 0010, rem = 0011
a = 00111100, b = 0111, quot = 1000, rem = 0100
a = 01010010, b = 0110, quot = 1000, rem = 0100
a = 01010010, b = 0110, quot = 1101, rem = 0100
a = 00111000, b = 0111, quot = 1101, rem = 0100
a = 00111000, b = 0111, quot = 1000, rem = 0000
a = 01100100, b = 0111, quot = 1000, rem = 0000
a = 01100100, b = 0111, quot = 1110, rem = 0010
a = 01101110, b = 0111, quot = 1110, rem = 0010
a = 01101110, b = 0111, quot = 1111, rem = 0101
```

Figure 7.7 Outputs for sequential shift-add/subtract restoring division.

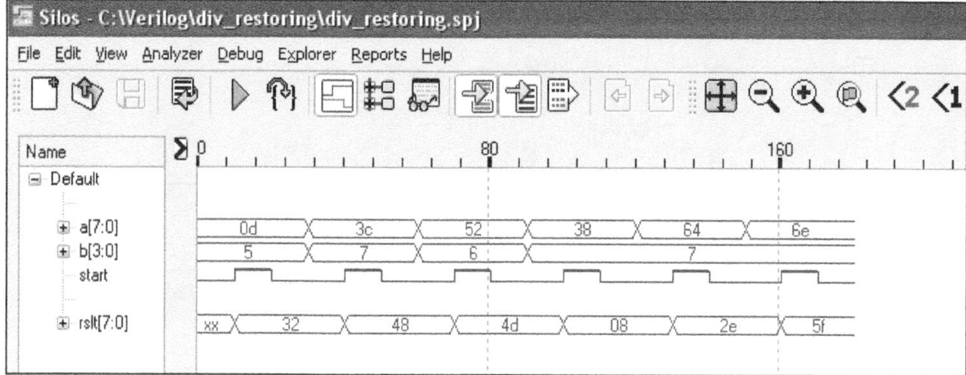

Figure 7.8 Waveforms for sequential shift-add/subtract restoring division.

7.1.2 Restoring Division — Version 2

This version of restoring division represents a faster method than the previous method presented in Section 7.1.1. The divisor is not added to the partial remainder to restore the previous partial remainder; instead the previous partial remainder — which is unchanged — is loaded into register *A*. This avoids the time required for addition. Additional hardware is required in the form of a 2:1 multiplexer which selects the appropriate version of the partial remainder to be loaded into register *A*. Either the previous partial remainder is selected, or the current partial remainder (the sum) is selected, controlled by the state of the carry-out, *cout*, from the adder.

If *cout* = 0, then the previous partial remainder (in register *A*) is selected and loaded into register *A*, while *cout* is loaded into the low-order quotient bit q_0. If *cout* = 1, then the sum output from the adder is selected and loaded into register *A*, and *cout* is loaded into the low-order quotient bit q_0. This sequence of operations is as follows:

$$\text{If } cout = 0, \text{ then } a_{n-1}\, a_{n-2} \ldots a_1\, a_0 \leftarrow a_{n-1}\, a_{n-2} \ldots a_1\, a_0$$
$$q_0 = 0$$

$$\text{If } cout = 1, \text{ then } a_{n-1}\, a_{n-2} \ldots a_1\, a_0 \leftarrow s_{n-1}\, s_{n-2} \ldots s_1\, s_0$$
$$q_0 = 1$$

A sequence counter is initialized to the number of bits in the divisor. When the sequence counter counts down to zero, the division operation is finished. The organization for this technique is shown in Figure 7.9. The 2s complement of the divisor is always added to the partial remainder by exclusive-ORing the divisor with a logic 1. Then, the appropriate partial remainder is selected by the multiplexer — either the previous partial remainder or the sum, which represents the current partial remainder. A numerical example that illustrates this method is shown in Figure 7.10 for a dividend of +13 and a divisor of +5, resulting in a quotient of +2 and a remainder of +3.

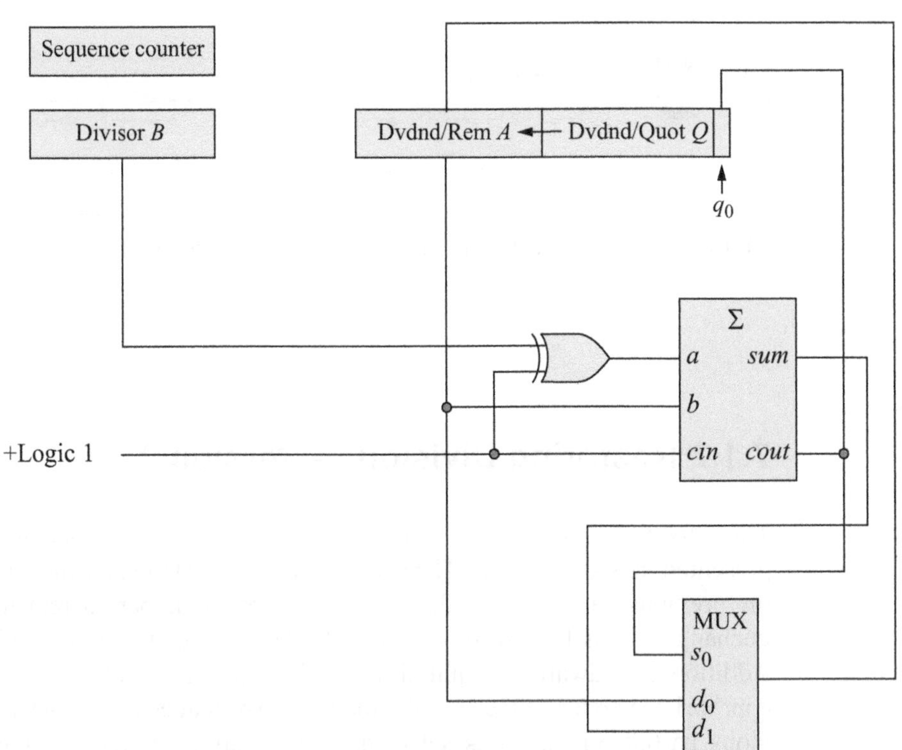

Figure 7.9 Restoring division using a multiplexer.

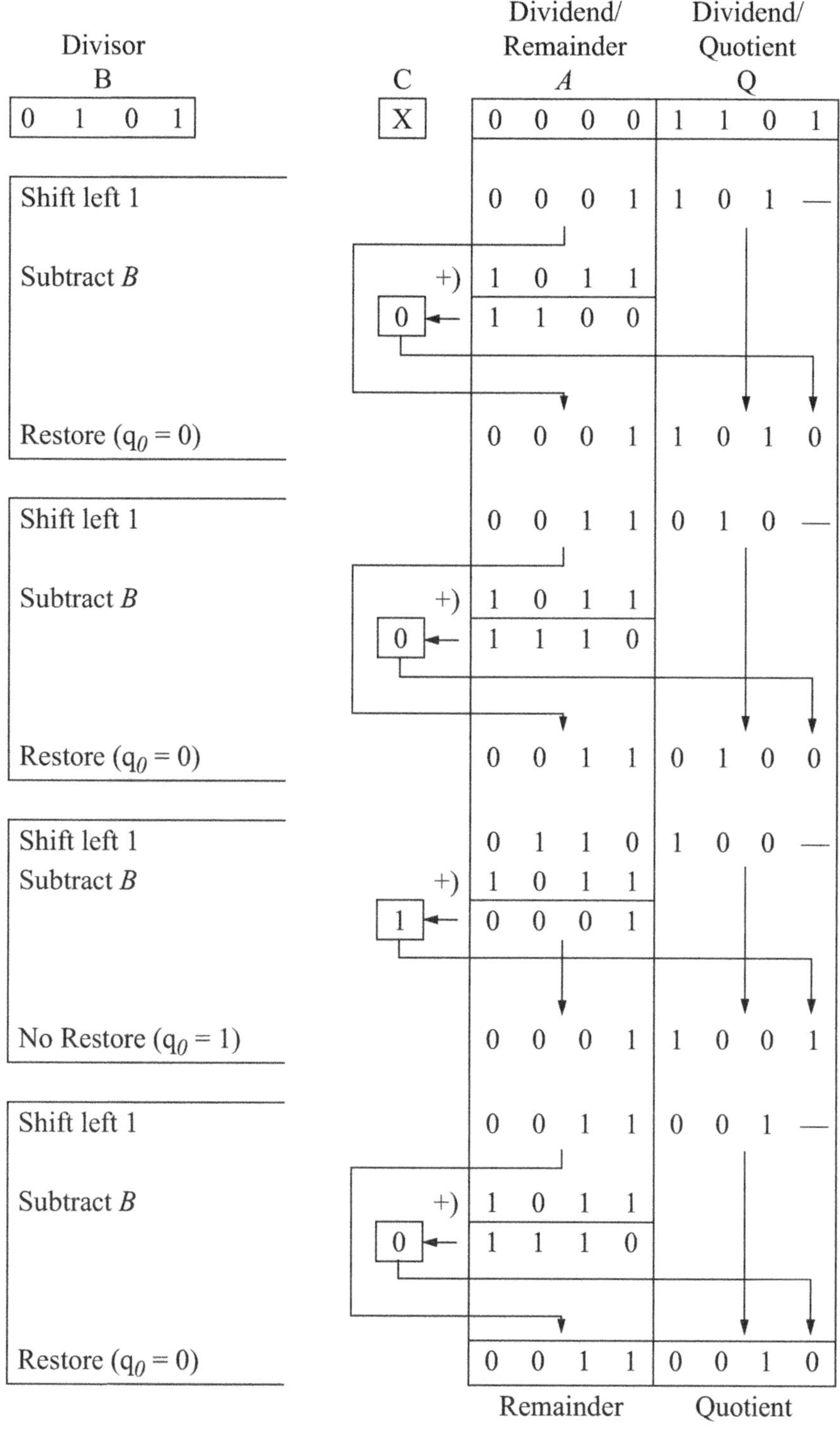

Figure 7.10 Numerical example illustrating restoring division using a multiplexer.

The behavioral/dataflow module for restoring division using a multiplexer is shown in Figure 7.11. Register *a[7:0]* is the dividend; register *b[3:0]* is the divisor. The result register, *rslt[7:0]*, represents the concatenation of the *A* register and the *Q* register in Figure 7.10, where *rslt[7:4]* depicts the remainder and *rslt[3:0]* depicts the quotient. The behavioral module conforms precisely to the algorithm shown in Figure 7.10.

The test bench is shown in Figure 7.12 and applies six input vectors to the design module for the dividend and divisor. The outputs are shown in Figure 7.13, and the waveforms are shown in Figure 7.14.

```
//behavioral/dataflow restoring division using a multiplexor
module div_restoring_vers2 (a, b, start, rslt);

input [7:0] a;              //dividend
input [3:0] b;              //divisor
input start;
output [7:0] rslt;          //rslt[7:4] is rem; rslt[3:0] is quot

wire [3:0] b_bar;

//define internal registers
reg [3:0] b_neg;
reg [7:0] rslt;
reg [3:0] count;
reg [4:0] sum;
reg cout;

assign b_bar = ~b;          //1s complement of divisor

always @ (b_bar)
   b_neg = b_bar + 1;       //2s complement of divisor

always @ (posedge start)
begin
   rslt = a;
   count = 4'b0100;

      if ((a!=0) && (b!=0))    //if a or b = 0, exit
         while (count)         //else do while loop
         begin
            rslt = rslt << 1;
            sum = rslt[7:4] + b_neg;
            cout = sum[4];

//continued on next page
```

Figure 7.11 Mixed-design module for restoring division using a multiplexer.

```
            if (cout == 0)
               begin
                  rslt[0] = cout;      //q0 = cout
                  rslt[7:4] = rslt[7:4];
                  count = count -1;
               end

            else
               begin
                  rslt[0] = cout;      //q0 = cout
                  rslt[7:4] = sum[3:0];
                  count = count - 1;
               end
         end
end
endmodule
```

Figure 7.11 (Continued)

```
//test bench for restoring division version 2
module div_restoring_vers2_tb;

reg [7:0] a;
reg [3:0] b;
reg start;
wire [7:0] rslt;

//display variables
initial
$monitor ("a = %b, b = %b, quot = %b, rem = %b",
            a, b, rslt[3:0], rslt[7:4]);

//apply input vectors
initial
begin
   #0    start = 1'b0;
         a = 8'b0000_1101;    b = 4'b0101;
   #10   start = 1'b1;
   #10   start = 1'b0;

   #10   a = 8'b0011_1100;    b = 4'b0111;
   #10   start = 1'b1;
   #10   start = 1'b0;
//continued on next page
```

Figure 7.12 Test bench for restoring division using a multiplexer.

```
   #10    a = 8'b0101_0010;     b = 4'b0110;
   #10    start = 1'b1;
   #10    start = 1'b0;

   #10    a = 8'b0011_1000;     b = 4'b0111;
   #10    start = 1'b1;
   #10    start = 1'b0;

   #10    a = 8'b0110_0100;     b = 4'b0111;
   #10    start = 1'b1;
   #10    start = 1'b0;

   #10    a = 8'b0110_1110;     b = 4'b0111;
   #10    start = 1'b1;
   #10    start = 1'b0;

   #10    $stop;
end

//instantiate the module into the test bench
div_restoring_vers2 inst1 (
   .a(a),
   .b(b),
   .start(start),
   .rslt(rslt)
   );

endmodule
```

Figure 7.12 (Continued)

```
a = 00001101, b = 0101, quot = xxxx, rem = xxxx
a = 00001101, b = 0101, quot = 0010, rem = 0011
a = 00111100, b = 0111, quot = 0010, rem = 0011
a = 00111100, b = 0111, quot = 1000, rem = 0100
a = 01010010, b = 0110, quot = 1000, rem = 0100
a = 01010010, b = 0110, quot = 1101, rem = 0100
a = 00111000, b = 0111, quot = 1101, rem = 0100
a = 00111000, b = 0111, quot = 1000, rem = 0000
a = 01100100, b = 0111, quot = 1000, rem = 0000
a = 01100100, b = 0111, quot = 1110, rem = 0010
a = 01101110, b = 0111, quot = 1110, rem = 0010
a = 01101110, b = 0111, quot = 1111, rem = 0101
```

Figure 7.13 Outputs for restoring division using a multiplexer.

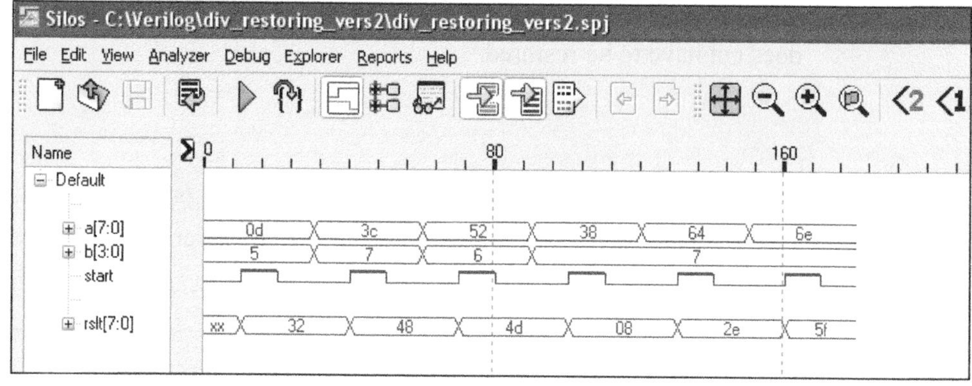

Figure 7.14 Waveforms for restoring division using a multiplexer.

7.2 Sequential Shift-Add/Subtract Nonrestoring Division

The speed of the sequential shift-add/subtract division algorithm can be increased by modifying the algorithm to avoid restoring the partial remainder in the event that a negative partial remainder occurs. This method of *nonrestoring division* allows both a positive partial remainder and a negative partial remainder to be utilized in the division process. The final partial remainder may require restoration if the sign is 1 (negative). This is required in order to have a final positive remainder.

In nonrestoring division, a negative partial remainder is not restored to the previous value but is used unchanged in the following cycle. If the value of the partial remainder is negative after subtracting the divisor B, then the dividend A is shifted left in the next cycle and the divisor is added to the partial remainder — that is, the operation is $2A + B$. If the value of the partial remainder is positive after subtracting the divisor, then the dividend is shifted left in the next cycle and the divisor is subtracted from the partial remainder — that is, the operation is $2A - B$.

Thus, in nonrestoring division, only the operations $2A + B$ and $2A - B$ are used in each cycle. Accordingly, if the sign of the partial remainder $a_{n-1} = 1$ (negative), then A is shifted left one bit position and B is added without restoring the partial remainder to its previous value. Similarly, if the sign of the partial remainder $a_{n-1} = 0$ (positive), then A is shifted left one bit position and B is subtracted. This sequence attempts to reduce the partial remainder toward zero for each cycle. Since the final remainder may have to be restored to its previous value, there can be either n or $n + 1$ shift-add/subtract cycles.

Two examples will now be presented to illustrate the algorithm for nonrestoring division. The first example is shown in Figure 7.15 and uses dividend A = 0011 0011

(+51) represented by dividend/remainder A and dividend/quotient Q; the divisor B = 0111 (+7). This examples requires only four cycles, since the final partial remainder does not have to be restored.

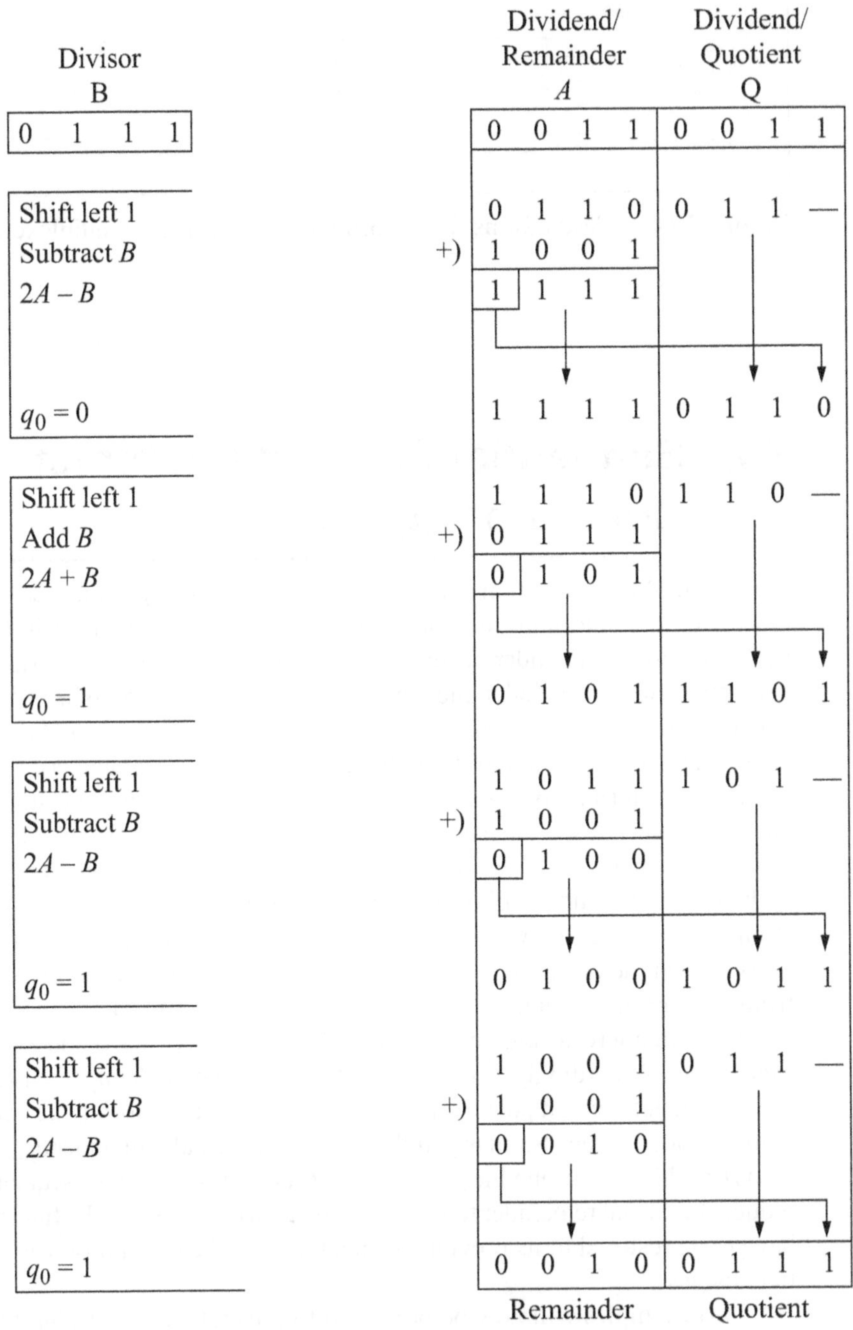

Figure 7.15 Example of nonrestoring division with no final restore cycle.

Figure 7.16 illustrates a nonrestoring division example where the dividend A = 0000 0111 (+7) and the divisor B = 0011 (+3). A final restore cycle is required in this case, because the sign of the final remainder is 1.

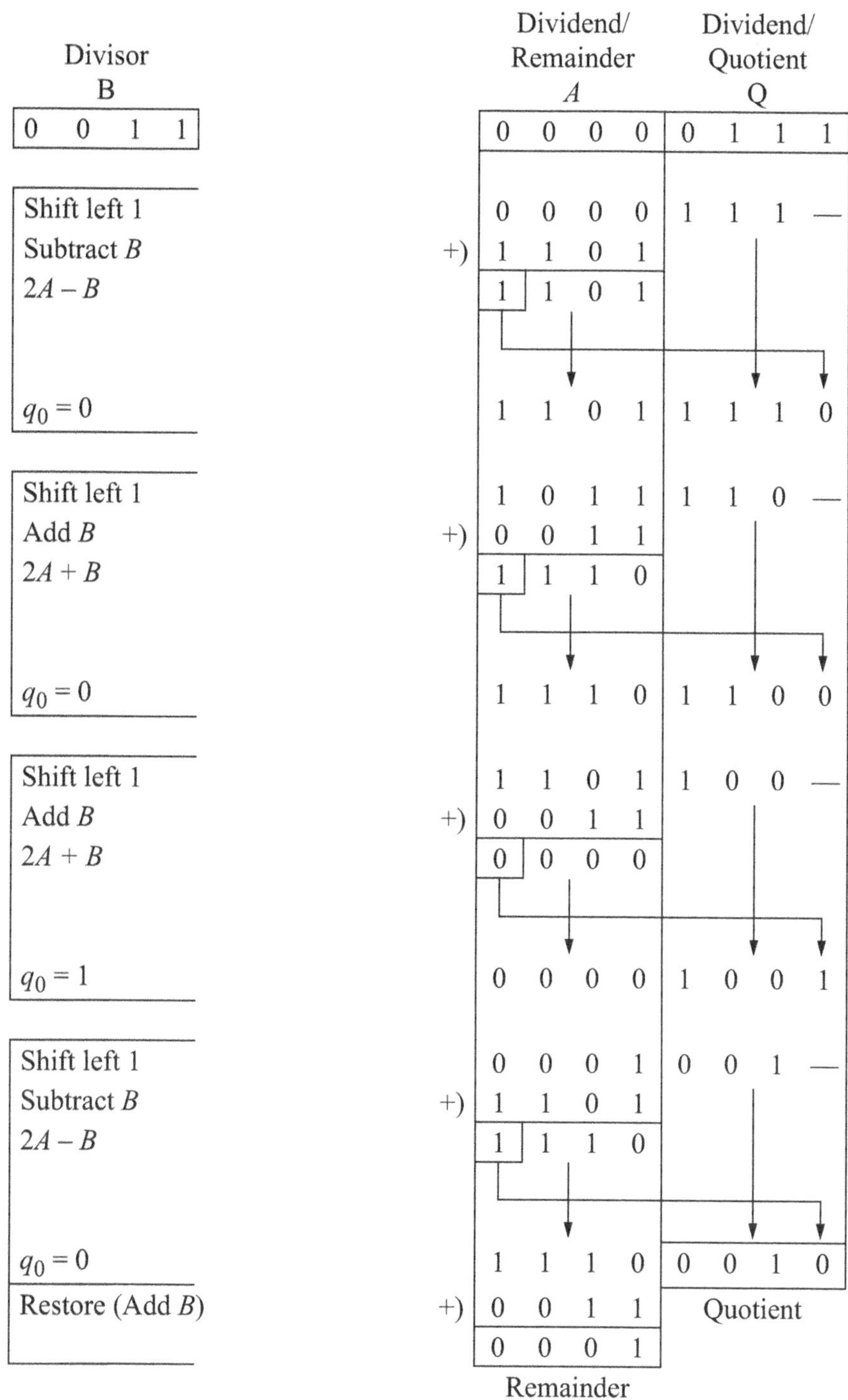

Figure 7.16 Example of nonrestoring division with a final restore cycle.

The sequential shift-add/subtract nonrestoring division algorithm will now be designed with Verilog HDL using behavioral and dataflow modeling, as shown in Figure 7.17. The inputs are an 8-bit dividend, *a[7:0]*; a 4-bit divisor, *b[3:0]*; and a *start* signal — the division operation begins on the positive edge of the *start* signal. The resulting output is *rslt[7:0]*, which contains the quotient, *rslt[3:0]*, and the remainder, *rslt[7:4]*.

```
//behavioral/dataflow nonrestoring division
module div_nonrestoring (a, b, start, rslt);

input [7:0] a;
input [3:0] b;
input start ;
output [7:0] rslt;

wire [3:0] b_bar;

//define internal registers
reg [3:0] b_neg;
reg [7:0] rslt;
reg [3:0] count;
reg part_rem_7;

assign b_bar = ~b;

always @ (b_bar)
   b_neg = b_bar + 1;

always @ (posedge start)
begin
   rslt = a;
   count = 4'b0100;
   part_rem_7 = 1'b0;

   if ((a!=0) && (b!=0))
      begin
         rslt = rslt << 1;

         rslt = {(rslt[7:4] + b_neg), rslt[3:0]};//2A-B
            if (rslt[7] == 1)
               begin
                  rslt = {rslt[7:1], 1'b0};
                  part_rem_7 = 1'b1;
                  count = count - 1;
               end            //continued on next page
```

Figure 7.17 Mixed-design module for nonrestoring division.

```
            else
               begin
                  rslt = {rslt[7:1], 1'b1};
                  part_rem_7 = 1'b0;
                  count = count - 1;
               end
      end

   while (count)
      begin
         rslt = rslt << 1;
         if (part_rem_7 == 1)
            begin
               rslt = {(rslt[7:4] + b), rslt[3:0]};//2A+B
               if (rslt[7] == 1)
                  begin
                     rslt = {rslt[7:1], 1'b0};
                     part_rem_7 = 1'b1;
                     count = count - 1;
                  end

               else
                  begin
                     rslt = {rslt[7:1], 1'b1};
                     part_rem_7 = 1'b0;
                     count = count - 1;
                  end
            end

         else
            begin
               rslt = {(rslt[7:4] + b_neg), rslt[3:0]}; //2A-B
               if (rslt[7] == 1)
                  begin
                     rslt = {rslt[7:1], 1'b0};
                     part_rem_7 = 1'b1;
                     count = count - 1;
                  end

               else
                  begin
                     rslt = {rslt[7:1], 1'b1};
                     part_rem_7 = 1'b0;
                     count = count - 1;
                  end
            end
      end            //continued on next page
```

Figure 7.17 (Continued)

```
   if (rslt[7] == 1)
      rslt = {(rslt[7:4] + b), rslt[3:0]};//restore
end

endmodule
```

Figure 7.17 (Continued)

In Figure 7.17, the 2s complement of the divisor B is obtained by the following statements:

```
assign b_bar = ~b;
always @ (b_bar)
    b_neg = b_bar + 1;
```

The first operation is always a shift-left and subtract operation ($2A - B$), as shown below. The second statement is the concatenation of the partial remainder plus the negation (2s complement) of the divisor and the low-order four bits of the dividend.

```
rslt = rslt << 1;
rslt = {(rslt[7:4] + b_neg), rslt[3:0]};
```

An internal register is declared for the sign bit of the partial remainder and initialized to zero, as shown below. This register is examined to determine if the next operation is an addition or a subtraction of the divisor.

```
reg part_rem_7;
part_rem_7 = 1'b0;
```

If the result of the first shift-left and subtract operation produces a negative difference (rslt[7] = = 1), then a 0 is inserted in the low-order quotient bit q_0, the partial remainder bit is set to 1, and the counter is decremented, as shown in the following statements:

```
rslt = {rslt[7:1], 1'b0};
part_rem_7 = 1'b1;
count = count – 1;
```

If the first shift-left and subtract operation produces a positive difference, then a 1 is inserted in the low-order quotient bit q_0, the partial remainder bit is set to 0, and the counter is decremented.

The remaining code segments use a sequence counter as a control variable. The dividend/remainder, A, and the dividend/quotient, Q — specified by *rslt* — are shifted left one bit position and the partial remainder bit is examined to determine if the next operation is an addition or a subtraction of the divisor. If an addition is required after the shift-left operation, then this is accomplished by the statement shown below, which is the concatenation of the partial remainder plus the divisor and the low-order four bits of the dividend. The counter is then decremented.

rslt = {(rst[7:4] + b), rslt[3:0]};

If a subtraction is required after the shift-left operation, then this is accomplished by the statement shown below, which is the concatenation of the partial remainder plus the negation (2s complement) of the divisor and the low-order four bits of the dividend. The counter is then decremented.

rslt = {(rslt[7:4] + b_neg), rslt[3:0]};

If a final restore operation is required — due to a negative sign for the partial remainder — then this is performed by the statements shown below.

if (rslt[7] = = 1)
rslt = {(rslt[7:4] + b), rslt[3:0]};

At the completion of the division operation, the quotient resides in *rslt[3:0]* and the remainder resides in *rslt[7:4]*.

The test bench is shown in Figure 7.18, in which several vectors are applied to the dividend and divisor. The outputs are shown in Figure 7.19, and the waveforms are shown in Figure 7.20.

```
//test bench for nonrestoring division
module div_nonrestoring_tb;

//define inputs
reg [7:0] a;
reg [3:0] b;
reg start;

//define output
wire [7:0] rslt;

//display variables
initial
$monitor ("a = %b, b = %b, quot = %b, rem = %b",
            a, b, rslt[3:0], rslt[7:4]);

//continued on next page
```

Figure 7.18 Test bench for sequential shift-add/subtract nonrestoring division.

```
//apply input vectors
initial
begin
   #0    start = 1'b0;
         a = 8'b0000_0111;    b = 4'b0011;
   #10   start = 1'b1;
   #10   start = 1'b0;

   #10   a = 8'b0000_1101;    b = 4'b0101;
   #10   start = 1'b1;
   #10   start = 1'b0;

   #10   a = 8'b0011_1100;    b = 4'b0111;
   #10   start = 1'b1;
   #10   start = 1'b0;

   #10   a = 8'b0101_0010;    b = 4'b0110;
   #10   start = 1'b1;
   #10   start = 1'b0;

   #10   a = 8'b0011_1000;    b = 4'b0111;
   #10   start = 1'b1;
   #10   start = 1'b0;

   #10   a = 8'b0110_0100;    b = 4'b0111;
   #10   start = 1'b1;
   #10   start = 1'b0;

   #10   a = 8'b0110_1110;    b = 4'b0111;
   #10   start = 1'b1;
   #10   start = 1'b0;

   #10   $stop;

end

//instantiate the module into the test bench
div_nonrestoring inst1 (
   .a(a),
   .b(b),
   .start(start),
   .rslt(rslt)
   );

endmodule
```

Figure 7.18 (Continued)

```
a = 00000111, b = 0011, quot = xxxx, rem = xxxx
a = 00000111, b = 0011, quot = 0010, rem = 0001
a = 00001101, b = 0101, quot = 0010, rem = 0001
a = 00001101, b = 0101, quot = 0010, rem = 0011
a = 00111100, b = 0111, quot = 0010, rem = 0011
a = 00111100, b = 0111, quot = 1000, rem = 0100
a = 01010010, b = 0110, quot = 1000, rem = 0100
a = 01010010, b = 0110, quot = 1101, rem = 0100
a = 00111000, b = 0111, quot = 1101, rem = 0100
a = 00111000, b = 0111, quot = 1000, rem = 0000
a = 01100100, b = 0111, quot = 1000, rem = 0000
a = 01100100, b = 0111, quot = 1110, rem = 0010
a = 01101110, b = 0111, quot = 1110, rem = 0010
a = 01101110, b = 0111, quot = 1111, rem = 0101
```

Figure 7.19 Outputs for sequential shift-add/subtract nonrestoring division.

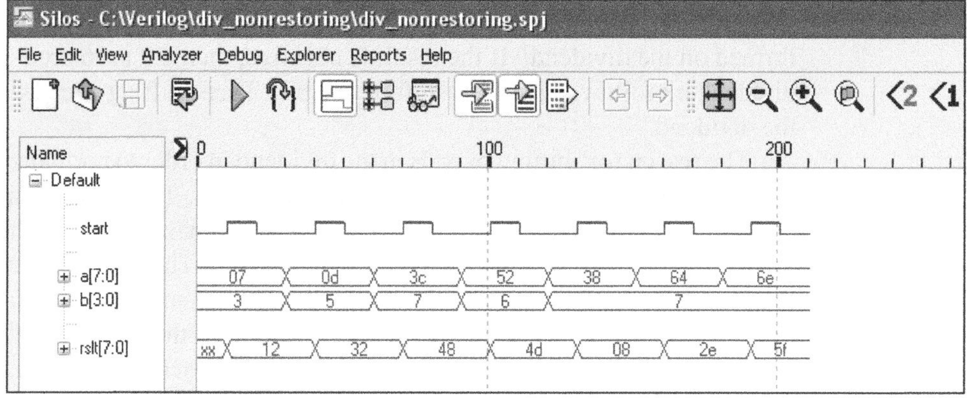

Figure 7.20 Waveforms for sequential shift-add/subtract nonrestoring division.

7.3 SRT Division

SRT division was developed independently by Sweeney, Robertson, and Tocher at approximately the same time as a way to increase the speed of a divide operation. It was intended to improve radix-2 floating-point arithmetic by shifting over strings of 0s or 1s in much the same way as the Booth algorithm shifts over strings of 0s in a multiply operation. The dividend and divisor are binary fractions with the radix point to the immediate left of the high-order significand bit. A positive divisor is normalized before beginning the divide operation by shifting the divisor left until there is a 1 bit in the high-order bit position. Thus, for a positive n-bit divisor

$$B = b_{n-1}\, b_{n-2}\, b_{n-3} \ldots b_1\, b_0$$

a normalized divisor is of the form

$$B = 0\ .\ 1\ b_{n-2}\, b_{n-3} \ldots b_1\, b_0$$

where 0 . 1 represent the sign bit and the high-order significand bit, respectively and $b_{n-2}\, b_{n-3} \ldots b_1\, b_0$ are 0s or 1s. The dividend is also normalized, but this is done during the divide operation. Normalizing the dividend and divisor provides a fixed reference point for both operands at the location of the radix point.

The dividend A, divisor B, quotient Q, and remainder R can be characterized by Equation 7.1.

$$\text{Dividend } A = (\text{Divisor } B \times \text{Quotient } Q) + \text{Remainder } R \qquad (7.1)$$

When the operands are normalized, the aligned operands are compared by subtracting the divisor from the dividend. If the result is positive, then a 1 is entered in the low-order dividend bit followed by a left shift of one bit. Then shifting over 0s is performed on the dividend. If the result is negative, then a 0 is entered in the low-order dividend bit followed by a left shift of one bit. Then shifting over 1s is performed on the dividend.

The reason for shifting over 0s in the dividend in order to normalize a positive partial remainder is explained in the following sentences. If there is a 0 in the high-order bit position of the significand, then the value is less than 0.5. Since the divisor is normalized, it has a value greater than or equal to 0.5. Therefore, the divisor is greater than the dividend which results in the quotient bit having a value of 0. For each shift of the dividend, a 0 is entered as the low-order bit of the dividend. When both operands are normalized, a subtraction is performed to determine if the dividend is larger than the divisor. A subtraction is always performed after shifting over 0s, because the partial remainder is to be diminished toward zero.

The reason for shifting over 1s in the dividend in order to normalize a negative partial remainder is explained in the following sentences. A subtraction may result in a negative partial remainder, such that

$$A = 1\ .\ 1\ a_i\, a_{i-1}\, a_{i-2} \ldots a_1\, a_0$$

Therefore, the value of A is greater than or equal to – 0.5 and less than zero. To normalize a negative partial remainder, the partial remainder is shifted left over 1s until a 0 appears in the high-order bit position. For each shift of the dividend, a 1 is entered as the low-order bit of the dividend. An addition is always performed after shifting over 1s, because the partial remainder is to be diminished toward zero.

The time required to perform a division operation is a function of the requisite number of additions to perform the operation. During the division operation when both operands are normalized, an addition or subtraction must take place in order to diminish the partial remainder toward zero and determine the next quotient bit. If the partial remainder is positive, then the normalized divisor is subtracted from the positive partial remainder. If the partial remainder is negative, then the normalized divisor is added to the negative partial remainder.

If the signs of the quotient and remainder are different, then the remainder must be corrected by shifting the remainder right one bit position and adding the normalized divisor. The operands are represented as n-bit significands in sign-magnitude representation. The sequence of steps to perform SRT division on positive operands are shown below.

1. Normalize divisor B.

2. Adjust dividend $A.Q$.

3. Shift over 0s in dividend $A.Q$.

4. Subtract normalized divisor B.

5. Load A, shift $A.Q$ left one bit position, set low-order quotient bit q_0.
   Set $q_0 = 1$ if partial remainder is positive.
   Set $q_0 = 0$ if partial remainder is negative.

6. Shift over 0s to normalize a positive partial remainder.
   Subtract normalized divisor B.
   Load A, shift $A.Q$ left one bit position, set low-order quotient bit $q_0 = 0$.

or

7. Shift over 1s to normalize a negative partial remainder.
   Add normalized divisor B.
   Load A, shift $A.Q$ left one bit position, set low-order quotient bit $q_0 = 1$.

8. If counter is not zero, then repeat step 6 or 7.

9. Remainder correction.
   If sign of remainder is different than sign of quotient, then
   shift remainder right one bit position, and
   add normalized divisor B.

For the examples that follow, it is assumed that a test for overflow has already been completed. The number of cycles required for SRT division is $n + 1$, where n is the number of bits in the divisor. Since SRT division is more complex than other methods previously presented, several examples will be given that exemplify the procedure. The examples include illustrations where both operands are positive and where the dividend is negative and the divisor is positive. The asterisks (*) are merely delimiters between the high-order quotient bit and the low-order remainder bit. A counter is used to control the division process and is initially set to a value of $n + 1$, then decremented by one for each left shift operation; when the counter contains a value of zero, the operation is finished.

Example 7.1 Let the dividend $A.Q = 0.0000\ 0110$ (+6) and the divisor $B = 0.0001$ (+1) to produce a quotient $Q = +6$ and a remainder $R = 0$. The divisor is normalized by shifting it left three bit positions with a corresponding shift to adjust the dividend. Then registers $A.Q$ shift left two bit positions to shift over 0s. The first add/subtract operation is always a subtract to determine the relative magnitude of the operands. If the sign is 0 after the subtraction, then a 1 is inserted in the low-order bit position of register pair $A.Q$. The operation continues until five shift-left operations have been completed, at which time the counter has decremented to zero.

$A.Q = 0.00000110$

$B = 0.0001$

Normalized $B = 0.1000$

2s complement of normalized $B = 1.1000$

Operation		8		7	6	5	4	3	2	1	0
				A				*Q*			
		0	.	0	0	0	0	0	1	1	0
Adjust dividend ← 3		0	.	0	0	1	1	0	*	*	*
Shift over 0s ← 2		0	.	1	1	0	*	*	*	0	0
Subtract normalized B	+)	1	.	1	0	0	0				
		0	.	0	1	0	*				
Load A, shift ← 1, $q_0 = 1$		0	.	1	0	*	*	*	0	0	1
Subtract normalized B	+)	1	.	1	0	0	0				
		0	.	0	0	*	*				
Load A, shift ← 1, $q_0 = 1$		0	.	0	*	*	*	0	0	1	1
Shift over 0s into termination ← 1		0	.	*	*	*	0	0	1	1	0
								Q			
Leftmost * is at bit 7, ∴ shift A 4 →				0	0	0	0				
				R							

Example 7.2 Let the dividend $A.Q = 0.0001\ 0001$ (+17) and the divisor $B = 0.0011$ (+3) to produce a quotient $Q = +5$ and a remainder $R = +2$. This example is similar to the previous example, but has conditions where the operation shifts over 0s and 1s. Also, the remainder must be corrected, because the sign of the remainder (1) is different than the sign of the quotient (0).

$A.Q = 0.00010001$

$B = 0.0011$

Normalized $B = 0.1100$

2s complement of normalized $B = 1.0100$

		8		7	6	5	4	3	2	1	0
					A				*Q*		
		0	.	0	0	0	1	0	0	0	1
Adjust dividend ← 2		0	.	0	1	0	0	0	1	*	*
Shift over 0s ← 1		0	.	1	0	0	0	1	*	*	0
Subtract normalized B	+)	1	.	0	1	0	0				
		1	.	1	1	0	0				
Load A, shift ← 1, $q_0 = 0$		1	.	1	0	0	1	*	*	0	0
Shift over 1s ← 1, $q_0 = 1$		1	.	0	0	1	*	*	0	0	1
Add normalized B	+)	0	.	1	1	0	0				
		1	.	1	1	1	*				
Load A, shift ← 1, $q_0 = 0$		1	.	1	1	*	*	0	0	1	0
Shift over 1s into termination ← 1		1	.	1	*	*	0	0	1	0	1
									Q		
Correct remainder											
Shift A 1 →		1	.	1	1	*	*				
Add normalized B	+)	0	.	1	1	0	0				
		0	.	1	0	*	*				
Leftmost * is at bit 5, ∴ shift A 2 →				0	0	1	0				
					R						

Example 7.3 Let the dividend $A.Q = 0.0010\ 1001$ (+41) and the divisor $B = 0.0011$ (+3) to produce a quotient $Q = +13$ and a remainder $R = +2$. The dividend is adjusted by shifting it left two bit positions, at which time it is normalized. After the first subtraction, the sign is negative (1); therefore, a 0 is inserted in the low-order bit position of register pair $A.Q$. After five shift-left operations, the counter has decremented to zero — note that not all 1s have been shifted over as the operation shifts into termination. As before, the remainder must be corrected, because the sign of the remainder (1) is different than the sign of the quotient (0).

$A.Q = 0.00101001$

$B = 0.0011$

Normalized $B = 0.1100$

2s complement of normalized $B = 1.0100$

	8		7	6	5	4	3	2	1	0
			A				*Q*			
	0	.	0	0	1	0	1	0	0	1
Adjust dividend ← 2	0	.	1	0	1	0	0	1	*	*
No shift over 0s	0	.	1	0	1	0	0	1	*	*
Subtract normalized B	+) 1	.	0	1	0	0				
	1	.	1	1	1	0				
Load A, shift ← 1, $q_0 = 0$	1	.	1	1	0	0	1	*	*	0
Shift over 1s ← 2, $q_1\ q_0 = 11$	1	.	0	0	1	*	*	0	1	1
Add normalized B	+) 0	.	1	1	0	0				
	1	.	1	1	1	*				
Load A, shift ← 1, $q_0 = 0$	1	.	1	1	*	*	0	1	1	0
Shift over 1s into termination ← 1	1	.	1	*	*	0	1	1	0	1
							Q			
Correct remainder										
Shift A 1 →	1	.	1	1	*	*				
Add normalized B	+) 0	.	1	1	0	0				
	0	.	1	0	*	*				
Leftmost * is at bit 5, ∴ shift A 2 →			0	0	1	0				
			R							

Example 7.4 Let the dividend $A.Q = 0.0100\ 1100$ (+76) and the divisor $B = 0.0110$ (+6) to produce a quotient $Q = +12$ and a remainder $R = +4$. After five shift-left operations, the counter has decremented to zero and the remainder is corrected.

$A.Q = 0.01001100$

$B = 0.0110$

Normalized $B = 0.1100$

2s complement of normalized $B = 1.0100$		8		7	6	5	4	3	2	1	0
					A				Q		
		0	.	0	1	0	0	1	1	0	0
Adjust dividend ← 1		0	.	1	0	0	1	1	0	0	*
No shift over 0s		0	.	1	0	0	1	1	0	0	*
Subtract normalized B	+)	1	.	0	1	0	0				
		[1]	.	1	1	0	1				
Load A, shift ← 1, $q_0 = 0$		1	.	1	0	1	1	0	0	*	0
Shift over 1s ← 1, $q_0 = 1$		1	.	0	1	1	0	0	*	0	1
Add normalized B	+)	0	.	1	1	0	0				
		[0]	.	0	0	1	0				
Load A, shift ← 1, $q_0 = 1$		0	.	0	1	0	0	*	0	1	1
Shift over 0s ← 1, $q_0 = 0$		0	.	1	0	0	*	0	1	1	0
Subtract normalized B	+)	1	.	0	1	0	0				
		[1]	.	1	1	0	*				
Load A, shift ← 1, $q_0 = 0$		1	.	1	0	*	0	1	1	0	0
									Q		
Correct remainder											
Shift A 1 →		1	.	1	1	0	*				
Add normalized B	+)	0	.	1	1	0	0				
		0	.	1	0	0	*				
Leftmost * is at bit 4, ∴ shift A 1 →				0	1	0	0				
					R						

Example 7.5 Let the dividend $A.Q = 0.0110\ 1110\ (+110)$ and the divisor $B = 0.0111$ $(+7)$ to produce a quotient $Q = +15$ and a remainder $R = +5$. The dividend is adjusted by a left shift of one bit position to match the left shift required to normalize the divisor.

After the first subtraction, the sign is negative (1); therefore, a 0 is inserted in the low-order bit position of register pair $A.Q$. Since the result is negative, register pair $A.Q$ shifts left over 1s into termination, because the counter has decremented to zero — note that not all 1s have been shifted over as the operation shifts into termination. As before, the remainder must be corrected, because the sign of the remainder (1) is different than the sign of the quotient (0).

$A.Q = 0.01101110$

$B = 0.0111$

Normalized $B = 0.1110$

			8		7	6	5	4	3	2	1	0
2s complement of normalized $B = 1.0010$						A				Q		
			0	.	0	1	1	0	1	1	1	0
Adjust dividend ← 1			0	.	1	1	0	1	1	1	0	*
No shift over 0s			0	.	1	1	0	1	1	1	0	*
Subtract normalized B		+)	1	.	0	0	1	0				
			1	.	1	1	1	1				
Load A, shift ← 1, $q_0 = 0$			1	.	1	1	1	1	1	0	*	0
Shift over 1s into termination ← 4			1	.	1	0	*	0	1	1	1	1
										Q		
Correct remainder												
Shift A 1 →			1	.	1	1	0	*				
Add normalized B		+)	0	.	1	1	1	0				
			0	.	1	0	1	*				
Leftmost * is at bit 4, ∴ shift A 1 →					0	1	0	1				
						R						

Example 7.6 Let the dividend $A.Q = 0.1110\ 1111$ (+239) and the divisor $B = 0.1111$ (+15) to produce a quotient $Q = +15$ and a remainder $R = +14$. The divisor is already normalized; therefore, the dividend does not require adjustment. There is also no shifting over 0s for register pair $A.Q$.

After the first subtraction, the sign is negative (1); therefore, a 0 is inserted in the low-order bit position of register pair $A.Q$. Since the result is negative, register pair $A.Q$ shifts left over 1s into termination, because the counter has decremented to zero — note that not all 1s have been shifted over as the operation shifts into termination. In this example, the remainder does not require correction, because the sign of the remainder is 1 and the sign of the quotient 1.

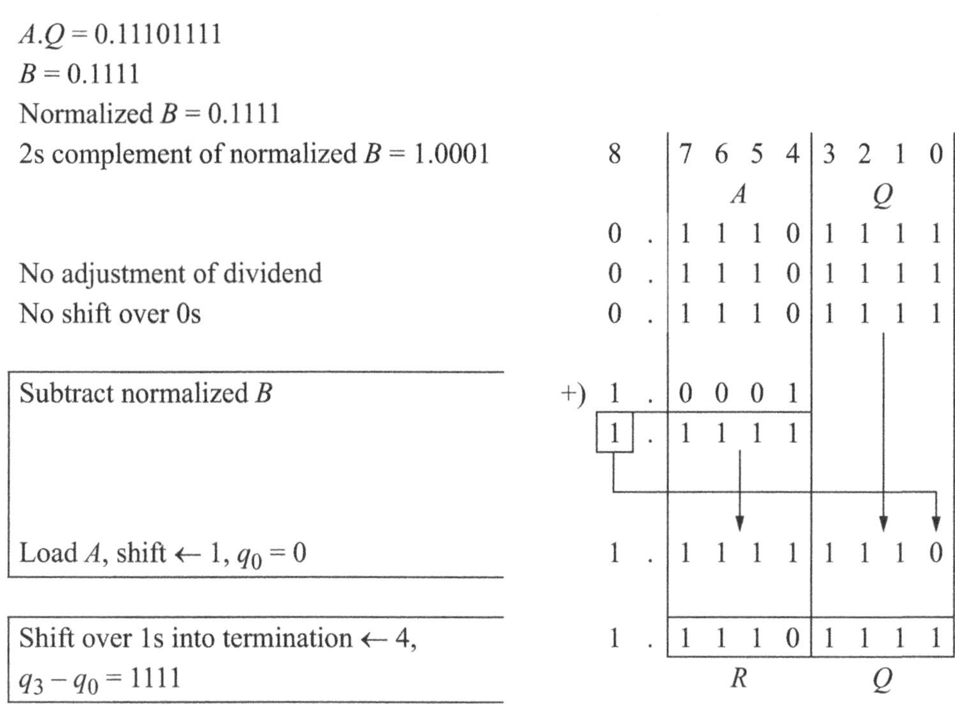

A.Q = 0.11101111
B = 0.1111
Normalized B = 0.1111
2s complement of normalized B = 1.0001
8
7 6 5 4
3 2 1 0
A
Q
0 . 1 1 1 0 1 1 1 1
No adjustment of dividend
0 . 1 1 1 0 1 1 1 1
No shift over 0s
0 . 1 1 1 0 1 1 1 1
Subtract normalized B
+) 1 . 0 0 0 1
1 . 1 1 1 1
Load A, shift ← 1, q0 = 0
1 . 1 1 1 1 1 1 1 0
Shift over 1s into termination ← 4,
q3 – q0 = 1111
1 . 1 1 1 0 1 1 1 1
R
Q

Example 7.7 This example will use a negative dividend $A.Q = 1.1110\ 0101$ (–27) and a positive divisor $B = 0.0110$ (+6) to produce a negative quotient $Q = –4$ and a negative remainder $R = –3$. Recall that the equation for the dividend is

$$\text{Dividend } A = (\text{Divisor } B \times \text{Quotient } Q) + \text{Remainder } R$$

Therefore, using the above values, the equation is

$$\text{Dividend } (-27) = [\text{Divisor } (+6) \times \text{Quotient } (-4)] + \text{Remainder } (-3)]$$

The divisor is normalized by a left shift of one bit position and the dividend is adjusted accordingly. Then the dividend $A.Q$ is shifted left over 1s until it is normalized with a 0 in the high-order bit position. Shifting a negative unnormalized number left over 1s does not destroy any information, because the 1s represent only positional information. Since the register pair $A.Q$ contains a negative number, an add operation takes place by adding the normalized divisor. After five left-shift operations, the counter contains a value of zero and the division terminates. The resulting quotient is positive and must be 2s complemented.

$A.Q = 1.11100101$

$B = 0.0110$

Normalized $B = 0.1100$

		8		7	6	5	4	3	2	1	0
2s complement of normalized $B = 1.0100$						A				Q	
		1	.	1	1	1	0	0	1	0	1
Adjust dividend ← 1		1	.	1	1	0	0	1	0	1	*
Shift over 1s ← 2		1	.	0	0	1	0	1	*	0	0
Add normalized B	+)	0	.	1	1	0	0				
		1	.	1	1	1	0				
Load A, shift ← 1, $q_0 = 1$		1	.	1	1	0	1	*	0	0	1
Shift over 1s ← 2; $q_1 - q_0 = 00$		1	.	0	1	*	0	0	1	0	0
2s complement Q							1	1	1	0	0
									Q		
* is at bit 5, ∴ shift A 2 →				1	1	0	1				
					R						

SRT division can be implemented in a variety of ways using behavioral modeling. The Verilog HDL code segment shown in Figure 7.21 illustrates one method of normalizing the divisor, adjusting the dividend, and shifting over 0s for a 9-bit dividend (sign plus 8-bit significand) and a 5-bit divisor (sign plus 4-bit significand). This code segment checks the high-order three bits of the divisor's significand and adjusts dividend accordingly. The **while** loop then shifts over 0s. This method can then be extended to check the remaining high-order bits of the divisor. However, a simpler method for division — one that is used by several companies — is to use the table lookup technique as described in the following section.

```
//behavioral/dataflow srt division segment
module srt_division (a, b, start, rslt);

input [8:0] a;
input [4:0] b;
input start;
output [8:0] rslt;

wire [4:0] b_norm_bar;

//define internal registers
reg [4:0] b_norm;
reg [4:0] b_norm_neg;
reg [8:0] rslt;
reg [3:0] count;

assign b_norm_bar = ~b_norm;

always @ (b_norm_bar)
   b_norm_neg = b_norm_bar + 1;

//normalize divisor, adjust dividend, shift over 0s
always @ (a or b)
begin
   count = 4'b0101;
   rslt = a;

   if (b[3:1] == 3'b000)
      begin
         b_norm = b << 3;       //normalize divisor
         rslt = rslt << 3;      //adjust dividend
      end

   while (count)                //shift over 0s
      begin
         if (rslt[7] == 1'b0)
            begin
               rslt[8:0] = {rslt[7:0], 1'b0};
               count = count - 1;
            end
. . .
```

Figure 7.21 Verilog code segment to illustrate normalizing the divisor, adjusting the dividend, and shifting over 0s.

7.3.1 SRT Division Using Table Lookup

The table lookup method uses the concatenated dividend and divisor to address a memory that contains the quotient and remainder at each memory address for the corresponding dividend and divisor. In order to demonstrate the feasibility of the table lookup method, only a select number of addresses will contain valid contents representing the quotient and remainder, thus keeping the address space to a reasonable size. The operands will consist of 6-bit dividends and 3-bit divisors. Memories can be represented in Verilog by an array of registers and are declared using a **reg** data type as follows:

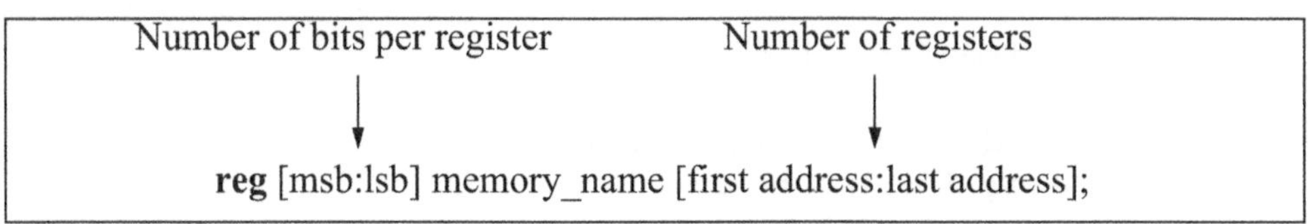

Figure 7.22 shows the contents of the memory called *opnds.srt*, where the 9-bit address is shown for reference. The leftmost six digits of the address represent the dividend; the rightmost three digits represent the divisor — the space is shown only for clarity. The address bits are not part of the memory contents — only the six bits for the resulting quotient and remainder are entered into the *opnds.srt* file.

Figure 7.23 contains the behavioral module for the table lookup method. The dividend and divisor are contained in the 9-bit variable called *opnds[8:0]* — the dividend is specified as *opnds[8:3]*, the divisor is specified as *opnds[2:0]*. Recall that overflow occurs if the high-order half of the dividend is greater than or equal to the divisor. Thus, overflow is checked by a behavioral construct using the **always** statement as shown below. Whenever the operands change value, overflow is checked.

```
always @ (opnds)
begin
    if (opnds[8:6] >= opnds[2:0]
        ovfl = 1'b1;
    else
        ovfl = 1'b0;
end
```

Since the concatenated operands constitute a 9-bit variable, the memory consists of an array of 512 six-bit registers, where the register contents represent the 3-bit quotient concatenated with the 3-bit remainder. The memory is called *mem_srt* and is loaded from a file called *opnds.srt* that contains the memory contents. The memory contents are saved in the project file with no **.v** extension. When the operands change value — thus changing the memory address — the 6-bit register *quot_rem* is loaded from memory *mem_srt* at the location of the corresponding operands.

```
dvdnd   dvsr quot rem                dvdnd   dvsr quot rem

000000 000   000000                  011101 011   000000
000000 001   000000                  011101 100   111001
000000 010   000000                  011101 101   000000
 . . .                                . . .
000110 000   000000                  100111 100   000000
000110 001   110000                  100111 101   111100
000110 010   000000                  100111 110   000000
 . . .                                . . .
000111 001   000000                  101010 101   000000
000111 010   011001                  101010 110   111000
000111 011   000000                  101010 111   000000
 . . .                                . . .
010001 010   000000                  110011 110   000000
010001 011   101010                  110011 111   111010
010001 100   000000                  110100 000   000000
 . . .                                . . .
010100 010   000000
010100 011   110010
010100 100   000000
 . . .
```

Figure 7.22 Contents of *opnds.srt* that contains the resulting quotient and remainder.

```
//behavioral srt division using table lookup
module srt_div_tbl_lookup (opnds, quot_rem, ovfl);

input [8:0] opnds;      //dvdnd 6 bits; dvsr 3 bits
output [5:0] quot_rem;
output ovfl;

wire [8:0] opnds;
reg [5:0] quot_rem;
reg ovfl;

//check for overflow
always @ (opnds)
begin
   if (opnds[8:6] >= opnds[2:0])
      ovfl = 1'b1;
   else
      ovfl = 1'b0;
end
//continued on next page
```

Figure 7.23 Behavioral module for SRT division using table lookup.

```
//define memory size
//mem_srt is an array of 512 six-bit registers
reg [5:0] mem_srt[0:511];

//define memory contents
//load mem_srt from file opnds.srt
initial
begin
   $readmemb ("opnds.srt", mem_srt);
end

//use the operands to access memory
always @ (opnds)
begin
   quot_rem = mem_srt [opnds];
end

endmodule
```

Figure 7.23 (Continued)

The test bench is shown in Figure 7.24 in which ten different combinations of dividends and divisors are applied to the behavioral module, including those that will generate an overflow. The outputs are shown in Figure 7.25, and the waveforms in Figure 7.26.

```
//test bench for srt division using table lookup
module srt_div_tbl_lookup_tb;


reg [8:0] opnds;
wire [5:0] quot_rem;
wire ovfl;


//display variables
initial
$monitor ("dvdnd=%b, dvsr=%b, quot=%b, rem=%b, ovfl=%b",
            opnds[8:3], opnds[2:0],
            quot_rem[5:3], quot_rem[2:0], ovfl);

//continued on next page
```

Figure 7.24 Test bench for SRT division using table lookup.

```
//apply stimulus
initial
begin
   #0    opnds = 9'b000110001;
   #10   opnds = 9'b000111010;
   #10   opnds = 9'b010001011;
   #10   opnds = 9'b011101100;
   #10   opnds = 9'b010100011;
   #10   opnds = 9'b100111101;
   #10   opnds = 9'b110011111;
   #10   opnds = 9'b101010110;
   #10   opnds = 9'b110001101;   //overflow
   #10   opnds = 9'b101000101;   //overflow

   #10   $stop;
end


//instantiate the module into the test bench
srt_div_tbl_lookup inst1 (
   .opnds(opnds),
   .quot_rem(quot_rem),
   .ovfl(ovfl)
   );


endmodule
```

Figure 7.24 (Continued)

```
dvdnd=000110, dvsr=001, quot=110, rem=000, ovfl=0
dvdnd=000111, dvsr=010, quot=011, rem=001, ovfl=0
dvdnd=010001, dvsr=011, quot=101, rem=010, ovfl=0
dvdnd=011101, dvsr=100, quot=111, rem=001, ovfl=0
dvdnd=010100, dvsr=011, quot=110, rem=010, ovfl=0
dvdnd=100111, dvsr=101, quot=111, rem=100, ovfl=0
dvdnd=110011, dvsr=111, quot=111, rem=010, ovfl=0
dvdnd=101010, dvsr=110, quot=111, rem=000, ovfl=0
dvdnd=110001, dvsr=101, quot=000, rem=000, ovfl=1
dvdnd=101000, dvsr=101, quot=000, rem=000, ovfl=1
```

Figure 7.25 Outputs for SRT division using table lookup.

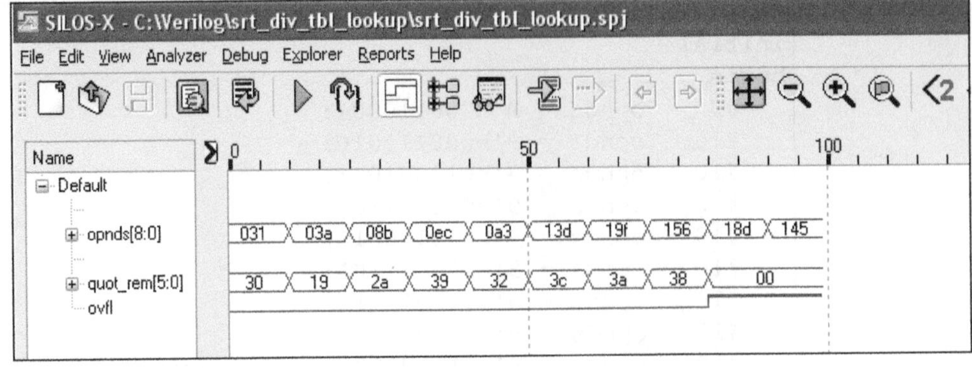

Figure 7.26 Waveforms for SRT division using table lookup.

7.3.2 SRT Division Using the Case Statement

SRT division — and any division technique — can also be accomplished using the **case** statement, which is a method that is analogous to the table lookup method, in that the resulting quotient and remainder are the direct result of a concatenated dividend and divisor. The combined dividends and divisors form case items which are compared with a case expression on a bit-by-bit basis. The first case item that matches the case expression is selected — in this case the quotient and remainder to the right of the colon. It is appropriate at this time to briefly review the **case** statement.

The **case** statement is an alternative to the **if . . . else if** construct and is a multiple-way conditional branch. It executes one of several different procedural statements depending on the comparison of a case expression with a case item. The expression and the case item are compared bit-by-bit and must match exactly. The statement that is associated with a case item may be a single procedural statement or a block of statements delimited by the keywords **begin . . . end.** The keywords **case**, **endcase**, and **default** are used in the **case** statement, which has the following syntax:

```
case (expression)
    case_item1 : procedural_statement1;
    case_item2 : procedural_statement2;
    case_item3 : procedural_statement3;
                 .
                 .
                 .
    case_itemn : procedural_statementn;
    default : default_statement;
endcase
```

The case expression may be an expression or a constant. The case items are evaluated in the order in which they are listed. If a match occurs between the case expression and a case item, then the corresponding procedural statement, or block of statements, is executed. If no match occurs, then the optional **default** statement is executed.

For SRT division, the case expression is the concatenated dividend and divisor expressed as the input operands *opnds*. The case items are the concatenated dividend and remainder and the procedural statements are the corresponding quotient and remainder.

The behavioral module to implement SRT division is shown in Figure 7.27 using 8-bit dividends and 4-bit divisors. Since the **case** statement would contain $2^{12} = 4096$ case items and corresponding procedural statements if all 4096 entries were used, only ten entries will be considered. The test bench is shown in Figure 7.28. The outputs and waveforms are shown in Figure 7.29 and Figure 7.30, respectively.

```
//behavioral srt division using the case statement
module srt_div_case2 (opnds, quot, rem, ovfl);

input [11:0] opnds;   //dividend 8 bits; divisor 4 bits
output [3:0] quot, rem;
output ovfl;

wire [11:0] opnds;
reg [3:0] quot, rem;
reg ovfl;

//check for overflow
always @ (opnds)
begin
   if (opnds[11:8] >= opnds[3:0])
      ovfl = 1'b1;
   else
      ovfl = 1'b0;
end

always @ (opnds)
begin
   case (opnds)
      //dvdnd = +6;dvsr = +1;quot = 6;rem = 0
      12'b000001100001 :begin
                           quot= 4'b0110;
                           rem = 4'b0000;
                        end

//continued on next page
```

Figure 7.27 Behavioral module for SRT division using the **case** statement.

```
      //dvdnd = +7;dvsr = +2;quot = 3;rem = 1
      12'b000001110010 :begin
                         quot= 4'b0011;
                         rem = 4'b0001;
                      end

      //dvdnd = +17;dvsr = +3;quot = 5;rem = 2
      12'b000100010011 :begin
                         quot= 4'b0101;
                         rem = 4'b0010;
                      end

      //dvdnd = +41;dvsr = +3;quot = 13;rem = 2
      12'b001010010011 :begin
                         quot= 4'b1101;
                         rem = 4'b0010;
                      end

      //dvdnd = +51;dvsr = +8;quot = 6;rem = 3
      12'b001100111000 :begin
                         quot= 4'b0110;
                         rem = 4'b0011;
                      end

      //dvdnd = +72;dvsr = +5;quot = 14;rem = 2
      12'b010010000101 :begin
                         quot= 4'b1110;
                         rem = 4'b0010;
                      end

      //dvdnd = +76;dvsr = +6;quot = 12;rem = 4
      12'b010011000110 :begin
                         quot= 4'b1100;
                         rem = 4'b0100;
                      end

      //dvdnd = +110;dvsr = +7;quot = 15;rem = 5
      12'b011011100111 :begin
                         quot= 4'b1111;
                         rem = 4'b0101;
                      end

//continued on next page
```

Figure 7.27 (Continued)

```
        //dvdnd = +97;dvsr = +5;quot = 19;rem = 2
        //overflow occurs
        12'b011000010101 :begin
                              quot= 4'bxxxx;
                              rem = 4'bxxxx;
                           end

        //dvdnd = +70;dvsr = +4;quot = 17;rem = 2
        //overflow occurs
        12'b010001100100 :begin
                              quot= 4'bxxxx;
                              rem = 4'bxxxx;
                           end

        default          :begin
                              quot= 4'b0000;
                              rem   = 4'b0000;
                           end
    endcase
end

endmodule
```

Figure 7.27 (Continued)

```
//test bench for srt division using the case statement
module srt_div_case_tb;

reg [11:0] opnds;
wire [3:0] quot, rem;
wire ovfl;

//display variables
initial
$monitor ("opnds= %b, quot = %b, rem = %b, ovfl = %b",
            opnds, quot, rem, ovfl);

//apply stimulus
initial
begin
        //dvdnd = +6;dvsr = +1;quot = 6;rem = 0
   #0   opnds = 12'b000001100001;

//continued on next page
```

Figure 7.28 Test bench for SRT division using the **case** statement.

```
        //dvdnd = +7;dvsr = +2;quot = 3;rem = 1
   #10  opnds = 12'b000001110010;

        //dvdnd = +17;dvsr = +3;quot = 5;rem = 2
   #10  opnds = 12'b000100010011;

        //dvdnd = +41;dvsr = +3;quot = 13;rem = 2
   #10  opnds = 12'b001010010011;

        //dvdnd = +51;dvsr = +8;quot = 6;rem = 3
   #10  opnds = 12'b001100111000;

        //dvdnd = +72;dvsr = +5;quot = 14;rem = 2
   #10  opnds = 12'b010010000101;

        //dvdnd = +76;dvsr = +6;quot = 12;rem = 4
   #10  opnds = 12'b010011000110;

        //dvdnd = +110;dvsr = +7;quot = 15;rem = 5
   #10  opnds = 12'b011011100111;

        //dvdnd = +97;dvsr = +5;quot = 19;rem = 2
        //overflow occurs
   #10  opnds = 12'b011000010101;

        //dvdnd = +70;dvsr = +4;quot = 17;rem = 2
        //overflow occurs
   #10  opnds = 12'b010001100100;


   #10  $stop;

end


//instantiate the module into the test bench
srt_div_case2 inst1 (
   .opnds(opnds),
   .quot(quot),
   .rem(rem),
   .ovfl(ovfl)
   );

endmodule
```

Figure 7.28 (Continued)

```
opnds= 000001100001, quot = 0110, rem = 0000, ovfl = 0
opnds= 000001110010, quot = 0011, rem = 0001, ovfl = 0
opnds= 000100010011, quot = 0101, rem = 0010, ovfl = 0
opnds= 001010010011, quot = 1101, rem = 0010, ovfl = 0
opnds= 001100111000, quot = 0110, rem = 0011, ovfl = 0
opnds= 010010000101, quot = 1110, rem = 0010, ovfl = 0
opnds= 010011000110, quot = 1100, rem = 0100, ovfl = 0
opnds= 011011100111, quot = 1111, rem = 0101, ovfl = 0
opnds= 011000010101, quot = xxxx, rem = xxxx, ovfl = 1
opnds= 010001100100, quot = xxxx, rem = xxxx, ovfl = 1
```

Figure 7.29 Outputs for SRT division using the **case** statement.

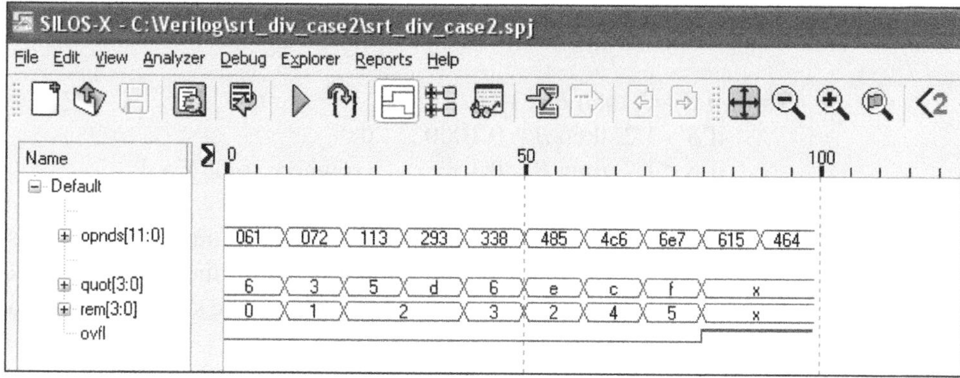

Figure 7.30 Waveforms for SRT division using the **case** statement.

7.4 Multiplicative Division

Multiplicative division is one type of convergence division in which a multiplier is used in the division process. This method uses a factor F_i for $i = 0, 1, 2, \ldots n$ to multiply the dividend A and divisor B without changing the value of the ratio $Q = A/B$. The division process is achieved by finding a factor F, such that $A \times F$ approaches Q while $B \times F$ approaches 1. For each iteration a factor F_i multiplies both the numerator A (dividend) and the denominator B (divisor) so that the numerator converges quadratically toward the quotient as the denominator converges quadratically toward 1. The equation that specifies the convergence technique is shown in Equation 7.2.

$$Q = \frac{A \times F_0 \times F_1 \times F_2 \times \ldots \times F_n}{\underbrace{\underbrace{\underbrace{B \times F_0}_{B_0} \times F_1}_{B_1} \times F_2}_{B_2} \times \ldots \times F_n} \quad \begin{matrix} \to Q \\ \to 1 \end{matrix} \tag{7.2}$$

Both operands are positive fractions; the divisor is normalized and the dividend is shifted accordingly. The convergence speed is a function of the multiplying factor F_i. Since the divisor B is a normalized positive fraction ($B = 0.1xxx \ldots x$), B can be defined by Equation 7.3, where $0 < \delta \leq 1/2$.

$$B = 1 - \delta \tag{7.3}$$

Thus, if $\delta > 0$, then $B < 1$
if $\delta = 1/2$, then $B = 0.1000 \ldots 0$
if $\delta < 1/2$, then $B = 0.1xxx \ldots x$, where some $x = 1$

Successive multipliers should be chosen for F_i, such that each B_i is greater than the previous B_i. Recall that B is a fraction less than 1; therefore, as B_i becomes greater, it approaches 1 while A approaches Q. Therefore, the sequence of denominators can be as shown in Equation 7.4.

$$\begin{aligned}
B_0 &= B \times F_0 & & \\
B_1 &= B \times F_0 \times F_1 & &= B_0 \times F_1 \\
B_2 &= B \times F_0 \times F_1 \times F_2 & &= B_1 \times F_2 \\
&\ldots \qquad \ldots & & \\
B_i &= B_{i-1} \times F_i & & \\
B_{i+1} &= B_i \times F_{i+1} & & \\
&\ldots \qquad \ldots & & \\
B_n &= B_{n-1} \times F_n & &
\end{aligned} \tag{7.4}$$

The iteration continues with each B_i being multiplied by F_{i+1} for $i = 0, 1, 2, \ldots n$ until for some n, $B_n \to 1$. It was shown that $B = 1 - \delta$. Now let F_0 be defined as $F_0 = 1 + \delta$. This is not an arbitrary choice as will be shown. From Equation 7.4, B_0 can now be written as shown in Equation 7.5 for $i = 0$, where δ is a fraction.

$$
\begin{aligned}
B_0 &= B \times F_0 \\
&= (1 - \delta)\,(1 + \delta) \\
B_0 &= 1 - \delta^2 \qquad (7.5)
\end{aligned}
$$

Since δ is a fraction, B_0 is closer to 1 than B. Now define F_1 for the next iteration as shown in Equation 7.6, which yields the equation for B_1, as shown in Equation 7.7 for $i = 1$. It is obvious that B_1 is closer to 1 than B_0.

$$F_1 = 1 + \delta^2 \qquad (7.6)$$

$$
\begin{aligned}
B_1 &= B \times F_0 \times F_1 \\
&= B_0 \times F_1 \\
&= (1 - \delta^2)\,(1 + \delta^2) \\
B_1 &= 1 - \delta^4 \qquad (7.7)
\end{aligned}
$$

Now F_2 is selected for the next iteration, as shown in Equation 7.8. The ith iteration for F_i is shown in Equation 7.9.

$$F_2 = 1 + \delta^4 \qquad (7.8)$$

$$F_i = 1 + \delta^{2^i} \qquad (7.9)$$

This provides the equation for F_i; however, it will be shown that it is advantageous to obtain F_i in terms of one of the operands, for example the divisor B. From Equation 7.7,

$$
\begin{aligned}
B_1 &= 1 - \delta^4 \\
&= 1 - \delta^{2^2} \\
&= 1 - \delta^{2^i} \qquad \text{for } i = 2 \qquad (7.10)
\end{aligned}
$$

Since $i = 2$, then $B_1 = B_{i-1}$. Thus,

$$B_{i-1} = 1 - \delta^{2^i} \tag{7.11}$$

Equation 7.11 is also true for any i; for example, $i = 3$, as shown in Equation 7.12. From Equation 7.4,

$$\begin{aligned} B_2 &= B_1 \times F_2 \\ &= (1 - \delta^4)(1 + \delta^4) \end{aligned} \tag{7.12}$$

Therefore,

$$\begin{aligned} B_2 &= 1 - \delta^8 \\ &= 1 - \delta^{2^i} \qquad \text{for } i = 3 \end{aligned} \tag{7.13}$$

Since $i = 3$, then $B_2 = B_{i-1}$. Thus,

$$B_{i-1} = 1 - \delta^{2^i} \tag{7.14}$$

Which is identical to Equation 7.11. Also,

$$\begin{aligned} F_i &= 1 + \delta^{2^i} \qquad \text{from Equation 7.9} \\ &= 2 - (1 - \delta^{2^i}) \\ F_i &= 2 - B_{i-1} \qquad \text{from Equation 7.14} \end{aligned} \tag{7.15}$$

It will now be shown that $2 - B_{i-1}$ is the 2s complement of B_{i-1}, because the division method uses binary fractions. For example, let $B_{i-1} = 0.100$. The 2s complement of 0.100 is 1.100 = 1.5 and $2 - B_{i-1} = 2 - 0.5 = 1.5$. As another example, let $B_{i-1} =$ 0.110. The 2s complement of 0.110 = 1.010 = 1.25 and $2 - B_{i-1} = 2 - 0.75 = 1.25$.

Therefore, the factor F_i can be derived by simply obtaining the 2s complement of the previous B_{i-1} term. It is this ease of obtaining F_i from B_{i-1} that makes multiplicative division so appealing. The paragraphs that follow show examples of different divisors.

Since $0 < \delta \leq 1/2$, therefore $0 < \delta^2 \leq 1/4$. And since $B_0 = 1 - \delta^2$ for $i = 0$, therefore $B_0 \geq 3/4$; that is, $B_0 = 0.11\,xxxx\ldots x$, where x = 0 or 1. Similarly, $0 < \delta^4 \leq 1/16$; therefore, $B_1 \geq 15/16$; that is, $B_1 = 0.1111\,xxxx\ldots x = 1/2 + 1/4 + 1/8 + 1/16 = 15/16$, where x = 0 or 1. Thus, as i increases, the denominator B_i approaches 1, where

$$B_i = 0.11111111\ldots 1111$$

which is the value nearest 1 for the word size of the machine. It is evident that B_i corresponds to the following bit configurations for increasing values of δ:

First δ gives	$B_0 =\ 0.11\,xxxx\ldots xx$
Second δ gives	$B_1 =\ 0.1111\,xxxx\ldots xx$
Third δ gives	$B_2 =\ 0.1111\ 1111\,xxxx\ldots xx$
Fourth δ gives	$B_3 =\ 0.1111\ 1111\ 1111\ 1111\,xxxx\ldots xx$
Fifth δ gives	$B_4 =\ 0.1111\ 1111\ 1111\ 1111\ 1111\ 1111\ 1111\ 1111\,xxxx\ldots xx$
Sixth δ gives	$B_5 =$ 64 1s

The procedure for obtaining the value of B_2 will now be explained, where $i = 3$. Since,

$$B_i = B_{i-1} \times F_i \qquad \text{from Equation 7.4}$$

$$B_i = (1 - \delta^{2^i})(1 + \delta^{2^i})$$

$$B_2 = 1 - \delta^{2^3}$$

$$B_2 = 1 - \delta^8$$

Since $0 < \delta^8 \leq 1/256$, therefore $B_2 \geq 255/256$ yielding the value for B_2 to be

$$B_2 = 0.1111\ 1111\,xxxx\ldots xx$$

Two multiplications are required for each iteration of multiplicative division — one to generate the next denominator and one to generate the next numerator that converges toward the quotient. The quotient can be expressed as

$$Q = \frac{A \times F_0 \times F_1 \times F_2 \times \ldots \times F_n}{B \times F_0 \times F_1 \times F_2 \times \ldots \times F_n} \begin{matrix} \rightarrow Q \\ \rightarrow 1 \end{matrix}$$

$$= A \times (1 + \delta) \times (1 + \delta^2) \times (1 + \delta^4) \times \ldots \times (1 + \delta^{2^n})$$

A small value for δ means that B is closer to 1 initially; therefore, convergence occurs more rapidly. The initial multiplying factor of $F_0 = 1 + \delta$ can be obtained using a table lookup procedure where the table resides in read-only memory (ROM). The high-order bits of the divisor can be used as the address inputs for the ROM, such that $B_0 = 0.11\,xxxx \ldots xx$. Different high-order divisor bits will produce different δ, such that $B_0 = 0.11\,xxxx \ldots xx$.

The output of the ROM is multiplied by the divisor B to generate B_0; the dividend is also multiplied by the ROM output F_0 to produce A_0. The process continues according to the sequences shown below.

$B_0 = B \times F_0$ F_0 is the output of the ROM

$A_0 = A \times F_0$

$F_1 =$ 2s complement of B_0

$B_1 = B_0 \times F_1$

$A_1 = A_0 \times F_1$

$F_2 =$ 2s complement of B_1

. . .   . . .

If $B_i = 0.1111 \ldots 11$, then $A_i = Q$ and the process terminates

If $B_i \neq 0.1111 \ldots 11$, then $F_{i+1} =$ 2s complement of B_i and the process continues

Examples of different divisors will now be shown. Let $B = 0001\ 1001$, where the high-order bits 0001 are the address inputs to the ROM whose output is F_0. Ideally, B_0 should be at least 0.1100 for this example.

Since $B_0 = B \times F_0$

$$\text{Therefore, } F_0 = \frac{B_0}{B} = \frac{0.1100}{0001\ 1001}$$

$$F_0 = \frac{0.75}{25} = 0.03$$

$$F_0 \approx 0.0000011111xx \ldots xx = .03$$

Now that F_0 has been established, B_0 can be determined.

$$\begin{aligned} B_0 &= B \times F_0 \\ &= (0001\ 1001) \times (0.0000011111\ldots) \\ &= 0.1100000\ldots \end{aligned}$$

F_1 is the 2s complement of B_0, therefore,

$$\begin{aligned} F_1 &= 0.0100000 \\ \text{Therefore } B_1 &= B_0 \times F_1 && \text{from Equation 7.4} \\ &= (0.1100000\ldots) \times (0.0100000) \\ &= 0.11110000 && \text{sign is extended} \end{aligned}$$

F_2 is the 2s complement of B_1, therefore,

$$\begin{aligned} F_2 &= 0.00010000 \\ \text{Therefore } B_2 &= B_1 \times F_2 && \text{from Equation 7.4} \\ &= (0.11110000\ldots) \times (0.00010000) \\ &= 0.1111111100000000 && \text{sign is extended} \end{aligned}$$

The hardware necessary for multiplicative division includes two high-speed multipliers: one for the numerator and one for the denominator. The 2s complement of the divisor can be realized with a carry-lookahead adder. A circuit is also required to detect a divisor of 0.1111 . . . 11 that indicates the end of the multiplicative division operation. There is a drawback with multiplicative division: the algorithm does not directly generate a remainder. The next section describes a method that produces both a quotient and a remainder.

7.5 Array Division

A combinational array can be used for division in much the same way as an array was used for multiplication as presented in the previous chapter. This is an extremely fast division operation, because the array is entirely combinational — the only delay is the propagation delay through the gates. Overflow can be detected in the usual manner: if the high-order half of the dividend is greater than or equal to the divisor, then overflow occurs.

This method uses the nonrestoring division algorithm, which is ideal for iterative arrays. As in the sequential nonrestoring division method, the divide operation is accomplished by a sequence of shifts, and additions or subtractions depending on the sign of the previous partial remainder. Recall that only two operations were required in the shift-subtract/add nonrestoring division method: $2A - B$ and $2A + B$ for a $2n$-bit dividend and an n-bit divisor.

The carry-out determines the quotient bit at any level and also the next operation, either $2A - B$ or $2A + B$, according to the following criteria:

$$q_0 = \begin{cases} 0, \text{ if carry-out} = 0; \text{ next operation is } 2A + B \\ 1, \text{ if carry-out} = 1; \text{ next operation is } 2A - B \end{cases}$$

The above criteria is analogous to the following statement: if the sign is 0, then the operation at the next row of the array is a shift-add ($2A + B$); if the sign is 1, then the operation at the next row of the array is a shift-subtract ($2A - B$). Shifting is accomplished by the placement of the cells along the diagonal of the array. The array consists of rows of identical cells incorporating a full adder in each cell with the ability to add or subtract. Subtraction or addition is determined by the state of the mode input, as shown in Figure 7.31. If the mode line is a logic 1, then the operation is subtraction; if the mode line is logic 0, then the operation is addition.

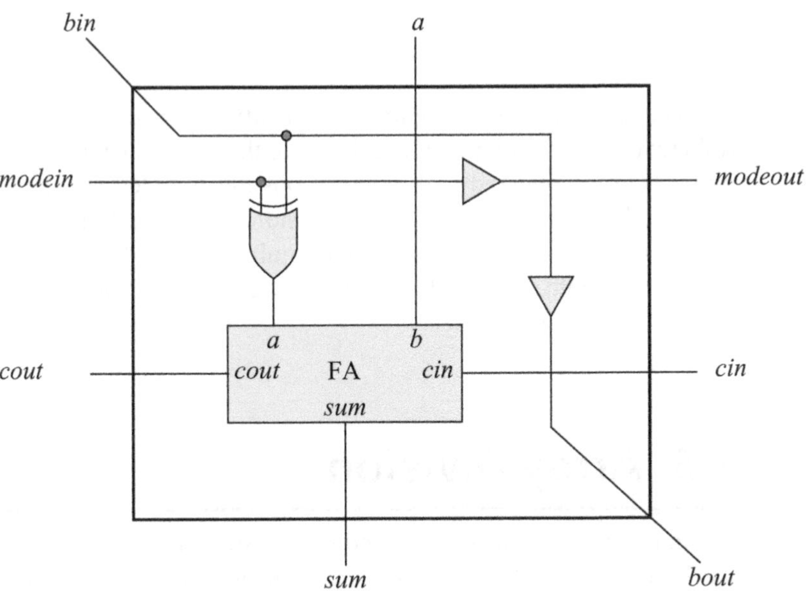

Figure 7.31 Typical cell for nonrestoring array division.

The organization for the nonrestoring division array is shown in Figure 7.32 complete with instantiation names and net names. The carry-out of the high-order cell in each row represents the quotient bit for that row. The carry-out connects to the mode input of the high-order cell in the row immediately below, which then propagates through all cells in the row and connects to the carry-in of the low-order cell. If the carry-out is 0, then the operation at the next lower level is addition; if the carry-out is 1, then the operation at the next lower level is subtraction. This directly corresponds to the nonrestoring division algorithm.

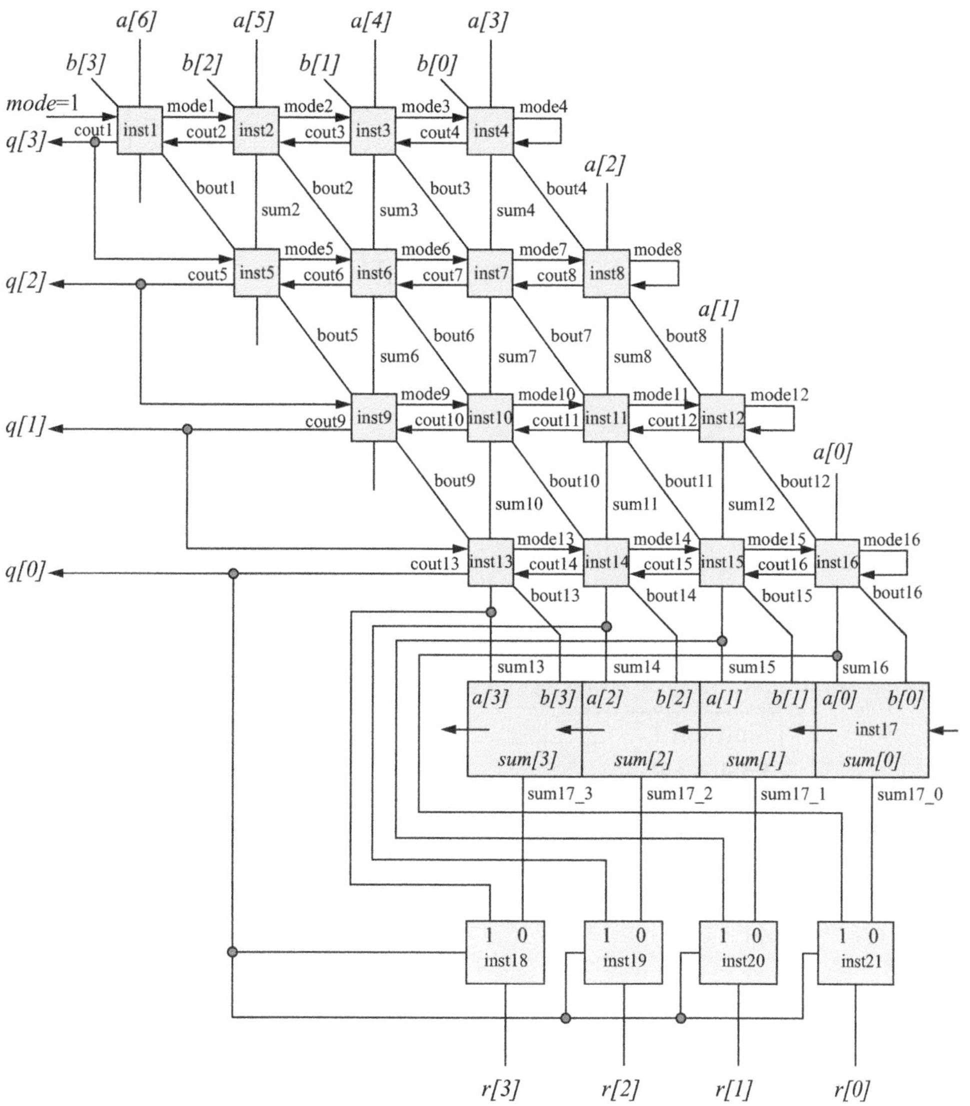

Figure 7.32 Organization of an array divider for a 7-bit dividend and a 4-bit divisor.

The shift operation is achieved by moving the divisor to the right along the diagonal; that is, the entire row of cells is positioned one column to the right, thus providing the requisite shift of the divisor relative to the previous partial remainder.

Restoring the final remainder in the array implementation is similar to that in the sequential implementation — the divisor is added to the previous partial remainder. Two examples that illustrate the array division technique will now be given in Figure 7.33 and Figure 7.34 using the paper-and-pencil method .

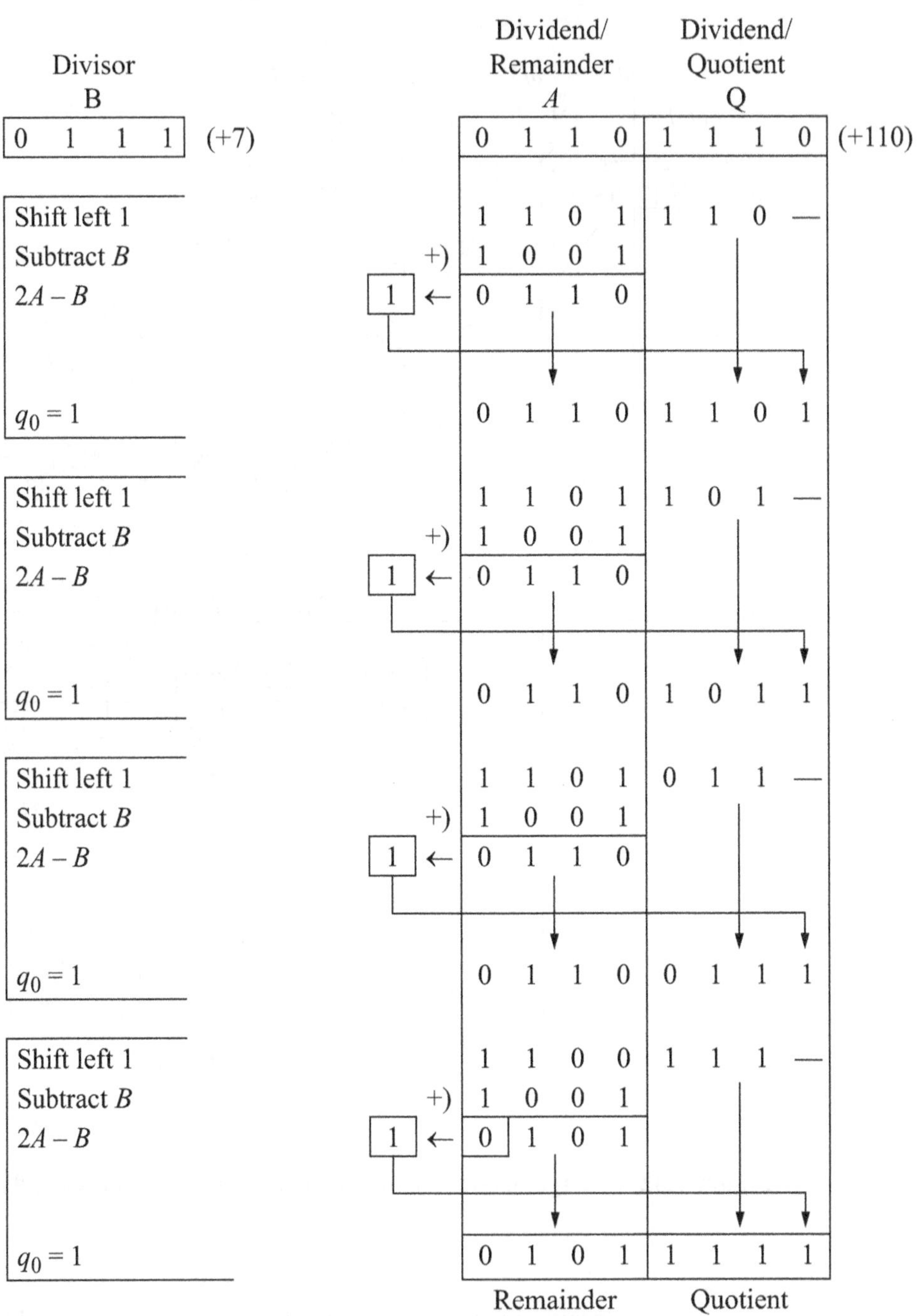

Figure 7.33 Example of array division with no restoration using the paper-and-pencil method.

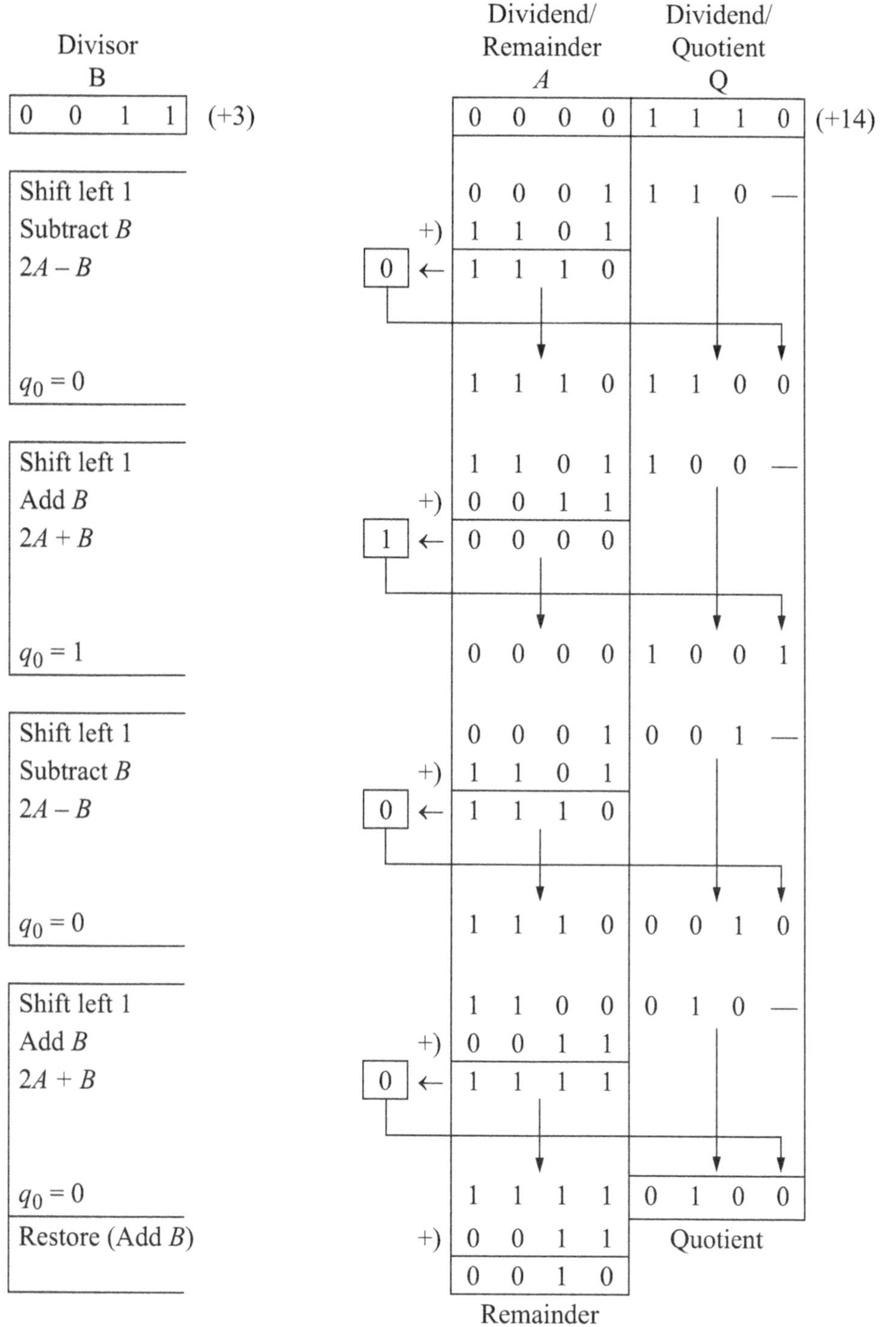

Figure 7.34 Example of array division with remainder restoration using the paper-and-pencil method.

The full adder that is used in the typical cell of Figure 7.31 is shown in the dataflow module of Figure 7.35. The typical cell is shown in the structural module of Figure 7.36. The test bench and outputs for the typical cell are shown in Figure 7.37 and Figure 7.38, respectively.

```
//dataflow full adder
module full_adder (a, b, cin, sum, cout);

input a, b, cin;
output sum, cout;

assign sum = (a ^ b) ^ cin;
assign cout = cin & (a ^ b) | (a & b);
endmodule
```

Figure 7.35 Full adder that is used in the typical cell of the divider array.

```
//structural full adder for an array divide
module full_adder_div_array (a, bin, cin, modein,
                              sum, cout, bout, modeout);

input a, bin, cin, modein;
output sum, cout, bout, modeout;

wire net1;          //define internal net

//instantiate the logic
xor2_df inst1 (
   .x1(modein),
   .x2(bin),
   .z1(net1)
   );

full_adder inst2 (
   .a(net1),
   .b(a),
   .cin(cin),
   .sum(sum),
   .cout(cout)
   );

buf inst3 (modeout, modein);
buf inst4 (bout, bin);

endmodule
```

Figure 7.36 Typical cell used in the divider array.

```
//test bench for the full adder cell used in the array divider
module full_adder_div_array_tb;
reg a, bin, cin, modein;
wire sum, cout, bout, modeout;

initial         //apply stimulus
begin : apply_stimulus
   reg [4:0] invect;
   for (invect = 0; invect < 16; invect = invect + 1)
   begin
      {a, bin, cin, modein} = invect [4:0];
      #10 $display ("a=%b, bin=%b, cin=%b, modein=%b,
                   sum=%b, cout=%b, bout=%b, modeout=%b",
               a, bin, cin, modein, sum, cout, bout, modeout);
   end
end
full_adder_div_array inst1 (      //instantiate the module
   .a(a),
   .bin(bin),
   .cin(cin),
   .modein(modein),
   .sum(sum),
   .cout(cout),
   .bout(bout),
   .modeout(modeout)
   );
endmodule
```

Figure 7.37 Test bench for the typical cell used in the divider array.

```
a=0, bin=0, cin=0, modein=0, sum=0, cout=0, bout=0, modeout=0
a=0, bin=0, cin=0, modein=1, sum=1, cout=0, bout=0, modeout=1
a=0, bin=0, cin=1, modein=0, sum=1, cout=0, bout=0, modeout=0
a=0, bin=0, cin=1, modein=1, sum=0, cout=1, bout=0, modeout=1
a=0, bin=1, cin=0, modein=0, sum=1, cout=0, bout=1, modeout=0
a=0, bin=1, cin=0, modein=1, sum=0, cout=0, bout=1, modeout=1
a=0, bin=1, cin=1, modein=0, sum=0, cout=1, bout=1, modeout=0
a=0, bin=1, cin=1, modein=1, sum=1, cout=0, bout=1, modeout=1
a=1, bin=0, cin=0, modein=0, sum=1, cout=0, bout=0, modeout=0
a=1, bin=0, cin=0, modein=1, sum=0, cout=1, bout=0, modeout=1
a=1, bin=0, cin=1, modein=0, sum=0, cout=1, bout=0, modeout=0
a=1, bin=0, cin=1, modein=1, sum=1, cout=1, bout=0, modeout=1
a=1, bin=1, cin=0, modein=0, sum=0, cout=1, bout=1, modeout=0
a=1, bin=1, cin=0, modein=1, sum=1, cout=0, bout=1, modeout=1
a=1, bin=1, cin=1, modein=0, sum=1, cout=1, bout=1, modeout=0
a=1, bin=1, cin=1, modein=1, sum=0, cout=1, bout=1, modeout=1
```

Figure 7.38 Outputs for the typical cell used in the divider array.

The 4-bit adder used in the divider array to restore the previous partial remainder is shown in the dataflow module of Figure 7.39. This adder adds the divisor to the previous partial remainder. The dataflow module for the 2:1 multiplexer that is used to select the final remainder or the restored partial remainder is shown in Figure 7.40.

```
//dataflow for a 4-bit adder
module adder4_df (a, b, cin, sum, cout);

//list inputs and outputs
input [3:0] a, b;
input cin;
output [3:0] sum;
output cout;


//define signals as wire for dataflow (or default to wire)
wire [3:0] a, b;
wire cin, cout;
wire [3:0] sum;


//continuous assignment for dataflow
//implement the 4-bit adder as a logic equation
//...concatenating cout and sum
assign {cout, sum} = a + b + cin;

endmodule
```

Figure 7.39 Four-bit divider used to restore the previous partial remainder.

```
//dataflow 2:1 multiplexer
module mux2_df (sel, data, z1);

//define inputs and output
input sel;
input [1:0] data;
output z1;

assign z1 = (~sel & data[0]) | (sel & data[1]);

endmodule
```

Figure 7.40 Dataflow module for a 2:1 multiplexer.

The array divider for an 8-bit dividend (with an implied high-order 0) and a 4-bit divisor is shown in the structural module of Figure 7.41 using the organization shown in Figure 7.32. The inputs are a 7-bit dividend, *a[6:0]*; a 4-bit divisor, *b[3:0]*; and a scalar mode control line, *mode*, to permit the first operation to be a subtraction. The outputs are a 4-bit quotient, *q[3:0]*, and a 4-bit remainder, *r[3:0]*.

As described previously, the carry-out of the high-order cell in each row represents the quotient bit for that row and is fed back to the mode input of the next lower row of cells. The array can be easily expanded to accommodate any size operands. The speed of the adder can be increased by using a carry-lookahead adder. The test bench that applies several operands to the design module is shown in Figure 7.42. The outputs and waveforms are shown in Figure 7.43 and Figure 7.44, respectively.

```
//structural array divide
module div_array2 (a, b, mode, q, r);

input [6:0] a;
input [3:0] b;
input mode;
output [3:0] q, r;

wire [6:0] a;
wire [3:0] b;
wire mode;
wire [3:0] q, r;

//define internal nets
wire   mode1, mode2, mode3, mode4, mode5, mode6, mode7, mode8,
       mode9, mode10, mode11, mode12, mode13, mode14, mode15,
       mode16,

       cout1, cout2, cout3, cout4, cout5, cout6, cout7, cout8,
       cout9, cout10, cout11, cout12, cout13, cout14, cout15,
       cout16,

       bout1, bout2, bout3, bout4, bout5, bout6, bout7, bout8,
       bout9, bout10, bout11, bout12, bout13, bout14, bout15,
       bout16,

       sum2, sum3, sum4, sum6, sum7, sum8, sum10,
       sum11, sum12, sum13, sum14, sum15, sum16,
       sum17_0, sum17_1, sum17_2, sum17_3;

//continued on next page
```

Figure 7.41 Structural module for the array divider of Figure 7.32.

```
//instantiate the logic for the array divider
//instantiate the array of full adders
full_adder_div_array inst1 (
   .a(a[6]),
   .bin(b[3]),
   .cin(cout2),
   .modein(mode),
   .cout(cout1),
   .bout(bout1),
   .modeout(mode1)
   );

full_adder_div_array inst2 (
   .a(a[5]),
   .bin(b[2]),
   .cin(cout3),
   .modein(mode1),
   .sum(sum2),
   .cout(cout2),
   .bout(bout2),
   .modeout(mode2)
   );

full_adder_div_array inst3 (
   .a(a[4]),
   .bin(b[1]),
   .cin(cout4),
   .modein(mode2),
   .sum(sum3),
   .cout(cout3),
   .bout(bout3),
   .modeout(mode3)
   );

full_adder_div_array inst4 (
   .a(a[3]),
   .bin(b[0]),
   .cin(mode4),
   .modein(mode3),
   .sum(sum4),
   .cout(cout4),
   .bout(bout4),
   .modeout(mode4)
   );

//continued on next page
```

Figure 7.41 (Continued)

```
full_adder_div_array inst5 (
    .a(sum2),
    .bin(bout1),
    .cin(cout6),
    .modein(cout1),
    .cout(cout5),
    .bout(bout5),
    .modeout(mode5)
    );

full_adder_div_array inst6 (
    .a(sum3),
    .bin(bout2),
    .cin(cout7),
    .modein(mode5),
    .sum(sum6),
    .cout(cout6),
    .bout(bout6),
    .modeout(mode6)
    );

full_adder_div_array inst7 (
    .a(sum4),
    .bin(bout3),
    .cin(cout8),
    .modein(mode6),
    .sum(sum7),
    .cout(cout7),
    .bout(bout7),
    .modeout(mode7)
    );

full_adder_div_array inst8 (
    .a(a[2]),
    .bin(bout4),
    .cin(mode8),
    .modein(mode7),
    .sum(sum8),
    .cout(cout8),
    .bout(bout8),
    .modeout(mode8)
    );

//continued on next page
```

Figure 7.41 (Continued)

```
full_adder_div_array inst9 (
   .a(sum6),
   .bin(bout5),
   .cin(cout10),
   .modein(cout5),
   .cout(cout9),
   .bout(bout9),
   .modeout(mode9)
   );

full_adder_div_array inst10 (
   .a(sum7),
   .bin(bout6),
   .cin(cout11),
   .modein(mode9),
   .sum(sum10),
   .cout(cout10),
   .bout(bout10),
   .modeout(mode10)
   );

full_adder_div_array inst11 (
   .a(sum8),
   .bin(bout7),
   .cin(cout12),
   .modein(mode10),
   .sum(sum11),
   .cout(cout11),
   .bout(bout11),
   .modeout(mode11)
   );

full_adder_div_array inst12 (
   .a(a[1]),
   .bin(bout8),
   .cin(mode12),
   .modein(mode11),
   .sum(sum12),
   .cout(cout12),
   .bout(bout12),
   .modeout(mode12)
   );

//continued on next page
```

Figure 7.41 (Continued)

```
full_adder_div_array inst13 (
   .a(sum10),
   .bin(bout9),
   .cin(cout14),
   .modein(cout9),
   .sum(sum13),
   .cout(cout13),
   .bout(bout13),
   .modeout(mode13)
   );

full_adder_div_array inst14 (
   .a(sum11),
   .bin(bout10),
   .cin(cout15),
   .modein(mode13),
   .sum(sum14),
   .cout(cout14),
   .bout(bout14),
   .modeout(mode14)
   );

full_adder_div_array inst15 (
   .a(sum12),
   .bin(bout11),
   .cin(cout16),
   .modein(mode14),
   .sum(sum15),
   .cout(cout15),
   .bout(bout15),
   .modeout(mode15)
   );

full_adder_div_array inst16 (
   .a(a[0]),
   .bin(bout12),
   .cin(mode16),
   .modein(mode15),
   .sum(sum16),
   .cout(cout16),
   .bout(bout16),
   .modeout(mode16)
   );

//continued on next page
```

Figure 7.41 (Continued)

```
//instantiate the 4-input adder
adder4_df inst17 (
   .a({sum13, sum14, sum15, sum16}),
   .b({bout13, bout14, bout15, bout16}),
   .cin(1'b0),
   .sum({sum17_3, sum17_2, sum17_1, sum17_0})
   );


//instantiate the 2:1 multiplexers
mux2_df inst18 (
   .sel(cout13),
   .data({sum13, sum17_3}),
   .z1(r[3])
   );

mux2_df inst19 (
   .sel(cout13),
   .data({sum14, sum17_2}),
   .z1(r[2])
   );

mux2_df inst20 (
   .sel(cout13),
   .data({sum15, sum17_1}),
   .z1(r[1])
   );

mux2_df inst21 (
   .sel(cout13),
   .data({sum16, sum17_0}),
   .z1(r[0])
   );


//assign the quotient outputs
assign    q[3] = cout1,
          q[2] = cout5,
          q[1] = cout9,
          q[0] = cout13;

endmodule
```

Figure 7.41 (Continued)

```
//test bench for array divider
module div_array2_tb;


reg [6:0] a;
reg [3:0] b;
reg mode;
wire [3:0] q, r;


//display inputs and outputs
initial
$monitor ("dvdnd = %b, dvsr = %b, quot = %b, rem = %b",
            a, b, q, r);


//apply input vectors
initial
begin
   #0    a = 7'b1101110;    b = 4'b0111;    mode = 1'b1;
   #10   a = 7'b0001110;    b = 4'b0011;    mode = 1'b1;
   #10   a = 7'b0000111;    b = 4'b0011;    mode = 1'b1;
   #10   a = 7'b0110011;    b = 4'b0111;    mode = 1'b1;
   #10   a = 7'b0000110;    b = 4'b0001;    mode = 1'b1;
   #10   a = 7'b0000111;    b = 4'b0010;    mode = 1'b1;
   #10   a = 7'b0010001;    b = 4'b0011;    mode = 1'b1;
   #10   a = 7'b0101001;    b = 4'b0011;    mode = 1'b1;
   #10   a = 7'b1001100;    b = 4'b0110;    mode = 1'b1;
   #10   a = 7'b1101110;    b = 4'b0111;    mode = 1'b1;
   #10   a = 7'b0111111;    b = 4'b0100;    mode = 1'b1;
   #10   a = 7'b0100101;    b = 4'b0101;    mode = 1'b1;

   #10   $stop;
end


//instantiate the module into the test bench
div_array2 inst1 (
   .a(a),
   .b(b),
   .mode(mode),
   .q(q),
   .r(r)
   );

endmodule
```

Figure 7.42 Test bench for the array divider.

```
dvdnd = 1101110, dvsr = 0111, quot = 1111, rem = 0101
dvdnd = 0001110, dvsr = 0011, quot = 0100, rem = 0010
dvdnd = 0000111, dvsr = 0011, quot = 0010, rem = 0001
dvdnd = 0110011, dvsr = 0111, quot = 0111, rem = 0010
dvdnd = 0000110, dvsr = 0001, quot = 0110, rem = 0000
dvdnd = 0000111, dvsr = 0010, quot = 0011, rem = 0001
dvdnd = 0010001, dvsr = 0011, quot = 0101, rem = 0010
dvdnd = 0101001, dvsr = 0011, quot = 1101, rem = 0010
dvdnd = 1001100, dvsr = 0110, quot = 1100, rem = 0100
dvdnd = 1101110, dvsr = 0111, quot = 1111, rem = 0101
dvdnd = 0111111, dvsr = 0100, quot = 1111, rem = 0011
dvdnd = 0100101, dvsr = 0101, quot = 0111, rem = 0010
```

Figure 7.43 Outputs for the array divider.

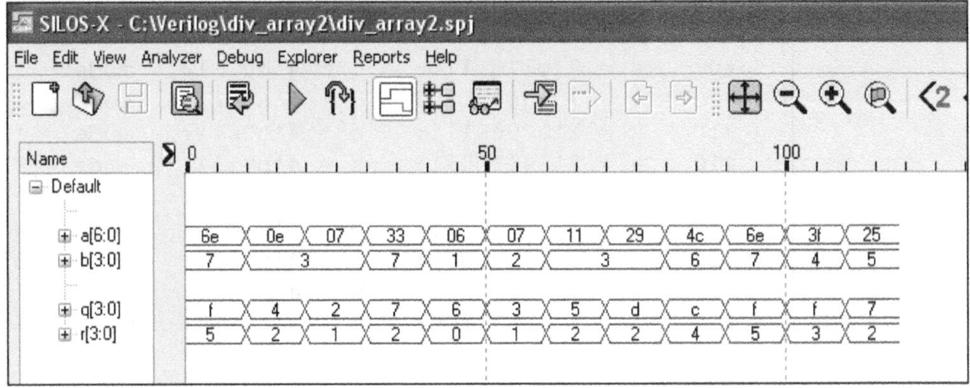

Figure 7.44 Waveforms for the array divider.

7.6 Problems

7.1 Determine whether the following operands produce an overflow for fixed-point binary division.

(a) Dividend = 0001 1111
Divisor = 0001

(b) Dividend = 0000 1111
Divisor = 0001

(c) Dividend = 0100 0001
Divisor = 0011

7.2 Determine whether the following operands produce an overflow for fixed-point binary division.

(a) Dividend = 001 0110
Divisor = 0011

(b) Dividend = 0110 1100
Divisor = 0101

(c) Dividend = 0010 1010
Divisor = 0011

7.3 Use the sequential shift-add/subtract restoring paper-and-pencil method to divide the operands shown below. The sign bit of the partial remainder determines whether to restore the previous partial remainder.

Dividend A = 0010 1111
Divisor B = 0011

7.4 Use the sequential shift-add/subtract restoring paper-and-pencil method to divide the operands shown below. The sign bit of the partial remainder determines whether to restore the previous partial remainder.

Dividend A = 0110 0101
Divisor B = 0111

7.5 Use sequential nonrestoring division to perform the following divide operation:

$A.Q$ (dividend) = 00100 1110
B (divisor) = 0101

7.6 Use sequential nonrestoring division to perform the following divide operation:

$A.Q$ (dividend) = 0110 0011
B (divisor) = 0111

7.7 Use sequential nonrestoring division to perform the following divide operation:

$A.Q$ (dividend) = 0000 0111
B (divisor) = 0010

7.8 Use sequential nonrestoring division to perform the following divide operation:

$$A.Q \text{ (dividend)} = 0101\ 0111$$
$$B \text{ (divisor)} = 0111$$

7.9 Use sequential nonrestoring division to perform the following divide operation:

$$A.Q \text{ (dividend)} = 0010\ 1111$$
$$B \text{ (divisor)} = 0101$$

7.10 Use the paper-and-pencil method to perform SRT division on the following operands: dividend $A.Q = 0.0000\ 0111$ and divisor $B = 0.0010$. Obtain the result as a 4-bit quotient and a 4-bit remainder.

7.11 Use the paper-and-pencil method to perform SRT division on the following operands: dividend $A.Q = 0.0011\ 0011$ and divisor $B = 0.1000$. Obtain the result as a 4-bit quotient and a 4-bit remainder.

7.12 Use the paper-and-pencil method to perform SRT division on the following operands: dividend $A.Q = 0.0100\ 1000$ and divisor $B = 0.0101$. Obtain the result as a 4-bit quotient and a 4-bit remainder.

7.13 Use the paper-and-pencil method to perform SRT division on the following operands: dividend $A.Q = 0.011\ 101$ and divisor $B = 0.100$. Obtain the result as a 3-bit quotient and a 3-bit remainder.

7.14 Use the paper-and-pencil method to perform SRT division on the following operands: dividend $A.Q = 0.100\ 111$ and divisor $B = 0.101$. Obtain the result as a 3-bit quotient and a 3-bit remainder.

7.15 Use the paper-and-pencil method to perform SRT division on the following operands: dividend $A.Q = 0.110\ 011$ and divisor $B = 0.111$. Obtain the result as a 3-bit quotient and a 3-bit remainder.

7.16 Use behavioral modeling to design an SRT divider using the table lookup method for a 4-bit dividend and a 2-bit divisor. Obtain the result as a 2-bit quotient and a 2-bit remainder. Obtain the behavioral module, the test bench module, the outputs, and the waveforms. Include two cases where overflow occurs.

7.17 Determine whether the following operands will generate a divide overflow:

(a) Dividend = 00111010
Divisor = 0011

(b) Dividend = 01101110
Divisor = 0101

(c) Dividend = 01001110
Divisor = 0110

7.18 Write the equation for F_i in terms of the divisor B for multiplicative division.

7.19 Obtain the bit configuration for the final value of the denominator using multiplicative division for a 32-bit processor.

7.20 Draw an array divider to perform division on the operands shown below. Show the bits for the quotient, all partial remainders, and the final remainder on the drawing.

$A = 0010001$ $B = 0011$

$A = 1100111$ $B = 0111$

7.21 Design an array divider using structural modeling to perform a divide operation on 5-bit dividends (with an implied high-order 0) and 3-bit divisors. Obtain the structural module, the test bench module that applies several input vectors for the dividend and divisor, the outputs, and the waveforms.

8.1 Addition with Sum Correction
8.2 Addition Using Multiplexers
8.3 Addition with Memory-Based Correction
8.4 Addition with Biased Augend
8.5 Problems

8

Decimal Addition

Many computers receive input data in decimal form and produce results in decimal form. The data can be converted to binary and calculations can then be performed in either the fixed-point or the floating-point number representation, then converted back to decimal. However, if the amount of input data is extensive, then it is more efficient to perform the operations in the decimal number representation also referred to as binary-coded decimal (BCD). A single BCD digit is represented by four binary bits. The most common code in BCD arithmetic is the 8421 code, as shown below.

			8421	8421
	15		0001	0101
+)	23	+)	0010	0011
	38		0011	1000

8.1 Addition with Sum Correction

Since four binary bits have $2^4 = 16$ combinations (0000 – 1111) and the range for a single decimal digit is 0 – 9, six of the sixteen combinations (1010 – 1111) are invalid for BCD. These invalid BCD digits must be converted to valid digits by adding six to the digit. Assume that the result of an operation is 10_{10} (1010). If 6_{10} (0110) is added to 1010, then the result is 0000 with a carry of 1; that is, 0001 0000 in BCD which is 10_{10}

in decimal. The reason 6 is added is because $10_{10} + 6_{10} = 16_{10}$; however, 16_{10} modulo-16 equals 0 with a carry of 1, or 0001 0000 in BCD.

Now assume that the result of an operation is 14_{10} (1110), then 0110 is added to the result as shown below. This yields a value of 0100 with a carry of 1, or 0001 0100 in BCD which is 14_{10} in decimal.

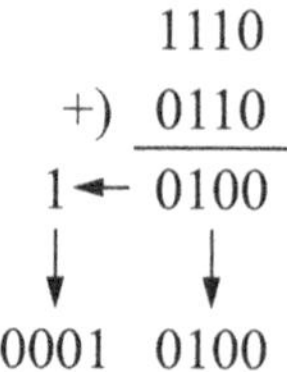

The reason 6 is added is because $14_{10} + 6_{10} = 20_{10}$. However, 20_{10} modulo-16 equals 4 with a carry of 1, or 0001 0100 in BCD. The process of adding 6 to the intermediate result is shown in Table 8.1, which enumerates the valid BCD numbers (0 to 9) and the invalid BCD numbers (10 to 15) that are adjusted to obtain a valid BCD number. For any binary number that is invalid for BCD, count six numbers beyond the invalid number to obtain the correct BCD number accompanied by a carry. If a carry is generated out of a digit position, then 0110 is added to the intermediate sum of the digit position, as shown in Table 8.1.

Table 8.1 Binary-Coded Decimal Digits Showing Valid Digits and Invalid Digits with Correction

	BCD Result 8421	Decimal Value			Carry	8421	Valid BCD Result
	0000	0	Valid				0000
	0001	1	Valid				0001
	0010	2	Valid				0010
	0011	3	Valid				0011
	0100	4	Valid				0100
	0101	5	Valid				0101
	0110	6	Valid				0110
	1110	7	Valid				0111
	1000	8	Valid				1000
	1001	9	Valid				1001
	1010	10	Invalid	+ 0110 =	1	0000 =	0001 0000
	1011	11	Invalid	+ 0110 =	1	0001 =	0001 0001
	1100	12	Invalid	+ 0110 =	1	0010 =	0001 0010
	1101	13	Invalid	+ 0110 =	1	0011 =	0001 0011
	1110	14	Invalid	+ 0110 =	1	0100 =	0001 0100
	1111	15	Invalid	+ 0110 =	1	0101 =	0001 0101
1	0000	16	Invalid	+ 0110 =	1	0110 =	0001 0110

Example 8.1 The numbers 73_{10} and 58_{10} will be added in BCD to yield a result of 131_{10} as shown below. Both *intermediate sums* (1011 and 1101) are invalid for BCD; therefore, 0110 must be added to the intermediate sums. Any carry that results from adding six to the intermediate sum is ignored because it provides no new information. The result of the BCD add operation is 0001 0011 0001. The carry produced from the low-order decade is also referred to as the *auxiliary carry*.

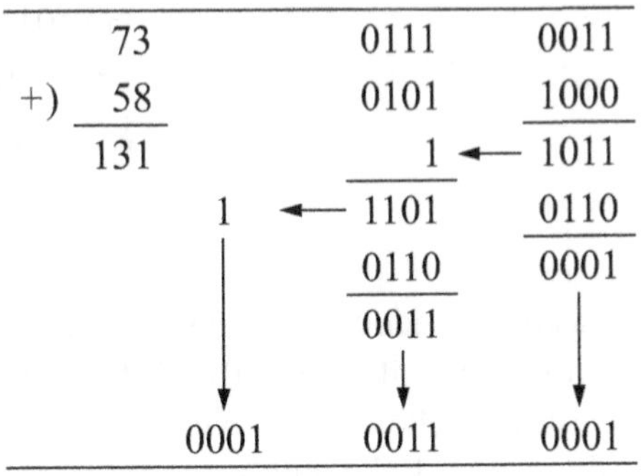

```
      73                     0111       0011
  +)  58                     0101       1000
     ---                               -----
     131                        1  ←    1011
                             -----
                 1    ←      1101       0110
                 |                     -----
                 |           0110       0001
                 |           -----        |
                 |           0011         |
                 |             ↓          ↓
                 ↓
               0001          0011       0001
```

Example 8.2 Another example of BCD addition is shown below, in which the intermediate sums are valid BCD numbers; however, there is a carry-out of the high-order decade. Whenever the unadjusted sum produces a carry-out, the intermediate sum must be corrected by adding six.

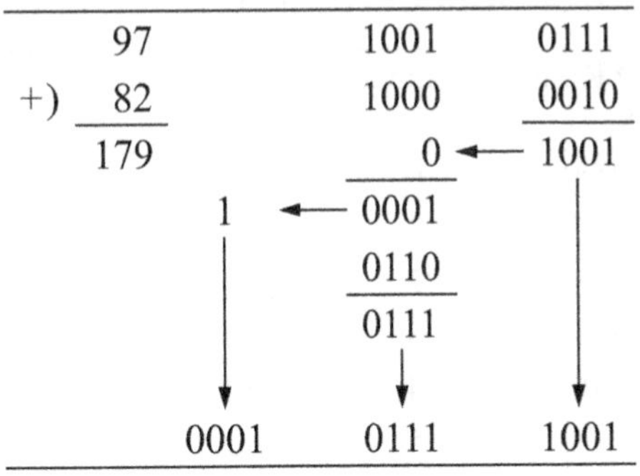

```
      97                     1001       0111
  +)  82                     1000       0010
     ---                               -----
     179                        0  ←    1001
                             -----        |
                 1    ←      0001         |
                 |                        |
                 |           0110         |
                 |           -----        |
                 |           0111         |
                 |             ↓          ↓
                 ↓
               0001          0111       1001
```

The condition for a correction (adjustment) that also produces a carry-out is shown in Equation 8.1. This specifies that a carry-out will be generated whenever bit position b_8 is a 1 in both decades ($1xxx + 1xxx$), when bit positions b_8 and b_4 are both 1s in the intermediate sum, or when bit positions b_8 and b_2 are both 1s in the intermediate sum.

$$\text{Carry} = c_8 + b_8 b_4 + b_8 b_2 \tag{8.1}$$

Example 8.3 This example illustrates all three conditions listed in Equation 8.1. The numbers 876 and 854 will be added using BCD arithmetic. The intermediate sums are: 0001 for the hundreds decade, 1101 for the tens decade, and 1010 for the units decade.

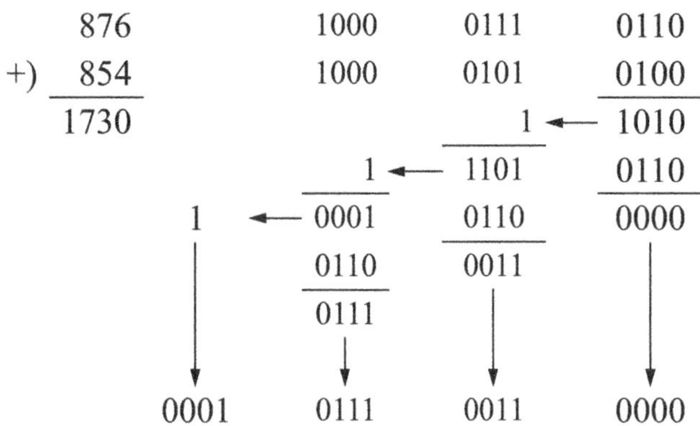

There are two forms of binary-coded decimal numbers: packed and unpacked. The packed format allows two BCD digits packed into one byte. The unpacked format allows only one BCD digit per byte — the high-order four bits contain zeroes; the low-order four bits contain the BCD digit. To convert unpacked digits to packed digits, simply shift one digit four bits to the left, then OR it with the other digit. Binary-coded decimal numbers may also have a sign associated with them. This usually occurs in the *packed* BCD format, as shown below, where the sign digits are the low-order bytes 1100 (+) and 1101 (–). Comparing signs is straightforward. This sign notation is similar to the sign notation used in binary. Note that the low-order bit of the sign digit is 0 (+) or 1 (–).

+53	0 0 0 0	0 1 0 1	0 0 1 1	1 1 0 0	Sign is +
–25	0 0 0 0	0 0 1 0	0 1 0 1	1 1 0 1	Sign is –

The algorithms used for BCD arithmetic are basically the same as those used for fixed-point arithmetic for radix 2. The main difference is that BCD arithmetic treats each digit as four bits, whereas fixed-point arithmetic treats each digit as a bit. Shifting operations are also different — decimal shifting is performed on 4-bit increments, whereas, fixed-point shifting is performed on individual bits. Thus, to shift the BCD number 1001 0110 0101 0011 two digits to the left with 0000 filing in the vacated positions results in 0101 0011 0000 0000.

The single element in decimal arithmetic has nine inputs and five outputs, as shown in Figure 8.1. Each of the two decimal operands is represented by a 4-bit BCD digit. A carry-in is also provided from the previous lower-order decimal element. The outputs of the decimal element are a 4-bit valid BCD digit and a carry-out.

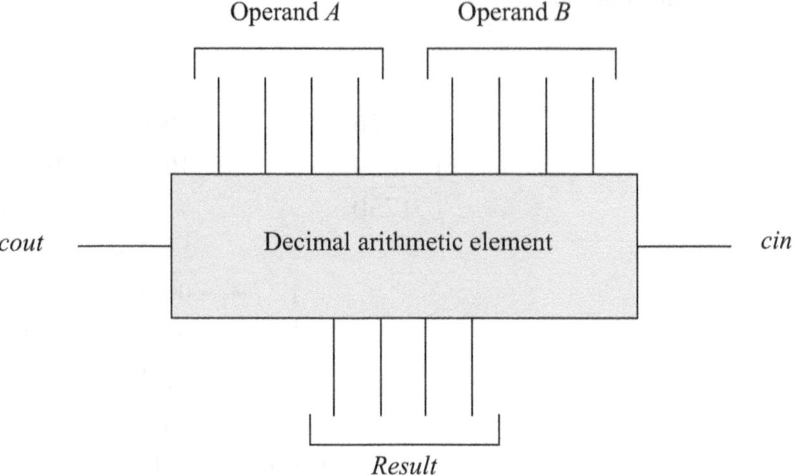

Figure 8.1 A 1-digit binary-coded decimal arithmetic element.

The decimal arithmetic element includes correction logic to generate a valid decimal digit when the intermediate sum exceeds 1001, according to Equation 8.1. Two adders are used for each decade of an n-digit decimal adder. These adders are 4-bit fixed-point binary adders of the type described in Chapter 4. The augend A and the addend B are defined as shown in Equation 8.2 for n-bit operands.

$$
\begin{aligned}
A = \; & a(n-1)_8 \, a(n-1)_4 \, a(n-1)_2 \, a(n-1)_1 \\
& a(n-2)_8 \, a(n-2)_4 \, a(n-2)_2 \, a(n-2)_1 \\
& \quad \ldots \\
& a(1)_8 \, a(1)_4 \, a(1)_2 \, a(1)_1 \\
& a(0)_8 \, a(0)_4 \, a(0)_2 \, a(0)_1 \\
\\
B = \; & b(n-1)_8 \, b(n-1)_4 \, b(n-1)_2 \, b(n-1)_1 \\
& b(n-2)_8 \, b(n-2)_4 \, b(n-2)_2 \, b(n-2)_1 \\
& \quad \ldots \\
& b(1)_8 \, b(1)_4 \, b(1)_2 \, b(1)_1 \\
& b(0)_8 \, b(0)_4 \, b(0)_2 \, b(0)_1
\end{aligned}
\qquad (8.2)
$$

A single stage of a decimal adder is shown in Figure 8.2 complete with instantiation names and net names. The carry-out of the decade directly corresponds to Equation 8.1. The carry-out of *adder_1* — in conjunction with the logic indicated by Equation 8.1 — specifies the carry-out of the decade and is connected to inputs b_4 and b_2 of *adder_2* with $b_8\,b_1 = 00$. This corrects an invalid decimal digit. The carry-out of *adder_2* can be ignored, because it provides no new information.

This decimal adder stage can be used in conjunction with other identical stages to design an n-digit parallel decimal adder. The carry-out of stage$_i$ connects to the carry-in of stage$_{i+1}$; therefore, this is a ripple adder for decimal operands. If a higher speed is required, then the carry-lookahead technique can be used. Since there is no carry-in to the low-order stage of the decimal adder, this decade can be minimized slightly by using a half adder as the low-order adder rather than a full adder — the remaining three adders would be full adders.

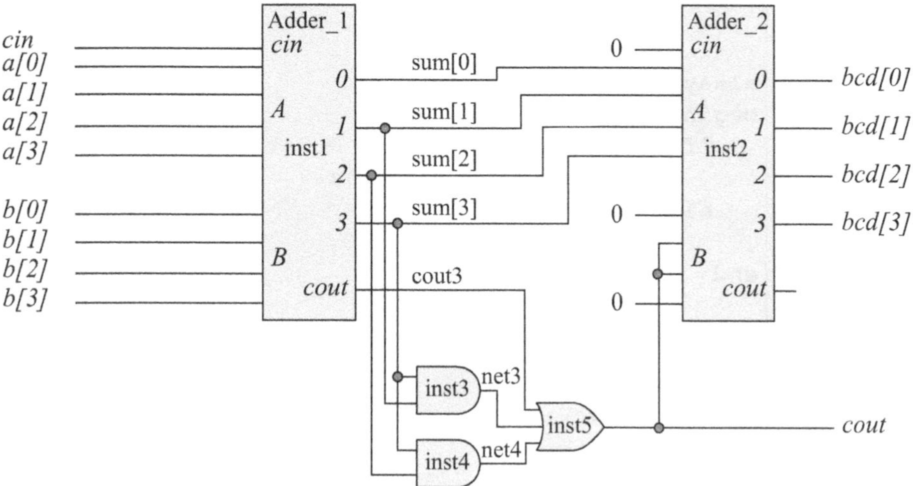

Figure 8.2 Typical stage of an n-digit decimal adder.

If the intermediate sum inputs *sum[0]*, *sum[1]*, *sum[2]*, and *sum[3]* to *adder_2* are invalid for BCD, then they are adjusted by adding six (0110) to the intermediate sum. Instantiations *inst3*, *inst4*, and *inst5* represent the logic expressed in Equation 8.1. The mixed-design module is shown in Figure 8.3 in which the instantiation names and the net names correspond to those in the logic diagram. A check is also made to ensure that the input operands contain valid BCD digits.

The test bench is shown in Figure 8.4, which applies several BCD numbers to the design module together with the carry-in to the module. The outputs and waveforms are shown in Figure 8.5 and Figure 8.6, respectively. If the *invalid_inputs* signal is set to 1, then the results of the operation are also invalid.

```
//mixed-design bcd adder
module add_bcd (a, b, cin, bcd, cout, invalid_inputs);

//define inputs and outputs
input [3:0] a, b;
input cin;
output [3:0] bcd;
output cout, invalid_inputs;

reg invalid_inputs;     //reg if used in always statement

//define internal nets
wire [3:0] sum;
wire cout3, net3, net4;


//check for invalid inputs
always @ (a or b)
begin
   if ((a > 4'b1001) || (b > 4'b1001))     //|| is logical or
      invalid_inputs = 1'b1;
   else
      invalid_inputs = 1'b0;
end


//instantiate the logic for adder_1
adder4 inst1 (
   .a(a[3:0]),
   .b(b[3:0]),
   .cin(cin),
   .sum(sum[3:0]),
   .cout(cout3)
   );

//instantiate the logic for adder_2
adder4 inst2 (
   .a(sum[3:0]),
   .b({1'b0, cout, cout, 1'b0}),
   .cin(1'b0),
   .sum(bcd[3:0])
   );


//continued on next page
```

Figure 8.3 Mixed-design module for one stage of a BCD adder.

```
//instantiate the logic for intermediate sum adjustment
and2_df inst3 (
   .x1(sum[3]),
   .x2(sum[1]),
   .z1(net3)
   );


and2_df inst4 (
   .x1(sum[3]),
   .x2(sum[2]),
   .z1(net4)
   );


or3_df inst5 (
   .x1(cout3),
   .x2(net3),
   .x3(net4),
   .z1(cout)
   );


endmodule
```

Figure 8.3 (Continued)

```
//test bench for mixed-design add_bcd
module add_bcd_tb;


reg [3:0] a, b;
reg cin;
wire [3:0] bcd;
wire cout, invalid_inputs;


//display variables
initial
$monitor ("a=%b, b=%b, cin=%b,
            cout=%b, bcd=%b, invalid_inputs=%b",
               a, b, cin, cout, bcd, invalid_inputs);


//continued on next page
```

Figure 8.4 Test bench for the BCD adder stage.

```
//apply input vectors
initial
begin
   #0     a = 4'b0011;    b = 4'b0011;    cin = 1'b0;
   #10    a = 4'b0101;    b = 4'b0110;    cin = 1'b0;
   #10    a = 4'b0101;    b = 4'b0100;    cin = 1'b0;
   #10    a = 4'b0111;    b = 4'b1000;    cin = 1'b0;
   #10    a = 4'b0111;    b = 4'b0111;    cin = 1'b0;
   #10    a = 4'b1000;    b = 4'b1001;    cin = 1'b0;
   #10    a = 4'b1001;    b = 4'b1001;    cin = 1'b0;
   #10    a = 4'b0101;    b = 4'b0110;    cin = 1'b1;
   #10    a = 4'b0111;    b = 4'b1000;    cin = 1'b1;
   #10    a = 4'b1001;    b = 4'b1001;    cin = 1'b1;
   #10    a = 4'b1000;    b = 4'b1000;    cin = 1'b0;
   #10    a = 4'b1000;    b = 4'b1000;    cin = 1'b1;
   #10    a = 4'b1001;    b = 4'b0111;    cin = 1'b0;
   #10    a = 4'b0111;    b = 4'b0010;    cin = 1'b1;
   #10    a = 4'b0011;    b = 4'b1000;    cin = 1'b0;

          //invalid inputs
   #10    a = 4'b1010;    b = 4'b0001;    cin = 1'b0;

          //invalid inputs
   #10    a = 4'b0011;    b = 4'b1100;    cin = 1'b0;

          //invalid inputs
   #10    a = 4'b1011;    b = 4'b1110;    cin = 1'b1;

   #10    $stop;

end


//instantiate the module into the test bench
add_bcd inst1 (
   .a(a),
   .b(b),
   .cin(cin),
   .bcd(bcd),
   .cout(cout),
   .invalid_inputs(invalid_inputs)
   );


endmodule
```

Figure 8.4 (Continued)

```
a=0011, b=0011, cin=0, cout=0, bcd=0110, invalid_inputs=0
a=0101, b=0110, cin=0, cout=1, bcd=0001, invalid_inputs=0
a=0101, b=0100, cin=0, cout=0, bcd=1001, invalid_inputs=0
a=0111, b=1000, cin=0, cout=1, bcd=0101, invalid_inputs=0
a=0111, b=0111, cin=0, cout=1, bcd=0100, invalid_inputs=0
a=1000, b=1001, cin=0, cout=1, bcd=0111, invalid_inputs=0
a=1001, b=1001, cin=0, cout=1, bcd=1000, invalid_inputs=0
a=0101, b=0110, cin=1, cout=1, bcd=0010, invalid_inputs=0
a=0111, b=1000, cin=1, cout=1, bcd=0110, invalid_inputs=0
a=1001, b=1001, cin=1, cout=1, bcd=1001, invalid_inputs=0
a=1000, b=1000, cin=0, cout=1, bcd=0110, invalid_inputs=0
a=1000, b=1000, cin=1, cout=1, bcd=0111, invalid_inputs=0
a=1001, b=0111, cin=0, cout=1, bcd=0110, invalid_inputs=0
a=0111, b=0010, cin=1, cout=1, bcd=0000, invalid_inputs=0
a=0011, b=1000, cin=0, cout=1, bcd=0001, invalid_inputs=0

a=1010, b=0001, cin=0, cout=1, bcd=0001, invalid_inputs=1
a=0011, b=1100, cin=0, cout=1, bcd=0101, invalid_inputs=1
a=1011, b=1110, cin=1, cout=1, bcd=0000, invalid_inputs=1
```

Figure 8.5 Outputs for the BCD adder stage.

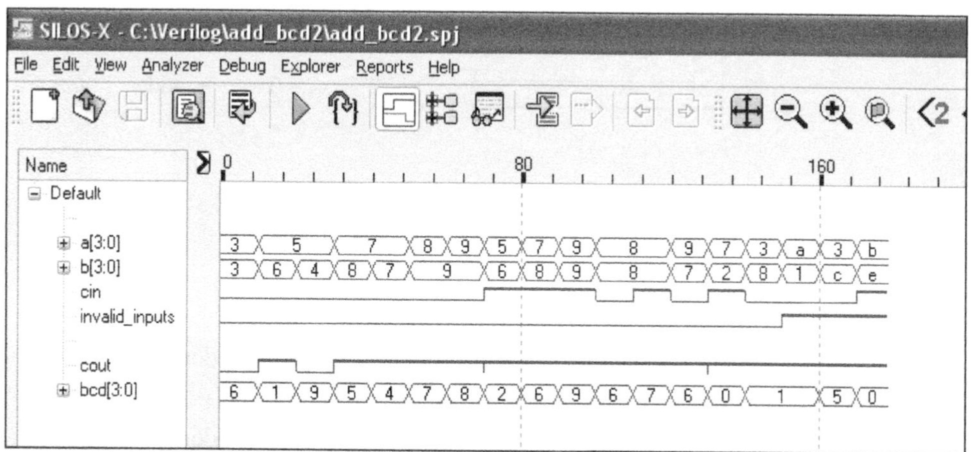

Figure 8.6 Waveforms for the BCD adder stage.

8.2 Addition Using Multiplexers

An alternative approach to determining whether to add six to correct an invalid decimal number is to use a multiplexer. The two operands are added in *adder_1* as before; however, a value of six is always added to this intermediate sum in *adder_2*, as shown in Figure 8.7. The sums from *adder_1* and *adder_2* are then applied to four 2:1 multiplexers. Selection of the *adder_1* sum or the *adder_2* sum is determined by ORing together the carry-out of both adders — *cout3* and *cout8* — to generate a select input *cout* to the multiplexers, as shown below. The logic diagram is shown in Figure 8.8, which shows the instantiation names and net names that will be used in the structural design module.

$$\text{Multiplexer select} = \begin{cases} \text{Adder\_1 sum if } cout \text{ is } 0 \\ \text{Adder\_2 sum if } cout \text{ is } 1 \end{cases}$$

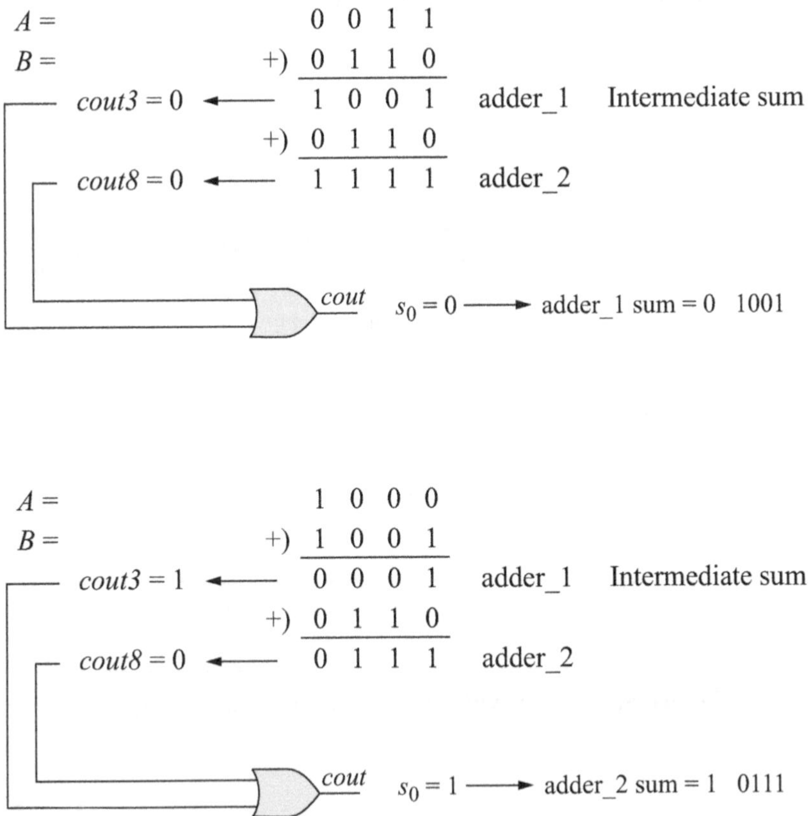

Continued on next page

Figure 8.7 Examples using a decimal adder with multiplexers.

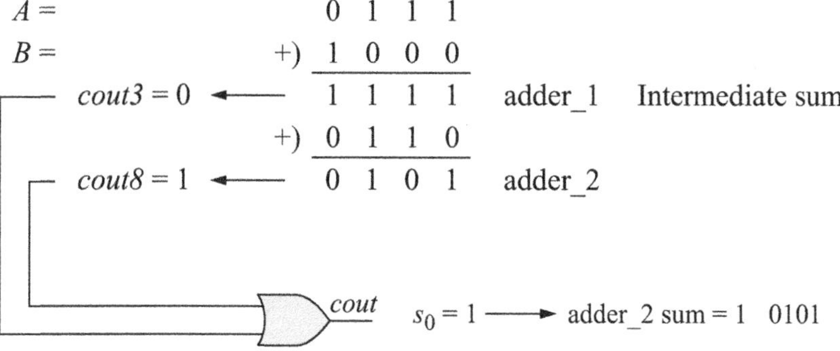

Figure 8.7 (Continued)

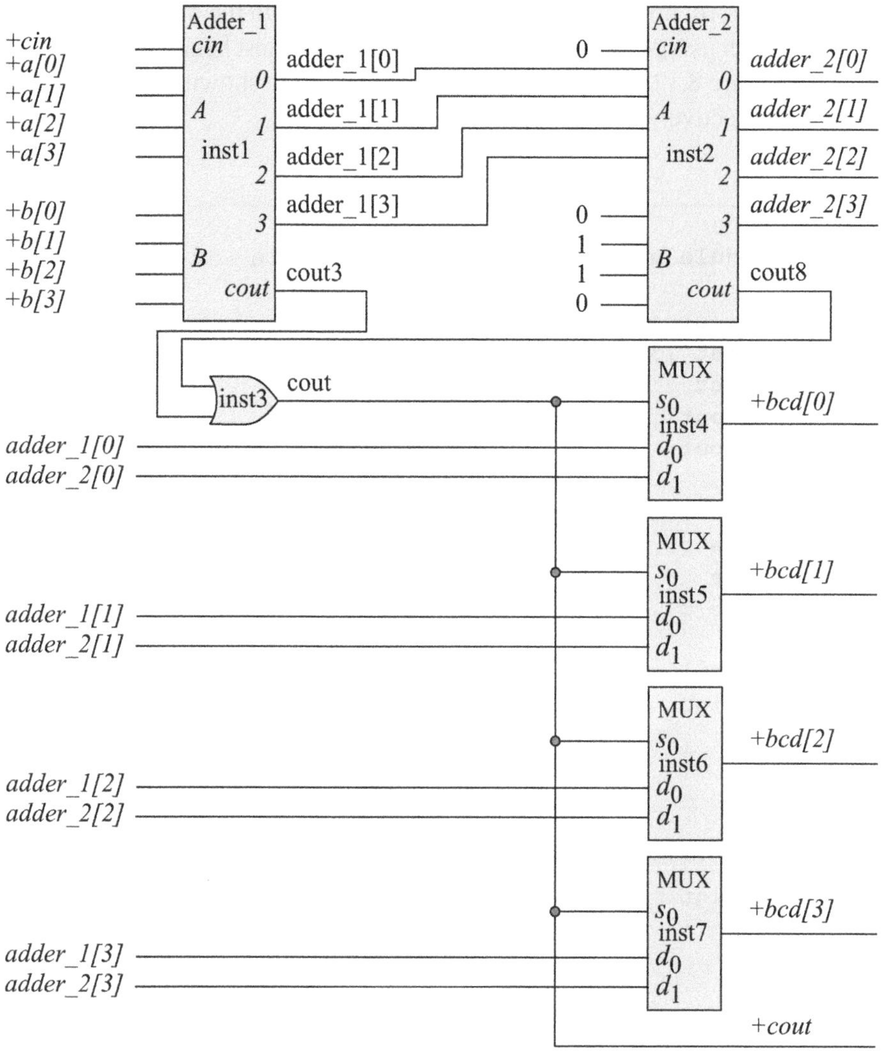

Figure 8.8 Decimal addition using multiplexers to obtain a valid decimal digit.

The propagation delay can be minimized if the carry lookahead (CLA) technique is utilized. Two CLA circuits are required — one for the upper adder (*adder_1*) and one for the lower adder (*adder_2*). The carry-out *cout* that is applied to the multiplexer is described by Equation 8.3 in terms of the group generate (GG) and group propagate (GP) functions.

$$\begin{aligned} cout &= cout3 \text{ (upper carry-out)} + cout8 \text{ (lower carry-out)} \\ &= [\text{GG} + (\text{GP} \bullet cin)]_{\text{upper}} + [\text{GG} + (\text{GP} \bullet cin)]_{\text{lower}} \end{aligned} \tag{8.3}$$

The 4-bit adder used in this design is shown in the dataflow module of Figure 8.9 using the CLA method, and the dataflow module for a 2:1 multiplexer is shown in Figure 8.10, both of which will be instantiated into the mixed-design module of the BCD adder using multiplexers of Figure 8.11. The test bench for the BCD adder is shown in Figure 8.12. The outputs and waveforms are shown in Figure 8.13 and Figure 8.14, respectively.

```
//dataflow carry lookahead adder
module adder4_cla (a, b, cin, sum, cout);

//i/o port declaration
input [3:0] a, b;
input cin;
output [3:0] sum;
output cout;

//define internal wires
wire g3, g2, g1, g0;
wire p3, p2, p1, p0;
wire c3, c2, c1, c0;

//define generate functions
assign   g0 = a[0] & b[0],//multiple statements with 1 assign
         g1 = a[1] & b[1],
         g2 = a[2] & b[2],
         g3 = a[3] & b[3];

//define propagate functions
assign   p0 = a[0] ^ b[0],//multiple statements with 1 assign
         p1 = a[1] ^ b[1],
         p2 = a[2] ^ b[2],
         p3 = a[3] ^ b[3];
//continued on next page
```

Figure 8.9 Carry lookahead adder used in the BCD adder.

```
//obtain the carry equations
assign   c0 = g0 | (p0 & cin),
         c1 = g1 | (p1 & g0) | (p1 & p0 & cin),
         c2 = g2 | (p2 & g1) | (p2 & p1 & g0) |
                  (p2 & p1 & p0 & cin),
         c3 = g3 | (p3 & g2) | (p3 & p2 & g1) |
                  (p3 & p2 & p1 & g0) |
                  (p3 & p2 & p1 & p0 & cin);

//obtain the sum equations
assign   sum[0] = p0 ^ cin,
         sum[1] = p1 ^ c0,
         sum[2] = p2 ^ c1,
         sum[3] = p3 ^ c2;

//obtain cout
assign   cout = c3;
endmodule
```

Figure 8.9 (Continued)

```
//dataflow 2:1 multiplexer
module mux2_df (sel, data, z1);

input sel;      //define inputs and output
input [1:0] data;
output z1;

assign z1 = (~sel & data[0]) | (sel & data[1]);
endmodule
```

Figure 8.10 Two-to-one multiplexer used in the BCD adder.

```
//mixed-design bcd adder using multiplexers
module add_bcd_mux3 (a, b, cin, bcd, cout, invalid_inputs);

//define inputs and outputs
input [3:0] a, b;
input cin;
output [3:0] bcd;
output cout, invalid_inputs;

//continued on next page
```

Figure 8.11 Mixed-design module for the BCD adder using multiplexers.

```
reg invalid_inputs;      //reg if used in always statement

//define internal nets
wire [3:0] adder_1, adder_2;
wire cout3, cout8;

//check for invalid inputs
always @ (a or b)
begin
   if ((a > 4'b1001) || (b > 4'b1001))
      invalid_inputs = 1'b1;
   else
      invalid_inputs = 1'b0;
end

//instantiate the adder for adder_1
adder4_cla inst1 (
   .a(a[3:0]),
   .b(b[3:0]),
   .cin(cin),
   .sum(adder_1),
   .cout(cout3)
   );

//instantiate the adder for adder_2
adder4_cla inst2 (
   .a(adder_1),
   .b({1'b0, 1'b1, 1'b1, 1'b0}),
   .cin(1'b0),
   .sum(adder_2),
   .cout(cout8)
   );

//instantiate the multiplexer select logic
or2_df inst3 (
   .x1(cout8),
   .x2(cout3),
   .z1(cout)
   );

//instantiate the 2:1 multiplexers
mux2_df inst4 (
   .sel(cout),
   .data({adder_2[0], adder_1[0]}),
   .z1(bcd[0])
   );
//continued on next page
```

Figure 8.11 (Continued)

```
mux2_df inst5 (
   .sel(cout),
   .data({adder_2[1], adder_1[1]}),
   .z1(bcd[1])
   );

mux2_df inst6 (
   .sel(cout),
   .data({adder_2[2], adder_1[2]}),
   .z1(bcd[2])
   );

mux2_df inst7 (
   .sel(cout),
   .data({adder_2[3], adder_1[3]}),
   .z1(bcd[3])
   );
endmodule
```

Figure 8.11 (Continued)

```
//test bench mixed-design bcd adder using multiplexers
module add_bcd_mux3_tb;

reg [3:0] a, b;
reg cin;
wire [3:0] bcd;
wire cout, invalid_inputs;

//display variables
initial
$monitor ("a=%b, b=%b, cin=%b,
            cout=%b, bcd=%b, invalid_inputs=%b",
            a, b, cin, cout, bcd, invalid_inputs);
//apply input vectors
initial
begin
   #0    a = 4'b0011;   b = 4'b0011;   cin = 1'b0;
   #10   a = 4'b0101;   b = 4'b0110;   cin = 1'b0;
   #10   a = 4'b0111;   b = 4'b1000;   cin = 1'b0;
   #10   a = 4'b0111;   b = 4'b0111;   cin = 1'b0;
   #10   a = 4'b1000;   b = 4'b1001;   cin = 1'b0;
   #10   a = 4'b1001;   b = 4'b1001;   cin = 1'b0;
   #10   a = 4'b0101;   b = 4'b0110;   cin = 1'b1;
//continued on next page
```

Figure 8.12 Test bench for the BCD adder using multiplexers.

```
    #10    a = 4'b0111;    b = 4'b1000;cin = 1'b1;
    #10    a = 4'b1001;    b = 4'b1001;cin = 1'b1;


           //invalid inputs
    #10    a = 4'b0111;    b = 4'b1011;cin = 1'b0;
           //invalid inputs
    #10    a = 4'b1111;    b = 4'b1000;cin = 1'b0;
           //invalid inputs
    #10    a = 4'b1101;    b = 4'b1010;cin = 1'b1;


    #10    $stop;
end


//instantiate the module into the test bench
add_bcd_mux3 inst1 (
    .a(a),
    .b(b),
    .cin(cin),
    .bcd(bcd),
    .cout(cout),
    .invalid_inputs(invalid_inputs)
    );

endmodule
```

Figure 8.12 (Continued)

```
a=0011, b=0011, cin=0, cout=0, bcd=0110, invalid_inputs=0
a=0101, b=0110, cin=0, cout=1, bcd=0001, invalid_inputs=0
a=0111, b=1000, cin=0, cout=1, bcd=0101, invalid_inputs=0
a=0111, b=0111, cin=0, cout=1, bcd=0100, invalid_inputs=0
a=1000, b=1001, cin=0, cout=1, bcd=0111, invalid_inputs=0
a=1001, b=1001, cin=0, cout=1, bcd=1000, invalid_inputs=0
a=0101, b=0110, cin=1, cout=1, bcd=0010, invalid_inputs=0
a=0111, b=1000, cin=1, cout=1, bcd=0110, invalid_inputs=0
a=1001, b=1001, cin=1, cout=1, bcd=1001, invalid_inputs=0
a=0111, b=1011, cin=0, cout=1, bcd=1000, invalid_inputs=1
a=1111, b=1000, cin=0, cout=1, bcd=1101, invalid_inputs=1
a=1101, b=1010, cin=1, cout=1, bcd=1110, invalid_inputs=1
```

Figure 8.13 Outputs for the BCD adder using multiplexers.

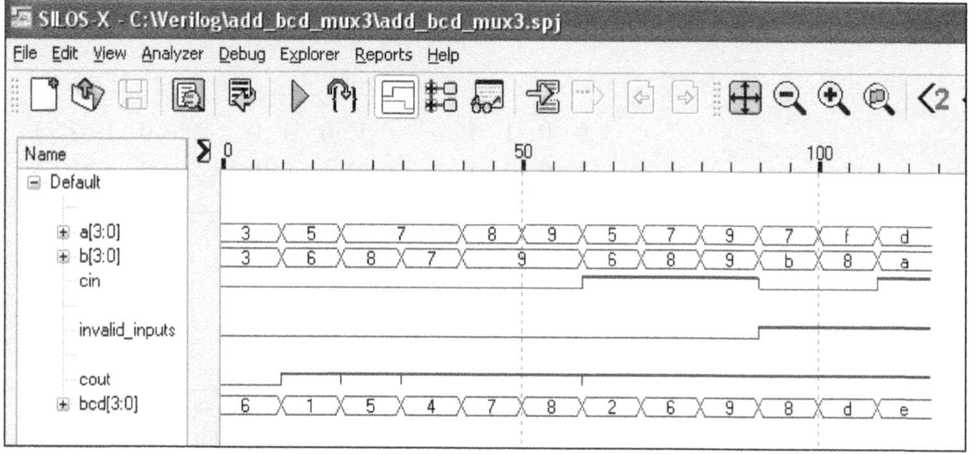

Figure 8.14 Waveforms for the BCD adder using multiplexers.

8.3 Addition with Memory-Based Correction

A memory can be used to correct the intermediate sum for a decimal add operation. This is a low-capacity memory containing only 32 four-bit words. One memory is required for each decade. This approach may require less hardware than using a second 4-bit fixed-point adder to correct the intermediate sum as presented in Section 8.1. Each memory is addressed by the concatenation of the carry-out of the stage with the 4-bit intermediate sum, as shown in Example 8.4 and Example 8.5.

Example 8.4 The decimal operands 76 and 53 will be added using memories to correct the intermediate sum to yield a result of 129.

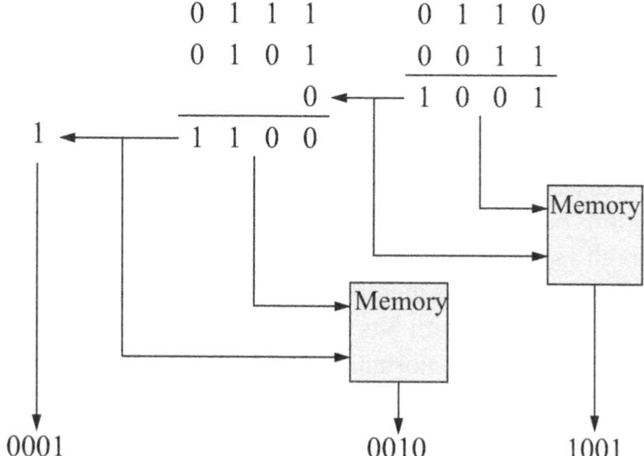

Example 8.5 The decimal operands 387 and 965 will be added using memories to correct the intermediate sum to yield a result of 1352.

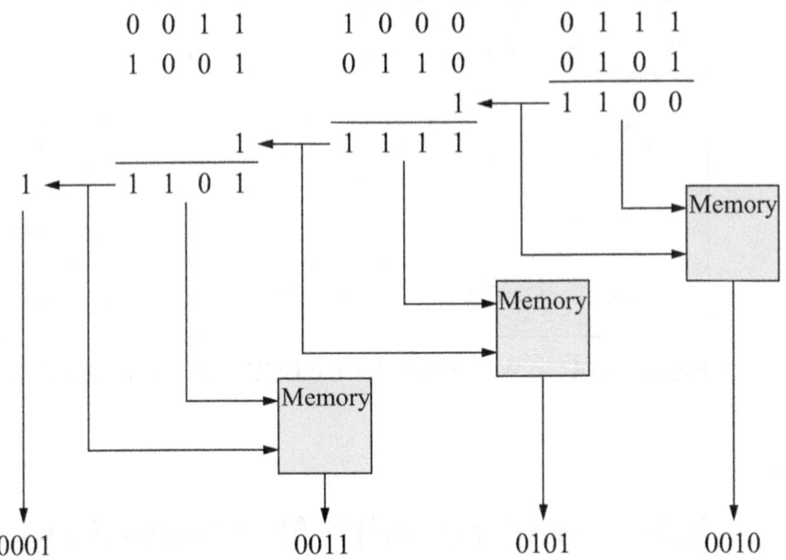

The 2-decade organization for this method is shown in Figure 8.15, which also indicates the instantiation names and net names. The memory for the units decade is addressed by the concatenated carry-out, *cout4*, and the intermediate sum, *adder_1[3:0]*; the memory for the tens decade is addressed by the concatenated carry-out, *cout8*, and the intermediate sum, *adder_2[3:0]*. The two carries are derived from the following general equation:

$$Carry = c_8 + b_8 b_4 + b_8 b_2 \tag{8.4}$$

Equation 8.4 specifies that a carry will be generated whenever there is a carry-out of the decade caused by a 1 bit in the high-order bit position of both operands, when bit positions b_8 and b_4 are both 1s in the intermediate sum, or when bit positions b_8 and b_2 are both 1s in the intermediate sum.

Table 8.2 lists the memory outputs for each address, which is a function of the concatenation of *cout* with bits 8 4 2 1. Note that only 20 of the 32 memory outputs are valid. There would normally be a carry generated for intermediate outputs 1010 – 1111, and no carry generated for intermediate outputs 0100 –1001.

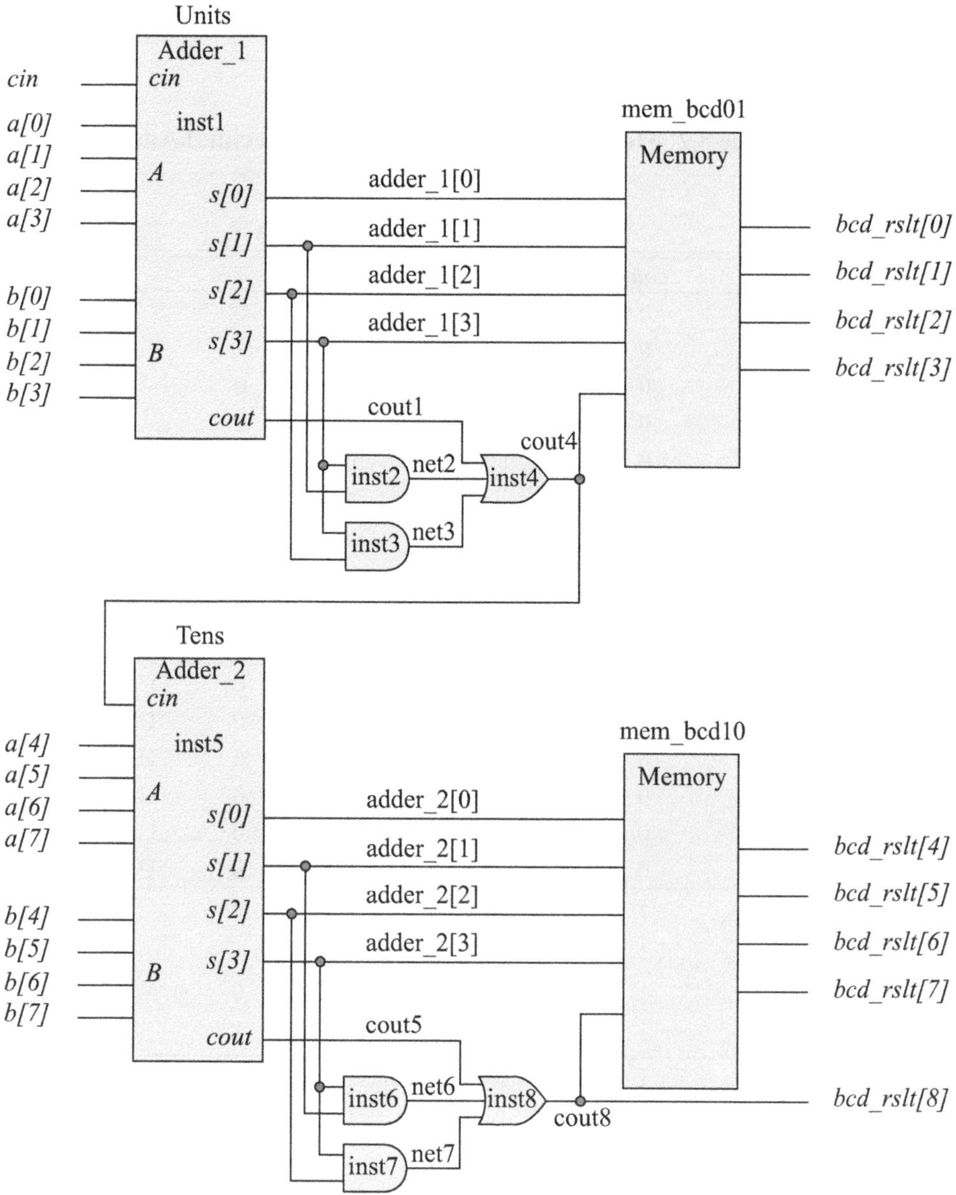

Figure 8.15 Organization of a 2-decade decimal adder using memories to correct the intermediate sum.

The mixed-design module for the BCD adder using memories to correct the intermediate sum is shown in Figure 8.16. A check is made to determine if any BCD input is invalid; that is, if the high-order four bits or the low-order four bits of either operand are greater than 1001, then the *invalid_inputs* output is set to a value of 1. The 4-bit adder used in this design utilizes the carry lookahead technique and is shown in Figure 8.9 of Section 8.2 for addition using multiplexers.

Table 8.2 Memory Correction Codes for Decimal Addition

		ROM Address				ROM Outputs				
	cout	8	4	2	1	8	4	2	1	
	0	0	0	0	0	0	0	0	0	
	0	0	0	0	1	0	0	0	1	
	0	0	0	1	0	0	0	1	0	
	0	0	0	1	1	0	0	1	1	
	0	0	1	0	0	0	1	0	0	
	0	0	1	0	1	0	1	0	1	
	0	0	1	1	0	0	1	1	0	
	0	0	1	1	1	0	1	1	1	
	0	1	0	0	0	1	0	0	0	
	0	1	0	0	1	1	0	0	1	
	0	1	0	1	0	0	0	0	0	Unused
	0	1	0	1	1	0	0	0	0	Unused
	0	1	1	0	0	0	0	0	0	Unused
	0	1	1	0	1	0	0	0	0	Unused
	0	1	1	1	0	0	0	0	0	Unused
	0	1	1	1	1	0	0	0	0	Unused
c_8	1	0	0	0	0	0	1	1	0	
c_8	1	0	0	0	1	0	1	1	1	
c_8	1	0	0	1	0	1	0	0	0	
c_8	1	0	0	1	1	1	0	0	1	
	1	0	1	0	0	0	0	0	0	Unused
	1	0	1	0	1	0	0	0	0	Unused
	1	0	1	1	0	0	0	0	0	Unused
	1	0	1	1	1	0	0	0	0	Unused
	1	1	0	0	0	0	0	0	0	Unused
	1	1	0	0	1	0	0	0	0	Unused
$b_8 \bullet b_2$	1	1	0	1	0	0	0	0	0	
$b_8 \bullet b_2$	1	1	0	1	1	0	0	0	1	
$b_8 \bullet b_4$	1	1	1	0	0	0	0	1	0	
$b_8 \bullet b_4$	1	1	1	0	1	0	0	1	1	
$b_8 \bullet b_4$	1	1	1	1	0	0	1	0	0	
$b_8 \bullet b_4$	1	1	1	1	1	0	1	0	1	

The 32-word memories are defined separately for each decade. For example, the memory for the units decade is defined as follows:

```
reg [3:0] mem_bcd01[0:31];
```

where *mem_bcd01* represents the memory for the units decade as an array of 32 four-bit registers. This memory is loaded from a file called *addbcd.rom01*, which is a replica of the contents shown in Table 8.2 and is saved in the project file without the **.v** extension. The memory is loaded by the following Verilog code:

```
initial
begin
        $readmemb ("addbcd.rom01", mem_bcd01);
end
```

The 32-word memory for the tens decade is defined and loaded in the same manner, as shown below.

```
reg [3:0] mem_bcd10[0:31];

initial
begin
    $readmemb ("addbcd.rom10", mem_bcd10);
end
```

When the operands change value, the combinational logic and memories produce the BCD results, as shown below. This Verilog code can be easily followed using the organization of Figure 8.15.

```
always @ (a or b)
begin
    bcd_rslt[3:0] = mem_bcd01[{cout4, adder_1}];
    bcd_rslt[7:4] = mem_bcd10[{cout8, adder_2}];
    bcd_rslt[8] = cout8;
end
```

The augend and addend are 8-bit operands to accommodate two BCD digits. Any intermediate sum that is invalid is corrected by the memories. The result is a 9-bit BCD sum which represents the sum obtained from the units and tens decades plus a carry-out of the tens decade. The concept shown in the organization of Figure 8.15 can be easily expanded to include any size of BCD operands.

The test bench is shown in Figure 8.17 using 8-bit operands for the augend *A* and the addend *B*. Also included are invalid inputs for some BCD decades that exceed a value of 1001. The outputs are shown in Figure 8.18 depicting the two operands and the resulting outputs, which are represented as hundreds *hun*, tens *ten*, and units *unit*. Invalid outputs are also indicated. The waveforms are shown in Figure 8.19.

```
//mixed-design for bcd addition using memories
module add_bcd_rom (a, b, cin, bcd_rslt, invalid_inputs);

input [7:0] a, b;
input cin;
output [8:0] bcd_rslt;
output invalid_inputs;

wire [7:0] a, b;
wire cin;
reg [8:0] bcd_rslt;
reg invalid_inputs;

//define internal nets
wire [3:0] adder_1, adder_2;
wire cout1, cout4, cout5, cout8;
wire net2, net3, net6, net7;

//check for invalid inputs
always @ (a or b)
begin
   if ((a[7:4] > 4'b1001) || (a[3:0] > 4'b1001) ||
       (b[7:4] > 4'b1001) || (b[3:0] > 4'b1001))
      invalid_inputs = 1'b1;
   else
      invalid_inputs = 1'b0;
end

//instantiate the 4-bit adder for the units decade
adder4_cla inst1 (
   .a(a[3:0]),
   .b(b[3:0]),
   .cin(1'b0),
   .sum(adder_1),
   .cout(cout1)
   );

//instantiate the logic for the carry-out from the units decade
and2_df inst2 (
   .x1(adder_1[3]),
   .x2(adder_1[1]),
   .z1(net2)
   );

//continued on next page
```

Figure 8.16 Mixed-design module for a BCD adder using memories to correct the intermediate sum.

```
and2_df inst3 (
   .x1(adder_1[3]),
   .x2(adder_1[2]),
   .z1(net3)
   );


or3_df inst4 (
   .x1(cout1),
   .x2(net2),
   .x3(net3),
   .z1(cout4)
   );


//instantiate the 4-bit adder for the tens decade
adder4_cla inst5 (
   .a(a[7:4]),
   .b(b[7:4]),
   .cin(cout4),
   .sum(adder_2),
   .cout(cout5)
   );


//instantiate the logic for the carry-out from the tens decade
and2_df inst6 (
   .x1(adder_2[3]),
   .x2(adder_2[1]),
   .z1(net6)
   );


and2_df inst7 (
   .x1(adder_2[3]),
   .x2(adder_2[2]),
   .z1(net7)
   );


or3_df inst8 (
   .x1(cout5),
   .x2(net6),
   .x3(net7),
   .z1(cout8)
   );

//continued on next page
```

Figure 8.16 (Continued)

```
//define memory size for units decade
//mem_bcd01 is an array of 32 four-bit registers
reg [3:0] mem_bcd01[0:31];

//define memory size for tens decade
//mem_bcd10 is an array of 32 four-bit registers
reg [3:0] mem_bcd10[0:31];

//define memory contents for units decade
//load mem_bcd01 from file addbcd.rom01
initial
begin
   $readmemb ("addbcd.rom01", mem_bcd01);
end

//define memory contents for tens decade
//load mem_bcd10 from file addbcd.rom10
initial
begin
   $readmemb ("addbcd.rom10", mem_bcd10);
end

//obtain the bcd result
always @ (a or b)
begin
   bcd_rslt[3:0] = mem_bcd01[{cout4, adder_1}];
   bcd_rslt[7:4] = mem_bcd10[{cout8, adder_2}];
   bcd_rslt[8] = cout8;
end

endmodule
```

Figure 8.16 (Continued)

```
//test bench for bcd addition using memories
module add_bcd_rom_tb;

reg [7:0] a, b;
reg cin;
wire [8:0] bcd_rslt;
wire invalid_inputs;

//continued on next page
```

Figure 8.17 Test bench for the BCD adder using memories to correct the intermediate sum.

```
//display variables
initial
$monitor ("a=%b, b=%b, bcd_hun=%b, bcd_ten=%b, bcd_unit=%b,
                invalid_inputs=%b",
         a, b, {{3{1'b0}}, bcd_rslt[8]}, bcd_rslt[7:4],
                bcd_rslt[3:0], invalid_inputs);
//replication operator {3{1'b0}} provides 4 bits for hundreds
initial      //apply input vectors
begin
         //98 + 98 = 196
   #0    a = 8'b1001_1000; b = 8'b1001_1000;

         //71 + 81 = 152
   #10   a = 8'b0111_0001; b = 8'b1000_0001;

         //62 + 74 = 136
   #10   a = 8'b0110_0010; b = 8'b0111_0100;

         //97 + 98 = 195
   #10   a = 8'b1001_0111; b = 8'b1001_1000;

         //23 + 75 = 098
   #10   a = 8'b0010_0011; b = 8'b0111_0101;

         //99 + 99 = 198
   #10   a = 8'b1001_1001; b = 8'b1001_1001;

         //65 + 38 = 103
   #10   a = 8'b0110_0101; b = 8'b0011_1000;

         //52 + 37 = 089
   #10   a = 8'b0101_0010; b = 8'b0011_0111;

   #10   a = 8'b0111_1011; b = 8'b1001_0011;//invalid inputs
   #10   a = 8'b1111_0110; b = 8'b1000_1010;//invalid inputs
   #10   a = 8'b1101_1100; b = 8'b1010_1100;//invalid inputs
   #10   $stop;
end

add_bcd_rom inst1 (   //instantiate the module
   .a(a),
   .b(b),
   .cin(cin),
   .bcd_rslt(bcd_rslt),
   .invalid_inputs(invalid_inputs)
   );
endmodule
```

Figure 8.17 (Continued)

```
a=10011000, b=10011000,
   bcd_hun=000x, bcd_ten=xxxx, bcd_unit=xxxx,
   invalid_inputs=0

a=01110001, b=10000001,
   bcd_hun=0001, bcd_ten=1001, bcd_unit=0110,
   invalid_inputs=0

a=01100010, b=01110100,
   bcd_hun=0001, bcd_ten=0101, bcd_unit=0010,
   invalid_inputs=0

a=10010111, b=10011000,
   bcd_hun=0001, bcd_ten=0011, bcd_unit=0110,
   invalid_inputs=0

a=00100011, b=01110101,
   bcd_hun=0001, bcd_ten=1001, bcd_unit=0101,
   invalid_inputs=0

a=10011001, b=10011001,
   bcd_hun=0000, bcd_ten=1001, bcd_unit=1000,
   invalid_inputs=0

a=01100101, b=00111000,
   bcd_hun=0001, bcd_ten=1001, bcd_unit=1000,
   invalid_inputs=0

a=01010010, b=00110111,
   bcd_hun=0001, bcd_ten=0000, bcd_unit=0011,
   invalid_inputs=0

a=01111011, b=10010011,
   bcd_hun=0000, bcd_ten=1000, bcd_unit=1001,
   invalid_inputs=1

a=11110110, b=10001010,
   bcd_hun=0001, bcd_ten=0111, bcd_unit=0100,
   invalid_inputs=1

a=11011100, b=10101100,
   bcd_hun=0001, bcd_ten=0000, bcd_unit=0110,
   invalid_inputs=1
```

Figure 8.18 Outputs for the BCD adder using memories to correct the intermediate sum.

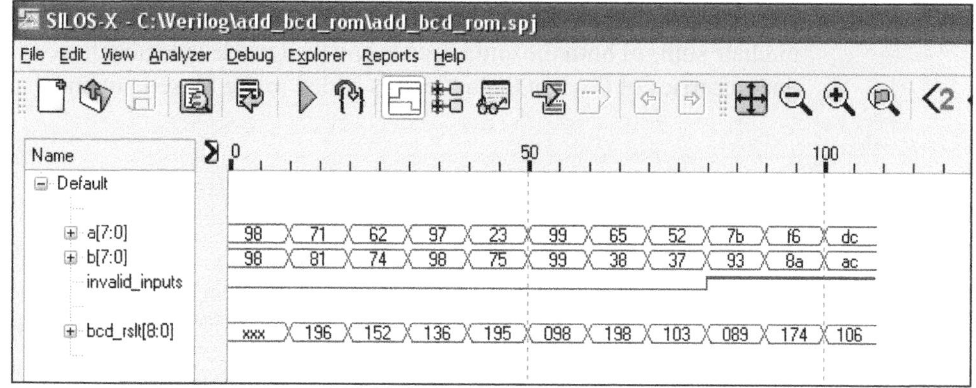

Figure 8.19 Waveforms for the BCD adder using memories to correct the intermediate sum.

8.4 Addition with Biased Augend

Another interesting approach to decimal addition is to bias one of the decimal operands prior to the add operation — in this case the augend — and then to remove the bias, if necessary, at the end of the operation depending on certain intermediate results. This is, in effect, preprocessing and postprocessing the decimal digits. The augend is biased by adding six to each BCD digit without generating a carry-out from the high-order bit position.

Then the addend digits are added to the biased augend digits and carries are allowed to propagate to the next higher-order BCD digit. Since a decimal digit cannot exceed nine (1001), when a bias of six (0110) is added to nine the result will be 15 (1111). Therefore, any nonzero digit that is added to nine — which was biased to 15 — will generate a carry-out of the high-order bit position for the corresponding digit. If a carry was produced in this manner, then the decimal digit is valid and the bias is not removed. Any BCD digit that did not generate a carry is corrected by removing the bias. This is accomplished by subtracting the bias; that is, by adding –6 (1010) in 2s complement.

Three examples are shown below using variable names that will be utilized in the behavioral module for a 2-digit BCD adder. Example 8.6 adds two 2-digit BCD numbers that generate intermediate sums that require correction by removing the bias, due to carries of zero. Example 8.7 also adds two 2-digit BCD numbers that generate intermediate sums that do not require correction, due to carries being generated between decades. Example 8.8 adds two 3-digit BCD numbers that generate carries of both zeroes (correction required) and ones (no correction required).

Example 8.6 Two BCD numbers, 73 and 26, will be added using a biased augend to yield a sum of 99. In this example, the bias of 0110 will be subtracted from the intermediate sums of both the units decade and the tens decade near the end of the operation due to carries of 0 when the addend is added to the biased augend.

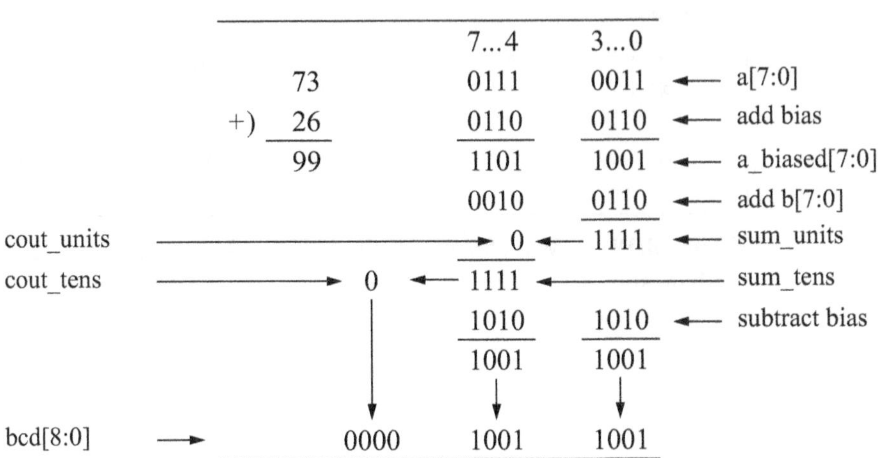

Example 8.7 Two BCD numbers, 86 and 88, will be added using a biased augend to yield a sum of 174. In this example, the bias of 0110 will not be subtracted from the units or tens decade near the end of the operation due to carries of 1 when the addend is added to the biased augend.

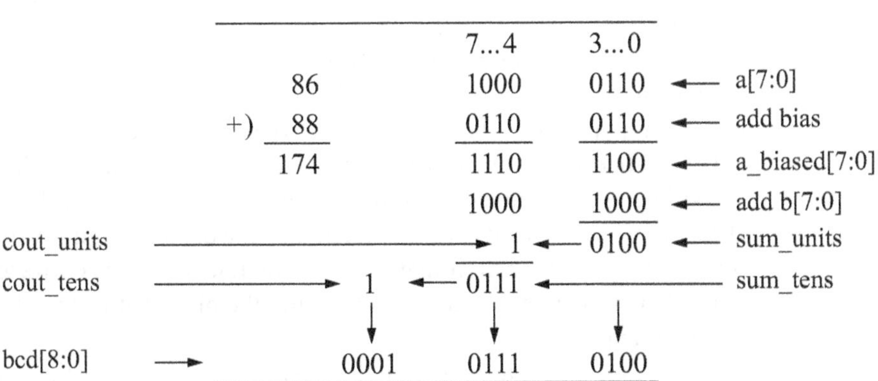

Example 8.8 Two BCD numbers, 597 and 872, will be added using a biased augend to yield a sum of 1469. In this example, the bias of 0110 will be subtracted from the units intermediate sum near the end of the operation due to a carry of 0 when the units decade of the addend is added to the units biased augend. The bias will not be subtracted from the tens or hundreds decades due to carries of 1.

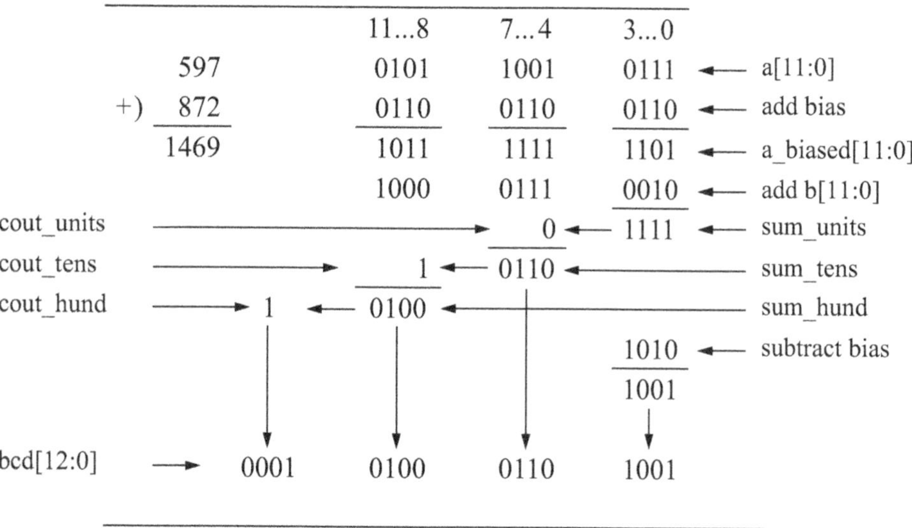

The behavioral module to add two 2-digit BCD numbers is shown in Figure 8.20. The inputs are an 8-bit augend vector, *a[7:0]*; an 8-bit addend vector, *b[7:0]*; and a scalar carry-in, *cin*. The outputs are a BCD result vector, *bcd[7:0]*, and a scalar carry-out, *cout_tens*. Target variables that are placed within the behavioral **always** construct are declared as type **reg**. The **always** statement executes the behavioral statements within the **always** block repeatedly in a looping manner and begins execution at time 0. Execution of the statements continues indefinitely until the simulation is terminated. Blocking statements are used in the **always** block so that a statement can complete execution before the next statement is executed.

The operation follows the algorithm previously described. The augend *a* is initially biased by adding 6 (0110), then the addend *b* and the carry-in *cin* are added to the biased augend. The carry-out that is generated determines whether the bias is subtracted or not subtracted.

The test bench is shown in Figure 8.21 in which several input vectors are applied to the behavioral design module. Note the use of the replication operator to provide a high-order digit of 0000 or 0001. The outputs and waveforms are shown in Figure 8.22 and Figure 8.23, respectively.

```
//behavioral for a 2-digit bcd adder with biased augend
module add_bcd_biased (a, b, cin, cout_tens, bcd);

input [7:0] a, b;
input cin;
output [7:0] bcd;
output cout_tens;
//continued on next page
```

Figure 8.20 Behavioral module for BCD addition using a biased augend.

```
//variables are declared as reg if used in always
reg [7:0] a_biased;
reg [4:0] sum_units;
reg [8:4] sum_tens;
reg cout_units, cout_tens;
reg bcd[7:0];


always @ (a or b)
begin
   a_biased[3:0] = (a[3:0] + 4'b0110);        //add bias
   sum_units = (a_biased[3:0] + b[3:0] + cin);
   cout_units = sum_units[4];

   a_biased[7:4] = (a[7:4] + 4'b0110);        //add bias
   sum_tens = (a_biased[7:4] + b[7:4]+ cout_units);
   cout_tens = sum_tens[8];

   if (cout_units == 1'b0)
      bcd[3:0] = sum_units[3:0] + 4'b1010;    //unbias
   else
      bcd[3:0] = sum_units[3:0];

   if (cout_tens == 1'b0)
      bcd[7:4] = sum_tens[7:4] + 4'b1010;     //unbias
   else
      bcd[7:4] = sum_tens[7:4];
end
endmodule
```

Figure 8.20 (Continued)

```
//test bench for bcd addition using a biased augend
module add_bcd_biased_tb;

reg [7:0] a, b;
reg cin;
wire [7:0] bcd;
wire cout_tens;

initial      //display variables
$monitor ("a_tens=%b, a_units=%b, b_tens=%b, b_units=%b,
             cin=%b, cout=%b, bcd_tens=%b, bcd_units=%b",
          a[7:4], a[3:0], b[7:4], b[3:0],
             cin, {{3{1'b0}}, cout_tens}, bcd[7:4], bcd[3:0]);
//continued on next page
```

Figure 8.21 Test bench for BCD addition using a biased augend.

```
//apply input vectors
initial
begin
         //63 + 36 + 0 = 099
   #0    a = 8'b0110_0011; b = 8'b0011_0110; cin = 1'b0;

         //25 + 52 + 0 = 077
   #10   a = 8'b0010_0101; b = 8'b0101_0010; cin = 1'b0;

         //63 + 75 + 1 = 139
   #10   a = 8'b0110_0011; b = 8'b0111_0101; cin = 1'b1;

         //79 + 63 + 0 = 142
   #10   a = 8'b0111_1001; b = 8'b0110_0011; cin = 1'b0;

         //99 + 99 + 1 = 199
   #10   a = 8'b1001_1001; b = 8'b1001_1001; cin = 1'b1;

         //27 + 63 + 0 = 090
   #10   a = 8'b0010_0111; b = 8'b0110_0011; cin = 1'b0;

         //63 + 27 + 1 = 091
   #10   a = 8'b0110_0011; b = 8'b0010_0111; cin = 1'b1;

         //44 + 85 + 0 = 129
   #10   a = 8'b0100_0100; b = 8'b1000_0101; cin = 1'b0;

         //86 + 88 + 0 = 174
   #10   a = 8'b1000_0110; b = 8'b1000_1000; cin = 1'b0;


   #10   $stop;
end


//instantiate the module into the test bench
add_bcd_biased inst1 (
   .a(a),
   .b(b),
   .cin(cin),
   .cout_tens(cout_tens),
   .bcd(bcd)
   );


endmodule
```

Figure 8.21 (Continued)

```
a_tens=0110, a_units=0011,
b_tens=0011, b_units=0110, cin=0,
 cout=0000, bcd_tens=1001, bcd_units=1001

a_tens=0010, a_units=0101,
b_tens=0101, b_units=0010, cin=0,
 cout=0000, bcd_tens=0111, bcd_units=0111

a_tens=0110, a_units=0011,
b_tens=0111, b_units=0101, cin=1,
 cout=0001, bcd_tens=0011, bcd_units=1001

a_tens=0111, a_units=1001,
b_tens=0110, b_units=0011, cin=0,
 cout=0001, bcd_tens=0100, bcd_units=0010

a_tens=1001, a_units=1001,
b_tens=1001, b_units=1001, cin=1,
 cout=0001, bcd_tens=1001, bcd_units=1001

a_tens=0010, a_units=0111,
b_tens=0110, b_units=0011, cin=0,
 cout=0000, bcd_tens=1001, bcd_units=0000

a_tens=0110, a_units=0011,
b_tens=0010, b_units=0111, cin=1,
 cout=0000, bcd_tens=1001, bcd_units=0001

a_tens=0100, a_units=0100,
b_tens=1000, b_units=0101, cin=0,
 cout=0001, bcd_tens=0010, bcd_units=1001

a_tens=1000, a_units=0110,
b_tens=1000, b_units=1000, cin=0,
 cout=0001, bcd_tens=0111, bcd_units=0100
```

Figure 8.22 Outputs for BCD addition using a biased augend.

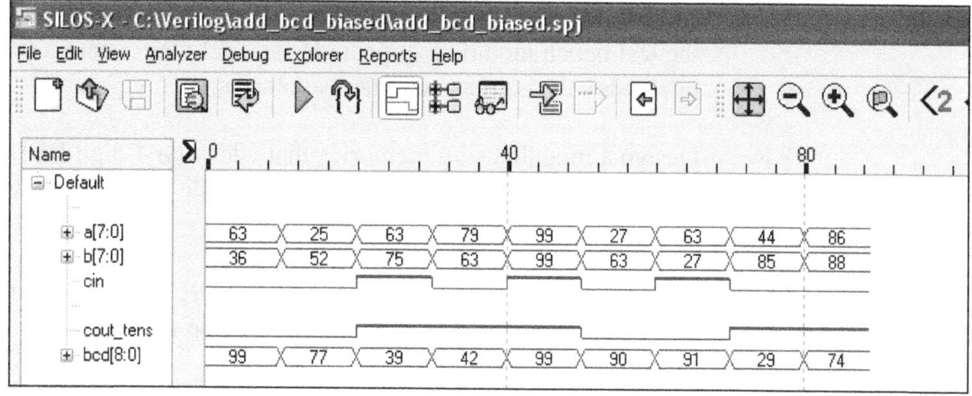

Figure 8.23 Waveforms for BCD addition using a biased augend.

8.5 Problems

8.1 Use the paper-and-pencil method to add the decimal digits 472 and 653.

8.2 Use the paper-and-pencil method to add the decimal digits 4673 and 9245.

8.3 Redesign the BCD adder of Figure 8.2 using a 4-bit adder for *Adder_1* and two half adders and one full adder for *Adder_2*. Use structural modeling and behavioral modeling in the same design module. Check for invalid inputs. Obtain the test bench that applies several inputs including invalid inputs, the outputs, and the waveforms.

8.4 Use mixed-design modeling — behavioral and structural — to design a BCD adder that adds two 8-bit BCD numbers using the sum correction method. Check for invalid inputs. Obtain the test bench that applies several inputs (including invalid inputs), the outputs, and the waveforms.

8.5 Add the decimal operands 786 and 956 using memories to correct the intermediate sum. Use the paper-and-pencil method to add the two 3-digit BCD numbers.

8.6 Add the decimal operands 999 and 999 using memories to correct the intermediate sum. Use the paper-and-pencil method to add the two 3-digit BCD numbers.

8.7 Design a mixed-design module using structural and behavioral modeling to add two 8-bit BCD operands using multiplexers. Obtain the design module, the test bench module that provides several input vectors (including invalid operands), the outputs, and the waveforms.

8.8 Design a module using memories that adds two 4-digit BCD operands. Obtain the design module, the test bench module that includes valid and invalid digits, the outputs, and the waveforms.

8.9 Use behavioral modeling to design a 1-digit BCD adder. Obtain the test bench for several input vectors, the outputs, and the waveforms.

8.10 Use behavioral modeling to design a 2-digit BCD adder. Obtain the test bench, outputs, and waveforms.

8.11 Use the paper-and-pencil method to add the following decimal numbers using a biased augend, augend = 8567 and addend = 9236, to generate a sum of 17803. Design a behavioral module to add two 4-digit BCD numbers using a biased augend. Obtain the test bench module that applies several 16-bit vectors for the augend and addend. Obtain the outputs, and the waveforms.

9.1 *Subtraction Examples*
9.2 *Two-Decade Addition/Subtraction Unit for A+B and A–B*
9.3 *Two-Decade Addition/Subtraction Unit for A+B, A–B, and B–A*
9.4 *Problems*

9

Decimal Subtraction

The binary-coded decimal (BCD) code is not self-complementing as is the radix 2 fixed-point number representation; that is, the $r-1$ complement cannot be acquired by inverting each bit of the 4-bit BCD digit. The rs complement for radix 2 was obtained by adding 1 to the $r-1$ complement; that is, adding 1 to the 1s complement. The same procedure is used in decimal notation; however, the $r-1$ complement is the 9s complement. Therefore, the rs complement for radix 10 is obtained by adding 1 to the 9s complement.

In fixed-point binary arithmetic for radix 2, subtraction is performed by adding the rs complement of the subtrahend B to the minuend A; that is, by adding the 2s complement of the subtrahend as shown in Equation 9.1, where B' is the 1s complement $(r-1)$ and 1 is added to the low-order bit position to form the 2s complement.

$$A - B = A + (B' + 1) \tag{9.1}$$

Subtraction in BCD is essentially the same as in fixed-point binary: Add the rs complement of the subtrahend B to the minuend A, as shown in Equation 9.1, where $(B' + 1)$ is the 10s complement for BCD. The examples which follow show subtraction operations using BCD numbers. Negative results can remain in 10s complement notation or be recomplemented to sign-magnitude notation with a negative sign.

9.1 Subtraction Examples

Several examples are given below that exemplify the principles of decimal subtraction. Negative results can either remain in 10s complement form to be used in further calculations or converted to sign magnitude notation with a negative sign. The result of each example is either true addition or true subtraction. *True addition* is where the result is the sum of the two numbers, regardless of the sign and corresponds to one of the following conditions:

$$
\begin{array}{ccc}
(+A) & + & (+B) \\
(-A) & + & (-B) \\
(+A) & - & (-B) \\
(-A) & - & (+B)
\end{array}
$$

True subtraction is where the result is the difference of the two numbers, regardless of the sign and corresponds to one of the following conditions:

$$
\begin{array}{ccc}
(+A) & - & (+B) \\
(-A) & - & (-B) \\
(+A) & + & (-B) \\
(-A) & + & (+B)
\end{array}
$$

Example 9.1 The number $+42_{10}$ will be subtracted from the number $+76_{10}$. This yields a result of $+34_{10}$. The 10s complement of the subtrahend is obtained as follows using radix 10 numbers: 9 – 4 = 5; 9 – 2 = 7 + 1 = 8, where 5 and 7 are the 9s complement of 4 and 2, respectively. A carry-out of the high-order decade indicates that the result is a positive number in BCD.

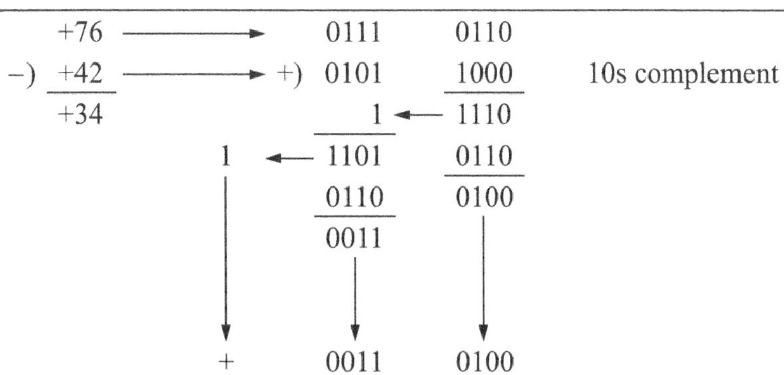

Example 9.2 The number $+87_{10}$ will be subtracted from the number $+76_{10}$. This yields a difference of -11_{10}, which is 1000 1001 represented as a negative BCD number in 10s complement. The 10s complement of the subtrahend is obtained as follows using radix 10 numbers: $9 - 8 = 1$; $9 - 7 = 2 + 1 = 3$, where 1 and 2 are the 9s complement of 8 and 7, respectively. A carry-out of 0 from the high-order decade indicates that the result is a negative BCD number in 10s complement. To obtain the result in radix 10, form the 10s complement of 89_{10}, which will yield 11_{10}. This is interpreted as a negative number.

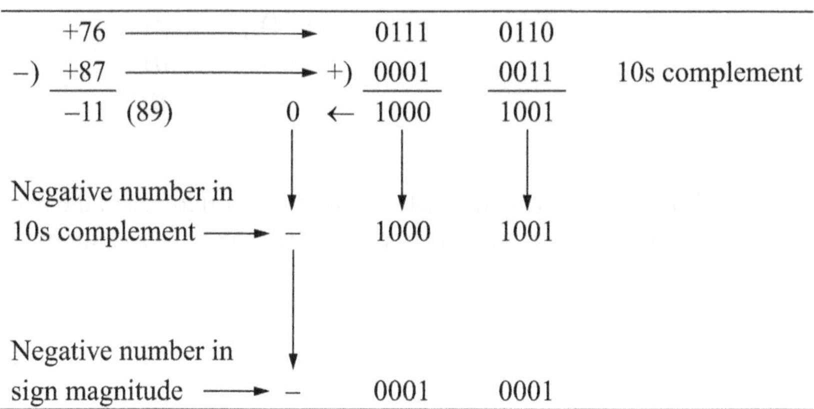

Example 9.3 The following decimal numbers will be added using BCD arithmetic: +28 and −18, as shown below. This can be considered as true subtraction, because the result is the difference of the two numbers, ignoring the signs. A carry of 1 from the high-order decade indicates a positive number.

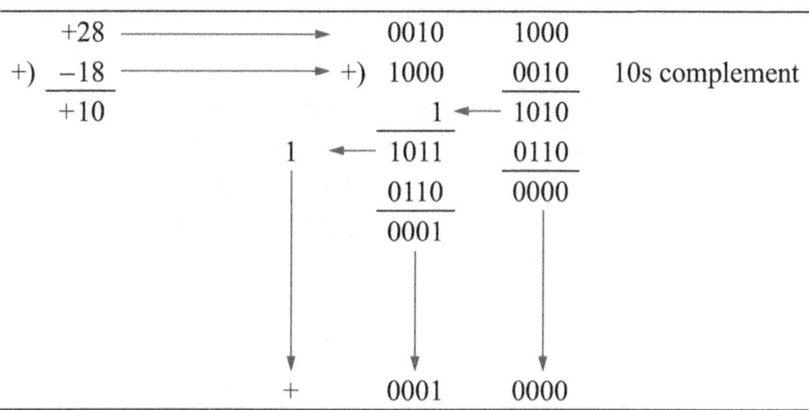

Example 9.4 The following decimal numbers will be subtracted using BCD arithmetic: +482 and +627, resulting in a difference of −145, as shown below. A carry of 0 from the high-order decade indicates a negative number in 10s complement notation.

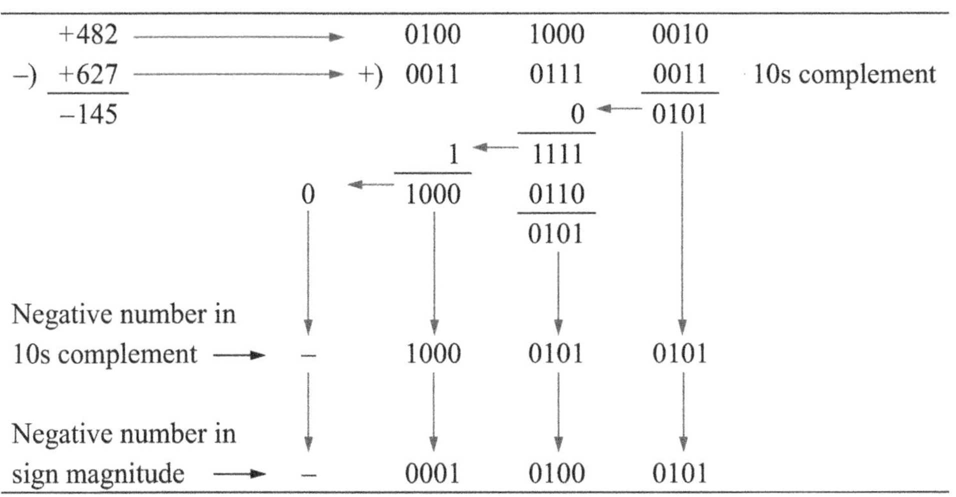

Example 9.5 The following decimal numbers will be subtracted using BCD arithmetic: +28 and +18, resulting in a difference of +10, as shown below. A carry of 1 from the high-order decade indicates a positive number.

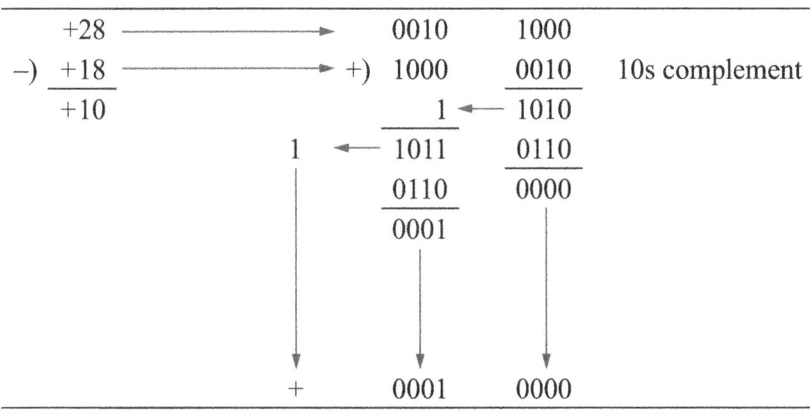

Example 9.6 The following decimal numbers will be subtracted using BCD arithmetic: +436 and +825, resulting in a difference of −389, as shown below. A carry of 0 from the high-order decade indicates a negative number in 10s complement.

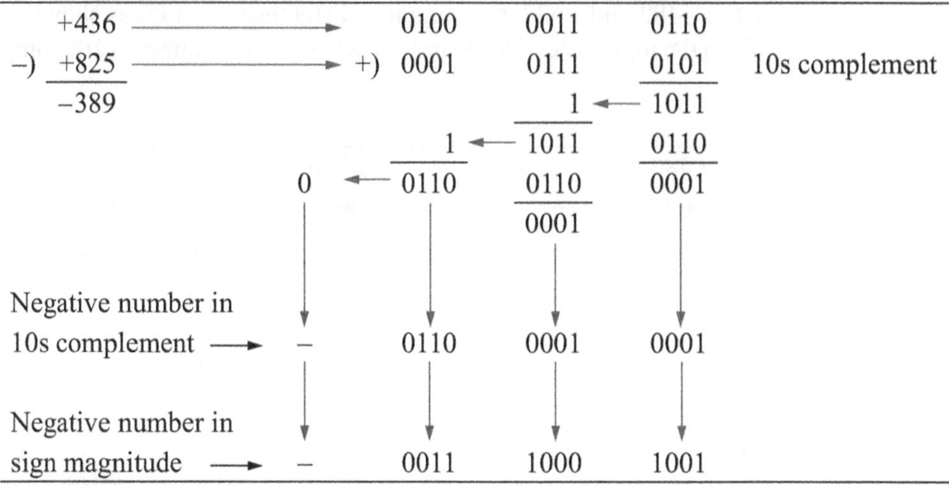

+436
–) +825
–389
0100 0011 0110
+) 0001 0111 0101
10s complement
1 1011
1 1011 0110
0 0110 0110 0001
0001
Negative number in
10s complement
– 0110 0001 0001
Negative number in
sign magnitude
– 0011 1000 1001

9.2 Two-Decade Addition/Subtraction Unit for A+B and A–B

As in fixed-point arithmetic, the decimal adder can be designed with minor modifications so that the operations of addition and subtraction can be performed by the same hardware. Refer to Appendix A for design modules of select logic functions to be instantiated into the designs of this chapter.

Before presenting the organization for the decimal adder/subtractor, a 9s complementer will be designed which will be used in the adder/subtractor module together with a carry-in to form the 10s complement of the subtrahend. As stated previously, a 9s complementer is required because BCD is not a self-complementing code; that is, it cannot form the $r - 1$ complement by inverting the four bits of each decade.

The truth table for a 9s complementer is shown in Table 9.1. A mode control input m will be used to determine whether operand *b[3:0]* will be added to operand *a[3:0]* or subtracted from operand *a[3:0]* as shown in the block diagram of the 9s complementer in Figure 9.1. The function *f[3:0]* is the 9s complement of the subtrahend *b[3:0]*.

Table 9.1 Nines Complementer

Subtrahend				9s Complement Subtraction (m = 1)				Addition (m = 0)			
b[3]	*b[2]*	*b[1]*	*b[0]*	*f[3]*	*f[2]*	*f[1]*	*f[0]*	*f[3]*	*f[2]*	*f[1]*	*f[0]*
0	0	0	0	1	0	0	1	0	0	0	0
0	0	0	1	1	0	0	0	0	0	0	1

Continued on next page

Table 9.1 Nines Complementer

Subtrahend				9s Complement Subtraction ($m = 1$)				Addition ($m = 0$)			
b[3]	*b[2]*	*b[1]*	*b[0]*	*f[3]*	*f[2]*	*f[1]*	*f[0]*	*f[3]*	*f[2]*	*f[1]*	*f[0]*
0	0	1	0	0	1	1	1	0	0	1	0
0	0	1	1	0	1	1	0	0	0	1	1
0	1	0	0	0	1	0	1	0	1	0	0
0	1	0	1	0	1	0	0	0	1	0	1
0	1	1	0	0	0	1	1	0	1	1	0
0	1	1	1	0	0	1	0	0	1	1	1
1	0	0	0	0	0	0	1	1	0	0	0
1	0	0	1	0	0	0	0	1	0	0	1

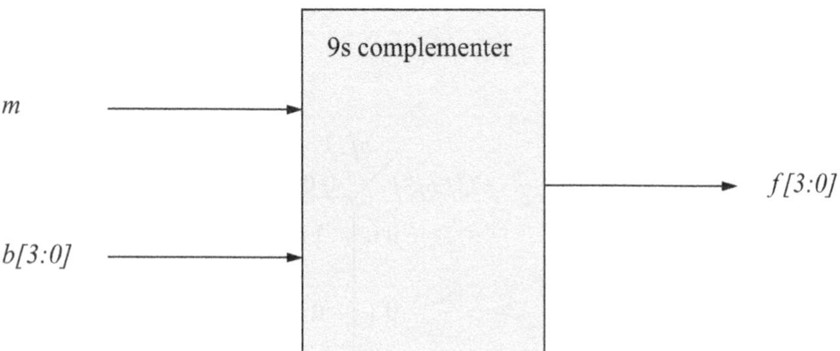

Figure 9.1 Block diagram for a 9s complementer.

If the operation is addition ($m = 0$), then the b operand is unchanged; that is, $f[i] = b[i]$. If the operation is subtraction ($m = 1$), then $b[0]$ is inverted; otherwise, $b[0]$ is unchanged. Therefore, the equation for $f[0]$ is

$$f[0] = b[0] \oplus m$$

The $b[1]$ variable is unchanged for both addition and subtraction; therefore, the equation for $f[1]$ is

$$f[1] = b[1]$$

The equations for $f[2]$ and $f[3]$ are less obvious. The Karnaugh maps representing $f[2]$ and $f[3]$ for a subtract operation are shown in Figure 9.2 and Figure 9.3, respectively. Since $f[2]$ does not depend upon $b[3]$, input $b[3]$ can be ignored; that is, *f[2]* = 1 whenever *b[1]* and *b[2]* are different. The equations for the 9s complement outputs are shown in Equation 9.2. The logic diagram for the 9s complementer is shown in Figure 9.4, including the instantiation names and net names for the dataflow/structural module of Figure 9.5.

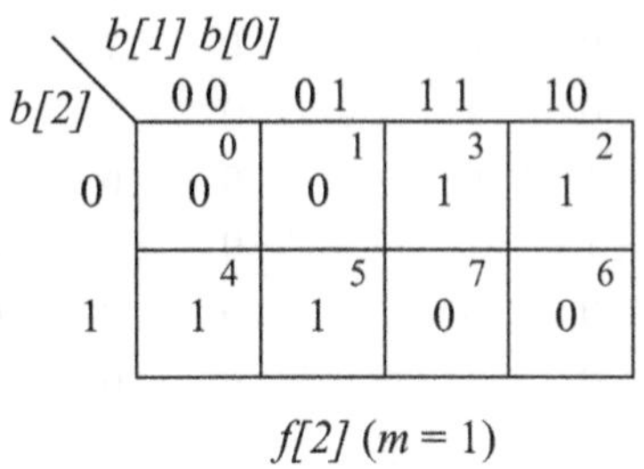

Figure 9.2 Karnaugh map for *f[2]* of the 9s complementer.

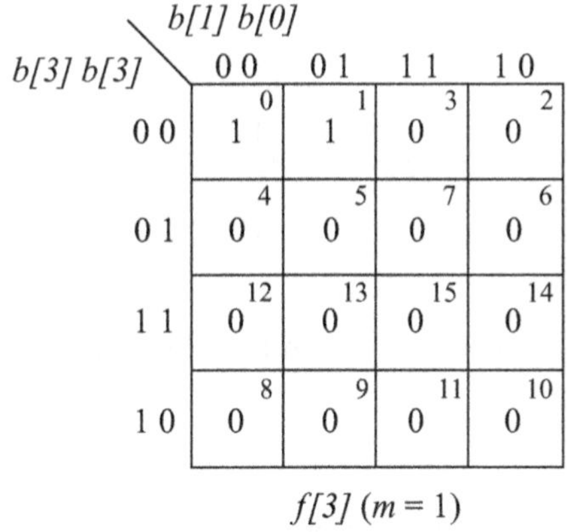

Figure 9.3 Karnaugh map for *f[3]* of the 9s complementer.

$$f[0] = b[0] \oplus m$$

$$f[1] = b[1]$$

$$f[2] = b[2]m' + (b[2] \oplus b[1])m$$

$$f[3] = b[3]m' + b[3]'b[2]'b[1]'m \qquad (9.2)$$

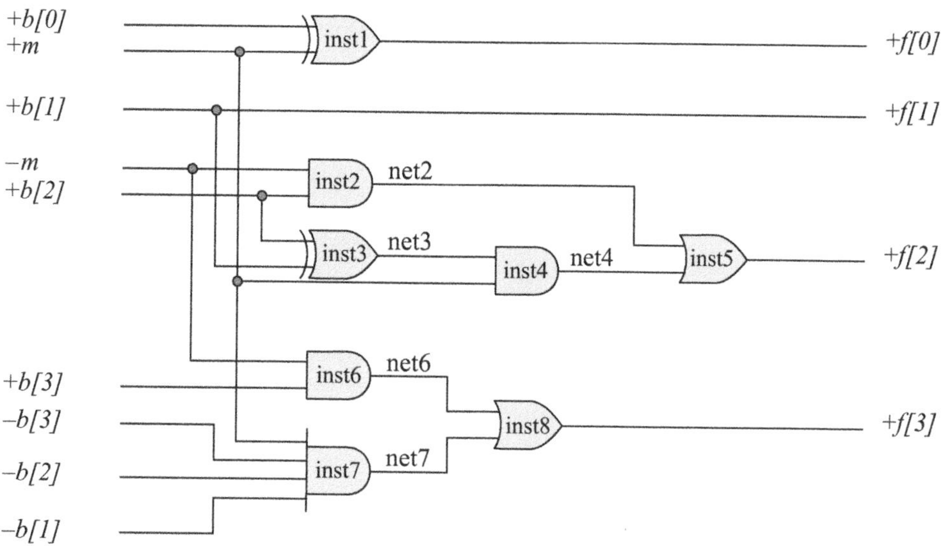

Figure 9.4 Logic diagram for a 9s complementer.

```
//dataflow/structural mixed-design module for 9s complementer
module nines_compl (m, b, f);

input m;
input [3:0] b;
output [3:0] f;


//define internal nets
wire net2, net3, net4, net6, net7;


//instantiate the logic gates for the 9s complementer
xor2_df inst1 (
   .x1(b[0]),
   .x2(m),
   .z1(f[0])
   );


assign f[1] = b[1];

//continued on next page
```

Figure 9.5 Mixed-design module for a 9s complementer.

```
and2_df inst2 (
    .x1(~m),
    .x2(b[2]),
    .z1(net2)
    );

xor2_df inst3 (
    .x1(b[2]),
    .x2(b[1]),
    .z1(net3)
    );

and2_df inst4 (
    .x1(net3),
    .x2(m),
    .z1(net4)
    );

or2_df inst5 (
    .x1(net2),
    .x2(net4),
    .z1(f[2])
    );

and2_df inst6 (
    .x1(~m),
    .x2(b[3]),
    .z1(net6)
    );

and4_df inst7 (
    .x1(m),
    .x2(~b[3]),
    .x3(~b[2]),
    .x4(~b[1]),
    .z1(net7)
    );

or2_df inst8 (
    .x1(net6),
    .x2(net7),
    .z1(f[3])
    );

endmodule
```

Figure 9.5 (Continued)

The test bench for the 9s complementer is shown in Figure 9.6, which applies input vectors for the subtrahend *b[3:0]* for addition ($m = 0$) and subtraction ($m = 1$). The outputs and waveforms are shown in Figure 9.7 and Figure 9.8, respectively.

```
//test bench for 9s complementer
module nines_compl_tb;

reg m;
reg [3:0] b;
wire [3:0] f;

initial          //display variables
$monitor ("m=%b, b=%b, f=%b", m, b, f);

initial          //apply input vectors
begin
//add -- do not complement
   #0     m = 1'b0;    b = 4'b0000;
   #10    m = 1'b0;    b = 4'b0001;
   #10    m = 1'b0;    b = 4'b0010;
   #10    m = 1'b0;    b = 4'b0011;
   #10    m = 1'b0;    b = 4'b0100;
   #10    m = 1'b0;    b = 4'b0101;
   #10    m = 1'b0;    b = 4'b0110;
   #10    m = 1'b0;    b = 4'b0111;
   #10    m = 1'b0;    b = 4'b1000;
   #10    m = 1'b0;    b = 4'b1001;

//subtract -- complement
   #10    m = 1'b1;    b = 4'b0000;
   #10    m = 1'b1;    b = 4'b0001;
   #10    m = 1'b1;    b = 4'b0010;
   #10    m = 1'b1;    b = 4'b0011;
   #10    m = 1'b1;    b = 4'b0100;
   #10    m = 1'b1;    b = 4'b0101;
   #10    m = 1'b1;    b = 4'b0110;
   #10    m = 1'b1;    b = 4'b0111;
   #10    m = 1'b1;    b = 4'b1000;
   #10    m = 1'b1;    b = 4'b1001;
   #10    $stop;
end

nines_compl inst1 (       //instantiate the module
   .m(m),
   .b(b),
   .f(f)
   );
endmodule
```

Figure 9.6 Test bench for the 9s complementer.

```
Add
m=0, b=0000, f=0000
m=0, b=0001, f=0001
m=0, b=0010, f=0010
m=0, b=0011, f=0011
m=0, b=0100, f=0100
m=0, b=0101, f=0101
m=0, b=0110, f=0110
m=0, b=0111, f=0111
m=0, b=1000, f=1000
m=0, b=1001, f=1001

Subtract
m=1, b=0000, f=1001
m=1, b=0001, f=1000
m=1, b=0010, f=0111
m=1, b=0011, f=0110
m=1, b=0100, f=0101
m=1, b=0101, f=0100
m=1, b=0110, f=0011
m=1, b=0111, f=0010
m=1, b=1000, f=0001
m=1, b=1001, f=0000
```

Figure 9.7 Outputs for the 9s complementer.

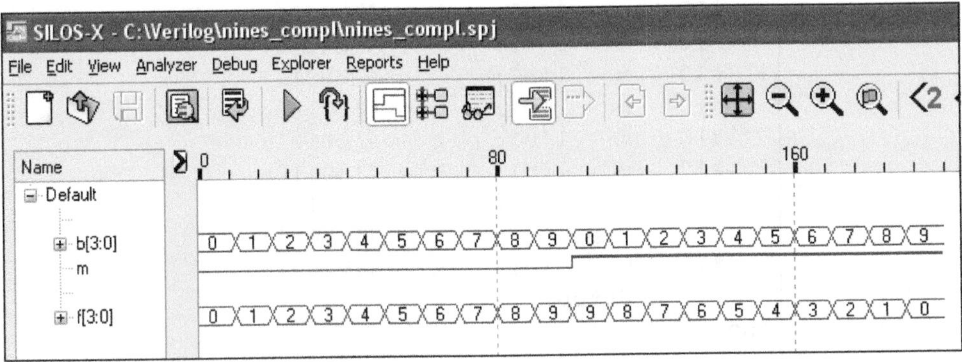

Figure 9.8 Waveforms for the 9s complementer.

The logic diagram for a BCD adder/subtractor is shown in Figure 9.9. The augend/minuend *a[3:0]* is connected directly to the *A* inputs of a fixed-point adder for the units decade; the addend/subtrahend *b[3:0]* is connected to the inputs of a 9s complementer whose outputs $f[3:0]$ connect to the B inputs of the adder. In a similar manner, the augend/minuend $a[7:4]$ connects to the A inputs of the adder for the tens decade; the addend/subtrahend $b[7:4]$ connects to a 9s complementer whose outputs $f[7:4]$ connect to the B inputs of the adder.

There is also a mode control input m to the 9s complementer circuits of the units decade and the tens decade. The mode control input specifies either an add operation ($m = 0$) or a subtract operation ($m = 1$). The mode control is also connected to the carry-in of the adder for the units decade.

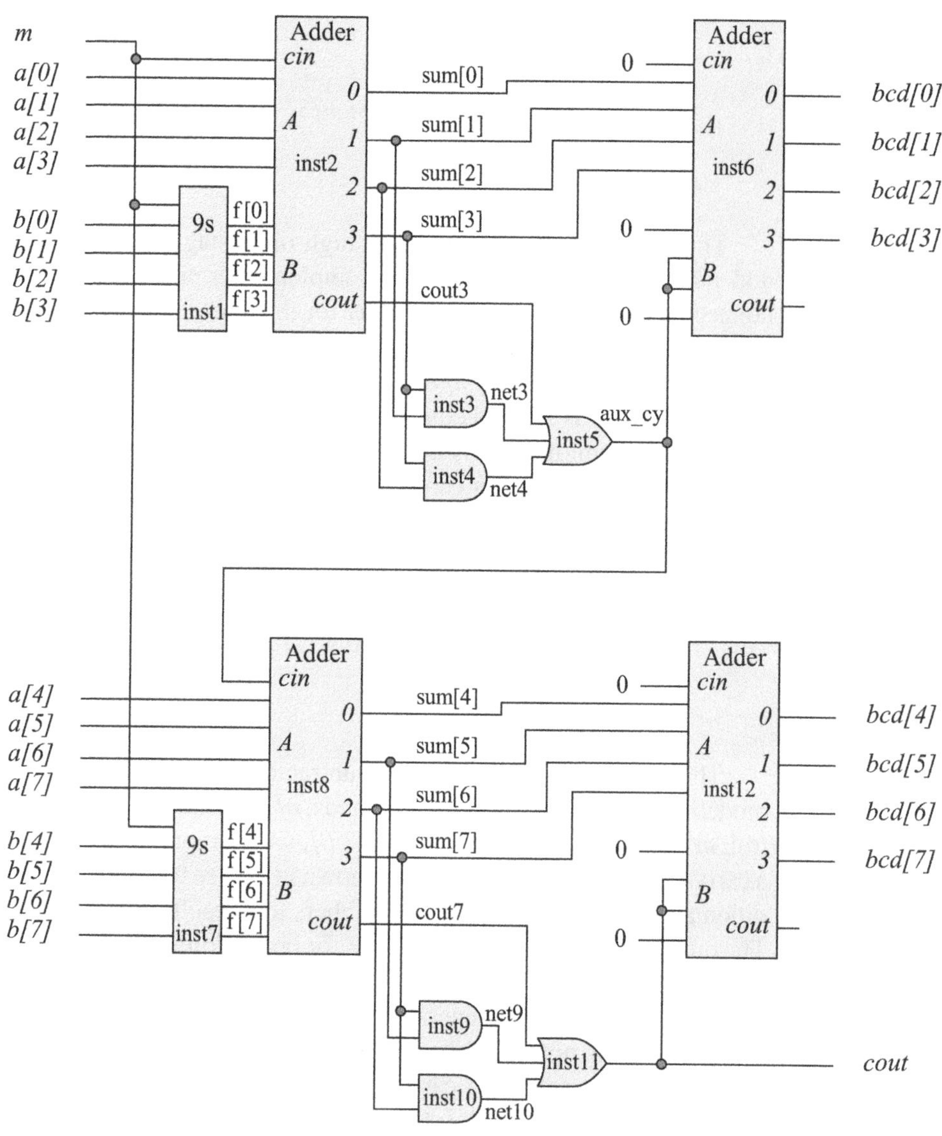

Figure 9.9 Logic diagram for a two-stage BCD adder/subtractor.

If the operation is addition, then the mode control input adds 0 to the uncomplemented version of operand *b[3:0]*. If the operation is subtraction, then the mode control adds a 1 to the 9s complement of operand *b[3:0]* to form the 10s complement. The outputs of the adder are a 4-bit intermediate sum, *sum[3:0]*, and a carry-out labeled *cout3*. The intermediate sum is connected to a second fixed-point adder in the units decade, which will add 0000 to the intermediate sum if no adjustment is required or add 0110 to the intermediate sum if adjustment is required.

The carry-out of the low-order stage is specified as an auxiliary carry, *aux_cy*, and is defined by Equation 9.3. The auxiliary carry determines whether 0000 or 0110 is added to the intermediate sum of the units decade to produce a valid BCD digit, as well as providing a carry-in to the tens decade.

$$aux\_cy = cout3 + sum[3]\ sum[1] + sum[3]\ sum[2] \tag{9.3}$$

The same rationale applies to the high-order stage for augend/minuend *a[7:4]* and addend/subtrahend *b[7:4]*. The 9s complementer produces outputs *f[7:4]* which are connected to the *B* inputs of the adder for the tens decade. The mode control input is also connected to the input of the 9s complementer. The fixed-point adder generates an intermediate sum of *sum[7:4]* and a carry-out, *cout7*. The carry-out of the BCD adder/subtractor is labeled *cout* and determines whether 0000 or 0110 is added to the intermediate sum of the tens decade. The equation for *cout* is shown in Equation 9.4. The BCD adder/subtractor can be extended to operands of any size by simply adding more stages.

$$cout = cout7 + sum[7]\ sum[5] + sum[7]\ sum[6] \tag{9.4}$$

The module for the BCD adder/subtractor is shown in Figure 9.10 using structural modeling. There are two input operands, *a[7:0]* and *b[7:0]*, and one input mode control, *m*. There are two outputs: *bcd[7:0]*, which represents a valid BCD number, and a carry-out, *cout*. The test bench is shown in Figure 9.11 and contains operands for addition and subtraction, including numbers that result in negative differences in BCD. The outputs are shown in Figure 9.12 for both addition and subtraction and the waveforms are shown in Figure 9.13.

The two adders that generate valid BCD digits each have a carry-out signal, *cout*; however *cout* is not used because it provides no new information. The result of the BCD adder/subtractor is a carry-out, *cout*, that represents the hundreds decade, and two valid BCD digits, *bcd[7:4]* and *bcd[3:0]*, that represent the tens decade and units decade, respectively.

```
//structural bcd adder/subtractor
module add_sub_bcd2 (a, b, m, bcd, cout);

input [7:0] a, b;
input m;
output [7:0] bcd;
output cout;

//define internal nets
wire [7:0] f;
wire [7:0] sum;
wire cout3, aux_cy, cout7;
wire net3, net4, net9, net10;

//instantiate the logic for the low-order (units) stage [3:0]
//instantiate the 9s complementer
nines_compl inst1 (
   .m(m),
   .b(b[3:0]),
   .f(f[3:0])
   );

//instantiate the adder for intermediate sum for units stage
adder4 inst2 (
   .a(a[3:0]),
   .b(f[3:0]),
   .cin(m),
   .sum(sum[3:0]),
   .cout(cout3)
   );

//instantiate the logic gates
and2_df inst3 (
   .x1(sum[3]),
   .x2(sum[1]),
   .z1(net3)
   );

and2_df inst4 (
   .x1(sum[2]),
   .x2(sum[3]),
   .z1(net4)
   );


//continued on next page
```

Figure 9.10 Structural module for the BCD adder/subtractor.

```
or3_df inst5 (
   .x1(cout3),
   .x2(net3),
   .x3(net4),
   .z1(aux_cy)
   );

//instantiate the adder for the bcd sum [3:0]
adder4 inst6 (
   .a(sum[3:0]),
   .b({1'b0, aux_cy, aux_cy, 1'b0}),
   .cin(1'b0),
   .sum(bcd[3:0])
   );

//instantiate the logic for the high-order (tens) stage [7:4]
//instantiate the 9s complementer
nines_compl inst7 (
   .m(m),
   .b(b[7:4]),
   .f(f[7:4])
   );

//instantiate the adder for intermediate sum for tens stage
adder4 inst8 (
   .a(a[7:4]),
   .b(f[7:4]),
   .cin(aux_cy),
   .sum(sum[7:4]),
   .cout(cout7)
   );

//instantiate the logic gates
and2_df inst9 (
   .x1(sum[7]),
   .x2(sum[5]),
   .z1(net9)
   );

and2_df inst10 (
   .x1(sum[6]),
   .x2(sum[7]),
   .z1(net10)
   );


//continued on next page
```

Figure 9.10 (Continued)

```
or3_df inst11 (
   .x1(cout7),
   .x2(net9),
   .x3(net10),
   .z1(cout)
   );

//instantiate the adder for the bcd sum [7:4]
adder4 inst12 (
   .a(sum[7:4]),
   .b({1'b0, cout, cout, 1'b0}),
   .cin(1'b0),
   .sum(bcd[7:4])
   );

endmodule
```

Figure 9.10 (Continued)

```
//test bench for the bcd adder subtractor
module add_sub_bcd2_tb;

reg [7:0] a, b;
reg m;

wire [7:0] bcd;
wire cout;

//display variables
initial
$monitor ("a=%b, b=%b, m=%b, bcd_hund=%b, bcd_tens=%b,
            bcd_units=%b",
         a, b, m, {{3{1'b0}}, cout}, bcd[7:4], bcd[3:0]);

//apply input vectors
initial
begin
//add bcd
   #0    a = 8'b1001_1001;    b = 8'b0110_0110;    m = 1'b0;
   #10   a = 8'b0010_0110;    b = 8'b0101_1001;    m = 1'b0;
   #10   a = 8'b0001_0001;    b = 8'b0011_0011;    m = 1'b0;

//continued on next page
```

Figure 9.11 Test bench for the BCD adder/subtractor.

```
      #10    a = 8'b0000_1000;     b = 8'b0000_0101;     m = 1'b0;
      #10    a = 8'b0110_1000;     b = 8'b0011_0101;     m = 1'b0;
      #10    a = 8'b1000_1001;     b = 8'b0101_1001;     m = 1'b0;
      #10    a = 8'b1001_0110;     b = 8'b1001_0011;     m = 1'b0;
      #10    a = 8'b1001_1001;     b = 8'b0000_0001;     m = 1'b0;
      #10    a = 8'b0111_0111;     b = 8'b0111_0111;     m = 1'b0;

   //subtract bcd
      #10    a = 8'b1001_1001;     b = 8'b0110_0110;     m = 1'b1;
      #10    a = 8'b1001_1001;     b = 8'b0110_0110;     m = 1'b1;
      #10    a = 8'b0011_0011;     b = 8'b0110_0110;     m = 1'b1;
      #10    a = 8'b0111_0110;     b = 8'b0100_0010;     m = 1'b1;
      #10    a = 8'b0111_0110;     b = 8'b1000_0111;     m = 1'b1;
      #10    a = 8'b0001_0001;     b = 8'b1001_1001;     m = 1'b1;
      #10    a = 8'b0001_1000;     b = 8'b0010_0110;     m = 1'b1;
      #10    a = 8'b0001_1000;     b = 8'b0010_1000;     m = 1'b1;
      #10    a = 8'b1001_0100;     b = 8'b0111_1000;     m = 1'b1;
      #10    $stop;
   end

   add_sub_bcd2 inst1 (          //instantiate the module
      .a(a),
      .b(b),
      .m(m),
      .bcd(bcd),
      .cout(cout)
      );
   endmodule
```

Figure 9.11 (Continued)

```
Addition
a=1001_1001, b=0110_0110, m=0,
   bcd_hund=0001, bcd_tens=0110, bcd_units=0101//+165

a=0010_0110, b=0101_1001, m=0,
   bcd_hund=0000, bcd_tens=1000, bcd_units=0101//+85

a=0001_0001, b=0011_0011, m=0,
   bcd_hund=0000, bcd_tens=0100, bcd_units=0100//+44

a=0000_1000, b=0000_0101, m=0,
   bcd_hund=0000, bcd_tens=0001, bcd_units=0011//+13

//continued on next page
```

Figure 9.12 Outputs for the BCD adder/subtractor.

```
a=0110_1000, b=0011_0101, m=0,
   bcd_hund=0001, bcd_tens=0000, bcd_units=0011//+103

a=1000_1001, b=0101_1001, m=0,
   bcd_hund=0001, bcd_tens=0100, bcd_units=1000//+148

a=1001_0110, b=1001_0011, m=0,
   bcd_hund=0001, bcd_tens=1000, bcd_units=1001//+189

a=1001_1001, b=0000_0001, m=0,
   bcd_hund=0001, bcd_tens=0000, bcd_units=0000//+100

a=0111_0111, b=0111_0111, m=0,
   bcd_hund=0001, bcd_tens=0101, bcd_units=0100//+154
-------------------------------------------------------

Subtraction
If bcd_hund = 0001, then the result is a positive number.
If bcd_hund = 0000, then the result is a negative number
in 10s complement.

a=1001_1001, b=0110_0110, m=1,
   bcd_hund=0001, bcd_tens=0011, bcd_units=0011//133 = +33

a=0011_0011, b=0110_0110, m=1,
   bcd_hund=0000, bcd_tens=0110, bcd_units=0111//067 = -33

a=0111_0110, b=0100_0010, m=1,
   bcd_hund=0001, bcd_tens=0011, bcd_units=0100//134 = +34

a=0111_0110, b=1000_0111, m=1,
   bcd_hund=0000, bcd_tens=1000, bcd_units=1001//089 = -11

a=0001_0001, b=1001_1001, m=1,
   bcd_hund=0000, bcd_tens=0001, bcd_units=0010//012 = -88

a=0001_1000, b=0010_0110, m=1,
   bcd_hund=0000, bcd_tens=1001, bcd_units=0010//092 = -08

a=0001_1000, b=0010_1000, m=1,
   bcd_hund=0000, bcd_tens=1001, bcd_units=0000//090 = -10

a=1001_0100, b=0111_1000, m=1,
   bcd_hund=0001, bcd_tens=0001, bcd_units=0110//116 = +16
```

Figure 9.12 (Continued)

The outputs in Figure 9.12 conform to the examples shown previously in this section. That is, if the carry-out of the high-order decade is a 1, then the result is a positive number in BCD. If the carry-out of the high-order decade is a 0, then the result is a negative BCD number in 10s complement notation, which can be changed to a negative number in sign-magnitude notation by forming the 10s complement of the result and preceding it with a negative sign.

For example, the result of Example 9.4 was 1000 0101 0101 with a carry-out of 0 from the high-order decade, indicating a negative BCD number in 10s complement notation. The 10s complement of the result is 9 – 8 = 1, 9 – 5 = 4, 9 – 5 = 4 + 1 = 5. Therefore, the result in sign-magnitude notation is –145 (– 0001 0100 0101).

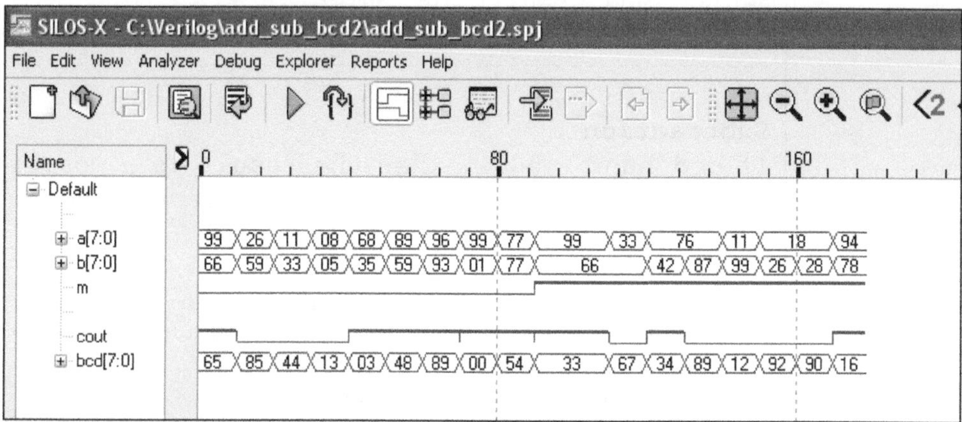

Figure 9.13 Waveforms for the BCD adder/subtractor.

9.3 Two-Decade Addition/Subtraction Unit for A+B, A–B, and B–A

It may sometimes be advantageous to have the BCD adder/subtractor perform the following operations: A+B, A–B, and B–A. This can be easily accomplished by utilizing a 9s complementer for both the augend/minuend inputs and the addend/subtrahend inputs to the adders that generate the intermediate sums. There must also be two mode control inputs: *m_a_minus_b* and *m_b_minus_a* that indicate whether the operation is A+B (both mode control inputs are deasserted); whether the operation is A–B (*m_a_minus_b* is asserted); or whether the operation is B–A (*m_b_minus_a* is asserted. Refer to Appendix A for design modules of select logic functions to be instantiated into the designs of this chapter.

In the examples that follow, recall that a carry-out of 1 from the high-order decade indicates a positive result; whereas, a carry-out of 0 from the high-order decade denotes a negative result in 10s complement representation. In this case, the result can be 10s complemented to obtain the result in sign-magnitude notation.

Example 9.7 The following decimal numbers will be subtracted using BCD arithmetic: +92 and +58, resulting in a difference of +34, as shown below. A carry-out of 1 from the high-order decade indicates a positive number.

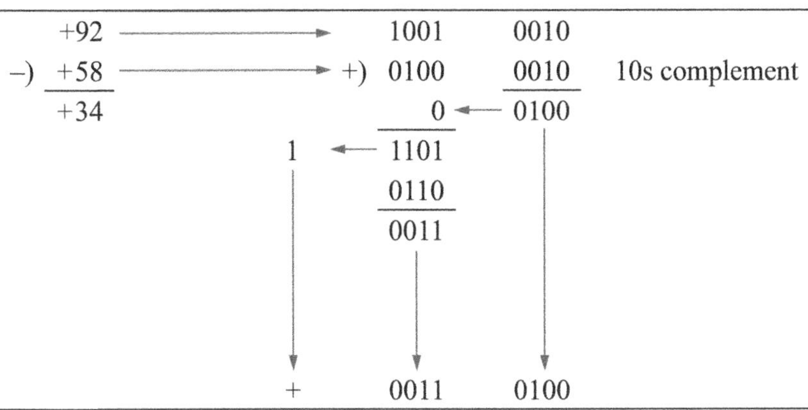

Example 9.8 The following decimal numbers will be subtracted using BCD arithmetic: +24 and +78, resulting in a difference of –54, as shown below. A carry-out of 0 from the high-order decade indicates a negative number in 10s complement representation.

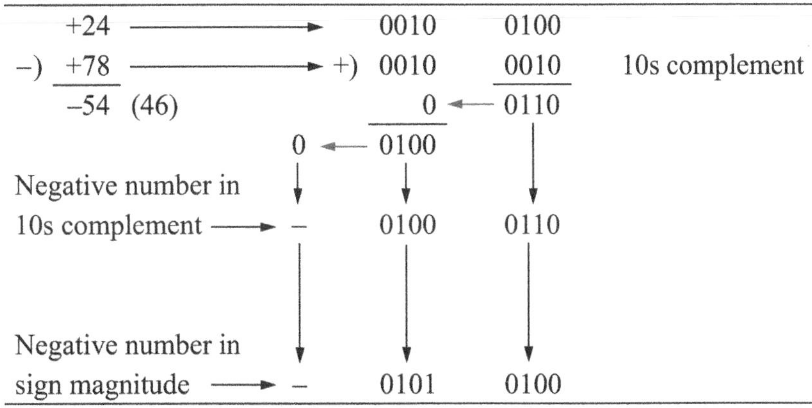

The organization and logic for this BCD adder/subtractor is shown in Figure 9.14, where the mode control inputs are ORed together and applied to the carry-in input of the units decade that generates the intermediate sum. The mode control inputs are connected to the appropriate 9s complementers as follows: if the operation is A+B, then both mode control inputs are deasserted; if the operation is A–B, then mode control input *m_a_minus_b* is asserted and connects to the 9s complementers for the B input of the units decade and tens decade of the adders; if the operation is B–A, then mode control input *m_b_minus_a* is asserted and connects to the 9s complementers for the A input of the adders.

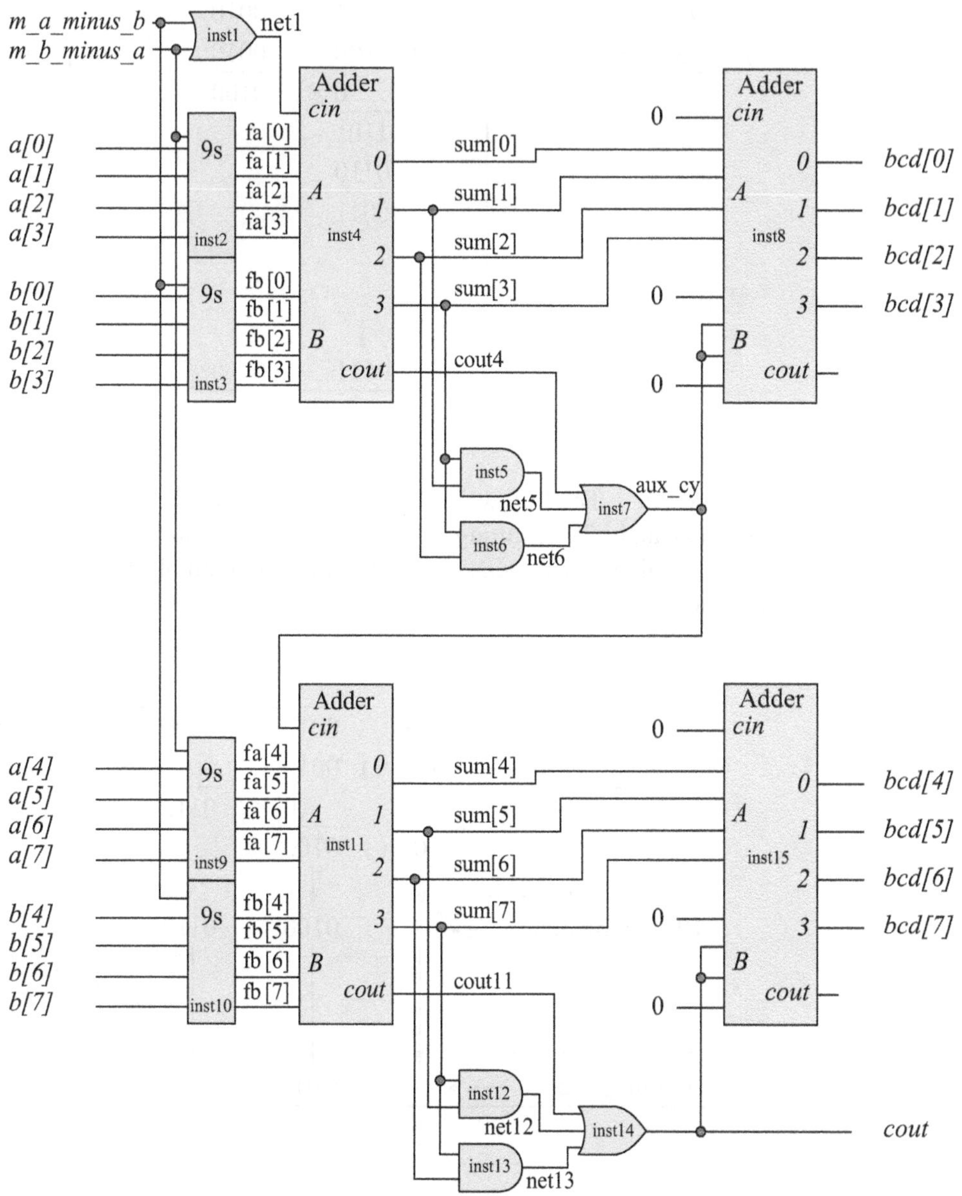

Figure 9.14 Logic diagram for a BCD adder/subtractor to implement the following operations: A+B, A–B, and B–A.

The structural module for the BCD adder/subtractor is shown in Figure 9.15 which instantiates the logic of Figure 9.14. The instantiation names and net names evinced in the structural module correspond to the names in the logic diagram. The test bench is shown in Figure 9.16 and applies input vectors for the operations of A+B, A–B, and B–A, including operands that produce negative results.

For example, the second set of input vectors for the operation A–B assigns the minuend a value of A = 0011 0011 (33) and the subtrahend a value of B = 0110 0110 (66) to yield a difference of 0000 0110 0111 (67) in 10s complement notation, or –33 in sign-magnitude notation. The outputs and waveforms are shown in Figure 9.17 and Figure 9.18, respectively.

```
//structural bcd adder subtractor
module add_sub_bcd3 (a, b, m_a_minus_b, m_b_minus_a,
                     bcd, cout);

input [7:0] a, b;
input m_a_minus_b, m_b_minus_a;
output [7:0] bcd;
output cout;

//define internal nets
wire [7:0] fa, fb;
wire [7:0] sum;
wire cout4, aux_cy, cout11;
wire net1, net5, net6, net12, net13;

//instantiate the logic for the low-order (units) stage [3:0]
or2_df inst1 (
   .x1(m_a_minus_b),
   .x2(m_b_minus_a),
   .z1(net1)
   );

//instantiate the 9s complementers
nines_compl inst2 (
   .m(m_b_minus_a),
   .b(a[3:0]),
   .f(fa[3:0])
   );

nines_compl inst3 (
   .m(m_a_minus_b),
   .b(b[3:0]),
   .f(fb[3:0])
   );             //continued on next page
```

Figure 9.15 Structural module for the adder/subtractor to execute the operations of A+B, A–B, and B–A.

```
//instantiate the adder for the intermediate sum
//for units stage
adder4 inst4 (
   .a(fa[3:0]),
   .b(fb[3:0]),
   .cin(net1),
   .sum(sum[3:0]),
   .cout(cout4)
   );

//instantiate the logic gates
and2_df inst5 (
   .x1(sum[3]),
   .x2(sum[1]),
   .z1(net5)
   );

and2_df inst6 (
   .x1(sum[3]),
   .x2(sum[2]),
   .z1(net6)
   );

or3_df inst7 (
   .x1(cout4),
   .x2(net5),
   .x3(net6),
   .z1(aux_cy)
   );

//instantiate the adder for the bcd sum [3:0]
adder4 inst8 (
   .a(sum[3:0]),
   .b({1'b0, aux_cy, aux_cy, 1'b0}),
   .cin(1'b0),
   .sum(bcd[3:0])
   );

//instantiate the logic for the high-order (tens) stage [7:4]
//instantiate the 9s complementers
nines_compl inst9 (
   .m(m_b_minus_a),
   .b(a[7:4]),
   .f(fa[7:4])
   );

//continued on next page
```

Figure 9.15 (Continued)

```
nines_compl inst10 (
    .m(m_a_minus_b),
    .b(b[7:4]),
    .f(fb[7:4])
    );

//instantiate the adder for the intermediate sum
//for tens stage
adder4 inst11 (
    .a(fa[7:4]),
    .b(fb[7:4]),
    .cin(aux_cy),
    .sum(sum[7:4]),
    .cout(cout11)
    );

//instantiate the logic gates
and2_df inst12 (
    .x1(sum[7]),
    .x2(sum[5]),
    .z1(net12)
    );

and2_df inst13 (
    .x1(sum[7]),
    .x2(sum[6]),
    .z1(net13)
    );

or3_df inst14 (
    .x1(cout11),
    .x2(net12),
    .x3(net13),
    .z1(cout)
    );

//instantiate the adder for the bcd sum [7:4]
adder4 inst15 (
    .a(sum[7:4]),
    .b({1'b0, cout, cout, 1'b0}),
    .cin(1'b0),
    .sum(bcd[7:4])
    );

endmodule
```

Figure 9.15 (Continued)

```
//test bench for the bcd adder subtractor
module add_sub_bcd3_tb;

reg [7:0] a, b;
reg m_a_minus_b, m_b_minus_a;

wire [7:0] bcd;
wire cout;

//display variables
initial
$monitor ("a=%b, b=%b, m_a_minus_b=%b, m_b_minus_a=%b,
            bcd_hund=%b, bcd_tens=%b, bcd_units=%b",
         a, b, m_a_minus_b, m_b_minus_a,
            {{3{1'b0}}, cout}, bcd[7:4], bcd[3:0]);


//apply input vectors
initial
begin
//add bcd
   #0     m_a_minus_b = 1'b0;  m_b_minus_a=1'b0;
          a = 8'b1001_1001;    b = 8'b0110_0110;
   #10    a = 8'b0010_0110;    b = 8'b0101_1001;
   #10    a = 8'b0001_0001;    b = 8'b0011_0011;
   #10    a = 8'b0000_1000;    b = 8'b0000_0101;
   #10    a = 8'b0110_1000;    b = 8'b0011_0101;
   #10    a = 8'b1000_1001;    b = 8'b0101_1001;
   #10    a = 8'b1001_0110;    b = 8'b1001_0011;
   #10    a = 8'b1001_1001;    b = 8'b0000_0001;
   #10    a = 8'b0111_0111;    b = 8'b0111_0111;



//subtract bcd a-b
   #10    m_a_minus_b = 1'b1;  m_b_minus_a=1'b0;
          a = 8'b1001_1001;    b = 8'b0110_0110;
   #10    a = 8'b0011_0011;    b = 8'b0110_0110;
   #10    a = 8'b0111_0110;    b = 8'b0100_0010;
   #10    a = 8'b0111_0110;    b = 8'b1000_0111;
   #10    a = 8'b0001_0001;    b = 8'b1001_1001;
   #10    a = 8'b0001_1000;    b = 8'b0010_0110;
   #10    a = 8'b0001_1000;    b = 8'b0010_1000;
   #10    a = 8'b1001_0100;    b = 8'b0111_1000;

//continued on next page
```

Figure 9.16 Test bench for the BCD adder/subtractor to execute the operations of A+B, A–B, and B–A.

```
//subtract bcd b-a
    #10     m_a_minus_b = 1'b0;   m_b_minus_a=1'b1;
            a = 8'b1001_1001;     b = 8'b0110_0110;
    #10     a = 8'b1001_1001;     b = 8'b0110_0110;
    #10     a = 8'b0011_0011;     b = 8'b0110_0110;
    #10     a = 8'b0111_0110;     b = 8'b0100_0010;
    #10     a = 8'b0111_0110;     b = 8'b1000_0111;
    #10     a = 8'b0001_0001;     b = 8'b1001_1001;
    #10     a = 8'b0001_1000;     b = 8'b0010_0110;
    #10     a = 8'b0001_1000;     b = 8'b0010_1000;
    #10     a = 8'b1001_0100;     b = 8'b0111_1000;

    #10     $stop;

end

//instantiate the module into the test bench
add_sub_bcd3 inst1 (
    .a(a),
    .b(b),
    .m_a_minus_b(m_a_minus_b),
    .m_b_minus_a(m_b_minus_a),
    .bcd(bcd),
    .cout(cout)
    );
endmodule
```

Figure 9.16 (Continued)

```
Addition a + b

a=1001_1001, b=0110_0110, m_a_minus_b=0, m_b_minus_a=0,
   bcd_hund=0001, bcd_tens=0110, bcd_units=0101

a=0010_0110, b=0101_1001, m_a_minus_b=0, m_b_minus_a=0,
   bcd_hund=0000, bcd_tens=1000, bcd_units=0101

a=0001_0001, b=0011_0011, m_a_minus_b=0, m_b_minus_a=0,
   bcd_hund=0000, bcd_tens=0100, bcd_units=0100

a=0000_1000, b=0000_0101, m_a_minus_b=0, m_b_minus_a=0,
   bcd_hund=0000, bcd_tens=0001, bcd_units=0011

a=0110_1000, b=0011_0101, m_a_minus_b=0, m_b_minus_a=0,
   bcd_hund=0001, bcd_tens=0000, bcd_units=0011 //next page
```

Figure 9.17 Outputs for the BCD adder/subtractor to execute the operations of A+B, A–B, and B–A.

```
a=0110_1000, b=0011_0101, m_a_minus_b=0, m_b_minus_a=0,
   bcd_hund=0001, bcd_tens=0000, bcd_units=0011

a=1000_1001, b=0101_1001, m_a_minus_b=0, m_b_minus_a=0,
   bcd_hund=0001, bcd_tens=0100, bcd_units=1000

a=1001_0110, b=1001_0011, m_a_minus_b=0, m_b_minus_a=0,
   bcd_hund=0001, bcd_tens=1000, bcd_units=1001

a=1001_1001, b=0000_0001, m_a_minus_b=0, m_b_minus_a=0,
   bcd_hund=0001, bcd_tens=0000, bcd_units=0000

a=0111_0111, b=0111_0111, m_a_minus_b=0, m_b_minus_a=0,
   bcd_hund=0001, bcd_tens=0101, bcd_units=0100
-------------------------------------------------------
```

Subtraction a - b

```
If bcd_hund = 0001, then the result is a positive number.
If bcd_hund = 0000, then the result is a negative number
in 10s complement.

a=1001_1001, b=0110_0110, m_a_minus_b=1, m_b_minus_a=0,
   bcd_hund=0001, bcd_tens=0011, bcd_units=0011

a=0011_0011, b=0110_0110, m_a_minus_b=1, m_b_minus_a=0,
   bcd_hund=0000, bcd_tens=0110, bcd_units=0111

a=0111_0110, b=0100_0010, m_a_minus_b=1, m_b_minus_a=0,
   bcd_hund=0001, bcd_tens=0011, bcd_units=0100

a=0111_0110, b=1000_0111, m_a_minus_b=1, m_b_minus_a=0,
   bcd_hund=0000, bcd_tens=1000, bcd_units=1001

a=0001_0001, b=1001_1001, m_a_minus_b=1, m_b_minus_a=0,
   bcd_hund=0000, bcd_tens=0001, bcd_units=0010

a=0001_1000, b=0010_0110, m_a_minus_b=1, m_b_minus_a=0,
   bcd_hund=0000, bcd_tens=1001, bcd_units=0010

a=0001_1000, b=0010_1000, m_a_minus_b=1, m_b_minus_a=0,
   bcd_hund=0000, bcd_tens=1001, bcd_units=0000

a=1001_0100, b=0111_1000, m_a_minus_b=1, m_b_minus_a=0,
   bcd_hund=0001, bcd_tens=0001, bcd_units=0110
-------------------------------------------------------
//continued on next page
```

Figure 9.17 (Continued)

```
Subtraction b - a

If bcd_hund = 0001, then the result is a positive number.
If bcd_hund = 0000, then the result is a negative number
in 10s complement.

a=1001_1001, b=0110_0110, m_a_minus_b=0, m_b_minus_a=1,
   bcd_hund=0000, bcd_tens=0110, bcd_units=0111

a=0011_0011, b=0110_0110, m_a_minus_b=0, m_b_minus_a=1,
   bcd_hund=0001, bcd_tens=0011, bcd_units=0011

a=0111_0110, b=0100_0010, m_a_minus_b=0, m_b_minus_a=1,
   bcd_hund=0000, bcd_tens=0110, bcd_units=0110

a=0111_0110, b=1000_0111, m_a_minus_b=0, m_b_minus_a=1,
   bcd_hund=0001, bcd_tens=0001, bcd_units=0001

a=0001_0001, b=1001_1001, m_a_minus_b=0, m_b_minus_a=1,
   bcd_hund=0001, bcd_tens=1000, bcd_units=1000

a=0001_1000, b=0010_0110, m_a_minus_b=0, m_b_minus_a=1,
   bcd_hund=0001, bcd_tens=0000, bcd_units=1000

a=0001_1000, b=0010_1000, m_a_minus_b=0, m_b_minus_a=1,
   bcd_hund=0001, bcd_tens=0001, bcd_units=0000

a=1001_0100, b=0111_1000, m_a_minus_b=0, m_b_minus_a=1,
   bcd_hund=0000, bcd_tens=1000, bcd_units=0100
```

Figure 9.17 (Continued)

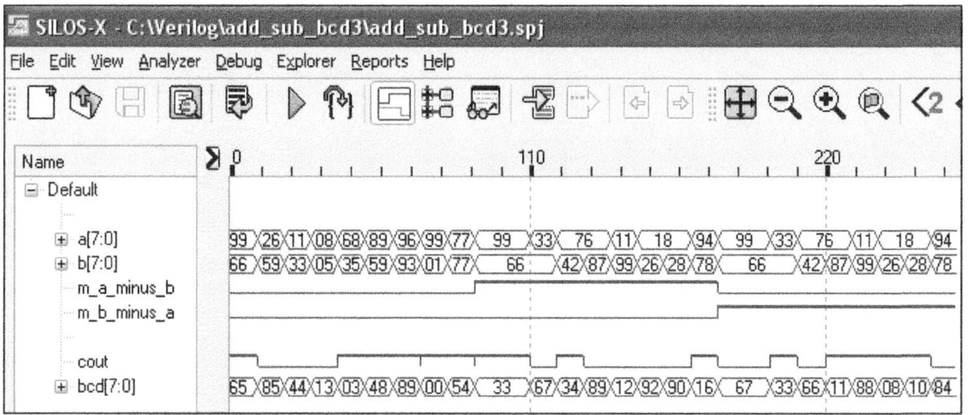

Figure 9.18 Waveforms for the BCD adder/subtractor to execute the operations of A+B, A–B, and B–A.

9.4 Problems

9.1 Perform the following decimal subtraction using the paper-and-pencil method: (+72) – (+56).

9.2 Perform the following decimal subtraction using the paper-and-pencil method: (+98) – (+120).

9.3 Perform the following decimal subtraction using the paper-and-pencil method: (+456) – (+230).

9.4 Perform the following decimal subtraction using the paper-and-pencil method: (+3476) – (+4532).

9.5 Perform the following decimal subtraction using the paper-and-pencil method: (+436) + (–825).

9.6 Use structural modeling to design a 2-digit BCD adder/subtractor using multiplexers. Obtain the test bench that provides several input vectors for the augend/minuend and the addend/subtrahend. Obtain the outputs and the waveforms.

9.7 Use structural modeling to design a 4-digit BCD adder/subtractor. Obtain the test bench that provides several input vectors for the augend/minuend and the addend/subtrahend. Obtain the outputs and the waveforms.

10.1 Binary-to-BCD Conversion
10.2 Multiplication Using Behavioral Modeling
10.3 Multiplication Using Structural Modeling
10.4 Multiplication Using Memory
10.5 Multiplication Using Table Lookup
10.6 Problems

10

Decimal Multiplication

The algorithms for decimal multiplication are more complex than those for fixed-point multiplication. This is because decimal digits consist of four bits and have values in the range of 0 to 9, whereas fixed-point digits have values of 0 or 1. Decimal multiplication can be achieved by a process of repeated addition in which the multiplicand is added to the previous partial product a number of times that is equal to the multiplier. This, however, is relatively slow process and will not be considered in this chapter.

It was stated in Chapter 8 that decimal arithmetic operations can be performed in the fixed-point number representation in which the result is converted to binary-coded decimal (BCD) if the amount of decimal data is not excessive. This is the method that is selected for the first two techniques, which are a fixed-point multiplication and conversion to BCD using behavioral modeling and structural modeling.

10.1 Binary-to-BCD Conversion

Converting from binary to BCD is accomplished by multiplying the BCD number by two repeatedly. Multiplying by two is accomplished by a left shift of one bit position followed by an adjustment, if necessary. For example, a left shift of BCD 1001 (9_{10}) results in 1 0010 which is 18 in binary, but only 12 in BCD. Adding six to the low-order BCD digit results in 1 1000, which is the required value of 18.

Instead of adding six after the shift, the same result can be obtained by adding three before the shift since a left shift multiplies any number by two. BCD digits in the range 0–4 do not require an adjustment after being shifted left, because the shifted number will be in the range 0–8, which can be contained in a 4-bit BCD digit. However, if the number to be shifted is in the range 5–9, then an adjustment will be required after the left shift, because the shifted number will be in the range 10–18, which requires two BCD digits. Therefore, three is added to the digit prior to being shifted left 1-bit position.

Figure 10.1 shows the procedure for converting from binary $0011\ 0111\ 1011_2$ (891_{10}) to BCD. Since there are 12 bits in the binary number, 12 left-shift operations are required, yielding the resulting BCD number of $1000\ 1001\ 0001_{BCD}$. Concatenated registers *A* and *B* are shifted left one bit position during each sequence. During the final left shift operation, no adjustment is performed.

	A Register (BCD)				*B* Register (Binary)		
	10^2	10^1	10^0				
	23...20	19...16	15...12		11 ... 8	7 ... 4	3 ... 0
Shift register = {*A* , *B*}	0000	0000	0000	←	0011	0111	1011
Shift left 1, no addition	0000	0000	0000		0110	1111	0110
Shift left 1, no addition	0000	0000	0000		1101	1110	1100
Shift left 1, no addition	0000	0000	0001		1011	1101	1000
Shift left 1, no addition	0000	0000	0011		0111	1011	0000
Shift left 1	0000	0000	0110		1111	0110	0000
Add 3			0011				
	0000	0000	1001		1111	0110	0000
Shift left 1, no addition	0000	0001	0011		1110	1100	0000
Shift left 1	0000	0010	0111		1101	1000	0000
Add 3			0011				
	0000	0010	1010		1101	1000	0000
Shift left 1	0000	0101	0101		1011	0000	0000
Add 3		0011	0011				
	0000	1000	1000		1011	0000	0000
Shift left 1, no addition	0001	0001	0001		0110	0000	0000
Shift left 1, no addition	0010	0010	0010		1100	0000	0000
Shift left 1	0100	0100	0101		1000	0000	0000
Add 3			0011				
	0100	0100	1000		1000	0000	0000
Shift left 1	1000	1001	0001		0000	0000	0000

Figure 10.1 Example of binary-to-BCD conversion.

10.2 Multiplication Using Behavioral Modeling

The algorithm shown in Figure 10.1 will be used for BCD multiplication by performing the multiply operation in the fixed-point number representation, and then converting the product to BCD notation. The design will be implemented using behavioral modeling. A 24-bit left-shift register — consisting of two 12-bit registers *A*, *a_reg*, and *B*, *b-reg*, in concatenation — is used for the shifting sequence.

A shift counter is used to determine the number of shift sequences to be executed. Since the final shift sequence is a left-shift operation only (no adjustment), the shift counter is set to a value of the binary length minus one; then a final left shift operation occurs. A **while** loop determines the number times that the procedural statements within the loop are executed and is a function of the shift counter value.

The **while** loop executes a procedural statement or a block of procedural statements as long as a Boolean expression returns a value of true (≥ 1). When the procedural statements are executed, the Boolean expression is reevaluated. The loop is executed until the expression returns a value of false, in this case a shift counter value of zero. If the evaluation of the expression is false, then the **while** loop is terminated and control is passed to the next statement in the module. If the expression is false before the loop is initially entered, then the **while** loop is never executed.

The Boolean expression may contain the following types of statements: arithmetic, logical, relational, equality, bitwise, reduction, shift, concatenation, replication, or conditional. If the **while** loop contains multiple procedural statements, then they are contained within the **begin** . . . **end** keywords. The syntax for a **while** statement is as follows:

while (expression)
procedural statement or block of procedural statements

Registers *A* and *B* are shifted left in concatenation by the following statement:

shift_reg = shift_reg << 1;

Then the three 4-bit segments of register *A* (*shift_reg[15:12]*, *shift_reg[19:16]*, and *shift_reg[23:20]*) are checked to determine if any of the three segments exceed a value of four (0100). If any segment is in the range 5–9, then a value of three is added to the segment and the shifting sequences continue.

The behavioral module is shown in Figure 10.2, where register *A* is reset to all zeroes and register *B* contains the product of the multiplicand and the multiplier. The test bench is shown in Figure 10.3, in which several input vectors are applied to the multiplicand and the multiplier, including binary and decimal values. The outputs are shown in Figure 10.4 and the waveforms are shown in Figure 10.5.

```
//behavioral bcd multiplier
module mul_bcd_behav (a, b, bcd);

input [7:0] a;
input [3:0] b;
output [11:0] bcd;

reg [11:0] a_reg, b_reg;
reg [23:0] shift_reg;
reg [3:0] shift_ctr;

always @ (a or b)
begin
   shift_ctr = 4'b1011;        //11 shift sequences
   a_reg = 12'b0000_0000_0000;//reset register a
   b_reg = a * b;              //register b contains product

   shift_reg = {a_reg, b_reg};

   while (shift_ctr)
      begin
         shift_reg = shift_reg << 1;
            if (shift_reg[15:12] > 4'b0100)
               shift_reg[15:12] = shift_reg[15:12] + 4'b0011;

            if (shift_reg[19:16] > 4'b0100)
               shift_reg[19:16] = shift_reg[19:16] + 4'b0011;

            if (shift_reg[23:20] > 4'b0100)
               shift_reg[23:20] = shift_reg[23:20] + 4'b0011;

         shift_ctr = shift_ctr - 1;

      end

         shift_reg = shift_reg << 1;
end

assign bcd = shift_reg[23:12];

endmodule
```

Figure 10.2 Behavioral module for BCD multiplication for an 8-bit multiplicand and a 4-bit multiplier using fixed-point multiplication, then converting the product to BCD.

```
//test bench for bcd multiplier
module mul_bcd_behav_tb;

reg [7:0] a;
reg [3:0] b;
wire [11:0] bcd;

//display variables
initial
$monitor ("a=%b, b=%b, bcd=%b", a, b, bcd);

//apply input vectors
initial
begin
          //99 x 9 = 891
   #0     a = 8'b0110_0011;     b = 4'b1001;

          //45 x 6 = 270
   #10    a = 8'b0010_1101;     b = 4'b0110;

          //77 x 7 = 539
   #10    a = 8'b0100_1101;     b = 4'b0111;

          //89 x 8 = 712
   #10    a = 8'b0101_1001;     b = 4'b1000;

          //37 x 6 = 222
   #10    a = 8'b0010_0101;     b = 4'b0110;

          //99 x 9 = 891
   #10    a = 8'd99;            b = 4'd9;

          //37 x 6 = 222
   #10    a = 8'd37;            b = 4'd6;

   #10    $stop;

end

//instantiate the module into the test bench
mul_bcd_behav inst1 (
   .a(a),
   .b(b),
   .bcd(bcd)
   );

endmodule
```

Figure 10.3 Test bench for the BCD multiplier.

```
a=0110_0011, b=1001, bcd=1000_1001_0001
a=0010_1101, b=0110, bcd=0010_0111_0000
a=0100_1101, b=0111, bcd=0101_0011_1001
a=0101_1001, b=1000, bcd=0111_0001_0010
a=0010_0101, b=0110, bcd=0010_0010_0010
a=0110_0011, b=1001, bcd=1000_1001_0001
a=0010_0101, b=0110, bcd=0010_0010_0010
```

Figure 10.4 Outputs for the BCD multiplier.

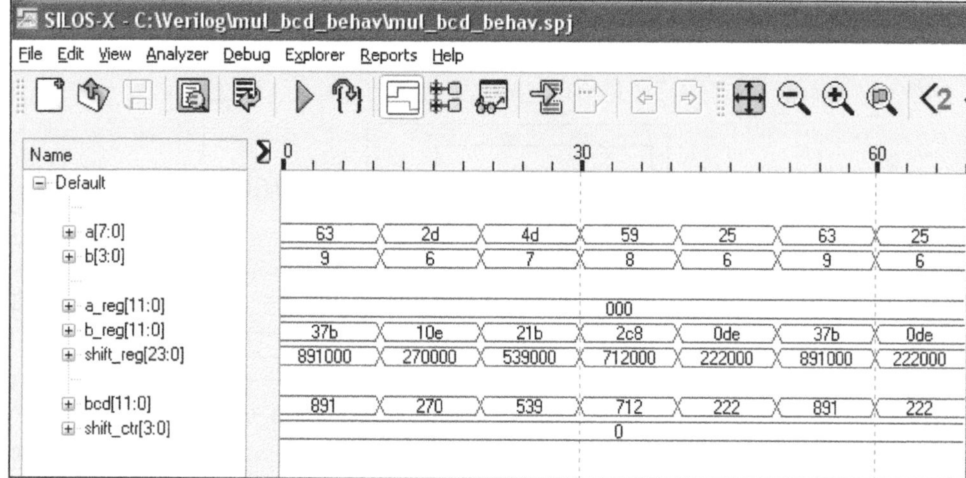

Figure 10.5 Waveforms for the BCD multiplier.

10.3 Multiplication Using Structural Modeling

Behavioral modeling does not require that the logic design of the project be completed before implementing it with Verilog HDL — simply specify the behavior of the machine and leave the logic design details to the synthesis tool. Structural modeling, however, requires that the project be designed in detail, including logic gates and other logic elements. These logic functions are then instantiated into a structural module.

This section describes a method to obtain a BCD product by multiplying the two operands using fixed-point multiplication and then changing the result to BCD, similar to the method described in Section 10.2, but using structural modeling. The logic diagram is shown in Figure 10.6.

The fixed-point multiplier is shown in instantiation *inst1* that multiplies an 8-bit multiplicand, *a[7:0]*, by a 4-bit multiplier, *b[3:0]*, to yield a 12-bit product, *prod[11:0]*. There are two 12-bit registers that shift left in concatenation. The binary product that is to be converted to BCD is loaded into register *B* (*b_reg[11:0]*). Register *A* (*a_reg[11:0]*) contains the data that is in the process of being converted to BCD. Instantiation *inst4*, labelled >4, contains the logic to detect a 4-bit value greater than four.

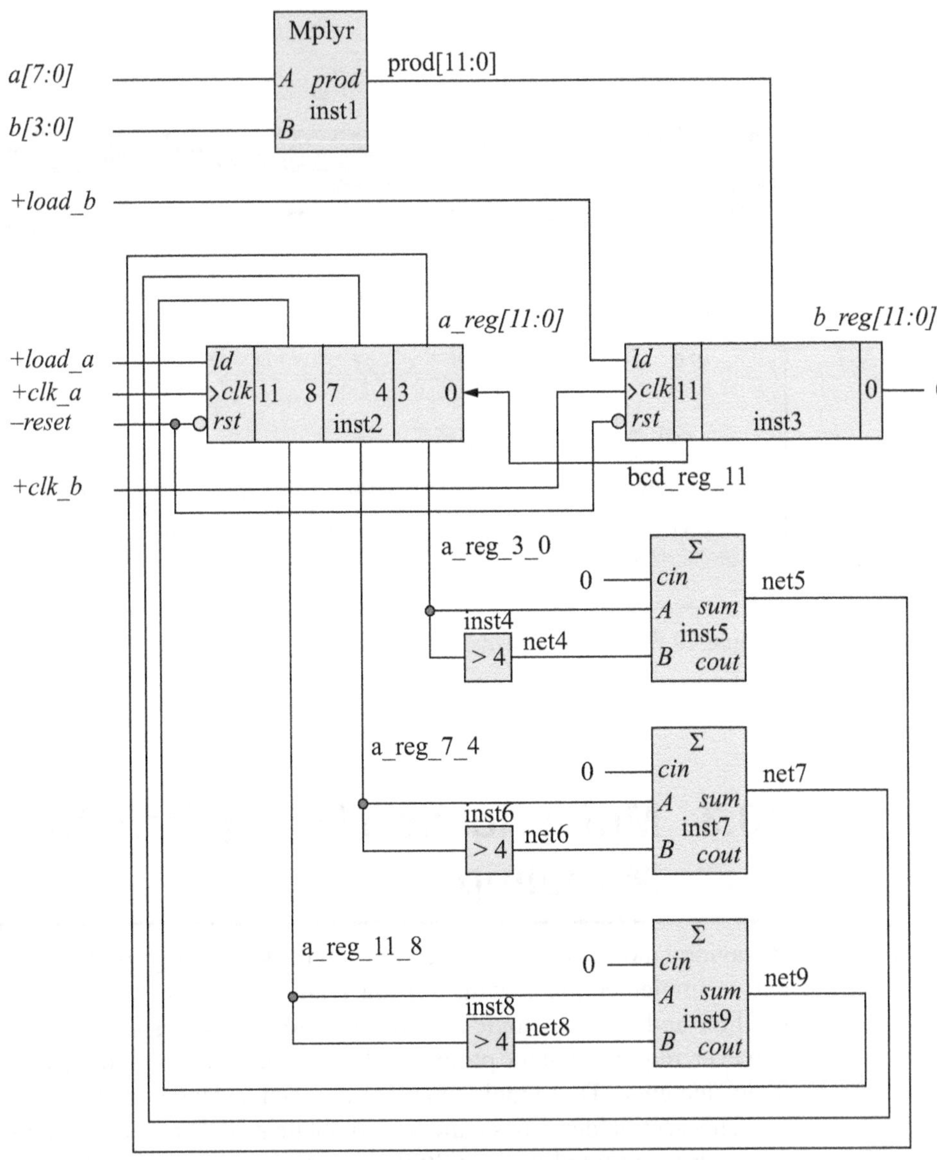

Figure 10.6 Logic diagram to obtain a BCD product from a fixed-point multiplication.

Two load signals are required: one to load register *B* (*load_b*) with the product at the beginning of the operation, and one to load register *A* (*load_a*) with the unadjusted or adjusted sum. The *load_b* signal occurs only once; the *load_a* signal occurs whenever register *A* is to be loaded with a new sum from the three adders. This sum is either a sum without adjustment — a value of three is not added because the sum is less than or equal to four; or the sum with adjustment — a value of three is added because the sum is greater than four.

There are also two clock signals: *clk_b* and *clk_a*. The *clk_b* signal is used to load register *B* with the product when the *load_b* signal is active, and also to shift register *B* left when the *load_b* signal is inactive. The *clk_a* signal is used to load register *A* with the sum from the three adders when the *load_a* signal is active, and also to shift register *A* when the *load_a* signal is inactive. Two load inputs are required, because registers *A* and *B* are loaded at different times. Two clock inputs are required, because if the same clock were applied to registers *A* and *B*, then when register *A* was loaded with a new sum, register *B* would be shifted left providing incorrect data for the following sequence.

Since a multiply operation provides a $2n$-bit product, there are 12 bits in register *B*. Therefore, there will be 12 shift sequences. The first 11 shift sequences consist of a left shift of registers *A* and *B* and an add operation to add zero or three to the segments in register *A*, depending on whether the values of the segments are greater than four. The twelfth and final shift occurs with no adjustment, regardless of the values of the individual segments. After the twelfth shift, the BCD product is in register *A*. Figure 10.1 can be used to follow the Verilog HDL code in conjunction with the logic diagram of Figure 10.6. Figure 10.7 shows the waveforms depicting the loading of register *B* and a representative sequence.

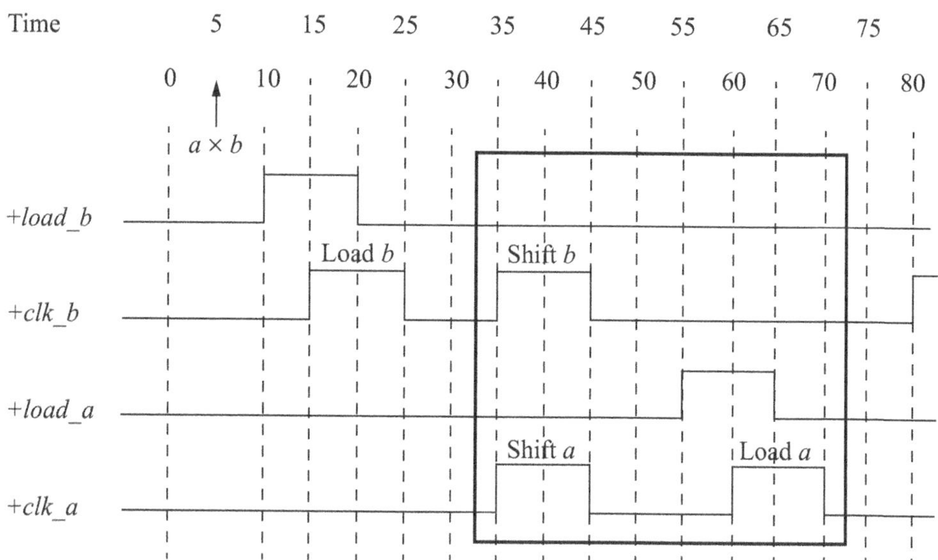

Figure 10.7 Representative sequence for structural BCD multiplication.

The behavioral module for the 8 × 4 multiplier is shown in Figure 10.8. The behavioral module for the 12-bit parallel-in, serial-out left shift register is shown in Figure 10.9, which will be used for registers *A* and *B*. The dataflow module for the greater-than-four (> 4) circuit is shown in Figure 10.10. The design of this circuit is obtained from a 4-variable Karnaugh map with an entry of 1 in all minterm locations that are greater than four. The test bench and outputs for the greater-than-four circuit are shown in Figure 10.11 and Figure 10.12, respectively. The behavioral module for the 4-bit adder is shown in Figure 10.13.

```
//behavioral 8 bits times 4 bits
module mul_8x4 (a, b, prod);

input [7:0] a;
input [3:0] b;
output [11:0] prod;

reg [11:0] prod;

always @ (a or b)
begin
   prod = a * b;
end
endmodule
```

Figure 10.8 Behavioral module for the 8 × 4 multiplier.

```
//behavioral parallel-in, serial-out left shift register
module shift_left_reg_piso12 (rst_n, clk, ser_in,
                              load, data_in, q);

input rst_n, clk, ser_in, load;
input [11:0] data_in;
output [11:0] q;

reg [11:0] q;

always @ (rst_n)
begin
   if (rst_n == 0)
      q <= 12'h000;
end
//continued on next page
```

Figure 10.9 Parallel-in, serial-out shift register.

```
always @ (posedge clk)
begin
   q[11] <= ((load && data_in[11]) || (~load && q[10]));
   q[10] <= ((load && data_in[10]) || (~load && q[9]));
   q[9]  <= ((load && data_in[9])  || (~load && q[8]));
   q[8]  <= ((load && data_in[8])  || (~load && q[7]));
   q[7]  <= ((load && data_in[7])  || (~load && q[6]));
   q[6]  <= ((load && data_in[6])  || (~load && q[5]));
   q[5]  <= ((load && data_in[5])  || (~load && q[4]));
   q[4]  <= ((load && data_in[4])  || (~load && q[3]));
   q[3]  <= ((load && data_in[3])  || (~load && q[2]));
   q[2]  <= ((load && data_in[2])  || (~load && q[1]));
   q[1]  <= ((load && data_in[1])  || (~load && q[0]));
   q[0]  <= ((load && data_in[0])  || (~load && ser_in));

end

endmodule
```

Figure 10.9 (Continued)

```
//dataflow greater than four detector
module gt_4_df (x, z);

input [3:0] x;
output [3:0] z;

wire [3:0] x;
wire [3:0] z;

assign   z[3] = 1'b0,
         z[2] = 1'b0,
         z[1] = x[3] | (x[2] & x[0]) | (x[2] & x[1]),
         z[0] = x[3] | (x[2] & x[0]) | (x[2] & x[1]);

endmodule
```

Figure 10.10 Dataflow module to detect a value greater than four.

```
//test bench for greater than four detector
module gt_4_df_tb;

reg [3:0] x;
wire [3:0] z;

//apply input vectors and display variables
initial
begin: apply_stimulus
   reg [4:0] invect;
      for (invect = 0; invect < 16; invect = invect + 1)
      begin
         x = invect [4:0];
         #10 $display ("x3 x2 x1 x0 = %b, z3 z2 z1 z0 = %b",
                        x, z);
      end
end

//instantiate the module into the test bench
gt_4_df inst1 (
   .x(x),
   .z(z)
   );

endmodule
```

Figure 10.11 Test bench for the greater-than-four circuit.

```
x3 x2 x1 x0 = 0000, z3 z2 z1 z0 = 0000
x3 x2 x1 x0 = 0001, z3 z2 z1 z0 = 0000
x3 x2 x1 x0 = 0010, z3 z2 z1 z0 = 0000
x3 x2 x1 x0 = 0011, z3 z2 z1 z0 = 0000
x3 x2 x1 x0 = 0100, z3 z2 z1 z0 = 0000
x3 x2 x1 x0 = 0101, z3 z2 z1 z0 = 0011
x3 x2 x1 x0 = 0110, z3 z2 z1 z0 = 0011
x3 x2 x1 x0 = 0111, z3 z2 z1 z0 = 0011
x3 x2 x1 x0 = 1000, z3 z2 z1 z0 = 0011
x3 x2 x1 x0 = 1001, z3 z2 z1 z0 = 0011
x3 x2 x1 x0 = 1010, z3 z2 z1 z0 = 0011
x3 x2 x1 x0 = 1011, z3 z2 z1 z0 = 0011
x3 x2 x1 x0 = 1100, z3 z2 z1 z0 = 0011
x3 x2 x1 x0 = 1101, z3 z2 z1 z0 = 0011
x3 x2 x1 x0 = 1110, z3 z2 z1 z0 = 0011
x3 x2 x1 x0 = 1111, z3 z2 z1 z0 = 0011
```

Figure 10.12 Outputs for the greater-than-four circuit.

```
//behavioral model for a 4-bit adder
module adder4 (a, b, cin, sum, cout);

input [3:0] a, b;
input cin;
output [3:0] sum;
output cout;

wire [3:0] a, b;
wire cin;
reg [3:0] sum;
reg cout;

always @ (a or b or cin)
begin
   sum  = a + b + cin;
   cout = (a[3] & b[3]) |
          ((a[3] | b[3]) & (a[2] & b[2])) |
          ((a[3] | b[3]) & (a[2] | b[2]) & (a[1] & b[1])) |
          ((a[3] | b[3]) & (a[2] | b[2]) & (a[1] | b[1]) &
                (a[0] & b[0])) |
          ((a[3] | b[3]) & (a[2] | b[2]) & (a[1] | b[1]) &
                (a[0] | b[0]) & cin);
end
endmodule
```

Figure 10.13 Behavioral module for a 4-bit adder.

The structural module for the multiplier is shown in Figure 10.14, which instantiates the design modules for the multiplier, the shift register, the greater-than-four circuit, and the 4-bit adder. The instantiation names and the net names correspond to those shown in the logic diagram of Figure 10.6.

```
//structural bcd multiplier
module mul_bcd_struc (rst_n, clk_a, clk_b, load_a, load_b,
                        a, b);
input [7:0] a;
input [3:0] b;
input rst_n, clk_a, clk_b, load_a, load_b;

//define internal nets
wire [11:0] prod;
wire [3:0] a_reg_11_8, a_reg_7_4, a_reg_3_0;
wire [3:0] net4, net5, net6, net7, net8, net9;
wire [11:0] b_reg;      //continued on next page
```

Figure 10.14 Structural module for BCD multiplier.

```
//define internal register
reg [11:0] a_reg;

assign b_reg_11 = b_reg[11];

//instantiate the multiplier
mul_8x4 inst1 (
   .a(a),
   .b(b),
   .prod(prod)
   );

//instantiate the parallel-in, serial-out
//11-bit left-shift register a_reg
shift_left_reg_piso12 inst2 (
   .rst_n(rst_n),
   .clk(clk_a),
   .ser_in(b_reg_11),
   .load(load_a),
   .data_in({net9, net7, net5}),
   .q({a_reg_11_8, a_reg_7_4, a_reg_3_0})
   );

//instantiate the parallel-in, serial-out
//11-bit left-shift register b_reg
shift_left_reg_piso12 inst3 (
   .rst_n(rst_n),
   .clk(clk_b),
   .ser_in(1'b0),
   .load(load_b),
   .data_in(prod),
   .q(b_reg)
   );

//instantiate the greater-than-four detectors
//and the 4-bit adders
gt_4_df inst4 (
   .x(a_reg_3_0),
   .z(net4)
   );

adder4 inst5 (
   .a(a_reg_3_0),
   .b(net4),
   .cin(1'b0),
   .sum(net5)
   );                  //continued on next page
```

Figure 10.14 (Continued)

```
gt_4_df inst6 (
    .x(a_reg_7_4),
    .z(net6)
    );

adder4 inst7 (
    .a(a_reg_7_4),
    .b(net6),
    .cin(1'b0),
    .sum(net7)
    );

gt_4_df inst8 (
    .x(a_reg_11_8),
    .z(net8)
    );

adder4 inst9 (
    .a(a_reg_11_8),
    .b(net8),
    .cin(1'b0),
    .sum(net9)
    );
endmodule
```

Figure 10.14 (Continued)

The test bench is shown in Figure 10.15 which applies the load and clock inputs to registers *A* and *B* at selected times to correspond to the loading and shifting of the registers. Alternatively, the load and clock pulses could be applied to the module by means of an **initial** statement using a **forever** loop.

An **initial** block begins at time zero, executes the statements in the block only once, then suspends execution. Multiple **initial** blocks in the same module all execute concurrently at time zero. If there is only one behavioral statement in the **initial** block, then the statements do not have to be grouped. If multiple behavioral statements are contained in the **initial** block, then the statements are enclosed within **begin** . . . **end** delimiters. Data types within an **initial** block must be declared as registers. The **initial** blocks are typically used for module initialization and monitoring.

The **forever** loop statement executes the procedural statements continuously. The loop is primarily used for timing control constructs, such as clock pulse generation. The **forever** procedural statement must be contained within an **initial** or an **always** block. An **always** statement executes at the beginning of simulation; the **forever** statement executes only when it is encountered in a procedural block. When using an **always** statement in this manner, care must be taken to assure that the pulses occur at the correct times; otherwise, the circuit will not operate correctly.

```
//test bench for structural bcd multiplier
module mul_bcd_struc_tb;

reg [7:0] a;
reg [3:0] b;
reg rst_n, clk_a, clk_b, load_a, load_b;

//apply input vectors
initial
begin
   #0    rst_n = 1'b0;
         a = 8'b0000_0000;    b = 8'b0000_0000;
         clk_a = 1'b0;        clk_b = 1'b0;
         load_a = 1'b0;       load_b = 1'b0;
   #5    rst_n = 1'b1;              //#5

         //99 x 9 = 891
         a = 8'b0110_0011;b = 4'b1001;

   #5    load_b = 1'b1;             //#10
   #5    clk_b = 1'b1;              //#15, ld b_reg with prod
   #5    load_b = 1'b0;             //#20
   #5    clk_b = 1'b0;              //#25

//sequence 1 -- shift a_reg & b_reg; load a_reg
   #10   clk_a = 1'b1;clk_b = 1'b1; //#35, shift registers
   #10   clk_a = 1'b0;clk_b = 1'b0; //#45
   #10   load_a = 1'b1;             //#55
   #5    clk_a = 1'b1;              //#60, load a_reg
   #5    load_a = 1'b0;             //#65
   #5    clk_a = 1'b0;              //#70

//sequence 2 -- shift a_reg & b_reg; load a_reg
   #10   clk_a = 1'b1;clk_b = 1'b1; //#80, shift registers
   #10   clk_a = 1'b0;clk_b = 1'b0; //#90
   #10   load_a = 1'b1;             //#100
   #5    clk_a = 1'b1;              //#105, load a_reg
   #5    load_a = 1'b0;             //#110
   #5    clk_a = 1'b0;              //#115

//sequence 3 -- shift a_reg & b_reg; load a_reg
   #10   clk_a = 1'b1;clk_b = 1'b1; //#35, shift registers
   #10   clk_a = 1'b0;clk_b = 1'b0; //#45
   #10   load_a = 1'b1;             //#55
   #5    clk_a = 1'b1;              //#60, load a_reg
   #5    load_a = 1'b0;             //#65
   #5    clk_a = 1'b0;              //#70  continued next pg
```

Figure 10.15 Test bench for the structural BCD multiplier.

```
//sequence 4 -- shift a_reg & b_reg; load a_reg
   #10   clk_a = 1'b1;clk_b = 1'b1; //#80, shift registers
   #10   clk_a = 1'b0;clk_b = 1'b0; //#90
   #10   load_a = 1'b1;              //#100
   #5    clk_a = 1'b1;               //#105, load a_reg
   #5    load_a = 1'b0;              //#110
   #5    clk_a = 1'b0;               //#115

//sequence 5 -- shift a_reg & b_reg; load a_reg
   #10   clk_a = 1'b1;clk_b = 1'b1; //#125, shift registers
   #10   clk_a = 1'b0;clk_b = 1'b0; //#135
   #10   load_a = 1'b1;              //#145
   #5    clk_a = 1'b1;               //#150, load a_reg
   #5    load_a = 1'b0;              //#155
   #5    clk_a = 1'b0;               //#160

//sequence 6 -- shift a_reg & b_reg; load a_reg
   #10   clk_a = 1'b1;clk_b = 1'b1; //#170, shift registers
   #10   clk_a = 1'b0;clk_b = 1'b0; //#180
   #10   load_a = 1'b1;              //#190
   #5    clk_a = 1'b1;               //#195, load a_reg
   #5    load_a = 1'b0;              //#200
   #5    clk_a = 1'b0;               //#205

//sequence 7 -- shift a_reg & b_reg; load a_reg
   #10   clk_a = 1'b1;clk_b = 1'b1; //#215, shift registers
   #10   clk_a = 1'b0;clk_b = 1'b0; //#225
   #10   load_a = 1'b1;              //#235
   #5    clk_a = 1'b1;               //#240, load a_reg
   #5    load_a = 1'b0;              //#245
   #5    clk_a = 1'b0;               //#250

//sequence 8 -- shift a_reg & b_reg; load a_reg
   #10   clk_a = 1'b1;clk_b = 1'b1; //#260, shift registers
   #10   clk_a = 1'b0;clk_b = 1'b0; //#270
   #10   load_a = 1'b1;              //#280
   #5    clk_a = 1'b1;               //#285, load a_reg
   #5    load_a = 1'b0;              //#290
   #5    clk_a = 1'b0;               //#295

//sequence 9 -- shift a_reg & b_reg; load a_reg
   #10   clk_a = 1'b1;clk_b = 1'b1; //#305, shift registers
   #10   clk_a = 1'b0;clk_b = 1'b0; //#315
   #10   load_a = 1'b1;              //#325
   #5    clk_a = 1'b1;               //#330, load a_reg
   #5    load_a = 1'b0;              //#335
   #5    clk_a = 1'b0;               //#340  continued next pg
```

Figure 10.15 (Continued)

```
//sequence 10 -- shift a_reg & b_reg; load a_reg
   #10   clk_a = 1'b1;clk_b = 1'b1; //#350, shift registers
   #10   clk_a = 1'b0;clk_b = 1'b0; //#360
   #10   load_a = 1'b1;             //#370
   #5    clk_a = 1'b1;              //#375, load a_reg
   #5    load_a = 1'b0;             //#380
   #5    clk_a = 1'b0;              //#385


//sequence 11 -- shift a_reg & b_reg; load a_reg
   #10   clk_a = 1'b1;clk_b = 1'b1; //#395, shift registers
   #10   clk_a = 1'b0;clk_b = 1'b0; //#405
   #10   load_a = 1'b1;             //#415
   #5    clk_a = 1'b1;              //#420, load a_reg
   #5    load_a = 1'b0;             //#425
   #5    clk_a = 1'b0;              //#430


//sequence 12 -- shift a_reg
   #10   clk_a = 1'b1;
   #10   clk_a = 1'b0;

   #10   $stop;

end



//instantiate the module into the test bench
mul_bcd_struc inst1 (
   .rst_n(rst_n),
   .clk_a(clk_a),
   .clk_b(clk_b),
   .load_a(load_a),
   .load_b(load_b),
   .a(a),
   .b(b)
   );

endmodule
```

Figure 10.15 (Continued)

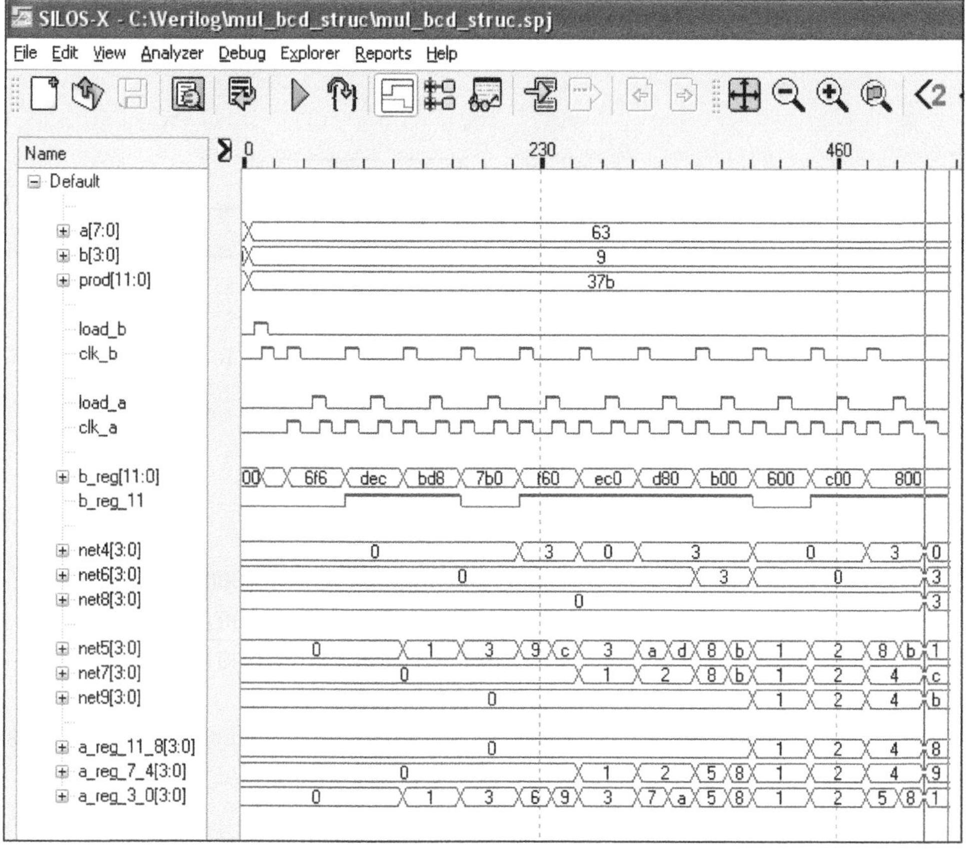

Figure 10.16 Waveforms for the BCD multiplier.

10.4 Multiplication Using Memory

With the advent of high-speed, high-capacity memories (including read-only memories), decimal multiplication can be easily implemented. Each digit of the multiplicand is multiplied by the low-order multiplier digit. This process repeats for all multiplier digits. The concatenated four bits of the multiplicand and the four bits of the multiplier constitute the memory address. The outputs from the memory are valid BCD digits — no correction is required. All corrections (or adjustments) are accomplished by the memory programming. A representative memory module is shown in Figure 10.17.

The decimal digits that are generated by multiplying each multiplicand digit by the low-order multiplier digit are aligned and added to form a product — or partial product if there is more than one multiplier digit. Two examples of multiplying two multiplicand digits by one multiplier digit are shown in Figure 10.18 and Figure 10.19, respectively. Figure 10.18 multiplies 99 by 9 to produce a product of 891; Figure 10.19 multiplies 46 by 7 to produce a product of 322.

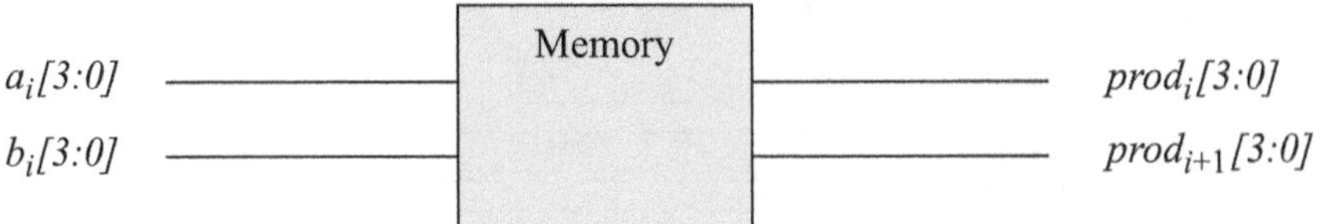

Figure 10.17 Representative memory module to form the product of two decimal digits.

	99		1000	0001
×)	9	1000	0001	↓
	891	1000	1001	0001

Figure 10.18 Example of multiplying two decimal digits by one decimal digit.

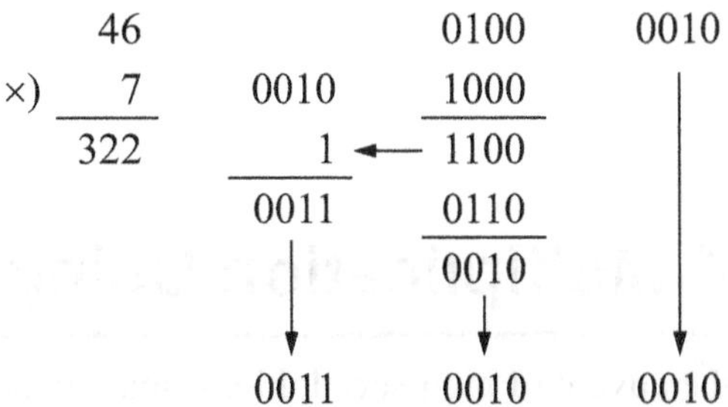

Figure 10.19 Example of multiplying two decimal digits by one decimal digit.

An example of multiplying three multiplicand digits 927 by two multiplier digits 54 to produce a product of 50058 is shown in Figure 10.20. This example yields two partial products *pp1* and *pp2* which are aligned and added together to generate a product of 50058. Any adjustment that is required to produce a valid BCD digit is accomplished by a decimal adder.

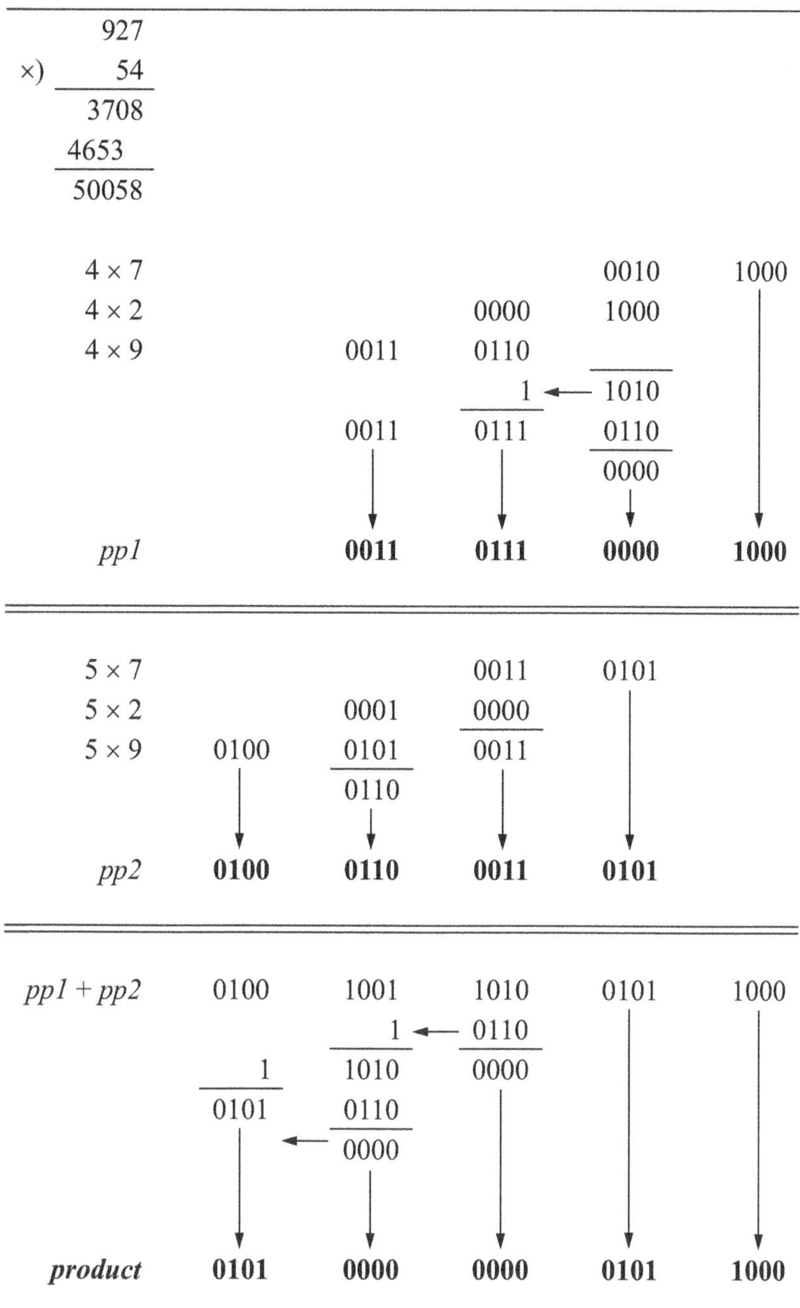

Figure 10.20 Example of multiplying three decimal digits by two decimal digits.

A BCD multiplier using memories will now be designed and implemented with a mixed-design Verilog HDL module. There are two multiplicand digits and one multiplier digit to yield a 3-digit product. The organization of the multiplier is shown in Figure 10.21. The 2-digit multiplicand is *a[7:4]* and *a[3:0]*; the one digit multiplier is *b[3:0]*; the 3-digit product is *bcd[11:8]*, *bcd[7:4]*, and *bcd[3:0]*.

There are two memories used in the design; one to multiply *a[3:0]* × *b[3:0]*, and one to multiply *a[7:4]* × *b[3:0]*. In order to incorporate all combinations of the concatenated *a[3:0]* and *b[3:0]* used for the address, each memory will consist of an array of 160 eight-bit registers — four bits for the tens decade and four bits for the units decade. A memory capacity of 160 bytes is required to accommodate all 16 combinations of each operand in conjunction with the other operand, with digits 1010 – 1111 yielding a result of 0000 0000. Table 10.1 lists the contents of the memories in which the multiplicand and the multiplier form the addresses that produce the product for the tens and the units decades for all combinations of the multiplicand and the multiplier.

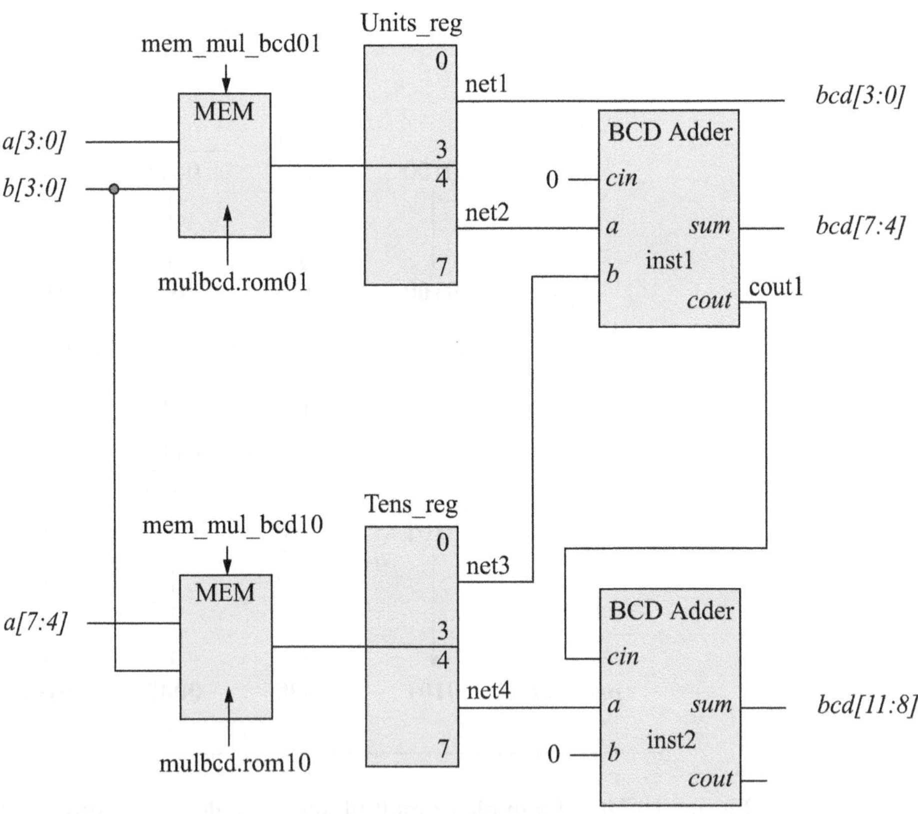

Figure 10.21 Organization of a decimal multiplication unit to multiply a 2-digit multiplicand by a 1-digit multiplier.

Table 10.1 Contents of the Memories for BCD Multiplication

Mpcnd	Mplyr	Tens	Units
0000	0000	0000	0000
0000	0001	0000	0000
0000	0010	0000	0000
0000	0011	0000	0000
0000	0100	0000	0000
0000	0101	0000	0000
0000	0110	0000	0000
0000	0111	0000	0000
0000	1000	0000	0000
0000	1001	0000	0000
0000	1010	0000	0000
0000	1011	0000	0000
0000	1100	0000	0000
0000	1101	0000	0000
0000	1110	0000	0000
0000	1111	0000	0000
0001	0000	0000	0000
0001	0001	0000	0001
0001	0010	0000	0010
0001	0011	0000	0011
0001	0100	0000	0100
0001	0101	0000	0101
0001	0110	0000	0110
0001	0111	0000	0111
0001	1000	0000	1000
0001	1001	0000	1001
0001	1010	0000	0000
0001	1011	0000	0000
0001	1100	0000	0000
0001	1101	0000	0000
0001	1110	0000	0000
0001	1111	0000	0000

Continued on next page

Table 10.1 Contents of the Memories for BCD Multiplication

Mpcnd	Mplyr	Tens	Units
0010	0000	0000	0000
0010	0001	0000	0001
0010	0010	0000	0100
0010	0011	0000	0110
0010	0100	0000	1000
0010	0101	0001	0000
0010	0110	0001	0010
0010	0111	0001	0100
0010	1000	0001	0110
0010	1001	0001	1000
0010	1010	0000	0000
0010	1011	0000	0000
0010	1100	0000	0000
0010	1101	0000	0000
0010	1110	0000	0000
0010	1111	0000	0000
0011	0000	0000	0000
0011	0001	0000	0011
0011	0010	0000	0110
0011	0011	0000	1001
0011	0100	0001	0010
0011	0101	0001	0101
0011	0110	0001	1000
0011	0111	0010	0001
0011	1000	0010	0100
0011	1001	0010	0111
0011	1010	0000	0000
0011	1011	0000	0000
0011	1100	0000	0000
0011	1101	0000	0000
0011	1110	0000	0000
0011	1111	0000	0000

Continued on next page

Table 10.1 Contents of the Memories for BCD Multiplication

Mpcnd	Mplyr	Tens	Units
0100	0000	0000	0000
0100	0001	0000	0100
0100	0010	0000	1000
0100	0011	0001	0010
0100	0100	0001	0110
0100	0101	0010	0000
0100	0110	0010	0100
0100	0111	0010	1000
0100	1000	0011	0010
0100	1001	0011	0110
0100	1010	0000	0000
0100	1011	0000	0000
0100	1100	0000	0000
0100	1101	0000	0000
0100	1110	0000	0000
0100	1111	0000	0000
0101	0000	0000	0000
0101	0001	0000	0101
0101	0010	0001	0000
0101	0011	0001	0101
0101	0100	0010	0000
0101	0101	0010	0101
0101	0110	0011	0000
0101	0111	0011	0101
0101	1000	0100	0000
0101	1001	0100	0101
0101	1010	0000	0000
0101	1011	0000	0000
0101	1100	0000	0000
0101	1101	0000	0000
0101	1110	0000	0000
0101	1111	0000	0000

Continued on next page

Table 10.1 Contents of the Memories for BCD Multiplication

Mpcnd	Mplyr	Tens	Units
0110	0000	0000	0000
0110	0001	0000	0110
0110	0010	0001	0010
0110	0011	0001	1000
0110	0100	0010	0100
0110	0101	0011	0000
0110	0110	0011	0110
0110	0111	0100	0010
0110	1000	0100	1000
0110	1001	0101	0100
0110	1010	0000	0000
0110	1011	0000	0000
0110	1100	0000	0000
0110	1101	0000	0000
0110	1110	0000	0000
0110	1111	0000	0000
0111	0000	0000	0000
0111	0001	0000	0111
0111	0010	0001	0100
0111	0011	0010	0001
0111	0100	0010	1000
0111	0101	0011	0101
0111	0110	0100	0010
0111	0111	0100	1001
0111	1000	0101	0110
0111	1001	0110	0011
0111	1010	0000	0000
0111	1011	0000	0000
0111	1100	0000	0000
0111	1101	0000	0000
0111	1110	0000	0000
0111	1111	0000	0000

Continued on next page

Table 10.1 **Contents of the Memories for BCD Multiplication**

Mpcnd	Mplyr	Tens	Units
1000	0000	0000	0000
1000	0001	0000	1000
1000	0010	0001	0110
1000	0011	0010	0100
1000	0100	0011	0010
1000	0101	0100	0000
1000	0110	0100	1000
1000	0111	0101	0110
1000	1000	0110	0100
1000	1001	0111	0010
1000	1010	0000	0000
1000	1011	0000	0000
1000	1100	0000	0000
1000	1101	0000	0000
1000	1110	0000	0000
1000	1111	0000	0000
1001	0000	0000	0000
1001	0001	0000	1001
1001	0010	0001	1000
1001	0011	0010	0111
1001	0100	0011	0110
1001	0101	0100	0101
1001	0110	0101	0100
1001	0111	0110	0011
1001	1000	0111	0010
1001	1001	1000	0001
1001	1010	0000	0000
1001	1011	0000	0000
1001	1100	0000	0000
1001	1101	0000	0000
1001	1110	0000	0000
1001	1111	0000	0000

The behavioral BCD adder that is instantiated into the mixed-design module is shown in Figure 10.22. The mixed-design module (behavioral and structural) is shown in Figure 10.23 with two vector inputs, *a[7:0]* and *b[3:0]*, one vector output, *bcd[11:0]*, and one scalar output, *invalid_outputs*, that is asserted whenever any 4-bit input is greater than 1001. The size of the units decade memory is defined as follows:

reg [7:0] mem_mul_bcd01 [0:159]

which is an array of 160 eight-bit registers. The contents of the units decade memory are loaded from the *mulbcd.rom01* file by the following behavioral statement:

$readmemb ("mulbcd.rom01", mem_mul_bcd01);

The *mulbcd.rom01* file contains the contents that are shown in Table 10.1 and is saved in the project folder without the **.v** extension. In a similar manner, the tens decade memory, *mem_mul_bcd10*, is defined for the same size as the units decade and is loaded from file *mulbcd.rom10*, which contains the same contents as the units decade file *mulbcd.rom01*.

```
//behavioral for a bcd adder
module add_bcd_behav (a, b, cin, bcd, cout);

input [3:0] a, b;
input cin;
output [3:0] bcd;
output cout;

reg [3:0] sum, bcd;
reg cout3, cout;

always @ (a or b)
begin
   {cout3, sum} = a + b + cin;
   cout = (cout3 || (sum[3] && sum[1]) || (sum[3] && sum[2]));
   bcd = sum + {1'b0, cout, cout, 1'b0};
end
endmodule
```

Figure 10.22 Behavioral module for a BCD adder.

```
//mixed-design for bcd multiplication using memory
module mul_bcd_rom2 (a, b, bcd, invalid_inputs);

input [7:0] a;
input [3:0] b;
output [11:0] bcd;
output invalid_inputs;

//continued on next page
```

Figure 10.23 Mixed-design module for the BCD multiplier.

```
wire [7:0] a;
wire [3:0] b;
wire [11:0] bcd;
reg invalid_inputs;

//define internal nets
wire [3:0] net1, net2, net3, net4;
wire cout1;

//define internal registers
reg [7:0] units_reg, tens_reg;

//check for invalid inputs
always @ (a or b)
begin
   if ((a[7:4] > 4'b1001) || (a[3:0] > 4'b1001) ||
         (b[3:0] > 4'b1001))
      invalid_inputs = 1'b1;
   else
      invalid_inputs = 1'b0;
end

//define memory size for units decade
//mem_mul_bcd01 is an array of 160 eight-bit registers
reg [7:0] mem_mul_bcd01 [0:159];

//define memory contents for units decade
//load file mulbcd.rom01 into memory mem_mul_bcd01
initial
begin
   $readmemb ("mulbcd.rom01", mem_mul_bcd01);
end

//define memory for tens decade
//mem_mul_bcd10 is an array of 160 eight-bit registers
reg [7:0] mem_mul_bcd10 [0:159];

//define memory contents for tens decade
//load file mulbcd.rom10 into memory mem_mul_bcd10
initial
begin
   $readmemb ("mulbcd.rom10", mem_mul_bcd10);
end

//continued on next page
```

Figure 10.23 (Continued)

```
//load units_reg and tens_reg
always @ (a or b)
begin
      units_reg = mem_mul_bcd01[{a[3:0], b[3:0]}];
      tens_reg = mem_mul_bcd10[{a[7:4], b[3:0]}];
end

assign      net1 = units_reg[3:0],
            net2 = units_reg[7:4],
            net3 = tens_reg[3:0];
assign   #2 net4 = tens_reg[7:4];

//instantiate the bcd adder for the units decade
add_bcd_behav inst1 (
   .a(net2),
   .b(net3),
   .cin(1'b0),
   .bcd(bcd[7:4]),
   .cout(cout1)
   );

//instantiate the bcd adder for the tens decade
add_bcd_behav inst2 (
   .a(net4),
   .b(4'b0000),
   .cin(cout1),
   .bcd(bcd[11:8])
   );

assign   bcd[3:0] = net1;
endmodule
```

Figure 10.23 (Continued)

The test bench, shown in Figure 10.24, applies several input vectors for the multiplicand and the multiplier, including an invalid input for the multiplicand. The outputs are shown in Figure 10.25, and the waveforms are shown in Figure 10.26.

The concept described in this section for decimal multiplication using memories can be easily extended to accommodate larger operands. For example, a 3-digit multiplicand multiplied by a 2-digit multiplier requires three memories and three decimal adders. When the partial product is obtained for the low-order multiplier digit, the partial product is shifted and stored. Then the next higher-order multiplier digit is applied to the memories and the resulting partial product is added to the previous partial product.

```
//test bench for bcd multiplication using memory
module mul_bcd_rom_tb;

reg [7:0] a;
reg [3:0] b;
wire [11:0] bcd;
wire invalid_inputs;

//display variables
initial
$monitor ("a=%b, b=%b, bcd=%b, invalid_inputs=%b",
            a, b, bcd, invalid_inputs);

//apply input vectors
initial
begin
         //99 x 9 = 891
   #15   a = 8'b1001_1001;b = 4'b1001;

         //2 x 3 = 6
   #15   a = 8'b0000_0010;b = 4'b0011;

         //77 x 6 = 442
   #15   a = 8'b0111_0111;b = 4'b0110;

         //97 x 5 = 485
   #15   a = 8'b1001_0111;b = 4'b0101;

         //46 x 7 = 322
   #15   a = 8'b0100_0110;b = 4'b0111;

         //55 x 8 = 440
   #15   a = 8'b0101_0101;b = 4'b1000;

         //55 x 6 = 330
   #15   a = 8'b0101_0101;b = 4'b0110;

         //56 x 9 = 504
   #15   a = 8'b0101_0110;b = 4'b1001;

   #15   a = 8'b1100_0011;b = 4'b1001;

   #15   $stop;

end

//continued on next page
```

Figure 10.24 Test bench for the decimal multiplier using memories.

```
//instantiate the module into the test bench
mul_bcd_rom2 inst1 (
   .a(a),
   .b(b),
   .bcd(bcd),
   .invalid_inputs(invalid_inputs)
   );
endmodule
```

Figure 10.24 (Continued)

```
a=1001_1001, b=1001, bcd=xxxx_1001_0001, invalid_inputs=0
a=1001_1001, b=1001, bcd=1000_1001_0001, invalid_inputs=0
a=0000_0010, b=0011, bcd=0000_0000_0110, invalid_inputs=0
a=0111_0111, b=0110, bcd=0100_0110_0010, invalid_inputs=0
a=1001_0111, b=0101, bcd=0100_1000_0101, invalid_inputs=0
a=0100_0110, b=0111, bcd=0011_0010_0010, invalid_inputs=0
a=0101_0101, b=1000, bcd=0100_0100_0000, invalid_inputs=0
a=0101_0101, b=0110, bcd=0100_0011_0000, invalid_inputs=0
a=0101_0110, b=1001, bcd=0101_0000_0100, invalid_inputs=0
a=1100_0011, b=1001, bcd=xxxx_xxxx_0111, invalid_inputs=1
```

Figure 10.25 Outputs for the decimal multiplier using memories.

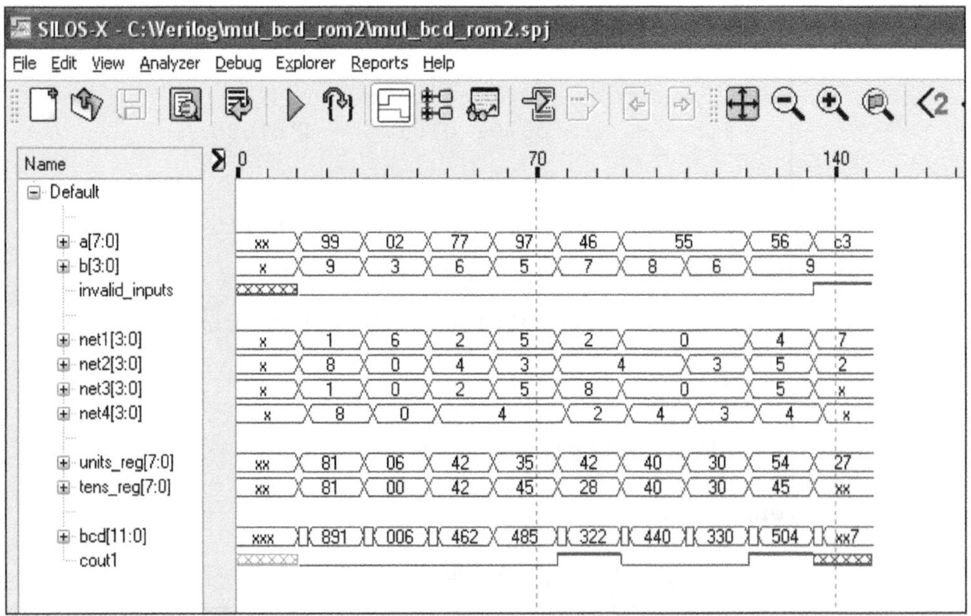

Figure 10.26 Waveforms for the decimal multiplier using memories.

10.5 Multiplication Using Table Lookup

The speed of decimal multiplication can be increased by using a table lookup method. This is similar to the table lookup method presented in Chapter 6 for fixed-point multiplication. The multiplicand table resides in memory and contains the following versions of the multiplicand:

Multiplicand × 0
Multiplicand × 1
Multiplicand × 2
Multiplicand × 4
Multiplicand × 8

Refer to Figure 10.27 for the procedure to initialize the multiplicand table, as shown below. The adder shown in Figure 10.27 is a BCD adder.

Multiplicand × 0 Location is reset.
Multiplicand × 1 is loaded into memory and register *D*.
Multiplicand × 1 and the contents of register *D* are added. The sum (multiplicand × 2) is loaded into memory location multiplicand × 2 and register *D*.
Multiplicand × 2 and the contents of register *D* are added.
The sum (multiplicand × 4) is loaded into memory location multiplicand × 4 and register *D*.
Multiplicand × 4 and the contents of register *D* are added.
The sum (multiplicand × 8) is loaded into memory location multiplicand × 8 and register *D*.
Multiplicand × 8 contains the value of multiplicand × 8.

The low-order digit of the multiplier is used to address the multiplicand table, as follows:

If $B(0)_1 =$	0	address multiplicand × 0 location
	1	address multiplicand × 1 location
If $B(0)_2 =$	0	address multiplicand × 0 location
	1	address multiplicand × 2 location
If $B(0)_4 =$	0	address multiplicand × 0 location
	1	address multiplicand × 4 location
If $B(0)_8 =$	0	address multiplicand × 0 location
	1	address multiplicand × 8 location

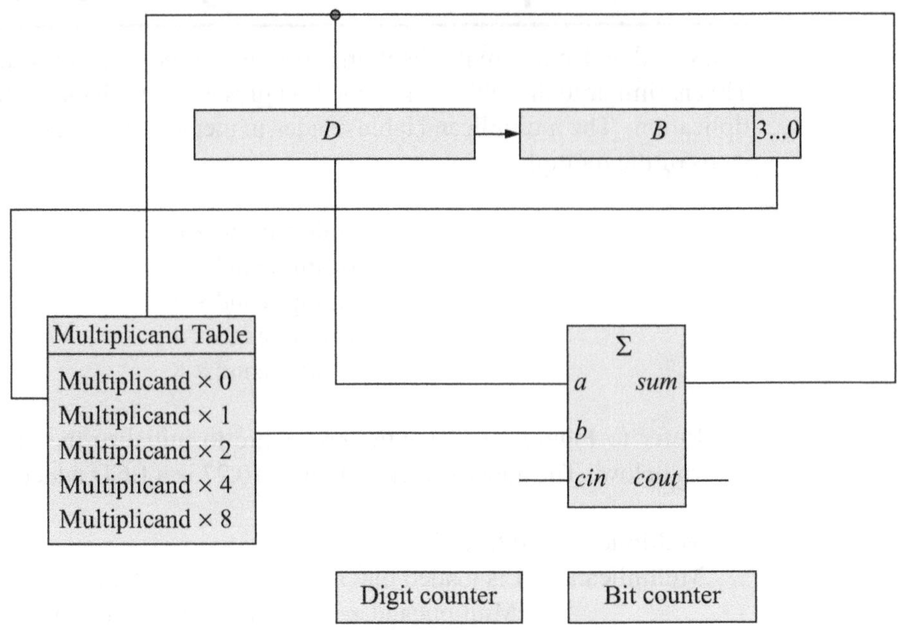

Figure 10.27 Organization of a decimal multiplier using the table lookup method.

Examples will best illustrate the technique involved in decimal multiplication using the table lookup method. In the examples which follow, the multiplicand and multiplier are both 2-digit operands. Register *D* is 12 bits and register *B* is eight bits. The bits in the low-order multiplier digit are scanned from right to left and are controlled by a 4-bit count-down counter labeled *Bit counter*. The end of the multiply operation is determined by a count-down counter labeled *Digit counter*, which is set to the number of digits in the multiplier.

Example 10.1 The decimal numbers 99 and 99 will be multiplied using the table lookup method to yield a product of 9801, as shown in Figure 10.28. The low-order multiplier digit is scanned from right to left to generate 1 × the multiplicand; 0 × the multiplicand; 0 × the multiplicand; and 8 × the multiplicand. See the next page for the detailed calculations. Each sum in boldface is added to the next versions of the multiplicand.

99	Multiplicand	1001 1001	
×) 99	Multiplier	1001 1001	
891	8 × multiplicand		1 × multiplicand
891	1 × multiplicand		8 × multiplicand
9801			

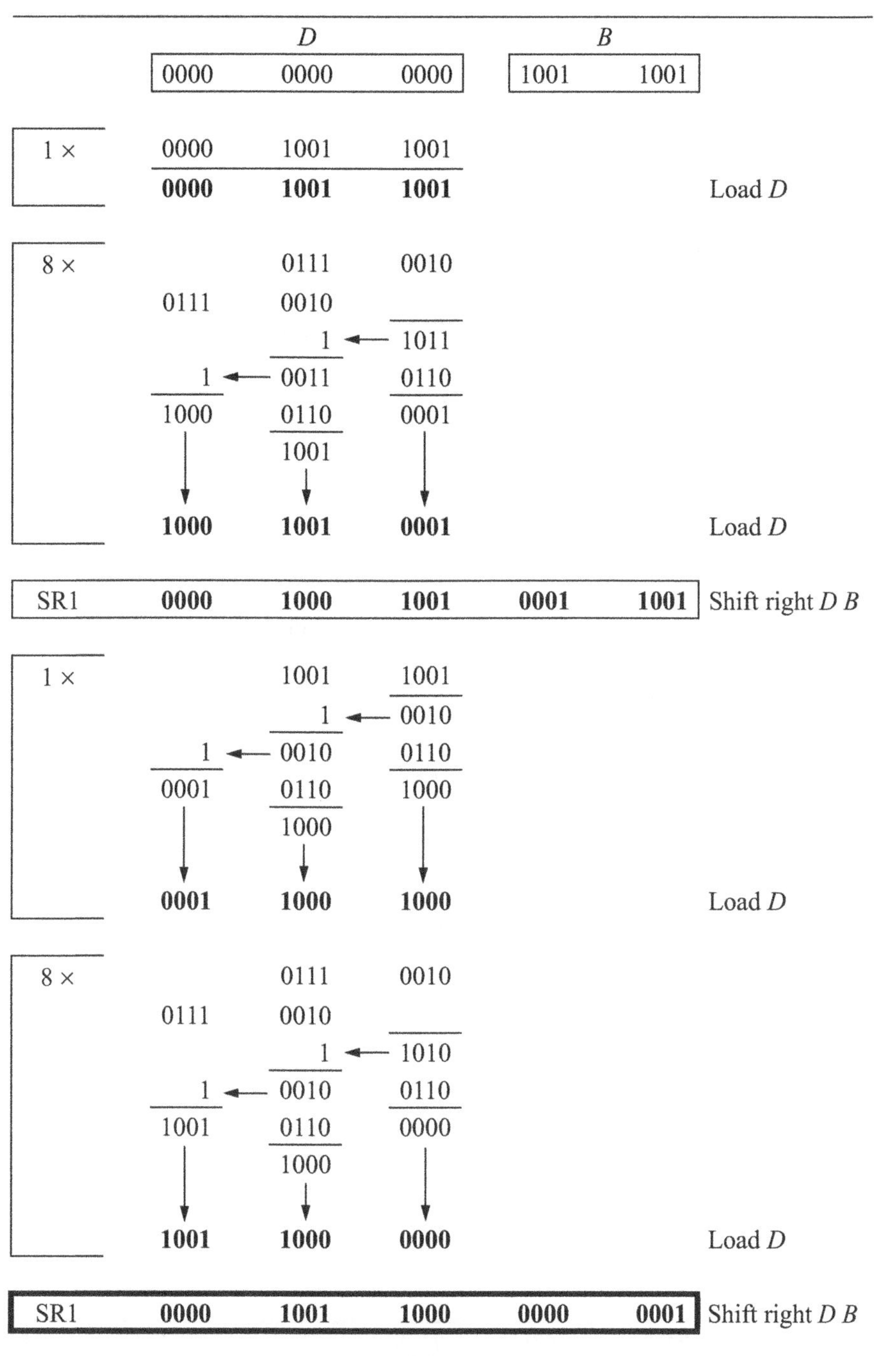

Figure 10.28 Example of decimal multiplication using the table lookup method.

Example 10.2 The decimal numbers 76 (0111 0110) and 63 (0110 0011) will be multiplied using the table lookup method to yield a product of 4788, as shown in Figure 10.29.

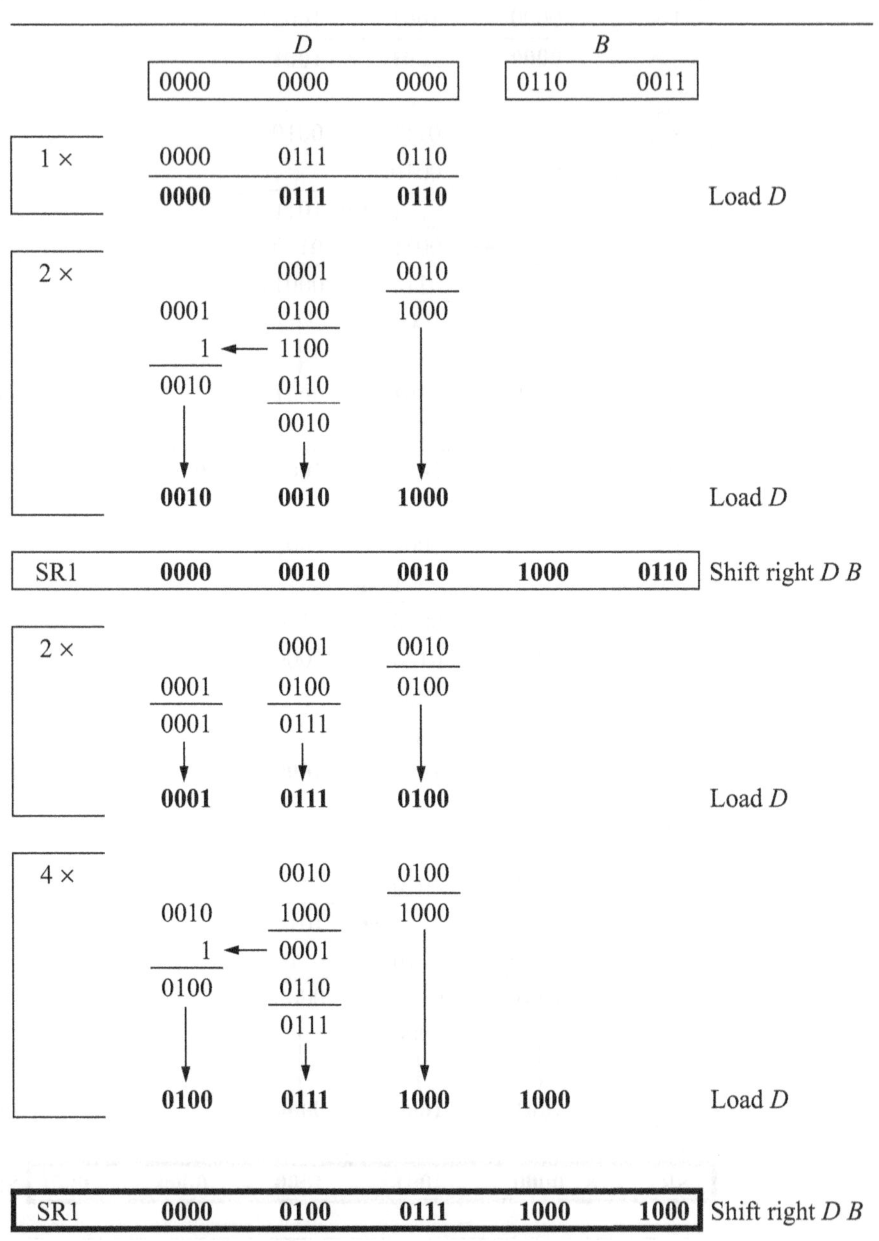

Figure 10.29 Example of decimal multiplication using the table lookup method.

10.6 Problems

10.1 Design a mixed-design (behavioral and structural) module to multiply the multiplicand *a[7:0]* by the multiplier *b[3:0]* using memories. Define the memory contents for all decimal combinations of the two operands. Use *reset* and *clock* as inputs, together with the two operands. The outputs are a 12-bit BCD result, *bcd[11:0]*, and an *invalid inputs* scalar. Obtain the design module, the test bench module for several input vectors, the outputs, and the waveforms.

10.2 Design a mixed-design (behavioral and structural) module to multiply the multiplicand *a[11:0]* by the multiplier *b[3:0]* using memories. Define the memory contents for all decimal combinations of the two operands. The outputs are a 16-bit BCD result, *bcd[15:0]*, and an *invalid inputs* scalar. There are no *reset* or *clock* inputs. Obtain the design module, the test bench module for several input vectors, the outputs, and the waveforms.

10.3 Using the paper-and-pencil method, multiply the decimal numbers 63 and 87 using the table lookup technique.

10.4 Using the paper-and-pencil method, multiply the decimal numbers 98 and 97 using the table lookup method.

10.5 Using the paper-and-pencil method, multiply the following decimal numbers utilizing memories: 934×76.

10.6 Design a 3-digit decimal adder using behavioral modeling. This adder can be used as part of the decimal multiply algorithm using the table lookup technique. Obtain the design module, the test bench module, the outputs, and the waveforms.

11.1 Restoring Division — Version 1
11.2 Restoring Division — Version 2
11.3 Division Using Table Lookup
11.4 Problems

11

Decimal Division

Decimal division can be considered as the inverse of decimal multiplication in which the dividend, divisor, and quotient correspond to the product, multiplicand, and multiplier, respectively. Decimal division is similar to fixed-point division, except for the digits used in the operation — 0 and 1 for fixed-point; 0 through 9 for decimal. Also, correct alignment must be established initially between the dividend and divisor. The divisor is subtracted from the dividend or previous partial remainder in much the same way as the restoring division method in fixed-point division. If the division is unsuccessful — producing a negative result — then the divisor is added to the negative partial remainder.

11.1 Restoring Division — Version 1

A straightforward method to perform binary-coded decimal (BCD) division is to implement the design using the fixed-point division algorithm and then convert the resulting quotient and remainder to BCD. This is a simple process and involves only a few lines of Verilog HDL code. Let A and B be the dividend and divisor, respectively, where

$$A = a_{2n-1}\, a_{2n-2} \ldots a_n\, a_{n-1} \ldots a_1\, a_0$$

$$B = b_{n-1}\, b_{n-2} \ldots b_1\, b_0$$

A review of the algorithm for fixed-point division is appropriate at this time. The dividend is initially shifted left 1 bit position. Then the divisor is subtracted from the dividend. Subtraction is accomplished by adding the 2s complement of the divisor. If the high-order bit of the subtract operation is 1, then a 0 is placed in the next lower-order bit position of the quotient; if the high-order bit is 0, then a 1 is placed in the next lower-order bit position of the quotient. The concatenated partial remainder and dividend are then shifted left 1 bit position. Fixed-point binary restoring division requires one subtraction for each quotient bit. Figure 11.1 illustrates the procedure.

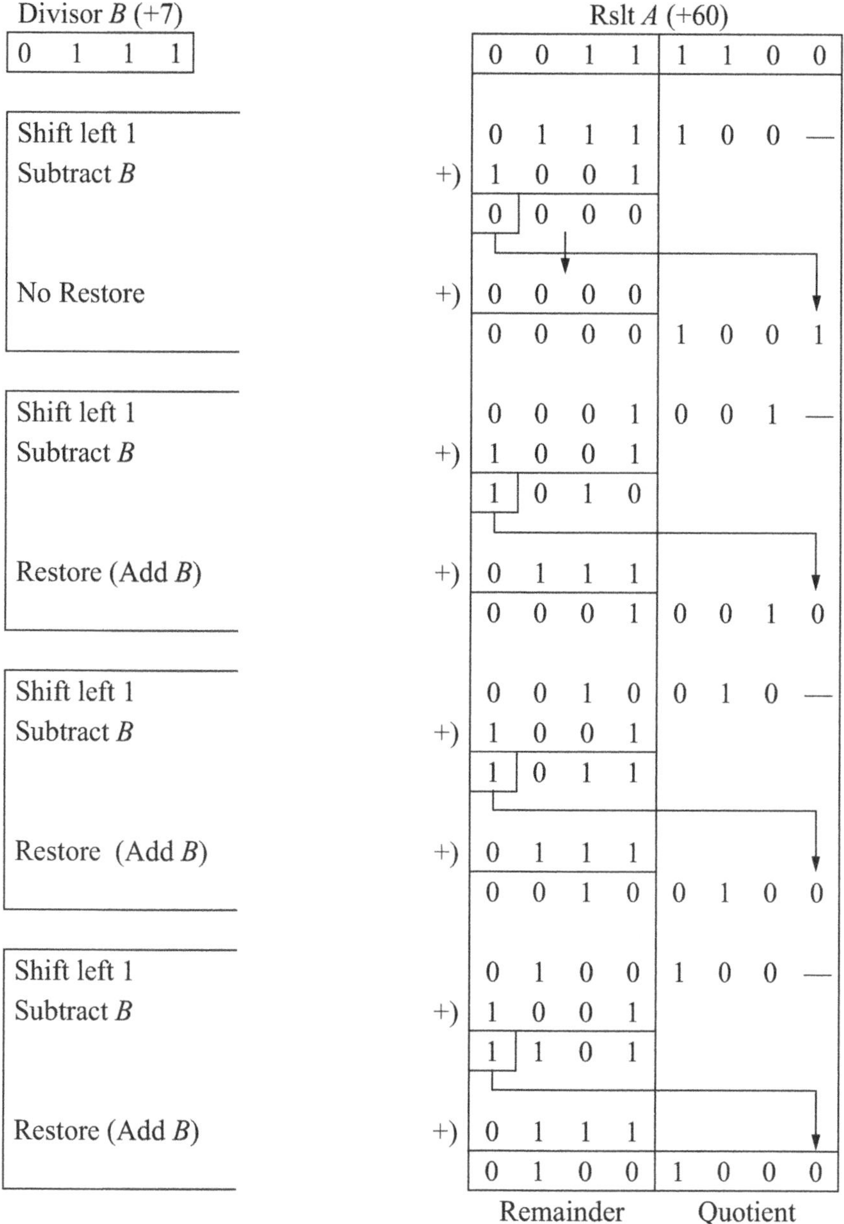

Figure 11.1 Example of fixed-point restoring division.

Since there are 4 bits in the divisor of Figure 11.1, there are four cycles. Each cycle begins with a left shift operation. The low-order quotient bit is left blank after each left shift operation —it will be set before the next left shift. Then the divisor is subtracted from the high-order half of the dividend. If the resulting difference is negative — indicated by a 1 in the high-order bit of the result — then the low-order quotient bit q_0 is set to 0 and the previous partial remainder is restored. If the high-order bit is 0, then q_0 is set to 1 and there is no restoration — the partial remainder thus obtained is loaded into the high-order half of the dividend. This sequence repeats for all four bits of the divisor. If the sign of the remainder is different than the sign of the dividend, then the previous partial remainder is restored.

Overflow can occur in division if the value of the dividend and the value of the divisor are disproportionate, yielding a quotient that exceeds the range of the machine's word size. Since the dividend has double the number of bits of the divisor, *overflow* can occur when the high-order half of the dividend has a value that is greater than or equal to that of the divisor.

If the high-order half of the dividend ($a_{2n-1} \ldots a_n$) and the divisor ($b_{n-1} \ldots b_0$) are equal, then the quotient will exceed *n* bits. If the value of the high-order half of the dividend is greater than the value of the divisor, then the value of the quotient will be even greater. Overflow can be detected by subtracting the divisor from the high-order half of the dividend before the first shift-subtract cycle. If the difference is positive, then an overflow has been detected; if the difference is negative, then the condition for overflow has not been met. Division by zero must also be avoided. This can be detected by the preceding method, since any high-order dividend bits will be greater than or equal to a divisor of all zeroes.

The mixed-design (behavioral and dataflow) module for this type of decimal division is shown in Figure 11.2. The dividend is an 8-bit vector, *a[7:0]*; the divisor is a 4-bit vector, *b[3:0]*; and the result is an 8-bit quotient/remainder vector, *rslt[7:0]*, with a carry-out of the high-order quotient bit, *cout_quot*. The operation begins on the positive assertion of a *start* pulse and follows the algorithm outlined in the previous discussion. The *rslt[7:0]* register is set to the value of the dividend, *a[7:0]*, and a sequence counter, *count*, is set to a value of four (0100), which is the size of the divisor. If the dividend and divisor are both nonzero, then the process continues until the count-down counter, *count*, reaches a value of zero, controlled by the **while** loop.

The **while** loop executes a procedural statement or a block of procedural statements — delimited by the keywords **begin . . . end** — as long as a Boolean expression returns a value of true. The loop is executed until the expression returns a value of false — the expression in this case being the value of the count-down counter, *count*. If the evaluation of the expression is false, then the **while** loop is terminated and control is passed to the next statement in the module. If the expression is false before the loop is initially entered, then the **while** loop is never executed.

The shift-left operation is accomplished by the following statement:

```
rslt = rslt << 1;
```

The divisor is then subtracted from the high-order half of the dividend, *rslt[7:4]*, by adding the negation of the divisor, *b_neg*, and concatenating the sum with the low-order half of the dividend, *rslt[3:0]*, by the following statement:

rslt = {(rslt[7:4] + b_neg), rslt[3:0]};

If the high-order bit, *rslt[7]*, of the sum is equal to a value of 1, then the previous partial remainder is restored by adding the divisor to the previous partial remainder and concatenating the result with *rslt[3:1]* together with a 0 in the low-order bit position of the result. The sequence counter is then decremented by 1.

If the high-order bit, *rslt[7]*, of the sum is equal to a value of 0, then the previous partial remainder is not restored and a 1 is placed in the low-order bit position of the result. The sequence counter is then decremented by 1.

The conversion to BCD is quite simple for a 2-digit dividend and a 1-digit divisor. Since the remainder can never be greater than 9 (1001) for the operands in this example, only the quotient need be considered, as shown below. If the quotient in *rslt[3:0]* is greater than 9, then 6 (0110) is added to the result and a carry-out may be generated on *cout_quot*.

```
if (rslt[3:0] > 4'b1001)
    {cout_quot, rslt[3:0]} = rslt[3:0] + 4'b0110;

else
    cout_quot = 1'b0;
```

```
//mixed-design for bcd restoring division
module div_bcd (a, b, start, rslt, cout_quot);

input [7:0] a;
input [3:0] b;
input start;
output [7:0] rslt;
output cout_quot;

wire [3:0] b_bar;

//define internal registers
reg [3:0] b_neg;
reg [7:0] rslt;
reg [3:0] count;
reg [3:0] quot;
reg cout_quot;

assign b_bar = ~b;

always @ (b_bar)
   b_neg = b_bar + 1;
//continued on next page
```

Figure 11.2 Mixed-design module for a BCD divider using a fixed-point division algorithm, then converting the result to BCD.

```
always @ (posedge start)
begin
   rslt = a;
   count = 4'b0100;

   if ((a!=0) && (b!=0))
      while (count)
         begin
            rslt = rslt << 1;
            rslt = {(rslt[7:4] + b_neg), rslt[3:0]};
               if (rslt[7] == 1)      //restore
                  begin
                     rslt = {(rslt[7:4]+b), rslt[3:1], 1'b0};
                     count = count - 1;
                  end

               else                   //no restore
                  begin
                     rslt = {rslt[7:1], 1'b1};
                     count = count - 1;
                  end
         end

   if (rslt[3:0] > 4'b1001)           //convert to bcd
      {cout_quot, rslt[3:0]} = rslt[3:0] + 4'b0110;

   else
      cout_quot = 1'b0;
end

endmodule
```

Figure 11.2 (Continued)

The test bench is shown in Figure 11.3, in which several input vectors are applied to both the dividend and the divisor. The first set of inputs represent those shown in Figure 11.1. Also included are vectors showing an invalid dividend (all zeroes) and a case where the high-order half of the dividend is equal to or greater than the divisor that yields an indeterminate result.

The outputs are shown in Figure 11.4, which lists the dividend and divisor as binary numbers and the quotient and remainder as BCD values. The waveforms are shown in Figure 11.5. For quotients and remainders that are larger than those utilized in this section, the procedure shown in Figure 10.2 can be used to convert the results to BCD.

```
//test bench for bcd restoring division
module div_bcd_tb;

reg [7:0] a;
reg [3:0] b;
reg start;
wire [7:0] rslt;
wire cout_quot;

//display variables
initial
$monitor ("a=%b, b=%b, quot_tens=%b, quot_units=%b, rem=%b",
        a, b, {{3{1'b0}}, cout_quot}, rslt[3:0], rslt[7:4]);

//apply input vectors
initial
begin
   #0    start = 1'b0;

         //60 / 7; quot = 8, rem = 4
         a = 8'b0011_1100; b = 4'b0111;
   #10   start = 1'b1;
   #10   start = 1'b0;

         //13 / 5; quot = 2, rem = 3
   #10   a = 8'b0000_1101; b = 4'b0101;
   #10   start = 1'b1;
   #10   start = 1'b0;

         //60 / 7; quot = 8, rem = 4
   #10   a = 8'b0011_1100; b = 4'b0111;
   #10   start = 1'b1;
   #10   start = 1'b0;

         //82 / 6; quot = 13, rem = 4
   #10   a = 8'b0101_0010; b = 4'b0110;
   #10   start = 1'b1;
   #10   start = 1'b0;

         //56 / 7; quot = 8, rem = 0
   #10   a = 8'b0011_1000; b = 4'b0111;
   #10   start = 1'b1;
   #10   start = 1'b0;

//continued on next page
```

Figure 11.3 Test bench for a BCD divider using a fixed-point division algorithm, then converting the result to BCD.

```
        //100 / 7; quot = 14, rem = 2
   #10  a = 8'b0110_0100; b = 4'b0111;
   #10  start = 1'b1;
   #10  start = 1'b0;

        //110 / 7; quot = 15, rem = 5
   #10  a = 8'b0110_1110; b = 4'b0111;
   #10  start = 1'b1;
   #10  start = 1'b0;

        //99 / 9; quot = 11, rem = 0
   #10  a = 8'b0110_0011; b = 4'b1001;
   #10  start = 1'b1;
   #10  start = 1'b0;

        //99 / 8; quot = 12, rem = 3
   #10  a = 8'b0110_0011; b = 4'b1000;
   #10  start = 1'b1;
   #10  start = 1'b0;

        //52 / 5; quot = 10, rem = 2
   #10  a = 8'b0011_0100; b = 4'b0101;
   #10  start = 1'b1;
   #10  start = 1'b0;

        //0 / 5; quot = 0, rem = 0; invalid dividend
   #10  a = 8'b0000_0000; b = 4'b0101;
   #10  start = 1'b1;
   #10  start = 1'b0;

        //86 / 5; overflow; results are indeterminate
   #10  a = 8'b0101_0110; b = 4'b0101;
   #10  start = 1'b1;
   #10  start = 1'b0;

   #10  $stop;
end

//instantiate the module into the test bench
div_bcd inst1 (
   .a(a),
   .b(b),
   .start(start),
   .rslt(rslt),
   .cout_quot(cout_quot)
   );
endmodule
```

Figure 11.3 (Continued)

```
Example of Figure 11.1
a = 0011_1100, b = 0111,
   quot_tens = 000x, quot_units = xxxx, rem = xxxx

a = 0011_1100, b = 0111,
   quot_tens = 0000, quot_units = 1000, rem = 0100
------------------------------------------------------------
a = 0000_1101, b = 0101,
   quot_tens = 0000, quot_units = 1000, rem = 0100

a = 0000_1101, b = 0101,
   quot_tens = 0000, quot_units = 0010, rem = 0011

a = 0011_1100, b = 0111,
   quot_tens = 0000, quot_units = 0010, rem = 0011

a = 0011_1100, b = 0111,
   quot_tens = 0000, quot_units = 1000, rem = 0100

a = 0101_0010, b = 0110,
   quot_tens = 0000, quot_units = 1000, rem = 0100

a = 0101_0010, b = 0110,
   quot_tens = 0001, quot_units = 0011, rem = 0100

a = 0011_1000, b = 0111,
   quot_tens = 0001, quot_units = 0011, rem = 0100

a = 0011_1000, b = 0111,
   quot_tens = 0000, quot_units = 1000, rem = 0000

a = 0110_0100, b = 0111,
   quot_tens = 0000, quot_units = 1000, rem = 0000

a = 0110_0100, b = 0111,
   quot_tens = 0001, quot_units = 0100, rem = 0010

a = 0110_1110, b = 0111,
   quot_tens = 0001, quot_units = 0100, rem = 0010

a = 0110_1110, b = 0111,
   quot_tens = 0001, quot_units = 0101, rem = 0101

a = 0110_0011, b = 1001,
   quot_tens = 0001, quot_units = 0101, rem = 0101
//continued on next page
```

Figure 11.4 Outputs for a BCD divider using a fixed-point division algorithm, then converting the result to BCD.

```
a = 0110_0011, b = 1001,
   quot_tens = 0001, quot_units = 0101, rem = 0101

a = 0110_0011, b = 1001,
   quot_tens = 0001, quot_units = 0001, rem = 0000

a = 0110_0011, b = 1000,
   quot_tens = 0001, quot_units = 0001, rem = 0000

a = 0110_0011, b = 1000,
   quot_tens = 0001, quot_units = 0010, rem = 0011

a = 0011_0100, b = 0101,
   quot_tens = 0001, quot_units = 0010, rem = 0011

a = 0011_0100, b = 0101,
   quot_tens = 0001, quot_units = 0000, rem = 0010


invalid dividend

a = 0000_0000, b = 0101,
   quot_tens = 0001, quot_units = 0000, rem = 0010

a = 0000_0000, b = 0101,
   quot_tens = 0000, quot_units = 0000, rem = 0000


overflow — results are indeterminate
high-order half of dividend is => divisor

a = 0101_0110, b = 0101,
   quot_tens = 0000, quot_units = 0000, rem = 0000

a = 0101_0110, b = 0101,
   quot_tens = 0001, quot_units = 0011, rem = 0101
```

Figure 11.4 (Continued)

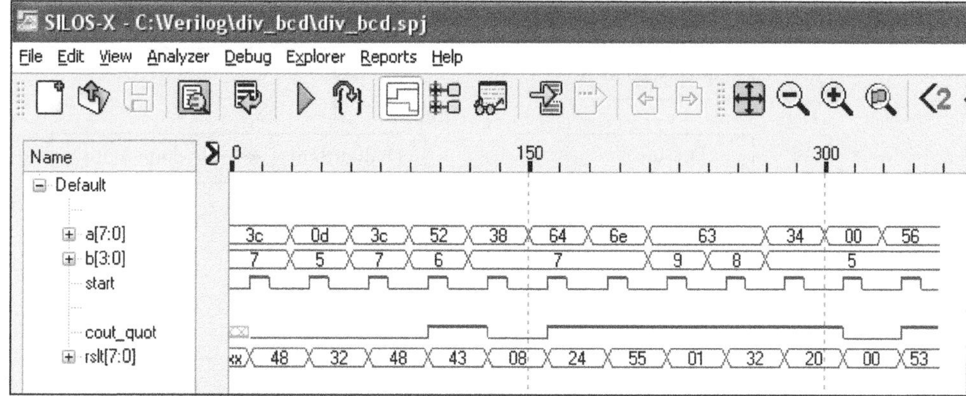

Figure 11.5 Waveforms for a BCD divider using a fixed-point division algorithm, then converting the result to BCD.

11.2 Restoring Division — Version 2

Recall that a multiplexer was used in one of the restoring division techniques in Chapter 7. A similar method can be used in decimal division; that is, obtain the fixed-point quotient and remainder, then convert the quotient and remainder (if applicable) to BCD.

This version of restoring division represents a faster method than the previous method presented in Section 11.1. The divisor is not added to the partial remainder to restore the previous partial remainder; instead the previous partial remainder — which is unchanged — is loaded into register A, as shown in Figure 11.6, which is reproduced from Chapter 7. This avoids the time required for addition. Additional hardware is required in the form of a 2:1 multiplexer which selects the appropriate version of the partial remainder to be loaded into register A. Either the previous partial remainder is selected or the current partial remainder (the sum) is selected, controlled by the state of the carry-out, *cout*, from the adder.

If *cout* = 0, then the previous partial remainder (in register A) is selected and loaded into register A, while *cout* is loaded into the low-order quotient bit q_0. If *cout* = 1, then the sum output from the adder is selected and loaded into register A, and *cout* is loaded into the low-order quotient bit q_0. This sequence of operations is as follows:

$$\text{If } cout = 0, \text{ then } a_{n-1}\ a_{n-2} \ldots a_1\ a_0 \leftarrow a_{n-1}\ a_{n-2} \ldots a_1\ a_0$$
$$q_0 = 0$$

$$\text{If } cout = 1, \text{ then } a_{n-1}\ a_{n-2} \ldots a_1\ a_0 \leftarrow s_{n-1}\ s_{n-2} \ldots s_1\ s_0$$
$$q_0 = 1$$

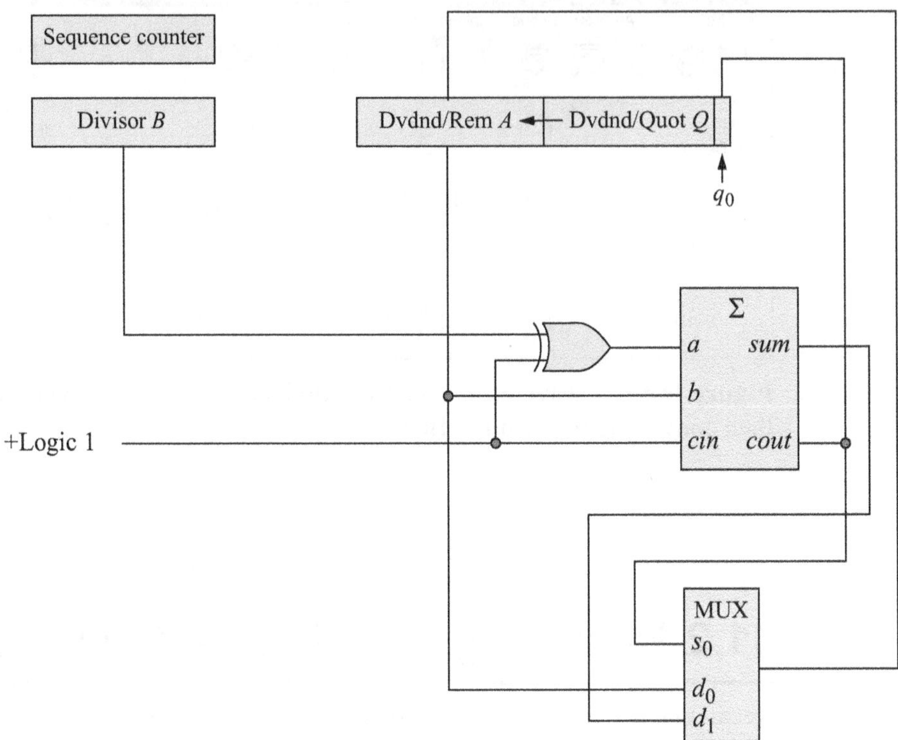

Figure 11.6 Restoring division using a multiplexer.

A sequence counter is initialized to the number of bits in the divisor. When the sequence counter counts down to zero, the division operation is finished. The 2s complement of the divisor is always added to the partial remainder by exclusive-ORing the divisor with a logic 1. Then the appropriate partial remainder is selected by the multiplexer — either the previous partial remainder or the sum, which represents the current partial remainder. A numerical example that illustrates this method is shown in Figure 11.7 for a dividend of +82 and a divisor of +6, resulting in a quotient of +13 and a remainder of +4.

The behavioral/dataflow module for restoring division using a multiplexer is shown in Figure 11.8. Register *a[7:0]* is the dividend; register *b[3:0]* is the divisor. The result register, *rslt[7:0]*, portrays the concatenation of the *A* register and the *Q* register in Figure 11.7, where *rslt[7:4]* represents the remainder, and *rslt[3:0]* represents the quotient. The quotient is converted to BCD in a manner identical to that described in the previous section. The behavioral module conforms precisely to the algorithm shown in Figure 11.7.

The test bench is shown in Figure 11.9 and applies several input vectors to the design module for the dividend and divisor. The outputs are shown in Figure 11.10 and the waveforms are shown in Figure 11.11.

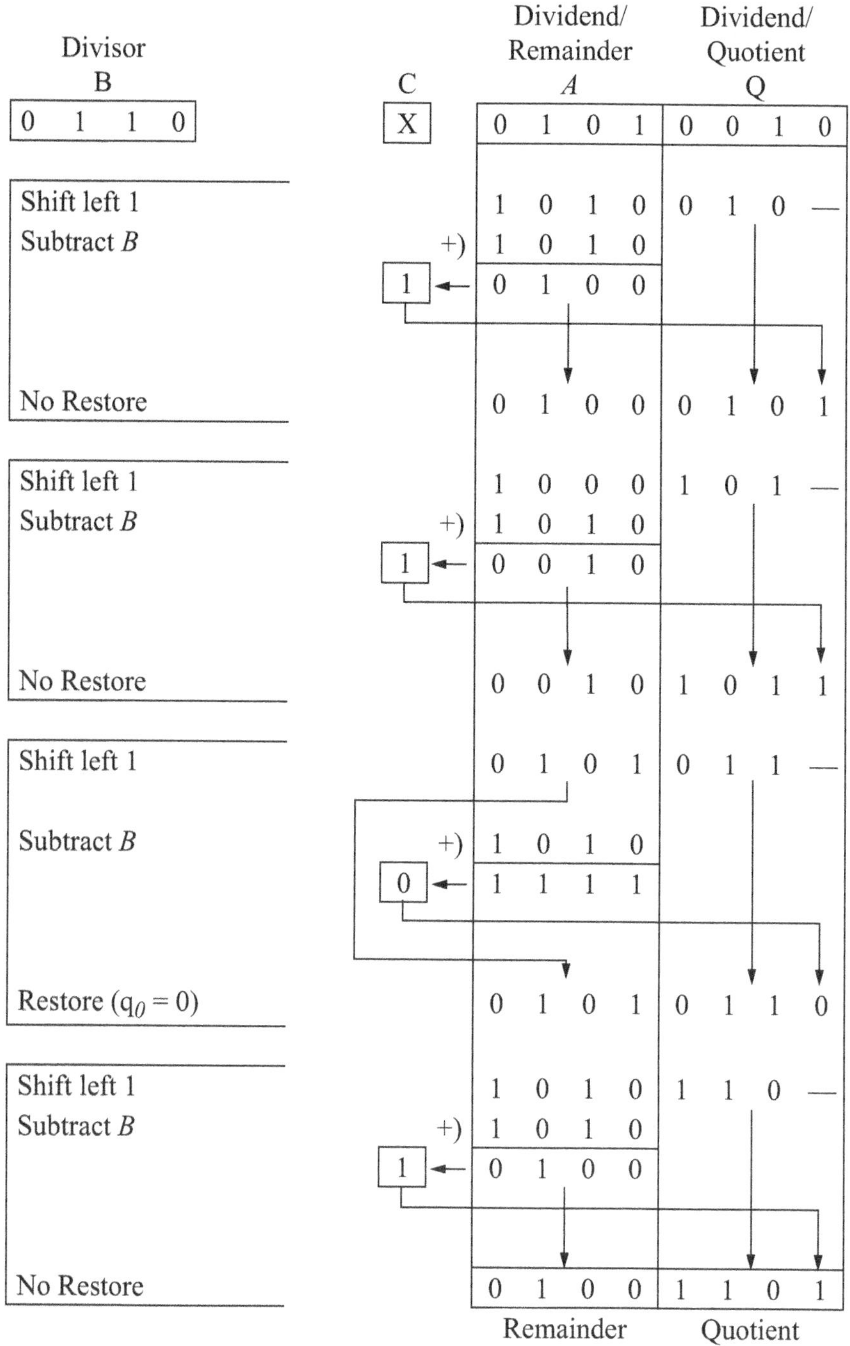

Figure 11.7 Numerical example illustrating restoring division using a multiplexer.

```
//mixed-design bcd restoring division using a multiplexor
module div_bcd_mux (a, b, start, rslt, cout_quot);

input [7:0] a;          //dividend
input [3:0] b;          //divisor
input start;
output [7:0] rslt;      //rslt[7:4] is rem; rslt[3:0] is quot
output cout_quot;
wire [3:0] b_bar;

reg [3:0] b_neg;        //define internal registers
reg [7:0] rslt;
reg [3:0] count;
reg [4:0] sum;
reg cout, cout_quot;

assign b_bar = ~b;      //1s complement of divisor
always @ (b_bar)
   b_neg = b_bar + 1;//2s complement of divisor

always @ (posedge start)
begin
   rslt = a;
   count = 4'b0100;

      if ((a!=0) && (b!=0))
         while (count)
         begin
            rslt = rslt << 1;
            sum = rslt[7:4] + b_neg;
            cout = sum[4];

            if (cout == 0)
               begin
                  rslt[0] = cout;    //q0 = cout
                  rslt[7:4] = rslt[7:4];
                  count = count -1;
               end

            else
               begin
                  rslt[0] = cout;    //q0 = cout
                  rslt[7:4] = sum[3:0];
                  count = count - 1;
               end
         end                  //continued on next page
```

Figure 11.8 Mixed-design module for fixed-point restoring division using a multiplexer, then converting the result to BCD.

```
//convert to bcd
   if (rslt[3:0] > 4'b1001)
      {cout_quot, rslt[3:0]} = rslt[3:0] + 4'b0110;

   else
      cout_quot = 1'b0;
end
endmodule
```

Figure 11.8 (Continued)

```
//test bench for restoring division
module div_bcd_mux_tb;

reg [7:0] a;
reg [3:0] b;
reg start;
wire [7:0] rslt;
wire cout_quot;

//display variables
initial
$monitor ("a=%b, b=%b, quot_tens=%b, quot_units=%b, rem=%b",
         a, b, {{3{1'b0}}, cout_quot}, rslt[3:0], rslt[7:4]);

//apply input vectors
initial
begin
   #0    start = 1'b0;

         //13 / 5; quot = 2, rem = 3
         a = 8'b0000_1101; b = 4'b0101;
   #10   start = 1'b1;
   #10   start = 1'b0;

         //60 / 7; quot = 8, rem = 4
   #10   a = 8'b0011_1100; b = 4'b0111;
   #10   start = 1'b1;
   #10   start = 1'b0;

//continued on next page
```

Figure 11.9 Test bench for fixed-point restoring division using a multiplexer, then converting the result to BCD.

```
        //82 / 6; quot = 13, rem = 4
   #10  a = 8'b0101_0010; b = 4'b0110;
   #10  start = 1'b1;
   #10  start = 1'b0;

        //56 / 7; quot = 8, rem = 0
   #10  a = 8'b0011_1000; b = 4'b0111;
   #10  start = 1'b1;
   #10  start = 1'b0;

        //100 / 7; quot = 14, rem = 2
   #10  a = 8'b0110_0100; b = 4'b0111;
   #10  start = 1'b1;
   #10  start = 1'b0;

        //110 / 7; quot = 15, rem = 5
   #10  a = 8'b0110_1110; b = 4'b0111;
   #10  start = 1'b1;
   #10  start = 1'b0;

        //99 / 9; quot = 11, rem = 0
   #10  a = 8'b0110_0011; b = 4'b1001;
   #10  start = 1'b1;
   #10  start = 1'b0;

        //0 / 5; quot = 0, rem = 0; invalid dividend
   #10  a = 8'b0000_0000; b = 4'b0101;
   #10  start = 1'b1;
   #10  start = 1'b0;

        //86 / 5; overflow; results are indeterminate
   #10  a = 8'b0101_0110; b = 4'b0101;
   #10  start = 1'b1;
   #10  start = 1'b0;

   #10  $stop;
end

div_bcd_mux inst1 (      //instantiate the module
   .a(a),
   .b(b),
   .start(start),
   .rslt(rslt),
   .cout_quot(cout_quot)
   );

endmodule
```

Figure 11.9 (Continued)

```
a = 00001101, b = 0101,
   quot_tens = 000x, quot_units = xxxx, rem = xxxx
a = 00001101, b = 0101,
   quot_tens = 0000, quot_units = 0010, rem = 0011
a = 00111100, b = 0111,
   quot_tens = 0000, quot_units = 0010, rem = 0011
a = 00111100, b = 0111,
   quot_tens = 0000, quot_units = 1000, rem = 0100
a = 01010010, b = 0110,
   quot_tens = 0000, quot_units = 1000, rem = 0100
a = 01010010, b = 0110,
   quot_tens = 0001, quot_units = 0011, rem = 0100
a = 00111000, b = 0111,
   quot_tens = 0001, quot_units = 0011, rem = 0100
a = 00111000, b = 0111,
   quot_tens = 0000, quot_units = 1000, rem = 0000
a = 01100100, b = 0111,
   quot_tens = 0000, quot_units = 1000, rem = 0000
a = 01100100, b = 0111,
   quot_tens = 0001, quot_units = 0100, rem = 0010
a = 01101110, b = 0111,
   quot_tens = 0001, quot_units = 0100, rem = 0010
a = 01101110, b = 0111,
   quot_tens = 0001, quot_units = 0101, rem = 0101
a = 01100011, b = 1001,
   quot_tens = 0001, quot_units = 0101, rem = 0101
a = 01100011, b = 1001,
   quot_tens = 0001, quot_units = 0001, rem = 0000

Invalid dividend

a = 00000000, b = 0101,
   quot_tens = 0001, quot_units = 0001, rem = 0000
a = 00000000, b = 0101,
   quot_tens = 0000, quot_units = 0000, rem = 0000


overflow - results are indeterminate
high-order half of dividend is => divisor

a = 01010110, b = 0101,
   quot_tens = 0000, quot_units = 0000, rem = 0000
a = 01010110, b = 0101,
   quot_tens = 0001, quot_units = 0100, rem = 0000
```

Figure 11.10 Outputs for fixed-point restoring division using a multiplexer, then converting the result to BCD.

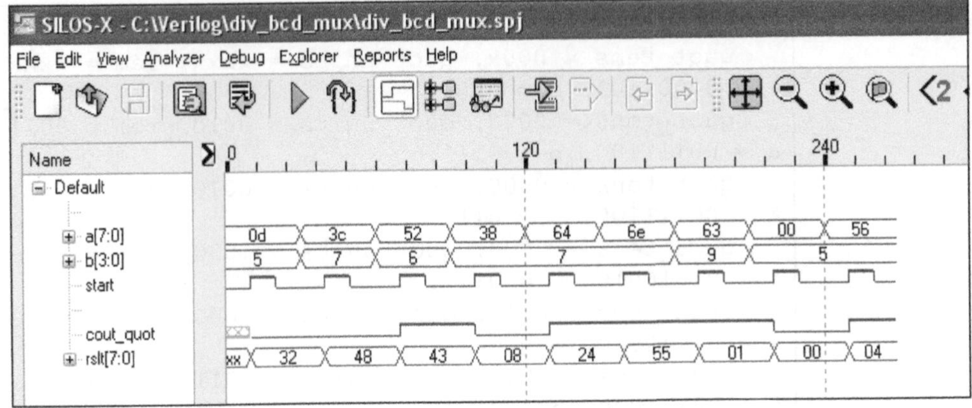

Figure 11.11 Waveforms for fixed-point restoring division using a multiplexer, then converting the result to BCD.

11.3 Division Using Table Lookup

Decimal division using the table lookup method is equivalent to the binary search technique used in programming; that is, it searches an ordered table for a particular entry beginning at the middle of the table. The entry is compared to a keyword. If the entry is less than the keyword, then the upper half of the table is searched; if the entry is equal to the keyword, then the desired entry is found; if the entry is greater than the keyword, then the lower half of the table is searched. This method essentially divides the table in half for each comparison until the desired entry is found. If the table has an even number of entries, then the search begins at the middle of the table ± 1.

The procedure first subtracts eight times the divisor from the dividend. Then, depending upon the sign of the partial remainder, performs the following operations: ± 4 × the divisor; ± 2 × the divisor; ± 1 × the divisor. If a partial remainder is negative, then a version of the divisor is added to the partial remainder; otherwise, a version of the divisor is subtracted from the partial remainder. Thus, there are always four operations per quotient digit.

Before the division operation begins, the divisor table is established. The divisor is loaded into memory location divisor × 1 and also into register *A*, as shown in the organization of Figure 11.12. Then the contents of register *A* are added to the contents of memory location × 1 to produce a sum of divisor × 2, which is loaded into memory location × 2 and register *A*. This process repeats until the divisor table contains the following contents: divisor × 8, divisor × 4, divisor × 2, and divisor × 1.

Next the actual division operation takes place. Eight times the divisor is subtracted from the dividend. If the difference is less than zero, then four times the divisor is added to the partial remainder. However, if the difference is greater than or equal to zero, then a value of eight is added to the quotient counter and four times the divisor is subtracted from the partial remainder. This process repeats until divisor × 1 is added or subtracted from the partial remainder. For every positive difference, the value of the divisor at that stage of the process is added to the quotient counter. Then registers *A*, *Q*, and the quotient counter are shifted left in concatenation with the quotient counter being shifted into the low-order digit position of register *Q*. The sequence counter — which was set to the number of digits in the divisor — is then decremented by one. When the sequence counter counts down to zero, the division operation is finished. Two examples are shown in Figure 11.13 and Figure 1.14 that illustrate this procedure.

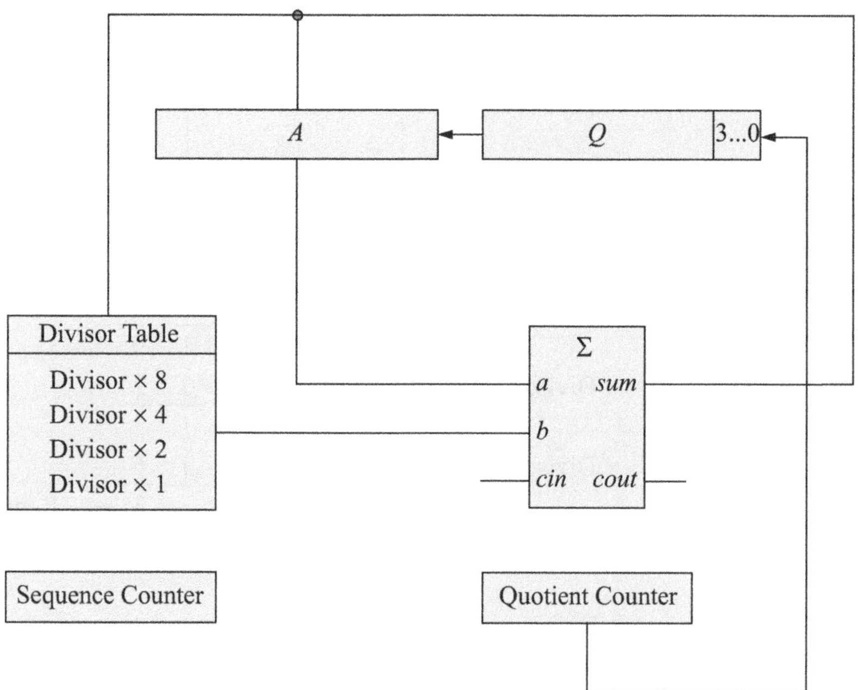

Figure 11.12 Organization of a decimal divider using the table lookup method.

```
      6 4
  4 | 2 5 9
      2 4
      ---
        1 9
        1 6
        ---
          3
```

Continued on next page

Figure 11.13 Example of decimal division using the table lookup method.

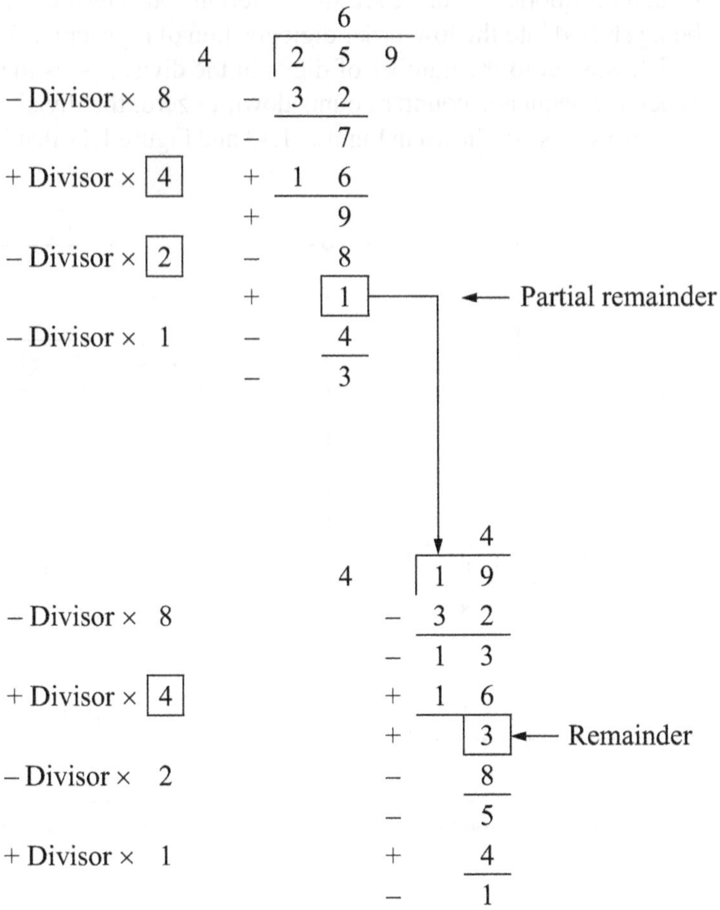

Figure 11.13 (Continued)

```
          9 7
    ---------
2 0 | 1 9 4 5
      1 8 0
      -----
        1 4 5
        1 4 0
        -----
            5
```

Continued on next page

Figure 11.14 Example of decimal division using the table lookup method.

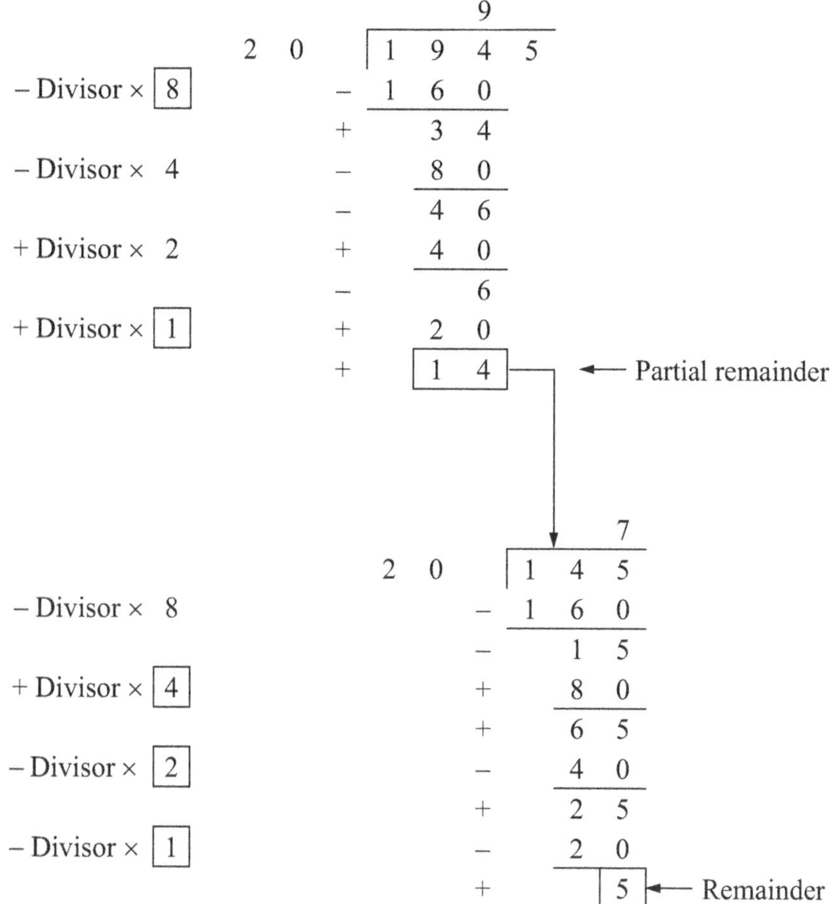

Figure 11.14 (Continued)

Whether an addition or subtraction occurs for the next version of the divisor depends upon the carry-out of the decimal adder/subtractor. For example, a carry-out of 1 indicates that the result of the addition/subtraction produced a positive partial remainder; therefore, the next operation is a subtraction of the next version of the divisor. This is illustrated in the segment shown in Figure 11.15, in which the first operation is – divisor × 8 for a dividend of 72 (0111 0010) and a divisor of 4 (0000 0100).

A carry-out of 0 indicates that the result of the addition/subtraction produced a negative partial remainder; therefore, the next operation is an addition of the next version of the divisor. This is illustrated in the segment of Figure 11.16, in which the first operation is – divisor × 8 for a dividend of 32 (0011 0010) and a divisor of 9 (0000 1001).

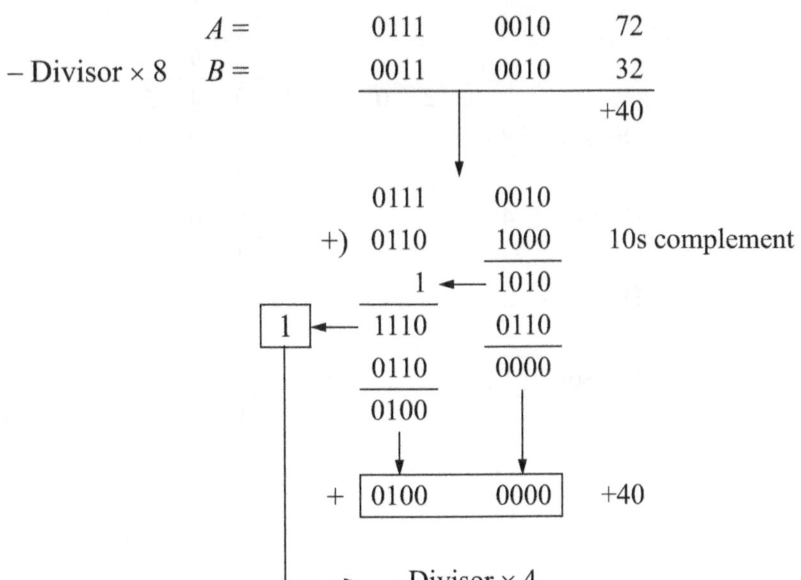

Figure 11.15 Table lookup division segment illustrating the use of the carry-out to determine the next version of the divisor to be used.

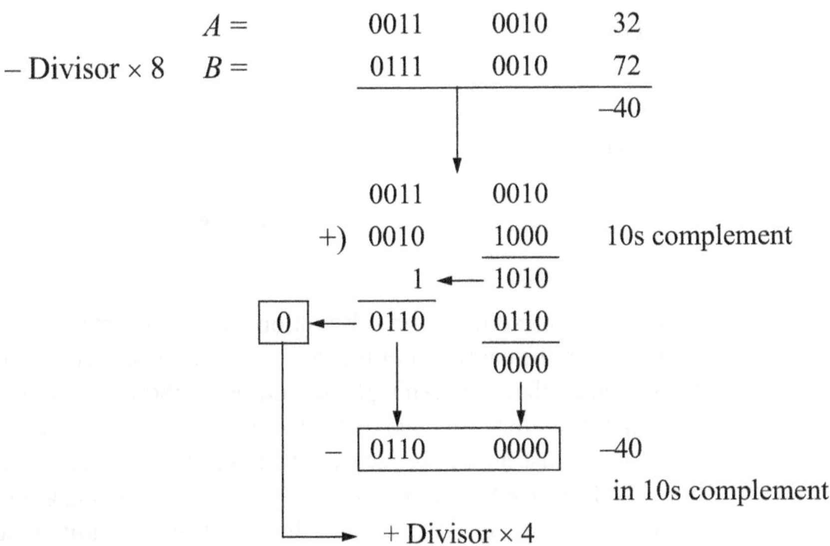

Figure 11.16 Table lookup division segment illustrating the use of the carry-out to determine the next version of the divisor to be used.

11.4 Problems

11.1 Use the decimal division restoring method version 1 to divide a dividend of +55 by a divisor of +6.

11.2 Use the decimal division restoring method version 2 to divide a dividend of +61 by a divisor of +7.

11.3 Use the table lookup method to perform the following divide operation: $+77 \div +5$.

11.4 Use the table lookup method to perform the following divide operation: $+1245 \div +13$.

11.5 Use the table lookup method to perform the following divide operation: $+2378 \div +42$.

12.1 *Floating-Point Format*
12.2 *Biased Exponents*
12.3 *Floating-Point Addition*
12.4 *Overflow and Underflow*
12.5 *General Floating-Point Organization*
12.6 *Verilog HDL Implementation*
12.7 *Problems*

12

Floating-Point Addition

The fixed-point number representation is appropriate for representing numbers with small numerical values that are considered as positive or negative integers; that is, the implied radix point is to the right of the low-order bit. The same algorithms for arithmetic operations can be employed if the implied radix point is to the immediate right of the sign bit, thus representing a signed fraction.

The range for a 16-bit fixed-point number is from (-2^{15}) to $(+2^{15} - 1)$, which is inadequate for some numbers; for example, the following operation:

$$28{,}400{,}000{,}000. \times 0.0000000546$$

This operation can also be written in scientific notation, as follows:

$$(0.284 \times 10^{11}) \times (0.546 \times 10^{-7})$$

where 10 is the *base* and 11 and –7 are the *exponents*. Floating-point notation is equivalent to scientific notation in which the radix point (or binary point) can be made to *float* around the fraction by changing the value of the exponent; thus, the term *floating-point*. In contrast, fixed-point numbers have the radix point located in a fixed position, usually to the immediate right of the low-order bit position, indicating an integer.

The base and exponent are called the *scaling factor*, which specify the position of the radix point relative to the *significand digits* (or fraction digits). Common bases are 2 for binary, 10 for decimal, and 16 for hexadecimal. The base in the scaling factor does not have to be explicitly specified in the floating-point number.

12.1 Floating-Point Format

The material presented in this chapter is based on the Institute of Electrical and Electronics Engineers (IEEE) Standard for Binary Floating-Point Arithmetic IEEE Std 754-1985 (Reaffirmed 1990). Floating-point numbers consist of the following three fields: a sign bit, s; an exponent, e; and a fraction, f. These parts represent a number that is obtained by multiplying the fraction, f, by a radix, r, raised to the power of the exponent, e, as shown in Equation 12.1 for the number A, where f and e are signed fixed-point numbers, and r is the radix (or base).

$$A = f \times r^e \tag{12.1}$$

The exponent is also referred to as the *characteristic*; the fraction is also referred to as the *significand* or *mantissa*. Although the fraction can be represented in sign-magnitude, diminished-radix complement, or radix complement, the fraction is predominantly expressed in sign-magnitude representation.

If the fraction is shifted left k bits, then the exponent is decremented by an amount equal to k; similarly, if the fraction is shifted right k bits, then the exponent is incremented by an amount equal to k. Consider an example in the radix 10 floating-point representation. Let $A = 0.0000074569 \times 10^{+3}$. This number can be rewritten as $A =$ 0.0000074569 +3 or $A = 0.74569$ –2, both with an implied base of 10.

Figure 12.1 shows the format for 32-bit single-precision and 64-bit double-precision floating-point numbers. The single-precision format consists of a sign bit that indicates the sign of the number, an 8-bit signed exponent, and a 24-bit fraction. The double-precision format consists of a sign bit, an 11-bit signed exponent, and a 52-bit fraction.

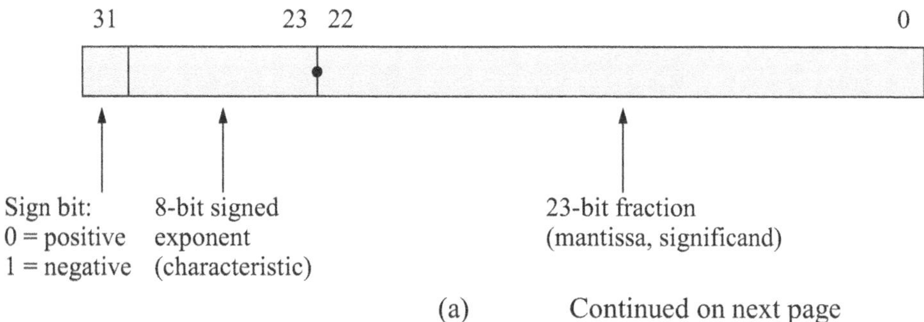

Continued on next page

Figure 12.1 IEEE floating-point formats: (a) 32-bit single precision and (b) 64-bit double precision.

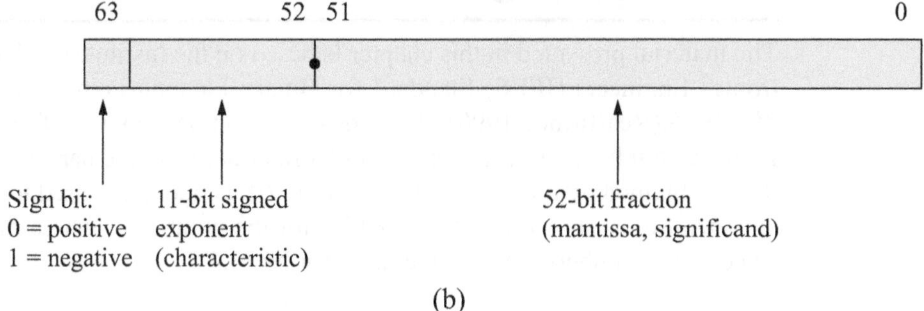

Figure 12.1 (Continued)

As mentioned earlier, the exponent is a signed number in 2s complement. When adding or subtracting floating-point numbers, the exponents are compared and made equal resulting in a right shift of the fraction with the smaller exponent. The comparison is easier if the exponents are unsigned — a simple comparator can be used for the comparison. As the exponents are being formed, a *bias* constant is added to the exponents such that all exponents are positive internally.

For the single-precision format, the bias constant is +127 — also called excess-127; therefore, the biased exponent has a range of

$$0 \leq e_{\text{biased}} \leq 255$$

The lower and upper limits of a biased exponent have special meanings. A value of $e_{\text{biased}} = 0$ with a fraction = 0 is used to represent a number with a value of zero. A value of $e_{\text{biased}} = 255$ with a fraction = 0 is used to represent a number with a value of infinity; that is, dividing a number by zero. The sign bit is still utilized providing values of ± 0 and $\pm \infty$.

Fractions in the IEEE format are normalized; that is, the leftmost significant bit is a 1. Figure 12.2 shows unnormalized and normalized numbers in the 32-bit format. Since there will always be a 1 to the immediate right of the radix point, the 1 bit is not explicitly shown — it is an *implied 1*.

	S	Exponent	Fraction
Unnormalized	0	0 0 0 0 0 1 1 1	• 0 0 1 1 x ... x

$+ \ .0011x \ldots x \times 2^7$

Continued on next page

Figure 12.2 Unnormalized and normalized floating-point numbers.

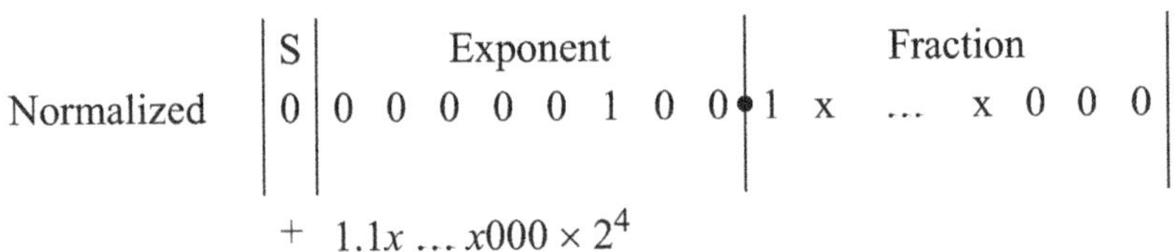

Figure 12.2 (Continued)

The scale factor for the 32-bit single-precision format has a range of 2^{-126} to 2^{+127}. The 64-bit double-precision format has an excess-1023 exponent providing a scale factor of 2^{-1022} to 2^{+1023}. Floating-point arithmetic operations may produce an overflow if the result is too large for the machine's word size or an underflow if the result is too small. Fraction overflow or underflow can normally be corrected by shifting the resulting fraction left or right and adjusting the exponent accordingly. Implementing a floating-point arithmetic processor can be accomplished entirely with hardware, entirely with software, or a combination of hardware and software.

12.2 Biased Exponents

Additional details on exponents are presented in this section. An unbiased exponent can be either a positive or a negative integer. During the addition or subtraction of two floating-point numbers, the exponents are compared and the fraction with the smaller exponent is shifted right by an amount equal to the difference of the two exponents. The comparison is simplified by adding a positive *bias constant* to each exponent during the formation of the numbers. This bias constant has a value that is equal to the most positive exponent. For example, if the exponents are represented by n bits, then the bias is $2^{n-1} - 1$. For $n = 4$, the most positive number is 0111 (+7). Therefore, all biased exponents are of the form shown in Equation 12.2.

$$e_{\text{biased}} = e_{\text{unbiased}} + 2^{n-1} - 1 \tag{12.2}$$

Unbiased exponents have the range shown in Equation 12.3. For four bits, this equates to 1000 to 0111.

$$-2^{n-1} \leq e_{\text{unbiased}} \leq +2^{n-1} - 1 \tag{12.3}$$

Biased exponents have the range shown in Equation 12.4. For four bits, this equates to 0000 to 1111.

$$0 \le e_{\text{biased}} \le 2^n - 1 \tag{12.4}$$

Comparing the biased exponents can be accomplished by means of a comparator, or by subtracting one from the other. If the exponents are e_A and e_B, then the shift amount is equal to the absolute value of $e_A - e_B$; that is, $|e_A - e_B|$. Subtraction can be accomplished by using the 1s complement or the 2s complement method.

Using the 1s complement method, if the subtraction produces a carry-out of the high-order bit position, then $e_A > e_B$ and fraction B is shifted right by the amount equal to the difference plus the end-around carry, as shown below.

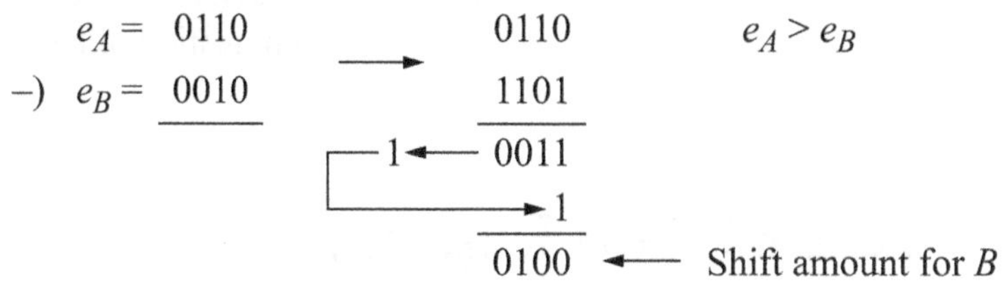

```
    eA = 0110          0110        eA > eB
-)  eB = 0010   →      1101
                     ------
               ┌─1 ← 0011
               └───→    1
                     ------
                       0100  ← Shift amount for B
```

Using the 1s complement method, if the subtraction produces no carry-out of the high-order bit position, then $e_A < e_B$ and fraction A is shifted right by the amount equal to the 1s complement of the difference, as shown below.

```
    eA = 0010          0010        eA < eB
-)  eB = 0110   →      1001
                     ------
                 0 ← 1011
                        ↓
1s complement  →      0100  ← Shift amount for A
```

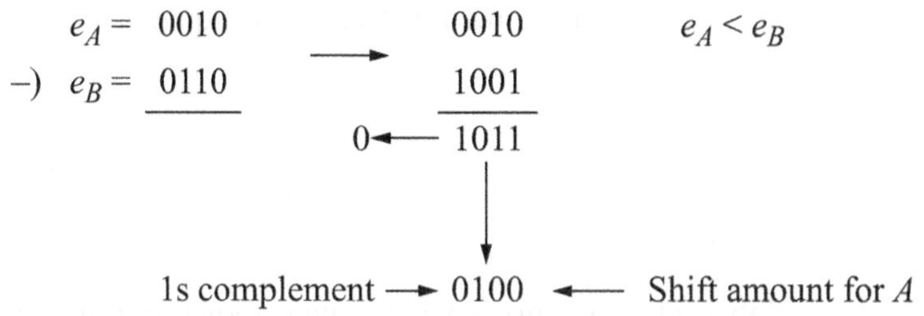

Using the 2s complement method, if the subtraction produces a carry-out of the high-order bit position, then $e_A > e_B$ and fraction B is shifted right by the amount equal to the difference, as shown below.

$$
\begin{array}{rrcrl}
 & e_A = 0110 & \longrightarrow & 0110 & e_A > e_B \\
-) & e_B = \underline{0010} & & \underline{1110} & \\
 & & 1 \longleftarrow & 0100 & \longleftarrow \text{Shift amount for } B
\end{array}
$$

Using the 2s complement method, if the subtraction produces no carry-out of the high-order bit position, then $e_A < e_B$ and fraction A is shifted right by the amount equal to the 2s complement of the difference , as shown below.

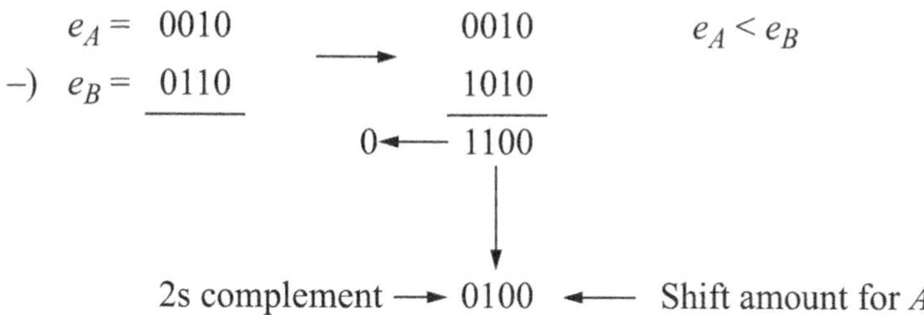

When adding two fractions, the exponents must be made equal. Although this is intuitively obvious, an example is shown below using radix 10 numbers. If the exponents are not equal the result will be incorrect.

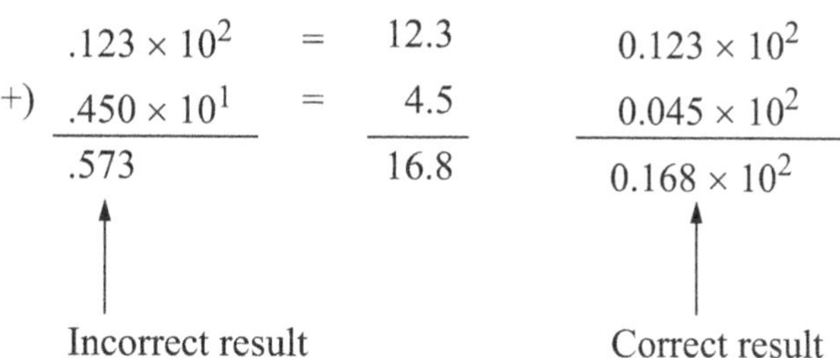

The advantage of using biased exponents is that they are easier to compare without having to consider the signs of the exponents. The main reason for biasing is to determine the correct alignment of the fractions by aligning the radix points, and to determine the number of bits to shift a fraction in order to obtain proper alignment. An additional advantage is that the smallest exponent contains only zeroes; therefore, the floating-point depiction of the number zero is an exponent of zero with a fraction of zero. If the biased exponent has a maximum value (255) and the fraction is nonzero, then this is interpreted as *Not a Number* (NaN), which is the result of zero divided by zero or the square root of -1.

Floating-point arithmetic operations are defined using normalized operands A and B, with $A = f_A \times r^{e_A}$ and $B = f_B \times r^{e_B}$, where f is the normalized fraction, e is the exponent, and r is the radix. The fraction, f, has the following range:

$$1/r \leq |f| \leq (1 - r^{-n}) \tag{12.5}$$

For $r = 2$, the fraction has a range from 1/2 to the term $1 - r^{-n}$, which gives the maximum value of the fraction. For $n = 4$, the range is from 1/2 (.1000) to 15/16 (.1111), as shown below.

$$\begin{aligned} 1 - r^{-4} &= 1 - 1/16 \\ &= 15/16 \end{aligned}$$

Therefore, the bit configuration of the fraction is

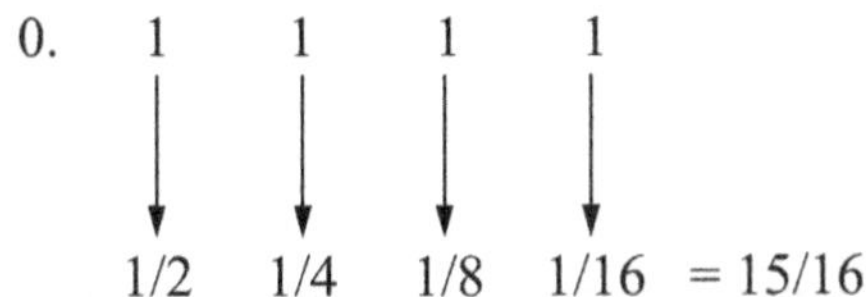

The biased exponent is a positive integer with n bits and has the following range: $0 \leq e_{\text{biased}} \leq r^n - 1$. For radix 2 and $n = 4$, the exponent has a maximum value of

$$r^n - 1 = 2^4 - 1 = 15$$

12.3 Floating-Point Addition

The addition of two fractions is identical to the addition algorithm presented in fixed-point addition. If the signs of the operands are the same ($A_{\text{sign}} \oplus B_{\text{sign}} = 0$), then this is referred to as *true addition* and the fractions are added. True addition corresponds to one of the following conditions:

$$\begin{aligned} &(+A) + (+B) \\ &(-A) + (-B) \\ &(+A) - (-B) \\ &(-A) - (+B) \end{aligned}$$

Floating-point addition is defined as shown in Equation 12.6 for two numbers A and B, where $A = f_A \times r^{e_A}$ and $B = f_B \times r^{e_B}$.

$$A + B = (f_A \times r^{e_A}) + (f_B \times r^{e_B})$$

$$= [f_A + (f_B \times r^{-(e_A - e_B)})] \times r^{e_A} \text{ for } e_A > e_B$$

$$= [(f_A \times r^{-(e_B - e_A)}) + f_B] \times r^{e_B} \text{ for } e_A \leq e_B \qquad (12.6)$$

The terms $r^{-(e_A - e_B)}$ and $r^{-(e_B - e_A)}$ are *shifting factors* to shift the fraction with the smaller exponent. This is analogous to a divide operation, since $r^{-(e_A - e_B)}$ is equivalent to $1/r^{(e_A - e_B)}$, which is a right shift. For $e_A > e_B$, fraction f_B is shifted right the number of bit positions specified by the absolute value of $| e_A - e_B |$. An example of using the shifting factor for addition is shown in Figure 12.3 for two operands $A = +26.5$ and $B = +4.375$. Since the implied 1 is part of the fractions, it must be considered when adding two normalized floating-point numbers.

The fractions must be properly aligned before addition can take place; therefore, the fraction with the smaller exponent is shifted right and the exponent is adjusted by increasing the exponent by one for each bit position shifted.

Before alignment

$A =$	$f_A \times r^5$		
$A =$	0 . 1 1 0 1 0 1 0 0	$\times 2^5$	+26.5
$B =$	$f_B \times r^3$		
$B =$	0 . 1 0 0 0 1 1 0 0	$\times 2^3$	+4.375

After alignment

$A =$	0 . 1 1 0 1 0 1 0 0	$\times 2^5$	+26.5
$B =$	0 . 0 0 1 0 0 0 1 1	$\times 2^5$	+4.375
$A + B =$	0 . 1 1 1 1 0 1 1 1	$\times 2^5$	+30.875

Figure 12.3 Example of adding two floating-point operands.

Operand B is shifted right by an amount equal to $| e_A - e_B | = | 5 - 3 | = 2$; that is, $B = f_B \times r^{-2}$, which is a divide by four operation accomplished by a right shift of 2 bit positions.

Figure 12.4 shows an example of floating-point addition when adding $A = +12$ and $B = +35$, in which the 8-bit fractions are not properly aligned initially and there is no postnormalization required. *Postnormalization* occurs when the resulting fraction overflows, requiring a right shift of one bit position with a corresponding increment of the exponent. The bit causing the overflow is shifted right into the high-order fraction bit position.

Before alignment

$A = f_A \times r^4$

$A = 0 \;.\; 1\;1\;0\;0\;0\;0\;0\;0 \quad \times 2^4 \qquad +12$

$B = f_B \times r^6$

$B = 0 \;.\; 1\;0\;0\;0\;1\;1\;0\;0 \quad \times 2^6 \qquad +35$

After alignment

$A = 0 \;.\; 0\;0\;1\;1\;0\;0\;0\;0 \quad \times 2^6 \qquad +12$

$B = 0 \;.\; 1\;0\;0\;0\;1\;1\;0\;0 \quad \times 2^6 \qquad +35$

$A + B = 0 \;.\; 1\;0\;1\;1\;1\;1\;0\;0 \quad \times 2^6 \qquad +47$

Figure 12.4 Example of floating-point addition in which the fractions are not properly aligned initially and there is no postnormalization.

Figure 12.5 shows an example of floating-point addition when adding $A = +26$ and $B = +20$, in which the 8-bit fractions are properly aligned and postnormalization is required. The sign of the augend A becomes the sign of the result.

Fractions are aligned

$A = 0 \;.\; 1\;1\;0\;1\;0\;0\;0\;0 \quad \times 2^5 \qquad +26$

$B = 0 \;.\; 1\;0\;1\;0\;0\;0\;0\;0 \quad \times 2^5 \qquad +20$

$A + B = 1 \leftarrow \;.\; 0\;1\;1\;1\;0\;0\;0\;0 \quad \times 2^5$

$A_{sign} \rightarrow 0 \;.\; 1\;0\;1\;1\;1\;0\;0\;0 \quad \times 2^6 \qquad +46$

Figure 12.5 Example of floating-point addition in which the fractions are properly aligned and there is postnormalization.

Figure 12.6 shows an example of floating-point addition when adding $A = -17.50$ and $B = -38.75$, in which the 8-bit fractions are not properly aligned and postnormalization is not required. This operation is called *true addition*, because the result is the sum of the two operands regardless of the signs.

Before alignment

$$A = f_A \times r^5$$

$$A = 1\ .\ 1\ 0\ 0\ 0\ 1\ 1\ 0\ 0 \qquad \times 2^5 \qquad -17.50$$

$$B = f_B \times r^6$$

$$B = 1\ .\ 1\ 0\ 0\ 1\ 1\ 0\ 1\ 1 \qquad \times 2^6 \qquad -38.75$$

After alignment

$$A = 1\ .\ 0\ 1\ 0\ 0\ 0\ 1\ 1\ 0 \qquad \times 2^6 \qquad -17.50$$

$$B = 1\ .\ 1\ 0\ 0\ 1\ 1\ 0\ 1\ 1 \qquad \times 2^6 \qquad -38.75$$

$$A + B = 1\ .\ 1\ 1\ 1\ 0\ 0\ 0\ 0\ 1 \qquad \times 2^6 \qquad -56.25$$

Figure 12.6 Example of floating-point addition in which the fractions are not properly aligned and there is no postnormalization.

The alignment and shifting of the fractions is now summarized. Equation 12.6 states that if $e_A > e_B$, then fraction f_A is added to the aligned fraction f_B with the exponent e_A assigned to the resulting sum. The radix points of the two operands must be aligned prior to the addition operation. This is achieved by comparing the relative magnitudes of the two exponents. The fraction with the smaller exponent is then shifted $| e_A - e_B |$ positions to the right.

The augend and addend are then added and the sum is characterized by the larger exponent. A carry-out of the high-order bit position may occur, yielding a result with an absolute value of $1 \leq | result | < 2$ before postnormalization.

12.4 Overflow and Underflow

The floating-point addition example of Figure 12.5 generated a carry-out of the high-order bit position, which caused a *fraction overflow*. When adding two numbers with the same sign, the absolute value of the result may be in the following range before postnormalization:

$$1 \leq | result | < 2$$

This indicates that the fraction is in the range of 1.000 . . . 0 to 1.111 . . .1. The overflow can be corrected by shifting the carry-out in concatenation with the fraction one bit position to the right and incrementing the exponent by 1. This operation is shown in Equation 12.7.

$$
\begin{aligned}
A + B &= (f_A \times r^{e_A}) + (f_B \times r^{e_B}) \\
&= \{[f_A + (f_B \times r^{-(e_A - e_B)})] \times r^{-1}\} \times r^{e_A + 1} \text{ for } e_A > e_B \\
&= \{[(f_A \times r^{-(e_B - e_A)}) + f_B] \times r^{-1}\} \times r^{e_B + 1} \text{ for } e_A \leq e_B \qquad (12.7)
\end{aligned}
$$

The term r^{-1} is the shifting factor that shifts the resulting fraction and the carry-out one bit position to the right. For radix 2, the shifting factor is 2^{-1} (or 1/2), which divides the result by 2 by executing a right shift of one bit position. The terms $r^{e_A + 1}$ and $r^{e_B + 1}$ increment the appropriate exponents by 1. Equation 12.6 is similar to Equation 12.7, but does not require a shift operation.

When aligning a fraction by shifting the fraction right and adjusting the exponent, bits may be lost off the right end of the fraction, resulting in a *fraction underflow*. This topic is discussed in Chapter 16 in the section on rounding.

12.5 General Floating-Point Organization

The same hardware that was utilized for fixed-point addition can also be used for the addition of floating-point fractions. However, additional hardware is necessary for fraction alignment, exponent comparison, and other characteristics that are unique to floating-point arithmetic, such as adding a bias to each exponent.

Floating-point addition can be implemented with two fixed-point adders: one for fraction addition and one for exponent processing. The organization for floating-point addition can also be used for floating-point subtraction. The hardware assigned to the fractions must perform addition and also have the ability to shift the appropriate fraction right for fraction alignment. The hardware required for the exponents must have the ability to compare exponents and to increment the exponents during fraction alignment. Exponent comparison can be accomplished with a simple comparator or by subtracting the exponents by adding the 2s complement of the subtrahend.

When the fractions are properly aligned, they are added as fixed-point numbers and the larger exponent is assigned to the result. A count-down counter contains the difference between the two exponents and decrements by one for each shift of the appropriate fraction.

The *addition algorithm* can be partitioned into seven steps, as shown below.

1. Normalize the operands.
2. Check for zero operands.
3. Bias the exponents.
4. Align the fractions by selecting the number with the smaller exponent and shifting the fraction (significand) to the right by an amount equal to the difference of the two exponents.
5. Add the fractions.
6. Set the exponent of the result equal to the larger exponent.
7. Normalize the result, if necessary.

The result of the addition operation may produce a sum that requires rounding. Rounding deletes one or more of the low-order bits and adjusts the remaining bits according to a specific rule. Rounding is presented in detail in Chapter 16. Figure 12.7 shows a flowchart representation of the above rules for floating-point addition.

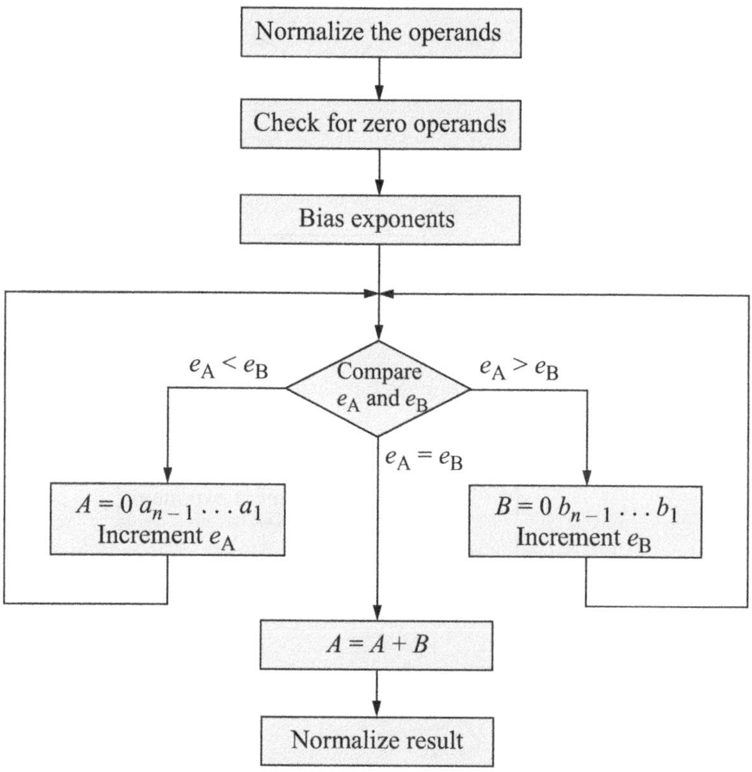

Figure 12.7 Flowchart depicting floating-point addition .

A floating-point number whose fraction is zero cannot be normalized; therefore, checking for a fraction with a value of zero is performed early in the algorithm — if a value of zero is detected, then the operation is terminated and the sum is equal to the nonzero operand. The bias generates exponents e_A and e_B as positive numbers that are compared for relative magnitudes. If the floating-point numbers are in single-precision format, then the *bias constant* has a value of +127; if the numbers are in double-precision format, then the bias constant has a value of +1023.

If the exponents are equal, then the fractions are already aligned and the addition operation can be executed. If the exponents are not equal, then the fraction with the smaller exponent is shifted right and its exponent is incremented by one. This process continues until the exponents are equal. Figure 12.8 shows a flowchart depicting postnormalization of the sum after floating-point addition.

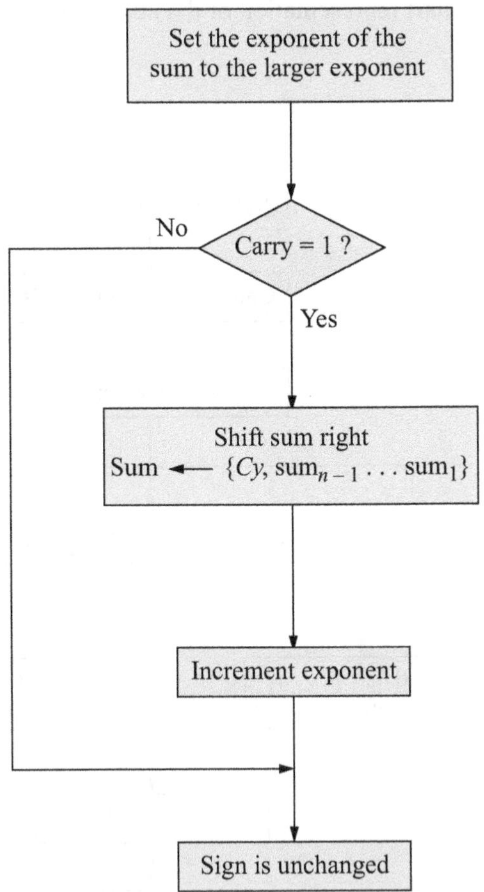

Figure 12.8 Postnormalization of the sum after floating-point addition.

When the sum of a true addition is equal to or greater than one, the carry-out and the sum are shifted right in concatenation, with the carry bit being shifted into the high-order bit position sum_{n-1} of the fraction. Therefore, no further normalization is necessary — the exponent is incremented by one and the sign bit is unchanged. During the right shift operation, the low-order bit may be lost, which may necessitate rounding. The carry-out from the high-order bit position of the sum does not necessarily indicate an overflow. Exponent overflow may occur during true addition when the value of the exponent exceeds the maximum size of the exponent field during alignment.

12.6 Verilog HDL Implementation

The Verilog HDL design of a floating-point adder will be implemented using behavioral modeling for the single-precision format of 32 bits, as shown in Figure 12.9. There are two inputs: the augend and the addend; there are three outputs: the sign of the floating-point number, the exponent, and the sum. The floating-point augend, *flp_a[31:0]*, consists of the sign bit, *sign_a*; the 8-bit exponent, *exp_a[7:0]*; and the 23-bit fraction, *fract_a[22:0]*. The addend, *flp_b[31:0]*, consists of the sign bit, *sign_b*; the 8-bit exponent, *exp_b[7:0]*; and the 23-bit fraction, *fract_b[22:0]*.

The exponents are biased by adding the bias constant of +127 (0111 1111) prior to the addition operation. Then the fractions are aligned by comparing the exponents, and a counter, *ctr_align[7:0]*, is set to the difference of the two exponents. The fraction with the smaller exponent is shifted right one bit position, the exponent is incremented by one, and the alignment counter is decremented by one. This process repeats until the alignment counter decrements to a value of zero at which point the fractions are aligned and the addition operation is performed. If the exponents are equal, then there is no alignment and the addition operation is performed immediately.

There may be a fraction overflow; therefore, the result is obtained by the following Verilog statement: *{cout, sum} = fract_a + fract_b;*, which allows for a carry-out concatenated with the sum. If there is a carry-out, then the carry-out and sum are shifted right one bit position in concatenation to postnormalize the result as follows: *{cout, sum} = {cout, sum} >> 1;* and the sign of the result is set equal to the sign of the augend.

The test bench is shown in Figure 12.10 and applies five sets of input vectors for the floating-point augend, *flp_a*, and the addend, *flp_b*, including a set of vectors that require postnormalization. In this example, the fractions are normalized without an implied 1 bit. The use of an implied 1 bit is left as a problem.

The outputs are shown in Figure 12.11 and indicate the sign, the biased exponent, and the sum for each set of input vectors. The exponents and sums can be easily verified by manual calculations.

The waveforms are shown in Figure 12.12 and indicate the value of the floating-point operands in hexadecimal. Although the sums shown in the waveforms appear to be incorrect, they are indeed correct, because the 23-bit fraction is partitioned into 4-bit segments while scanning the fraction from right to left with an implied zero in the high-order bit position. For example, the outputs indicate that the sum for the first vector is 1011 1100 0000 0000 0000 000. However, the waveforms show the sum as 5e0000. By scanning the sum in the outputs of Figure 12.11 from right to left and including a 0 bit in the high-order bit position to accommodate the hexadecimal digit of four bits, the waveforms are correct, as shown below.

0101	1110	0000	0000	0000	0000
5	e	0	0	0	0

The same is true for the fifth set of vectors, in which the sum in the outputs of Figure 12.11 is 1101 1101 1100 0000 0000 000, but translates to 6ee000 because of the right-to-left scan with a high-order 0 bit, as shown below.

0110	1110	1110	0000	0000	0000
6	e	e	0	0	0

```
//behavioral floating-point addition
module add_flp3 (flp_a, flp_b, sign, exponent, sum);

input [31:0] flp_a, flp_b;
output [22:0] sum;
output sign;
output [7:0] exponent;

//variables used in an always block
//are declared as registers
reg sign_a, sign_b;
reg [7:0] exp_a, exp_b;
reg [7:0] exp_a_bias, exp_b_bias;
reg [22:0] fract_a, fract_b;
reg [7:0] ctr_align;
reg [22:0] sum;
reg sign;
reg [7:0] exponent;
reg cout;

always @ (flp_a or flp_b)
begin
   sign_a = flp_a [31];
   sign_b = flp_b [31];
//continued on next page
```

Figure 12.9 Behavioral design module for floating-point addition.

```
   exp_a = flp_a [30:23];
   exp_b = flp_b [30:23];

   fract_a = flp_a [22:0];
   fract_b = flp_b [22:0];

//bias exponents
   exp_a_bias = exp_a + 8'b0111_1111;
   exp_b_bias = exp_b + 8'b0111_1111;

//align fractions
   if (exp_a_bias < exp_b_bias)
      ctr_align = exp_b_bias - exp_a_bias;

      while (ctr_align)
         begin
            fract_a = fract_a >> 1;
            exp_a_bias = exp_a_bias + 1;
            ctr_align = ctr_align - 1;
         end

   if (exp_b_bias < exp_a_bias)
      ctr_align = exp_a_bias - exp_b_bias;

      while (ctr_align)
         begin
            fract_b = fract_b >> 1;
            exp_b_bias = exp_b_bias + 1;
            ctr_align = ctr_align - 1;
         end

//add fractions
   {cout, sum} = fract_a + fract_b;

//normalize result
   if (cout == 1)
      begin
         {cout, sum} = {cout, sum} >> 1;
         exp_b_bias = exp_b_bias + 1;
      end

   sign = sign_a;
   exponent = exp_b_bias;
   exp_unbiased = exp_b_bias - 8'b0111_1111;

end
endmodule
```

Figure 12.9 (Continued)

```
//test bench for floating-point addition
module add_flp3_tb;
reg [31:0] flp_a, flp_b;
wire sign;
wire [7:0] exponent;
wire [22:0] sum;
initial     //display variables
$monitor ("sign=%b, exp_biased=%b, exp_unbiased=%b, sum=%b",
            sign, exponent, exp_unbiased, sum);
initial     //apply input vectors
begin
      //+12 + +35 = +47
      //          s ----e---- -------------f-------------
#0    flp_a = 32'b0_0000_0100_1100_0000_0000_0000_0000_000;
      flp_b = 32'b0_0000_0110_1000_1100_0000_0000_0000_000;

      //+26 + +20 = +46
      //          s ----e---- -------------f-------------
#10   flp_a = 32'b0_0000_0101_1101_0000_0000_0000_0000_000;
      flp_b = 32'b0_0000_0101_1010_0000_0000_0000_0000_000;

      //+26.5 + +4.375 = +30.875
      //          s ----e---- -------------f-------------
#10   flp_a = 32'b0_0000_0101_1101_0100_0000_0000_0000_000;
      flp_b = 32'b0_0000_0011_1000_1100_0000_0000_0000_000;

      //+11 + +34 = +45
      //          s ----e---- -------------f-------------
#10   flp_a = 32'b0_0000_0100_1011_0000_0000_0000_0000_000;
      flp_b = 32'b0_0000_0110_1000_1000_0000_0000_0000_000;

      //+23.75 + +87.125 = +110.875
      //          s ----e---- -------------f-------------
#10   flp_a = 32'b0_0000_0101_1011_1110_0000_0000_0000_000;
      flp_b = 32'b0_0000_0111_1010_1110_0100_0000_0000_000;
#10   $stop;
end
add_flp3 inst1 (  //instantiate the module
   .flp_a(flp_a),
   .flp_b(flp_b),
   .sign(sign),
   .exponent(exponent),
   .exp_unbiased(exp_unbiased),
   .sum(sum)
   );
endmodule
```

Figure 12.10 Test bench module for floating-point addition.

```
sign = 0, exp_biased = 1000_0101,
   exp_unbiased = 0000_0110,
   sum = 1011_1100_0000_0000_0000_000

sign = 0, exp_biased = 1000_0101,
   exp_unbiased = 0000_0110,
   sum = 1011_1000_0000_0000_0000_000

sign = 0, exp_biased = 1000_0100,
   exp_unbiased = 0000_0101,
   sum = 1111_0111_0000_0000_0000_000

sign = 0, exp_biased = 1000_0101,
   exp_unbiased = 0000_0110,
   sum = 1011_0100_0000_0000_0000_000

sign = 0, exp_biased = 1000_0110,
   exp_unbiased = 0000_0111,
   sum = 1101_1101_1100_0000_0000_000
```

Figure 12.11 Outputs for floating-point addition.

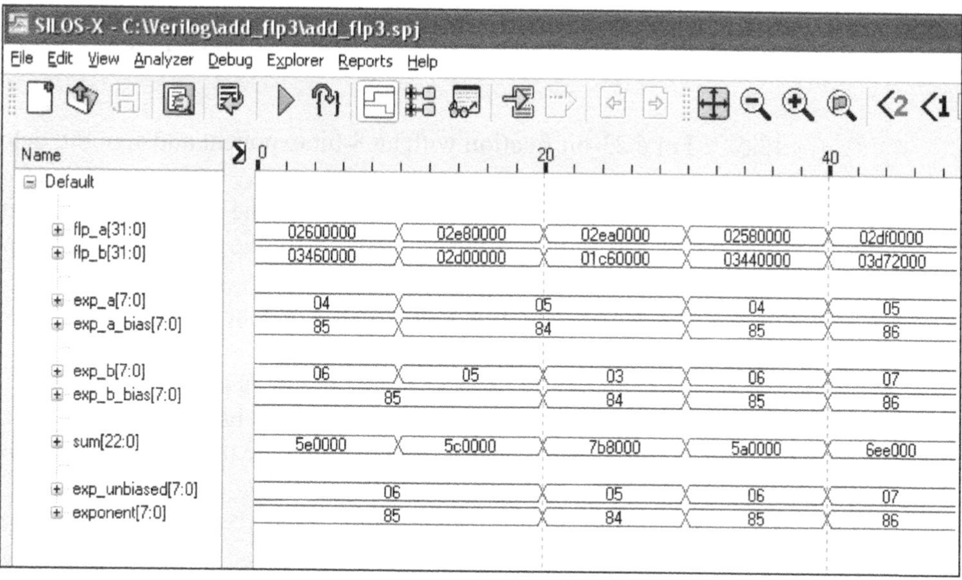

Figure 12.12 Waveforms for floating-point addition.

12.7 Problems

12.1 What is the binary bit configuration for the minimum and maximum value of an 11-bit biased exponent?

12.2 Add the two floating-point numbers shown below with no implied 1 bit.

$$A = 0.1100\ 1000\ 0000\ldots 0\ \times 2^6$$
$$B = 0.1000\ 1000\ 0000\ldots 0\ \times 2^3$$

12.3 Perform the following floating-point addition for positive operands with no implied 1 bit:

$$A = 0.1001\ 0000\ \times 2^6$$
$$B = 0.1111\ 0000\ \times 2^2$$

12.4 Indicate which of the equations shown below correctly represent floating-point addition.

(a) $A + B = (f_A \times r^{e_A}) \times (f_B \times r^{e_B})$

(b) $A + B = [f_A + (f_B \times r^{-(e_A - e_B)})] \times r^{e_A}$, for $e_A > e_B$

(c) $A + B = (f_A \times r^{e_A}) + (f_B \times r^{e_B})$

(d) $A + B = [(f_A \times r^{-(e_B - e_A)}) + f_B] \times r^{e_B}$, for $e_A > e_B$

12.5 For a 23-bit fraction with an 8-bit exponent and sign bit, determine:

(a) The largest positive number with the most negative unbiased exponent
(b) The largest positive number with the smallest biased exponent

12.6 For a 23-bit fraction with an 8-bit exponent and sign bit, determine:

(a) The largest positive number with an unbiased exponent
(b) The largest positive number with a biased exponent
(c) The most negative number with the most negative unbiased exponent

12.7 Perform floating-point addition on the fixed-point binary numbers shown below with no implied 1 bit.

$$A = 0000\ 1111$$
$$B = 0000\ 0110$$

12.8 Perform floating-point addition on the fixed-point binary numbers shown below with no implied 1 bit.

$A = 0011\ 1011$
$B = 0011\ 1101$

12.9 Perform floating-point addition on the numbers shown below with no implied 1 bit.

$A = 0.1000\ 1110 \times 2^6$
$B = 0.1101\ 0000 \times 2^3$

12.10 Obtain the normalized floating-point representation for the decimal number shown below using the single-precision format. Use a biased exponent and an implied 1 bit.

+9.75

12.11 Obtain the normalized floating-point representation for the decimal number shown below using the single-precision format. Use a biased exponent and an implied 1 bit.

–0.125

12.12 Add the decimal operands shown below using the single-precision floating-point format with biased exponents and an implied 1 bit.

$A = (-10),\ B = (-35)$

12.13 Design a floating-point adder using behavioral modeling to add two operands using the single-precision format. This implementation will not use biased exponents and will not have an implied 1 bit. Obtain the design module, the test bench module with at least three sets of input vectors, including one with an integer and fraction, and the outputs.

12.14 Design a floating-point adder using behavioral modeling to add two operands using the single-precision format. This implementation will use biased exponents and will have an implied 1 bit. Obtain the design module, the test bench module with several sets of input vectors, including at least one with an integer and fraction, the outputs, and the waveforms.

13.1 Numerical Examples
13.2 Flowcharts
13.3 Verilog HDL Implementations
13.4 Problems

13

Floating-Point Subtraction

Floating-point addition and subtraction are more complicated than multiplication and division because of the requirement for aligning the fractions before adding or subtracting. Fraction overflow can also occur in subtraction since subtraction is accomplished by adding the 2s complement of the subtrahend. The subtraction of two fractions is identical to the subtraction algorithm presented in fixed-point addition. If the signs of the operands are the same ($A_{sign} \oplus B_{sign} = 0$) and the operation is subtraction, then this is referred to as *true subtraction* and the fractions are subtracted. If the signs of the operands are different ($A_{sign} \oplus B_{sign} = 1$) and the operation is addition, then this is also specified as *true subtraction*. True subtraction corresponds to one of the following conditions:

$$\begin{array}{ccc} (+A) & - & (+B) \\ (-A) & - & (-B) \\ (+A) & + & (-B) \\ (-A) & + & (+B) \end{array}$$

As in fixed-point notation, the same hardware can be used for both floating-point addition and subtraction to add or subtract the fractions. Additional hardware is required for exponent processing. All operands will consist of normalized fractions properly aligned with biased exponents. Floating-point subtraction is defined as shown in Equation 13.1 for two numbers A and B, where $A = f_A \times r^{e_A}$ and $B = f_B \times r^{e_B}$.

$$A - B = (f_A \times r^{e_A}) - (f_B \times r^{e_B})$$

$$= [f_A - (f_B \times r^{-(e_A - e_B)})] \times r^{e_A} \text{ for } e_A > e_B$$

$$= [(f_A \times r^{-(e_B - e_A)}) - f_B] \times r^{e_B} \text{ for } e_A \leq e_B \qquad (13.1)$$

The terms $r^{-(e_A - e_B)}$ and $r^{-(e_B - e_A)}$ are *shifting factors* to shift the fraction with the smaller exponent. This is analogous to a divide operation, since $r^{-(e_A - e_B)}$ is equivalent to $1/r^{(e_A - e_B)}$, which is a right shift. For $e_A > e_B$, fraction f_B is shifted right the number of bit positions specified by the absolute value of $| e_A - e_B |$. An example of using the shifting factor for subtraction is shown in Figure 13.1 for two operands, $A = +36.5$ and $B = +5.75$. Since the *implied 1* is part of the fractions, it must be considered when subtracting two normalized floating-point numbers — the implied 1 is shown as the high-order bit in Figure 13.1. The equation for floating-point subtraction that corrects for fraction overflow is shown in Equation 13.2.

Before alignment				
$A =$	0 . 1 0 0 1 | 0 0 1 0	$\times 2^6$	+36.5	
$B =$	0 . 1 0 1 1 | 1 0 0 0	$\times 2^3$	+5.75	
After alignment				
$A =$	0 . 1 0 0 1 | 0 0 1 0	$\times 2^6$	+36.5	
$B =$	0 . 0 0 0 1 | 0 1 1 1	$\times 2^6$	+5.75	
Subtract fractions				
$A =$	0 . 1 0 0 1 | 0 0 1 0	$\times 2^6$		
+) $B' + 1 =$	0 . 1 1 1 0 | 1 0 0 1	$\times 2^6$		
$1 \leftarrow$	0 . 0 1 1 1 | 1 0 1 1	$\times 2^6$	+30.75	
Postnormalize (SL1)	0 . 1 1 1 1 | 0 1 1 0	$\times 2^5$	+30.75	

Figure 13.1 Example of floating-point subtraction.

$$A - B = (f_A \times r^{e_A}) - (f_B \times r^{e_B})$$

$$= \{[f_A - (f_B \times r^{-(e_A - e_B)})] \times r^{-1}\} \times r^{e_A + 1} \text{ for } e_A > e_B$$

$$= \{[(f_A \times r^{-(e_B - e_A)}) - f_B] \times r^{-1}\} \times r^{e_B + 1} \text{ for } e_A \leq e_B \quad (13.2)$$

13.1 Numerical Examples

Subtraction can yield a result that is either true addition or true subtraction. *True addition* produces a result that is the sum of the two operands disregarding the signs; *true subtraction* produces a result that is the difference of the two operands disregarding the signs. There are four cases that yield a true addition, as shown in Figure 13.2, and eight cases that yield a true subtraction, as shown in Figure 13.3.

```
    – Small number          – Large number
–)  + Large number      –)  + Small number
    True addition           True addition

    + Large number          + Small number
–)  – Small number      –)  – Large number
    True addition           True addition
```

Figure 13.2 Examples of true addition.

```
    + Large number          + Small number
–)  + Small number      –)  + Large number
    True subtraction        True subtraction

    – Small number          – Large number
–)  – Large number      –)  – Small number
    True subtraction        True subtraction
```

Continued on next page

Figure 13.3 Examples of true subtraction.

+ Small number	– Small number
+) – Large number	+) + Large number
True subtraction	True subtraction

+ Large number	– Large number
+) – Small number	+) + Small number
True subtraction	True subtraction

Figure 13.3 (Continued)

True addition

Several numerical examples will be presented now to illustrate both true addition and true subtraction. Four examples of true addition are shown below in Examples 13.1, Example 13.2, Example 13.3, and Example 13.4. For true addition, the following rules apply, whether $|fract\_a| < |fract\_b|$ or $|fract\_a| > |fract\_b|$:

(1) Bias the exponents.
(2) Align the fractions
(4) Perform the addition.
(5) The sign of the result is the sign of the minuend.
(6) If carry-out =1, then {cout, rslt} >> 1
(7) Increment the exponent by 1.

Example 13.1 Subtraction will be performed on the decimal numbers (–22) – (+28) to yield a result of –50.

Before alignment

A =	1 . 1 0 1 1	0 0 0 0	$\times 2^5$	–22
B =	0 . 1 1 1 0	0 0 0 0	$\times 2^5$	+28

After alignment (already aligned)

A =	1 . 1 0 1 1	0 0 0 0	$\times 2^5$	–22
+) B =	1 . 1 1 1 0	0 0 0 0	$\times 2^5$	–28
1 ←	. 1 0 0 1	0 0 0 0	$\times 2^5$	
Postnormalize	1 . 1 1 0 0	1 0 0 0	$\times 2^6$	–50

Example 13.2 Subtraction will be performed on the decimal numbers (–40) – (+24) to yield a result of –64.

Before alignment

A = 1	.	1	0	1	0	0	0	0	0	$\times 2^6$	–40
B = 0	.	1	1	0	0	0	0	0	0	$\times 2^5$	+24

After alignment

A = 1	.	1	0	1	0	0	0	0	0	$\times 2^6$	–40
B = 0	.	0	1	1	0	0	0	0	0	$\times 2^6$	+24

Add fractions

A = 1	.	1	0	1	0	0	0	0	0	$\times 2^6$	–40
+) B = 1	.	0	1	1	0	0	0	0	0	$\times 2^6$	–24
1 ←	.	0	0	0	0	0	0	0	0	$\times 2^6$	
Postnormalize 1	.	1	0	0	0	0	0	0	0	$\times 2^7$	–64

Example 13.3 Subtraction will be performed on the decimal numbers (+34) – (–11) to yield a result of +45.

Before alignment

A = 0	.	1	0	0	0	1	0	0	0	$\times 2^6$	+34
B = 1	.	1	0	1	1	0	0	0	0	$\times 2^4$	–11

After alignment

A = 0	.	1	0	0	0	1	0	0	0	$\times 2^6$	+34
B = 1	.	0	0	1	0	1	1	0	0	$\times 2^6$	–11

Add fractions

A = 0	.	1	0	0	0	1	0	0	0	$\times 2^6$	+34
+) B = 1	.	0	0	1	0	1	1	0	0	$\times 2^6$	–11
0 ←	.	1	0	1	1	0	1	0	0	$\times 2^6$	+45
No postnormalize 0	.	1	0	1	1	0	1	0	0	$\times 2^6$	+45

Example 13.4 Subtraction will be performed on the decimal numbers (+16) – (–42) to yield a result of +58.

Before alignment					
	$A =$ 0 .	1 0 0 0	0 0 0 0	$\times 2^5$	+16
	$B =$ 1 .	1 0 1 0	1 0 0 0	$\times 2^6$	–42
After alignment					
	$A =$ 0 .	0 1 0 0	0 0 0 0	$\times 2^6$	+16
	$B =$ 1 .	1 0 1 0	1 0 0 0	$\times 2^6$	–42
Add fractions					
	$A =$ 0 .	0 1 0 0	0 0 0 0	$\times 2^6$	+16
	+) $B =$ 1 .	1 0 1 0	1 0 0 0	$\times 2^6$	–42
	0 ← .	1 1 1 0	1 0 0 0	$\times 2^6$	+58
No postnormalize	0 .	1 1 1 0	1 0 0 0	$\times 2^6$	+58

Examples of true subtraction will now be presented that illustrate the remaining categories shown in Figure 13.3. For true subtraction, the rules shown below apply depending on whether $|fract\_a| < |fract\_b|$ or $|fract\_a| > |fract\_b|$, on the state of a *mode* control input, and on the signs of the operands. If $mode = 0$, then the operation is addition; if $mode = 1$, then the operation is subtraction.

True subtraction

- Bias the exponents.
- Align the fractions

- If $|fract\_a| < |fract\_b|$ and $mode = 1$ and sign of $fract\_a$ = sign of $fract\_b$.
  - (1) 2s complement *fract_b*.
  - (2) Perform the addition.
  - (3) Sign of the result = A_{sign}'.

  If $sum_{n-1} = 1$
  - (4) Then 2s complement the result. There is only one sign bit for the resulting floating-point number; therefore, the fraction cannot also be signed. The number must be in sign-magnitude notation.
  - (5) Postnormalize, if necessary (shift left 1 and decrement the exponent).

- If $|fract\_a| < |fract\_b|$ and $mode = 0$ and sign of $fract\_a \neq$ sign of $fract\_b$.
  - (1) 2s complement $fract\_b$.
  - (2) Perform the addition.
  - (3) Sign of the result = A_{sign}'

  If $sum_{n-1} = 1$
  - (4) Then 2s complement the result. There is only one sign bit for the resulting floating-point number; therefore, the fraction cannot also be signed. The number must be in sign-magnitude notation.
  - (5) Postnormalize, if necessary (shift left 1 and decrement the exponent).

- If $|fract\_a| > |fract\_b|$ and $mode = 0$ and sign of $fract\_a \neq$ sign of $fract\_b$.
  - (1) 2s complement $fract\_b$.
  - (2) Perform the addition.
  - (3) Sign of the result = A_{sign}.
  - (4) Postnormalize, if necessary (shift left 1 and decrement the exponent).

- If $|fract\_a| > |fract\_b|$ and $mode = 1$ and sign of $fract\_a =$ sign of $fract\_b$.
  - (1) 2s complement $fract\_b$.
  - (2) Perform the addition.
  - (3) Sign of the result = A_{sign}.
  - (4) Postnormalize, if necessary (shift left 1 and decrement the exponent).

Example 13.5 Subtraction will be performed on the decimal numbers (+11) – (+34) to yield a result of –23.

Before alignment

$A =$	0	.	1 0 1 1	0 0 0 0	$\times 2^4$	+11
$B =$	0	.	1 0 0 0	1 0 0 0	$\times 2^6$	+34

After alignment

$A =$	0	.	0 0 1 0	1 1 0 0	$\times 2^6$
+) $B' + 1 =$	0	.	0 1 1 1	1 0 0 0	$\times 2^6$

Add fractions

$A =$	0	.	0 0 1 0	1 1 0 0	$\times 2^6$	+11
+) $B =$	0	.	0 1 1 1	1 0 0 0	$\times 2^6$	+34
	0 ⟵	.	1 0 1 0	0 1 0 0	$\times 2^6$	

2s complement	1	.	0 1 0 1	1 1 0 0	$\times 2^6$	
Postnormalize	1	.	1 0 1 1	1 0 0 0	$\times 2^5$	–23

Example 13.6 Subtraction will be performed on the decimal numbers (–23) – (–36) to yield a result of +13.

	Sign		Fraction	Exponent	Decimal
Before alignment					
A =	1	.	1 0 1 1 | 1 0 0 0	$\times 2^5$	–23
B =	1	.	1 0 0 1 | 0 0 0 0	$\times 2^6$	–36
After alignment					
A =	1	.	0 1 0 1 | 1 1 0 0	$\times 2^6$	–23
2s complement B =	1	.	1 0 0 1 | 0 0 0 0	$\times 2^6$	–36
Add fractions					
A =	1	.	0 1 0 1 | 1 1 0 0	$\times 2^6$	
+) $B' + 1$ =	1	.	0 1 1 1 | 0 0 0 0	$\times 2^6$	
0 ←		.	1 1 0 0 | 1 1 0 0	$\times 2^6$	
2s complement	0	.	0 0 1 1 | 0 1 0 0	$\times 2^6$	+13
Postnormalize	0	.	1 1 0 1 | 0 0 0 0	$\times 2^4$	+13

Example 13.7 Subtraction will be performed on the decimal numbers (–47) – (–35) to yield a result of –12.

	Sign		Fraction	Exponent	Decimal
Before alignment					
A =	1	.	1 0 1 1 | 1 1 0 0	$\times 2^6$	–47
B =	1	.	1 0 0 0 | 1 1 0 0	$\times 2^6$	–35
After alignment (already aligned)					
A =	1	.	1 0 1 1 | 1 1 0 0	$\times 2^6$	
+) $B' + 1$ =	1	.	0 1 1 1 | 0 1 0 0	$\times 2^6$	
1 ←		.	0 0 1 1 | 0 0 0 0	$\times 2^6$	
	1	.	0 0 1 1 | 0 0 0 0	$\times 2^6$	–12
Postnormalize	1	.	1 1 0 0 | 0 1 0 0	$\times 2^4$	–12

Example 13.8 Subtraction will be performed on the decimal numbers (+28) + (–35) to yield a result of –7.

Before alignment						
$A =$	0	.	1 1 1 0	0 0 0 0	$\times 2^5$	+28
$B =$	1	.	1 0 0 0	1 1 0 0	$\times 2^6$	–35
After alignment						
$A =$	0	.	0 1 1 1	0 0 0 0	$\times 2^6$	+28
$B =$	1	.	1 0 0 0	1 1 0 0	$\times 2^6$	–35
Add fractions						
$A =$	0	.	0 1 1 1	0 0 0 0	$\times 2^6$	
+) $B' + 1 =$	1	.	0 1 1 1	0 1 0 0	$\times 2^6$	
0 ←		.	1 1 1 0	0 1 0 0	$\times 2^6$	
2s complement	1	.	0 0 0 1	1 1 0 0	$\times 2^6$	–7
Postnormalize	1	.	1 1 1 0	0 0 0 0	$\times 2^3$	–7

Example 13.9 Subtraction will be performed on the decimal numbers (–36) + (+48) to yield a result of +12.

Before alignment						
$A =$	1	.	1 0 0 1	0 0 0 0	$\times 2^6$	–36
$B =$	0	.	1 1 0 0	0 0 0 0	$\times 2^6$	+48
After alignment (already aligned)						
$A =$	1	.	1 0 0 1	0 0 0 0	$\times 2^6$	
+) $B' + 1 =$	0	.	0 1 0 0	0 0 0 0	$\times 2^6$	
0 ←		.	1 1 0 1	0 0 0 0	$\times 2^6$	
2s complement	0	.	0 0 1 1	0 0 0 0	$\times 2^6$	+12
Postnormalize	0	.	1 1 0 0	0 0 0 0	$\times 2^4$	+12

Example 13.10 Subtraction will be performed on the decimal numbers (+45) + (–13) to yield a result of +32.

Before alignment

$A =$	0	.	1 0 1 1	0 1 0 0	$\times 2^6$	+45
$B =$	1	.	1 1 0 1	0 0 0 0	$\times 2^4$	–13

After alignment

$A =$	0	.	1 0 1 1	0 1 0 0	$\times 2^6$	+45
$B =$	1	.	0 0 1 1	0 1 0 0	$\times 2^6$	–13

Add fractions

$A =$	0	.	1 0 1 1	0 1 0 0	$\times 2^6$	
+) $B' + 1 =$	1	.	1 1 0 0	1 1 0 0	$\times 2^6$	
	1 ⟵	.	1 0 0 0	0 0 0 0	$\times 2^6$	
	0	.	1 0 0 0	0 0 0 0	$\times 2^6$	+32

Example 13.11 Subtraction will be performed on the decimal numbers (–37) + (+17) to yield a result of –20.

Before alignment

$A =$	1	.	1 0 0 1	0 1 0 0	$\times 2^6$	–37
$B =$	0	.	1 0 0 0	1 0 0 0	$\times 2^5$	+17

After alignment

$A =$	1	.	1 0 0 1	0 1 0 0	$\times 2^6$	–37
$B =$	0	.	0 1 0 0	0 1 0 0	$\times 2^6$	+17

Add fractions

$A =$	1	.	1 0 0 1	0 1 0 0	$\times 2^6$	
+) $B' + 1 =$	0	.	1 0 1 1	1 1 0 0	$\times 2^6$	
	1 ⟵	.	0 1 0 1	0 0 0 0	$\times 2^6$	
Postnormalize	1	.	1 0 1 0	0 0 0 0	$\times 2^5$	–20

13.2 Flowcharts

The flowchart of Figure 13.4 graphically portrays the steps required for true addition and true subtraction for floating-point operands. It illustrates the alignment of the fractions and also contains the addition algorithm as well as the subtraction algorithm. If the exponents are not equal, then the fraction with the smaller exponent is shifted right and its exponent is increased by 1. A right shift of the operand A fraction is specified by the following equation: $A = 0\ a_{n-1}\ a_{n-2} \cdots a_1$

This process repeats until the exponents are equal. The fractions are then added or subtracted depending on the operation and on the signs of the operands. If the operation is addition and the signs are the same, that is, $A_{sign} \oplus B_{sign} = 0$, then the fractions are added. If the signs are different, that is, $A_{sign} \oplus B_{sign} = 1$, then the fractions are subtracted by adding the 2s complement of the addend to the augend. If the operation is subtraction and the signs are the same, that is, $A_{sign} \oplus B_{sign} = 0$, then the 2s complement of the subtrahend is added to the minuend. If the operation is subtraction and the signs are different, that is, $A_{sign} \oplus B_{sign} = 1$, then the fractions are added.

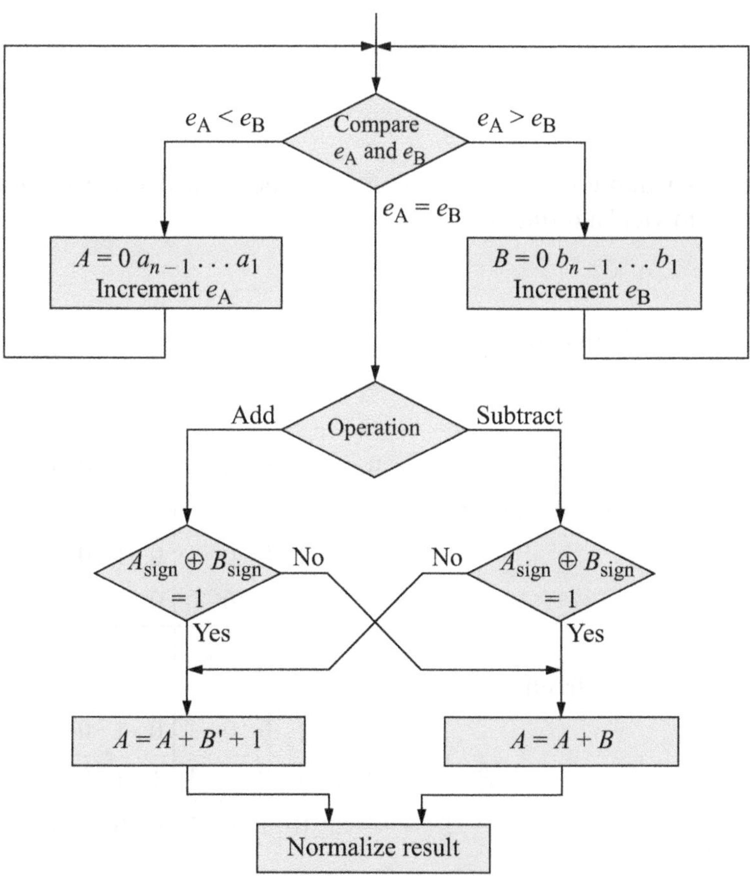

Figure 13.4 Flowchart showing alignment of fractions for addition/subtraction.

Postnormalization Figure 13.5 illustrates the normalization process for true subtraction; Figure 13.6 illustrates the normalization process for true addition. When the result of a subtraction produces a carry-out = 0 from the high-order bit position $n - 1$, the result must be 2s complemented and the sign bit inverted. This difference may be positive, negative, or zero. A negative result is in 2s complement notation and occurs when $| A < B |$. Since the floating-point number must be in sign-magnitude notation, the fraction cannot be a signed number — there cannot be two signs associated with the number — therefore, the result is 2s complemented in order to express the result in the correct sign-magnitude representation.

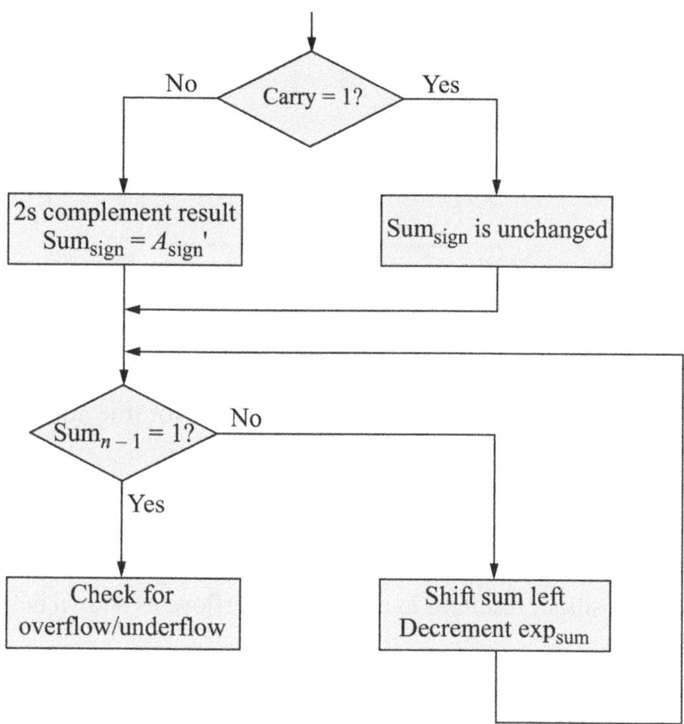

Figure 13.5 Normalization process for true subtraction.

If an operation produces a 0 in bit position $n - 1$ of the result, then the result is normalized by shifting the fraction left until a 1 bit appears in bit position $n - 1$ and the exponent is decreased accordingly. When the sum of a true addition is equal to or greater than 1.000 ... 00, the result is shifted right 1 bit position and the exponent is incremented by 1; that is, the carry bit is shifted right into the high-order bit position of the fraction and the sign is unchanged.

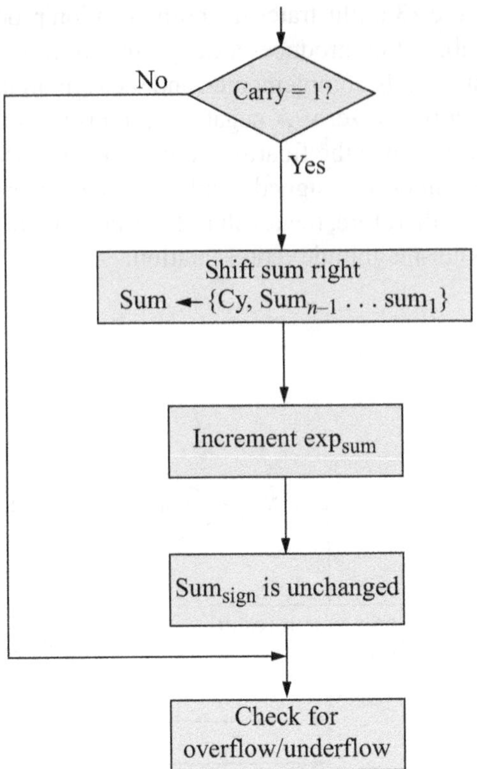

Figure 13.6 Normalization process for true addition.

A brief review of fraction overflow/underflow and exponent overflow/underflow will now be presented. An addition may generate a carry-out of the high-order bit position, resulting in a fraction overflow, as shown below.

$$1 \leq |\text{Resulting fraction}| < 2$$

1.000 . . . 0   1.111 . . . 1

This condition can be corrected by shifting the fraction and the carry-out one bit position to the right and increasing the exponent by one, as shown in Equation 13.3. Fraction underflow occurs during alignment when bits are lost off the right end of a fraction during a right shift for addition or subtraction.

Exponent overflow may occur during multiplication — covered in Chapter 14 — when the exponents are added, or during postnormalization during addition. Exponent underflow may occur during division — covered in Chapter 15 — when the exponents are subtracted, or during addition/subtraction when the shift count $| e_A - e_B |$ may be very large, indicating that one operand is very small when compared to the other operand.

$$A + B = (f_A \times r^{e_A}) + (f_B \times r^{e_B})$$

$$= \{[f_A + (f_B \times r^{-(e_A - e_B)})] \times r^{-1}\} \times r^{e_A + 1} \text{ for } e_A > e_B$$

$$= \{[(f_A \times r^{-(e_B - e_A)}) + f_B] \times r^{-1}\} \times r^{e_B + 1} \text{ for } e_A \leq e_B \qquad (13.3)$$

13.3 Verilog HDL Implementations

This section will design five behavioral modules: one for true addition; one for true subtraction where the minuend fraction is less than the subtrahend fraction and the signs are equal; one for true subtraction where the minuend fraction is less than the subtrahend fraction and the signs are unequal; one for true subtraction where the minuend fraction is greater than the subtrahend fraction and the signs are unequal; and one for true subtraction where the minuend fraction is greater than the subtrahend fraction and the signs are equal. These modules will not use the mode control input. The mode control input will be used in a behavioral design that combines all five of the above modules and is given as a problem.

The single-precision format is used in all three modules and is reproduced in Figure 13.7 for convenience. All exponents will be biased by adding a bias constant of +127 (0111 1111). The fractions will be aligned, if necessary, then added and the result postnormalized, if necessary.

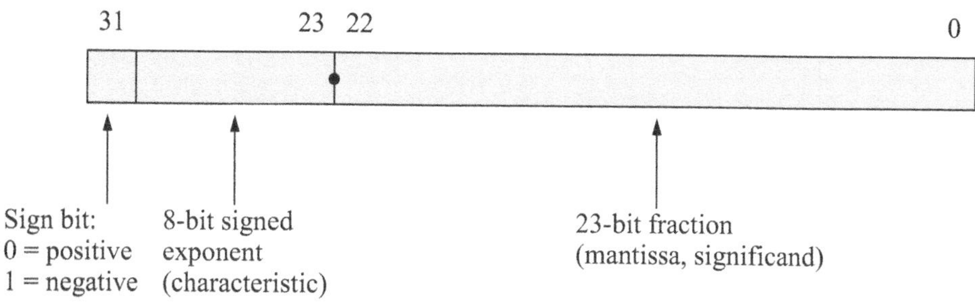

Figure 13.7 Single-precision floating-point format.

13.3.1 True Addition

The behavioral module for subtraction that results in true addition is shown in Figure 13.8. There are two 32-bit fractions: the minuend, *fract_a[31:0]*, and the subtrahend, *fract_b[31:0]*, each containing a sign bit, *sign_a* and *sign_b*; an 8-bit exponent, *exp_a[7:0]* and *exp_b[7:0]*; and a 23-bit fraction, *fract_a[22:0]* and *fract_b[22:0]*. The signs, exponents, and fractions are then defined, as they appear in Figure 13.7. The exponents are biased as follows: *exp_a_bias* = *exp_a* + *8'b0111_1111*, *exp_b_bias* = *exp_b* + *8'b0111_1111*, then the fractions are aligned.

There is a count-down counter, *ctr_align*, which is set to the difference of the two exponents. The test bench is shown in Figure 13.9, which applies twelve different input vectors for the operands according to Figure 13.2. The outputs shown in Figure 13.10, list the sign, the biased exponent, the unbiased exponent, and the resulting fraction.

```
//behavioral floating-point subtraction
//true addition: fract_a < fract_b or fract_a > fract_b
module sub_flp2 (flp_a, flp_b, sign, exponent,
                  exp_unbiased, rslt);

input [31:0] flp_a, flp_b;
output sign;
output [7:0] exponent, exp_unbiased;
output [22:0] rslt;

//variables used in an always block are declared as registers
reg sign_a, sign_b;
reg [7:0] exp_a, exp_b;
reg [7:0] exp_a_bias, exp_b_bias;
reg [22:0] fract_a, fract_b;
reg [7:0] ctr_align;
reg [22:0] rslt;
reg sign;
reg [7:0] exponent, exp_unbiased;
reg cout;

//==========================================================
//define sign, exponent, and fraction
always @ (flp_a or flp_b)
begin
   sign_a = flp_a[31];
   sign_b = flp_b[31];

   exp_a = flp_a[30:23];
   exp_b = flp_b[30:23];

   fract_a = flp_a[22:0];
   fract_b = flp_b[22:0];
//continued on next page
```

Figure 13.8 Behavioral module for subtraction that results in true addition.

```
//bias exponents
   exp_a_bias = exp_a + 8'b0111_1111;
   exp_b_bias = exp_b + 8'b0111_1111;


//align fractions
   if (exp_a_bias < exp_b_bias)
      ctr_align = exp_b_bias - exp_a_bias;

      while (ctr_align)
         begin
            fract_a = fract_a >> 1;
            exp_a_bias = exp_a_bias + 1;
            ctr_align = ctr_align - 1;
         end


   if (exp_b_bias < exp_a_bias)
      ctr_align = exp_a_bias - exp_b_bias;

      while (ctr_align)
         begin
            fract_b = fract_b >> 1;
            exp_b_bias = exp_b_bias + 1;
            ctr_align = ctr_align - 1;
         end

//============================================================
//obtain rslt
      {cout, rslt} = fract_a + fract_b;
      sign = sign_a;


//postnormalize
   if (cout == 1)
      begin
         {cout, rslt} = {cout, rslt} >> 1;
         exp_b_bias = exp_b_bias + 1;
      end

      exponent = exp_b_bias;
      exp_unbiased = exp_b_bias - 8'b0111_1111;

end

endmodule
```

Figure 13.8 (Continued)

```
//test bench for floating-point subtraction
module sub_flp2_tb;

reg [31:0] flp_a, flp_b;
wire sign;
wire [7:0] exponent, exp_unbiased;
wire [22:0] rslt;

//display variables
initial
$monitor ("sign = %b, exp_biased = %b, exp_unbiased = %b,
            rslt = %b",
         sign, exponent, exp_unbiased, rslt);

//apply input vectors
initial
begin
      //(-19) - (+25) = -44
      //            s ----e---- --------------f-------------
#0    flp_a = 32'b1_0000_0101_1001_1000_0000_0000_0000_000;
      flp_b = 32'b0_0000_0101_1100_1000_0000_0000_0000_000;

      //(-22) - (+28) = -50
      //            s ----e---- --------------f-------------
#10   flp_a = 32'b1_0000_0101_1011_0000_0000_0000_0000_000;
      flp_b = 32'b0_0000_0101_1110_0000_0000_0000_0000_000;

      //(-16) - (+23) = -39
      //            s ----e---- --------------f-------------
#10   flp_a = 32'b1_0000_0101_1000_0000_0000_0000_0000_000;
      flp_b = 32'b0_0000_0101_1011_1000_0000_0000_0000_000;

      //(-13) - (+54) = -67
      //            s ----e---- --------------f-------------
#10   flp_a = 32'b1_0000_0100_1101_0000_0000_0000_0000_000;
      flp_b = 32'b0_0000_0110_1101_1000_0000_0000_0000_000;

      //(-40) - (+24) = -64
      //            s ----e---- --------------f-------------
#10   flp_a = 32'b1_0000_0110_1010_0000_0000_0000_0000_000;
      flp_b = 32'b0_0000_0101_1100_0000_0000_0000_0000_000;

      //(+34) - (-11) = +45
      //            s ----e---- --------------f-------------
#10   flp_a = 32'b0_0000_0110_1000_1000_0000_0000_0000_000;
      flp_b = 32'b1_0000_0100_1011_0000_0000_0000_0000_000;
//continued on next page
```

Figure 13.9 Test bench for subtraction that results in true addition.

```
        //(+16) - (-42) = +58
        //          s ----e---- --------------f-------------
#10     flp_a = 32'b0_0000_0101_1000_0000_0000_0000_0000_000;
        flp_b = 32'b1_0000_0110_1010_1000_0000_0000_0000_000;

        //(-86) - (+127) = -213
        //          s ----e---- --------------f-------------
#10     flp_a = 32'b1_0000_0111_1010_1100_0000_0000_0000_000;
        flp_b = 32'b0_0000_0111_1111_1110_0000_0000_0000_000;

        //(-127) - (+76) = -202
        //          s ----e---- --------------f-------------
#10     flp_a = 32'b1_0000_0111_1111_1100_0000_0000_0000_000;
        flp_b = 32'b0_0000_0111_1001_1000_0000_0000_0000_000;

        //(-127) - (+76) = -203
        //          s ----e---- --------------f-------------
#10     flp_a = 32'b1_0000_0111_1111_1110_0000_0000_0000_000;
        flp_b = 32'b0_0000_0111_1001_1000_0000_0000_0000_000;

        //(+76) - (-127) = +203
        //          s ----e---- --------------f-------------
#10     flp_a = 32'b0_0000_0111_1001_1000_0000_0000_0000_000;
        flp_b = 32'b1_0000_0111_1111_1110_0000_0000_0000_000;

        //(+127) - (-77) = +204
        //          s ----e---- --------------f-------------
#10     flp_a = 32'b0_0000_0111_1111_1110_0000_0000_0000_000;
        flp_b = 32'b1_0000_0111_1001_1010_0000_0000_0000_000;

#10     $stop;

end

//instantiate the module into the test bench
sub_flp2 inst1 (
   .flp_a(flp_a),
   .flp_b(flp_b),
   .sign(sign),
   .exponent(exponent),
   .exp_unbiased(exp_unbiased),
   .rslt(rslt)
   );

endmodule
```

Figure 13.9 (Continued)

```
sign = 1, exp_biased = 10000101,
   exp_unbiased = 00000110, rslt = 10110000000000000000000

sign = 1, exp_biased = 10000101,
   exp_unbiased = 00000110, rslt = 11001000000000000000000

sign = 1, exp_biased = 10000101,
   exp_unbiased = 00000110, rslt = 10011100000000000000000

sign = 1, exp_biased = 10000110,
   exp_unbiased = 00000111, rslt = 10000110000000000000000

sign = 1, exp_biased = 10000110,
   exp_unbiased = 00000111, rslt = 10000000000000000000000

sign = 0, exp_biased = 10000101,
   exp_unbiased = 00000110, rslt = 10110100000000000000000

sign = 0, exp_biased = 10000101,
   exp_unbiased = 00000110, rslt = 11101000000000000000000

sign = 1, exp_biased = 10000111,
   exp_unbiased = 00001000, rslt = 11010101000000000000000

sign = 1, exp_biased = 10000111,
   exp_unbiased = 00001000, rslt = 11001010000000000000000

sign = 1, exp_biased = 10000111,
   exp_unbiased = 00001000, rslt = 11001011000000000000000

sign = 0, exp_biased = 10000111,
   exp_unbiased = 00001000, rslt = 11001011000000000000000

sign = 0, exp_biased = 10000111,
   exp_unbiased = 00001000, rslt = 11001100000000000000000
```

Figure 13.10 Outputs for subtraction that results in true addition.

13.3.2 True Subtraction — Version 1

This version of true subtraction contains the following attributes: *fract_a* < *fract_b* and *sign_a* = *sign_b*. The exponents are biased and the fractions are aligned as before. In this version, *fract_a* is 2s complemented instead of *fract_b*; that is, the smaller fraction is subtracted from the larger fraction. Then the addition occurs, and the concatenated *cout* and *rslt* are replaced by the sum of the two fractions. The sign of the result is made equal to the complement of the sign of the operand that represents *fract_a*. If the high-order bit of the result is equal to 0, then the result is shifted left and the biased exponent is decremented by one — this repeats until the high-order bit of the result is a 1.

The behavioral module is shown in Figure 13.11. The test bench module, shown in Figure 13.12, applies six input vectors for the two operands. In each case, *fract_a* is less than *fract_b* and the operation is subtraction. The outputs are shown in Figure 13.13 displaying the sign of the floating-point number, the biased exponent, the unbiased exponent, and the resulting fraction. The results can be verified by placing the radix point in the appropriate position as indicated by the unbiased exponent.

```
//behavioral floating-point subtraction
//true subtraction: fract_a < fract_b, sign_a=sign_b
module sub_flp7 (flp_a, flp_b, sign, exponent,
                 exp_unbiased, rslt);

input [31:0] flp_a, flp_b;
output sign;
output [7:0] exponent, exp_unbiased;
output [22:0] rslt;

//variables used in an always block are declared as registers
reg sign_a, sign_b;
reg [7:0] exp_a, exp_b;
reg [7:0] exp_a_bias, exp_b_bias;
reg [22:0] fract_a, fract_b;
reg [7:0] ctr_align;
reg [22:0] rslt;
reg sign;
reg [7:0] exponent, exp_unbiased;
reg cout;

//=========================================================
//define sign, exponent, and fraction
always @ (flp_a or flp_b)
begin
   sign_a = flp_a[31];
   sign_b = flp_b[31];

   exp_a = flp_a[30:23];
   exp_b = flp_b[30:23];

   fract_a = flp_a[22:0];
   fract_b = flp_b[22:0];
//continued on next page
```

Figure 13.11 Behavioral module for true subtraction, where *fract_a* is less than *fract_b* and *sign_a* = *sign_b*.

```
//bias exponents
   exp_a_bias = exp_a + 8'b0111_1111;
   exp_b_bias = exp_b + 8'b0111_1111;

//align fractions
   if (exp_a_bias < exp_b_bias)
      ctr_align = exp_b_bias - exp_a_bias;

      while (ctr_align)
         begin
            fract_a = fract_a >> 1;
            exp_a_bias = exp_a_bias + 1;
            ctr_align = ctr_align - 1;
         end

   if (exp_b_bias < exp_a_bias)
      ctr_align = exp_a_bias - exp_b_bias;

      while (ctr_align)
         begin
            fract_b = fract_b >> 1;
            exp_b_bias = exp_b_bias + 1;
            ctr_align = ctr_align - 1;
         end

//==============================================================
//obtain rslt
   if (fract_a < fract_b)
      begin
         fract_a = ~fract_a + 1;
         {cout, rslt} = fract_a + fract_b;
         sign = ~sign_a;
      end

//postnormalize
      while (rslt[22] == 0)
         begin
            rslt = rslt << 1;
            exp_b_bias = exp_b_bias - 1;
         end

   exponent = exp_b_bias;
   exp_unbiased = exp_b_bias - 8'b0111_1111;

end

endmodule
```

Figure 13.11 (Continued)

```
//test bench for floating-point subtraction
module sub_flp7_tb;

reg [31:0] flp_a, flp_b;
wire sign;
wire [7:0] exponent, exp_unbiased;
wire [22:0] rslt;

initial     //display variables
$monitor ("sign=%b, exp_biased=%b, exp_unbiased=%b, rslt=%b",
         sign, exponent, exp_unbiased, rslt);

initial     //apply input vectors
begin
      //(-60) - (-127) = +67
      //          s ----e---- --------------f-------------
#0    flp_a = 32'b1_0000_0110_1111_0000_0000_0000_0000_000;
      flp_b = 32'b1_0000_0111_1111_1110_0000_0000_0000_000;

      //(+11) - (+34) = -23
      //          s ----e---- --------------f-------------
#10   flp_a = 32'b0_0000_0100_1011_0000_0000_0000_0000_000;
      flp_b = 32'b0_0000_0110_1000_1000_0000_0000_0000_000;

      //(-23) - (-36) = +13
      //          s ----e---- --------------f-------------
#10   flp_a = 32'b1_0000_0101_1011_1000_0000_0000_0000_000;
      flp_b = 32'b1_0000_0110_1001_0000_0000_0000_0000_000;

      //(-7) - (-38) = +31
      //          s ----e---- --------------f-------------
#10   flp_a = 32'b1_0000_0011_1110_0000_0000_0000_0000_000;
      flp_b = 32'b1_0000_0110_1001_1000_0000_0000_0000_000;

      //(+127) - (+255) = -128
      //          s ----e---- --------------f-------------
#10   flp_a = 32'b0_0000_0111_1111_1110_0000_0000_0000_000;
      flp_b = 32'b0_0000_1000_1111_1111_0000_0000_0000_000;

      //(+47) - (+72) = -25
      //          s ----e---- --------------f-------------
#10   flp_a = 32'b0_0000_0110_1011_1100_0000_0000_0000_000;
      flp_b = 32'b0_0000_0111_1001_0000_0000_0000_0000_000;

#10   $stop;
end         //continued on next page
```

Figure 13.12 Test bench for true subtraction, where *fract_a* is less than *fract_b* and *sign_a* = *sign_b*.

```
//instantiate the module into the test bench
sub_flp7 inst1 (
   .flp_a(flp_a),
   .flp_b(flp_b),
   .sign(sign),
   .exponent(exponent),
   .exp_unbiased(exp_unbiased),
   .rslt(rslt)
   );
endmodule
```

Figure 13.12 (Continued)

```
sign = 0, exp_biased = 10000110,
   exp_unbiased = 00000111, rslt = 1000011000000000000000

sign = 1, exp_biased = 10000100,
   exp_unbiased = 00000101, rslt = 1011100000000000000000

sign = 0, exp_biased = 10000011,
   exp_unbiased = 00000100, rslt = 1101000000000000000000

sign = 0, exp_biased = 10000100,
   exp_unbiased = 00000101, rslt = 1111100000000000000000

sign = 1, exp_biased = 10000111,
   exp_unbiased = 00001000, rslt = 1000000000000000000000

sign = 1, exp_biased = 10000100,
   exp_unbiased = 00000101, rslt = 1100100000000000000000
```

Figure 13.13 Outputs for true subtraction, where *fract_a* is less than *fract_b* and *sign_a* = *sign_b*.

13.3.3 True Subtraction — Version 2

This version of true subtraction contains the following attributes: *fract_a* < *fract_b* and *sign_a* ≠ *sign_b*. The exponents are biased and the fractions are aligned as before. In this version, *fract_a* is 2s complemented instead of *fract_b*; that is, the smaller fraction is subtracted from the larger fraction. Then the addition occurs, and the concatenated *cout* and *rslt* are replaced by the sum of the two fractions. The sign of the result is made equal to the complement of the sign of the operand that represents *fract_a*. If the high-order bit of the result is equal to 0, then the result is shifted left and the biased exponent is decremented by one — this repeats until the high-order bit of the result is a 1.

The behavioral module is shown in Figure 13.14. The test bench module, shown in Figure 13.15, applies nine input vectors for the two operands. In each case, *fract_a* is less than *fratc_b* and the operation is addition, which results in a subtraction because the signs of the two operands are different. The outputs are shown in Figure 13.16 displaying the sign of the floating-point number, the biased exponent, the unbiased exponent, and the resulting fraction. The results can be verified by placing the radix point in the appropriate position as indicated by the unbiased exponent.

```
//behavioral floating-point subtraction
//true subtraction: fract_a < fract_b, sign_a!=sign_b
module sub_flp8 (flp_a, flp_b, sign, exponent,
                 exp_unbiased, rslt);

input [31:0] flp_a, flp_b;
output sign;
output [7:0] exponent, exp_unbiased;
output [22:0] rslt;

//variables used in an always block are declared as registers
reg sign_a, sign_b;
reg [7:0] exp_a, exp_b;
reg [7:0] exp_a_bias, exp_b_bias;
reg [22:0] fract_a, fract_b;
reg [7:0] ctr_align;
reg [22:0] rslt;
reg sign;
reg [7:0] exponent, exp_unbiased;
reg cout;

//=============================================================
//define sign, exponent, and fraction
always @ (flp_a or flp_b)
begin
   sign_a = flp_a[31];
   sign_b = flp_b[31];

   exp_a = flp_a[30:23];
   exp_b = flp_b[30:23];

   fract_a = flp_a[22:0];
   fract_b = flp_b[22:0];
//continued on next page
```

Figure 13.14 Behavioral module for true subtraction, where *fract_a* is less than *fract_b* and *sign_a* ≠ *sign_b*.

```
//bias exponents
   exp_a_bias = exp_a + 8'b0111_1111;
   exp_b_bias = exp_b + 8'b0111_1111;

//align fractions
   if (exp_a_bias < exp_b_bias)
      ctr_align = exp_b_bias - exp_a_bias;

      while (ctr_align)
         begin
            fract_a = fract_a >> 1;
            exp_a_bias = exp_a_bias + 1;
            ctr_align = ctr_align - 1;
         end

   if (exp_b_bias < exp_a_bias)
      ctr_align = exp_a_bias - exp_b_bias;

      while (ctr_align)
         begin
            fract_b = fract_b >> 1;
            exp_b_bias = exp_b_bias + 1;
            ctr_align = ctr_align - 1;
         end

//=====================================================================
//obtain rslt
   if (fract_a < fract_b)
      begin
         fract_a = ~fract_a + 1;
         {cout, rslt} = fract_a + fract_b;
         sign = ~sign_a;
      end

//postnormalize
      while (rslt[22] == 0)
         begin
            rslt = rslt << 1;
            exp_b_bias = exp_b_bias - 1;
         end

   exponent = exp_b_bias;
   exp_unbiased = exp_b_bias - 8'b0111_1111;

end

endmodule
```

Figure 13.14 (Continued)

```
//test bench for floating-point subtraction
module sub_flp8_tb;

reg [31:0] flp_a, flp_b;
wire sign;
wire [7:0] exponent, exp_unbiased;
wire [22:0] rslt;

initial          //display variables
$monitor ("sign=%b, exp_biased=%b, exp_unbiased=%b, rslt=%b",
            sign, exponent, exp_unbiased, rslt);

//apply input vectors
initial
begin
      //(+36) + (-140) = -104
      //             s ----e---- --------------f-------------
#0    flp_a = 32'b0_0000_0110_1001_0000_0000_0000_0000_000;
      flp_b = 32'b1_0000_1000_1000_1100_0000_0000_0000_000;

      //(-85) + (+127) = +42
      //             s ----e---- --------------f-------------
#10   flp_a = 32'b1_0000_0111_1010_1010_0000_0000_0000_000;
      flp_b = 32'b0_0000_0111_1111_1110_0000_0000_0000_000;

      //(+76) + (-127) = -51
      //             s ----e---- --------------f-------------
#10   flp_a = 32'b0_0000_0111_1001_1000_0000_0000_0000_000;
      flp_b = 32'b1_0000_0111_1111_1110_0000_0000_0000_000;

      //(+25) + (-120) = -95
      //             s ----e---- --------------f-------------
#10   flp_a = 32'b0_0000_0101_1100_1000_0000_0000_0000_000;
      flp_b = 32'b1_0000_0111_1111_0000_0000_0000_0000_000;

      //(-36) + (+48) = +12
      //             s ----e---- --------------f-------------
#10   flp_a = 32'b1_0000_0110_1001_0000_0000_0000_0000_000;
      flp_b = 32'b0_0000_0110_1100_0000_0000_0000_0000_000;

      //(+28) + (-35) = -7
      //             s ----e---- --------------f-------------
#10   flp_a = 32'b0_0000_0101_1110_0000_0000_0000_0000_000;
      flp_b = 32'b1_0000_0110_1000_1100_0000_0000_0000_000;

//continued on next page
```

Figure 13.15 Test bench for true subtraction, where *fract_a* is less than *fract_b* and *sign_a* ≠ *sign_b*.

```
        //(+11) + (-34) = -23
        //            s ----e---- --------------f-------------
#10     flp_a = 32'b0_0000_0100_1011_0000_0000_0000_0000_000;
        flp_b = 32'b1_0000_0110_1000_1000_0000_0000_0000_000;

        //(-127) + (+150) = +23
        //            s ----e---- --------------f-------------
#10     flp_a = 32'b1_0000_0111_1111_1110_0000_0000_0000_000;
        flp_b = 32'b0_0000_1000_1001_0110_0000_0000_0000_000;

        //(+127) + (-128) = -1
        //            s ----e---- --------------f-------------
#10     flp_a = 32'b0_0000_0111_1111_1110_0000_0000_0000_000;
        flp_b = 32'b1_0000_1000_1000_0000_0000_0000_0000_000;

#10     $stop;

end

//instantiate the module into the test bench
sub_flp8 inst1 (
   .flp_a(flp_a),
   .flp_b(flp_b),
   .sign(sign),
   .exponent(exponent),
   .exp_unbiased(exp_unbiased),
   .rslt(rslt)
   );
endmodule
```

Figure 13.15 (Continued)

```
sign = 1, exp_biased = 10000110,
   exp_unbiased = 00000111, rslt = 1101000000000000000000

sign = 0, exp_biased = 10000101,
   exp_unbiased = 00000110, rslt = 1010100000000000000000

sign = 1, exp_biased = 10000101,
   exp_unbiased = 00000110, rslt = 1100110000000000000000

sign = 1, exp_biased = 10000110,
   exp_unbiased = 00000111, rslt = 1011111000000000000000
//continued on next page
```

Figure 13.16 Outputs for true subtraction, where *fract_a* is less than *fract_b* and *sign_a* ≠ *sign_b*.

```
sign = 0, exp_biased = 10000011,
   exp_unbiased = 00000100, rslt = 1100000000000000000000

sign = 1, exp_biased = 10000010,
   exp_unbiased = 00000011, rslt = 1110000000000000000000

sign = 1, exp_biased = 10000100,
   exp_unbiased = 00000101, rslt = 1011100000000000000000

sign = 0, exp_biased = 10000100,
   exp_unbiased = 00000101, rslt = 1011100000000000000000

sign = 1, exp_biased = 10000000,
   exp_unbiased = 00000001, rslt = 1000000000000000000000
```

Figure 13.16 (Continued)

13.3.4 True Subtraction — Version 3

This version of true subtraction contains the following attributes: *fract_a* > *fract_b* and *sign_a* ≠ *sign_b*. The exponents are biased and the fractions are aligned as before. In this version, *fract_b* is 2s complemented instead of *fract_a*; that is, the smaller fraction is subtracted from the larger fraction. Then the addition occurs, and the concatenated *cout* and *rslt* are replaced by the sum of the two fractions. The sign of the result is made equal to the sign of the operand that represents *fract_a*. If the high-order bit of the result is equal to 0, then the result is shifted left one bit position and the biased exponent is decremented by one — this repeats until the high-order bit of the result is a 1.

The behavioral module is shown in Figure 13.17. The test bench module, shown in Figure 13.18, applies ten input vectors for the two operands. In each case, *fract_a* is greater than *fratc_b* and the operation is addition, which results in a subtraction because the signs of the two operands are different. The outputs are shown in Figure 13.19 displaying the sign of the floating-point number, the biased exponent, the unbiased exponent, and the resulting fraction. The results can be verified by placing the radix point in the appropriate position as indicated by the unbiased exponent.

```
//behavioral floating-point subtraction
//true subtraction: fract_a > fract_b, sign_a != sign_b
module sub_flp9 (flp_a, flp_b, sign, exponent,
                  exp_unbiased, rslt);
//continued on next page
```

Figure 13.17 Behavioral module for true subtraction, where *fract_a* is greater than *fract_b* and *sign_a* ≠ *sign_b*.

```
input [31:0] flp_a, flp_b;
output sign;
output [7:0] exponent, exp_unbiased;
output [22:0] rslt;

//variables used in an always block are declared as registers
reg sign_a, sign_b;
reg [7:0] exp_a, exp_b;
reg [7:0] exp_a_bias, exp_b_bias;
reg [22:0] fract_a, fract_b;
reg [7:0] ctr_align;
reg [22:0] rslt;
reg sign;
reg [7:0] exponent, exp_unbiased;
reg cout;

//==============================================================
//define sign, exponent, and fraction
always @ (flp_a or flp_b)
begin
   sign_a = flp_a[31];
   sign_b = flp_b[31];

   exp_a = flp_a[30:23];
   exp_b = flp_b[30:23];

   fract_a = flp_a[22:0];
   fract_b = flp_b[22:0];

//bias exponents
   exp_a_bias = exp_a + 8'b0111_1111;
   exp_b_bias = exp_b + 8'b0111_1111;

//align fractions
   if (exp_a_bias < exp_b_bias)
      ctr_align = exp_b_bias - exp_a_bias;

      while (ctr_align)
         begin
            fract_a = fract_a >> 1;
            exp_a_bias = exp_a_bias + 1;
            ctr_align = ctr_align - 1;
         end

//continued on next page
```

Figure 13.17 (Continued)

```
   if (exp_b_bias < exp_a_bias)
      ctr_align = exp_a_bias - exp_b_bias;

      while (ctr_align)
         begin
            fract_b = fract_b >> 1;
            exp_b_bias = exp_b_bias + 1;
            ctr_align = ctr_align - 1;
         end

//===========================================================
//obtain rslt
   if (fract_a > fract_b)
      begin
         fract_b = ~fract_b + 1;
         {cout, rslt} = fract_a + fract_b;
         sign = sign_a;
      end

//postnormalize
      while (rslt[22] == 0)
         begin
            rslt = rslt << 1;
            exp_b_bias = exp_b_bias - 1;
         end

   exponent = exp_b_bias;
   exp_unbiased = exp_b_bias - 8'b0111_1111;
end
endmodule
```

Figure 13.17 (Continued)

```
//test bench for floating-point subtraction
module sub_flp9_tb;

reg [31:0] flp_a, flp_b;
wire sign;
wire [7:0] exponent, exp_unbiased;
wire [22:0] rslt;

initial      //display variables
$monitor ("sign=%b, exp_biased=%b, exp_unbiased=%b, rslt=%b",
            sign, exponent, exp_unbiased, rslt);
//continued on next page
```

Figure 13.18 Test bench for true subtraction, where *fract_a* is greater than *fract_b* and *sign_a* ≠ *sign_b*.

```
initial          //apply input vectors
begin
      //(+72) + (-46) = +26
      //          s ----e---- --------------f-------------
#0    flp_a = 32'b0_0000_0111_1001_0000_0000_0000_0000_000;
      flp_b = 32'b1_0000_0110_1011_1000_0000_0000_0000_000;

      //(-126) + (+25) = -101
      //          s ----e---- --------------f-------------
#10   flp_a = 32'b1_0000_0111_1111_1100_0000_0000_0000_000;
      flp_b = 32'b0_0000_0101_1100_1000_0000_0000_0000_000;

      //(+172) + (-100) = +72
      //          s ----e---- --------------f-------------
#10   flp_a = 32'b0_0000_1000_1010_1100_0000_0000_0000_000;
      flp_b = 32'b1_0000_0111_1100_1000_0000_0000_0000_000;

      //(+117) + (-10) = +107
      //          s ----e---- --------------f-------------
#10   flp_a = 32'b0_0000_0111_1110_1010_0000_0000_0000_000;
      flp_b = 32'b1_0000_0100_1010_0000_0000_0000_0000_000;

      //(+127) + (-75) = +52
      //          s ----e---- --------------f-------------
#10   flp_a = 32'b0_0000_0111_1111_1110_0000_0000_0000_000;
      flp_b = 32'b1_0000_0111_1001_0110_0000_0000_0000_000;

      //(-127) + (+120) = -7
      //          s ----e---- --------------f-------------
#10   flp_a = 32'b1_0000_0111_1111_1110_0000_0000_0000_000;
      flp_b = 32'b0_0000_0111_1111_0000_0000_0000_0000_000;

      //(+135) + (-127) = +8
      //          s ----e---- --------------f-------------
#10   flp_a = 32'b0_0000_1000_1000_0111_0000_0000_0000_000;
      flp_b = 32'b1_0000_0111_1111_1110_0000_0000_0000_000;

      //(-180) + (+127) = -53
      //          s ----e---- --------------f-------------
#10   flp_a = 32'b1_0000_1000_1011_0100_0000_0000_0000_000;
      flp_b = 32'b0_0000_0111_1111_1110_0000_0000_0000_000;

      //(+85.75) + (-70.50) = +15.25
      //          s ----e---- --------------f-------------
#10   flp_a = 32'b0_0000_0111_1010_1011_1000_0000_0000_000;
      flp_b = 32'b1_0000_0111_1000_1101_0000_0000_0000_000;
//continued on next page
```

Figure 13.18 (Continued)

```
      //(-96.50) + (+30.25) = -66.25
      //          s ----e---- -------------f-------------
#10   flp_a = 32'b1_0000_0111_1100_0001_0000_0000_0000_000;
      flp_b = 32'b0_0000_0101_1111_0010_0000_0000_0000_000;

#10   $stop;

end

//instantiate the module into the test bench
sub_flp9 inst1 (
   .flp_a(flp_a),
   .flp_b(flp_b),
   .sign(sign),
   .exponent(exponent),
   .exp_unbiased(exp_unbiased),
   .rslt(rslt)
   );
endmodule
```

Figure 13.18 (Continued)

```
sign = 0, exp_biased = 10000100,
   exp_unbiased = 00000101, rslt = 1101000000000000000000

sign = 1, exp_biased = 10000110,
   exp_unbiased = 00000111, rslt = 1100101000000000000000

sign = 0, exp_biased = 10000110,
   exp_unbiased = 00000111, rslt = 1001000000000000000000

sign = 0, exp_biased = 10000110,
   exp_unbiased = 00000111, rslt = 1101011000000000000000

sign = 0, exp_biased = 10000101,
   exp_unbiased = 00000110, rslt = 1101000000000000000000

sign = 1, exp_biased = 10000010,
   exp_unbiased = 00000011, rslt = 1110000000000000000000

sign = 0, exp_biased = 10000011,
   exp_unbiased = 00000100, rslt = 1000000000000000000000
//continued on next page
```

Figure 13.19 Outputs for true subtraction, where *fract_a* is greater than *fract_b* and *sign_a* ≠ *sign_b*.

```
sign = 1, exp_biased = 10000101,
   exp_unbiased = 00000110, rslt = 11010100000000000000000

sign = 0, exp_biased = 10000011,
   exp_unbiased = 00000100, rslt = 11110100000000000000000

sign = 1, exp_biased = 10000110,
   exp_unbiased = 00000111, rslt = 10000100100000000000000
```

Figure 13.19 (Continued)

13.3.5 True Subtraction — Version 4

This version of true subtraction contains the following attributes: *fract_a* > *fract_b* and *sign_a* = *sign_b*. The exponents are biased and the fractions are aligned as before. In this version, *fract_b* is 2s complemented instead of *fract_a*; that is, the smaller fraction is subtracted from the larger fraction. Then the addition occurs, and the concatenated *cout* and *rslt* are replaced by the sum of the two fractions. The sign of the result is made equal to the sign of the operand that represents *fract_a*. If the high-order bit of the result is equal to 0, then the result is shifted left one bit position and the biased exponent is decremented by one — this repeats until the high-order bit of the result is a 1.

The behavioral module is shown in Figure 13.20. The test bench module, shown in Figure 13.21, applies eight input vectors for the two operands. In each case, *fract_a* is greater than *fratc_b* and the operation is subtraction. This results in a true subtraction because the signs of the two operands are the same. The outputs are shown in Figure 13.22 displaying the sign of the floating-point number, the biased exponent, the unbiased exponent, and the resulting fraction. The results can be verified by placing the radix point in the appropriate position as indicated by the unbiased exponent.

```
//behavioral floating-point subtraction
//true subtraction: fract_a > fract_b, sign_a= sign_b
module sub_flp10 (flp_a, flp_b, sign, exponent,
                  exp_unbiased, rslt);

input [31:0] flp_a, flp_b;
output sign;
output [7:0] exponent, exp_unbiased;
output [22:0] rslt;
//continued on next page
```

Figure 13.20 Behavioral module for true subtraction, where *fract_a* is greater than *fract_b* and *sign_a* = *sign_b*.

```
//variables used in an always block are declared as registers
reg sign_a, sign_b;
reg [7:0] exp_a, exp_b;
reg [7:0] exp_a_bias, exp_b_bias;
reg [22:0] fract_a, fract_b;
reg [7:0] ctr_align;
reg [22:0] rslt;
reg sign;
reg [7:0] exponent, exp_unbiased;
reg cout;

//=========================================================
//define sign, exponent, and fraction
always @ (flp_a or flp_b)
begin
   sign_a = flp_a[31];
   sign_b = flp_b[31];

   exp_a = flp_a[30:23];
   exp_b = flp_b[30:23];

   fract_a = flp_a[22:0];
   fract_b = flp_b[22:0];

   exp_a_bias = exp_a + 8'b0111_1111;  //bias exponents
   exp_b_bias = exp_b + 8'b0111_1111;

//align fractions
   if (exp_a_bias < exp_b_bias)
      ctr_align = exp_b_bias - exp_a_bias;

      while (ctr_align)
         begin
            fract_a = fract_a >> 1;
            exp_a_bias = exp_a_bias + 1;
            ctr_align = ctr_align - 1;
         end

   if (exp_b_bias < exp_a_bias)
      ctr_align = exp_a_bias - exp_b_bias;

      while (ctr_align)
         begin
            fract_b = fract_b >> 1;
            exp_b_bias = exp_b_bias + 1;
            ctr_align = ctr_align - 1;
         end      //continued on next page
```

Figure 13.20 (Continued)

```
//=====================================================================
//obtain rslt
   if (fract_a > fract_b)
      begin
         fract_b = ~fract_b + 1;
         {cout, rslt} = fract_a + fract_b;
         sign = sign_a;
      end

//postnormalize
      while (rslt[22] == 0)
         begin
            rslt = rslt << 1;
            exp_b_bias = exp_b_bias - 1;
         end

   exponent = exp_b_bias;
   exp_unbiased = exp_b_bias - 8'b0111_1111;

end

endmodule
```

Figure 13.20 (Continued)

```
//test bench for floating-point subtraction
module sub_flp10_tb;


reg [31:0] flp_a, flp_b;
wire sign;
wire [7:0] exponent, exp_unbiased;
wire [22:0] rslt;


//display variables
initial
$monitor ("sign=%b, exp_biased=%b, exp_unbiased=%b, rslt=%b",
            sign, exponent, exp_unbiased, rslt);

//continued on next page
```

Figure 13.21 Test bench for true subtraction, where *fract_a* is greater than *fract_b* and *sign_a* = *sign_b*.

```
//apply input vectors
initial
begin
      //(-130) - (-25) = -105
      //          s ----e---- --------------f-------------
#0    flp_a = 32'b1_0000_1000_1000_0010_0000_0000_0000_000;
      flp_b = 32'b1_0000_0101_1100_1000_0000_0000_0000_000;

      //(+105) - (+5) = +100
      //          s ----e---- --------------f-------------
#10   flp_a = 32'b0_0000_0111_1101_0010_0000_0000_0000_000;
      flp_b = 32'b0_0000_0011_1010_0000_0000_0000_0000_000;

      //(+72) - (+47) = +25
      //          s ----e---- --------------f-------------
#10   flp_a = 32'b0_0000_0111_1001_0000_0000_0000_0000_000;
      flp_b = 32'b0_0000_0110_1011_1100_0000_0000_0000_000;

      //(-127) - (-60) = -67
      //          s ----e---- --------------f-------------
#10   flp_a = 32'b1_0000_0111_1111_1110_0000_0000_0000_000;
      flp_b = 32'b1_0000_0110_1111_0000_0000_0000_0000_000;

      //(+36.5) - (+5.75) = +30.75
      //          s ----e---- --------------f-------------
#10   flp_a = 32'b0_0000_0110_1001_0010_0000_0000_0000_000;
      flp_b = 32'b0_0000_0011_1011_1000_0000_0000_0000_000;

      //(-720.75) - (-700.25) = -20.50
      //          s ----e---- --------------f-------------
#10   flp_a = 32'b1_0000_1010_1011_0100_0011_0000_0000_000;
      flp_b = 32'b1_0000_1010_1010_1111_0001_0000_0000_000;

      //(+963.50) - (+520.25) = +443.25
      //          s ----e---- --------------f-------------
#10   flp_a = 32'b0_0000_1010_1111_0000_1110_0000_0000_000;
      flp_b = 32'b0_0000_1010_1000_0010_0001_0000_0000_000;

      //(+5276) - (+4528) = +748
      //          s ----e---- --------------f-------------
#10   flp_a = 32'b0_0000_1101_1010_0100_1110_0000_0000_000;
      flp_b = 32'b0_0000_1101_1000_1101_1000_0000_0000_000;

#10   $stop;

end
//continued on next page
```

Figure 13.21 (Continued)

```
//instantiate the module into the test bench
sub_flp10 inst1 (
   .flp_a(flp_a),
   .flp_b(flp_b),
   .sign(sign),
   .exponent(exponent),
   .exp_unbiased(exp_unbiased),
   .rslt(rslt)
   );

endmodule
```

Figure 13.21 (Continued)

```
sign = 1, exp_biased = 10000110,
   exp_unbiased = 00000111, rslt = 11010010000000000000000

sign = 0, exp_biased = 10000110,
   exp_unbiased = 00000111, rslt = 11001000000000000000000

sign = 0, exp_biased = 10000100,
   exp_unbiased = 00000101, rslt = 11001000000000000000000

sign = 1, exp_biased = 10000110,
   exp_unbiased = 00000111, rslt = 10000110000000000000000

sign = 0, exp_biased = 10000100,
   exp_unbiased = 00000101, rslt = 11110110000000000000000

sign = 1, exp_biased = 10000100,
   exp_unbiased = 00000101, rslt = 10100100000000000000000

sign = 0, exp_biased = 10001000,
   exp_unbiased = 00001001, rslt = 11011101101000000000000

sign = 0, exp_biased = 10001001,
   exp_unbiased = 00001010, rslt = 10111011000000000000000
```

Figure 13.22 Outputs for true subtraction, where *fract_a* is greater than *fract_b* and *sign_a* = *sign_b*.

13.4 Problems

13.1 Perform the following operation on the two operands: (–86) – (+127).

13.2 Perform the following operation on the two operands: (–13) – (+54).

13.3 Perform the following operation on the two operands: (+127) – (–77).

13.4 Perform the following operation on the two operands: (+76) – (–127).

13.5 Perform the following operation on the two operands: (–60) – (–127).

13.6 Perform the following operation on the two operands: (–7) – (–38).

13.7 Perform the following operation on the two operands: (+127) – (+255).

13.8 Perform the following operation on the two operands: (+47) – (+72).

13.9 Perform the following operation on the two operands: (+36) + (–140).

13.10 Perform the following operation on the two operands: (–85) + (+127).

13.11 Perform the following operation on the two operands: (–127) + (+150).

13.12 Perform the following operation on the two operands: (+127) + (–128).

13.13 Perform the following operation on the two operands: (+135) + (–127).

13.14 Perform the following operation on the two operands: (–180) + (+127).

13.15 Perform the following operation on the two operands: (+85.5) + (–70.5).

13.16 Perform the following operation on the two operands: (–35.75) + (+26.50).

13.17 Perform the following operation on the two operands: (–47.25) – (–18.75).

13.18 Perform the following operation on the two operands: (+36.50) – (+5.75).

13.19 Perform the following operation on the two operands: (–130) – (–25).

13.20 Perform the following operation on the two operands: (+105) – (+5).

13.21 Design a behavioral module that incorporates all of the methods used to implement true addition and true subtraction — one method for true addition and four methods for true subtraction as described in Section 13.1 under the headings of *true addition* and *true subtraction*.

Obtain the behavioral design module, the test bench module that applies several input vectors for the two floating-point operands, and the outputs.

14.1 Double Bias
14.2 Flowcharts
14.3 Numerical Examples
14.4 Verilog HDL Implementations
14.5 Problems

14

Floating-Point Multiplication

Floating-point multiplication is slightly easier than addition or subtraction, because the exponents do not have to be compared and the fractions do not have to be aligned. For floating-point multiplication, the exponents are added and the fractions are multiplied. Both operations can be done in parallel. Any of the algorithms presented in fixed-point multiplication can be used to multiply the floating-point fractions.

Multiplication is performed on normalized floating-point operands A and B using biased exponents, such that

$$A = f_A \times r^{e_A}$$
$$B = f_B \times r^{e_B}$$

where f is the normalized fraction, e is the exponent, and r is the radix. Floating-point multiplication is defined as shown in Equation 14.1, which shows fraction multiplication and exponent addition performed simultaneously.

$$\begin{aligned} A \times B &= (f_A \times r^{e_A}) \times (f_B \times r^{e_B}) \\ &= (f_A \times f_B) \times r^{e_A + e_B} \end{aligned} \tag{14.1}$$

The multiplicand and multiplier are represented in sign-magnitude notation in either the single-precision format or the double-precision format, as shown in Figure 14.1. The single-precision format will be the primary format used in this chapter.

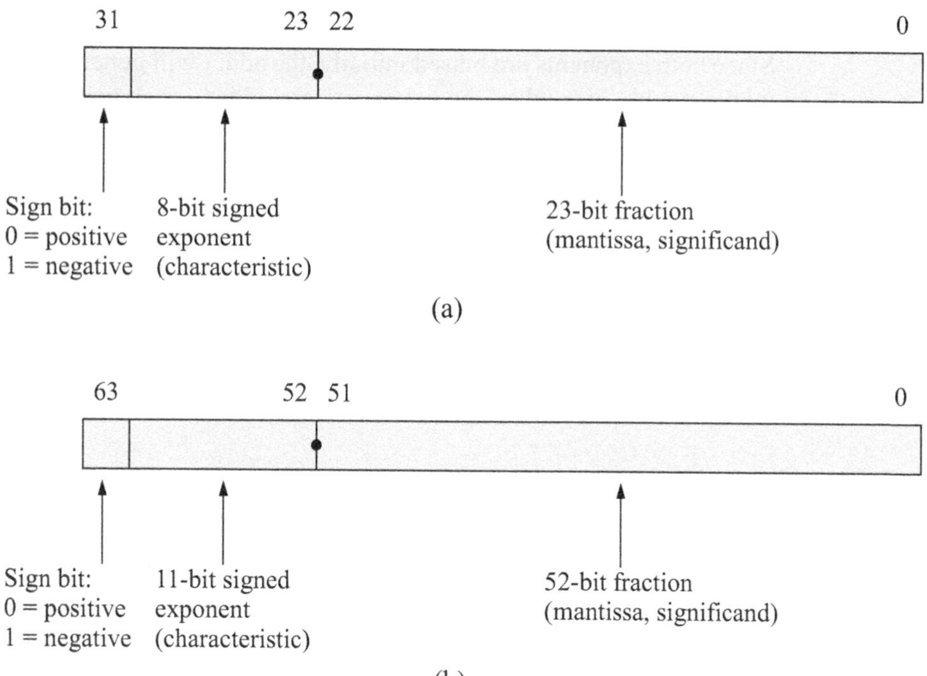

Figure 14.1 IEEE floating-point formats: (a) 32-bit single precision and (b) 64-bit double precision.

Multiplication using the single-precision format generates a double-precision $2n$-bit product; therefore, the resulting fraction is 46 bits. However, the range of a single-precision fraction in conjunction with the exponent is sufficiently accurate so that a single-precision result is usually adequate. Therefore, the low-order half of the fraction can be truncated, which may require a rounding procedure to be performed — rounding techniques are covered in Chapter 16.

The sign of the product is determined by the signs of the floating-point numbers. If the signs are the same, then the sign of the product is positive; if the signs are different, then the sign of the product is negative. This can be determined by the exclusive-OR of the two signs, as shown in Equation 14.2.

$$\text{Product sign} = A_{\text{sign}} \oplus B_{\text{sign}} \tag{14.2}$$

If the resulting product fraction contains all zeroes, then the sign of the result is positive, regardless of the signs of the operands. Exponent overflow may occur when adding two positive exponents, indicating that the magnitude of the product is greater than the capacity of the floating-point format. The product can then be shifted left and the exponent decremented by one for each shift.

14.1 Double Bias

Since both exponents are biased initially, the adder will generate a resulting exponent with a double bias when the exponents are added, as shown below. Therefore, the exponent must be restored to a single bias by subtracting the bias.

$$(e_A + \text{bias}) + (e_B + \text{bias}) = (e_A + e_B) + 2\ \text{bias}$$

An example will illustrate this concept. Let the exponents be e_A = 0000 1010 (10) and e_B = 0000 0101 (5). Each exponent will be biased, then added to produce a double bias. The bias will then be subtracted to produce a single bias, then subtracted again to produce the sum of the two unbiased exponent: e_A = 0000 1010 (10) and e_B = 0000 0101 (5) = $(e_A + e_B)_{\text{unbiased}}$ = 0000 1111 (15).

$$\begin{aligned} e_{A(\text{unbiased})} &= 0000\_1010 \\ \text{+) bias} &= \underline{0111\_1111} \\ e_{A(\text{biased})} &= 1000\_1001 \end{aligned}$$

$$\begin{aligned} e_{B(\text{unbiased})} &= 0000\_0101 \\ \text{+) bias} &= \underline{0111\_1111} \\ e_{B(\text{biased})} &= 1000\_0100 \end{aligned}$$

$$\begin{aligned} e_{A(\text{biased})} &= \quad 1000\_1001 \\ \text{+) } e_{B(\text{biased})} &= \quad \underline{1000\_0100} \\ \text{Double bias} &= 1 \leftarrow 0000\_1101 \end{aligned}$$

Restore to single bias by subtracting the bias; that is, by adding the 2s complement of 0111 1111 (1000 0001).

$$\begin{aligned} e_{A(\text{biased})} + e_{B(\text{biased})} &= \quad 0000\_1101 \\ \text{+) 2s complement of bias} &= \quad \underline{1000\_0001} \\ (e_A + e_B)_{\text{single bias}} &= 1 \leftarrow 1000\_1110 \end{aligned}$$

$$\begin{aligned} (e_A + e_B)_{\text{single bias}} &= \quad 1000\_1110 \\ \text{+) 2s complement of bias} &= \quad \underline{1000\_0001} \\ (e_A + e_B)_{\text{no bias}} &= 1 \leftarrow 0000\_1111 \end{aligned}$$

14.2 Flowcharts

The multiplication algorithm can be partitioned into eight steps, as shown below.

1. Normalize operands
2. Check for zero operands.
3. Determine the product sign.
4. Bias the exponents.
5. Add the exponents.
6. Restore the exponent sum to a single bias by subtracting the bias.
7. Multiply the fractions.
8. Normalize the product, if necessary.

Figure 14.2 shows a flowchart representation of the above rules for floating-point multiplication.

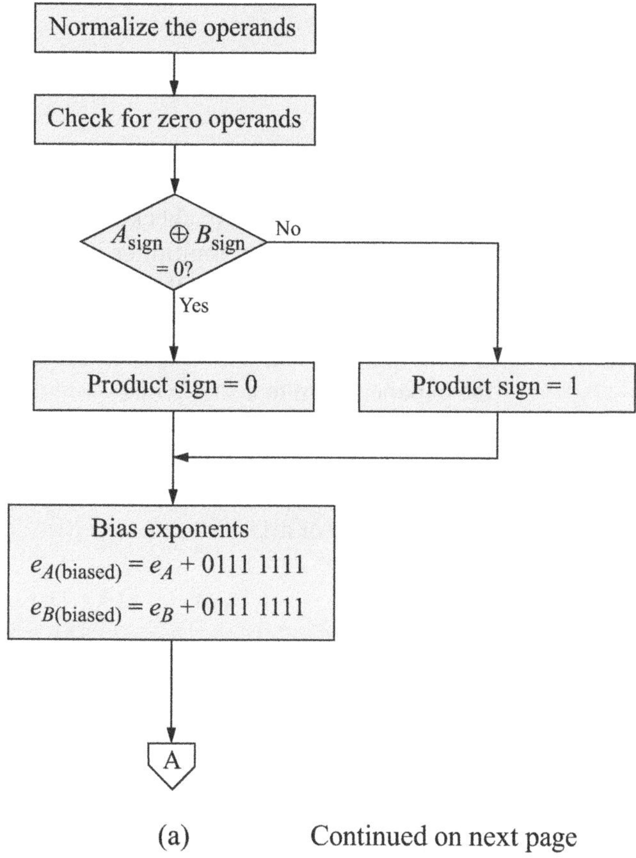

(a)

Continued on next page

Figure 14.2 Flowchart illustrating floating-point multiplication: (a) check for zero operands, determine product sign, bias exponents and (b) add exponents, restore to single bias, multiply fractions, and normalize the result.

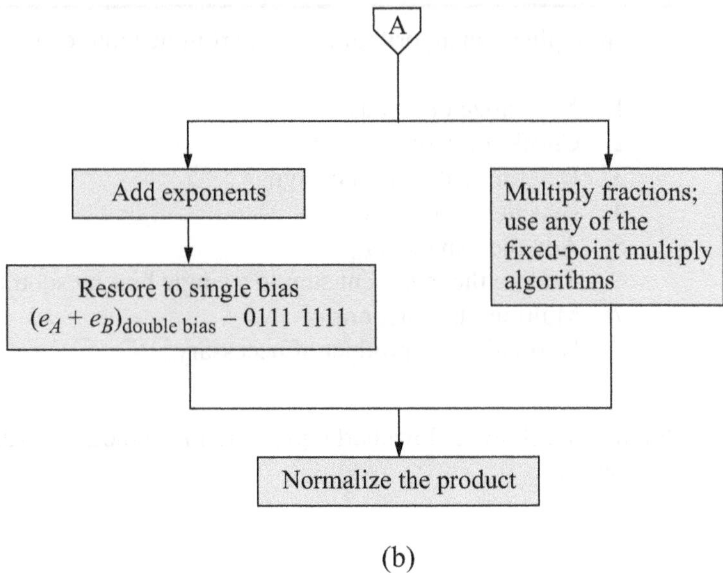

(b)

Figure 14.2 (Continued)

The two operands are checked to determine if either operand is equal to zero. If the multiplicand or the multiplier is equal to zero, then the product is set to zero. If both operands are nonzero, then the sign of the product is determined. The product is positive if both operands have the same sign; otherwise, the product is negative. This is followed by biasing the exponents, then adding the biased exponents and restoring the exponent sum to a single bias.

The next two operations are executed in parallel: exponent addition and fraction multiplication, as shown in Figure 14.2(b). The value of the resulting fraction has the range shown in Equation 14.3, which equates to the following numerical range, where x indicates a 0 or a 1:

$$(.1xxx\ldots x) \times (.1xxx\ldots x) \geq .01xxx\ldots x)$$

$$1/r^2 \leq |f_A \times f_B| < 1 \tag{14.3}$$

The multiplication process is accomplished using any of the algorithms described in the chapter on fixed-point multiplication. One of the Verilog HDL implementations presented in this chapter uses the sequential add-shift technique for floating-point multiplication.

14.3 Numerical Examples

This section presents numerical examples using the sequential add-shift method with 8-bit operands. Register *A* contains the normalized multiplicand fraction, *fract_a*; register *prod* contains the high-order *n* bits of the partial product (initially set to all zeroes); and register *B* contains the normalized multiplier fraction, *fract_b*.

Since the multiplication involves two *n*-bit operands, a count-down sequence counter, *count*, is set to a value that represents the number of bits in one of the operands. The counter is decremented by one for each step of the add-shift sequence. When the counter reaches a value of zero, the operation is finished and the product is normalized, if necessary.

If the low-order bit of register *fract_b* is equal to zero, then zeroes are added to the partial product and the sum is loaded into register *prod*. In this case, it is not necessary to perform an add operation — a right shift can accomplish the same result. However, it may require less logic if the same add-shift sequence occurs for each cycle. The sequence counter is then decremented by one. If the low-order bit of register *fract_b* is equal to one, then the multiplicand is added to the partial product. The sum is loaded into register *prod* and the sequence counter is decremented.

Example 14.1 A multiplicand fraction *fract_a* = 0.1010 0000 $\times 2^3$ (+5) is multiplied by a multiplier *fract_b* = 0.1100 0000 $\times 2^2$ (+3) with partial product *D* = 0000 0000 to produce a product of *prod* = 0.1111 0000 0000 0000 $\times 2^4$ (+15).

fract_a (+5)		*prod*	*fract_b* (+3)	
1010 0000	*prod*	0000 0000	1100 0000	
Shift right 6		0000 0000	0000 0011	
	+)	1010 0000	↓	Add-shift
	0 ←	1010 0000	0000 0011	
Shift right 1		0101 0000	0000 0001	
	+)	1010 0000	↓	Add-shift
	0 ←	1111 0000	0000 0001	
Shift right 1		0111 1000	0000 0000	
8 cycles (count = 0)				
Postnormalize		1111 0000	0000 0000	
Product =	0.	1111 0000	0000 0000	$\times 2^{(3+2)-1} = 2^4$

Example 14.2 This example shows more detail of the individual steps. Each step will always be an add-shift-right sequence. A multiplicand fraction *fract_a* = 0.1100 1000 × 2^5 (+25) is multiplied by a multiplier *fract_b* = 0.1100 1000 × 2^5 (+25) with partial product *prod* = 0000 0000 to produce a product of *prod* = 0.1001 1100 0100 0000 × 2^{10} (+625).

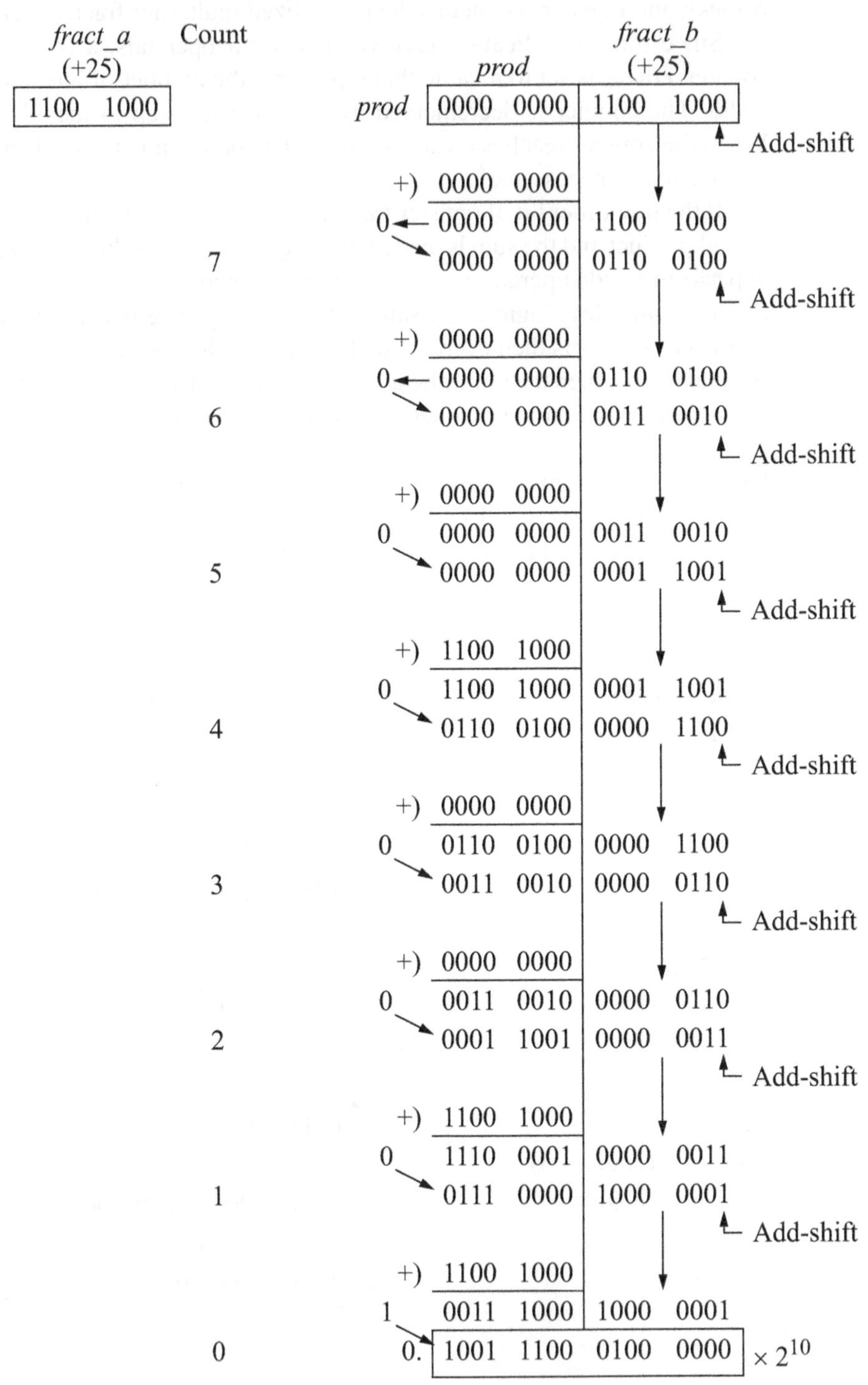

14.4 Verilog HDL Implementations

This section will design two behavioral modules: one using the Verilog HDL multiplication operator (*), and one using the sequential add-shift method. Arithmetic operations are performed on one (unary) or two (binary) operands in various radices. The result of an arithmetic operation is interpreted as an unsigned value or as a signed value in 2s complement representation on scalar nets, vector nets, and registers.

The single-precision format is used in both modules and is reproduced in Figure 13.7 for convenience. All exponents will be biased by adding a bias constant of +127 (0111 1111). The exponents will then be added and one bias constant will be subtracted. Then the fractions will be multiplied and the product postnormalized, if necessary.

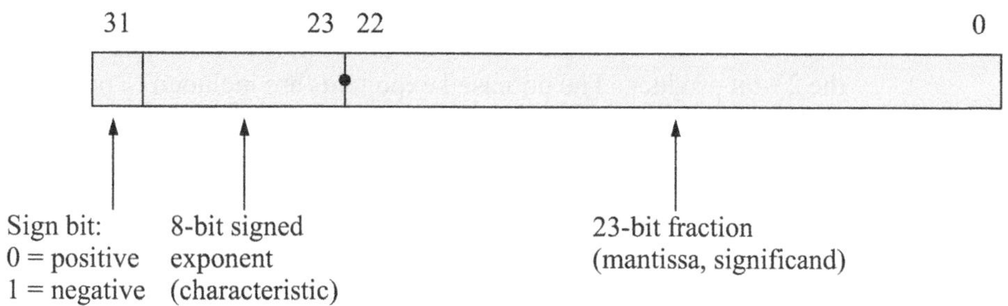

Figure 14.3 Single-precision floating-point format.

14.4.1 Floating-Point Multiplication — Version 1

The behavioral module in this section has two inputs that represent the floating-point multiplicand *flp_a[31:0]* and the floating-point multiplier *flp_b[31:0]*. There are five outputs: a scalar output indicating the *sign* of the number; a biased exponent, *exponent[7:0]*, with a single bias; an unbiased exponent, *exp_unbiased[7:0]*, that indicates the true value of the resulting exponent with no bias; the sum of the two exponents, *exp_sum[8:0]*, which includes a high-order bit for a carry; and the 23-bit product, *prod[22:0]*.

Target variables used in an **always** statement are declared as type **reg**. This is shown in the behavioral module of Figure 14.4. The **always** statement is a behavioral construct that begins at time 0 and executes the statements in the associated block continuously in a looping fashion until the simulation is terminated. The **always** construct never exits the corresponding block. There can be more than one **always** statement in a behavioral module.

An **always** statement is often used with an *event control list* — or *sensitivity list* — to execute a sequential block. When a change occurs to a variable in the sensitivity list, the statement or block of statements in the **always** block is executed. The keyword **or** is used to indicate multiple events as shown in Figure 14.4. When either one of the inputs *flp_a* or *flp_b* changes state, the statements in the **always** block — delimited by the keywords **begin** . . . **end** — are executed.

The three fields of the floating-point numbers are then defined for both the multiplicand and the multiplier. The signs of the numbers, *sign_a* and *sign_b*, are declared as the high-order bit of each operand; the exponents, *exp_a* and *exp_b*, are declared as 8-bit values, *flp_a[30:23]* and *flp_b[30:23]*, as defined in the single-precision format; and the fractions are declared as 23-bit variables, *fract_a[22:0]* and *fract_b[22:0]*, as defined in the single-precision format.

The exponents are then processed and the fractions are multiplied. The product may then be normalized, depending on the state of the high-order fraction bit *prod[22]*. The sign of the product is determined by comparing the signs of both operands using the exclusive-OR function. If the exclusive-OR generates an output of zero, then the sign of the product is positive; otherwise, the sign is negative.

The test bench, shown in Figure 14.5, provides eight input vectors for the two operands ranging from small numbers to large numbers. The outputs are shown in Figure 14.6 and show the sign of the product, the biased and unbiased exponents, and the 23-bit product. The unbiased exponents are included to permit easy interpretation of the product. The waveforms are shown in Figure 14.7 in hexadecimal notation. The hexadecimal product is formed by scanning the product from right to left; therefore, there will be an implied zero to the immediate left of the product shown in the outputs.

```
//behavioral floating-point multiplication
module mul_flp (flp_a, flp_b, sign, exponent,
               exp_unbiased, exp_sum, prod);

input [31:0] flp_a, flp_b;
output sign;
output [7:0] exponent, exp_unbiased;
output [8:0] exp_sum;
output [22:0] prod;

//variables used in an always block are declared as registers
reg sign_a, sign_b;
reg [7:0] exp_a, exp_b;
reg [7:0] exp_a_bias, exp_b_bias;
reg [8:0] exp_sum;
reg [22:0] fract_a, fract_b;
reg [45:0] prod_dbl;
reg [22:0] prod;
reg sign;
reg [7:0] exponent, exp_unbiased;
//continued on next page
```

Figure 14.4 Behavioral module for floating-point multiplication using the multiply operator.

```
//define sign, exponent, and fraction
always @ (flp_a or flp_b)
begin
   sign_a = flp_a[31];
   sign_b = flp_b[31];


   exp_a = flp_a[30:23];
   exp_b = flp_b[30:23];


   fract_a = flp_a[22:0];
   fract_b = flp_b[22:0];


//bias exponents
   exp_a_bias = exp_a + 8'b0111_1111;
   exp_b_bias = exp_b + 8'b0111_1111;


//add exponents
   exp_sum = exp_a_bias + exp_b_bias;


//remove one bias
   exponent = exp_sum - 8'b0111_1111;

   exp_unbiased = exponent - 8'b0111_1111;


//multiply fractions
   prod_dbl = fract_a * fract_b;
   prod = prod_dbl[45:23];


//postnormalize product
   while (prod[22] == 0)
      begin
         prod = prod << 1;
         exp_unbiased = exp_unbiased - 1;
      end

   sign = sign_a ^ sign_b;

end

endmodule
```

Figure 14.4 (Continued)

```
//test bench for floating-point multiplication
module mul_flp_tb;

reg [31:0] flp_a, flp_b;
wire sign;
wire [7:0] exponent, exp_unbiased;
wire [8:0] exp_sum;
wire [22:0] prod;

//display variables
initial
$monitor ("sign = %b, exp_biased = %b, exp_unbiased = %b,
         prod = %b",
         sign, exp_sum, exp_unbiased, prod);

//apply input vectors
initial
begin
      //+5 x +3 = +15
      //            s ----e---- --------------f-------------
#0    flp_a = 32'b0_0000_0011_1010_0000_0000_0000_0000_000;
      flp_b = 32'b0_0000_0010_1100_0000_0000_0000_0000_000;

      //+6 x +4 = +24
      //            s ----e---- --------------f-------------
#10   flp_a = 32'b0_0000_0011_1100_0000_0000_0000_0000_000;
      flp_b = 32'b0_0000_0011_1000_0000_0000_0000_0000_000;

      //-5 x +5 = -25
      //            s ----e---- --------------f-------------
#10   flp_a = 32'b1_0000_0011_1010_0000_0000_0000_0000_000;
      flp_b = 32'b0_0000_0011_1010_0000_0000_0000_0000_000;

      //+7 x -5 = -35
      //            s ----e---- --------------f-------------
#10   flp_a = 32'b0_0000_0011_1110_0000_0000_0000_0000_000;
      flp_b = 32'b1_0000_0011_1010_0000_0000_0000_0000_000;

      //+25 x +25 = +625
      //            s ----e---- --------------f-------------
#10   flp_a = 32'b0_0000_0101_1100_1000_0000_0000_0000_000;
      flp_b = 32'b0_0000_0101_1100_1000_0000_0000_0000_000;

//continued on next page
```

Figure 14.5 Test bench module for floating-point multiplication using the multiply operator.

```
        //+76 x +55 = +4180
        //            s ----e---- --------------f-------------
#10     flp_a = 32'b0_0000_0111_1001_1000_0000_0000_0000_000;
        flp_b = 32'b0_0000_0110_1101_1100_0000_0000_0000_000;

        //-48 x -17 = +816
        //            s ----e---- --------------f-------------
#10     flp_a = 32'b1_0000_0110_1100_0000_0000_0000_0000_000;
        flp_b = 32'b1_0000_0101_1000_1000_0000_0000_0000_000;

        //+3724 x +853 = +3,176,572
        //            s ----e---- --------------f-------------
#10     flp_a = 32'b0_0000_1100_1110_1000_1100_0000_0000_000;
        flp_b = 32'b0_0000_1010_1101_0101_0100_0000_0000_000;

#10     $stop;

end

//instantiate the module into the test bench
mul_flp inst1 (
   .flp_a(flp_a),
   .flp_b(flp_b),
   .sign(sign),
   .exponent(exponent),
   .exp_unbiased(exp_unbiased),
   .exp_sum(exp_sum),
   .prod(prod)
   );
endmodule
```

Figure 14.5 (Continued)

```
sign = 0, exp_biased = 100000011,
   exp_unbiased = 00000100,
   prod = 1111_0000_0000_0000_0000_000

sign = 0, exp_biased = 100000100,
   exp_unbiased = 00000101,
   prod = 1100_0000_0000_0000_0000_000

//continued on next page
```

Figure 14.6 Outputs for floating-point multiplication using the multiply operator.

```
sign = 1, exp_biased = 100000100,
   exp_unbiased = 00000101,
   prod = 1100_1000_0000_0000_0000_000

sign = 1, exp_biased = 100000100,
   exp_unbiased = 00000110,
   prod = 1000_1100_0000_0000_0000_000

sign = 0, exp_biased = 100001000,
   exp_unbiased = 00001010,
   prod = 1001_1100_0100_0000_0000_000

sign = 0, exp_biased = 100001011,
   exp_unbiased = 00001101,
   prod = 1000_0010_1010_0000_0000_000

sign = 0, exp_biased = 100001001,
   exp_unbiased = 00001010,
   prod = 1100_1100_0000_0000_0000_000

sign = 0, exp_biased = 100010100,
   exp_unbiased = 00010110,
   prod = 1100_0001_1110_0001_1111_000
```

Figure 14.6 (Continued)

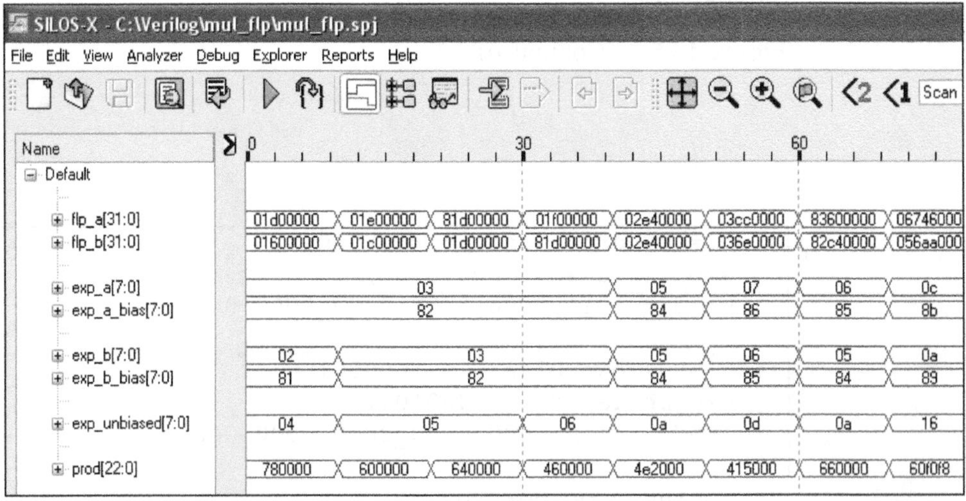

Figure 14.7 Waveforms for floating-point multiplication using the multiply operator.

14.4.2 Floating-Point Multiplication — Version 2

This version of floating-point multiplication is similar to the previous version except for the use of a different method for multiplying the fractions. The previous version used the multiplication operator to multiply the fractions, whereas, this version uses the sequential add-shift technique. The behavioral module of Figure 14.8 employs two **always** statements — one to define the fields of the single-precision format when the floating-point operands change value, and one to perform the multiplication when a *start* signal is asserted.

Since the fractions contain 23 bits, a count-down counter is set to a value of 10111 (23) to accommodate the add-shift procedure. The fractions are then checked to determine if either fraction is zero. If both fractions are nonzero, then the multiplication begins; if either fractions is zero, then the product is assigned a value of zero.

Since multiplication produces a $2n$-bit product, the product contains 46 bits: *prod[45:0]*. The low-order bit of the multiplier, *fract_b_reg[0]*, is extended to 23 bits by the replication statement *{23{fract_b_reg[0]}}*, then ANDed with the 23-bit multiplicand. This result — either zero or the multiplicand — is then added to the high-order half the partial product, *prod[45:23]*.

The product is then replaced by the concatenation of *cout*, the high-order half of the product, and the low-order 22 bits of the product, as follows: *prod = {cout, prod[45:23], prod[22:1]};*, which shifts *cout* and the product one bit position to the right to provide a new 46-bit product. The multiplier is then shifted right one bit position and the counter is decremented by one. This process repeats until the counter decrements to a value of zero. The product is then normalized, if necessary. The sign of the product is determined by the exclusive-OR of the signs of the multiplicand and the multiplier.

The test bench is shown in Figure 14.9 and applies several input vectors to the behavioral design module, including an all-zeroes fraction. The outputs and waveforms are shown in Figure 14.10 and Figure 14.11, respectively.

```
//behavioral floating-point multiplication
module mul_flp3 (flp_a, flp_b, start, sign, exponent,
                    exp_unbiased, cout, prod);

input [31:0] flp_a, flp_b;
input start;
output sign;
output [7:0] exponent, exp_unbiased;
output cout;
output [45:0] prod;
//continued on next page
```

Figure 14.8 Behavioral module for floating-point multiplication using the sequential add-shift method.

```
//variables used in an always block are declared as registers
reg sign_a, sign_b;
reg [7:0] exp_a, exp_b;
reg [7:0] exp_a_bias, exp_b_bias;
reg [7:0] exp_sum;
reg [22:0] fract_a, fract_b;
reg [22:0] fract_b_reg;
reg sign;
reg [7:0] exponent, exp_unbiased;
reg cout;
reg [45:0] prod;
reg [22:0] product;
reg [4:0] count;

//define sign, exponent, and fraction
always @ (flp_a or flp_b)
begin
   sign_a = flp_a[31];
   sign_b = flp_b[31];

   exp_a = flp_a[30:23];
   exp_b = flp_b[30:23];

   fract_a = flp_a[22:0];
   fract_b = flp_b[22:0];

//bias exponents
   exp_a_bias = exp_a + 8'b0111_1111;
   exp_b_bias = exp_b + 8'b0111_1111;

//add exponents
   exp_sum = exp_a_bias + exp_b_bias;

//remove one bias
   exponent = exp_sum - 8'b0111_1111;

   exp_unbiased = exponent - 8'b0111_1111;
end

//multiply fractions
always @ (posedge start)
   begin
      fract_b_reg = fract_b;
      prod = 0;
      count = 5'b10111;

//continued on next page
```

Figure 14.8 (Continued)

```
      if ((fract_a != 0) && (fract_b != 0))
         while (count)
            begin
               {cout, prod[45:23]} = (({23{fract_b_reg[0]}} &
                                          fract_a) + prod[45:23]);
               prod = {cout, prod[45:23], prod[22:1]};
               fract_b_reg = fract_b_reg >> 1;
               count = count - 1;
            end

      else
         begin
            sign = 1'b0;
            exp_unbiased = 8'h00;
            prod = 0;
         end

//postnormalize result
   if (prod != 0)
      while (prod[45] == 0)
         begin
            prod = prod << 1;
            exp_unbiased = exp_unbiased - 1;
         end

   sign = sign_a ^ sign_b;
end
endmodule
```

Figure 14.8 (Continued)

```
//test bench for floating-point multiplication
module mul_flp3_tb;

reg [31:0] flp_a, flp_b;
reg start;
wire sign;
wire [7:0] exponent, exp_unbiased;
wire [8:0] exp_sum;
wire [45:0] prod;

initial        //display variables
$monitor ("sign = %b, exp_unbiased = %b, prod = %b",
            sign, exp_unbiased, prod);
//continued on next page
```

Figure 14.9 Test bench for floating-point multiplication using the sequential add-shift method.

```
//apply input vectors
initial
begin
#0      start = 1'b0;
        //+5 x +3 = +15
        //           s ----e---- -------------f-------------
        flp_a = 32'b0_0000_0011_1010_0000_0000_0000_0000_000;
        flp_b = 32'b0_0000_0010_1100_0000_0000_0000_0000_000;
#10     start = 1'b1;
#10     start = 1'b0;


        //+6 x +4 = +24
        //           s ----e---- -------------f-------------
#10     flp_a = 32'b0_0000_0011_1100_0000_0000_0000_0000_000;
        flp_b = 32'b0_0000_0011_1000_0000_0000_0000_0000_000;
#10     start = 1'b1;
#10     start = 1'b0;


        //-5 x +5 = -25
        //           s ----e---- -------------f-------------
#10     flp_a = 32'b1_0000_0011_1010_0000_0000_0000_0000_000;
        flp_b = 32'b0_0000_0011_1010_0000_0000_0000_0000_000;
#10     start = 1'b1;
#10     start = 1'b0;


        //+7 x -5 = -35
        //           s ----e---- -------------f-------------
#10     flp_a = 32'b0_0000_0011_1110_0000_0000_0000_0000_000;
        flp_b = 32'b1_0000_0011_1010_0000_0000_0000_0000_000;
#10     start = 1'b1;
#10     start = 1'b0;


        //+25 x +25 = +625
        //           s ----e---- -------------f-------------
#10     flp_a = 32'b0_0000_0101_1100_1000_0000_0000_0000_000;
        flp_b = 32'b0_0000_0101_1100_1000_0000_0000_0000_000;
#10     start = 1'b1;
#10     start = 1'b0;

//continued on next page
```

Figure 14.9 (Continued)

```
        //+76 x +55 = +4180
        //          s ----e---- --------------f-------------
#10     flp_a = 32'b0_0000_0111_1001_1000_0000_0000_0000_000;
        flp_b = 32'b0_0000_0110_1101_1100_0000_0000_0000_000;
#10     start = 1'b1;
#10     start = 1'b0;


        //-48 x -17 = +816
        //          s ----e---- --------------f-------------
#10     flp_a = 32'b1_0000_0110_1100_0000_0000_0000_0000_000;
        flp_b = 32'b1_0000_0101_1000_1000_0000_0000_0000_000;
#10     start = 1'b1;
#10     start = 1'b0;


        //+3724 x +853 = +3,176,572
        //          s ----e---- --------------f-------------
#10     flp_a = 32'b0_0000_1100_1110_1000_1100_0000_0000_000;
        flp_b = 32'b0_0000_1010_1101_0101_0100_0000_0000_000;
#10     start = 1'b1;
#10     start = 1'b0;


        //0 x +853 = +0
        //          s ----e---- --------------f-------------
#10     flp_a = 32'b0_0000_1100_0000_0000_0000_0000_0000_000;
        flp_b = 32'b0_0000_1010_1101_0101_0100_0000_0000_000;
#10     start = 1'b1;
#10     start = 1'b0;

#20     $stop;

end

//instantiate the module into the test bench
mul_flp3 inst1 (
    .flp_a(flp_a),
    .flp_b(flp_b),
    .start(start),
    .sign(sign),
    .exponent(exponent),
    .exp_unbiased(exp_unbiased),
    .prod(prod)
    );

endmodule
```

Figure 14.9 (Continued)

```
(+5) × (+3) = +15
sign = 0,
   exp_unbiased = 00000100,
   prod = 111100000000000000000000000000000000000000000000

(+6) × (+4) = +24
sign = 0,
   exp_unbiased = 00000101,
   prod = 110000000000000000000000000000000000000000000000

(-5 × (+5) = -25
sign = 1,
   exp_unbiased = 00000101,
   prod = 110010000000000000000000000000000000000000000000

(+7) × (-5) = -35
sign = 1,
   exp_unbiased = 00000110,
   prod = 100011000000000000000000000000000000000000000000

(+25) × (+25) = +625
sign = 0,
   exp_unbiased = 00001010,
   prod = 100111000100000000000000000000000000000000000000

(+76) × (+55) = +4180
sign = 0,
   exp_unbiased = 00001101,
   prod = 100000101010000000000000000000000000000000000000

(-48) × (-17) = +816
sign = 0,
   exp_unbiased = 00001010,
   prod = 110011000000000000000000000000000000000000000000

(+3724) × (+853) = +3,176,572
sign = 0,
   exp_unbiased = 00010110,
   prod = 110000011110000111110000000000000000000000000000

(+0) × (+853) = +0
sign = 0,
   exp_unbiased = 00000000,
   prod = 000000000000000000000000000000000000000000000000
```

Figure 14.10 Outputs for floating-point multiplication using the sequential add-shift method.

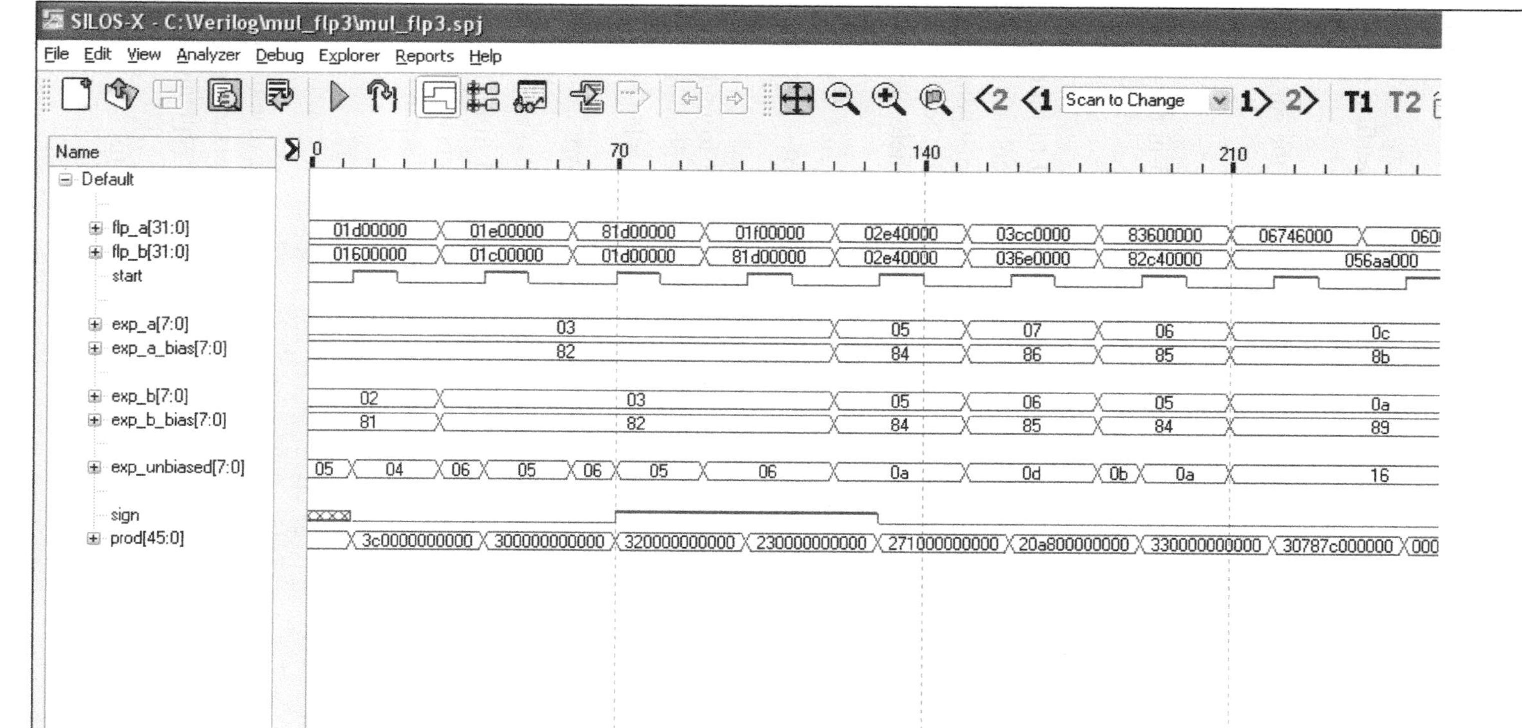

Figure 14.11 Waveforms for floating-point multiplication.

14.5 Problems

14.1 For a 32-bit floating-point format with a 7-bit exponent and a 24-bit fraction, determine the largest positive number with an unbiased exponent and the most negative number with the most negative unbiased exponent.

14.2 Show that there can be no fraction overflow in floating-point multiplication.

14.3 Represent the decimal number –5 in the single-precision format with an implied 1 bit and a biased exponent.

14.4 Represent the decimal number –1.75 in the single-precision format with an implied 1 bit and a biased exponent.

14.5 Use the sequential add-shift method to multiply the two decimal numbers shown below as floating-point numbers using 8-bit fractions. Show multiple shifts as a single shift amount, not as separate shifts. For example, to shift over three low-order zeroes, perform the shift operation in one statement. There are eight bits in each fraction; therefore, the sequence counter is initialized to a value of eight.

$$-7 \times -15$$

14.6 Use the sequential add-shift method to multiply the two decimal numbers shown below as floating-point numbers using 8-bit fractions. Show multiple shifts as a single shift amount, not as separate shifts.

$$-35 \times +72$$

14.7 Use the sequential add-shift method to multiply the two decimal numbers shown below as floating-point numbers using 8-bit fractions. Show multiple shifts as a single shift amount, not as separate shifts.

$$+80 \times +37$$

14.8 Use the sequential add-shift method to multiply the two decimal numbers shown below as floating-point numbers using 8-bit fractions. Show multiple shifts as a single shift amount, not as separate shifts.

$$+34 \times -68$$

14.9 Use the sequential add-shift method to multiply the two decimal numbers shown below as floating-point numbers using 8-bit fractions. Show each shift as a separate step; therefore, each step will always be an add-shift-right sequence providing eight separate cycles. The sequence counter is initialized to a value of eight.

$$+7 \times -5$$

14.10 A floating-point format is shown below for 14 bits. Design a behavioral module for floating-point multiplication using the sequential add-shift method for two operands: *flp_a[13:0]* and *flp_b[13:0]*. Obtain the test bench for several different input vectors, including both positive and negative operands. Obtain the outputs and waveforms.

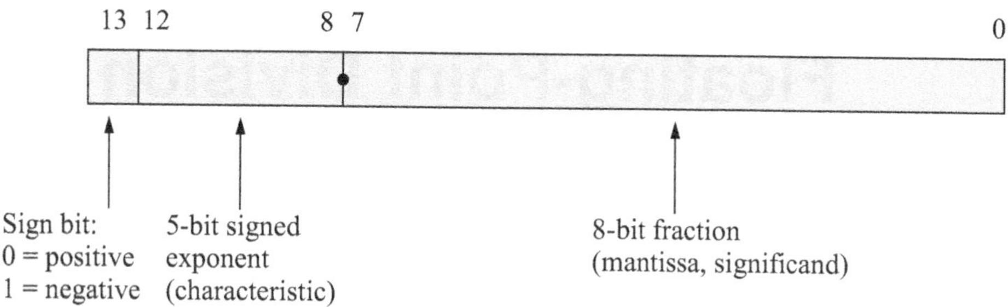

14.11 The single-precision floating-point format is shown below for 32 bits. Design a behavioral module for floating-point multiplication using the multiplication operator for two operands: *flp_a[31:0]* and *flp_b[31:0]*. Check for exponent overflow and zero-valued operands. Obtain the test bench for several different input vectors, including positive operands, negative operands, exponent overflow, and zero-valued operands. Obtain the outputs and waveforms.

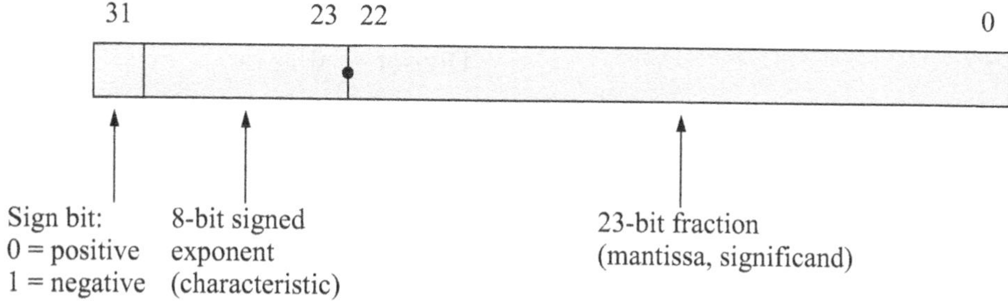

15.1 Zero Bias
15.2 Exponent Overflow/Underflow
15.3 Flowcharts
15.4 Numerical Examples
15.5 Problems

15

Floating-Point Division

Floating-point division performs two operations in parallel: fraction division and exponent subtraction. Fraction division can be accomplished using any of the methods presented in the chapter on fixed-point division. The dividend is usually $2n$ bits and the divisor is n bits; that is, the dividend conforms to the double-precision format and the divisor conforms to the single-precision format.

Divide overflow is determined in the same way as in fixed-point division; that is, if the high-order half of the dividend is greater than or equal to the divisor, then divide overflow occurs. The problem is resolved by shifting the dividend right one bit position and incrementing the exponent by one. Since both operands were normalized, this assures that the entire dividend is smaller than the divisor, as shown below.

$$\begin{aligned} \text{High-order half of Dividend} \quad & 0.01xxx\ldots xx \\ = \quad & \\ \text{Divisor} = \quad & 0.1xxx\ldots xx \end{aligned}$$

This is referred to as *dividend alignment*, providing the ranges for the two operands, as shown below.

$$\begin{aligned} 1/4 \ \le \ & \text{Dividend} \ < \ 1/2 \\ 1/2 \ \le \ & \text{Divisor} \ < \ 1 \end{aligned}$$

Both operands are checked for a value of zero. If the dividend is zero, then the exponent, quotient, and remainder are set to zero. If the divisor is zero, then the result is infinity and the operation is terminated. Division is performed on normalized floating-point operands *A* and *B* using biased exponents, such that

$$A = f_A \times r^{e_A}$$

$$B = f_B \times r^{e_B}$$

where *f* is the normalized fraction, *e* is the exponent, and *r* is the radix. Floating-point division is defined as shown in Equation 15.1, which shows fraction division and exponent subtraction performed simultaneously.

$$\begin{aligned} A / B &= (f_A \times r^{e_A}) / (f_B \times r^{e_B}) \\ &= (f_A / f_B) \times r^{e_A - e_B} \end{aligned} \tag{15.1}$$

The dividend is usually represented in sign-magnitude notation in the double-precision format since the dividend is 2*n* bits; the divisor is usually represented in sign-magnitude notation in single-precision format since the divisor is *n* bits, as shown in Figure 15.1.

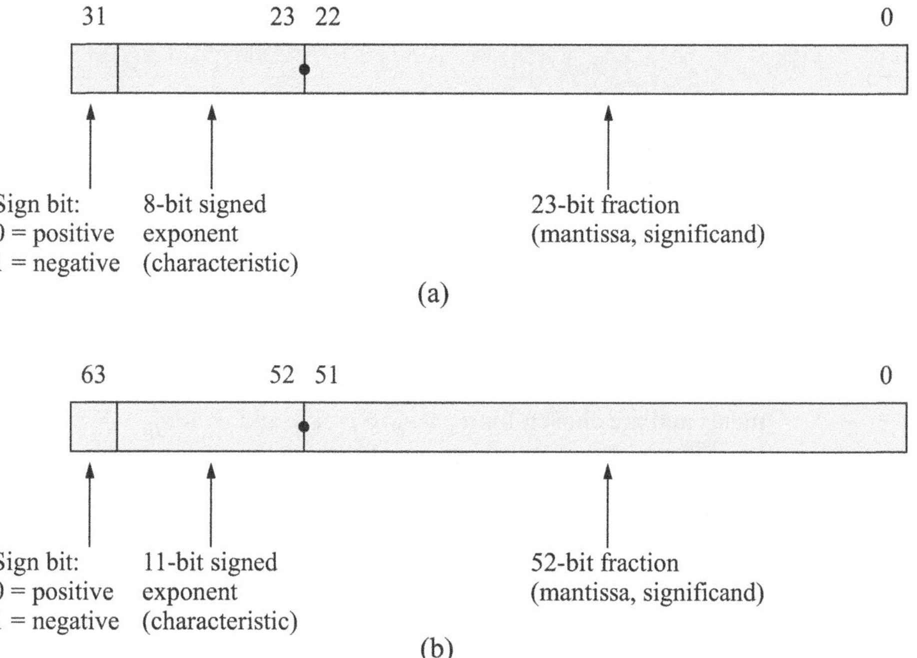

Figure 15.1 IEEE floating-point formats: (a) 32-bit single precision and (b) 64-bit double precision.

The sign of the quotient is determined by the signs of the floating-point numbers. If the signs are the same, then the sign of the quotient is positive; if the signs are different, then the sign of the quotient is negative. This can be determined by the exclusive-OR of the two signs, as shown in Equation 15.2. The sign of the remainder is the same as the sign of the dividend.

$$\text{Quotient sign} = A_{\text{sign}} \oplus B_{\text{sign}} \tag{15.2}$$

15.1 Zero Bias

As was stated previously, the divisor exponent is subtracted from the dividend exponent in parallel with fraction division. The exponents are subtracted and the carry-out is examined. If the carry-out = 1, then the dividend exponent was greater than or equal to the divisor exponent ($e_A \geq e_B$). If the carry-out = 0, then the dividend exponent was less than the divisor exponent ($e_A < e_B$). Since both exponents were initially biased, the difference generates a result with no bias, as shown in Equation 15.3.

$$\begin{aligned} e_A - e_B &= (e_A + \text{bias}) - (e_B + \text{bias}) \\ &= e_A + \text{bias} - e_B - \text{bias} \\ &= (e_A - e_B)_{\text{unbiased}} \end{aligned} \tag{15.3}$$

Therefore, the bias must be added to the difference so that the resulting exponent is properly biased. Thus, for the single-precision format:

$$(e_A - e_B)_{\text{biased}} = (e_A - e_B)_{\text{unbiased}} + 0111\ 1111$$

Restoring the bias may result in an exponent overflow, in which case the division operation is terminated. Examples will now be presented that illustrate the previous statements and are chosen for $e_A > e_B$, $e_A = e_B$, and $e_A < e_B$.

Example 15.1 Let $e_A > e_B$, where $e_{A(\text{unbiased})} = 0001\ 0110\ (22)$ and $e_{B(\text{unbiased})} = 0000\ 1010\ (10)$. Therefore, $e_A - e_B = 22 - 10 = 12$.

$$\begin{aligned} e_{A(\text{unbiased})} &= 0001\ \ 0110 \\ \text{Add bias} +)\ &\underline{\ 0111\ \ 1111\ } \\ e_{A(\text{biased})} &= 1001\ \ 0101 \end{aligned}$$

$$\begin{array}{rl} e_{B(\text{unbiased})} = & 0000\ \ 1010 \\ \text{Add bias +)} & \underline{0111\ \ 1111} \\ e_{B(\text{biased})} = & 1000\ \ 1001 \end{array}$$

$e_{A(\text{biased})} - e_{B(\text{biased})}$

$$\begin{array}{rll} e_{A(\text{biased})} = & 1001\ \ 0101 & \\ \text{+) 2s complement of } e_{B(\text{biased})} = & \underline{0111\ \ 0111} & \\ 1 \leftarrow & 0000\ \ 1100 & 12 \end{array}$$

Restore to single bias by adding the bias.

$$\begin{array}{rl} (e_A - e_B)_{\text{unbiased}} = & 0000\ \ 1100 \\ \text{Add bias} = & \underline{0111\ \ 0111} \\ (e_A - e_B)_{\text{biased}} = & 1001\ \ 1011 \end{array}$$

Example 15.2 Let $e_A = e_B$, where $e_{A(\text{unbiased})}$ = 0001 0101 (21) and $e_{B(\text{unbiased})}$ = 0001 0101 (21). Therefore, $e_A - e_B = 21 - 21 = 0$.

$$\begin{array}{rl} e_{A(\text{unbiased})} = & 0001\ \ 0101 \\ \text{Add bias +)} & \underline{0111\ \ 1111} \\ e_{A(\text{biased})} = & 1001\ \ 0100 \end{array}$$

$$\begin{array}{rl} e_{B(\text{unbiased})} = & 0001\ \ 0101 \\ \text{Add bias +)} & \underline{0111\ \ 1111} \\ e_{B(\text{biased})} = & 1001\ \ 0100 \end{array}$$

$e_{A(\text{biased})} - e_{B(\text{biased})}$

$$\begin{array}{rll} e_{A(\text{biased})} = & 1001\ \ 0100 & \\ \text{+) 2s complement of } e_{B(\text{biased})} = & \underline{0110\ \ 1100} & \\ 1 \leftarrow & 0000\ \ 0000 & 0 \end{array}$$

Restore to single bias by adding the bias.

$$\begin{array}{rl} (e_A - e_B)_{\text{unbiased}} = & 0000\ \ 0000 \\ \text{Add bias} = & \underline{0111\ \ 0111} \\ (e_A - e_B)_{\text{biased}} = & 0111\ \ 1111 \end{array}$$

Example 15.3 Let $e_A < e_B$, where $e_{A(\text{unbiased})} = 0000\ 1001$ (9) and $e_{B(\text{unbiased})} = 0001\ 0011$ (19). Therefore, $e_A - e_B = 9 - 19 = -10$.

$$\begin{array}{r r}
e_{A(\text{unbiased})} = & 0000\ \ 1001 \\
\text{Add bias +)} & \underline{0111\ \ 1111} \\
e_{A(\text{biased})} = & 1000\ \ 1000
\end{array}$$

$$\begin{array}{r r}
e_{B(\text{unbiased})} = & 0001\ \ 0011 \\
\text{Add bias +)} & \underline{0111\ \ 1111} \\
e_{B(\text{biased})} = & 1001\ \ 0010
\end{array}$$

$e_{A(\text{biased})} - e_{B(\text{biased})}$

$$\begin{array}{r r l}
e_{A(\text{biased})} = & 1000\ \ 1000 & \\
\text{+) 2s complement of } e_{B(\text{biased})} = & \underline{0110\ \ 1110} & \\
0 \leftarrow & 1111\ \ 0110 & -10
\end{array}$$

If cout = 0, then 2s complement to obtain the difference of 0000 1010 (10).

Restore to single bias by adding the bias.

$$\begin{array}{r r}
(e_A - e_B)_{\text{unbiased}} = & 1111\ \ 0110 \\
\text{Add bias} = & \underline{0111\ \ 1111} \\
(e_A - e_B)_{\text{biased}} = & 0111\ \ 0101
\end{array}$$

Example 15.4 Let $e_A > e_B$, where $e_{A(\text{unbiased})} = 0001\ 0011$ (19) and $e_{B(\text{unbiased})} = 0000\ 1001$ (9). Therefore, $e_A - e_B = 19 - 9 = 10$.

$$\begin{array}{r r}
e_{A(\text{unbiased})} = & 0001\ \ 0011 \\
\text{Add bias +)} & \underline{0111\ \ 1111} \\
e_{A(\text{biased})} = & 1001\ \ 0010
\end{array}$$

$$\begin{array}{r r}
e_{B(\text{unbiased})} = & 0000\ \ 1001 \\
\text{Add bias +)} & \underline{0111\ \ 1111} \\
e_{B(\text{biased})} = & 1000\ \ 1000
\end{array}$$

$e_{A(\text{biased})} - e_{B(\text{biased})}$

$$\begin{array}{rrrr} e_{A(\text{biased})} = & 1001\ \ 0010 & \\ +)\ \text{2s complement of } e_{B(\text{biased})} = & \underline{0111\ \ 1000} & \\ 1 \leftarrow & 0000\ \ 1010 & 10 \end{array}$$

Restore to single bias by adding the bias.

$$\begin{array}{rr} (e_A - e_B)_{\text{unbiased}} = & 0000\ \ 1010 \\ \text{Add bias} = & \underline{0111\ \ 1111} \\ (e_A - e_B)_{\text{biased}} = & 1000\ \ 1001 \end{array}$$

15.2 Exponent Overflow/Underflow

For the single-precision format using 8-bit exponents, the most positive signed unbiased exponent is 0111 1111 (+127). A bias of 0111 1111 is added to the unbiased exponent to form a biased unsigned exponent, such that $\text{exponent}_{\text{biased}} = \text{exponent}_{\text{unbiased}} +$ 0111 1111. This is referred to as the excess-127 format. The low-order value (0) and the high-order value (255) represent special values. If $\text{exponent}_{\text{biased}} = 0$ and the mantissa = 0, then the value of the floating-point number = 0. If $\text{exponent}_{\text{biased}} = 255$ and the mantissa = 0, then the value of the floating-point number = ∞. The sign bit then represents ± 0 and $\pm \infty$.

Therefore, the normal range for a biased exponent is $1 \leq \text{exponent}_{\text{biased}} \leq 254$, which provides a range for an unbiased exponent of $-126 \leq \text{exponent}_{\text{unbiased}} \leq +127$. Examples of exponent overflow and underflow will now be presented for $e_A > e_B$ and $e_A < e_B$.

Example 15.5 For $e_A > e_B$, let $e_{A(\text{unbiased})} = 0111\ 1111$ (+127) and $e_{B(\text{unbiased})} = 0111\ 1110$ (+126).

$$\begin{array}{rrr} e_{A(\text{unbiased})} = & 0111\ \ 1111 & (+127) \\ \text{Add bias} +) & \underline{0111\ \ 1111} & \\ e_{A(\text{biased})} = & 1111\ \ 1110 & \end{array}$$

$$\begin{array}{rrr} e_{B(\text{unbiased})} = & 0111\ \ 1110 & (+126) \\ \text{Add bias} +) & \underline{0111\ \ 1111} & \\ e_{B(\text{biased})} = & 1111\ \ 1101 & \end{array}$$

$e_A - e_B$

$$\begin{array}{rr}
e_{A(\text{biased})} = & 1111\ \ 1110 \\
\text{Add 2s complement of } e_{B(\text{biased})}\ +) & \underline{0000\ \ 0011} \\
(e_A - e_B)_{(\text{unbiased})} = & 0000\ \ 0001 \\
\text{Add bias } +) & \underline{0111\ \ 1111} \\
\text{Biased exponent} = & 1000\ \ 0000 \quad 128 \text{ (No overflow)}
\end{array}$$

Example 15.6 For $e_A > e_B$, let $e_{A(\text{unbiased})} = 0100\ 0000\ (+64)$ and $e_{B(\text{unbiased})} = 1100\ 0000\ (-64)$.

$$\begin{array}{rr}
e_{A(\text{unbiased})} = & 0100\ \ 0000 \quad (+64) \\
\text{Add bias } +) & \underline{0111\ \ 1111} \\
e_{A(\text{biased})} = & 1011\ \ 1111
\end{array}$$

$$\begin{array}{rr}
e_{B(\text{unbiased})} = & 1100\ \ 0000 \quad (-64) \\
\text{Add bias } +) & \underline{0111\ \ 1111} \\
e_{B(\text{biased})} = & 0011\ \ 1111
\end{array}$$

$e_A - e_B$

$$\begin{array}{rr}
e_{A(\text{biased})} = & 1011\ \ 1111 \\
\text{Add 2s complement of } e_{B(\text{biased})}\ +) & \underline{1100\ \ 0001} \\
(e_A - e_B)_{(\text{unbiased})} = & 1000\ \ 0000 \\
\text{Add bias } +) & \underline{0111\ \ 1111} \\
\text{Biased exponent} = & 1111\ \ 1111 \quad 255 \text{ (Overflow)}
\end{array}$$

Example 15.7 For $e_A < e_B$, let $e_{A(\text{unbiased})} = 1000\ 0010\ (-126)$ and $e_{B(\text{unbiased})} = 1111\ 1110\ (-2)$.

$$\begin{array}{rr}
e_{A(\text{unbiased})} = & 1000\ \ 0010 \quad (-126) \\
\text{Add bias } +) & \underline{0111\ \ 1111} \\
e_{A(\text{biased})} = & 0000\ \ 0001
\end{array}$$

$$\begin{array}{rr} e_{B(\text{unbiased})} = & 1111\ \ 1110 \\ \text{Add bias} +) & \underline{0111\ \ 1111} \\ e_{B(\text{biased})} = & 0111\ \ 1101 \end{array} \qquad (-2)$$

$e_A - e_B$

$$\begin{array}{rr} e_{A(\text{biased})} = & 0000\ \ 0001 \\ \text{Add 2s complement of } e_{B(\text{biased})} +) & \underline{1000\ \ 0011} \\ (e_A - e_B)_{(\text{unbiased})} = & 1000\ \ 0100 \\ \text{Add bias} +) & \underline{0111\ \ 1111} \\ \text{Biased exponent} = & 0000\ \ 0011 \end{array} \qquad 3 \text{ (No Underflow)}$$

Example 15.8 For $e_A < e_B$, let $e_{A(\text{unbiased})} = 1000\ 0010\ (-126)$ and $e_{B(\text{unbiased})} = 0000\ 0010\ (+2)$.

$$\begin{array}{rr} e_{A(\text{unbiased})} = & 1000\ \ 0010 \\ \text{Add bias} +) & \underline{0111\ \ 1111} \\ e_{A(\text{biased})} = & 0000\ \ 0001 \end{array} \qquad (-126)$$

$$\begin{array}{rr} e_{B(\text{unbiased})} = & 0000\ \ 0010 \\ \text{Add bias} +) & \underline{0111\ \ 1111} \\ e_{B(\text{biased})} = & 1000\ \ 0001 \end{array} \qquad (+2)$$

$e_A - e_B$

$$\begin{array}{rr} e_{A(\text{biased})} = & 0000\ \ 0001 \\ \text{Add 2s complement of } e_{B(\text{biased})} +) & \underline{0111\ \ 1111} \\ (e_A - e_B)_{(\text{unbiased})} = & 1000\ \ 0000 \\ \text{Add bias} +) & \underline{0111\ \ 1111} \\ \text{Biased exponent} = & 1111\ \ 1111 \end{array} \qquad 255 \text{ (Underflow)}$$

15.3 Flowcharts

The division algorithm can be partitioned into nine steps, as shown below.

1. Normalize the operands
2. Check for zero operands.
3. Determine the quotient sign.
4. Bias the exponents.
5. Align dividend
6. Subtract the exponents.
7. Restore the bias by adding the bias.
8. Divide the fractions.
9. Normalize the product, if necessary.

Figure 15.2 shows a flowchart representation of the above rules for floating-point division.

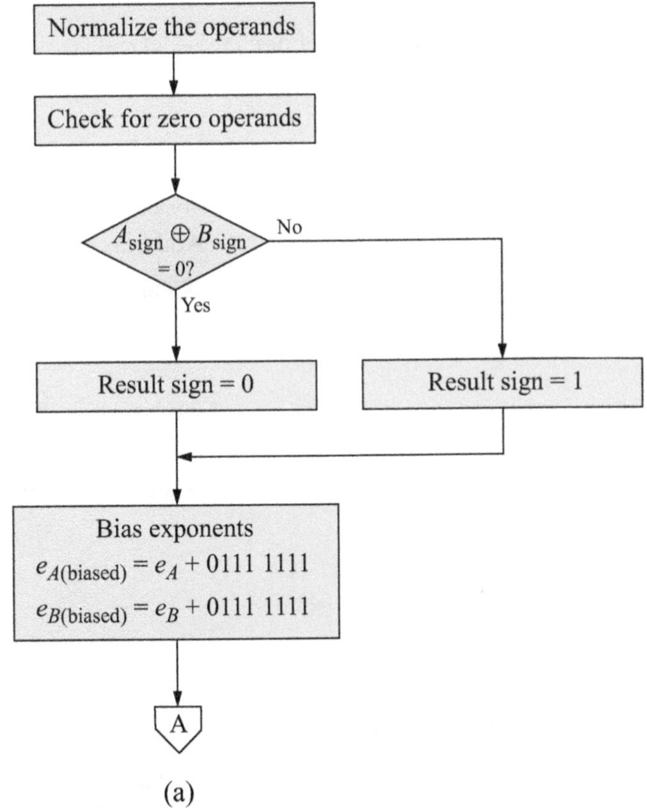

(a)

Figure 15.2 Flowchart illustrating floating-point division: (a) check for zero operands, determine product sign, bias exponents and (b) dividend alignment, check for exponent overflow, subtract exponents, restore bias, divide fractions, and normalize the result.

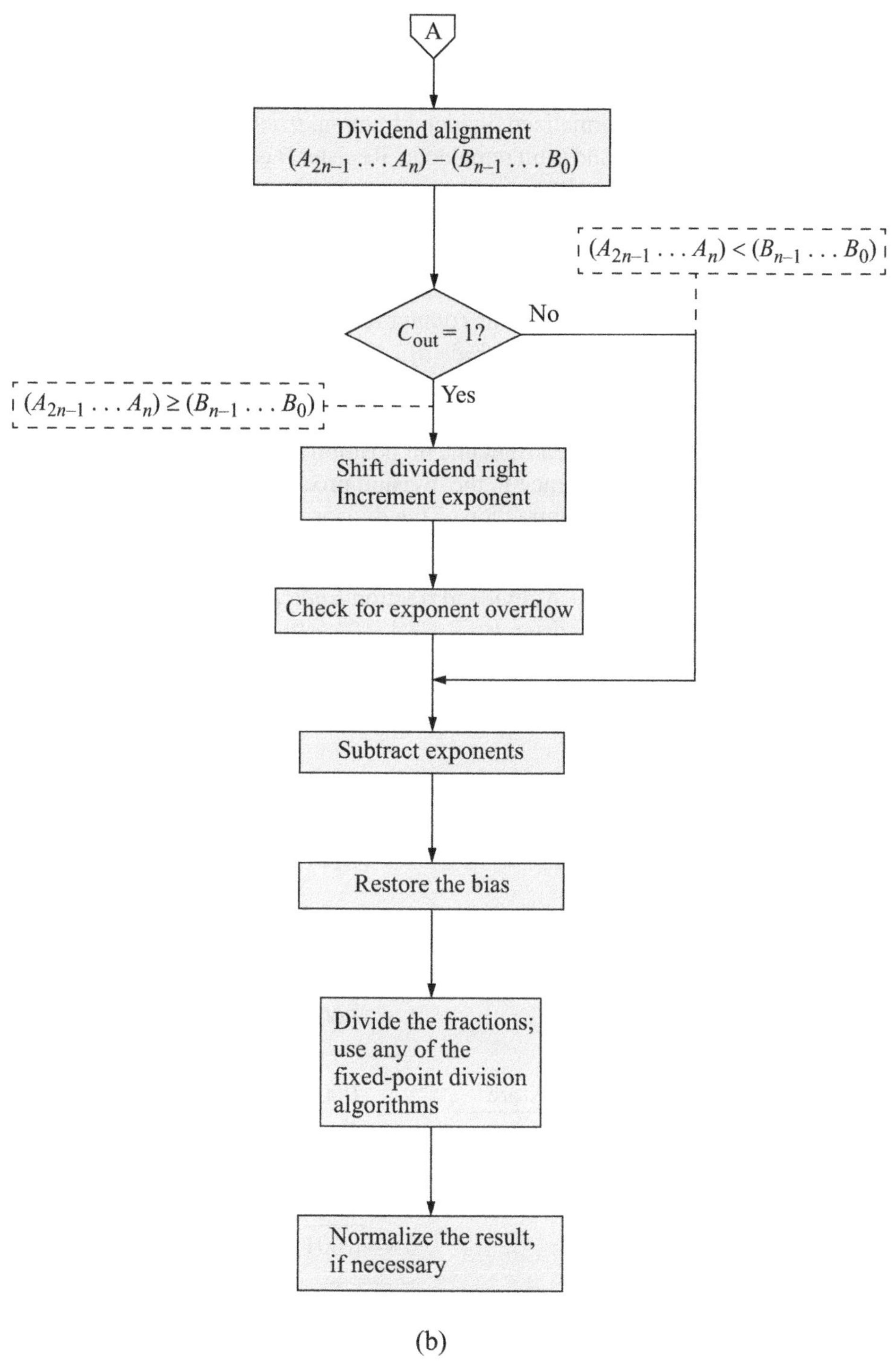

Figure 15.2 (Continued)

15.4 Numerical Examples

This section presents numerical examples using the sequential shift-subtract/add restoring division method with 4-bit divisors and 8-bit dividends. Register A contains the $2n$-bit normalized dividend fraction, *fract_a*, which will eventually contain the n-bit quotient and n-bit remainder. Register B contains the n-bit normalized divisor fraction, *fract_b*.

Since the division process involves one n-bit divisor and one $2n$-bit dividend, a count-down sequence counter, *count*, is set to a value that represents the number of bits in the divisor. The counter is decremented by one for each step of the shift-subtract/add sequence. When the counter reaches a value of zero, the operation is finished and the quotient resides in *fract_a[3:0]* and the remainder resides in *fract_a[7:4]*.

If the value of the high-order half of the dividend is greater than or equal to the value of the divisor, then an overflow condition exists. To resolve this problem, the dividend is shifted right one bit position and the dividend exponent is incremented by one. Each sequence in the division process consists of a shift left of one bit position followed by a subtraction of the divisor.

Example 15.9 A dividend fraction *fract_a* $= 0.1010\ 0100 \times 2^7$ (+82) is divided by a divisor fraction *fract_b* $= 0.1001 \times 2^4$ (+9) to yield a quotient of 1001×2^4 (+9) and a remainder of 0001×2^4 (+1).

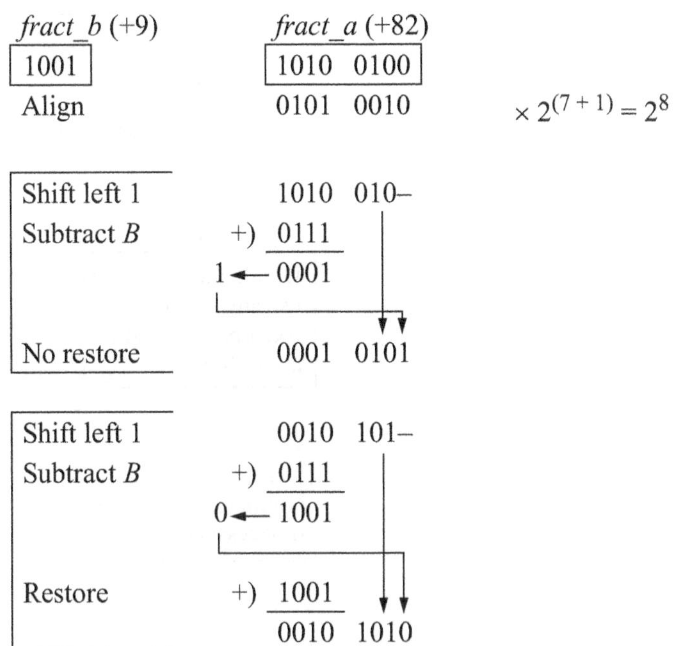

Continued on next page

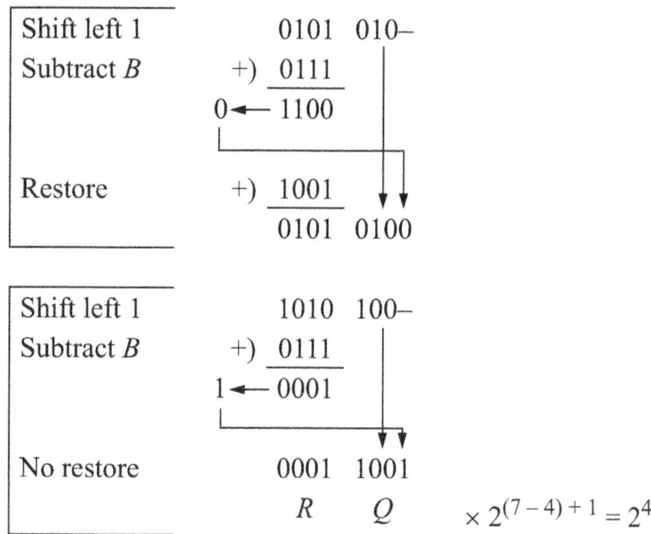

Shift left 1
Subtract B
Restore
0101 010–
+) 0111
0 ← 1100
+) 1001
0101 0100
Shift left 1
Subtract B
No restore
1010 100–
+) 0111
1 ← 0001
0001 1001
R Q
× 2^(7 – 4) + 1 = 2^4

Example 15.10 A dividend fraction *fract_a* = 0.1000 0110 × 2^7 (+67) is divided by a divisor fraction *fract_b* = 0.1000 × 2^4 (+8) to yield a quotient of 1000 × 2^4 (+8) and a remainder of 0011 × 2^4 (+3).

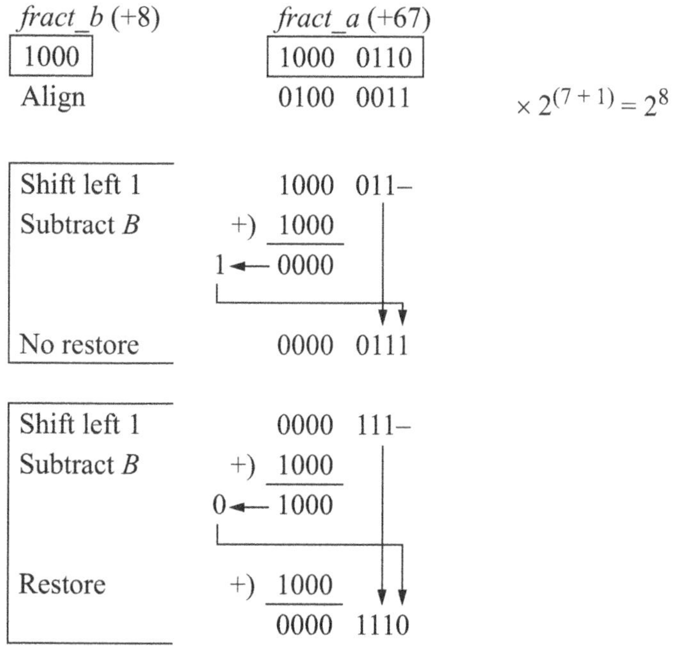

fract_b (+8)
1000
Align
fract_a (+67)
1000 0110
0100 0011
× 2^(7 + 1) = 2^8
Shift left 1
Subtract B
No restore
1000 011–
+) 1000
1 ← 0000
0000 0111
Shift left 1
Subtract B
Restore
0000 111–
+) 1000
0 ← 1000
+) 1000
0000 1110

Continued on next page

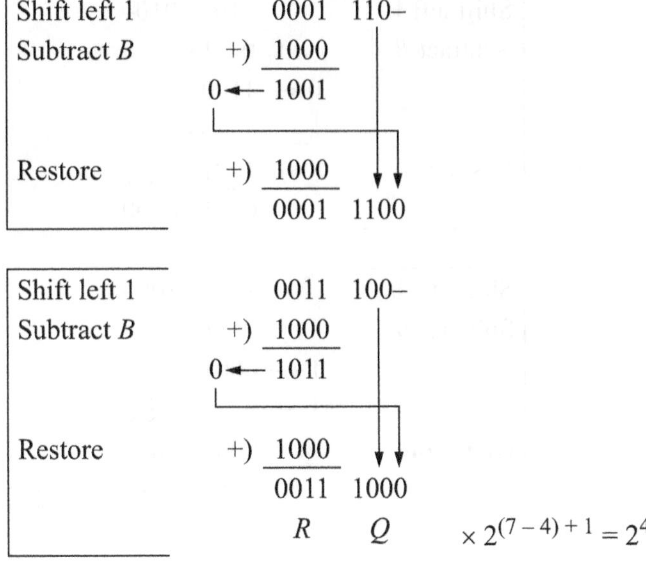

Example 15.11 A dividend fraction *fract_a* = 0.1100 1000 × 2^5 (+25) is divided by a divisor fraction *fract_b* = 0.1010 × 2^3 (+5) to yield a quotient of 1010 × 2^3 (+5) and a remainder of 0000 × 2^3 (+0).

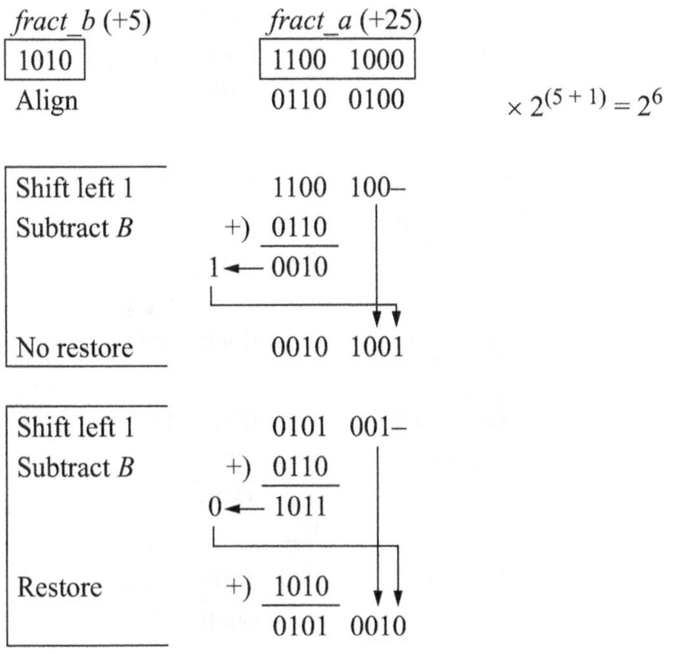

Continued on next page

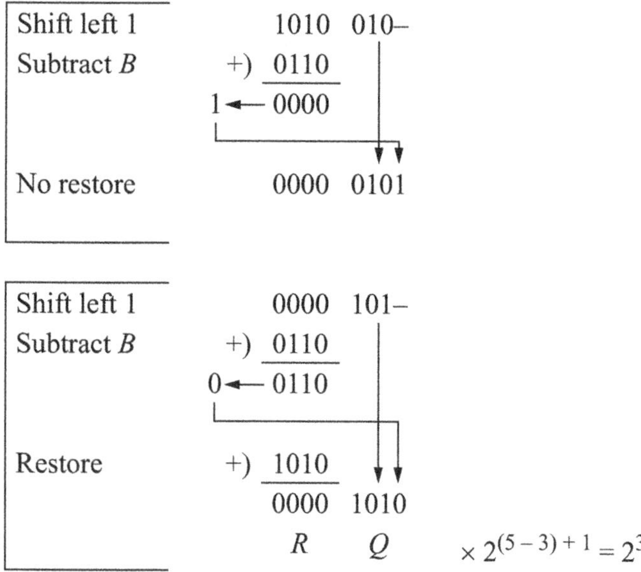

Shift left 1
Subtract B
1010 010–
+) 0110
1 ← 0000
No restore
0000 0101
Shift left 1
Subtract B
0000 101–
+) 0110
0 ← 0110
Restore
+) 1010
0000 1010
R Q
× 2^(5 – 3) + 1 = 2^3

15.5 Problems

15.1 How is quotient overflow established in floating-point division? How is the problem resolved? What is the range of the dividend and divisor after the quotient overflow issue has been resolved?

15.2 Write the equation that adjusts for a fraction overflow in floating-point division.

15.3 Comment on the biasing problem when the exponents are operated on during floating-point division.

15.4 Let the dividend fraction $fract\_a = 0.1111\ 0000 \times 2^4$ (+15) be divided by a divisor fraction $fract\_b = 0.1010 \times 2^3$ (+5). Use the paper-and-pencil technique to obtain the quotient and remainder.

15.5 Let the dividend fraction $fract\_a = 0.1010\ 0010 \times 2^7$ (+81) be divided by a divisor fraction $fract\_b = 0.1001 \times 2^4$ (+9). Use the paper-and-pencil technique to obtain the quotient and remainder.

15.6 Let the dividend fraction *fract_a* = $0.1110\ 1000 \times 2^7$ (+116) be divided by a divisor fraction *fract_b* = 0.1101×2^4 (+13). Use the paper-and-pencil technique to obtain the quotient and remainder.

15.7 It was stated in this chapter that division for floating-point fractions can be implemented using any of the methods described in fixed-point division. Therefore, using fixed-point restoring division, design a Verilog HDL behavioral and dataflow module to perform division on 16-bit dividends and 8-bit divisors, which represent the fractions. Obtain the test bench, outputs, and waveforms.

15.8 Using the floating-point formats shown below, design a behavioral module to divide a 14-bit dividend by a 10-bit divisor. Obtain the test bench, outputs, and waveforms.

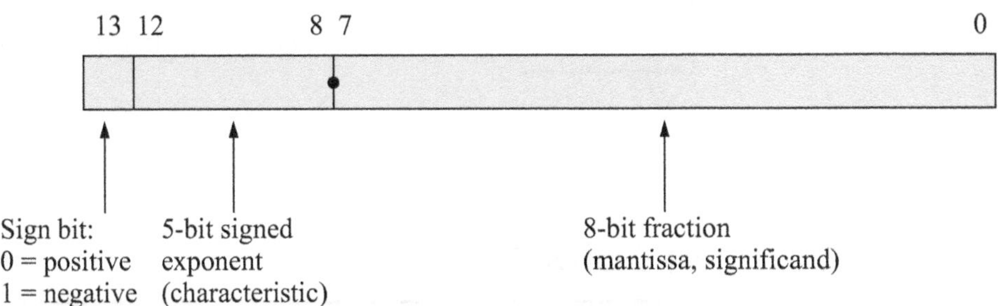

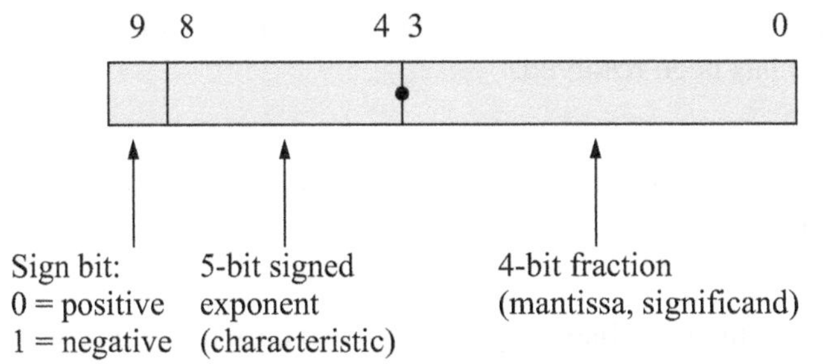

16.1 Rounding Methods
16.2 Guard Bits
16.3 Verilog HDL Implementations
16.4 Problems

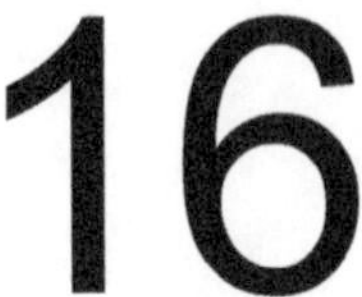

Additional Floating-Point Topics

The results of a floating-point operation should be as precise as possible. However, arithmetic calculations may produce results that are imprecise due to fraction alignment and rounding techniques. Fraction alignment may result in low-order significand bits being lost off the right end of the fraction. Rounding reduces the number of bits in the significand, thus limiting the precision of the result. Guard bits are also discussed, which retain extra bits during the floating-point operation to maintain maximum accuracy.

16.1 Rounding Methods

Rounding deletes one or more low-order bits of the significand and adjusts the retained bits according to a particular rounding technique. The reason for rounding is to reduce the number of bits in the result in order to conform to the size of the significand; that is, in order to be retained within the word size of the machine. Since bits are deleted, this limits the precision of the result.

In some floating-point operations, the result may exceed the number of bits of the significand. For example, rounding can occur when adding two n-bit numbers that result in a sum of $n + 1$ bits. The overflow is handled by shifting the fraction right 1 bit position, resulting in the low-order bit being lost unless it is saved. Rounding attempts to dispose of the extra bits and yet preserve a high degree of accuracy. This section presents three common techniques for rounding that still maintain a high degree of accuracy.

16.1.1 Truncation Rounding

This method of rounding is also called *chopping*. Truncation deletes extra bits and makes no changes to the retained bits. Aligning fractions during addition or subtraction could result is losing several low-order bits, so there is obviously an error associated with truncation. Assume that the following fraction is to be truncated to four bits:

$$0.b_{-1}\, b_{-2}\, b_{-3}\, b_{-4}\, b_{-5}\, b_{-6}\, b_{-7}\, b_{-8}$$

Then all fractions in the range $0.b_{-1}\, b_{-2}\, b_{-3}\, b_{-4}\, 0000$ to $0.b_{-1}\, b_{-2}\, b_{-3}\, b_{-4}\, 1111$ will be truncated to $0.b_{-1}\, b_{-2}\, b_{-3}\, b_{-4}$. The error ranges from 0 to .00001111. In general, the error ranges from 0 to approximately 1 in the low-order position of the retained bits.

Truncation is a fast and easy method for deleting bits resulting from a fraction underflow and requires no additional hardware. There is one disadvantage in that a significant error may result. Recall that fraction underflow can occur when aligning fractions during addition or subtraction when one of the fractions is shifted to the right.

A graph indicating the truncation function is shown in Figure 16.1, which truncates 4-bit fractions to three bits. Note that the function lies entirely below the ideal truncation line and contacts the line only where there is no truncation error. For example, fraction 0.0010 truncates to 0.001 and fraction 0.0110 truncates to 0.011. An error is introduced when fractions 0.0010 and 0.0011 truncate to 0.001 and fractions 0.0110 and 0.0111 truncate to 0.011.

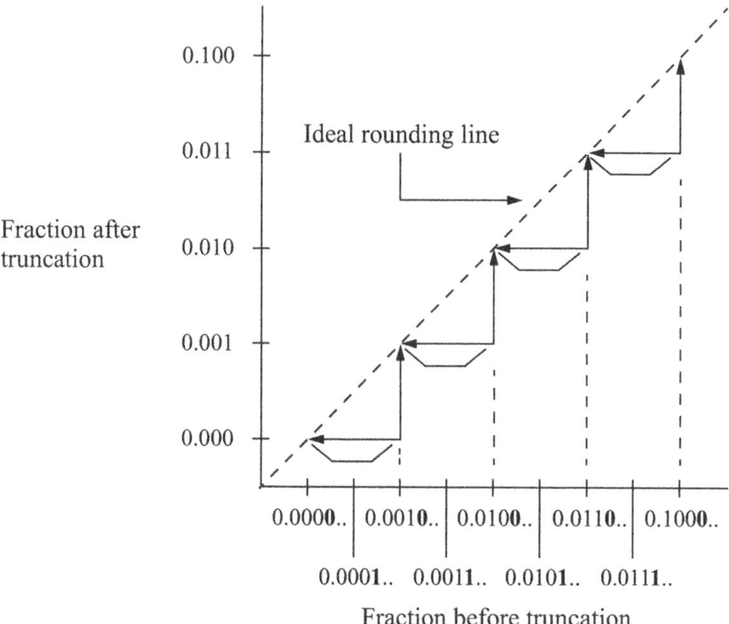

Figure 16.1 Graph illustrating the truncation function for rounding.

Any fraction below the ideal truncation line represents an error, because the error range is asymmetrical. Truncation does not round up or round down, but simply deletes a specified number of the low-order significand bits.

16.1.2 Adder-Based Rounding

The result of a floating-point arithmetic operation can be rounded to the nearest number that contains n bits. This method is called *adder-based rounding* and rounds the result to the nearest approximation that contains n bits. The operation is as follows: The bits to be deleted are truncated and a 1 is added to the retained bits if the high-order bit of the deleted bits is a 1. When a 1 is added to the retained bits, the carry is propagated to the higher-order bits. If the addition results in a carry out of the high-order bit position, then the fraction is shifted right one bit position and the exponent is incremented.

Consider the fraction $0.b_{-1}\,b_{-2}\,b_{-3}\,b_{-4}\,1\,x\,x\,x$ — where the xs are 0s or 1s — which is to be truncated and rounded to four bits. Using adder-based rounding, this rounds to $0.b_{-1}\,b_{-2}\,b_{-3}\,b_{-4} + 0.0001$ and the retained bits of fraction $0.b_{-1}\,b_{-2}\,b_{-3}\,b_{-4}\,0\,x\,x\,x$ round to $0.b_{-1}\,b_{-2}\,b_{-3}\,b_{-4}$. The first fraction approaches the true value from above; the second fraction approaches the true value from below. Examples of adder-based rounding are shown in Figure 16.2 and illustrate approaching the true value both from above and approaching the true value from below.

		Delete		
	0. 0 1 0 1	1 0 0 0	$\times 2^8 = +88$	True value
+)	0. 0 0 0 1			
Rounded result	0. 0 1 1 0		$\times 2^8 = +96$	Approach from above

(a)

		Delete		
	0. 0 1 1 1	0 1 1 1	$\times 2^8 = +119$	True value
	↓			
Rounded result	0. 0 1 1 1		$\times 2^8 = +112$	Approach from below

(b)

Figure 16.2 Adder-based rounding: (a) true value approached from above and (b) true value approached from below.

In Figure 16.2(a), the part of the fraction to be rounded has a value that is greater than or equal to 1/2 its maximum value of 15. Therefore, a 1 is added to the retained bits, which results in the true value being approached from above. That is, the part being deleted has a maximum value of 1111 (15), while its actual value is 1000 (8). Since a value of $8 \geq 7.5$, a 1 is added to the retained bits. A similar reasoning is used for Figure 16.2(b); however, the actual value of the part to be deleted is 0111 (7). Since $7 < 7.5$, the low-order four bits are deleted and a 1 is not added to the retained bits, which results in the true value being approached from below.

Figure 16.3 shows a graph that represents adder-based rounding. Note that the graph is symmetrical around the ideal rounding line. The odd-valued fractions — with a 1 in the high-order bit position of the bits being deleted — are rounded up and approach the true value from above. The even-valued fractions — with a 0 in the high-order bit position of the bits being deleted — are rounded down and approach the true value from below.

Adder-based rounding is an unbiased method that generates the nearest approximation to the number being rounded. Although adder-based rounding is obviously a better method of rounding than truncation, additional hardware is required to accommodate the addition cycle, thus adding more delay to the rounding operation.

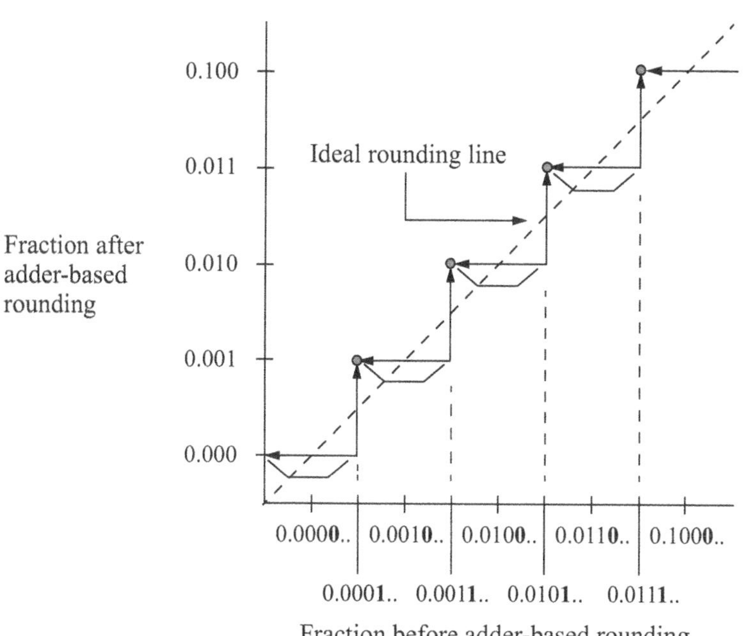

Figure 16.3 Graph illustrating the adder-based rounding method.

16.1.3 Von Neumann Rounding

The von Neumann rounding method is also referred to as *jamming* and is similar to truncation. If the bits to be deleted are all zeroes, then the bits are truncated and there is no change to the retained bits. However, if the bits to be deleted are not all zeroes, then the bits are deleted and the low-order bit of the retained bits is set to 1.

Thus, when 8-bit fractions are rounded to four bits, fractions in the range

$$0.b_{-1}\ b_{-2}\ b_{-3}\ b_{-4}\ 0001 \text{ to } 0.b_{-1}\ b_{-2}\ b_{-3}\ b_{-4}\ 1111$$

will all be rounded to $0.b_{-1}\ b_{-2}\ b_{-3}\ 1$. Therefore, the error ranges from approximately −1 to +1 in the low-order bit of the retained bits when

$$0.b_{-1}\ b_{-2}\ b_{-3}\ b_{-4}\ 0001 \text{ is rounded to } 0.b_{-1}\ b_{-2}\ b_{-3}\ 1$$

and when

$$0.b_{-1}\ b_{-2}\ b_{-3}\ b_{-4}\ 1111 \text{ is rounded to } 0.b_{-1}\ b_{-2}\ b_{-3}\ 1$$

Although the error range is larger in von Neumann rounding than with truncation rounding, the error range is symmetrical about the ideal rounding line and is an unbiased approximation, as shown in Figure 16.4. Assuming that individual errors are evenly distributed over the error range, then positive errors will be inclined to offset negative errors for long sequences of floating-point calculations involving rounding. The von Neumann rounding method has the same total bias as adder-based rounding; however, it requires no more time than truncation.

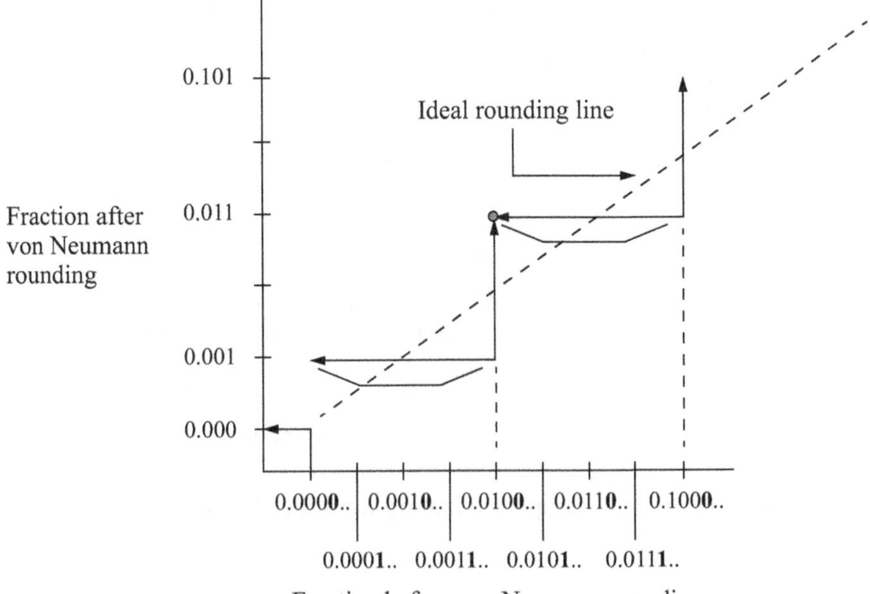

Figure 16.4 Graph illustrating the von Neumann rounding method.

16.2 Guard Bits

In order to achieve maximum accuracy in the result of a floating-point operation, additional bits must be maintained. These bits are called *guard bits* and are positioned to the right of the low-order fraction bit and may consist of one or more bits. The guard bits are the result of fraction alignment during exponent alignment and can be used with any rounding method.

When a fraction is shifted right during floating-point addition when the exponent values are different, the bits that are shifted off the low-order end of the fraction are retained in a guard-digit register that is concatenated with the significand fraction. If the rounding scheme is truncation, then the guard bits do not have to be retained.

During floating-point multiplication, two n-bit fractions generate a $2n$-bit product, which may contain a 0 bit in the high-order bit position. If the rounding method is truncation, then at least one of the high-order bits of the lower n-bit portion of the result must be retained for use as a guard bit. This bit becomes the low-order bit of the product when the product is postnormalized.

If a rounding method other than truncation is used, then at least two of the high-order bits of the lower n-bit portion of the result must be retained for use as guard bits. One bit is shifted left during postnormalization and one bit is used for rounding.

There may also be a third guard bit used in the rounding method, which is the low-order bit of the three guard bits and is initialized to 0. If a 1 bit is shifted into this bit position during a floating-point operation, then it remains in that state; otherwise, it remains in a 0 state; thus, it is often referred to as the *sticky bit*. This is advantageous in some rounding schemes, because it indicates if a 1 bit was shifted right into this position during any of the intermediate steps in a floating-point operation. Guard bits (or digits) are used in other radices; for example, radix 16. In this case, the guard digits consist of four bits. Figure 16.5 shows a typical guard digit register containing three bits, which is concatenated to the right of the significand register.

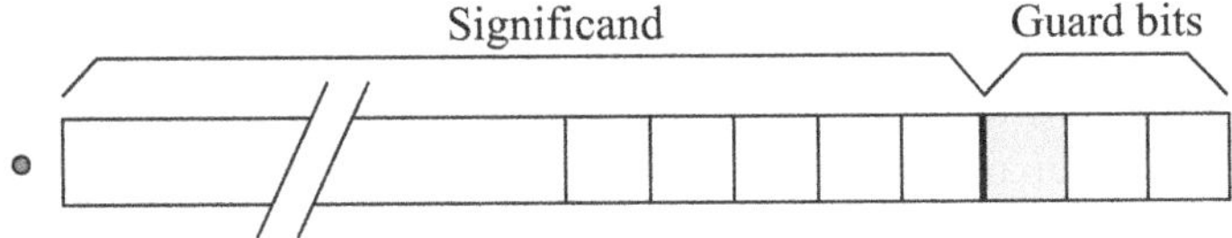

Figure 16.5 Typical significand with guard bits.

16.3 Verilog HDL Implementations

This section uses Verilog HDL to design three methods for implementing adder-based rounding. The first method uses memory to generate the desired rounded fraction in a behavioral module. Memory provides a faster rounding operation because it negates the necessity of utilizing an adder with accompanying propagation delays.

The second method uses dataflow modeling as the implementation technique for designing combinational logic to generate the desired rounded fraction. The *continuous assignment* statement models dataflow behavior and is used to design combinational logic without using gates and interconnecting nets. Continuous assignment statements provide a Boolean correspondence between the right-hand side expression and the left-hand side target. The continuous assignment statement uses the keyword **assign** and has the following syntax with optional drive strength and delay:

assign [drive_strength] [delay] left-hand side target = right-hand side expression

The continuous assignment statement assigns a value to a net (**wire**) that has been previously declared — it cannot be used to assign a value to a register. Therefore, the left-hand target must be a scalar or vector net or a concatenation of scalar and vector nets. The operands on the right-hand side can be registers, nets, or function calls. The registers and nets can be declared as either scalars or vectors.

The third method uses behavioral modeling without the advantage of high-speed memory. This design adds two floating-point operands and adjusts the sum based upon the adder-based rounding algorithm previously discussed.

16.3.1 Adder-Based Rounding Using Memory

The adder-based rounding method requires an additional operation to add 1 to the retained bits if there is a 1 in the high-order bit position of the bits being deleted. The additional time required to perform this operation can be eliminated by using high-speed memory to obtain the desired rounded fraction. The memory is programmed so that the addressed contents generate the required rounded fraction. The memory address consists of the fraction that is right-concatenated with the high-order bit of the bits being deleted.

The memory produces the correct rounded fraction for all addresses, except when the highest addressed memory location is all 1s. In this case, truncation rounding is used or the result is left unchanged. In either case, the overall error is small because the remaining memory outputs produce correctly rounded results. Figure 16.6 shows a general organization for this method, utilizing a 5-bit fraction with three guard bits, that will be used in the Verilog HDL implementation. This organization can be easily expanded to accommodate larger fractions with a corresponding larger memory.

The behavioral module is shown in Figure 16.7 with a 6-bit input fraction named *fract_a[5:0]*, and a 5-bit rounded output result named *rslt[4:0]*. The memory, called *mem_round*, is defined as an array of 64 five-bit registers by the following statement:

reg [4:0] mem_round [0:63];

The memory is loaded from a file called *fract.round* by the following statement:

$readmemb ("fract.round", mem_load);

The contents of memory are saved in the project file as *fract.round* without the **.v** extension. Whenever the fraction, *frcat_a*, changes value in the sensitivity list of the **always** statement, the result, *rslt*, is set equal to the contents of memory at the location specified by *fract_a*, as shown below.

```
always @ (fract_a)
begin
    rslt = mem_round [fract_a];
end
```

In order to keep the contents of *fract.round* to a minimum yet still illustrate the technique for memory-based rounding, the memory will consist of 64 words with each word containing five bits. The contents of *fract.round* are shown in Figure 16.8. The memory address, *mem addr*, is shown for reference; the memory contents consist only of the memory outputs, *mem out*, which represent the correctly rounded fractions. The test bench is shown in Figure 16.9 providing 16 input vectors for the fraction. The outputs and waveforms are shown in Figure 16.10 and Figure 16.11, respectively.

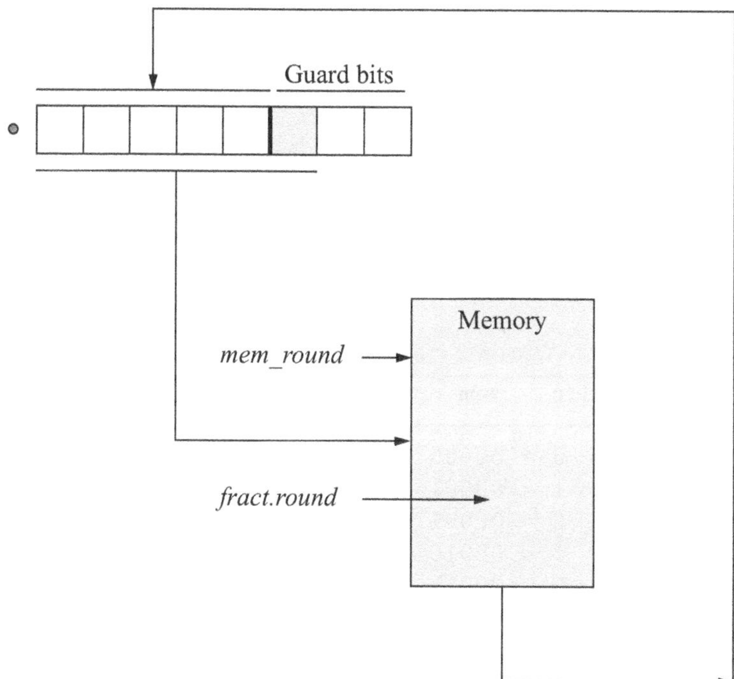

Figure 16.6 General organization for adder-based rounding using memory.

```
//behavioral floating-point adder-based rounding
module mem_flp_add_based_round (fract_a, rslt);

//list inputs and outputs
input [5:0] fract_a;
output [4:0] rslt;

//list wire and reg
wire [5:0] fract_a;
reg [4:0] rslt;

//define memory size
//mem_round is an array of 64 5-bit registers
reg [4:0] mem_round [0:63];

//define memory contents
//load mem_round from file fract.round
initial
begin
   $readmemb ("fract.round", mem_round);
end

//use the fraction fract_a to access the memory
always @ (fract_a)
begin
   rslt = mem_round [fract_a];
end

endmodule
```

Figure 16.7 Behavioral module for adder-based rounding using memory.

```
mem addr   mem out (rounded fraction)
------------------
00000  0 = 00000
00000  1 = 00001
00001  0 = 00001
00001  1 = 00010
00010  0 = 00010
00010  1 = 00011
00011  0 = 00011
00011  1 = 00100
------------------
//continued on next page
```

Figure 16.8 Contents of file *fract.round* for memory *mem_round*.

```
mem addr   mem out                    mem addr   mem out
------------------                    ------------------
00100  0 = 00100                      11000  0 = 11000
00100  1 = 00101                      11000  1 = 11001
00101  0 = 00101                      11001  0 = 11001
00101  1 = 00110                      11001  1 = 11010
00110  0 = 00110                      11010  0 = 11010
00110  1 = 00111                      11010  1 = 11011
00111  0 = 00111                      11011  0 = 11011
00111  1 = 01000                      11011  1 = 11100
------------------                    ------------------
01000  0 = 01000                      11100  0 = 11100
01000  1 = 01001                      11100  1 = 11101
01001  0 = 01001                      11101  0 = 11101
01001  1 = 01010                      11101  1 = 11110
01010  0 = 01010                      11110  0 = 11110
01010  1 = 01011                      11110  1 = 11111
01011  0 = 01011                      11111  0 = 11111
01011  1 = 01100                      11111  1 = 11111
------------------                    ------------------
01100  0 = 01100
01100  1 = 01101
01101  0 = 01101
01101  1 = 01110
01110  0 = 01110
01110  1 = 01111
01111  0 = 01111
01111  1 = 10000
------------------
10000  0 = 10000
10000  1 = 10001
10001  0 = 10001
10001  1 = 10010
10010  0 = 10010
10010  1 = 10011
10011  0 = 10011
10011  1 = 10100
------------------
10100  0 = 10100
10100  1 = 10101
10101  0 = 10101
10101  1 = 10110
10110  0 = 10110
10110  1 = 10111
10111  0 = 10111
10111  1 = 11000
------------------
```

Figure 16.8 (Continued)

```
//test bench for floating-point adder-based rounding
module mem_flp_add_based_round_tb;


//inputs are reg and outputs are wire for test benches
reg [5:0] fract_a;
wire [4:0] rslt;


//display variables
initial
$monitor ("fraction = %b, rounded result = %b",
            fract_a, rslt);


//apply input vectors for fraction
initial
begin
   #0    fract_a = 6'b00000_1;
   #10   fract_a = 6'b00000_0;
   #10   fract_a = 6'b00110_0;
   #10   fract_a = 6'b00110_1;
   #10   fract_a = 6'b01011_0;
   #10   fract_a = 6'b01011_1;
   #10   fract_a = 6'b01111_0;
   #10   fract_a = 6'b01111_1;
   #10   fract_a = 6'b10011_0;
   #10   fract_a = 6'b10011_1;
   #10   fract_a = 6'b10111_0;
   #10   fract_a = 6'b10111_1;
   #10   fract_a = 6'b11011_0;
   #10   fract_a = 6'b11011_1;
   #10   fract_a = 6'b11110_0;
   #10   fract_a = 6'b11110_1;

   #10   $stop;

end


//instantiate the module into the test bench
mem_flp_add_based_round inst1 (
   .fract_a(fract_a),
   .rslt(rslt)
   );

endmodule
```

Figure 16.9 Test bench for adder-based rounding using memory.

```
fraction = 00000_1, rounded result = 00001
fraction = 00000_0, rounded result = 00000
fraction = 00110_0, rounded result = 00110
fraction = 00110_1, rounded result = 00111
fraction = 01011_0, rounded result = 01011
fraction = 01011_1, rounded result = 01100
fraction = 01111_0, rounded result = 01111
fraction = 01111_1, rounded result = 10000
fraction = 10011_0, rounded result = 10011
fraction = 10011_1, rounded result = 10100
fraction = 10111_0, rounded result = 10111
fraction = 10111_1, rounded result = 11000
fraction = 11011_0, rounded result = 11011
fraction = 11011_1, rounded result = 11100
fraction = 11110_0, rounded result = 11110
fraction = 11110_1, rounded result = 11111
```

Figure 16.10 Outputs for adder-based rounding using memory.

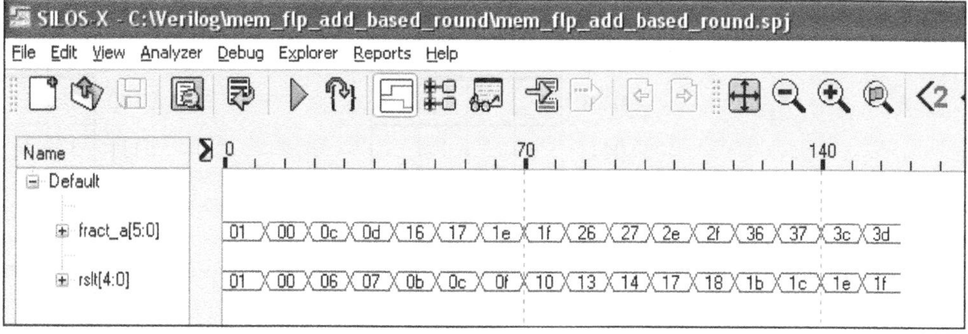

Figure 16.11 Waveforms for adder-based rounding using memory.

16.3.2 Adder-Based Rounding Using Combinational Logic

The operation of an adder-based rounding scheme can be implemented directly with combinational logic using dataflow modeling. Dataflow modeling uses the continuous assignment statement. The **assign** statement continuously monitors the right-hand side expression. If a variable changes value, then the expression is evaluated and the result is assigned to the target after any specified delay. If no delay is specified, then the default delay is zero. The continuous assignment statement can be considered to be a form of behavioral modeling, because the circuit is designed by specifying its behavior, not by instantiating logic circuits, as is the case with structural modeling.

This design uses operands of the same size as in the previous section. The fraction is designated by the variable *X[1:6]* and the rounded result is designated as *rslt[4:0]*. The fraction is of the form $X = x_1 x_2 x_3 x_4 x_5\_x_6$, where $x_1 x_2 x_3 x_4 x_5$ represents the significand, and x_6 is the high-order bit of the guard bits, which determines whether a 1 is added to the significand.

Since there are six variables in the input vector, the design uses five 6-variable Karnaugh maps — one map for each bit of the result. There are 64 minterms in a 6-variable Karnaugh map — 0 through 63 — which accommodate all combinations of the input fraction and guard bit. If $x_6 = 0$, then the fraction is unchanged; if $x_6 = 1$, then the fraction is incremented by 1. This is evident in each minterm location in the Karnaugh maps. The Karnaugh maps are shown in Figure 6.12.

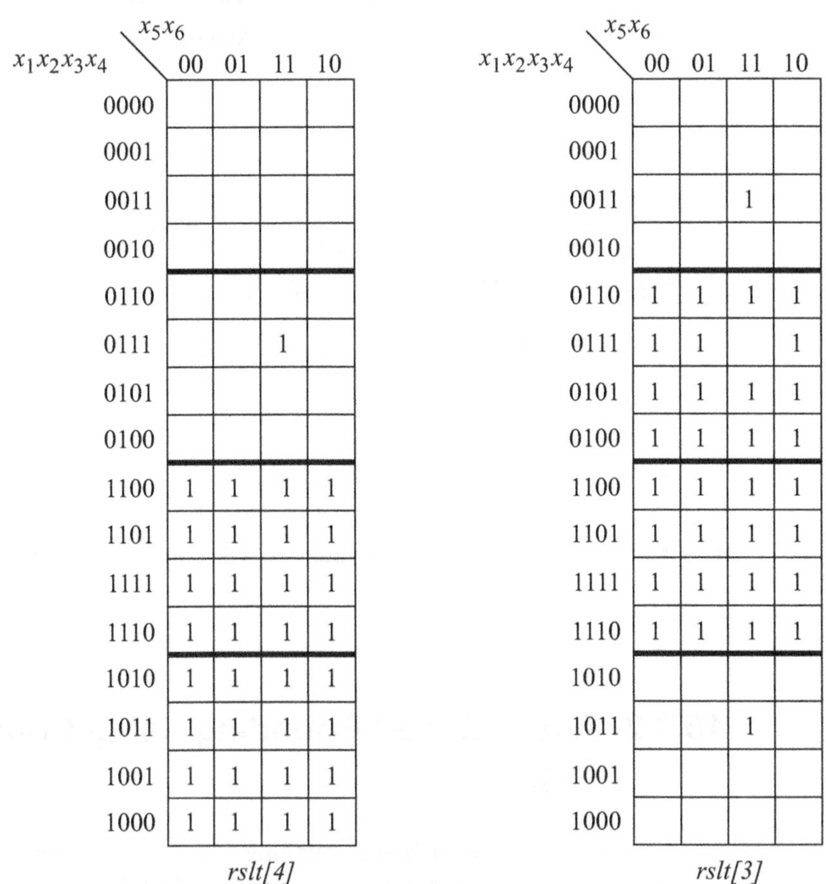

rslt[4]

$x_1x_2x_3x_4$ \ x_5x_6	00	01	11	10
0000				
0001				
0011				
0010				
0110				
0111			1	
0101				
0100				
1100	1	1	1	1
1101	1	1	1	1
1111	1	1	1	1
1110	1	1	1	1
1010	1	1	1	1
1011	1	1	1	1
1001	1	1	1	1
1000	1	1	1	1

rslt[3]

$x_1x_2x_3x_4$ \ x_5x_6	00	01	11	10
0000				
0001				
0011			1	
0010				
0110	1	1	1	1
0111	1	1		1
0101	1	1	1	1
0100	1	1	1	1
1100	1	1	1	1
1101	1	1	1	1
1111	1	1	1	1
1110	1	1	1	1
1010				
1011			1	
1001				
1000				

Continued on next page

Figure 16.12 Karnaugh maps for adder-based rounding using combinational logic.

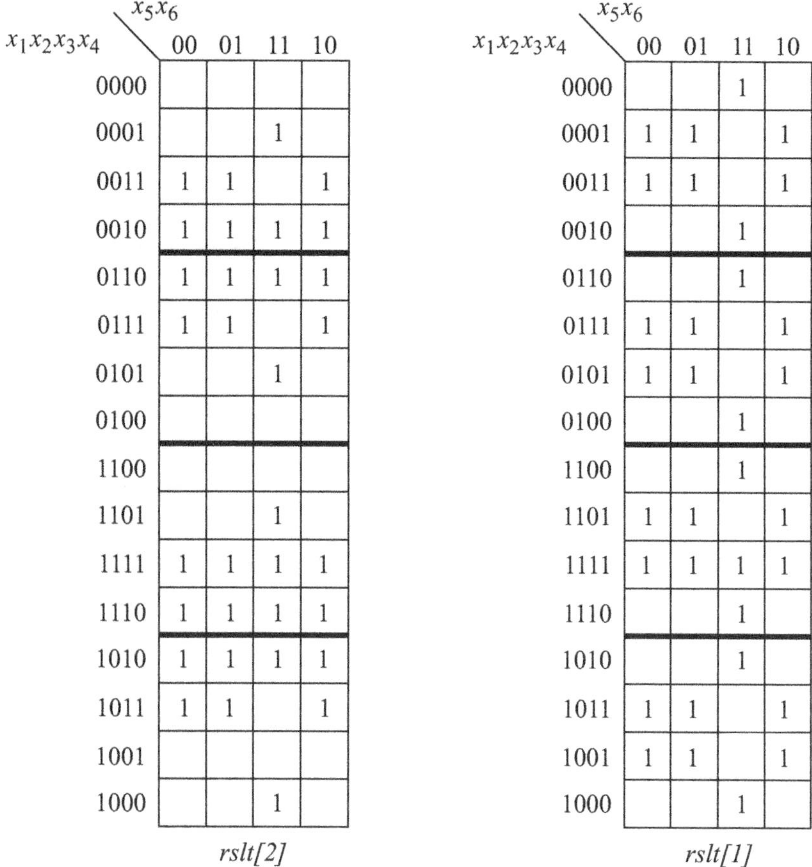

Continued on next page

Figure 16.12 (Continued)

The equations for *RSLT[4:0]* are shown in Equation 16.1 and are obtained from the Karnaugh maps of Figure 16.12.

$$
\begin{aligned}
rslt[4] &= x_1 + x_2 x_3 x_4 x_5 x_6 \\
rslt[3] &= x_1 x_2 + x_2 x_3' + x_2 x_5' + x_2 x_6' + x_2 x_4' + x_2' x_3 x_4 x_5 x_6 \\
rslt[2] &= x_3 x_4' + x_3 x_6' + x_3 x_5' + x_1 x_2 x_3 + x_3' x_4 x_5 x_6 \\
rslt[1] &= x_4 x_5' + x_4 x_6' + x_4' x_5 x_6 + x_1 x_2 x_3 x_4 \\
rslt[0] &= x_5' x_6 + x_5 x_6' + x_1 x_2 x_3 x_4 x_5
\end{aligned}
\qquad (16.1)
$$

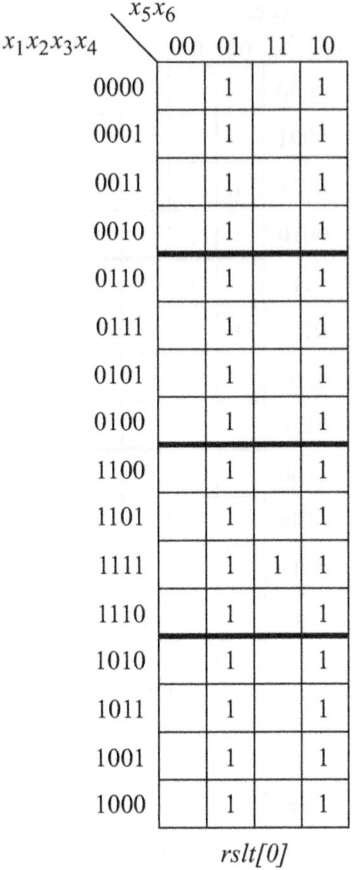

x_5x_6
$x_1x_2x_3x_4$
00 01 11 10
0000
0001
0011
0010
0110
0111
0101
0100
1100
1101
1111
1110
1010
1011
1001
1000
1
rslt[0]

Figure 16.12 (Continued)

The logic diagram is shown in Figure 16.13 and is designed from the equations of Equation 16.1.

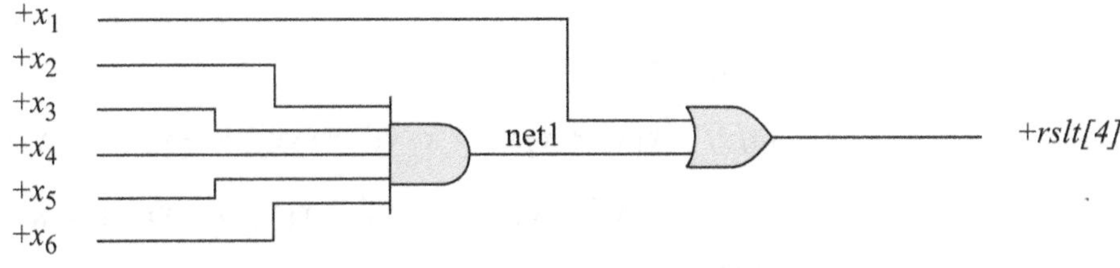

Continued on next page

Figure 16.13 Logic diagram for adder-based rounding using combinational logic.

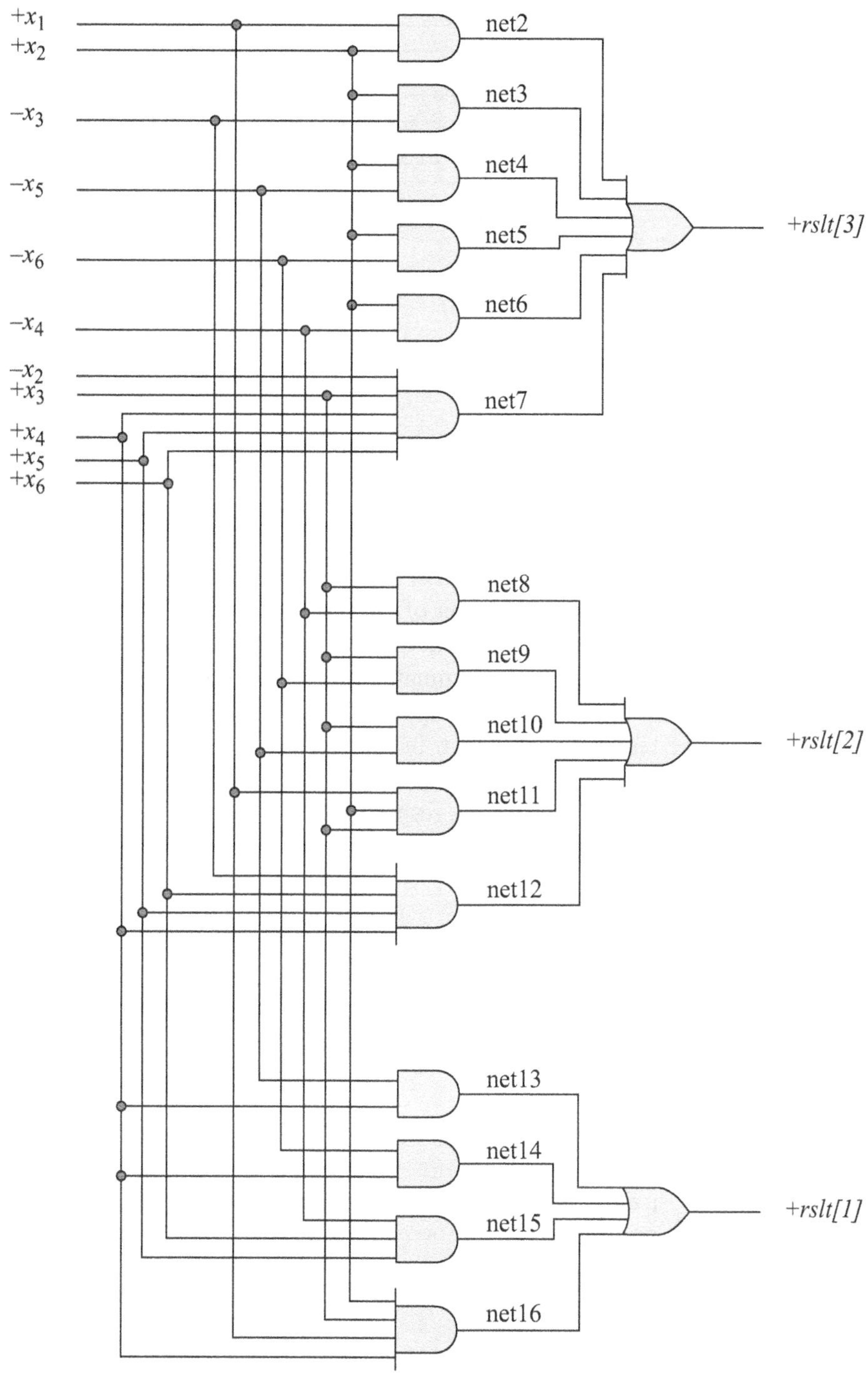

Continued on next page

Figure 16.13 (Continued)

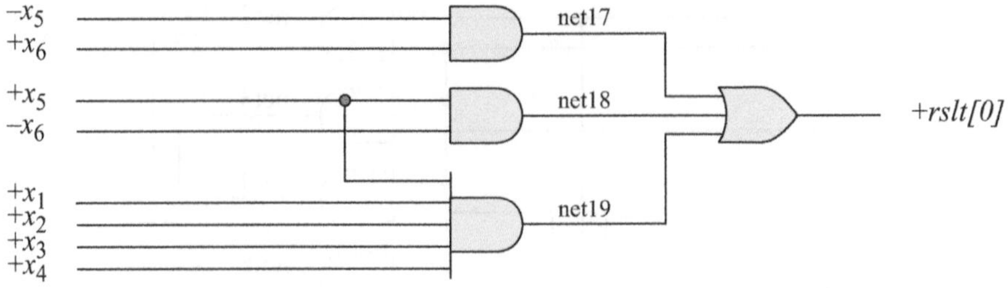

Figure 16.13 (Continued)

The dataflow module is shown in Figure 16.14 and shows the input vector, *x[1:6]*, that represents the fraction, and the output vector, *rslt[4:0]*, that represents the rounded fraction. The internal nets are defined in accordance with the net names shown in Figure 16.13. Each of the five outputs, *rslt[4:0]*, are designed using a single **assign** statement for each bit of the result vector. Using the **assign** statement in this manner allows each statement in the block to be terminated by a comma, except the last statement, which is terminated by a semicolon. The equations for the result bits conform to those shown in Equation 16.1.

The test bench is shown in Figure 16.15 and applies the same input vectors as shown in Figure 16.9 for comparison. The outputs and waveforms are shown in Figure 16.16 and Figure 16.17, respectively.

```
//dataflow for floating-point adder-based rounding
module adder_based_round_df (x, rslt);

input [1:6] x;
output [4:0] rslt;

//define internal wires
wire   net1, net2, net3, net4, net5, net6, net7,
       net8, net9, net10, net11, net12, net13,
       net14, net15, net16, net17, net18, net19;

//define logic for rslt[4]
assign    net1 = x[2] & x[3] & x[4] & x[5] & x[6],
          rslt[4] = x[1] | net1;
//continued on next page
```

Figure 16.14 Dataflow module for adder-based rounding using combinational logic.

```
//define logic for rslt[3]
assign    net2 = x[1] & x[2],
          net3 = x[2] & ~x[3],
          net4 = x[2] & ~x[5],
          net5 = x[2] & ~x[6],
          net6 = x[2] & ~x[4],
          net7 = ~x[2] & x[3] & x[4] & x[5] & x[6],
          rslt[3] = net2 | net3 | net4 | net5 | net6 | net7;

//define logic for rslt[2]
assign    net8 = x[3] & ~x[4],
          net9 = x[3] & ~x[6],
          net10 = x[3] & ~x[5],
          net11 = x[1] & x[2] & x[3],
          net12 = ~x[3] & x[4] & x[5] & x[6],
          rslt[2] = net8 | net9 | net10 | net11 | net12;

//define logic for rslt[1]
assign    net13 = x[4] & ~x[5],
          net14 = x[4] & ~x[6],
          net15 = ~x[4] & x[5] & x[6],
          net16 = x[1] & x[2] & x[3] & x[4],
          rslt[1] = net13 | net14 | net15 | net16;

//define logic for rslt[0]
assign    net17 = ~x[5] & x[6],
          net18 = x[5] & ~x[6],
          net19 = x[1] & x[2] & x[3] & x[4] & x[5],
          rslt[0] = net17 | net18 | net19;

endmodule
```

Figure 16.14 (Continued)

```
//test bench for floating-point adder-based rounding
module adder_based_round_df_tb;

reg [1:6] x;
wire [4:0] rslt;

//display variables
initial
$monitor ("fraction = %b, rounded result = %b", x, rslt);
//continued on next page
```

Figure 16.15 Test bench for adder-based rounding using combinational logic.

```
initial  //apply input vectors for fraction
begin
   #0     x = 6'b00000_1;
   #10    x = 6'b00000_0;
   #10    x = 6'b00110_0;
   #10    x = 6'b00110_1;
   #10    x = 6'b01011_0;
   #10    x = 6'b01011_1;
   #10    x = 6'b01111_0;
   #10    x = 6'b01111_1;
   #10    x = 6'b10011_0;
   #10    x = 6'b10011_1;
   #10    x = 6'b10111_0;
   #10    x = 6'b10111_1;
   #10    x = 6'b11011_0;
   #10    x = 6'b11011_1;
   #10    x = 6'b11110_0;
   #10    x = 6'b11110_1;
   #10    $stop;
end

adder_based_round_df inst1 (  //instantiate the module
   .x(x),
   .rslt(rslt)
   );
endmodule
```

Figure 16.15 (Continued)

```
fraction = 00000_1, rounded result = 00001
fraction = 00000_0, rounded result = 00000
fraction = 00110_0, rounded result = 00110
fraction = 00110_1, rounded result = 00111
fraction = 01011_0, rounded result = 01011
fraction = 01011_1, rounded result = 01100
fraction = 01111_0, rounded result = 01111
fraction = 01111_1, rounded result = 10000
fraction = 10011_0, rounded result = 10011
fraction = 10011_1, rounded result = 10100
fraction = 10111_0, rounded result = 10111
fraction = 10111_1, rounded result = 11000
fraction = 11011_0, rounded result = 11011
fraction = 11011_1, rounded result = 11100
fraction = 11110_0, rounded result = 11110
fraction = 11110_1, rounded result = 11111
```

Figure 16.16 Outputs for adder-based rounding using combinational logic.

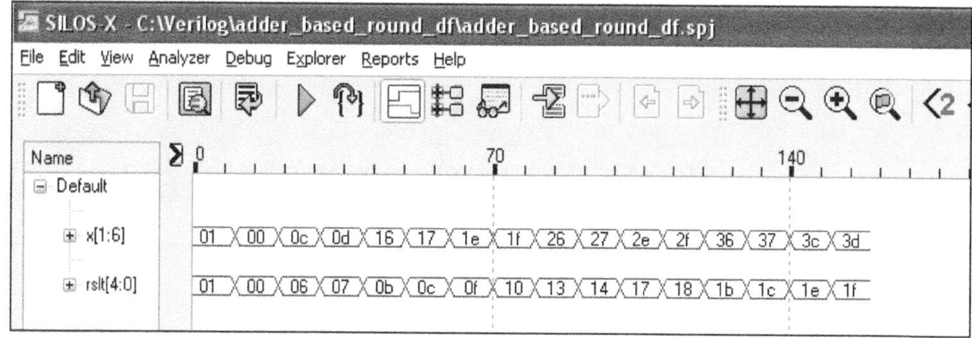

Figure 16.17 Waveforms for adder-based rounding using combinational logic.

16.3.3 Adder-Based Rounding Using Behavioral Modeling

Adder-based rounding can also be accomplished using behavioral modeling without utilizing memory. This modeling technique describes the *behavior* of a digital system and is not concerned with the direct implementation of logic gates, but rather the architecture of the system. This is an algorithmic approach to hardware implementation and represents a higher level of abstraction than other modeling methods.

Recall that Verilog HDL contains two structured procedure statements or behaviors: **initial** and **always**, which are briefly reviewed here. A behavior may consist of a single statement or a block of statements delimited by the keywords **begin** . . . **end**. A module may contain multiple **initial** and **always** statements. These statements are the basic statements used in behavioral modeling and execute concurrently starting at time zero in which the order of execution is not important. All other behavioral statements are contained inside these structured procedure statements.

Consider the floating-point number format shown in Figure 16.18 with a sign bit, a 4-bit exponent, an 8-bit fraction, and three guard bits. This is the format that will be used to model adder-based rounding when adding two floating-point numbers. The behavioral module is shown in Figure 16.19 containing two floating-point numbers: *flp_a[15:0]* and *flp_b[15:0]*.

The fractions are labelled *fract_a[10:0]*, where *fract_a[2]* is the guard bit, and *fract_b[10:0]*, where *fract_b[2]* is the guard bit. The bias for this 16-bit format is 0111. One of the two fractions may have to be aligned before the add operation; therefore, if the high-order guard bit in either floating-point number is a 1 after alignment, then a 1 is added to the resulting sum. This is accomplished by the following two statements:

```
if ((guard_a[2] | guard_b[2]) == 1)
    sum[10:3] = sum[10:3] + 8'b0000_0001;
```

The test bench is shown in Figure 16.20 and applies eight input vectors to the augend and addend, including integers and integers with fractions. The outputs and waveforms are shown in Figure 16.21 and Figure 16.22, respectively.

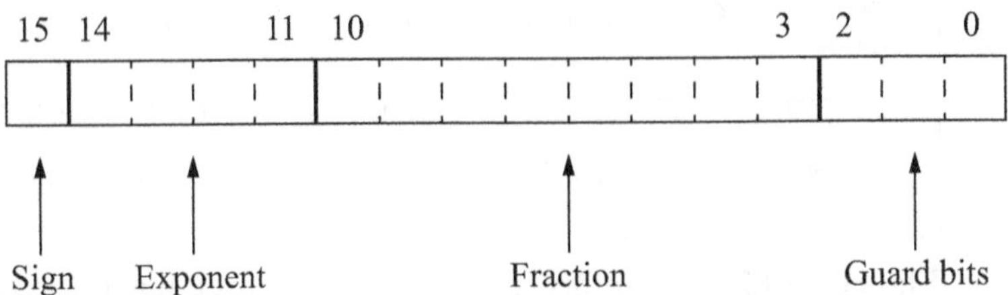

Figure 16.18 A 16-bit floating-point number used in a behavioral module for adder-based rounding.

```
//behavioral adder-based rounding
module adder_based_round_bh2 (flp_a, flp_b, sign, exponent,
                              exp_unbiased, sum);

input [15:0] flp_a, flp_b;
output [10:3] sum;
output sign;
output [3:0] exponent, exp_unbiased;

//variables used in an always block are declared as registers
reg sign_a, sign_b;
reg [3:0] exp_a, exp_b;
reg [3:0] exp_a_bias, exp_b_bias;
reg [10:3] fract_a, fract_b;
reg [2:0] guard_a, guard_b;
reg [3:0] ctr_align;
reg [10:3] sum;
reg sign;
reg [3:0] exponent, exp_unbiased;
reg cout;

always @ (flp_a or flp_b)
begin
   sign_a = flp_a[15];
   sign_b = flp_b[15];

   exp_a = flp_a[14:11];
   exp_b = flp_b[14:11];
//continued on next page
```

Figure 16.19 Behavioral module for adder-based rounding.

```
   fract_a = flp_a[10:3];
   fract_b = flp_b[10:3];

   guard_a = flp_a[2:0];
   guard_b = flp_b[2:0];

//bias exponents
   exp_a_bias = exp_a + 4'b0111;
   exp_b_bias = exp_b + 4'b0111;

//align fractions
   if (exp_a_bias < exp_b_bias)
      ctr_align = exp_b_bias - exp_a_bias;

      while (ctr_align)
         begin
            {fract_a, guard_a} = {fract_a, guard_a} >> 1;
            exp_a_bias = exp_a_bias + 1;
            ctr_align = ctr_align - 1;
         end

   if (exp_b_bias < exp_a_bias)
      ctr_align = exp_a_bias - exp_b_bias;

      while (ctr_align)
         begin
            {fract_b, guard_b} = {fract_b, guard_b} >> 1;
            exp_b_bias = exp_b_bias + 1;
            ctr_align = ctr_align - 1;
         end

//add fractions
   {cout, sum[10:3]} = fract_a[10:3] + fract_b[10:3];

//execute adder-based rounding
   if ((guard_a[2] | guard_b[2]) == 1)
      sum[10:3] = sum[10:3] + 8'b0000_0001;

//normalize result
   if (cout == 1)
      begin
         {cout, sum[10:3]} = {cout, sum[10:3]} >> 1;
         exp_b_bias = exp_b_bias + 1;
      end

//continued on next page
```

Figure 16.19 (Continued)

```
   sign = sign_a;
   exponent = exp_b_bias;
   exp_unbiased = exp_b_bias - 4'b0111;

end
endmodule
```

Figure 16.19 (Continued)

```
//test bench for behavioral adder-based rounding
module adder_based_round_bh2_tb;

reg [15:0] flp_a, flp_b;
wire [10:3] sum;
wire sign;
wire [3:0] exponent, exp_unbiased;

//display variables
initial
$monitor ("sign=%b, exp_biased=%b, exp_unbiased=%b, sum=%b",
            sign, exponent, exp_unbiased, sum[10:3]);

//apply input vectors
initial
begin
         //+12 + +35 = +47
         //          s --e- ------f------
   #0    flp_a = 16'b0_0100_1100_0000_000;
         flp_b = 16'b0_0110_1000_1100_000;


         //+26 + +20 = +46
         //          s --e- ------f------
   #10   flp_a = 16'b0_0101_1101_0000_000;
         flp_b = 16'b0_0101_1010_0000_000;


         //+26.5 + +4.375 = +30.875
         //          s --e- ------f------
   #10   flp_a = 16'b0_0101_1101_0100_000;
         flp_b = 16'b0_0011_1000_1100_000;

//continued on next page
```

Figure 16.20 Test bench for behavioral adder-based rounding.

```
        //+11 + +34 = +45
        //          s --e- ------f------
   #10  flp_a = 16'b0_0100_1011_0000_000;
        flp_b = 16'b0_0110_1000_1000_000;


        //+58.2500 + +9.4375 = +67.6875
        //          s --e- ------f------
   #10  flp_a = 16'b0_0110_1110_1001_000;
        flp_b = 16'b0_0100_1001_0111_000;


        //+50.2500 + +9.4375 = +59.6875
        //          s --e- ------f------
   #10  flp_a = 16'b0_0110_1100_1001_000;
        flp_b = 16'b0_0100_1001_0111_000;


        //+51.2500 + +11.0625 = +62.3125
        //          s --e- ------f------
   #10  flp_a = 16'b0_0110_1100_1101_000;
        flp_b = 16'b0_0100_1011_0001_000;


        //+12.75000 + +4.34375 = +17.09375
        //          s --e- ------f------
   #10  flp_a = 16'b0_0100_1100_1100_000;
        flp_b = 16'b0_0011_1000_1011_000;

   #10  $stop;

end


//instantiate the module into the test bench
adder_based_round_bh2 inst1 (
   .flp_a(flp_a),
   .flp_b(flp_b),
   .sum(sum),
   .sign(sign),
   .exponent(exponent),
   .exp_unbiased(exp_unbiased)
   );

endmodule
```

Figure 16.20 (Continued)

```
(+12) + (+35) = +47
sign=0, exp_biased=1101, exp_unbiased=0110, sum=1011_1100

(+26) + (+20) = +46
sign=0, exp_biased=1101, exp_unbiased=0110, sum=1011_1000

(+26.5) + (4.375) = +30.875
sign=0, exp_biased=1100, exp_unbiased=0101, sum=1111_0111

(+11) + (+34) = +45
sign=0, exp_biased=1101, exp_unbiased=0110, sum=1011_0100

(+58.2500) + (+9.4375) = +67.6875
sign=0, exp_biased=1110, exp_unbiased=0111, sum=1000_0111

(+50.2500) + (+9.4375) = +59.6875
sign=0, exp_biased=1101, exp_unbiased=0110, sum=1110_1111

(+51.2500) + (+11.0625) = +62.3125
sign=0, exp_biased=1101, exp_unbiased=0110, sum=1111_1001

(+12.75000) + (+4.34375) = +17.09375
sign=0, exp_biased=1100, exp_unbiased=0101, sum=1000_1001
```

Figure 16.21 Outputs for behavioral adder-based rounding.

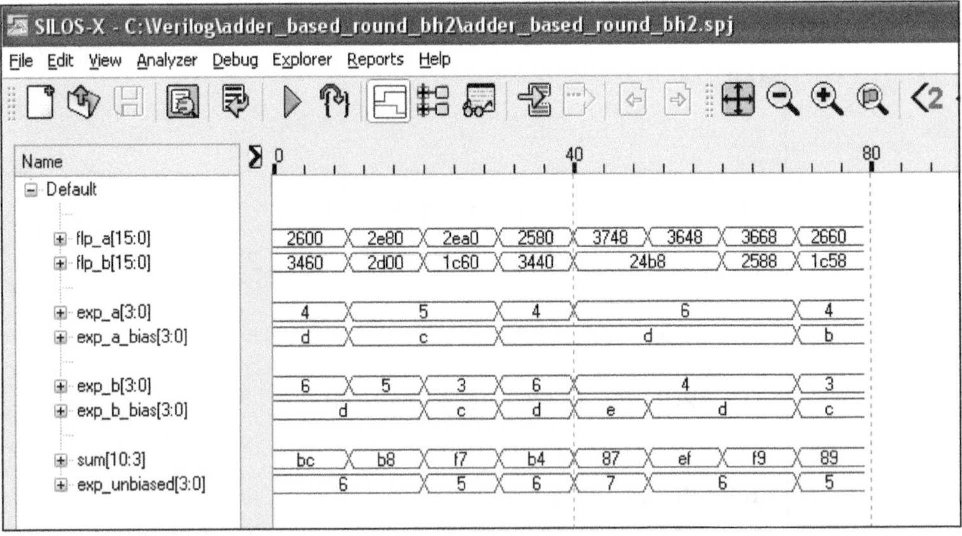

Figure 16.22 Waveforms for behavioral adder-based rounding.

16.3.4 Combined Truncation, Adder-Based and von Neumann Rounding

This final Verilog HDL design incorporates all three rounding methods in one behavioral module so that the precision of each method can be compared relative to the other two rounding techniques. The operation is addition and the format for the 16-bit floating-point number is shown in Figure 16.23. After the fields of the format are defined, the exponents are biased by adding the bias constant of 0111. The fractions are then aligned and added, after which postnormalization is performed, if necessary.

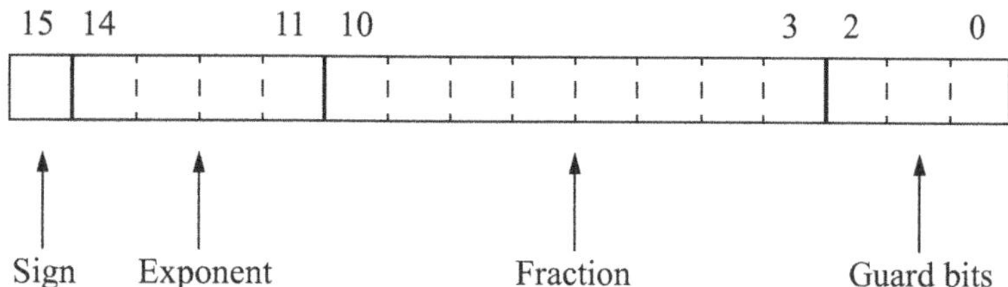

Figure 16.23 Format used for a 16-bit floating-point add operation in which all three rounding methods are employed.

The inputs are two floating-point numbers labeled *flp_a[15:0]* and *flp_b[15:0]*. The sum outputs for truncation, adder-based, and von Neumann rounding are defined as *sum_trunc[10:3]*, *sum_add[10:3]*, and *sum_von[10:3]*, respectively.

Truncation rounding is performed by the following statement, which simply truncates the resulting sum without considering the guard digits:

```
sum_trunc = sum[10:3];
```

Adder-based rounding is performed by the statements shown below, based on the state of the high-order guard bits, *guard_a[2]* and *guard_b[2]*. If either guard bit is equal to 1, then 1 is added to *sum[10:3]*, which is then assigned to the adder-based sum, *sum_add*; otherwise, *sum[10:3]* is assigned to the adder-based sum unchanged.

```
if ((guard_a[2] == 1) | (guard_b[2] == 1))
    sum_add = sum[10:3] + 8'b0000_0001;
else
    sum_add = sum[10:3];
```

Von Neumann rounding is performed by the statements shown below, based on the value of *guard_a[2:0]* and *guard_b[2:0]*. If either guard field is not equal to zero, then the low-order bit of *sum[10:3]* is set to a value of 1. Then *sum[10:3]* is assigned to the von Neumann sum, *sum_von*; otherwise, *sum[10:3]* is assigned to the von Neumann sum unchanged.

```
if ((guard_a != 3'b000) | (guard_b != 3'b000))
    sum_von = {sum[10:4], 1'b1};
else
    sum_von = sum[10:3];
```

The behavioral module is shown in Figure 16.24. The test bench module, shown in Figure 16.25, applies several input vectors to the behavioral module. The input vectors consist of both integers and integers with fractions. The outputs are shown in Figure 16.26, which clearly illustrates the precision achieved by each rounding technique for the same input vectors. The waveforms are shown in Figure 16.27.

```
//behavioral truncation, adder-based, von neumann rounding
module trunc_add_von_round (flp_a, flp_b, sign, exponent,
            exp_unbiased, sum, sum_trunc, sum_add, sum_von);

input [15:0] flp_a, flp_b;
output [10:3] sum, sum_trunc, sum_add, sum_von;
output sign;
output [3:0] exponent, exp_unbiased;

//variables used in an always block are declared as registers
reg sign_a, sign_b;
reg [3:0] exp_a, exp_b;
reg [3:0] exp_a_bias, exp_b_bias;
reg [10:3] fract_a, fract_b;
reg [2:0] guard_a, guard_b;
reg [3:0] ctr_align;
reg [10:3] sum, sum_trunc, sum_add, sum_von;
reg sign;
reg [3:0] exponent, exp_unbiased;
reg cout;

always @ (flp_a or flp_b)
begin
   sign_a = flp_a[15];
   sign_b = flp_b[15];

   exp_a = flp_a[14:11];
   exp_b = flp_b[14:11];
//continued on next page
```

Figure 16.24 Behavioral module illustrating the three rounding techniques of truncation, adder-based and von Neumann rounding.

```
   fract_a = flp_a[10:3];
   fract_b = flp_b[10:3];

   guard_a = flp_a[2:0];
   guard_b = flp_b[2:0];

//bias exponents
   exp_a_bias = exp_a + 4'b0111;
   exp_b_bias = exp_b + 4'b0111;

//align fractions
   if (exp_a_bias < exp_b_bias)
      ctr_align = exp_b_bias - exp_a_bias;

      while (ctr_align)
         begin
            {fract_a, guard_a} = {fract_a, guard_a} >> 1;
            exp_a_bias = exp_a_bias + 1;
            ctr_align = ctr_align - 1;
         end

   if (exp_b_bias < exp_a_bias)
      ctr_align = exp_a_bias - exp_b_bias;

      while (ctr_align)
         begin
            {fract_b, guard_b} = {fract_b, guard_b} >> 1;
            exp_b_bias = exp_b_bias + 1;
            ctr_align = ctr_align - 1;
         end


//add fractions
   {cout, sum[10:3]} = fract_a[10:3] + fract_b[10:3];


//postnormalize result
   if (cout == 1)
      begin
         {cout, sum[10:3]} = {cout, sum[10:3]} >> 1;
         exp_b_bias = exp_b_bias + 1;
      end


//execute truncation rounding
   sum_trunc = sum[10:3];

//continued on next page
```

Figure 16.24 (Continued)

```
//execute adder-based rounding
   if ((guard_a[2] == 1) | (guard_b[2] == 1))
      sum_add = sum[10:3] + 8'b0000_0001;
   else
      sum_add = sum[10:3];

//execute von neumann rounding
   if ((guard_a != 3'b000) | (guard_b != 3'b000))
      sum_von = {sum[10:4], 1'b1};
   else
      sum_von = sum[10:3];

   sign = sign_a;
   exponent = exp_b_bias;
   exp_unbiased = exp_b_bias - 4'b0111;
end
endmodule
```

Figure 16.24 (Continued)

```
//test bench for truncation, adder-based, von neumann rounding
module trunc_add_von_round_tb;

reg [15:0] flp_a, flp_b;
wire [10:3] sum, sum_trunc, sum_add, sum_von;
wire sign;
wire [3:0] exponent, exp_unbiased;

initial      //display variables
$monitor ("sign=%b, exp_unbiased=%b, sum_trunc=%b,
            sum_add=%b, sum_von=%b",
            sign, exp_unbiased, sum_trunc, sum_add, sum_von);

initial      //apply input vectors
begin
         //(+63.500) + (+7.875) = +71.375
         //          s --e- ------f------
   #0    flp_a = 16'b0_0110_1111_1110_000;
         flp_b = 16'b0_0011_1111_1100_000;

         //(+35) + (+12) = +47
         //          s --e- ------f------
   #10   flp_a = 16'b0_0110_1000_1100_000;
         flp_b = 16'b0_0100_1100_0000_000;
//continued on next page
```

Figure 16.25 Test bench for the three rounding techniques of truncation, adder-based and von Neumann rounding.

```
        //(+26.5) + (+4.375) = +30.875
        //           s --e- ------f------
    #10 flp_a = 16'b0_0101_1101_0100_000;
        flp_b = 16'b0_0011_1000_1100_000;

        //(+58.2500) + (+9.4375) = +67.6875
        //           s --e- ------f------
    #10 flp_a = 16'b0_0110_1110_1001_000;
        flp_b = 16'b0_0100_1001_0111_000;

        //(+50.2500) + (+9.4375) = +59.6875
        //           s --e- ------f------
    #10 flp_a = 16'b0_0110_1100_1001_000;
        flp_b = 16'b0_0100_1001_0111_000;

        //(+12.75000) + (+4.34375) = +17.09375
        //           s --e- ------f------
    #10 flp_a = 16'b0_0100_1100_1100_000;
        flp_b = 16'b0_0011_1000_1011_000;

        //(+51.2500) + (+11.0625) = +62.3125
        //           s --e- ------f------
    #10 flp_a = 16'b0_0110_1100_1101_000;
        flp_b = 16'b0_0100_1011_0001_000;

        //(+42.250) + (+9.875) = +52.125
        //           s --e- ------f------
    #10 flp_a = 16'b0_0110_1010_1001_000;
        flp_b = 16'b0_0100_1001_0111_000;

    #10 $stop;

end

//instantiate the module into the test bench
trunc_add_von_round inst1 (
    .flp_a(flp_a),
    .flp_b(flp_b),
    .sum(sum),
    .sum_trunc(sum_trunc),
    .sum_add(sum_add),
    .sum_von(sum_von),
    .sign(sign),
    .exponent(exponent),
    .exp_unbiased(exp_unbiased)
    );

endmodule
```

Figure 16.25 (Continued)

```
//(+63.500) + (+7.875) = +71.375
sign=0, exp_unbiased=0111,
   sum_trunc=1000_1110, sum_add=1000_1111, sum_von=1000_1111


//(+35) + (+12) = +47
sign=0, exp_unbiased=0110,
   sum_trunc=1011_1100, sum_add=1011_1100, sum_von=1011_1100


//(+26.5) + (+4.375) = +30.875
sign=0, exp_unbiased=0101,
   sum_trunc=1111_0111, sum_add=1111_0111, sum_von=1111_0111


//(+58.2500) + (+9.4375) = +67.6875
sign=0, exp_unbiased=0111,
   sum_trunc=1000_0111, sum_add=1000_1000, sum_von=1000_0111


//(+50.2500) + (+9.4375) = +59.6875
sign=0, exp_unbiased=0110,
   sum_trunc=1110_1110, sum_add=1110_1111, sum_von=1110_1111


//(+12.75000) + (+4.34375) = +17.09375
sign=0, exp_unbiased=0101,
   sum_trunc=1000_1000, sum_add=1000_1001, sum_von=1000_1001


//(+51.2500) + (+11.0625) = +62.3125
sign=0, exp_unbiased=0110,
   sum_trunc=1111_1001, sum_add=1111_1001, sum_von=1111_1001


//(+42.250) + (+9.875) = +52.125
sign=0, exp_unbiased=0110,
   sum_trunc=1100_1110, sum_add=1100_1111, sum_von=1100_1111
```

Figure 16.26 Outputs for the three rounding techniques of truncation, adder-based and von Neumann rounding.

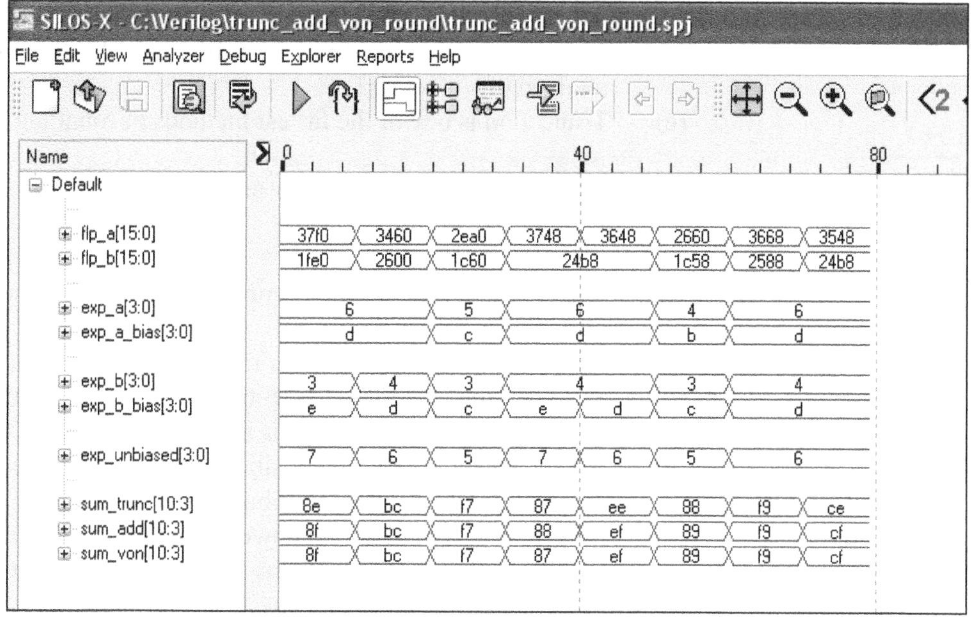

Figure 16.27 Waveforms for the three rounding techniques of truncation, adder-based and von Neumann rounding.

16.4 Problems

16.1 What is the purpose of rounding?
How is precision affected?
What situations can cause rounding to occur?

16.2 Name three methods of rounding and discuss their different attributes regarding:

(a) How rounding is accomplished
(b) Symmetry in relation to the ideal rounding line
(c) Relative speeds of the rounding methods
(d) Additional hardware required to perform the rounding

16.3 Which of the statements shown below are true for truncation. Indicate all correct answers.

(a) Truncation is also referred to as jamming.
(b) Remove the extra bits and make no change to the retained bits.

(c) The truncation function lies entirely below the ideal rounding line, touching it only where there is no error.
(d) The bits to be deleted are truncated and the low-order bit of the retained bits is set to 1.
(e) Truncation is one of the fastest methods of rounding.

16.4 Which of the statements shown below are true for adder-based rounding. Indicate all correct answers.

(a) The rounding function is symmetric with respect to the ideal rounding line.
(b) Adder-base rounding requires no more time than truncation.
(c) The rounding function lies entirely below the ideal rounding line, touching it only where there is an error.
(d) Add 1 to the low-order bit position of the retained bits if there is a 1 in the high-order bit position of bits being removed.
(e) Adder-based rounding is slower than truncation or von Neumann rounding because it requires an extra addition operation.

16.5 Which of the statements shown below are true for von Neumann rounding. Indicate all correct answers.

(a) The von Neumann method is also referred to as chopping.
(b) The bits to be deleted are truncated and the low-order bit of the retained bits is set to 1 if the bits to be deleted are not all 0s.
(c) Von Neumann rounding is also referred to as jamming.
(d) The function is symmetrical with respect to the ideal rounding line.
(e) Von Neumann rounding requires no more time than truncation.

16.6 A memory address consists of the three low-order bits of the significand right-concatenated with the high-order bit of the three guard bits. The memory contents contain the rounded results of the adder-based rounding technique. Obtain the memory contents.

16.7 A fraction and associated guard bits are shown below. Determine the rounded results using truncation, adder-based rounding, and von Neumann rounding.

Fraction	Guard
. 0 0 1 0 1 0 0 1	1 0 1

16.8 Let the augend be $A = 0.1100\ 1101 \times 2^6$.
Let the addend be $B = 0.1011\ 0001 \times 2^4$.
Perform the addition operation and round the result using all three rounding methods.

16.9 Let the augend be $A = 0.1110\ 1001 \times 2^6$.
Let the addend be $B = 0.1001\ 0111 \times 2^4$.
Perform the addition operation and round the result using all three rounding methods.

16.10 Let the augend be $A = 0.1100\ 1001 \times 2^6$.
Let the addend be $B = 0.1001\ 0111 \times 2^4$.
Perform the addition operation and round the result using all three rounding methods.

16.11 Using the floating-point format shown below, design a Verilog HDL module to perform addition on two operands, then round the result using truncation. Let the fractions be *fract_a[10:0]* and *fract_b[10:0]*. Truncate the result to eight bits labeled *trunc_sum[7:0]*. Obtain the behavioral module, the test bench module containing integers and integers with fractions, and the outputs.

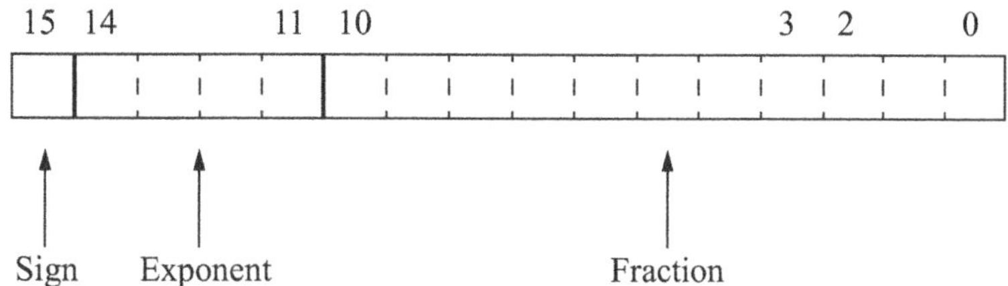

16.12 Using the floating-point format shown below, design a Verilog HDL module to perform addition on two operands, then round the result using adder-based rounding. Let the fractions be *fract_a[10:0]* and *fract_b[10:0]*. Round the result to eight bits labeled *sum[10:3]*. Obtain the behavioral module, the test bench module containing integers and integers with fractions, and the outputs.

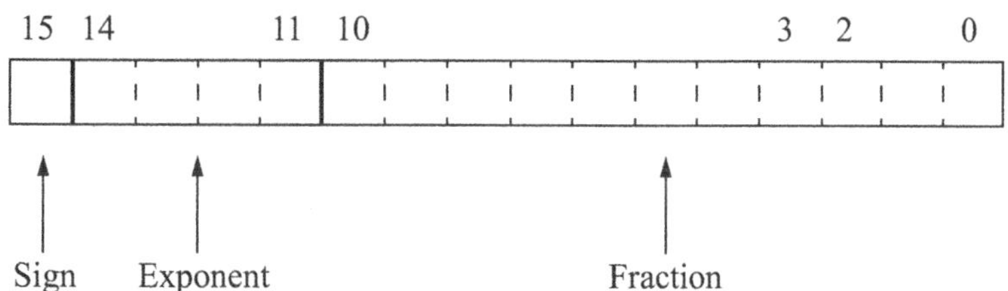

16.13 Using the floating-point format shown below, design a Verilog HDL module to perform addition on two operands, then round the result using von Neumann rounding. Let the fractions be *fract_a[10:0]* and *fract_b[10:0]*. Round the result to eight bits labeled *sum[10:3]*. Obtain the behavioral module, the test bench module containing integers and integers with fractions, and the outputs.

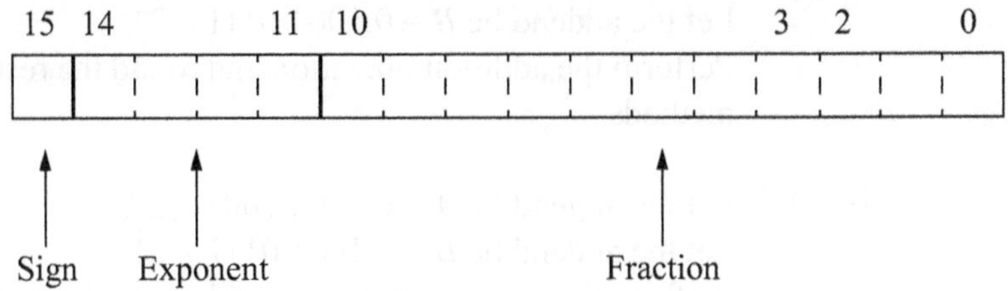
15 14 11 10 3 2 0
Sign
Exponent
Fraction

17

17.1 *Residue Checking*
17.2 *Parity-Checked Shift Register*
17.3 *Parity Prediction*
17.4 *Condition Codes for Addition*
17.5 *Logical and Algebraic Shifters*
17.6 *Arithmetic and Logic Units*
17.7 *Count-Down Counter*
17.8 *Shift Registers*
17.9 *Problems*

Additional Topics in Computer Arithmetic

This chapter presents several additional topics in computer arithmetic, including residue checking. Residue checking is a significant application that is especially suitable for addition, subtraction, and multiplication. This is a method of checking the accuracy of an arithmetic operation and is done in parallel with the operation. Residue checking will be implemented for an add operation. The adder and residue checker are implemented as independent logic functions. Checking is accomplished by using a residue for the operands based on a particular modulus.

Parity prediction is an additional way to check for errors in addition and generally requires less hardware than residue checking. Condition codes are covered which allow the condition codes of sum < 0, sum = 0, sum > 0, and overflow to be obtained concurrently with the add operation.

Shifting operations are frequently utilized in many arithmetic operations; for example, the sequential add-shift multiply operation and the sequential shift-subtract/add restoring division operation, among others. Therefore, various types of shifting devices will be presented and designed. Different categories of arithmetic and logic units (ALUs) will be presented for a wide variety of functions. Conversion between different number representations is also presented; for example, binary-to-binary-coded decimal (BCD) and BCD-to-binary.

Counters are also an important logic macro device used in several arithmetic applications to determine the duration of an operation and other specific functions. Different types of counters are presented for various moduli, together with unique counters that generate specific pulses for certain applications.

17.1 Residue Checking

This section presents a method of detecting errors in addition operations. When an integer A is divided by an integer m (modulus), a unique remainder R (residue) is generated, where $R < m$. If two integers A and B are divided by the same modulus, m, and produce the same remainder, R, then A and B are *congruent*, as shown below.

$$A \equiv B \quad \text{mod-}m$$

Congruence is indicated either by the symbol ≡ or by the symbol ⇒. The residue, R, of a number A modulo-m is specified by Equation 17.1. The integers A and B are of the form shown in Equation 17.2, where a_i and b_i are the digit values and r is the radix.

$$R(A) \equiv A \text{ mod-}m \tag{17.1}$$

$$\begin{aligned} A &= a_{n-1} r^{n-1} \; a_{n-2} r^{n-2} \ldots a_1 r^1 \; a_0 r^0 \\ B &= b_{n-1} r^{n-1} \; b_{n-2} r^{n-2} \ldots b_1 r^1 \; b_0 r^0 \end{aligned} \tag{17.2}$$

The congruence relationship can be obtained by dividing the integer A by an integer m (modulus) to produce Equation 17.3, where $0 \leq R < m$.

$$A = (quotient \times m) + R \tag{17.3}$$

Numbers that are congruent to the same modulus can be added and the result is a valid congruence, as shown in Equation 17.4.

$$\begin{aligned} R(A) &\equiv A \text{ mod-}m \\ R(B) &\equiv B \text{ mod-}m \\ R(A) + R(B) &\equiv (A \text{ mod-}m) + (B \text{ mod-}m) \\ &\equiv (A + B) \text{ mod-}m \\ &\equiv R(A + B) \end{aligned} \tag{17.4}$$

The operations of subtraction and multiplication yield similar results; that is, the difference of the residues of two numbers is congruent to the residue of their difference, and the product of the residues of two numbers is congruent to the residue of their products, as shown in Equation 17.5. A similar procedure is not available for division.

$$R(A) - R(B) \equiv R(A - B)$$
$$R(A) \times R(B) \equiv R(A \times B) \tag{17.5}$$

For the radix 2 number system, modulo-3 residue can detect single-bit errors using only two check bits. The modulo-3 results are shown in Table 17.1 for binary data 0000 through 1111.

Table 17.1 Modulo-3 Residue

Binary Data				Modulo-3 Residue	
8	4	2	1	2	1
0	0	0	0	0	0
0	0	0	1	0	1
0	0	1	0	1	0
0	0	1	1	0	0
0	1	0	0	0	1
0	1	0	1	1	0
0	1	1	0	0	0
0	1	1	1	0	1
1	0	0	0	1	0
1	0	0	1	0	0
1	0	1	0	0	1
1	0	1	1	1	0
1	1	0	0	0	0
1	1	0	1	0	1
1	1	1	0	1	0
1	1	1	1	0	0

Examples will now be presented to illustrate error detection in addition using Equation 17.4. The *carry-in* and *carry-out* will also be considered.

Example 17.1 Let the augend A and the addend B be decimal values of 42 and 87, respectively, to yield a sum of 129. The *carry-in* = 0; the *carry-out* = 0.

$A =$	0010 1010 →	mod-3 = 0	→ mod-3 = **0**	
$B =$	0101 0111 →	mod-3 = 0		
	0 ←			No error
Sum = 0 ←	1000 0001 →	→	→ mod-3 = **0**	

Example 17.2 Let the augend *A* and the addend *B* be decimal values of 42 and 87, respectively, with a *carry-in* = 1 and a *carry-out* = 0 to yield a sum of 130.

```
  A =       0010  1010  →  mod-3 = 0
                                       → mod-3 = 0 + cin = 1
  B =       0101  0111  →  mod-3 = 0
                    1                                         No error
 ----------------------
 Sum =      1000  0010  →      →       → mod-3 = 1
```

Example 17.3 Let the augend *A* and the addend *B* be decimal values of 122 and 149, respectively, with a *carry-in* = 0 and a *carry-out* = 1 to yield a sum of 271.

```
  A =       0111  1010  →  mod-3 = 2
                                       → mod-3 = 1 + cin = 1
  B =       1001  0101  →  mod-3 = 2
                    0                                         No error
 ----------------------
 Sum =  1 ← 0000  1111  →      →       → mod-3 = 1
```

Example 17.4 Let the augend *A* and the addend *B* be decimal values of 122 and 148, respectively, with a *carry-in* = 1 and a *carry-out* = 1 to yield a sum of 271.

```
  A =       0111  1010  →  mod-3 = 2
                                       → mod-3 = 0 + cin = 1
  B =       1001  0100  →  mod-3 = 1
                    1                                         No error
 ----------------------
 Sum =  1 ← 0000  1111  →      →       → mod-3 = 1
```

Example 17.5 Let the augend *A* and the addend *B* be decimal values of 122 and 148, respectively, with a *carry-in* = 1 and a *carry-out* = 1 to yield a sum of 271. Assume that an error occurred in adding the low-order four bits, as shown below. The modulo-3 sum of the residues of the augend and addend is 1; however, the modulo-3 residue of the sum is 0, indicating an error.

```
  A =       0111  1010  →  mod-3 = 2
                                       → mod-3 = 0 + cin = 1
  B =       1001  0100  →  mod-3 = 1
                    1                                         Error
 ----------------------
 Sum =  1 ← 0000  1110  →      →       → mod-3 = 0
```

The same technique can be used for detecting errors in binary-coded decimal (BCD) addition. However, each digit is treated separately for each operand. Then the modulo-3 sum of the residues of each digit determines the residue of the operand. Examples are shown below.

Example 17.6 The following two decimal numbers will be added using BCD addition: 452 and 120, to yield a sum of 572.

				Residue of segments					
A =	0100	0101	0010	1	2	2	→ 2	→ mod-3 = **2**	No error
B =	0001	0010	0000	1	2	0	→ 0		
Sum =	0101	0111	0010	2	1	2	→	→ mod-3 = **2**	

Example 17.7 The following two decimal numbers will be added using BCD addition: 676 and 738, to yield a sum of 1414.

					Residue of segments					
A =		0110	0111	0110		0	1	0	→ 1	→ mod-3 = **1**
B =		0111	0011	1000		1	0	2	→ 0	
			1 ←	1110						
		1 ←	1011	0110						
	1 ←	1110	0110	0100						
		0110	0001							
		0100								No error
	↓	↓	↓	↓						
Sum =	0001	0100	0001	0100	1	1	1	1	→	→ mod-3 = **1**

Example 17.8 The following two decimal numbers will be added using BCD addition: 732 and 125, to yield a sum of 857. If no error occurred in the addition operation, then the sum of the residues of the augend and the addend is 2 and the residue of the sum is also 2. Assume that an error occurred in adding the middle four bits, as shown below. The modulo-3 sum of the residues of the augend and addend is 2; however, the modulo-3 residue of the sum is 1, indicating an error.

				Residue of segments					
A =	0111	0011	0010	1	0	2	→ 0	→ mod-3 = **2**	Error
B =	0001	0010	0101	1	2	2	→ 2		
Sum =	1000	010**0**	0111	2	1	1	→	→ mod-3 = **1**	

Example 17.9 The following two decimal numbers will be added using BCD addition: 769 and 476, to yield a sum of 1245.

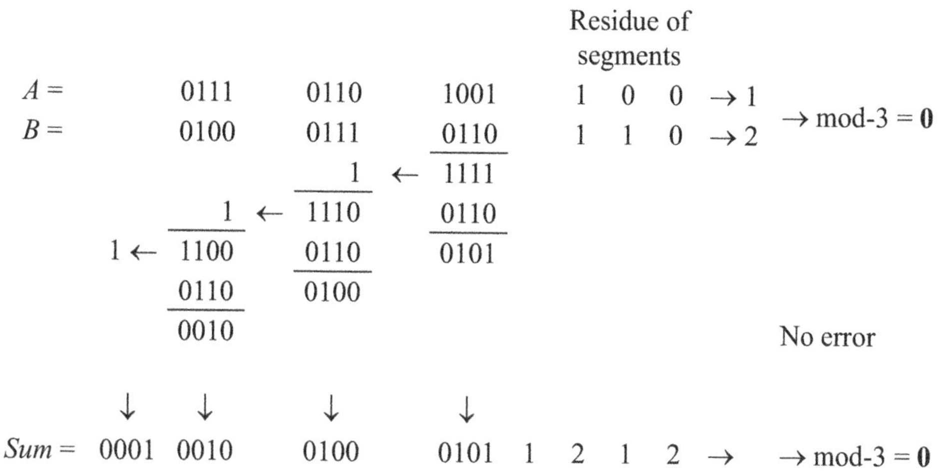

17.1.1 Dataflow Modeling

The Verilog HDL dataflow module for modulo-3 residue checking is shown in Figure 17.1 using the concept of Equation 17.4, where the arithmetic operator % indicates the modulus (remainder/residue) operation. The test bench is shown in Figure 17.2 and the outputs are shown in Figure 17.3. The outputs show the residue of augend *a* (*residue_a*) and the addend *b* (*residue_b*), the residue of the residues of *a* and *b* (*residue_a_b*), and the residue of the sum (*residue_sum*). The waveforms are shown in Figure 17.4.

```
//dataflow modulo-3 residue checking
module residue_chkg (a, b, cin, sum, cout,
            residue_a, residue_b, residue_a_b, residue_sum);

input [7:0] a, b;
input cin;
output [7:0] sum;
output cout;
output [1:0] residue_a, residue_b, residue_a_b, residue_sum;

assign {cout, sum} = a + b + cin;     //obtain sum

//obtain residues for a and b
assign    residue_a = a % 3,
          residue_b = b % 3;              //continued on next page
```

Figure 17.1 Dataflow module for modulo-3 residue checking.

```
//obtain the residue of the residues
assign   residue_a_b = (residue_a + residue_b + cin) % 3;

//obtain the residue of the sum
assign   residue_sum = {cout, sum}  % 3;

endmodule
```

Figure 17.1 (Continued)

```
//test bench for modulo-3 residue checking
module residue_chkg_tb;

reg [7:0] a, b;
reg cin;
wire [7:0] sum;
wire cout;

wire [1:0] residue_a, residue_b, residue_a_b, residue_sum;

//display variables
initial
$monitor ("a=%b, b=%b, cin=%b, sum=%b, cout=%b, residue_a=%b,
            residue_b=%b, residue_a_b=%b, residue_sum=%b",
         a, b, cin, sum, cout, residue_a, residue_b,
         residue_a_b, residue_sum);

//apply input vectors
initial
begin
   #0    a = 8'b0111_1010;
         b = 8'b0011_0101;
         cin = 1'b0;

   #10   a = 8'b0111_1010;
         b = 8'b1001_0100;
         cin = 1'b1;

   #10   a = 8'b0010_1010;
         b = 8'b0101_0111;
         cin = 1'b1;

   #10   $stop;
end                     //continued on next page
```

Figure 17.2 Test bench for modulo-3 residue checking.

```
//instantiate the module into the test bench
residue_chkg inst1 (
   .a(a),
   .b(b),
   .cin(cin),
   .sum(sum),
   .cout(cout),
   .residue_a(residue_a),
   .residue_b(residue_b),
   .residue_a_b(residue_a_b),
   .residue_sum(residue_sum)
   );

endmodule
```

Figure 17.2 (Continued)

```
a=0111_1010, b=0011_0101, cin=0,
   sum=1010_1111, cout=0,
   residue_a=10, residue_b=10,
   residue_a_b=01, residue_sum=01


a=0111_1010, b=1001_0100, cin=1,
   sum=0000_1111, cout=1,
   residue_a=10, residue_b=01,
   residue_a_b=01, residue_sum=01


a=0010_1010, b=0101_0111, cin=1,
   sum=1000_0010, cout=0,
   residue_a=00, residue_b=00,
   residue_a_b=01, residue_sum=01
```

Figure 17.3 Outputs for modulo-3 residue checking.

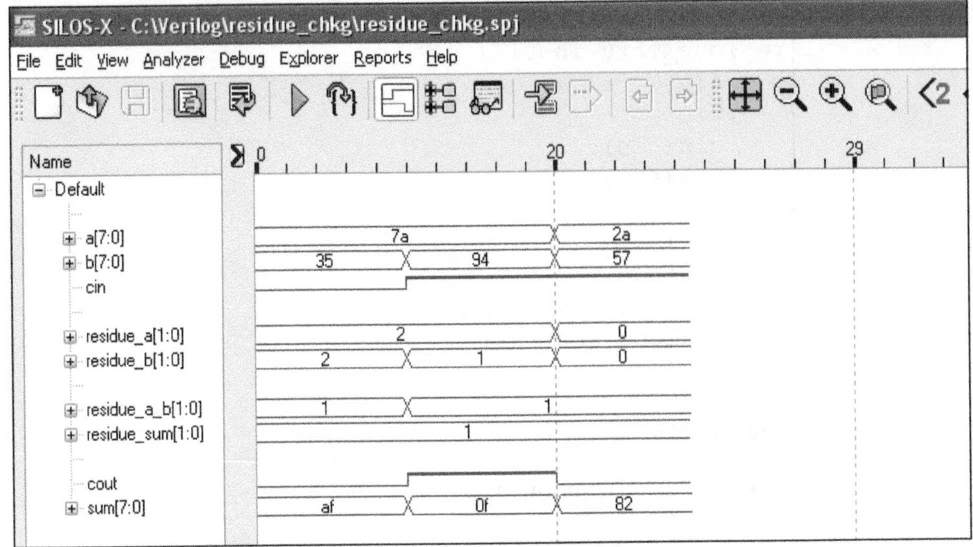

Figure 17.4 Waveforms for modulo-3 residue checking.

17.1.2 Structural Modeling

A combinational array can be used for residue checking in much the same way as an array was used for fixed-point division. The divisor, however, is a constant value of three that represents the modulus. The array generates a remainder which is the residue and has binary values of 00, 01, or 10. The quotient is unused. This is an extremely fast division operation, because the array is entirely combinational — the only delay is the propagation delay through the gates.

This method uses the nonrestoring division algorithm, which is ideal for iterative arrays. The divide operation is accomplished by a sequence of shifts and additions or subtractions depending on the state of the previous high-order carry-out. The carry-out determines the quotient bit at any level and also the next operation, either $2A - B$ or $2A + B$, according to the following criteria:

$$q_0 = \begin{cases} 0, \text{ if carry-out} = 0\text{; next operation is } 2A + B \\ 1, \text{ if carry-out} = 1\text{; next operation is } 2A - B \end{cases}$$

The above criteria is analogous to the following statement: if the sign is 0, then the operation at the next row of the array is a shift-add ($2A + B$); if the sign is 1, then the operation at the next row of the array is a shift-subtract ($2A - B$).

Two examples are shown in Figure 17.5 and Figure 17.6 using the paper-and-pencil method that illustrate the residue generation technique for a 6-bit dividend and a 3-bit divisor to yield a 3-bit quotient and remainder (residue).

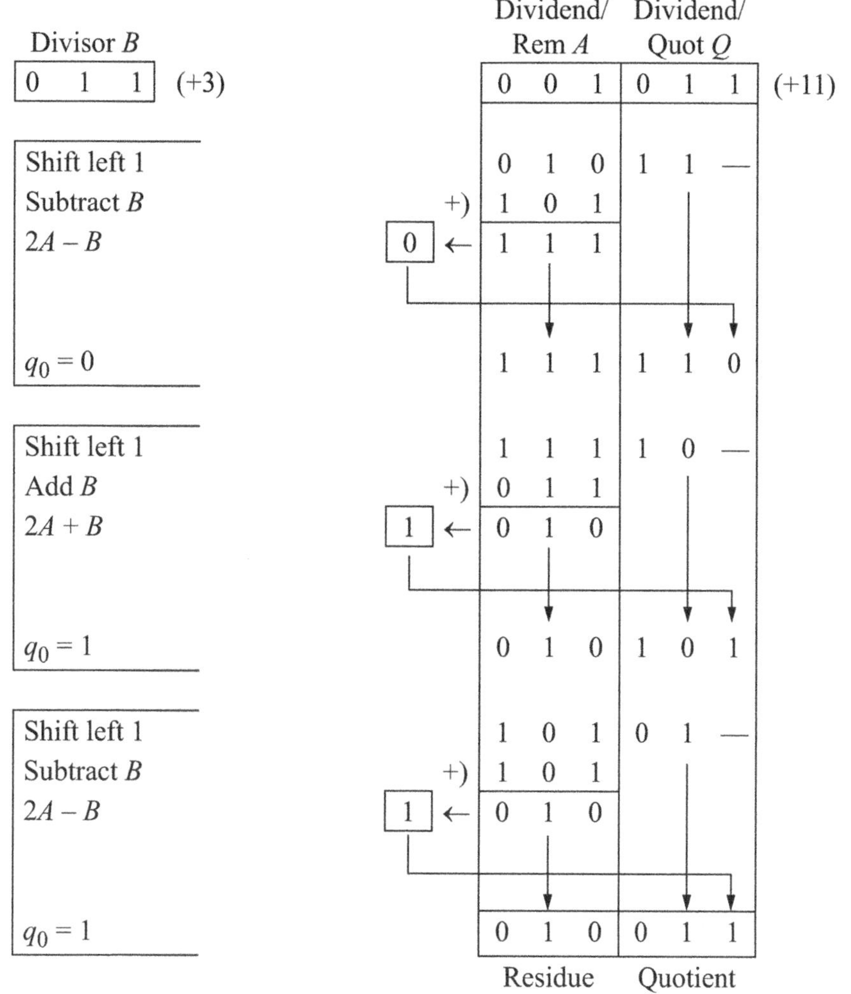

Figure 17.5 Example of residue generation using the paper-and-pencil method.

Shifting is accomplished by the placement of the cells along the diagonal of the array. The array consists of rows of identical cells incorporating a full adder in each cell with the ability to add or subtract. Subtraction or addition is determined by the state of the mode input, as shown in the typical full adder cell of Figure 17.7. If the mode line is a logic 1, then the operation is subtraction; if the mode line is logic 0, then the operation is addition.

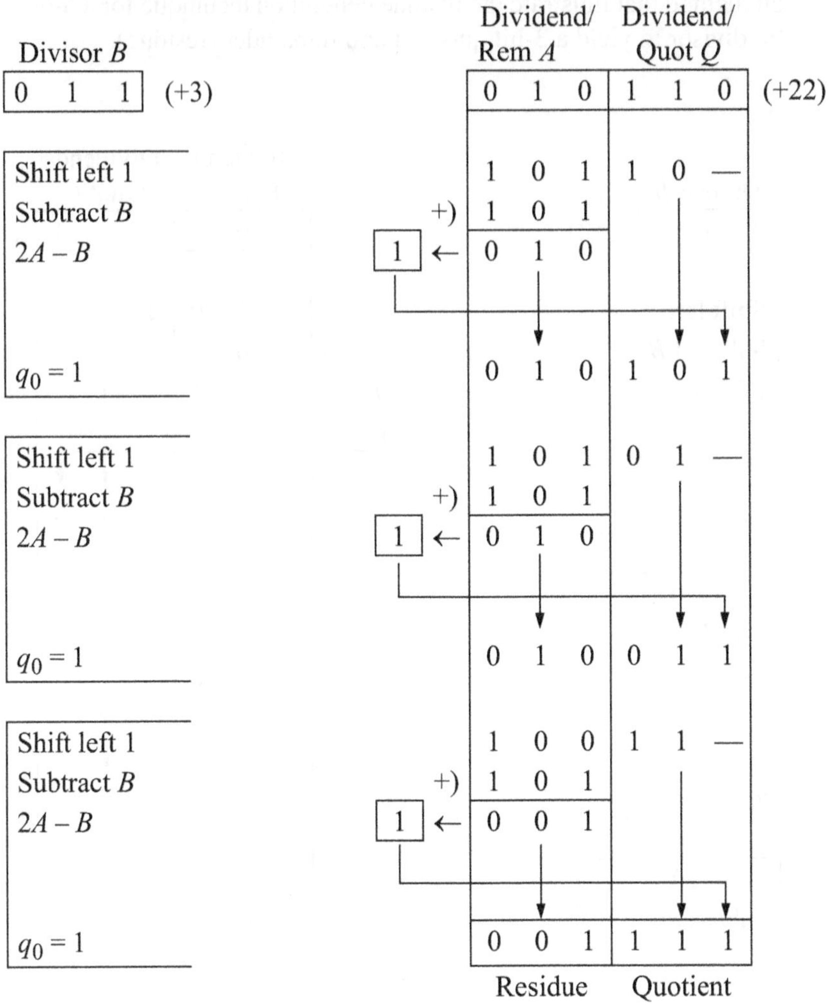

Operation		Dividend/ Rem A	Dividend/ Quot Q	
Divisor B: 0 1 1 (+3)		0 1 0	1 1 0	(+22)
Shift left 1		1 0 1	1 0 —	
Subtract B	+)	1 0 1		
$2A - B$	1 ←	0 1 0		
$q_0 = 1$		0 1 0	1 0 1	
Shift left 1		1 0 1	0 1 —	
Subtract B	+)	1 0 1		
$2A - B$	1 ←	0 1 0		
$q_0 = 1$		0 1 0	0 1 1	
Shift left 1		1 0 0	1 1 —	
Subtract B	+)	1 0 1		
$2A - B$	1 ←	0 0 1		
$q_0 = 1$		0 0 1	1 1 1	
		Residue	Quotient	

Figure 17.6 Example of residue generation using the paper-and-pencil method.

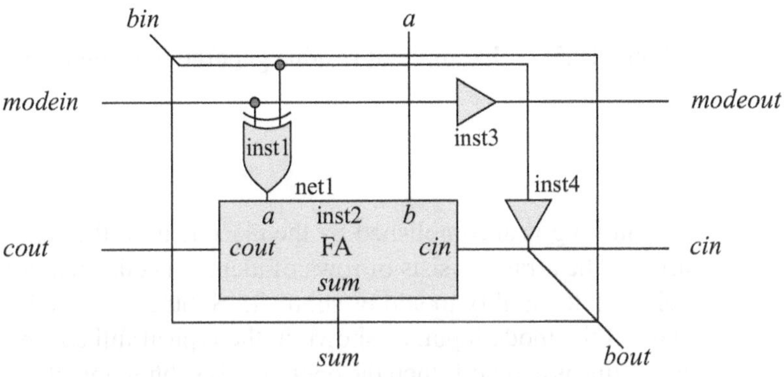

Figure 17.7 Typical cell for a residue generator.

The structural module for the typical cell of the residue generator is shown in Figure 17.8. The test bench and outputs are shown in Figure 17.9 and Figure 17.10, respectively.

```
//structural full adder for an array divide
module full_adder_div_array (a, bin, cin, modein,
                    sum, cout, bout, modeout);

input a, bin, cin, modein;
output sum, cout, bout, modeout;

wire a, bin, cin, modein;
wire sum, cout, bout, modeout;

wire net1;          //define internal net

//instantiate the logic
xor2_df inst1 (
   .x1(modein),
   .x2(bin),
   .z1(net1)
   );

full_adder inst2 (
   .a(net1),
   .b(a),
   .cin(cin),
   .sum(sum),
   .cout(cout)
   );

buf inst3 (modeout, modein);
buf inst4 (bout, bin);
endmodule
```

Figure 17.8 Structural module for the typical cell of the residue generator.

```
//test bench for the full adder used in the array divider
module full_adder_div_array_tb;

reg a, bin, cin, modein;
wire sum, cout, bout, modeout;

//continued on next page
```

Figure 17.9 Test bench for the typical cell of the residue generator.

```
//apply stimulus
initial
begin : apply_stimulus
   reg [4:0] invect;
   for (invect = 0; invect < 16; invect = invect + 1)
   begin
      {a, bin, cin, modein} = invect [4:0];
      #10 $display ("a=%b, bin=%b, cin=%b, modein=%b,
         sum=%b, cout=%b, bout=%b, modeout=%b",
               a, bin, cin, modein, sum, cout, bout, modeout);
   end
end

//instantiate the module into the test bench
full_adder_div_array inst1 (
   .a(a),
   .bin(bin),
   .cin(cin),
   .modein(modein),
   .sum(sum),
   .cout(cout),
   .bout(bout),
   .modeout(modeout)
   );
endmodule
```

Figure 17.9 (Continued)

```
a=0, bin=0, cin=0, modein=0, sum=0, cout=0, bout=0, modeout=0
a=0, bin=0, cin=0, modein=1, sum=1, cout=0, bout=0, modeout=1
a=0, bin=0, cin=1, modein=0, sum=1, cout=0, bout=0, modeout=0
a=0, bin=0, cin=1, modein=1, sum=0, cout=1, bout=0, modeout=1
a=0, bin=1, cin=0, modein=0, sum=1, cout=0, bout=1, modeout=0
a=0, bin=1, cin=0, modein=1, sum=0, cout=0, bout=1, modeout=1
a=0, bin=1, cin=1, modein=0, sum=0, cout=1, bout=1, modeout=0
a=0, bin=1, cin=1, modein=1, sum=1, cout=0, bout=1, modeout=1
a=1, bin=0, cin=0, modein=0, sum=1, cout=0, bout=0, modeout=0
a=1, bin=0, cin=0, modein=1, sum=0, cout=1, bout=0, modeout=1
a=1, bin=0, cin=1, modein=0, sum=0, cout=1, bout=0, modeout=0
a=1, bin=0, cin=1, modein=1, sum=1, cout=1, bout=0, modeout=1
a=1, bin=1, cin=0, modein=0, sum=0, cout=1, bout=1, modeout=0
a=1, bin=1, cin=0, modein=1, sum=1, cout=0, bout=1, modeout=1
a=1, bin=1, cin=1, modein=0, sum=1, cout=1, bout=1, modeout=0
a=1, bin=1, cin=1, modein=1, sum=0, cout=1, bout=1, modeout=1
```

Figure 17.10 Outputs for the typical cell of the residue generator.

The organization for the residue generator using a division array is shown in Figure 17.11 complete with instantiation names and net names. The carry-out of the high-order cell in each row represents the quotient bit for that row. The carry-out connects to the mode input of the high-order cell in the row immediately below, which then propagates through all cells in the row and connects to the carry-in of the low-order cell. If the carry-out is 0, then the operation at the next lower level is addition; if the carry-out is 1, then the operation at the next lower level is subtraction.

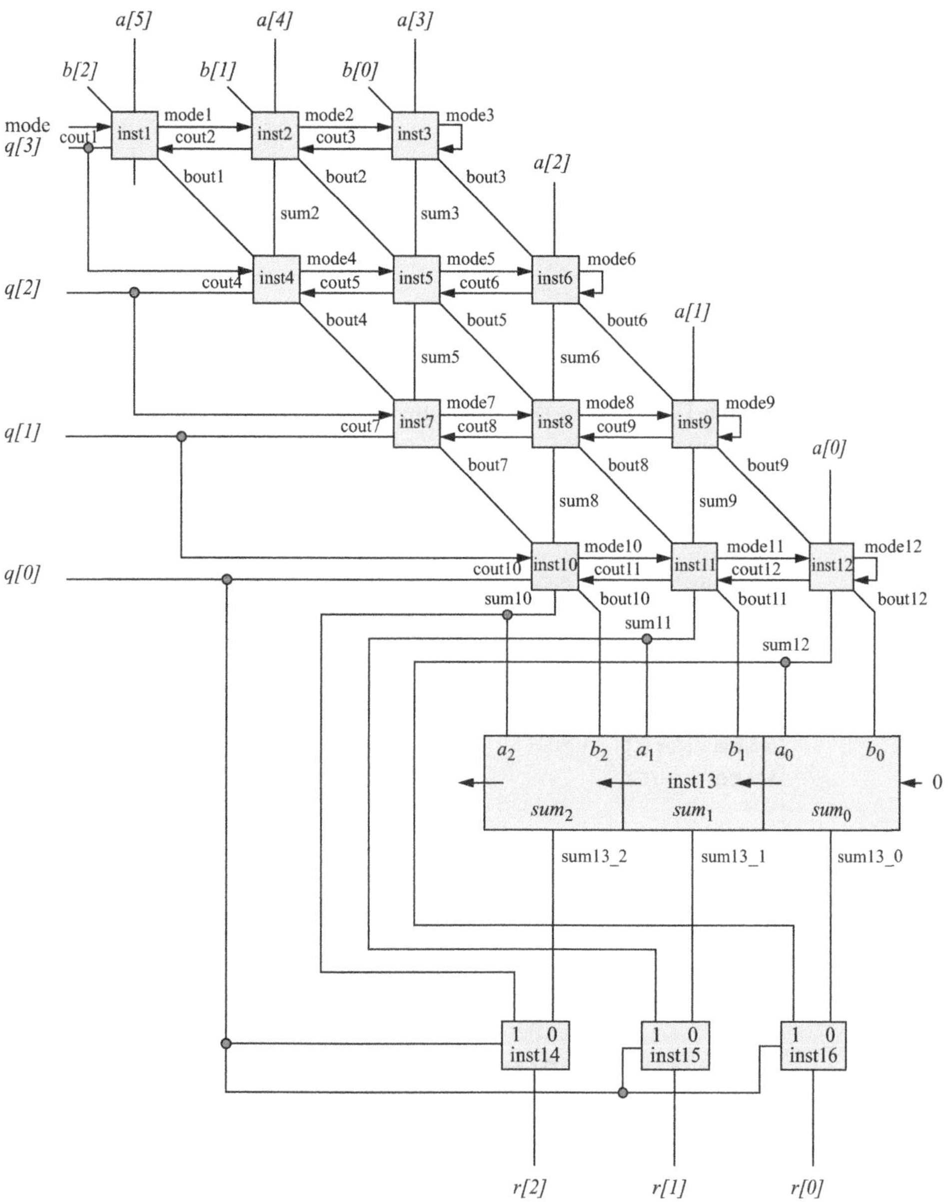

Figure 17.11 Organization of the array for a modulo-3 residue generator.

The shift operation is achieved by moving the divisor to the right along the diagonal; that is, the entire row of cells is positioned one column to the right, thus providing the requisite shift of the divisor relative to the previous partial remainder. Restoring the final remainder in the array implementation is similar to that in the sequential implementation — the divisor is added to the previous partial remainder.

The dataflow module for the 3-input adder that is instantiated into the array as *inst13* is shown in Figure 17.12. The test bench, which applies all 128 inputs to the adder, is shown in Figure 17.13, and a selection of the outputs is shown in Figure 17.14.

```
//dataflow for a 3-bit adder
module adder3_df (a, b, cin, sum, cout);

input [2:0] a, b;     //list inputs and outputs
input cin;
output [2:0] sum;
output cout;

assign {cout, sum} = a + b + cin;
endmodule
```

Figure 17.12 Dataflow module for the 3-input adder that is instantiated into the residue generator array.

```
//test bench for 3-bit dataflow adder
module adder3_df_tb;

reg [2:0] a, b;
reg cin;
wire [2:0] sum;
wire cout;

//apply stimulus
initial
begin : apply_stimulus
   reg [7:0] invect;
   for (invect = 0; invect < 128; invect = invect + 1)
   begin
      {a, b, cin} = invect [7:0];
      #10 $display ("a=%b, b=%b, cin=%b, cout=%b, sum=%b",
                        a, b, cin, cout, sum);
   end
end
//continued on next page
```

Figure 17.13 Test bench for the 3-input adder of Figure 17.12.

```
//instantiate the module into the test bench
adder3_df inst1 (
   .a(a),
   .b(b),
   .cin(cin),
   .sum(sum),
   .cout(cout)
   );
endmodule
```

Figure 17.13 (Continued)

```
a=000, b=000, cin=0,
   cout=0, sum=000
a=000, b=000, cin=1,
   cout=0, sum=001
a=000, b=001, cin=0,
   cout=0, sum=001
a=000, b=001, cin=1,
   cout=0, sum=010
a=000, b=010, cin=0,
   cout=0, sum=010
a=000, b=010, cin=1,
   cout=0, sum=011
a=000, b=011, cin=0,
   cout=0, sum=011
a=000, b=011, cin=1,
   cout=0, sum=100
a=000, b=100, cin=0,
   cout=0, sum=100
a=000, b=100, cin=1,
   cout=0, sum=101
a=000, b=101, cin=0,
   cout=0, sum=101
a=000, b=101, cin=1,
   cout=0, sum=110
a=000, b=110, cin=0,
   cout=0, sum=110
a=000, b=110, cin=1,
   cout=0, sum=111
a=000, b=111, cin=0,
   cout=0, sum=111
      . . .
a=111, b=000, cin=0,
   cout=0, sum=111
a=111, b=000, cin=1,
   cout=1, sum=000
a=111, b=001, cin=0,
   cout=1, sum=000
a=111, b=001, cin=1,
   cout=1, sum=001
a=111, b=010, cin=0,
   cout=1, sum=001
a=111, b=010, cin=1,
   cout=1, sum=010
a=111, b=011, cin=0,
   cout=1, sum=010
a=111, b=011, cin=1,
   cout=1, sum=011
a=111, b=100, cin=0,
   cout=1, sum=011
a=111, b=100, cin=1,
   cout=1, sum=100
a=111, b=101, cin=0,
   cout=1, sum=100
a=111, b=101, cin=1,
   cout=1, sum=101
a=111, b=110, cin=0,
   cout=1, sum=101
a=111, b=110, cin=1,
   cout=1, sum=110
a=111, b=111, cin=0,
   cout=1, sum=110
a=111, b=111, cin=1,
   cout=1, sum=111
```

Figure 17.14 Outputs for the 3-bit adder of Figure 17.12.

The structural module for the modulo-3 residue generator is shown in Figure 17.15. Instantiations *inst14*, *inst15*, and *inst16* are 2:1 multiplexers. This module is the first of two modules that will be used to design a modulo-3 residue checker — the second structural module is the actual residue checker that will instantiate the residue generator. The test bench is shown in Figure 17.16 and applies all valid vectors that generate a 4-bit quotient. The outputs are shown in Figure 17.17.

```
//structural mod-3 residue generator
module residue_gen (a, b, mode, q, r);

input [5:0] a;
input [2:0] b;
input mode;
output [3:0] q;
output [2:0] r;

//define internal nets
wire  mode1, mode2, mode3, mode4, mode5, mode6,
      mode7, mode8, mode9, mode10, mode11, mode12,
      cout1, cout2, cout3, cout4, cout5, cout6,
      cout7, cout8, cout9, cout10, cout11, cout12,
      bout1, bout2, bout3, bout4, bout5, bout6,
      bout7, bout8, bout9, bout10, bout11, bout12,
      sum2, sum3, sum5, sum6, sum8, sum9, sum10, sum11, sum12,
      sum13_0, sum13_1, sum13_2;

//instantiate the logic for the array divider and full adders
full_adder_div_array inst1 (
   .a(a[5]),
   .bin(b[2]),
   .cin(cout2),
   .modein(mode),
   .cout(cout1),
   .bout(bout1),
   .modeout(mode1)
   );

full_adder_div_array inst2 (
   .a(a[4]),
   .bin(b[1]),
   .cin(cout3),
   .modein(mode1),
   .sum(sum2),
   .cout(cout2),
   .bout(bout2),
   .modeout(mode2)
   );                    //continued on next page
```

Figure 17.15 Structural module for the residue generator.

```
full_adder_div_array inst3 (
   .a(a[3]),
   .bin(b[0]),
   .cin(mode3),
   .modein(mode2),
   .sum(sum3),
   .cout(cout3),
   .bout(bout3),
   .modeout(mode3)
   );

full_adder_div_array inst4 (
   .a(sum2),
   .bin(bout1),
   .cin(cout5),
   .modein(cout1),
   .cout(cout4),
   .bout(bout4),
   .modeout(mode4)
   );

full_adder_div_array inst5 (
   .a(sum3),
   .bin(bout2),
   .cin(cout6),
   .modein(mode4),
   .sum(sum5),
   .cout(cout5),
   .bout(bout5),
   .modeout(mode5)
   );

full_adder_div_array inst6 (
   .a(a[2]),
   .bin(bout3),
   .cin(mode6),
   .modein(mode5),
   .sum(sum6),
   .cout(cout6),
   .bout(bout6),
   .modeout(mode6)
   );

//continued on next page
```

Figure 17.15 (Continued)

```
full_adder_div_array inst7 (
   .a(sum5),
   .bin(bout4),
   .cin(cout8),
   .modein(cout4),
   .cout(cout7),
   .bout(bout7),
   .modeout(mode7)
   );


full_adder_div_array inst8 (
   .a(sum6),
   .bin(bout5),
   .cin(cout9),
   .modein(mode7),
   .sum(sum8),
   .cout(cout8),
   .bout(bout8),
   .modeout(mode8)
   );


full_adder_div_array inst9 (
   .a(a[1]),
   .bin(bout6),
   .cin(mode9),
   .modein(mode8),
   .sum(sum9),
   .cout(cout9),
   .bout(bout9),
   .modeout(mode9)
   );



full_adder_div_array inst10 (
   .a(sum8),
   .bin(bout7),
   .cin(cout11),
   .modein(cout7),
   .sum(sum10),
   .cout(cout10),
   .bout(bout10),
   .modeout(mode10)
   );

//continued on next page
```

Figure 17.15 (Continued)

```
full_adder_div_array inst11 (
   .a(sum9),
   .bin(bout8),
   .cin(cout12),
   .modein(mode10),
   .sum(sum11),
   .cout(cout11),
   .bout(bout11),
   .modeout(mode11)
   );

full_adder_div_array inst12 (
   .a(a[0]),
   .bin(bout9),
   .cin(mode12),
   .modein(mode11),
   .sum(sum12),
   .cout(cout12),
   .bout(bout12),
   .modeout(mode12)
   );

adder3_df inst13 (         //instantiate the 3-input adder
   .a({sum10, sum11, sum12}),
   .b({bout10, bout11, bout12}),
   .cin(1'b0),
   .sum({sum13_2, sum13_1, sum13_0})
   );

mux2_df inst14 (           //instantiate the 2:1 multiplexers
   .sel(cout10),
   .data({sum10, sum13_2}),
   .z1(r[2])
   );

mux2_df inst15 (
   .sel(cout10),
   .data({sum11, sum13_1}),
   .z1(r[1])
   );

mux2_df inst16 (
   .sel(cout10),
   .data({sum12, sum13_0}),
   .z1(r[0])
   );
//continued on next page
```

Figure 17.15 (Continued)

```
//assign the quotient outputs
assign   q[3] = cout1,
         q[2] = cout4,
         q[1] = cout7,
         q[0] = cout10;
endmodule
```

Figure 17.15 (Continued)

```
//test bench for the residue generator
module residue_gen_tb;

reg [5:0] a;
reg [2:0] b;
reg mode;
wire [3:0] q;
wire [2:0] r;

initial        //display inputs and outputs
$monitor ("dividend=%b, divisor=%b, quotient=%b, residue=%b",
           a, b, q, r[1:0]);

initial        //apply input vectors
begin
   #0    a = 6'b00_0000; b = 3'b011; mode =1'b1;
   #10   a = 6'b00_0001; b = 3'b011; mode =1'b1;   //1
   #10   a = 6'b00_0010; b = 3'b011; mode =1'b1;
   #10   a = 6'b00_0011; b = 3'b011; mode =1'b1;
   #10   a = 6'b00_0100; b = 3'b011; mode =1'b1;   //4

   #10   a = 6'b00_0101; b = 3'b011; mode =1'b1;   //5
   #10   a = 6'b00_0110; b = 3'b011; mode =1'b1;
   #10   a = 6'b00_0111; b = 3'b011; mode =1'b1;
   #10   a = 6'b00_1000; b = 3'b011; mode =1'b1;   //8

   #10   a = 6'b00_1001; b = 3'b011; mode =1'b1;   //9
   #10   a = 6'b00_1010; b = 3'b011; mode =1'b1;
   #10   a = 6'b00_1011; b = 3'b011; mode =1'b1;
   #10   a = 6'b00_1100; b = 3'b011; mode =1'b1;   //12

   #10   a = 6'b00_1101; b = 3'b011; mode =1'b1;   //13
   #10   a = 6'b00_1110; b = 3'b011; mode =1'b1;
   #10   a = 6'b00_1111; b = 3'b011; mode =1'b1;
   #10   a = 6'b01_0000; b = 3'b011; mode =1'b1;   //16
//continued on next page
```

Figure 17.16 Test bench for the residue generator.

```
    #10   a = 6'b01_0001; b = 3'b011; mode =1'b1;    //17
    #10   a = 6'b01_0010; b = 3'b011; mode =1'b1;
    #10   a = 6'b01_0011; b = 3'b011; mode =1'b1;
    #10   a = 6'b01_0100; b = 3'b011; mode =1'b1;    //20

    #10   a = 6'b01_0101; b = 3'b011; mode =1'b1;    //21
    #10   a = 6'b01_0110; b = 3'b011; mode =1'b1;
    #10   a = 6'b01_0111; b = 3'b011; mode =1'b1;
    #10   a = 6'b01_1000; b = 3'b011; mode =1'b1;    //24

    #10   a = 6'b01_1001; b = 3'b011; mode =1'b1;    //25
    #10   a = 6'b01_1010; b = 3'b011; mode =1'b1;
    #10   a = 6'b01_1011; b = 3'b011; mode =1'b1;
    #10   a = 6'b01_1100; b = 3'b011; mode =1'b1;    //28

    #10   a = 6'b01_1101; b = 3'b011; mode =1'b1;    //29
    #10   a = 6'b01_1110; b = 3'b011; mode =1'b1;
    #10   a = 6'b01_1111; b = 3'b011; mode =1'b1;
    #10   a = 6'b10_0000; b = 3'b011; mode =1'b1;    //32

    #10   a = 6'b10_0001; b = 3'b011; mode =1'b1;    //33
    #10   a = 6'b10_0010; b = 3'b011; mode =1'b1;
    #10   a = 6'b10_0011; b = 3'b011; mode =1'b1;
    #10   a = 6'b10_0100; b = 3'b011; mode =1'b1;    //36

    #10   a = 6'b10_0101; b = 3'b011; mode =1'b1;    //37
    #10   a = 6'b10_0110; b = 3'b011; mode =1'b1;
    #10   a = 6'b10_0111; b = 3'b011; mode =1'b1;
    #10   a = 6'b10_1000; b = 3'b011; mode =1'b1;    //40

    #10   a = 6'b10_1001; b = 3'b011; mode =1'b1;    //41
    #10   a = 6'b10_1010; b = 3'b011; mode =1'b1;
    #10   a = 6'b10_1011; b = 3'b011; mode =1'b1;
    #10   a = 6'b10_1100; b = 3'b011; mode =1'b1;    //44

    #10   a = 6'b10_1101; b = 3'b011; mode =1'b1;    //45
    #10   a = 6'b10_1110; b = 3'b011; mode =1'b1;
    #10   a = 6'b10_1111; b = 3'b011; mode =1'b1;    //47

    #10   $stop;
end

//instantiate the module into the test bench
residue_gen inst1 (
    .a(a),
    .b(b),
    .mode(mode),
    .q(q),
    .r(r)
    );
endmodule
```

Figure 17.16 (Continued)

```
dividend=000000, divisor=011, quotient=0000, residue=00
dividend=000001, divisor=011, quotient=0000, residue=01
dividend=000010, divisor=011, quotient=0000, residue=10
dividend=000011, divisor=011, quotient=0001, residue=00
dividend=000100, divisor=011, quotient=0001, residue=01
dividend=000101, divisor=011, quotient=0001, residue=10
dividend=000110, divisor=011, quotient=0010, residue=00
dividend=000111, divisor=011, quotient=0010, residue=01
dividend=001000, divisor=011, quotient=0010, residue=10
dividend=001001, divisor=011, quotient=0011, residue=00
dividend=001010, divisor=011, quotient=0011, residue=01
dividend=001011, divisor=011, quotient=0011, residue=10
dividend=001100, divisor=011, quotient=0100, residue=00
dividend=001101, divisor=011, quotient=0100, residue=01
dividend=001110, divisor=011, quotient=0100, residue=10
dividend=001111, divisor=011, quotient=0101, residue=00
dividend=010000, divisor=011, quotient=0101, residue=01
dividend=010001, divisor=011, quotient=0101, residue=10
dividend=010010, divisor=011, quotient=0110, residue=00
dividend=010011, divisor=011, quotient=0110, residue=01
dividend=010100, divisor=011, quotient=0110, residue=10
dividend=010101, divisor=011, quotient=0111, residue=00
dividend=010110, divisor=011, quotient=0111, residue=01
dividend=010111, divisor=011, quotient=0111, residue=10
dividend=011000, divisor=011, quotient=1000, residue=00
dividend=011001, divisor=011, quotient=1000, residue=01
dividend=011010, divisor=011, quotient=1000, residue=10
dividend=011011, divisor=011, quotient=1001, residue=00
dividend=011100, divisor=011, quotient=1001, residue=01
dividend=011101, divisor=011, quotient=1001, residue=10
dividend=011110, divisor=011, quotient=1010, residue=00
dividend=011111, divisor=011, quotient=1010, residue=01
dividend=100000, divisor=011, quotient=1010, residue=10
dividend=100001, divisor=011, quotient=1011, residue=00
dividend=100010, divisor=011, quotient=1011, residue=01
dividend=100011, divisor=011, quotient=1011, residue=10
dividend=100100, divisor=011, quotient=1100, residue=00
dividend=100101, divisor=011, quotient=1100, residue=01
dividend=100110, divisor=011, quotient=1100, residue=10
dividend=100111, divisor=011, quotient=1101, residue=00
dividend=101000, divisor=011, quotient=1101, residue=01
dividend=101001, divisor=011, quotient=1101, residue=10
dividend=101010, divisor=011, quotient=1110, residue=00
dividend=101011, divisor=011, quotient=1110, residue=01
dividend=101100, divisor=011, quotient=1110, residue=10
dividend=101101, divisor=011, quotient=1111, residue=00
dividend=101110, divisor=011, quotient=1111, residue=01
dividend=101111, divisor=011, quotient=1111, residue=10
```

Figure 17.17 Outputs for the residue generator.

The second structural module that checks the residue of the sum of two operands will now be presented. Recall that the residue of the sum is equal to the sum of the residues as specified in Equation 17.4, which is replicated in Equation 17.6 for convenience. This states that the sum of the augend A and the addend B is applied to the inputs of a residue generator.

In parallel with the add operation, separate residues for A and B are obtained using two residue generators. These residues are then added in a 2-input adder, and the resulting sum is applied to another residue generator. The residue of the sum is then compared to the residue of the sum of the residues. If the two residues are equal, then there was no error in the addition operation; if the two residues are unequal, then a miscalculation occured in the addition operation and an error is indicated. Residue checking can detect multiple bit errors providing that the error result is not a multiple of three.

$$
\begin{aligned}
R(A) &\equiv A \text{ mod-}m \\
R(B) &\equiv B \text{ mod-}m \\
R(A) + R(B) &\equiv (A \text{ mod-}m) + (B \text{ mod-}m) \\
&\equiv (A + B) \text{ mod-}m \\
&\equiv R(A + B)
\end{aligned}
\tag{17.6}
$$

The organization for the residue checker is shown in Figure 17.18, which instantiates the residue generator that was previously designed, together with a 4-bit adder, a 2-bit adder, and the necessary logic gates to implement a 2-bit comparator. The carry-in to the 2-bit adder that generates the sum of the individual operand residues is connected to an input labeled *gen_err*. This input modifies the addition operation of the augend and addend so that an error is generated.

The behavioral module for the 4-bit adder is shown in Figure 17.19. In this implementation, there is no carry-out — the high-order bit, *sum[4]*, of the adder represents the carry out of the adder. The test bench, shown in Figure 17.20, applies several input vectors for the augend and addend. The outputs are shown in Figure 17.21.

The dataflow module for the 2-bit adder is shown in Figure 17.22, also with no carry-out. The test bench, shown in Figure 17.23, applies eleven input vectors for the operands. The outputs are shown in Figure 17.24.

The structural module for the residue checker is shown in Figure 17.25 with the internal nets defined to correspond with those declared in the organization of Figure 17.18. The residue generator is instantiated four times into the structural module together with the modules for the adders and logic gates.

The test bench is shown in Figure 17.26, which applies several input vectors for the augend and addend in which no error occurs, and one input vector in which a residue error occurs due to the *gen_err* input being active. The outputs and waveforms are shown in Figure 17.27 and Figure 17.28, respectively.

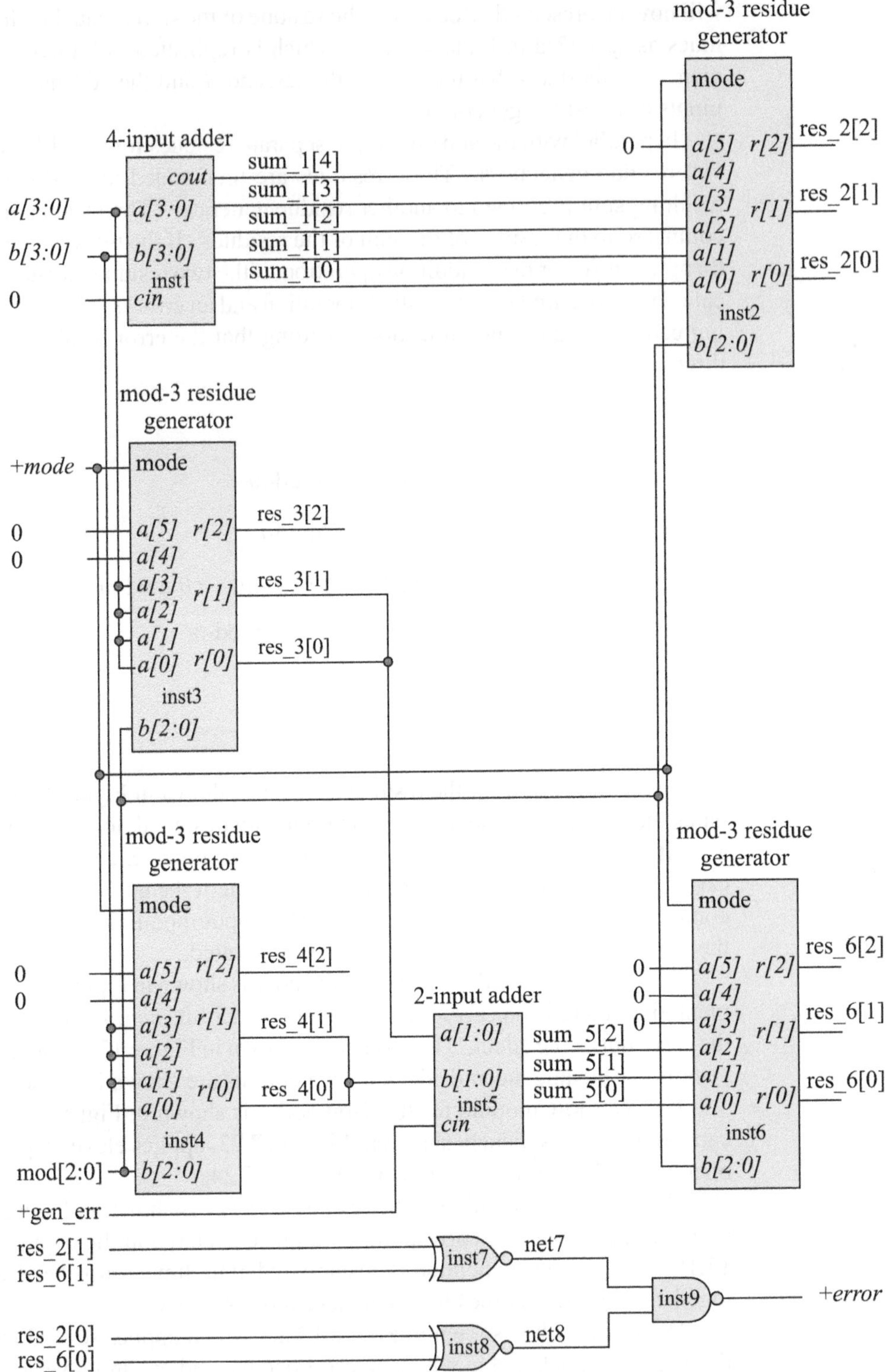

Figure 17.18 Organization for the modulo-3 residue checker.

```
//behavioral 4-bit adder
module adder_4_behav (a, b, cin, sum);

input [3:0] a, b;
input cin;
output [4:0] sum;

reg [4:0] sum;    //used in always statement

always @ (a or b or cin)
begin
   sum = a + b + cin;
end
endmodule
```

Figure 17.19 Behavioral module for the 4-bit adder.

```
//behavioral 4-bit adder test bench
module adder_4_behav_tb;

reg [3:0] a, b;
reg cin;
wire [4:0] sum;

initial        //display variables
$monitor ("a b cin = %b_%b_%b, sum = %b", a, b, cin, sum);

initial        //apply input vectors
begin
   #0    a = 4'b0011;   b = 4'b0100;   cin = 1'b0;
   #10   a = 4'b1100;   b = 4'b0011;   cin = 1'b0;
   #10   a = 4'b0111;   b = 4'b0110;   cin = 1'b1;
   #10   a = 4'b1001;   b = 4'b0111;   cin = 1'b1;
   #10   a = 4'b1101;   b = 4'b0111;   cin = 1'b1;
   #10   a = 4'b1111;   b = 4'b0110;   cin = 1'b1;
   #10   $stop;
end

adder_4_behav inst1 (      //instantiate the module
   .a(a),
   .b(b),
   .cin(cin),
   .sum(sum)
   );
endmodule
```

Figure 17.20 Test bench for the 4-bit adder.

```
a b cin = 0011_0100_0, sum = 0_0111
a b cin = 1100_0011_0, sum = 0_1111
a b cin = 0111_0110_1, sum = 0_1110
a b cin = 1001_0111_1, sum = 1_0001
a b cin = 1101_0111_1, sum = 1_0101
a b cin = 1111_0110_1, sum = 1_0110
```

Figure 17.21 Outputs for the 4-bit adder.

```
//dataflow for a 2-bit adder
module adder_2_df (a, b, cin, sum);

//list inputs and outputs
input [1:0] a, b;
input cin;
output [2:0] sum;

assign sum = a + b + cin;
endmodule
```

Figure 17.22 Dataflow module for the 2-bit adder.

```
//test bench for 2-input adder
module adder_2_df_tb;

reg [1:0] a, b;
reg cin;
wire [2:0] sum;

//display signals
initial
$monitor ("a=%b, b=%b, cin=%b, sum=%b", a, b, cin, sum);

//apply stimulus
initial
begin
   #0    a = 2'b00;  b = 2'b00;   cin = 1'b0; //sum = 000
   #10   a = 2'b01;  b = 2'b10;   cin = 1'b0; //sum = 011
   #10   a = 2'b10;  b = 2'b11;   cin = 1'b0; //sum = 101
   #10   a = 2'b11;  b = 2'b11;   cin = 1'b0; //sum = 110
   #10   a = 2'b10;  b = 2'b10;   cin = 1'b0; //sum = 100
   #10   a = 2'b10;  b = 2'b01;   cin = 1'b0; //sum = 011
//continued on next page
```

Figure 17.23 Test bench for the 2-bit adder.

```
    #10   a = 2'b11;  b = 2'b00;  cin = 1'b0; //sum = 011
    #10   a = 2'b11;  b = 2'b01;  cin = 1'b1; //sum = 101
    #10   a = 2'b01;  b = 2'b10;  cin = 1'b1; //sum = 100
    #10   a = 2'b11;  b = 2'b10;  cin = 1'b1; //sum = 110
    #10   a = 2'b11;  b = 2'b11;  cin = 1'b1; //sum = 111
    #10   $stop;
end

//instantiate the module into the test bench
adder_2_df inst1 (
    .a(a),
    .b(b),
    .cin(cin),
    .sum(sum)
    );
endmodule
```

Figure 17.23 (Continued)

```
a=00, b=00, cin=0, sum=000
a=01, b=10, cin=0, sum=011
a=10, b=11, cin=0, sum=101
a=11, b=11, cin=0, sum=110
a=10, b=10, cin=0, sum=100
a=10, b=01, cin=0, sum=011
a=11, b=00, cin=0, sum=011
a=11, b=01, cin=1, sum=101
a=01, b=10, cin=1, sum=100
a=11, b=10, cin=1, sum=110
a=11, b=11, cin=1, sum=111
```

Figure 17.24 Outputs for the 2-bit adder.

```
//structural residue checking for modulo-3
module residue_chk2 (a, b, mod3, mode, gen_err, sum_1, res_2,
                    res_3, res_4, res_6, error);

input [3:0] a, b;
input mode, gen_err;
input [2:0] mod3;
output [4:0] sum_1;
output [2:0] res_2, res_3, res_4, res_6;
output error;         //continued on next page
```

Figure 17.25 Structural module for the modulo-3 residue checker.

```
//define internal nets
wire [4:0] sum_1;
wire [2:0] res_2, res_3, res_4, res_6;
wire [2:0] sum_5;
wire net7, net8;


//instantiate the 4-bit adder
adder_4_behav inst1 (
   .a(a),
   .b(b),
   .cin(1'b0),
   .sum(sum_1)
   );


//instantiate the modulo-3 residue generator
//to obtain the residue of the sum
residue_gen inst2 (
   .a({1'b0, sum_1}),
   .b(mod3),
   .mode(mode),
   .r(res_2)
   );


//instantiate the modulo-3 residue generator
//to obtain the residue of operand a (augend)
residue_gen inst3 (
   .a({1'b0, 1'b0, a}),
   .b(mod3),
   .mode(mode),
   .r(res_3)
   );


//instantiate the modulo-3 residue generator
//to obtain the residue of operand b (addend)
residue_gen inst4 (
   .a({1'b0, 1'b0, b}),
   .b(mod3),
   .mode(mode),
   .r(res_4)
   );

//continued on next page
```

Figure 17.25 (Continued)

```
//instantiate the 2-bit adder
adder_2_df inst5 (
    .a({res_3[1], res_3[0]}),
    .b({res_4[1], res_4[0]}),
    .cin(gen_err),
    .sum(sum_5)
    );


//instantiate the modulo-3 residue generator
//to obtain the residue of the sum of the
//residues of operands a and b
residue_gen inst6 (
    .a({1'b0, 1'b0, 1'b0, sum_5}),
    .b(mod3),
    .mode(mode),
    .r(res_6)
    );


//instantiate the logic for the comparator
xnor2_df inst7 (
    .x1(res_2[1]),
    .x2(res_6[1]),
    .z1(net7)
    );


xnor2_df inst8 (
    .x1(res_2[0]),
    .x2(res_6[0]),
    .z1(net8)
    );


nand2_df inst9 (
    .x1(net7),
    .x2(net8),
    .z1(error)
    );


endmodule
```

Figure 17.25 (Continued)

```
//test bench for the mod-3 residue checker
module residue_chk2_tb;

reg [3:0] a, b;
reg mode, gen_err;
reg [2:0] mod3;
wire [4:0] sum_1;
wire [2:0] res_2, res_3, res_4, res_6;
wire error;

initial      //display inputs and output
$monitor ("a=%b, b=%b, sum=%b, res_sum=%b, res_a=%b, res_b=%b,
            res_sum_res=%b, error=%b",
         a, b, sum_1, {res_2[1], res_2[0]}, {res_3[1],
            res_3[0]}, {res_4[1], res_4[0]},
            {res_6[1], res_6[0]}, error);

initial      //apply input vectors
begin
   #0    a = 4'b0110;   b = 4'b0101;
         mod3 = 3'b011; mode = 1'b1;   gen_err = 1'b0;

   #10   a = 4'b0101;   b = 4'b0101;
         mod3 = 3'b011; mode = 1'b1;   gen_err = 1'b0;

   #10   a = 4'b1010;   b = 4'b0111;
         mod3 = 3'b011; mode = 1'b1;   gen_err = 1'b0;

   #10   a = 4'b1111;   b = 4'b1101;
         mod3 = 3'b011; mode = 1'b1;   gen_err = 1'b0;

   #10   a = 4'b1011;   b = 4'b0111;
         mod3 = 3'b011; mode = 1'b1;   gen_err = 1'b0;

   #10   a = 4'b1111;   b = 4'b1111;
         mod3 = 3'b011; mode = 1'b1;   gen_err = 1'b0;

   #10   a = 4'b0111;   b = 4'b1001;
         mod3 = 3'b011; mode = 1'b1;   gen_err = 1'b0;

   #10   a = 4'b1101;   b = 4'b0111;
         mod3 = 3'b011; mode = 1'b1;   gen_err = 1'b0;

   #10   a = 4'b1101;   b = 4'b0111;
         mod3 = 3'b011; mode = 1'b1;   gen_err = 1'b1;
   #10   $stop;
end                 //continued on next page
```

Figure 17.26 Test bench for the modulo-3 residue checker.

```
residue_chk2 inst1 (     //instantiate the module
   .a(a),
   .b(b),
   .mode(mode),
   .gen_err(gen_err),
   .mod3(mod3),
   .sum_1(sum_1),
   .res_2(res_2),
   .res_3(res_3),
   .res_4(res_4),
   .res_6(res_6),
   .error(error)
   );
endmodule
```

Figure 17.26 (Continued)

```
a=0110, b=0101, sum=01011, res_sum=10,
   res_a=00, res_b=10, res_sum_res=10, error=0

a=0101, b=0101, sum=01010, res_sum=01
   res_a=10, res_b=10, res_sum_res=01, error=0

a=1010, b=0111, sum=10001, res_sum=10,
   res_a=01, res_b=01, res_sum_res=10, error=0

a=1111, b=1101, sum=11100, res_sum=01,
   res_a=00, res_b=01, res_sum_res=01, error=0

a=1011, b=0111, sum=10010, res_sum=00,
   res_a=10, res_b=01, res_sum_res=00, error=0

a=1111, b=1111, sum=11110, res_sum=00,
   res_a=00, res_b=00, res_sum_res=00, error=0

a=0111, b=1001, sum=10000, res_sum=01,
   res_a=01, res_b=00, res_sum_res=01, error=0

a=1101, b=0111, sum=10100, res_sum=10,
   res_a=01, res_b=01, res_sum_res=10, error=0

a=1101, b=0111, sum=10100, res_sum=10,
   res_a=01, res_b=01, res_sum_res=00, error=1
```

Figure 17.27 Outputs for the modulo-3 residue checker.

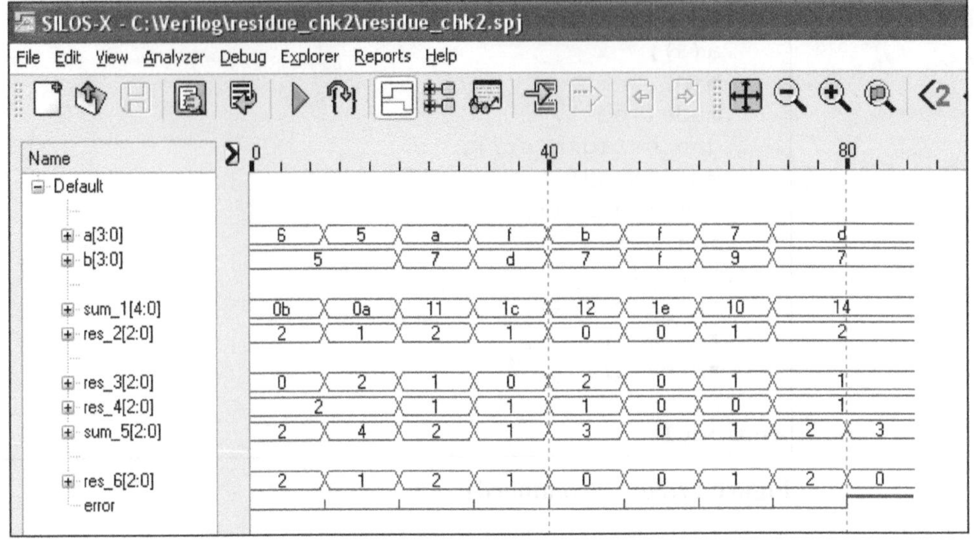

Figure 17.28 Waveforms for the modulo-3 residue checker.

17.2 Parity-Checked Shift Register

Shift registers that perform the operations of shift left logical (SLL), shift left algebraic (SLA), shift right logical (SRL), or shift right algebraic (SRA) can be checked using parity bits.

Shift left logical (SLL) The logical shift operations are much simpler to implement than the arithmetic shift operations. For SLL, the high-order bit of the unsigned operand is shifted out of the left end of the shifter for each shift cycle. Zeroes are entered from the right and fill the vacated low-order bit positions.

Shift left algebraic (SLA) SLA operates on signed operands in 2s complement representation for radix 2. The numeric part of the operand is shifted left the number of bit positions specified in the shift count field. The sign remains unchanged and does not participate in the shift operation. All remaining bits participate in the left shift. Bits are shifted out of the high-order numeric position. Zeroes are entered from the right and fill the vacated low-order bit positions.

Shift right logical (SRL) For SRL, the low-order bit of the unsigned operand is shifted out of the right end of the shifter for each shift cycle. Zeroes are entered from the left and fill the vacated high-order bit positions.

Shift right algebraic (SRA) The numeric part of the signed operand is shifted right the number of bits specified by the shift count. The sign of the operand remains unchanged. All numeric bits participate in the right shift. The sign bit propagates right to fill in the vacated high-order numeric bit positions.

Consider the 9-bit shift register shown in Figure 17.29. For an SLL shift operation, the bit shifted out is compared with the bit shifted in. The shift operation may or may not change the parity of the register that was generated after the shift-left operation. The parity of the shift register before the shift operation is contained in the *byte parity* block. The bit shifted out is contained in the *shift_reg[8]* block. This bit is compared with the bit that was shifted in. After the left shift operation, the new parity of the register is obtained and compared with the exclusive-OR in *inst3*. A final comparison is made with the exclusive-OR in *inst6* by comparing the output of the exclusive-OR in *inst5* with the original parity bit of the shift register.

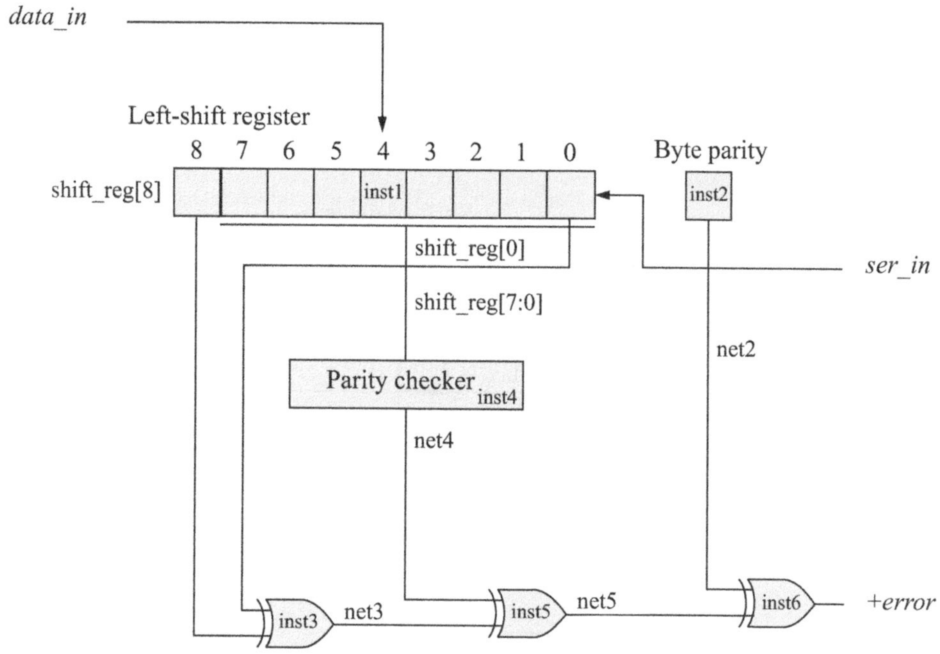

Figure 17.29 Shift register checking using parity.

Two examples are shown in Figure 17.30, one depicting no error and one depicting an error. The block labeled *parity chk* is implemented using the reduction exclusive-NOR operator, which operates as follows: If there are an odd number of 1s in the operand, then the result is 0; otherwise, the result is 1. For a vector x_1, the reduction exclusive-NOR ($\wedge\sim x_1$) is the inverse of the reduction exclusive-OR operator. The reduction exclusive-NOR operator can be used as an odd parity checker. Assume that bit position 4 has a *stuck-at-1* fault caused by a hardware malfunction, as shown in Figure 17.30(b). Then the parity of the shifted contents will be altered, causing an error to be generated.

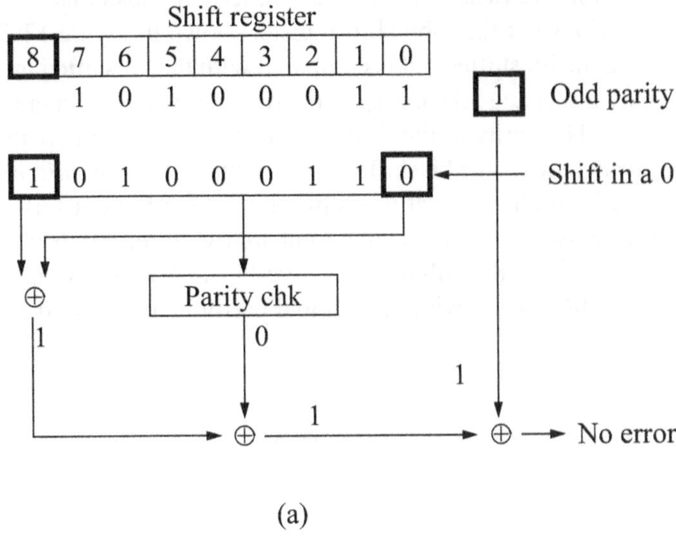

(a)

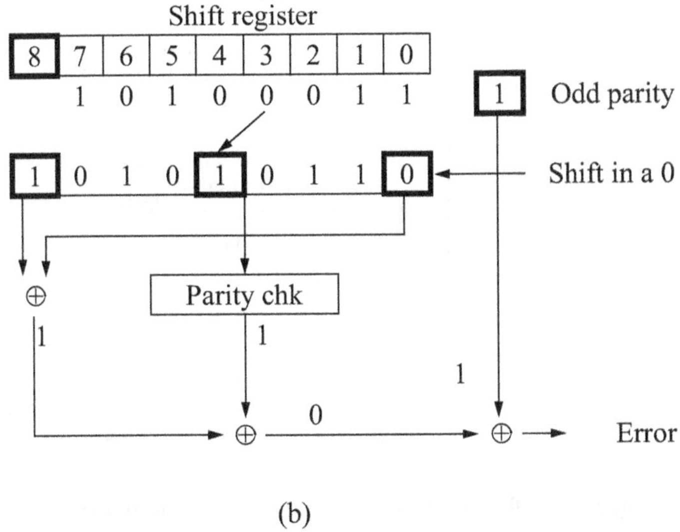

(b)

Figure 17.30 Examples of checking a shift register by means of parity.

Figure 17.31 shows the behavioral module for a parity checked shift register. The statement labeled *net4* is the exclusive-NOR operation. If the contents of the shifted register contain an odd number of 1s, then *net4* is assigned a value of 0; otherwise, it is assigned a value of 1. The test bench is shown in Figure 17.32, in which different *data-in* vectors are assigned together with applicable values for the serial input, *ser-in*, and the parity bit, *par-in*, to provide odd parity to the shift register. In the final set of inputs, the parity bit provides even parity to the shift register in order to generate an error. The outputs are shown in Figure 17.33 displaying the data input vector, the serial input value, the resulting shift register contents, and the error output. The waveforms are shown in Figure 17.34.

```
//behavioral shift register check using parity
module shift_reg_par_chk (data_in, ser_in, par_in, start,
                          shift_reg, error);

input [8:0] data_in;
input ser_in, par_in, start;
output [8:0] shift_reg;
output error;

reg [8:0] shift_reg;
reg byte_par;
reg net2, net3, net4, net5;
reg error;

always @ (start)
begin
   shift_reg = data_in;
   shift_reg = shift_reg << 1;
   byte_par = par_in;
   shift_reg[0] = ser_in;
   net2 = par_in;
   net3 = shift_reg[8] ^ shift_reg[0];
   net4 = ~(shift_reg[0] ^ shift_reg[1] ^ shift_reg[2]
         ^ shift_reg[3] ^ shift_reg[4] ^ shift_reg[5]
         ^ shift_reg[6] ^ shift_reg[7]);
   net5 = net3 ^ net4;
   error = net2 ^ net5;
end
endmodule
```

Figure 17.31 Behavioral module for the parity-checked shift register.

```
//test bench for shift register parity check
module shift_reg_par_chk_tb;

reg [8:0] data_in;
reg ser_in, par_in, start;
wire [8:0] shift_reg;
wire error;
//continued on next page
```

Figure 17.32 Test bench for the parity-checked shift register.

```
//display variables
initial
$monitor ("data_in=%b, serial_in=%b, shift_reg=%b, error=%b",
        data_in, ser_in, shift_reg, error);

//apply input vectors
initial
begin
   #0     start = 1'b0;

   #10    data_in = 9'b0_1010_0011;
          ser_in = 1'b0; par_in = 1'b1;
   #10    start = 1'b1;
   #10    start = 1'b0;

   #10    data_in = 9'b0_0111_1110;
          ser_in = 1'b1; par_in = 1'b1;
   #10    start = 1'b1;
   #10    start = 1'b0;

   #10    data_in = 9'b0_1001_1101;
          ser_in = 1'b1; par_in = 1'b0;
   #10    start = 1'b1;
   #10    start = 1'b0;

   #10    data_in = 9'b0_1111_1111;
          ser_in = 1'b1; par_in = 1'b1;
   #10    start = 1'b1;
   #10    start = 1'b0;

   #10    data_in = 9'b0_0000_0001;
          ser_in = 1'b1; par_in = 1'b0;
   #10    start = 1'b1;
   #10    start = 1'b0;

   #10    data_in = 9'b0_1010_0011;
          ser_in = 1'b0; par_in = 1'b0;
   #10    start = 1'b1;
   #10    start = 1'b0;

   #10    $stop;

end

//continued on next page
```

Figure 17.32 (Continued)

```
//instantiate the module into the test bench
shift_reg_par_chk inst1 (
   .data_in(data_in),
   .ser_in(ser_in),
   .par_in(par_in),
   .start(start),
   .shift_reg(shift_reg),
   .error(error)
   );

endmodule
```

Figure 17.32 (Continued)

```
data_in = 0_1010_0011, serial_in = 0,
   shift_reg = 1_0100_0110, error = 0


data_in = 0_0111_1110, serial_in = 1,
   shift_reg = 0_1111_1101, error = 0


data_in = 0_1001_1101, serial_in = 1,
   shift_reg = 1_0011_1011, error = 0


data_in = 0_1111_1111, serial_in = 1,
   shift_reg = 1_1111_1111, error = 0


data_in = 0_0000_0001, serial_in = 1,
   shift_reg = 0_0000_0011, error = 0


data_in = 0_1010_0011, serial_in = 0,
   shift_reg = 1_0100_0110, error = 1
```

Figure 17.33 Outputs for the parity-checked shift register.

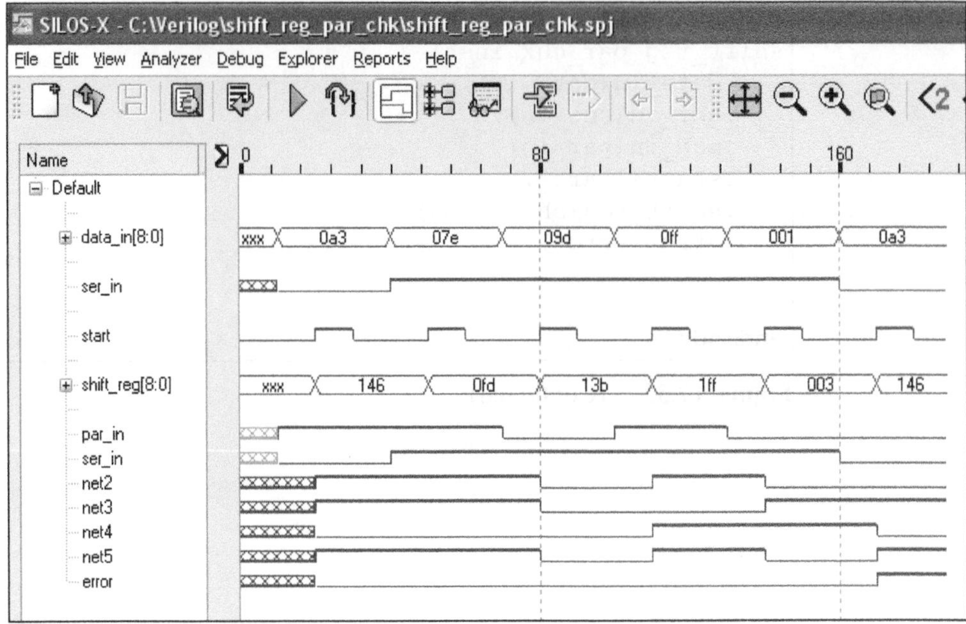

Figure 17.34 Waveforms for the parity-checked shift register.

17.3 Parity Prediction

Parity prediction can be considered a form of checksum error detection. For digital transmission applications, the *checksum* is the sum derived from the application of an algorithm that is calculated before and after transmission to ensure that the data is free from errors. The checksum character is a numerical value that is based on the number of asserted bits in the message and is appended to the end of the message. The receiving unit then applies the same algorithm to the message and compares the results with the appended checksum character.

There are many versions of checksum algorithms. The version presented in this section is used for addition and is based on the parities of the augend, the addend, and the sum. The parity of the sum does not necessarily reflect errors that are generated during addition. However, it is possible to detect addition errors by utilizing the parities of the input operands in conjunction with the parity of the carries and the parity of the sum.

This method is referred to as *parity prediction*. Parity prediction is an effective way to check for errors in addition. It usually requires less hardware than the modulo-3 residue checking method, but is slightly slower. The parity of the sum is obtained as shown in Equation 17.7 for n-bit operands, where P_{sum} is the parity of the sum, P_a is the parity of the augend, P_b is the parity of the addend, and P_{cy} is the parity of the carries that are generated during the addition operation.

Parity prediction compares the predicted parity of the sum with the actual parity of the sum, as shown in Equation 17.7, where c_{-1} is the carry-in to the adder. The actual parity of the sum is denoted by P_{sum}. The predicted parity of the sum is specified by $P_a \oplus P_b \oplus P_{cy}$. The block diagram for a parity-checked adder using parity prediction is shown in Figure 17.35.

$$P_{sum} = sum_{n-1} \oplus sum_{n-2} \oplus \ldots \oplus sum_1 \oplus sum_0$$

$$= (a_{n-1} \oplus b_{n-1} \oplus cy_{n-2}) \oplus (a_{n-2} \oplus b_{n-2} \oplus cy_{n-3})$$
$$\oplus \ldots \oplus (a_1 \oplus b_1 \oplus cy_0) \oplus (a_0 \oplus b_0 \oplus c_{-1})$$

$$= (a_{n-1} \oplus a_{n-2} \oplus \ldots \oplus a_1 \oplus a_0) \oplus (b_{n-1} \oplus b_{n-2} \oplus \ldots \oplus b_1 \oplus b_0)$$
$$\oplus (cy_{n-2} \oplus cy_{n-3} \oplus \ldots \oplus cy_0 \oplus c_{-1})$$

$$P_{sum} = P_a \oplus P_b \oplus P_{cy} \tag{17.7}$$

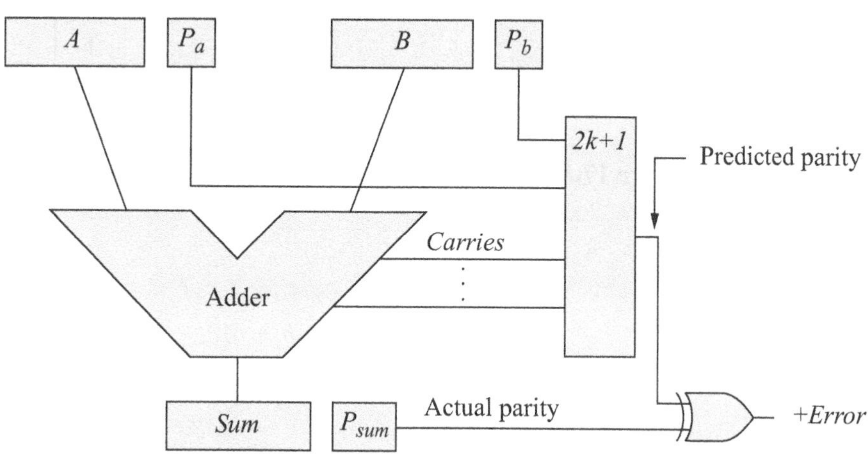

Figure 17.35 Block diagram for a parity-checked adder.

The block labeled *2k+1* is an odd parity element, whose output is a logic 1 if the number of logic 1s on the inputs is odd. The output of the *2k+1* predicted parity circuit is compared to the actual parity of the sum. An error will be undetected if there is an even number of errors in the operands. The carries are obtained from a carry-look-ahead technique using the *generate* and *propagate* functions. The truth table for stage $_i$ of a full adder is shown in Table 17.2, where the carry-in c_{i-1} to stage $_i$ is the carry-out from the previous lower-order stage$_{i-1}$. The Karnaugh map for the carry-out from stage$_i$ (c_i) is shown in Figure 17.36 and the equation for c_i is shown in Equation 17.8.

Table 17.2 Truth Table for a Full Adder

a_i	b_i	c_{i-1}	c_i	sum_i
0	0	0	0	0
0	0	1	0	1
0	1	0	0	1
0	1	1	1	0
1	0	0	0	1
1	0	1	1	0
1	1	0	1	0
1	1	1	1	1

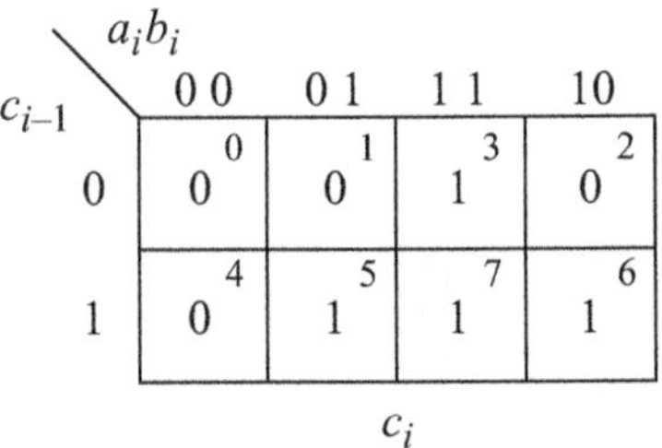

Figure 17.36 Karnaugh map for the carry-out of stage$_i$ of a full adder.

$$
\begin{aligned}
c_i &= a_ib_i + a_ic_{i-1} + b_ic_{i-1} \\
&= a_ib_i + (a_i + b_i)\,c_{i-1}
\end{aligned}
\qquad (17.8)
$$

Two auxiliary functions can now be defined: *generate* and *propagate*, as shown below.

Generate: $G_i = a_ib_i$

Propagate: $P_i = a_i + b_i$

Equation 17.8 can now be rewritten in terms of the generate and propagate functions, as shown in Equation 17.9 for adder stage$_i$. Equation 17.9 can be used recursively to obtain the carries for parity prediction, as shown in Equation 17.10, where c_{-1} is the carry-in to the low-order stage of the adder.

$$c_i = G_i + P_i c_{i-1} \tag{17.9}$$

$$\begin{aligned}
c_0 &= G_0 + P_0 c_{-1} \\
c_1 &= G_1 + P_1 c_0 \\
&= G_1 + P_1(G_0 + P_0 c_{-1}) \\
&= G_1 + P_1 G_0 + P_1 P_0 c_{-1} \\
c_2 &= G_2 + P_2 c_1 \\
&= G_2 + P_2(G_1 + P_1 G_0 + P_1 P_0 c_{-1}) \\
&= G_2 + P_2 G_1 + P_2 P_1 G_0 + P_2 P_1 P_0 c_{-1} \\
&\quad \ldots
\end{aligned} \tag{17.10}$$

Examples for parity prediction are shown below.

	$a =$	0	1	1	1		$P_a =$	0	
+)	$b =$	0	0	1	1		$P_b =$	1	$\oplus = \mathbf{1}$
	Carries =	1	1	1	0	← (c_{in})	$P_{cy} =$	0	
	Sum =	1	0	1	0		$P_{sum} =$	**1**	

	$a =$	0	1	0	1		$P_a =$	1	
+)	$b =$	0	0	1	0		$P_b =$	0	$\oplus = \mathbf{0}$
	Carries =	0	0	0	0	← (c_{in})	$P_{cy} =$	1	
	Sum =	0	1	1	1		$P_{sum} =$	**0**	

	$a =$	0	1	1	0		$P_a = 1$
+)	$b =$	0	1	1	0		$P_b = 1$ $\oplus = \mathbf{1}$
	Carries =	1	1	0	0	← (c_{in})	$P_{cy} = 1$
	Sum =	1	1	0	0		$P_{sum} = \mathbf{1}$

	$a =$	0	1	1	0		$P_a = 1$
+)	$b =$	0	1	1	0		$P_b = 1$ $\oplus = \mathbf{0}$
	Carries =	1	1	0	1	← (c_{in})	$P_{cy} = 0$
	Sum =	1	1	0	1		$P_{sum} = \mathbf{0}$

	$a =$	1	1	1	1		$P_a = 1$
+)	$b =$	1	1	1	1		$P_b = 1$ $\oplus = \mathbf{1}$
	Carries =	1	1	1	1	← (c_{in})	$P_{cy} = 1$
	Sum =	1	1	1	1		$P_{sum} = \mathbf{1}$

	$a =$	1	1	1	1		$P_a = 1$
+)	$b =$	1	1	1	1		$P_b = 1$ $\oplus = \mathbf{0}$
	Carries =	1	1	1	0	← (c_{in})	$P_{cy} = 0$
	Sum =	1	1	1	0		$P_{sum} = \mathbf{0}$

As a final example, a byte adder will be examined together with the carries and the carry-in to the adder, as shown below. Then the logic diagram will be designed to illustrate the operation of the parity prediction hardware. The logic diagram is shown in Figure 17.37. The logic blocks labeled *2k+1* are the general symbol to generate appropriate outputs if there are an odd number of logic 1s on the input.

The inputs c_{in}, c_0, c_1, c_2, c_3, c_4, c_5, c_6 represent the carries. The outputs of the logic block labeled 1 represents the parity of the carries P_{cy}, which, in conjunction with the parity of operand $a(P_a)$ and the parity of operand $b(P_b)$, denote the predicted parity. The predicted parity is then compared with the actual parity of the sum to indicate whether an error occurred. Assume that an error occurred during addition and bits c_{in} c_0 c_1 c_2 c_3 c_4 c_5 c_6 = 1 0 0 1 0 0 1 1 making the parity of the sum a logic 1. The inputs to the exclusive-OR circuit would then be 0 and 1, generating a logic 1 output indicating an error.

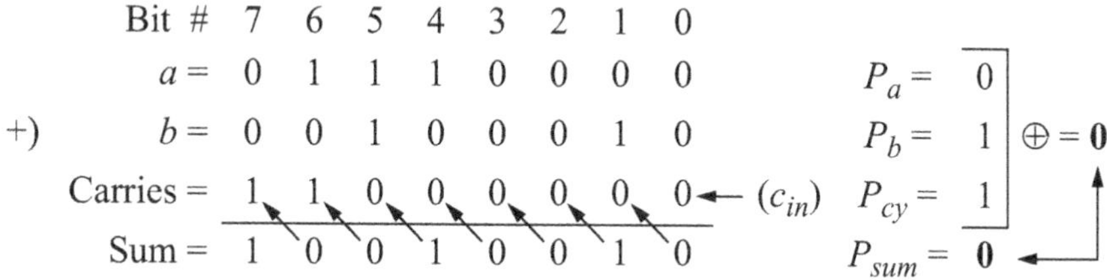

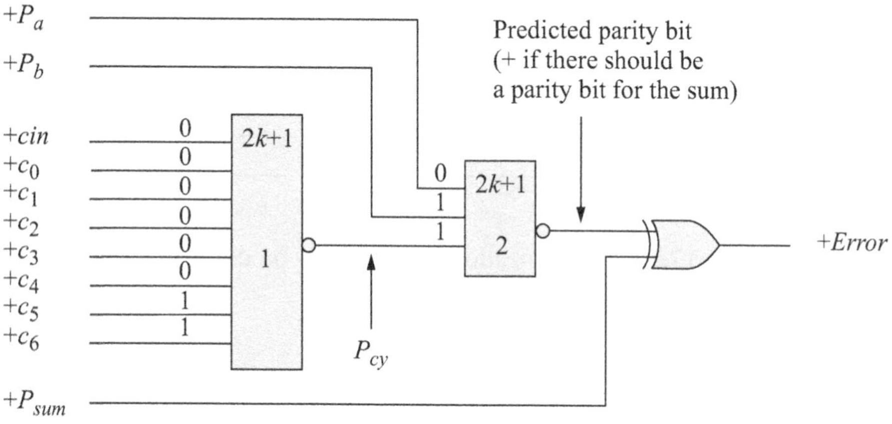

Figure 17.37 Logic diagram for parity prediction.

The organization for the parity prediction hardware is shown in Figure 17.38 and includes the instantiation names and net names. The carries are generated by the carry lookahead adder as $c_0\ c_1\ c_2\ c_3\ c_4\ c_5\ c_6$. The exclusive-OR circuit in instantiation *inst3* is used to generate an error. When the *gen_err* input is active at a high voltage level, the carry-out c_6 from the adder is inverted, which results in an error.

The adder in *inst1* is an 8-bit carry lookahead adder as shown in Figure 17.39. The dataflow module for this adder is shown in Figure 17.40 using the generate and propagate functions. The test bench is shown in Figure 17.41, which applies several input vectors for the augend and addend, together with a carry-in. In this test bench, the values of the augend, the addend, and the sum are to be displayed as decimal values, as shown in the outputs of Figure 17.42. The dataflow module for the 8-bit odd parity generator is shown in Figure 17.43.

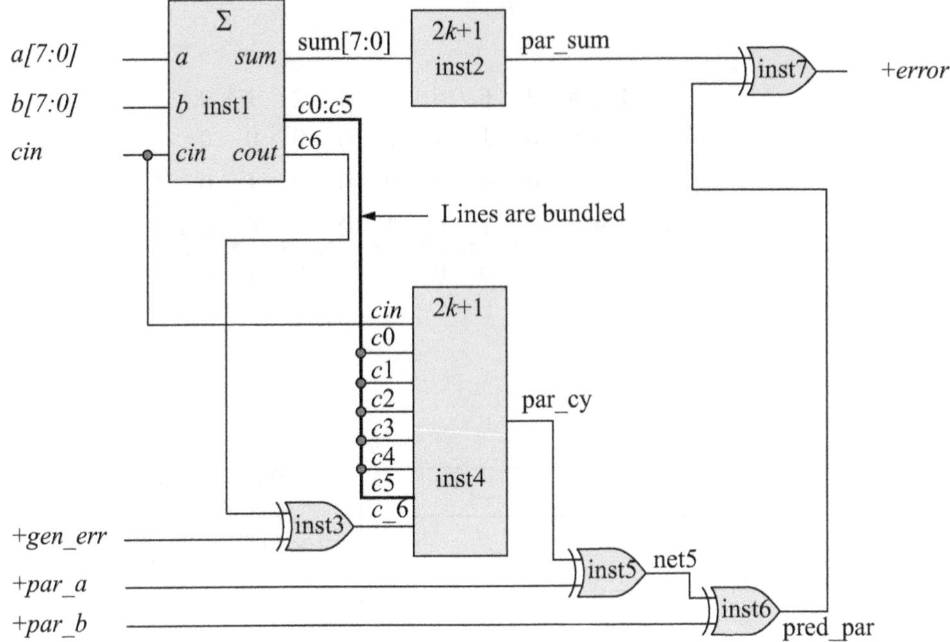

Figure 17.38 Organization for the parity prediction hardware.

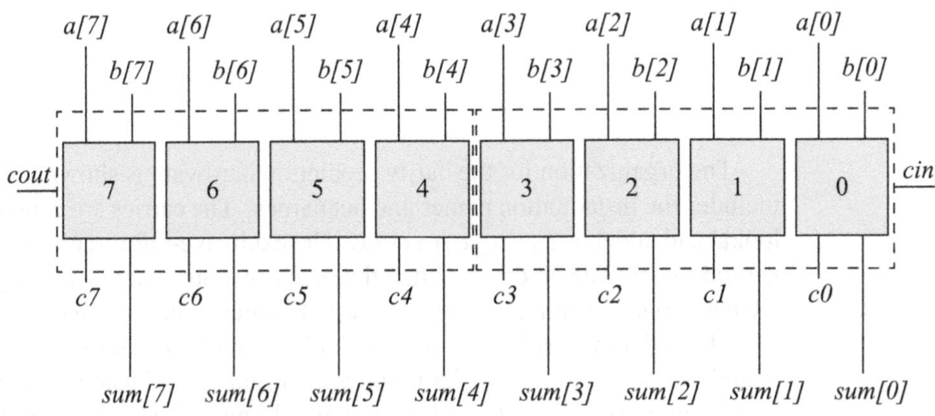

Figure 17.39 An 8-bit carry lookahead adder.

```
//dataflow 8-bit carry lookahead adder
module adder_cla8 (a, b, cin, sum, cout);

//input and output port declaration
input [7:0] a, b;
input cin;
output [7:0] sum;
output cout;

//define internal wires
wire g7, g6, g5, g4, g3, g2, g1, g0;
wire p7, p6, p5, p4, p3, p2, p1, p0;
wire c7, c6, c5, c4, c3, c2, c1, c0;

//define generate functions
assign  g7 = a[7] & b[7],//multiple statements using 1 assign
        g6 = a[6] & b[6],
        g5 = a[5] & b[5],
        g4 = a[4] & b[4],
        g3 = a[3] & b[3],
        g2 = a[2] & b[2],
        g1 = a[1] & b[1],
        g0 = a[0] & b[0];

//define propagate functions
assign  p7 = a[7] ^ b[7],//multiple statements using 1 assign
        p6 = a[6] ^ b[6],
        p5 = a[5] ^ b[5],
        p4 = a[4] ^ b[4],
        p3 = a[3] ^ b[3],
        p2 = a[2] ^ b[2],
        p1 = a[1] ^ b[1],
        p0 = a[0] ^ b[0];

//obtain the carry equations for low order
assign  c0 = g0 | (p0 & cin),
        c1 = g1 | (p1 & g0) | (p1 & p0 & cin),
        c2 = g2 | (p2 & g1) | (p2 & p1 & g0)
              | (p2 & p1 & p0 & cin),
        c3 = g3 | (p3 & g2) | (p3 & p2 & g1)
              | (p3 & p2 & p1 & g0) | (p3 & p2 & p1 & p0 & cin);

//continued on next page
```

Figure 17.40 Dataflow module for the 8-bit carry lookahead adder to be used in the parity prediction design.

```
//obtain the carry equations for high order
assign   c4 = g4 | (p4 & c3),
         c5 = g5 | (p5 & g4) | (p5 & p4 & c3),
         c6 = g6 | (p6 & g5) | (p6 & p5 & g4)
            | (p6 & p5 & p4 & c3),
         c7 = g7 | (p7 & g6) | (p7 & p6 & g5)
            | (p7 & p6 & p5 & g4) | (p7 & p6 & p5 & p4 & c3);

//obtain the sum equations
assign   sum[0] = p0 ^ cin,
         sum[1] = p1 ^ c0,
         sum[2] = p2 ^ c1,
         sum[3] = p3 ^ c2,
         sum[4] = p4 ^ c3,
         sum[5] = p5 ^ c4,
         sum[6] = p6 ^ c5,
         sum[7] = p7 ^ c6;

//obtain cout
assign   cout = c7;
endmodule
```

Figure 17.40 (Continued)

```
//test bench for dataflow 8-bit carry lookahead adder
module adder_cla8_tb;

reg [7:0] a, b;
reg cin;
wire [7:0] sum;
wire cout;

//display signals
initial
$monitor ("a = %d, b = %d, cin = %b, cout = %b, sum = %d",
            a, b, cin, cout, sum);

initial      //apply stimulus
begin
   #0    a = 8'b0000_0000;
         b = 8'b0000_0000;
         cin = 1'b0;        //cout = 0, sum = 0000_0000

//continued on next page
```

Figure 17.41 Test bench for the 8-bit carry lookahead adder to be used in the parity prediction design.

```
    #10    a = 8'b0000_0001;
           b = 8'b0000_0010;
           cin = 1'b0;          //cout = 0, sum = 0000_0011

    #10    a = 8'b0000_0010;
           b = 8'b0000_0110;
           cin = 1'b0;          //cout = 0, sum = 0000_1000

    #10    a = 8'b0000_0111;
           b = 8'b0000_0111;
           cin = 1'b0;          //cout = 0, sum = 0000_110

    #10    a = 8'b0000_1001;
           b = 8'b0000_0110;
           cin = 1'b0;          //cout = 0, sum = 0000_1111

    #10    a = 8'b0000_1100;
           b = 8'b0000_1100;
           cin = 1'b0;          //cout = 0, sum = 0001_1000

    #10    a = 8'b0000_1111;
           b = 8'b0000_1110;
           cin = 1'b0;          //cout = 0, sum = 0001_1101

    #10    a = 8'b0000_1110;
           b = 8'b0000_1110;
           cin = 1'b1;          //cout = 0, sum = 0001_1101

    #10    a = 8'b0000_1111;
           b = 8'b0000_1111;
           cin = 1'b1;          //cout = 0, sum = 0001_1111

    #10    a = 8'b1111_0000;
           b = 8'b0000_1111;
           cin = 1'b1;          //cout = 1, sum = 0000_0000

    #10    a = 8'b0111_0000;
           b = 8'b0000_1111;
           cin = 1'b1;          //cout = 0, sum = 1000_0000

    #10    a = 8'b1011_1000;
           b = 8'b0100_1111;
           cin = 1'b1;          //cout = 1, sum = 0000_1000

    #10    $stop;
end
//continued on next page
```

Figure 17.41 (Continued)

```
//instantiate the module into the test bench
adder_cla8 inst1 (
   .a(a),
   .b(b),
   .cin(cin),
   .sum(sum),
   .cout(cout)
   );

endmodule
```

Figure 17.41 (Continued)

```
a = 0,     b = 0,    cin = 0,   cout = 0,   sum = 0
a = 1,     b = 2,    cin = 0,   cout = 0,   sum = 3
a = 2,     b = 6,    cin = 0,   cout = 0,   sum = 8
a = 7,     b = 7,    cin = 0,   cout = 0,   sum = 14
a = 9,     b = 6,    cin = 0,   cout = 0,   sum = 15
a = 12,    b = 12,   cin = 0,   cout = 0,   sum = 24
a = 15,    b = 14,   cin = 0,   cout = 0,   sum = 29
a = 14,    b = 14,   cin = 1,   cout = 0,   sum = 29
a = 15,    b = 15,   cin = 1,   cout = 0,   sum = 31
a = 240,   b = 15,   cin = 1,   cout = 1,   sum = 0
a = 112,   b = 15,   cin = 1,   cout = 0,   sum = 128
a = 184,   b = 79,   cin = 1,   cout = 1,   sum = 8
```

Figure 17.42 Outputs for the 8-bit carry lookahead adder to be used in the parity prediction design.

```
//8-bit odd parity generator using dataflow modeling
module par_gen8_df (x, par_bit);

input [7:0] x;
output par_bit;

//use continuous assignment
assign par_bit = ~{(x[0] ^ x[1]) ^ (x[2]^x[3])
                  ^ (x[4] ^ x[5]) ^ (x[6]^x[7])};

endmodule
```

Figure 17.43 Dataflow module for the 8-bit parity generator to be used in the parity prediction design.

The structural module for the parity prediction hardware is shown in Figure 17.44, which instantiates the modules for the logic blocks depicted in Figure 17.38 and presented in the design modules of Figures 17.40 (8-bit carry lookahead adder) and Figure 17.43 (8-bit parity generator) plus the four exclusive-OR functions.

The test bench is shown in Figure 17.45, which applies input vectors for the augend and addend plus appropriate values for the scalar inputs. The second and fifth set of vectors cause an error to be generated by asserting the *gen_err* input. The outputs are shown in Figure 17.46 and the waveforms are shown in Figure 17.47 that display the input values of the operands plus the values of the internal nets.

```
//structural parity prediction
module parity_predict (a, b, par_a, par_b, cin,
                        gen_err, error);

input [7:0] a, b;
input par_a, par_b, cin, gen_err;
output error;

//define internal nets
wire [7:0] sum;
wire c0, c1, c2, c3, c4, c5, c6, c_6;
wire par_sum, par_cy, net5, pred_par;

//instantiate the 8-bit carry-lookahead adder
adder8_cla inst1 (
   .a(a),
   .b(b),
   .cin(cin),
   .sum(sum),
   .c0(c0),
   .c1(c1),
   .c2(c2),
   .c3(c3),
   .c4(c4),
   .c5(c5),
   .c6(c6),
   .c7(c7)
   );

//instantiate the 8-bit parity generator
//to obtain the actual parity of the sum
par_gen8_df inst2 (
   .x(sum),
   .par_bit(par_sum)
   );
//continued on next page
```

Figure 17.44 Structural module for the parity prediction implementation.

```
//instantiate the exclusive-or circuit
//to generate an error
xor2_df inst3 (
   .x1(c6),
   .x2(gen_err),
   .z1(c_6)
   );

//instantiate the 8-bit parity generator
//to obtain the parity of the carries
par_gen8_df inst4 (
   .x({cin, c0, c1, c2, c3, c4, c5, c_6}),
   .par_bit(par_cy)
   );

//instantiate the logic to obtain the predicted parity
xor2_df inst5 (
   .x1(par_cy),
   .x2(par_a),
   .z1(net5)
   );

xor2_df inst6 (
   .x1(net5),
   .x2(par_b),
   .z1(pred_par)
   );

//instantiate the logic to obtain the error output
xor2_df inst7 (
   .x1(par_sum),
   .x2(pred_par),
   .z1(error)
   );
endmodule
```

Figure 17.44 (Continued)

```
//test bench for parity prediction
module parity_predict_tb;

reg [7:0] a, b;
reg par_a, par_b, cin, gen_err;
wire error;
//continued on next page
```

Figure 17.45 Test bench for the parity prediction implementation.

```
//display variables
initial
$monitor ("a=%b, b=%b, cin=%b, par_a=%b, par_b=%b,
            gen_err=%b, error=%b",
         a, b, cin, par_a, par_b, gen_err, error);

//apply input vectors
initial
begin
   #0    a = 8'b0001_1101;    b = 8'b0010_0110;
         par_a = 1'b1;  par_b = 1'b0;  cin = 1'b0;
         gen_err = 1'b0;

   #10   a = 8'b0001_1101;    b = 8'b0010_0110;
         par_a = 1'b1;  par_b = 1'b0;  cin = 1'b0;
         gen_err = 1'b1;

   #10   a = 8'b0111_1110;    b = 8'b0001_0001;
         par_a = 1'b1;  par_b = 1'b1;  cin = 1'b0;
         gen_err = 1'b0;

   #10   a = 8'b1000_1001;    b = 8'b0110_0110;
         par_a = 1'b0;  par_b = 1'b1;  cin = 1'b1;
         gen_err = 1'b0;

   #10   a = 8'b0011_1110;    b = 8'b0010_1101;
         par_a = 1'b0;  par_b = 1'b1;  cin = 1'b0;
         gen_err = 1'b1;

   #10   $stop;

end

//instantiate the module into the test bench
parity_predict inst1 (
   .a(a),
   .b(b),
   .par_a(par_a),
   .par_b(par_b),
   .cin(cin),
   .gen_err(gen_err),
   .error(error)
   );

endmodule
```

Figure 17.45 (Continued)

```
a = 0001_1101,     b = 0010_0110,     cin = 0,
   par_a = 1, par_b = 0, gen_err = 0, error = 0

a = 0001_1101,     b = 0010_0110,     cin = 0,
   par_a = 1, par_b = 0, gen_err = 1, error = 1

a = 0111_1110,     b = 0001_0001,     cin = 0,
   par_a = 1, par_b = 1, gen_err = 0, error = 0

a = 1000_1001,     b = 0110_0110,     cin = 1,
   par_a = 0, par_b = 1, gen_err = 0, error=0

a = 0011_1110,     b = 0010_1101,     cin = 0,
   par_a = 0, par_b = 1, gen_err = 1, error = 1
```

Figure 17.46 Outputs for the parity prediction implementation.

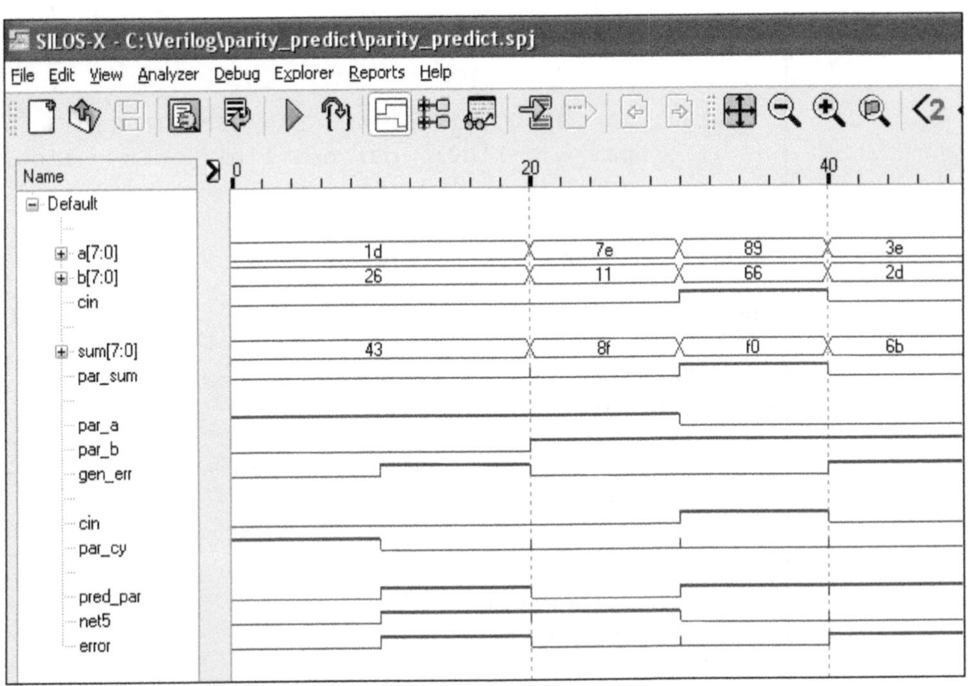

Figure 17.47 Waveforms for the parity prediction implementation.

17.4 Condition Codes for Addition

For high-speed addition, it is desirable to have the following condition codes generated in parallel with the add operation: sum = 0; sum < 0; sum > 0; and overflow. This negates the necessity of waiting to obtain the condition codes until after the sum is available.

The symbol for a full adder for stage$_i$ is shown in Figure 17.48, and the truth table for the sum s_i is shown in Table 17.3, indicating that s_i has a value of 0 in four rows. The Karnaugh map for the sum s_i is shown in Figure 17.49, and the equation for s_i is shown in Equation 17.11.

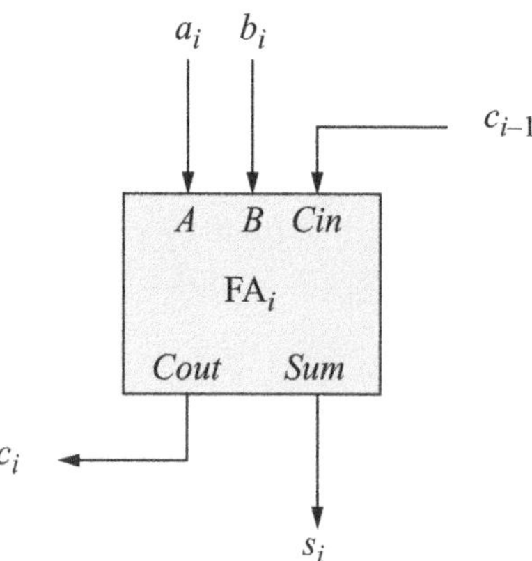

Figure 17.48 Symbol for a full adder.

Table 17.3 Truth Table for the Sum and Carry of a Full Adder

a_i	b_i	c_{i-1}	c_i	s_i
0	0	0	0	**0**
0	0	1	0	1
0	1	0	0	1
0	1	1	1	**0**
1	0	0	0	1
1	0	1	1	**0**
1	1	0	1	**0**
1	1	1	1	1

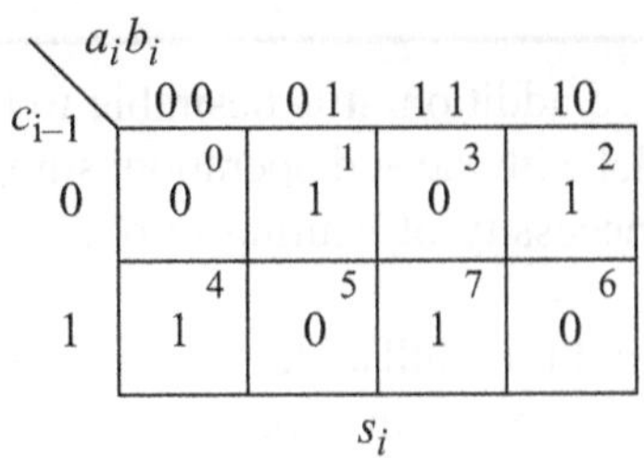

Figure 17.49 Karnaugh map for the sum of a full adder.

$$
\begin{aligned}
s_i &= a_i' b_i c_{i-1}' + a_i b_i' c_{i-1}' + a_i' b_i' c_{i-1} + a_i b_i c_{i-1} \\
&= c_{i-1}'(a_i \oplus b_i) + c_{i-1}(a_i \oplus b_i)' \\
&= (a_i \oplus b_i) \oplus c_{i-1} \qquad (+ \text{ if } s_i \text{ is } 1; - \text{ if } s_i \text{ is } 0) \qquad (17.11)
\end{aligned}
$$

Assume that the operands consist of 4-bytes (bits 0 through 31, where bit 0 is the low-order bit), using the format shown in Figure 17.50 and that the addition operation is an algebraic operation; that is, the operands are signed numbers in 2s complement representation. The operands consist of two parts: the numeric part (bits 0 through 30) and the sign part (bit 31). The logic diagram for the sum s_i together with the logic to indicate that a byte contains all zeroes is shown in Figure 17.51. Two exclusive-OR circuits are required for each bit and 16 exclusive-OR circuits for each byte.

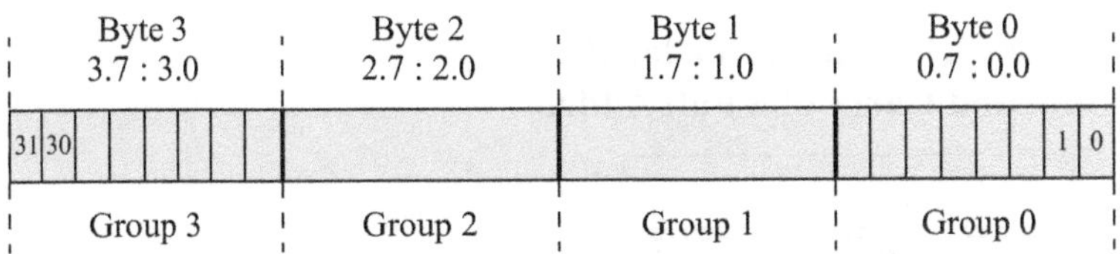

Figure 17.50 Format for the augend and the addend.

The logic to determine if the numeric part is zero or nonzero, the sum is zero or nonzero, and the sign is positive or negative is shown in Figure 17.52. The logic to determine the four condition codes is shown in Figure 17.53. Using this approach, the condition codes for the sum can be obtained independently and in parallel with the addition operation, by utilizing a carry lookahead adder.

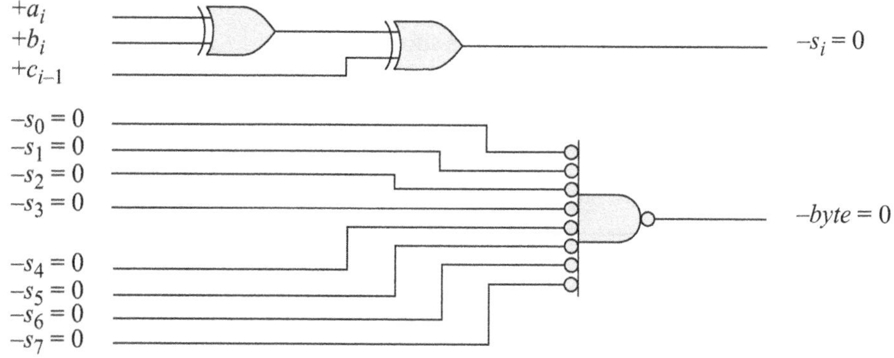

Figure 17.51 Logic diagram to indicate that the sum s_i is zero and that a byte is zero.

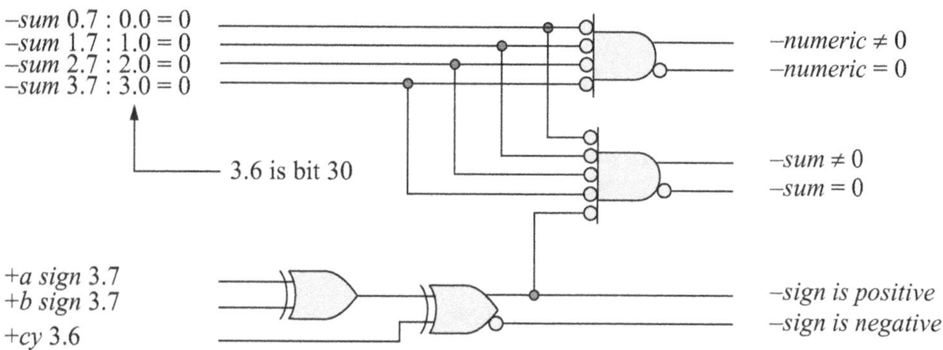

Figure 17.52 Logic diagram to determine whether the numeric part is zero or nonzero, the sum is zero or nonzero, and the sign is positive or negative.

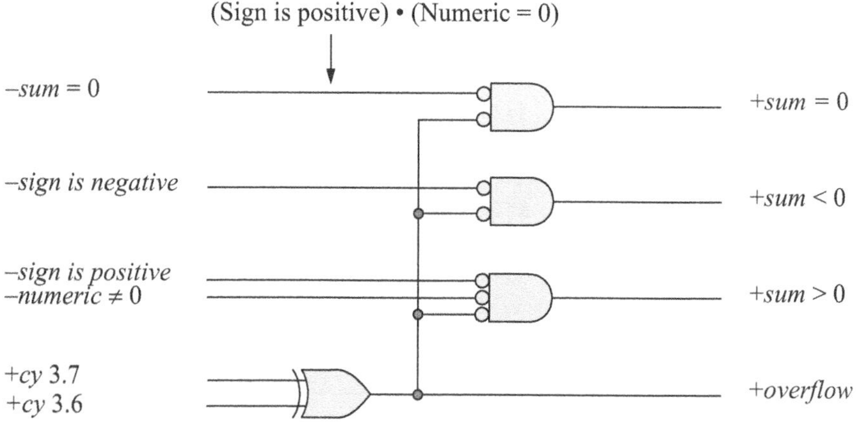

Figure 17.53 Logic diagram to generate the four condition codes.

A Verilog HDL design will now be presented that obtains the condition codes for single-byte operands using an 8-bit carry lookahead adder. The dataflow module for the carry lookahead adder is shown in Figure 17.54, which will be instantiated into the design module.

```
//dataflow 8-bit carry-lookahead adder
module adder8_cla (a, b, cin, sum, c0, c1, c2, c3,
                    c4, c5, c6, c7, cout);

//declare input and output ports
input [7:0] a, b;
input cin;
output [7:0] sum;
output c0, c1, c2, c3, c4, c5, c6, c7;
output cout;

//define internal wires
wire g7, g6, g5, g4, g3, g2, g1, g0;
wire p7, p6, p5, p4, p3, p2, p1, p0;
wire c7, c6, c5, c4, c3, c2, c1, c0;

//define generate functions using one assign
assign   g7 = a[7] & b[7],
         g6 = a[6] & b[6],
         g5 = a[5] & b[5],
         g4 = a[4] & b[4],
         g3 = a[3] & b[3],
         g2 = a[2] & b[2],
         g1 = a[1] & b[1],
         g0 = a[0] & b[0];

//define propagate functions using one assign
assign   p7 = a[7] ^ b[7],
         p6 = a[6] ^ b[6],
         p5 = a[5] ^ b[5],
         p4 = a[4] ^ b[4],
         p3 = a[3] ^ b[3],
         p2 = a[2] ^ b[2],
         p1 = a[1] ^ b[1],
         p0 = a[0] ^ b[0];


//continued on next page
```

Figure 17.54 Carry lookahead adder to be used for condition code generation.

```
//obtain the carry equations for low order
assign   c0 = g0 | (p0 & cin),
         c1 = g1 | (p1 & g0) | (p1 & p0 & cin),
         c2 = g2 | (p2 & g1) | (p2 & p1 & g0) |
               (p2 & p1 & p0 & cin),
         c3 = g3 | (p3 & g2) | (p3 & p2 & g1) |
               (p3 & p2 & p1 & g0) | (p3 & p2 & p1 & p0 & cin);

//obtain the carry equations for high order
assign   c4 = g4 | (p4 & c3),
         c5 = g5 | (p5 & g4) | (p5 & p4 & c3),
         c6 = g6 | (p6 & g5) | (p6 & p5 & g4) |
               (p6 & p5 & p4 & c3),
         c7 = g7 | (p7 & g6) | (p7 & p6 & g5) |
               (p7 & p6 & p5 & g4) | (p7 & p6 & p5 & p4 & c3);

//obtain the sum equations
assign   sum[0] = p0 ^ cin,
         sum[1] = p1 ^ c0,
         sum[2] = p2 ^ c1,
         sum[3] = p3 ^ c2,
         sum[4] = p4 ^ c3,
         sum[5] = p5 ^ c4,
         sum[6] = p6 ^ c5,
         sum[7] = p7 ^ c6;

//obtain cout
assign   cout = c7;

endmodule
```

Figure 17.54 (Continued)

The logical organization to obtain the condition codes for a one-byte augend and a one-byte addend is shown in Figure 17.55. The block labeled *detect num = 0* determines whether the numeric part — bit 0 through bit 6 — is equal to zero. This will then be used to determine if the sum is zero or greater than zero. Outputs *+sum_lt0*, *+sum_eq0*, and *+sum_gt0* are active at a high voltage level when the sum < 0, sum $= 0$, and sum > 0, respectively.

The concepts discussed previously in this section are combined to obtain the condition codes for a one-byte augend and a one-byte addend. This implementation can be easily expanded to accommodate any size operands. The dataflow module to determine the condition codes for single-byte operands is shown in Figure 17.56. The test bench is shown in Figure 17.57 and contains inputs to generate all four condition codes. The outputs and waveforms are shown in Figure 17.58 and Figure 17.59, respectively.

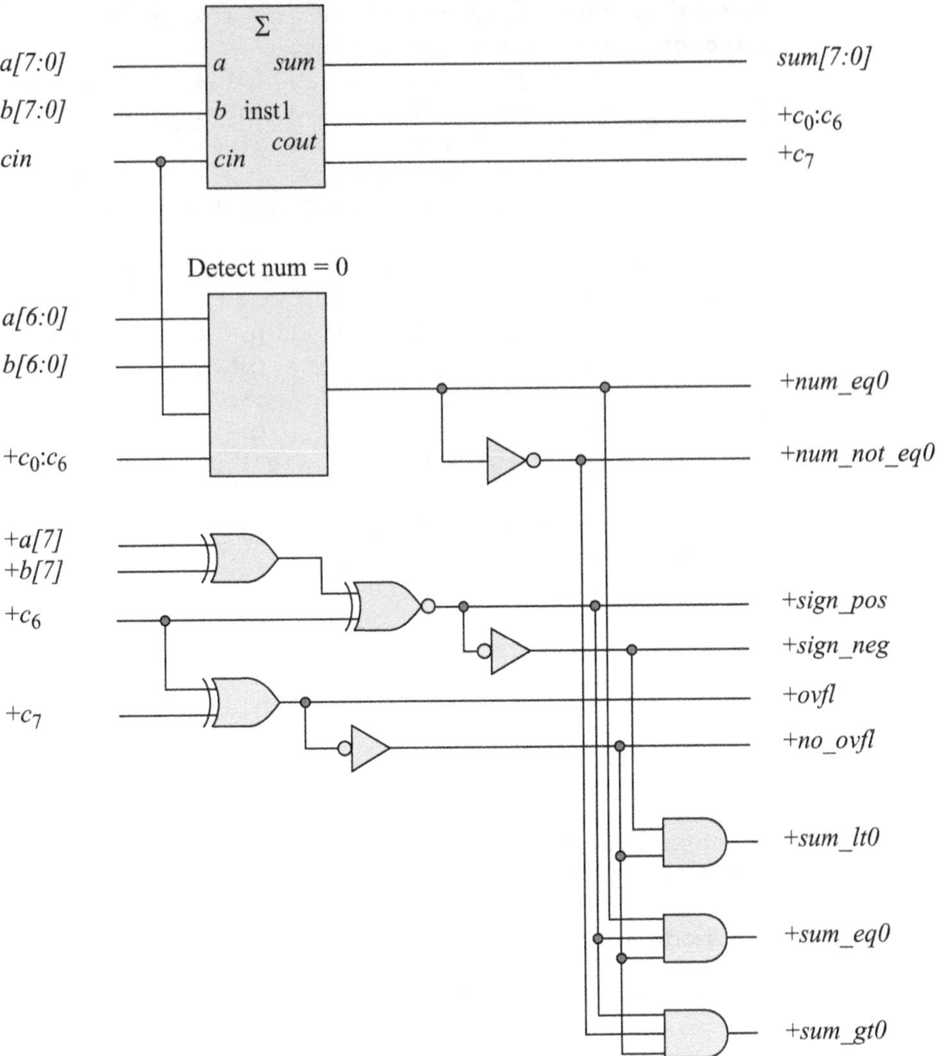

Figure 17.55 Logical organization to generate condition codes of sum < 0, sum = 0, sum > 0, and overflow.

```
//dataflow for condition codes
module cond_codes (a, b, cin, sum, sum_lt0, sum_eq0, sum_gt0,
                        ovfl);
input [7:0] a, b;
input cin;
output [7:0] sum;
output sum_lt0, sum_eq0, sum_gt0, ovfl;   //next page
```

Figure 17.56 Dataflow module for condition code generation for addition.

```
//define internal nets
wire [7:0] sum;
wire c0, c1, c2, c3, c4, c5, c6, c7;
wire s0_eq0, s1_eq0, s2_eq0, s3_eq0,
        s4_eq0, s5_eq0, s6_eq0, s7_eq0;

//instantiate the 8-bit carry lookahead adder
adder8_cla inst1 (
   .a(a[7:0]),
   .b(b[7:0]),
   .cin(cin),
   .sum(sum[7:0]),
   .c0(c0),
   .c1(c1),
   .c2(c2),
   .c3(c3),
   .c4(c4),
   .c5(c5),
   .c6(c6),
   .c7(c7)
   );

//detect numeric equal to zero
assign  s0_eq0 = ((a[0] ^ b[0]) ~^ cin),
        s1_eq0 = ((a[1] ^ b[1]) ~^ c0),
        s2_eq0 = ((a[2] ^ b[2]) ~^ c1),
        s3_eq0 = ((a[3] ^ b[3]) ~^ c2),
        s4_eq0 = ((a[4] ^ b[4]) ~^ c3),
        s5_eq0 = ((a[5] ^ b[5]) ~^ c4),
        s6_eq0 = ((a[6] ^ b[6]) ~^ c5);

assign  num_eq0 = s0_eq0 & s1_eq0 & s2_eq0 & s3_eq0 &
                  s4_eq0 & s5_eq0 & s6_eq0;

//determine sign (positive or negative)
assign  sign_pos = (a[7] ^ b[7]) ~^ c6;

//detect overflow
assign  ovfl = (c6 ^ c7);

//detect sum < 0, sum = 0, sum > 0
assign  sum_lt0 = (~sign_pos) & (~ovfl),
        sum_eq0 = (num_eq0) & (sign_pos) & (~ovfl),
        sum_gt0 = (sign_pos) & (~num_eq0) & (~ovfl);

endmodule
```

Figure 17.56 (Continued)

```
//test bench for condition codes
module cond_codes_tb;


reg [7:0] a, b;
reg cin;
wire [7:0] sum;
wire sum_lt0, sum_eq0, sum_gt0, ovfl;


//display variables
initial
$monitor ("a=%b, b=%b, cin=%b, sum=%b, sum_lt0=%b,
            sum_eq0=%b, sum_gt0=%b, ovfl=%b",
            a, b, cin, sum, sum_lt0, sum_eq0, sum_gt0, ovfl);


//apply stimulus
initial
begin
   #0    a=8'b0000_0000;    b=8'b0000_0000;    cin=1'b0;
   #10   a=8'b0000_0001;    b=8'b0000_0001;    cin=1'b0;
   #10   a=8'b1000_0010;    b=8'b0000_0001;    cin=1'b0;
   #10   a=8'b1100_0011;    b=8'b1001_1100;    cin=1'b0;

   #10   a=8'b0000_0000;    b=8'b0000_0000;    cin=1'b1;
   #10   a=8'b0000_0001;    b=8'b0000_0001;    cin=1'b1;
   #10   a=8'b1000_0010;    b=8'b0000_0001;    cin=1'b1;
   #10   a=8'b1100_0011;    b=8'b1001_1100;    cin=1'b1;

   #10   a=8'b0000_0111;    b=8'b1111_1001;    cin=1'b0;

   #10   $stop;
end


//instantiate the module into the test bench
cond_codes inst1 (
   .a(a),
   .b(b),
   .cin(cin),
   .sum(sum),
   .sum_lt0(sum_lt0),
   .sum_eq0(sum_eq0),
   .sum_gt0(sum_gt0),
   .ovfl(ovfl)
   );

endmodule
```

Figure 17.57 Test bench for the generation of condition codes for addition.

```
a=0000_0000, b=0000_0000, cin=0, sum=0000_0000,
   sum_lt0=0, sum_eq0=1, sum_gt0=0, ovfl=0

a=0000_0001, b=0000_0001, cin=0, sum=0000_0010,
   sum_lt0=0, sum_eq0=0, sum_gt0=1, ovfl=0

a=1000_0010, b=0000_0001, cin=0, sum=1000_0011,
   sum_lt0=1, sum_eq0=0, sum_gt0=0, ovfl=0

a=1100_0011, b=1001_1100, cin=0, sum=0101_1111,
   sum_lt0=0, sum_eq0=0, sum_gt0=0, ovfl=1

a=0000_0000, b=0000_0000, cin=1, sum=0000_0001,
   sum_lt0=0, sum_eq0=0, sum_gt0=1, ovfl=0

a=0000_0001, b=0000_0001, cin=1, sum=0000_0011,
   sum_lt0=0, sum_eq0=0, sum_gt0=1, ovfl=0

a=1000_0010, b=0000_0001, cin=1, sum=1000_0100,
   sum_lt0=1, sum_eq0=0, sum_gt0=0, ovfl=0

a=1100_0011, b=1001_1100, cin=1, sum=0110_0000,
   sum_lt0=0, sum_eq0=0, sum_gt0=0, ovfl=1

a=00000111, b=11111001, cin=0, sum=00000000,
   sum_lt0=0, sum_eq0=1, sum_gt0=0, ovfl=0
```

Figure 17.58 Outputs for condition code generation for addition.

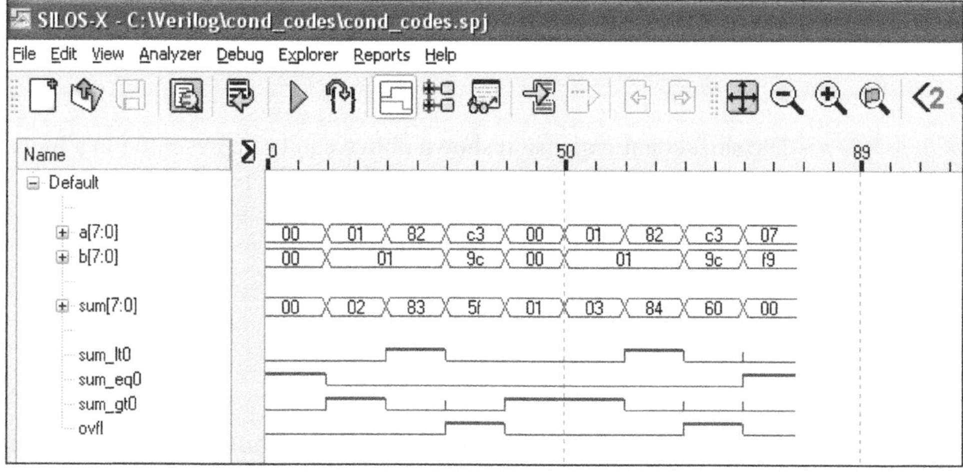

Figure 17.59 Waveforms for condition code generation for addition.

17.5 Logical and Algebraic Shifters

Since several arithmetic operations utilize a shifter, this section describes a shifter for both logical (unsigned operand) shifts and arithmetic (signed operand) shifts. Arithmetic algorithms, such as the sequential add-shift multiplication method, the sequential shift-subtract/add restoring division method, and floating-point operations all use some form of shift register.

As described in Section 17.2, there are four basic shift operations: *shift left logical* (SLL) and *shift left algebraic* (SLA) for unsigned and signed operands, respectively; *shift right logical* (SRL) and *shift right algebraic* (SRA) for unsigned and signed operands, respectively.

Shift count Assume a 6-bit shift count field for radix 2 as follows:

$$C = c_5 c_4 c_3 c_2 c_1 c_0 \tag{17.12}$$

where C is the shift count and c_i is a shift bit, either 0 or 1, with a binary weight of 2^i. For example, let the shift count be

$$\begin{array}{rcccccc} & 2^5 & 2^4 & 2^3 & 2^2 & 2^1 & 2^0 \\ C = & 0 & 0 & 1 & 1 & 0 & 0_2 \end{array}$$

The shift count can be expressed as

$$\begin{aligned} C = & \; c_5{\times}2^5 + c_4{\times}2^4 + c_3{\times}2^3 + c_2{\times}2^2 + c_1{\times}2^1 + c_0{\times}2^0 \\ = & \; 0{\times}2^5 + 0{\times}2^4 + 1{\times}2^3 + 1{\times}2^2 + 0{\times}2^1 + 0{\times}2^0 \\ = & \; 8 + 4 \\ = & \; 12 \end{aligned}$$

The shift count expansion shown above can be represented in a more compact mathematical form as a summation of terms as shown in Equation 17.13.

$$C = \sum_{i=0}^{n-1} c_i 2^i \tag{17.13}$$

where $c_i = 0$ or 1, and $2^n - 1$ is the maximum shift count. If $n = 6$, then the maximum shift count is $2^6 - 1 = 63$.

17.5.1 Behavioral Modeling

A shifter that can shift unsigned operands for *shift left logical* (SLL) and *shift right logical* (SRL), and can shift signed operands for *shift left algebraic* (SLA) and *shift right algebraic* (SRA), can be easily implemented in Verilog HDL using behavioral modeling. Constants, such as the code that defines the type of shifting, can be defined in a module by means of the keyword **parameter**. Parameter statements assign values to constants, which cannot be changed during simulation; however, the value of a constant can be changed during compilation. Examples of parameters are shown in Table 17.4.

Table 17.4 Examples of Parameters

Examples	Comments
parameter width = 8	Defines a bus width of 8 bits
parameter width = 16, depth = 512	Defines a memory with two bytes per word and 512 words
parameter out_port = 8	Defines an output port with an address of 8

When there are many paths from which to choose, nested **if-else if** statements can be cumbersome. Recall that the **case** statement is an alternative to the **if**. . . **else if** construct and may simplify the readability of the Verilog HDL code. The **case** statement is a multiple-way conditional branch. It executes one of several different procedural statements depending on the comparison of an expression with a case item. The expression and the case item are compared bit-by-bit and must match exactly. The statement that is associated with a case item may be a single procedural statement or a block of statements delimited by the keywords **begin** . . . **end**. The keywords **case**, **endcase**, and **default** are used in a **case** statement. The **case** statement has the following syntax:

```
case (expression)
    case_item1 : procedural_statement1;
    case_item2 : procedural_statement2;
    case_item3 : procedural_statement3;
                .
                .
                .
    case_itemn : procedural_statementn;
    default : default_statement;
endcase
```

The case_expression is evaluated, then compared to each alternative case_item in the order listed: case_item1 through case_item*n*. The **case** statement compares the case_expression (case_item) with each alternative on a bit-by-bit basis. The first alternative that matches the case_expression will cause the corresponding procedural statement of that alternative to be executed. If none of the alternatives match, then the **default** statement is executed.

The type of shifts are defined by keyword **parameter** as follows:

parameter	sll	2'b00,	(shift left logical)
	sla	2'b01,	(shift left algebraic)
	srl	2'b10,	(shift right logical)
	sra	2'b11;	(shift right algebraic)

The behavioral module is shown in Figure 17.60. For shift left logical, the high-order bit of the unsigned operand is shifted out of the left end of the shifter for each shift cycle. Zeroes are entered from the right and fill the vacated low-order bit positions.

For shift left algebraic, the numeric part of the operand is shifted left the number of bit positions specified by the shift amount. The sign remains unchanged and does not participate in the shift operation. The bits are shifted out of the high-order numeric bit position and 0s are shifted into the vacated register positions on the right. If a bit shifted out is different than the sign bit, then an overflow has occured.

For shift right logical, the low-order bit of the unsigned operand is shifted out of the right end of the shifter for each shift cycle. Zeroes are entered from the left and fill the vacated high-order bit positions.

For shift right algebraic, the numeric part of the signed operand is shifted right the number of bits specified by the shift amount. The sign of the operand remains unchanged. All numeric bits participate in the right shift. The sign bit propagates right to fill in the vacated high-order numeric bit positions.

```
//behavioral logical and algebraic shifter
module comb_shifter (a, shft_code, shft_amt, shft_rslt);

input [7:0] a;
input [1:0] shft_code;
input [3:0] shft_amt;
output [7:0] shft_rslt;

wire [7:0] a;
wire [3:0] shft_amt;

//variables used in always are declared as registers
reg [7:0] reg_a;
reg [7:0] shft_rslt;
reg [15:0] sra_reg;      //continued on next page
```

Figure 17.60 Behavioral module for the logical and algebraic shifter.

```
//define shift codes
parameter   sll = 2'b00,
            sla = 2'b01,
            srl = 2'b10,
            sra = 2'b11;


//perform the shift operations
always @ (a or shft_code)
begin
   case (shft_code)
      sll:
         begin
            reg_a = a << shft_amt;
            shft_rslt = reg_a;
         end


      sla:
         begin
            reg_a = a;
            reg_a = reg_a << shft_amt;
            reg_a[7] = a[7];
            shft_rslt = reg_a;
         end


      srl:
         begin
            reg_a = a >> shft_amt;
            shft_rslt = reg_a;
         end


      sra:
         begin
            sra_reg[15:8] = {8{a[7]}};
            sra_reg[7:0] = a;
            sra_reg = sra_reg >> shft_amt;
            shft_rslt = sra_reg[7:0];
         end

   endcase
end

endmodule
```

Figure 17.60 (Continued)

The test bench is shown in Figure 17.61, which applies twelve input vectors, shift code, and shift amount for the operand labeled *a*. The outputs and waveforms are shown in Figure 17.62 and Figure 17.63, respectively.

```
//test bench for logical and algebraic shifter
module comb_shifter_tb;

reg [7:0] a;
reg [1:0] shft_code;
reg [3:0] shft_amt;
wire [7:0] shft_rslt;

//display variables
initial
$monitor ("a=%b, shft_code=%b, shft_amt=%b, shft_rslt=%b",
            a, shft_code, shft_amt, shft_rslt);

//apply input vectors
initial
begin
         //shift left logical
   #0    a = 8'b0000_1111;
         shft_code = 2'b00;   shft_amt = 4'b0010;

         //shift left algebraic
   #10   a = 8'b1000_1111;
         shft_code = 2'b01;   shft_amt = 4'b0010;

         //shift right logical
   #10   a = 8'b0000_1111;
         shft_code = 2'b10;   shft_amt = 4'b0010;

         //shift right algebraic
   #10   a = 8'b1000_1111;
         shft_code = 2'b11;   shft_amt = 4'b0010;

         //shift left logical
   #10   a = 8'b1111_1111;
         shft_code = 2'b00;   shft_amt = 4'b0100;

         //shift left algebraic
   #10   a = 8'b1111_1111;
         shft_code = 2'b01;   shft_amt = 4'b0100;

//continued on next page
```

Figure 17.61 Test bench for the logical and algebraic shifter.

```
        //shift right logical
   #10  a = 8'b1111_1111;
        shft_code = 2'b10;   shft_amt = 4'b0100;

        //shift right algebraic
   #10  a = 8'b1111_1111;
        shft_code = 2'b11;   shft_amt = 4'b0100;

        //shift left logical
   #10  a = 8'b1100_0011;
        shft_code = 2'b00;   shft_amt = 4'b0101;

        //shift left algebraic
   #10  a = 8'b1100_0011;
        shft_code = 2'b01;   shft_amt = 4'b0101;

        //shift right logical
   #10  a = 8'b1100_0011;
        shft_code = 2'b10;   shft_amt = 4'b0101;

        //shift right algebraic
   #10  a = 8'b0111_1111;
        shft_code = 2'b11;   shft_amt = 4'b0101;

   #10  $stop;
end

comb_shifter inst1 (        //instantiate the module
   .a(a),
   .shft_code(shft_code),
   .shft_amt(shft_amt),
   .shft_rslt(shft_rslt)
   );
endmodule
```

Figure 17.61 (Continued)

```
sll = 00, sla = 01, srl = 10, sra = 11
a=0000_1111, shft_code=00, shft_amt=0010, shft_rslt=0011_1100

a=1000_1111, shft_code=01, shft_amt=0010, shft_rslt=1011_1100

a=0000_1111, shft_code=10, shft_amt=0010, shft_rslt=0000_0011

a=1000_1111, shft_code=11, shft_amt=0010, shft_rslt=1110_0011
//continued on next page
```

Figure 17.62 Outputs for the logical and algebraic shifter.

```
sll = 00, sla = 01, srl = 10, sra = 11
a=1111_1111, shft_code=00, shft_amt=0100, shft_rslt=1111_0000

a=1111_1111, shft_code=01, shft_amt=0100, shft_rslt=1111_0000

a=1111_1111, shft_code=10, shft_amt=0100, shft_rslt=0000_1111

a=1111_1111, shft_code=11, shft_amt=0100, shft_rslt=1111_1111


sll = 00, sla = 01, srl = 10, sra = 11
a=1100_0011, shft_code=00, shft_amt=0101, shft_rslt=0110_0000

a=1100_0011, shft_code=01, shft_amt=0101, shft_rslt=1110_0000

a=1100_0011, shft_code=10, shft_amt=0101, shft_rslt=0000_0110

a=0111_1111, shft_code=11, shft_amt=0101, shft_rslt=0000_0011
```

Figure 17.62 (Continued)

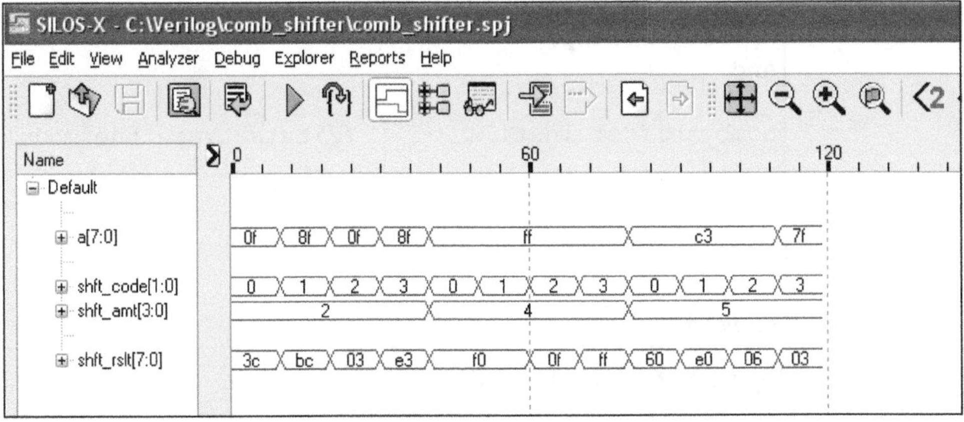

Figure 17.63 Waveforms for the logical and algebraic shifter.

17.5.2 Structural Modeling

This section describes a high-speed shift-left-logical shifter using combinational logic in the form of multiplexers. The high-speed in the shifting element is achieved due to the small delay in the logic gates of the multiplexers; that is, there is no clocked sequential operation. The structural design module in this section utilizes only one byte in the shifter; however, the concept can be easily extended for larger operands. Eight 4:1 multiplexers are used to perform the shifting operation, as shown in Figure 17.64.

If the shift amount is *shft_amt[00]*, then no shifting occurs — the input operand, *a[7:0]*, is passed through the 0 input of the multiplexers of the shifting element unchanged. If the shift amount is *shft_amt[01]*, then all bit positions of operand *a[7:0]* are shifted left one bit position by assigning a logic 0 to input 1 of the *inst0* multiplexer, bit *a[0]* to input 1 of the *inst1* multiplexer, and the remaining bits assigned to the appropriate data inputs of the remaining multiplexers. Bit *a[7]* is shifted off the left end of the shifting element, representing a logical left shift operation.

If the shift amount is *shft_amt[11]*, then operand *a[7:0]* is shifted left three bit positions with zeroes filling the vacated low-order bit positions. In this case, input 3 of multiplexers 7 through 0 are assigned the values *a[4] a[3] a[2] a[1] a[0]* 0 0 0.

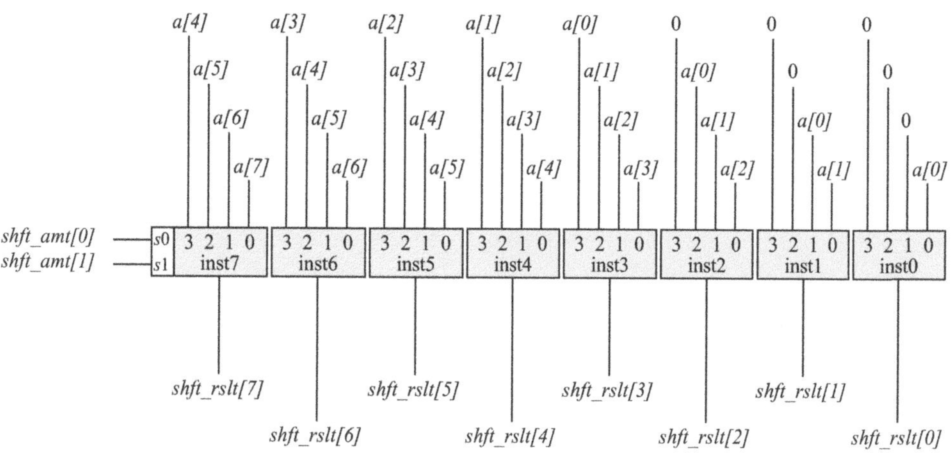

Figure 17.64 Logical organization for a high-speed shifter.

The 4:1 multiplexer is shown in the behavioral module of Figure 17.65 using the **case** statement. The structural module for the shifter is shown in Figure 17.66 that instantiates the 4:1 multiplexer eight times. The multiplexer has two select inputs, four data inputs, and one output. Consider instantiation *inst1* in the structural module. The data inputs labeled *data* correspond to the inputs shown in the logical organization for *inst1*. In this case, the data inputs are 0 0 *a[0] a[1]*, and are represented in the structural module as *({{2{1'b0}}, a[0], a[1]});*, where *{{2{1'b0}}* is the replication operator that provides two zeroes that are concatenated with *a[0] a[1]*.

The test bench, shown in Figure 17.67, applies eight input vectors that are used for all four shift amounts. The outputs and waveforms are shown in Figure 17.68 and Figure 17.69, respectively.

```
//example of a 4:1 multiplexer using a case statement
module mux_4_1_case (sel, data, out);

input [1:0] sel;
input [3:0] data;
output out;

reg out;

always @ (sel or data)
begin
   case (sel)
      (0) : out = data[0];
      (1) : out = data[1];
      (2) : out = data[2];
      (3) : out = data[3];
   endcase
end
endmodule
```

Figure 17.65 Four-to-one multiplexer to be used in the combinational shifter.

```
//structural combinational shifter using multiplexers
module shifter_usg_mux (a, shft_amt, shft_rslt);

input [7:0] a;
input [1:0] shft_amt;
output [7:0] shft_rslt;

//instantiate the multiplexers
mux_4_1_case inst0 (
   .sel(shft_amt),
   .data({{3{1'b0}}, a[0]}),
   .out(shft_rslt[0])
   );

mux_4_1_case inst1 (
   .sel(shft_amt),
   .data({{2{1'b0}}, a[0], a[1]}),
   .out(shft_rslt[1])
   );

//continued on next page
```

Figure 17.66 Structural module for the combinational shifter using multiplexers.

```
mux_4_1_case inst2 (
   .sel(shft_amt),
   .data({1'b0, a[0], a[1], a[2]}),
   .out(shft_rslt[2])
   );


mux_4_1_case inst3 (
   .sel(shft_amt),
   .data({a[0], a[1], a[2], a[3]}),
   .out(shft_rslt[3])
   );


mux_4_1_case inst4 (
   .sel(shft_amt),
   .data({a[1], a[2], a[3], a[4]}),
   .out(shft_rslt[4])
   );


mux_4_1_case inst5 (
   .sel(shft_amt),
   .data({a[2], a[3], a[4], a[5]}),
   .out(shft_rslt[5])
   );


mux_4_1_case inst6 (
   .sel(shft_amt),
   .data({a[3], a[4], a[5], a[6]}),
   .out(shft_rslt[6])
   );


mux_4_1_case inst7 (
   .sel(shft_amt),
   .data({a[4], a[5], a[6], a[7]}),
   .out(shft_rslt[7])
   );


endmodule
```

Figure 17.66 (Continued)

```
//test bench for shifter using multiplexers
module shifter_usg_mux_tb;

reg[7:0] a;
reg [1:0] shft_amt;
wire [7:0] shft_rslt;

//display variables
initial
$monitor ("a=%b, shft_amt=%b, shft_rslt=%b",
            a, shft_amt, shft_rslt);

//apply input vectors
initial
begin
   #0     a = 8'b0000_0000;
          shft_amt = 2'b00;

   #10    a = 8'b0000_1111;
          shft_amt = 2'b01;

   #10    a = 8'b0000_1111;
          shft_amt = 2'b10;

   #10    a = 8'b0000_1111;
          shft_amt = 2'b11;

   #10    a = 8'b1111_0000;
          shft_amt = 2'b00;

   #10    a = 8'b1111_0000;
          shft_amt = 2'b01;

   #10    a = 8'b1111_0000;
          shft_amt = 2'b10;

   #10    a = 8'b1111_0000;
          shft_amt = 2'b11;
   #10    $stop;
end

shifter_usg_mux inst1 (     //instantiate the module
   .a(a),
   .shft_amt(shft_amt),
   .shft_rslt(shft_rslt)
   );
endmodule
```

Figure 17.67 Test bench for the combinational shifter using multiplexers.

```
a=0000_0000, shft_amt=00, shft_rslt=0000_0000

a=0000_1111, shft_amt=01, shft_rslt=0001_1110

a=0000_1111, shft_amt=10, shft_rslt=0011_1100

a=0000_1111, shft_amt=11, shft_rslt=0111_1000

a=1111_0000, shft_amt=00, shft_rslt=1111_0000

a=1111_0000, shft_amt=01, shft_rslt=1110_0000

a=1111_0000, shft_amt=10, shft_rslt=1100_0000

a=1111_0000, shft_amt=11, shft_rslt=1000_0000
```

Figure 17.68 Outputs for the combinational shifter using multiplexers.

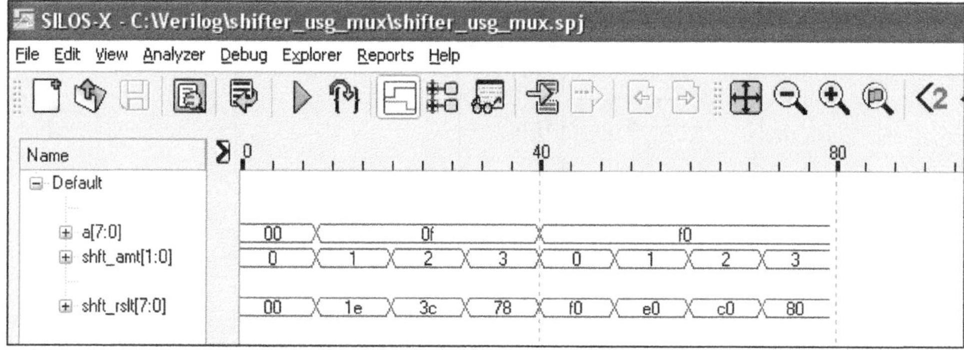

Figure 17.69 Waveforms for the combinational shifter using multiplexers.

The concept of using multiplexers for a combinational shifter can be easily expanded to accommodate larger operands and larger shift amounts simply by using multiplexers with more data inputs with a corresponding increase in the number of select inputs. For example, an 8:1 multiplexer can shift operands zero to seven bit positions.

An alternative approach to achieve the same shift amount as an 8:1 multiplexer is to use two 4:1 multiplexers, whose outputs connect to a 2:1 multiplexer, using the truth table of Table 17.5 and shown in Figure 17.70.

Table 17.5 Truth Table for Figure 17.70

Select Inputs			
s_2	s_1	s_0	Outputs
0	0	0	d_0
0	0	1	d_1
0	1	0	d_2
0	1	1	d_3
1	0	0	d_4
1	0	1	d_5
1	1	0	d_6
1	1	1	d_7

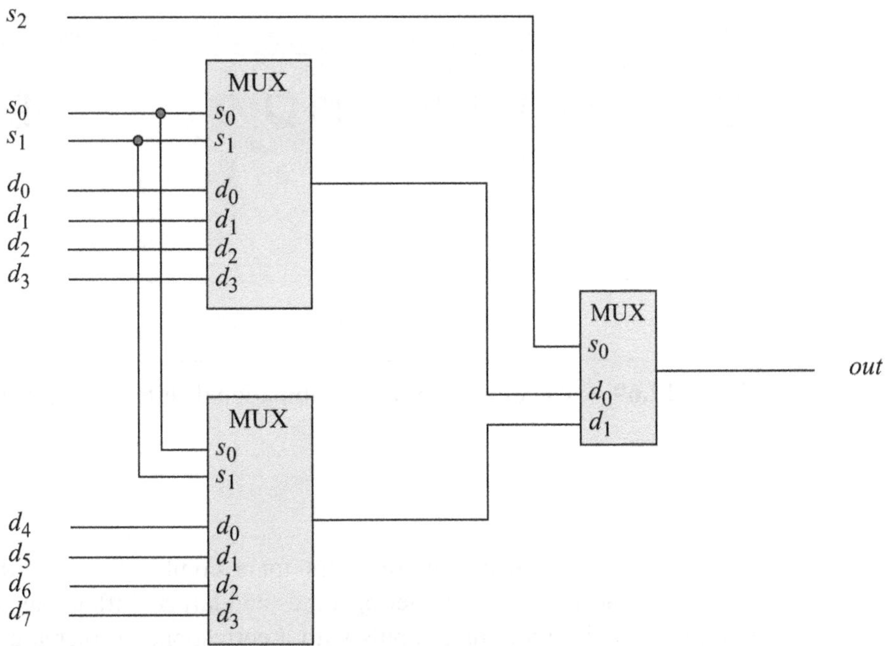

Figure 17.70 Logic to shift from zero to seven bit positions using two 4:1 multiplexers and one 2:1 multiplexer.

17.6 Arithmetic and Logic Units

Arithmetic and logic units (ALUs) perform the arithmetic operations of addition, subtraction, multiplication, and division in fixed-point, decimal, and floating-point number representations. They also perform various logical operations such as AND, OR, exclusive-OR, exclusive-NOR, and complementation. An ALU is one of five main units of a computer: input, output, memory (storage), arithmetic and logic unit, and control (or sequencer), which is usually microprogrammed. The ALU is the core of the computer and, together with the control unit, form the processor.

There are two types of information in a computer: instructions and data. The instructions perform arithmetic and logic operations, transfer data, and execute programs from memory. Data are numbers or coded characters. There are four general categories of storage hierarchy in a computer: internal registers, cache memory, main memory, and peripheral memory. Internal registers are very high speed, but have very small capacity; cache memory is high speed — but slower than registers — and has a larger capacity; main memory is lower speed, but has a larger capacity than cache; peripheral memory is the lowest speed, but has the largest capacity.

Most computer operations are executed by the ALU. For example, operands located in memory can be transferred to the ALU, where an operation is performed on the operands, then the result is stored in memory. The ALU performs the following: calculations in different number representations; logical operations; comparisons; and shift operations such as shift right algebraic, shift right logical, shift left algebraic, and shift left logical. The algebraic shift operations refer to signed operands in 2s complement representation; the logical shift operations refer to unsigned operands. ALUs also perform operations on packed and unpacked decimal operands. This section will present two examples of designing arithmetic and logic units using Verilog HDL.

17.6.1 Four-Function Arithmetic and Logic Unit

A 4-function arithmetic and logic unit will be designed using mixed-design modeling, (dataflow and behavioral) to execute the following four operations: add, multiply, AND, and OR. The operands are eight bits defined as *a[7:0]* and *b[7:0]*. The result of all operations will be 16 bits *rslt[15:0]*. The state of the following flags will be determined:

The parity flag *pf* = 1 if the result has an even number of 1s; otherwise, *pf* = 0.
The sign flag *sf* represents the state of the leftmost result bit.
The zero flag *zf* = 1 if the result of an operation is zero; otherwise, *zf* = 0.

There are three inputs: *a[7:0]*, *b[7:0]*, and the operation code *opcode[1:0]* that defines the operations to be performed, as shown in Table 17.6. There are four outputs: the 16-bit result *rslt[15:0]*, the parity flag *pf*, the sign flag *sf*, and the zero flag *zf*. The **case** statement will be used together with the **parameter** keyword to define the operation code for each operation.

Table 17.6 Operation Codes for the 4-Function ALU

Operation	Operation Code
Add	00
Multiply	01
AND	10
OR	11

The design module is shown in Figure 17.71, and the test bench module is shown in Figure 17.72 using several values for each of the four operations. The outputs and waveforms are shown in Figure 17.73 and Figure 17.74, respectively.

```
//mixed-design 4-function alu
module alu_4fctn3 (a, b, opcode, rslt, pf, sf, zf);

input [7:0] a, b;
input [1:0] opcode;
output [15:0] rslt;
output pf, sf, zf;

reg [15:0] rslt;

assign pf = ~^rslt;
assign sf = rslt[15];
assign zf = (rslt == 16'h0000);

parameter   add_op  = 2'b00,        //define the opcodes
            mul_op  = 2'b01,
            and_op  = 2'b10,
            or_op   = 2'b11;

always @ (a or b or opcode)
begin
   case (opcode)
         add_op: rslt = {8'h00, a + b};
         mul_op: rslt = a * b;
         and_op: rslt = {8'h00, a & b};
         or_op:  rslt = {8'h00, a | b};
         default: rslt = 16'hxxxx;
   endcase
end
endmodule
```

Figure 17.71 Mixed-design module for the 4-function arithmetic and logic unit.

```
//test bench for 4-function alu
module alu_4fctn3_tb;

reg [7:0] a, b;   //inputs are reg for test bench
reg [1:0] opcode;
wire [15:0] rslt; //outputs are wire for test bench
wire pf, sf, zf;

initial
$monitor ("a=%b, b=%b, op=%b, rslt=%b, pf_sf_zf=%b",
            a, b, opcode, rslt, {pf,sf,zf});

initial
begin

//add op ------------------------------------------------

   #0 a = 8'b0000_0001; b = 8'b0000_0010; opcode = 2'b00;
   #10a = 8'b1100_0101; b = 8'b0011_0110; opcode = 2'b00;
   #10a = 8'b0000_0101; b = 8'b1011_0110; opcode = 2'b00;
   #10a = 8'b1000_1010; b = 8'b0011_1010; opcode = 2'b00;
   #10a = 8'b1111_1110; b = 8'b0000_0001; opcode = 2'b00;
   #10a = 8'b0000_0000; b = 8'b0000_0000; opcode = 2'b00;

//mul_op ------------------------------------------------

   #10a = 8'b0000_0100; b = 8'b0000_0011; opcode = 2'b01;
   #10a = 8'b1000_0000; b = 8'b0000_0010; opcode = 2'b01;
   #10a = 8'b1010_1010; b = 8'b0101_0101; opcode = 2'b01;
   #10a = 8'b1111_1111; b = 8'b1111_1111; opcode = 2'b01;

//and op ------------------------------------------------

   #10a = 8'b1111_0000; b = 8'b0000_1111; opcode = 2'b10;
   #10a = 8'b1010_1010; b = 8'b1100_0011; opcode = 2'b10;
   #10a = 8'b1111_1111; b = 8'b1111_0000; opcode = 2'b10;

//or op -------------------------------------------------

   #10a = 8'b1111_0000; b = 8'b0000_1111; opcode = 2'b11;
   #10a = 8'b1010_1010; b = 8'b1100_0011; opcode = 2'b11;
   #10a = 8'b1111_1111; b = 8'b1111_0000; opcode = 2'b11;

   #10 $stop;
end

//continued on next page
```

Figure 17.72 Test bench for the 4-function arithmetic and logic unit.

```
alu_4fctn3 inst1 (        //instantiate the module
   .a(a),
   .b(b),
   .opcode(opcode),
   .rslt(rslt),
   .pf(pf),
   .sf(sf),
   .zf(zf)
   );
endmodule
```

Figure 17.72 (Continued)

```
a=0000_0001, b=0000_0010, op=00,
   rslt=0000_0000_0000_0011, pf_sf_zf=100
a=1100_0101, b=0011_0110, op=00,
   rslt=0000_0000_1111_1011, pf_sf_zf=000
a=0000_0101, b=1011_0110, op=00,
   rslt=0000_0000_1011_1011, pf_sf_zf=100
a=1000_1010, b=0011_1010, op=00,
   rslt=0000_0000_1100_0100, pf_sf_zf=000
a=1111_1110, b=0000_0001, op=00,
   rslt=0000_0000_1111_1111, pf_sf_zf=100
a=0000_0000, b=0000_0000, op=00,
   rslt=0000_0000_0000_0000, pf_sf_zf=101
a=0000_0100, b=0000_0011, op=01,
   rslt=0000_0000_0000_1100, pf_sf_zf=100
a=1000_0000, b=0000_0010, op=01,
   rslt=0000_0001_0000_0000, pf_sf_zf=000
a=1010_1010, b=0101_0101, op=01,
   rslt=0011_1000_0111_0010, pf_sf_zf=000
a=1111_1111, b=1111_1111, op=01,
   rslt=1111_1110_0000_0001, pf_sf_zf=110
a=1111_0000, b=0000_1111, op=10,
   rslt=0000_0000_0000_0000, pf_sf_zf=101
a=1010_1010, b=1100_0011, op=10,
   rslt=0000_0000_1000_0010, pf_sf_zf=100
a=1111_1111, b=1111_0000, op=10,
   rslt=0000_0000_1111_0000, pf_sf_zf=100
a=1111_0000, b=0000_1111, op=11,
   rslt=0000_0000_1111_1111, pf_sf_zf=100
a=1010_1010, b=1100_0011, op=11,
   rslt=0000_0000_1110_1011, pf_sf_zf=100
a=1111_1111, b=1111_0000, op=11,
   rslt=0000_0000_1111_1111, pf_sf_zf=100
```

Figure 17.73 Outputs for the 4-function arithmetic and logic unit.

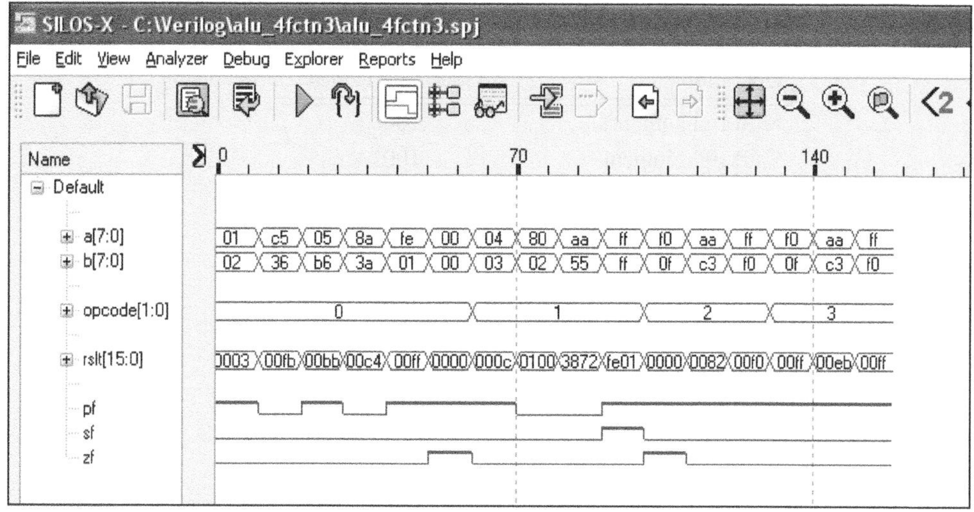

Figure 17.74 Waveforms for the 4-function arithmetic and logic unit.

17.6.2 Sixteen-Function Arithmetic and Logic Unit

A 16-function arithmetic and logic unit (ALU) will be designed in this section using behavioral modeling with the **case** construct. There are two 8-bit inputs: operands, *a[7:0]* and *b[7:0]*, and one 4-bit input, *opcode[3:0]*. There are two outputs *result[7:0]*, which contains the result of all operations, except multiplication, and *result_mul[15:0]* to accommodate the 2*n*-bit product. The **parameter** keyword will declare and assign values to the operation codes for use with the **case** statement. The operations and the operation codes are listed in Table 17.7.

Table 17.7 Operation codes for the 16-function ALU

Operation	Operation Codes
Add	0000
Subtract	0001
Multiply	0010
Divide	0011
AND	0100
OR	0101
Exclusive-OR	0110
NOT	0111

Continued on next page

Table 17.7 Operation codes for the 16-function ALU

Operation	Operation Codes
Shift right algebraic	1000
Shift right logical	1001
Shift left algebraic	1010
Shift left logical	1011
Rotate right	1100
Rotate left	1101
Increment	1110
Decrement	1111

The behavioral module is shown in Figure 17.75. Some operations require two operands, while some operations require only one operand; in that case, operand *a[7:0]* is used. The test bench is shown in Figure 17.76, providing input vectors for each operation. The outputs and waveforms are shown in Figure 17.77 and Figure 17.78, respectively.

```
//behavioral 16-function alu
module alu_16fctn (a, b, opcode, result, result_mul);

//list which are input or output
input [7:0] a, b;
input [3:0] opcode;
output [7:0] result;
output [15:0] result_mul;

//specify wire for input and reg for output
wire [7:0] a, b;      //inputs are wire
wire [3:0] opcode;
reg [7:0] result;     //outputs are reg
reg [15:0] result_mul;

//define the opcodes
parameter    add_op = 4'b0000,
             sub_op = 4'b0001,
             mul_op = 4'b0010,
             div_op = 4'b0011,
//-----------------------------
//continued on next page
```

Figure 17.75 Behavioral module for the 16-function arithmetic and logic unit.

```
        and_op = 4'b0100,
        or_op  = 4'b0101,
        xor_op = 4'b0110,
        not_op = 4'b0111,
//--------------------------------------------------
        sra_op = 4'b1000,
        srl_op = 4'b1001,
        sla_op = 4'b1010,
        sll_op = 4'b1011,
//--------------------------------------------------
        ror_op = 4'b1100,
        rol_op = 4'b1101,
        inc_op = 4'b1110,
        dec_op = 4'b1111;

always @ (a or b or opcode)
begin
   case (opcode)
      add_op : result = a + b;
      sub_op : result = a - b;
      mul_op : result_mul = a * b;
      div_op : result = a / b;
//--------------------------------------------------
      and_op : result = a & b;
      or_op  : result = a | b;
      xor_op : result = a ^ b;
      not_op : result = ~a;
//--------------------------------------------------
      sra_op : result = {a[7], a[7], a[6], a[5],
                         a[4], a[3], a[2], a[1]};
      srl_op : result = a >> 1;
      sla_op : result = {a[7], a[5], a[4], a[3],
                         a[2], a[1], a[0], 1'b0};
      sll_op : result = a << 1;
//--------------------------------------------------
      ror_op : result = {a[0], a[7], a[6], a[5],
                         a[4], a[3], a[2], a[1]};
      rol_op : result = {a[6], a[5], a[4], a[3],
                         a[2], a[1], a[0], a[7]};
      inc_op : result = a + 1;
      dec_op : result = a - 1;
      default : result = 0;

   endcase
end

endmodule
```

Figure 17.75 (Continued)

```
//alu_16fctn test bench
module alu_16fctn_tb;

reg [7:0] a, b;         //inputs are reg for test bench
reg [3:0] opcode;
wire [7:0] result;      //outputs are wire for test bench
wire [15:0] result_mul;

initial
$monitor ("a=%b, b=%b, opcode=%b, rslt=%b, rslt_mul=%h",
            a, b, opcode, result, result_mul);

initial
begin
//add op -------------------------------------------
   #0     a = 8'b00000000;
          b = 8'b00000000;
          opcode = 4'b0000;

   #10    a = 8'b01100010;      //a = 98d = 62h
          b = 8'b00011100;      //b = 28d = 1Ch
          opcode = 4'b0000;     //result = 126d = 007eh

   #10    a = 8'b11111111;      //a = 255d = ffh
          b = 8'b11111111;      //b = 255d = ffh
          opcode = 4'b0000;     //result = 510d = 01feh

//sub op -------------------------------------------
   #10    a = 8'b10000000;      //a = 128d = 80h
          b = 8'b01100011;      //b = 99d = 63h
          opcode = 4'b0001;     //result = 29d = 001dh

   #10    a = 8'b11111111;      //a = 255d = ffh
          b = 8'b00001111;      //b = 15d = fh
          opcode = 4'b0001;     //result = 240d = 00f0h

//mul op -------------------------------------------
   #10    a = 8'b00011001;      //a = 25d = 19h
          b = 8'b00011001;      //b = 25d = 19h
          opcode = 4'b0010;     //result_mul = 625d = 0271h

   #10    a = 8'b10000100;      //a = 132d = 84h
          b = 8'b11111010;      //b = 250d = fah
          opcode = 4'b0010;     //result_mul = 33000d = 80e8h

//continued on next page
```

Figure 17.76 Test bench for the 16-function arithmetic and logic unit.

```
   #10   a = 8'b11110000;      //a = 240d = f0h
         b = 8'b00010100;      //b = 20d = 14h
         opcode = 4'b0010;     //result_mul = 4800d = 12c0h

//div op ---------------------------------------
   #10   a = 8'b11110000;      //a = 240d = f0h
         b = 8'b00001111;      //b = 15d = 0fh
         opcode = 4'b0011;     //result = 16d = 10h

   #10   a = 8'b00010000;      //a = 16d = 10h
         b = 8'b00010010;      //b = 18d = 12h
         opcode = 4'b011;      //result = 0d = 0h

//and op ---------------------------------------
   #10   a = 8'b11111111;
         b = 8'b11110000;
         opcode = 4'b0100;     //result = 11110000

//or op ----------------------------------------
   #10   a = 8'b10101010;
         b = 8'b11110101;
         opcode = 4'b0101;     //result = 11111111

//xor op ---------------------------------------
   #10   a = 8'b00001111;
         b = 8'b10101010;
         opcode = 4'b0110;     //result = 10100101

//not op ---------------------------------------
   #10   a = 8'b00001111;
         opcode = 4'b0111;

//sra op ---------------------------------------
   #10   a = 8'b10001110;
         opcode = 4'b1000;

//srl op ---------------------------------------
   #10   a = 8'b11110011;
         opcode = 4'b1001;

//sla op ---------------------------------------
   #10   a = 8'b10001111;
         opcode = 4'b1010;

//sll op ---------------------------------------
   #10   a = 8'b01110111;
         opcode = 4'b1011;        //continued on next page
```

Figure 17.76 (Continued)

```
//ror op -------------------------------------------
   #10   a = 8'b01010101;
         opcode = 4'b1100;

//rol op -------------------------------------------
   #10   a = 8'b01010101;
         opcode = 4'b1101;

//inc op -------------------------------------------
   #10   a = 8'b11111101;
         opcode = 4'b1110;

//dec op -------------------------------------------
   #10   a = 8'b11111110;
         opcode = 4'b1111;

   #10   $stop;
end

//instantiate the module into the test bench
alu_16fctn inst1 (
   .a(a),
   .b(b),
   .opcode(opcode),
   .result(result),
   .result_mul(result_mul)
   );
endmodule
```

Figure 17.76 (Continued)

```
a=0000_0000, b=0000_0000, opcode=0000,
   rslt=0000_0000, rslt_mul=xxxx

a=0110_0010, b=0001_1100, opcode=0000,
   rslt=0111_1110, rslt_mul=xxxx

a=1111_1111, b=1111_1111, opcode=0000,
   rslt=1111_1110, rslt_mul=xxxx

a=1000_0000, b=0110_0011, opcode=0001,
   rslt=0001_1101, rslt_mul=xxxx

//continued on next page
```

Figure 17.77 Outputs for the 16-function arithmetic and logic unit.

```
a=1111_1111, b=0000_1111, opcode=0001,
   rslt=1111_0000, rslt_mul=xxxx

a=0001_1001, b=0001_1001, opcode=0010,
   rslt=1111_0000, rslt_mul=0271

a=1000_0100, b=1111_1010, opcode=0010,
   rslt=1111_0000, rslt_mul=80e8

a=1111_0000, b=0001_0100, opcode=0010,
   rslt=1111_0000, rslt_mul=12c0

a=1111_0000, b=0000_1111, opcode=0011,
   rslt=0001_0000, rslt_mul=12c0

a=0001_0000, b=0001_0010, opcode=0011,
   rslt=0000_0000, rslt_mul=12c0

a=1111_1111, b=1111_0000, opcode=0100,
   rslt=1111_0000, rslt_mul=12c0

a=1010_1010, b=1111_0101, opcode=0101,
   rslt=1111_1111, rslt_mul=12c0

a=0000_1111, b=1010_1010, opcode=0110,
   rslt=1010_0101, rslt_mul=12c0

a=0000_1111, b=1010_1010, opcode=0111,
   rslt=1111_0000, rslt_mul=12c0

a=1000_1110, b=1010_1010, opcode=1000,
   rslt=1100_0111, rslt_mul=12c0

a=1111_0011, b=1010_1010, opcode=1001,
   rslt=0111_1001, rslt_mul=12c0

a=1000_1111, b=1010_1010, opcode=1010,
   rslt=1001_1110, rslt_mul=12c0

a=0111_0111, b=1010_1010, opcode=1011,
   rslt=1110_1110, rslt_mul=12c0

a=0101_0101, b=1010_1010, opcode=1100,
   rslt=1010_1010, rslt_mul=12c0

a=0101_0101, b=1010_1010, opcode=1101,
   rslt=1010_1010, rslt_mul=12c0    //continued on next page
```

Figure 17.77 (Continued)

```
a=1111_1101, b=1010_1010, opcode=1110,
   rslt=1111_1110, rslt_mul=12c0

a=1111_1110, b=1010_1010, opcode=1111,
   rslt=1111_1101, rslt_mul=12c0
```

Figure 17.77 (Continued)

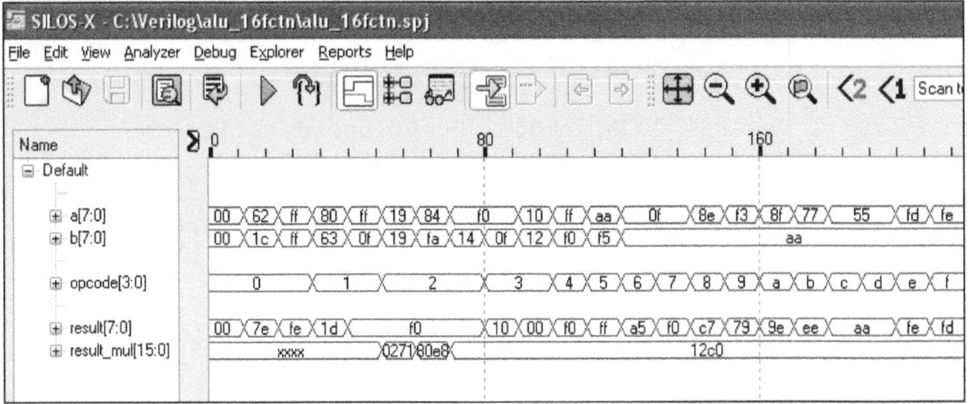

Figure 17.78 Waveforms for the 16-function arithmetic and logic unit.

17.7 Count-Down Counter

In many of the designs in this book, count-down counters were used to control the sequence of operations. The counter was set to a predetermined value and counted down to zero. When the counter reached a value of zero, the operation was terminated. One of the easiest way to design any counter is by using the **case** statement. A 4-bit count-down counter with a beginning count of 1111 will be designed using behavioral modeling.

The design module is shown in Figure 17.79. The statements *if (rst_n == 1'b0)* and *count <= 4'b1111;* specifies that if the reset input is active at a low voltage level, the counter is set to a value of 1111 using the nonblocking assignment symbol (<=). The assignment symbol is used to represent a nonblocking procedural assignment. *Nonblocking assignments* allow the scheduling of assignments without blocking execution of the following statements in a sequential procedural block. A nonblocking assignment is used to synchronize assignment statements so that they appear to execute at the same time.

The Verilog HDL simulator schedules a *nonblocking assignment* statement to execute, then proceeds to the next statement in the block without waiting for the previous nonblocking statement to complete execution. That is, the right-hand expression is evaluated and the value is stored in the event queue and is *scheduled* to be assigned to the left-hand target. The assignment is made at the end of the current time step if there are no intrastatement delays specified.

Nonblocking assignments are typically used to model several concurrent assignments that are caused by a common event, such as the low-to-high transition of a clock pulse or a change to any variable in a sensitivity list (event control list). The order of the assignments is irrelevant because the right-hand side evaluations are stored in the event queue before any assignments are made.

The test bench is shown in Figure 17.80 and defines the reset input, the clock input, and the length of simulation. The outputs and waveforms are shown in Figure 17.81 and Figure 17.82, respectively.

```
//behavioral 4-bit count-down counter
module ctr_down4_case (rst_n, clk, count);

input rst_n, clk;
output [3:0] count;

wire rst_n, clk;
reg [3:0] count;
reg [3:0] next_count;     //define internal reg

//set next count
always @ (posedge clk or negedge rst_n)
begin
   if (rst_n == 1'b0)
      count <= 4'b1111;
   else
      count <= next_count;
end

//determine next count
always @ (count)
begin
   case (count)
      4'b1111 : next_count = 4'b1110;
      4'b1110 : next_count = 4'b1101;
      4'b1101 : next_count = 4'b1100;
      4'b1100 : next_count = 4'b1011;

//continued on next page
```

Figure 17.79 Behavioral module for the 4-bit count-down counter.

```
        4'b1011 : next_count = 4'b1010;
        4'b1010 : next_count = 4'b1001;
        4'b1001 : next_count = 4'b1000;
        4'b1000 : next_count = 4'b0111;

        4'b0111 : next_count = 4'b0110;
        4'b0110 : next_count = 4'b0101;
        4'b0101 : next_count = 4'b0100;
        4'b0100 : next_count = 4'b0011;

        4'b0011 : next_count = 4'b0010;
        4'b0010 : next_count = 4'b0001;
        4'b0001 : next_count = 4'b0000;
        4'b0000 : next_count = 4'b0000;
        default : next_count = 4'b0000;
    endcase
end

endmodule
```

Figure 17.79 (Continued)

```
//test bench for 4-bit count-down counter
module ctr_down4_case_tb;

reg clk, rst_n;
wire [3:0] count;

initial
$monitor ("count = %b", count);

initial     //define reset
begin
   #0 rst_n = 1'b0;
   #5 rst_n = 1'b1;
end

initial     //define clock
begin
   clk = 1'b0;
   forever
      #10   clk = ~clk;
end
//continued on next page
```

Figure 17.80 Test bench for the 4-bit count-down counter.

```
//define length of simulation
initial
begin
   #330  $stop;
end

//instantiate the module into the test bench
ctr_down4_case inst1 (
   .clk(clk),
   .rst_n(rst_n),
   .count(count)
   );

endmodule
```

Figure 17.80 (Continued)

```
count = 1111                count = 0111
count = 1110                count = 0110
count = 1101                count = 0101
count = 1100                count = 0100
count = 1011                count = 0011
count = 1010                count = 0010
count = 1001                count = 0001
count = 1000                count = 0000
```

Figure 17.81 Outputs for the 4-bit count-down counter.

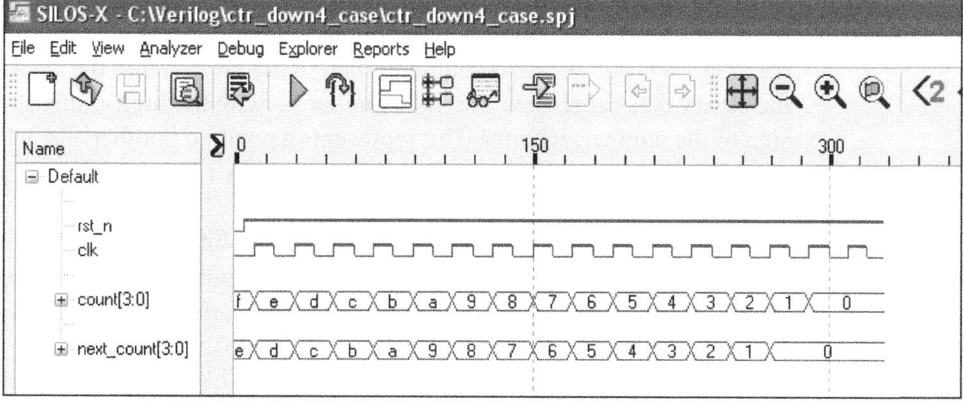

Figure 17.82 Waveforms for the 4-bit count-down counter.

17.8 Shift Registers

A shift register can be defined as an ordered set of storage elements and associated combinational logic, where each cell performs an identical or similar operation. Each cell of a register stores one bit of binary information. Different types of shift registers are also used in various arithmetic operations. There are five main categories of synchronous shift registers: parallel-in, parallel-out; parallel-in, serial-out; serial-in, parallel-out; serial-in, serial-out; and parallel-in, serial-in, serial-out registers.

The next state of a register is usually a direct correspondence to the input vector, whose binary variables connect to the flip-flop data inputs, either directly or through δ next-state logic. Most registers are used primarily for temporary storage of binary data, either signed or unsigned, and do not modify the data internally; that is, the state of the register is unchanged until the next active clock transition. Other registers may modify the data in some elementary manner, such as shifting left or shifting right, where a left shift of one bit corresponds to a multiply-by-two operation, and a right shift of one bit corresponds to a divide-by-two operation.

An n-bit register requires n storage elements, either *SR* latches, *D* flip-flops, or *JK* flip-flops. There are 2^n different states in an n-bit register, where each n-tuple corresponds to a unique state of the register.

This section describes four different types of shift registers: a parallel-in, serial-out (PISO) register; a serial-in, serial-out (SISO) register; a parallel-in, serial-in, serial-out (PISISO) register, and a serial-in, parallel-out (SIPO) register.

17.8.1 Parallel-In, Serial-Out Shift Register

A *parallel-in, serial-out* (PISO) register accepts binary input data in parallel and generates binary output data in serial form. The binary data can be shifted either left or right under control of a shift direction signal and a clock pulse, which is applied to all flip-flops simultaneously. The register shifts left or right 1 bit position at each active clock transition. Bits shifted out of one end of the register are lost unless the register is cyclic, in which case, the bits are shifted (or rotated) into the other end.

This section describes a PISO shift-right register in which two conditions determine the value of the bits shifted into the vacated positions on the left. If the binary data represents an unsigned number, then 0s are shifted into the vacated positions. If the binary data represents a signed number — with the high-order bit specified as the sign of the number, where a 0 bit represents a positive number and a 1 bit represents a negative number — then the sign bit extends right 1 bit position for each active clock transition.

Figure 17.83 shows the logic symbol for a 12-bit PISO register with four inputs: a reset input, *rst*; a clock input, *clk*; a load input, *ld*; and a 12-bit input vector, *x[11:0]*. In this design, the flip-flop outputs are also available, if needed, from the register and are labeled *piso_reg[11:0]*; the serial output is labeled *z*.

The behavioral module is shown in Figure 17.84 and the test bench is shown in Figure 17.85. It is common practice to label an active-low reset as *rst_n*. The outputs and waveforms are shown in Figure 17.86 and Figure 17.87, respectively.

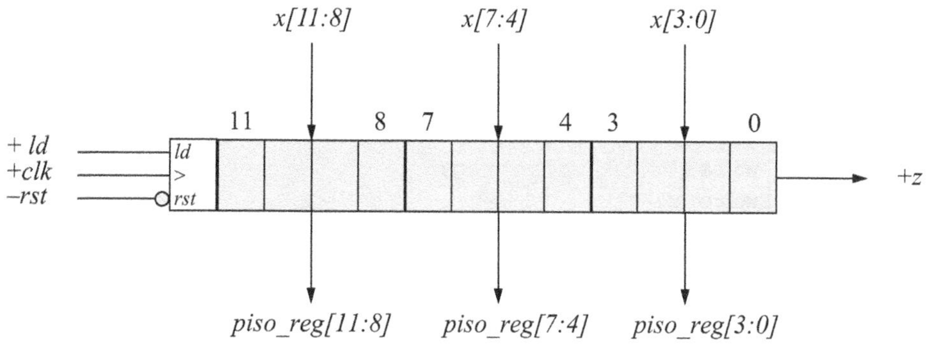

Figure 17.83 Logic symbol for the parallel-in, serial-out 12-bit shift register.

```
//behavioral 12-bit piso shift register
module shift_reg_piso12 (rst_n, clk, ld, x, piso_reg, z);

//define inputs and outputs
input rst_n, clk, ld;
input [11:0] x;
output [11:0] piso_reg;
output z;

//variables used in an always are declared as registers
reg [11:0] piso_reg;

assign z = piso_reg[0];

always @ (negedge rst_n or posedge clk)
begin
   if (rst_n == 1'b0)
      piso_reg <= 12'b0000_0000_0000;

   else if (ld == 1'b1)
      piso_reg[7:0] = x[7:0];

   else
      piso_reg <= {1'b0, piso_reg[11:1]};
end

endmodule
```

Figure 17.84 Behavioral module for the parallel-in, serial-out 12-bit register.

```
//test bench for the 12-bit piso shift register
module shift_reg_piso12_tb;

reg rst_n, clk, ld;
reg [11:0] x;
wire [11:0] piso_reg;
wire z;

//define clock
initial
begin
   clk = 1'b0;
   forever
      #10 clk = ~clk;
end

//display variables
initial
$monitor ("x=%b, piso_reg=%b, z=%b", x, piso_reg, z);

//apply inputs
initial
begin
   #0    rst_n = 1'b0;
         ld = 1'b0;
         x = 12'b0000_0000_0000;

   #3    rst_n = 1'b1;

   #5    x = 12'b1111_1111_1111;
         ld = 1'b1;

   #7    ld = 1'b0;

   #200  $stop;
end

//instantiate the module into the test bench
shift_reg_piso12 inst1 (
   .rst_n(rst_n),
   .clk(clk),
   .ld(ld),
   .x(x),
   .piso_reg(piso_reg),
   .z(z)
   );
endmodule
```

Figure 17.85 Test bench for the parallel-in, serial-out 12-bit register.

```
x=000000000000, piso_reg=000000000000, z=0
x=111111111111, piso_reg=000000000000, z=0
x=111111111111, piso_reg=000011111111, z=1
x=111111111111, piso_reg=000001111111, z=1
x=111111111111, piso_reg=000000111111, z=1
x=111111111111, piso_reg=000000011111, z=1
x=111111111111, piso_reg=000000001111, z=1
x=111111111111, piso_reg=000000000111, z=1
x=111111111111, piso_reg=000000000011, z=1
x=111111111111, piso_reg=000000000001, z=1
x=111111111111, piso_reg=000000000000, z=0
```

Figure 17.86 Outputs for the parallel-in, serial-out 12-bit register.

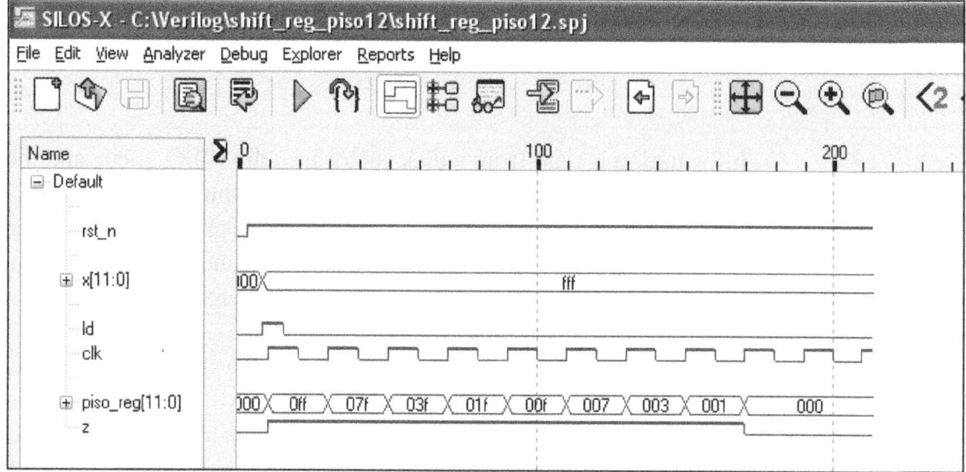

Figure 17.87 Waveforms for the parallel-in, serial-out 12-bit register.

17.8.2 Serial-In, Serial-Out Shift Register

Another type of shift register that is commonly used in arithmetic circuits is the *serial-in, serial-out* (SISO) shift register. The synthesis of a SISO register is identical to that of a SIPO register, with the exception that only one output is required. The rightmost flip-flop provides the output for the register as shown in Figure 17.88; however, the flip-flop outputs are also available as a parallel-out configuration.

An important application of a SISO register is to deserialize data from a disk drive in conjunction with a SIPO register. A serial bit stream is read from a disk drive and converted into parallel bits by means of a SIPO register. When eight bits have been shifted into the register, the bytes are shifted in parallel into a matrix of SISO registers, where each bit is shifted into a particular column. The SISO register, in this application, performs the function of a first-in, first-out (FIFO) queue and acts as a buffer between the disk drive and the system input/output (I/O) data bus.

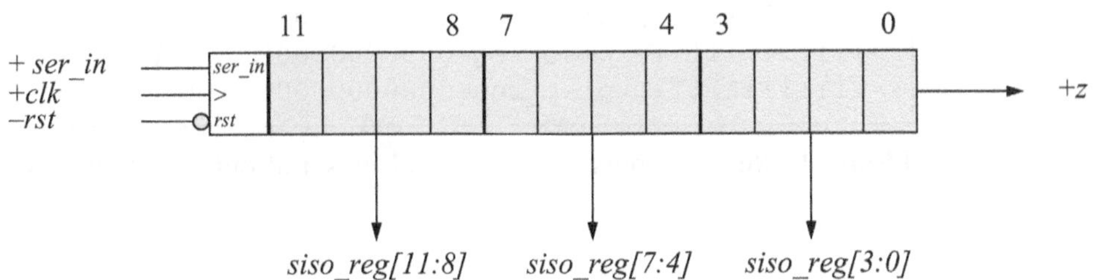

Figure 17.88 Organization of a serial-in, serial-out shift register.

Information is read from a disk sector into the FIFO queue and then transferred to a destination by means of the data bus. The destination may be a CPU register or a storage location if direct-memory access (DMA) is implemented. The mode of transfer is bit parallel, byte serial and is either synchronous, where the transfer rate is determined by a system clock, or asynchronous, where the transfer rate is determined by the disk control unit.

The mode of data transfer between a disk subsystem and a destination is usually in *burst mode*, in which the disk control unit remains logically connected to the system bus for the entire data transfer sequence; that is, until the complete sector has been transferred. Some systems, however, do not allow burst mode transfer, because this would prevent other peripheral devices from gaining control of the bus. This presents no problem when the disk control unit contains a FIFO queue. In this situation, data continues to be read from the disk and is transferred to the FIFO queue, where the bytes are retained until the disk control unit again gains control of the bus. The FIFO queue prevents data from being lost while the control unit is arbitrating for bus control.

The same implementation of a SISO register matrix can be used as an instruction queue in a CPU instruction pipeline. The CPU prefetches instructions from memory during unused memory cycles and stores the instructions in the FIFO queue. Thus, an instruction stream can be placed in the instruction queue to wait for decoding and execution by the processor. Instruction queueing provides an effective method to increase system throughput.

The behavioral module for a SISO shift register is shown in Figure 17.89. The test bench, shown in Figure 17.90, applies a sequence of several input vectors that are applied to the serial-in input. The outputs and waveforms are shown in Figure 17.91 and Figure 17.92, respectively.

```
//behavioral 12-bit siso shift register
module shift_reg_siso12 (rst_n, clk, ser_in, siso_reg, z);

input rst_n, clk, ser_in;
output [11:0] siso_reg;
output z;

//variable used in an always block are declared as registers
reg [11:0] siso_reg;

assign z = siso_reg[0];

always @ (negedge rst_n or posedge clk)
begin
   if (rst_n == 1'b0)
      siso_reg <= 12'b000_0000_0000;

   else
      siso_reg <= {ser_in, siso_reg[11:1]};
end
endmodule
```

Figure 17.89 Organization for the serial-in, serial-out shift register.

```
//test bench for siso shift register
module shift_reg_siso12_tb;

reg rst_n, clk, ser_in;
wire [11:0] siso_reg;
wire z;

//define clock
initial
begin
   clk = 1'b0;
   forever
      #10 clk = ~clk;
end

//display variables
initial
$monitor ("ser_in=%b, siso_reg=%b, z=%b",
            ser_in, siso_reg, z);
//continued on next page
```

Figure 17.90 Test bench for the serial-in, serial-out shift register.

```
initial  //apply inputs
begin
   #0    rst_n = 1'b0;
   #5    rst_n = 1'b1;
   #3    ser_in = 1'b1;
   #17   ser_in = 1'b0;
   #20   ser_in = 1'b1;
   #20   ser_in = 1'b1;
   #20   ser_in = 1'b1;
   #20   ser_in = 1'b1;
   #20   ser_in = 1'b1;
   #20   ser_in = 1'b1;
   #20   ser_in = 1'b0;
   #20   ser_in = 1'b0;
   #20   ser_in = 1'b0;
   #20   ser_in = 1'b0;
   #20   ser_in = 1'b1;
   #20   ser_in = 1'b1;
   #20   ser_in = 1'b1;
   #20   ser_in = 1'b1;
   #10   $stop;
end

shift_reg_siso12 inst1 (       //instantiate the module
   .rst_n(rst_n),
   .clk(clk),
   .ser_in(ser_in),
   .siso_reg(siso_reg),
   .z(z)
   );
endmodule
```

Figure 17.90 (Continued)

```
ser_in=x, siso_reg=000000000000, z=0
ser_in=1, siso_reg=000000000000, z=0
ser_in=1, siso_reg=100000000000, z=0
ser_in=0, siso_reg=100000000000, z=0
ser_in=0, siso_reg=010000000000, z=0
ser_in=1, siso_reg=010000000000, z=0
ser_in=1, siso_reg=101000000000, z=0
ser_in=1, siso_reg=110100000000, z=0
ser_in=1, siso_reg=111010000000, z=0
ser_in=1, siso_reg=111101000000, z=0
//continued on next page
```

Figure 17.91 Outputs for the serial-in, serial-out shift register.

```
ser_in=1, siso_reg=111110100000, z=0
ser_in=1, siso_reg=111111010000, z=0
ser_in=0, siso_reg=111111010000, z=0
ser_in=0, siso_reg=011111101000, z=0
ser_in=0, siso_reg=001111110100, z=0
ser_in=0, siso_reg=000111111010, z=0
ser_in=0, siso_reg=000011111101, z=1
ser_in=1, siso_reg=000011111101, z=1
ser_in=1, siso_reg=100001111110, z=0
ser_in=1, siso_reg=110000111111, z=1
ser_in=1, siso_reg=111000011111, z=1
ser_in=1, siso_reg=111100001111, z=1
```

Figure 17.91 (Continued)

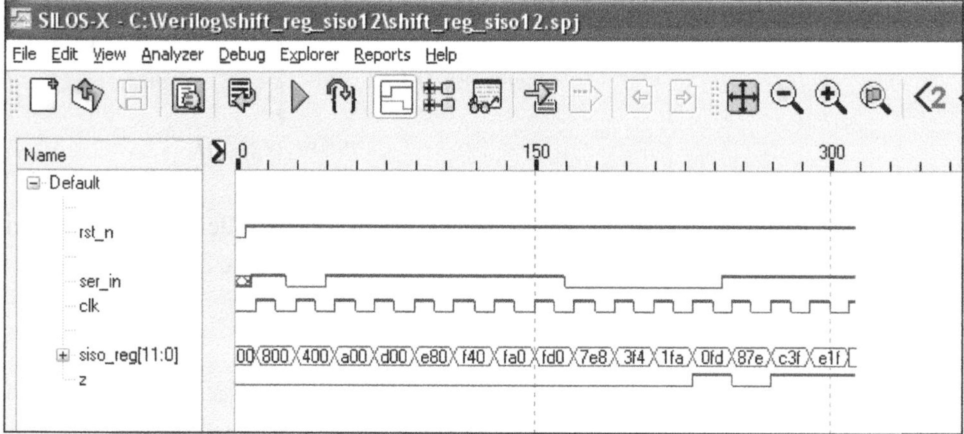

Figure 17.92 Waveforms for the serial-in, serial-out shift register.

17.8.3 Parallel-In, Serial-In, Serial-Out Shift Register

Another type of shift register that is sometimes used is one whose inputs consist of two sets of parallel inputs and a serial input, and generate a serial output. Two load control inputs selectively choose one set of parallel inputs. A logical organization of this type of shift register is shown in Figure 17.93 with two parallel input vectors: *x[11:0]* selected by load control input *ld1*, and *y[11:0]* selected by load control input *ld2*. There can also be a parallel output vector, *pisiso_reg[11:0]*, from the shift register flip-flop outputs. The serial output is labeled *z*. There is one set of AND-OR gates for each parallel input signal.

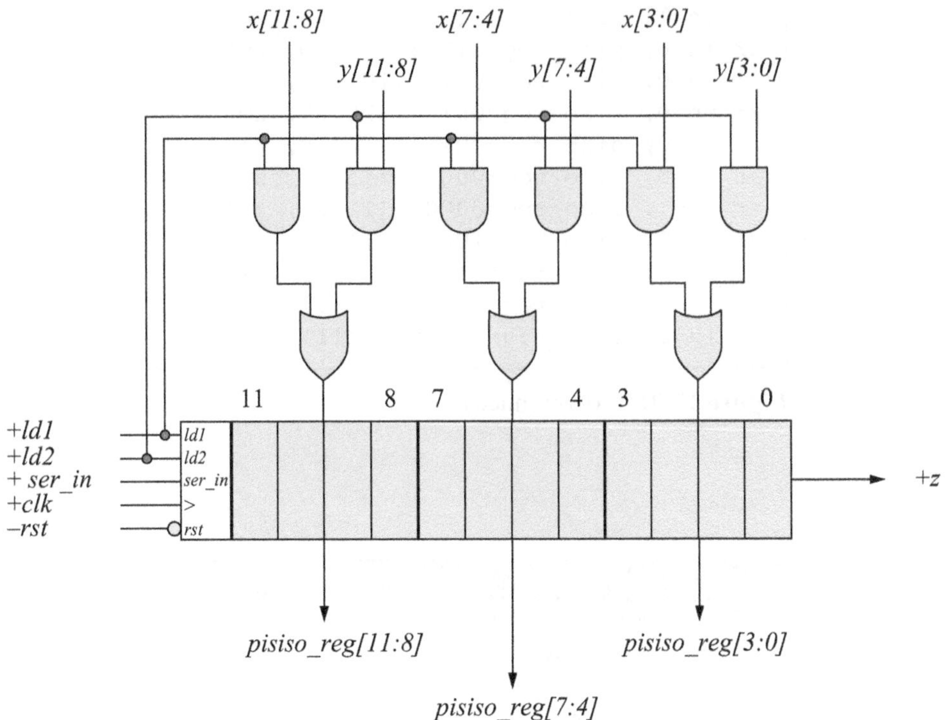

Figure 17.93 Logical organization for the parallel-in, serial-in, serial-out shift register.

The procedure for the synthesis of this type of register is relatively straightforward. This 12-bit register receives a parallel input vector from one of two sources depending on the asserted level of the two load control input lines. The input operand is stored in twelve flip-flops. Upon application of a clock signal, the operand shifts right one bit position. The serial output, z_1, is generated from the output of flip-flop *pisiso_reg[0]*. Zeroes fill the vacated positions on the left.

Upon completion of the load cycle, *pisiso_reg[11:0]* = *x[11:0]* or *y[11:0]*. During the shift sequence, *pisiso_reg[i]* = *pisiso_reg[i+1]*. After twelve shift cycles, the state of the register is *pisiso_reg[11:0]* = 0000 0000 0000, and the process repeats with a new input vector. One application of a PISO register is to convert data from a parallel bus into serial data for use by a single-track device, such as a disk drive. The serialization process occurs during a write operation.

The behavioral module for the 12-bit parallel-in, serial-in, serial-out shift register is shown in Figure 17.94, and the test bench, shown in Figure 17.95, applies input vectors for *x[11:0]* and *y[11:0]*. The outputs and waveforms are shown in Figure 17.96 and Figure 17.97, respectively.

```
//behavioral 12-bit pisiso shift register
module shift_reg_pisiso12 (rst_n, clk, ser_in, ld1, ld2, x, y,
                            pisiso_reg, z);

//define inputs and outputs
input rst_n, clk, ser_in, ld1, ld2;
input [11:0] x, y;
output [11:0] pisiso_reg;
output z;

//variables used in an always block are declared as registers
reg [11:0] pisiso_reg;

assign z = pisiso_reg[0];

always @ (negedge rst_n or posedge clk)
begin
   if (rst_n == 1'b0)
      pisiso_reg <= 12'b0000_0000_0000;

   else if ((ld1 == 1'b1) && (ld2 == 1'b0))
      pisiso_reg = x;

   else if ((ld1 == 1'b0)  && (ld2 == 1'b1))
      pisiso_reg = y;

   else
      pisiso_reg <= {ser_in, pisiso_reg[11:1]};
end

endmodule
```

Figure 17.94 Behavioral module for the 12-bit parallel-in, serial-in, serial-out shift register.

```
//test bench for 12-bit pisiso shift register
module shift_reg_pisiso12_tb;

reg rst_n, clk, ser_in, ld1, ld2;
reg [11:0] x, y;
wire [11:0] pisiso_reg;
wire z;

//continued on next page
```

Figure 17.95 Test bench for 12-bit the parallel-in, serial-in, serial-out shift register.

```
//define clock
initial
begin
   clk = 1'b0;
   forever
      #10clk = ~clk;
end

initial  //display variables
$monitor ("x=%b, y=%b, ser_in=%b, pisiso_reg=%b, z=%b",
            x, y, ser_in, pisiso_reg, z);

initial  //apply inputs
begin
   #0    rst_n = 1'b0;
         ser_in = 1'b0;
         ld1 = 1'b0;
         ld2 = 1'b0;
         x = 12'b0000_0000_0000;
         y = 12'b0000_0000_0000;

   #5    rst_n = 1'b1;
         x = 12'b1111_0000_1111;
         y = 12'b1111_0000_0000;
         ld1 = 1'b1;
         ld2 = 1'b0;

   #20   ld1 = 1'b0;

   #120  ld2 = 1'b1;
   #20   ld2 = 1'b0;
   #120  $stop;
end

//instantiate the module into the test bench
shift_reg_pisiso12 inst1 (
   .rst_n(rst_n),
   .clk(clk),
   .ser_in(ser_in),
   .ld1(ld1),
   .ld2(ld2),
   .x(x),
   .y(y),
   .pisiso_reg(pisiso_reg),
   .z(z)
   );
endmodule
```

Figure 17.95 (Continued)

```
x=000000000000, y=000000000000, ser_in=0,
   pisiso_reg=000000000000, z=0

x=111100001111, y=111100000000, ser_in=0,
   pisiso_reg=000000000000, z=0

x=111100001111, y=111100000000, ser_in=0,
   pisiso_reg=111100001111, z=1

x=111100001111, y=111100000000, ser_in=0,
   pisiso_reg=011110000111, z=1

x=111100001111, y=111100000000, ser_in=0,
   pisiso_reg=001111000011, z=1

x=111100001111, y=111100000000, ser_in=0,
   pisiso_reg=000111100001, z=1

x=111100001111, y=111100000000, ser_in=0,
   pisiso_reg=000011110000, z=0

x=111100001111, y=111100000000, ser_in=0,
   pisiso_reg=000001111000, z=0

x=111100001111, y=111100000000, ser_in=0,
   pisiso_reg=000000111100, z=0

x=111100001111, y=111100000000, ser_in=0,
   pisiso_reg=111100000000, z=0

x=111100001111, y=111100000000, ser_in=0,
   pisiso_reg=011110000000, z=0

x=111100001111, y=111100000000, ser_in=0,
   pisiso_reg=001111000000, z=0

x=111100001111, y=111100000000, ser_in=0,
   pisiso_reg=000111100000, z=0

x=111100001111, y=111100000000, ser_in=0,
   pisiso_reg=000011110000, z=0

x=111100001111, y=111100000000, ser_in=0,
   pisiso_reg=000001111000, z=0

x=111100001111, y=111100000000, ser_in=0,
   pisiso_reg=000000111100, z=0
```

Figure 17.96 Outputs for the 12-bit parallel-in, serial-in, serial-out shift register.

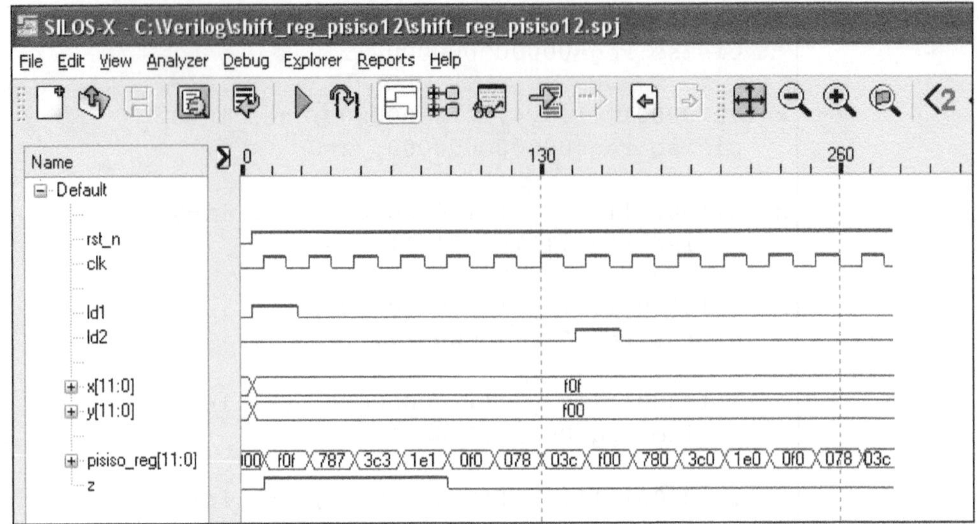

Figure 17.97 Waveforms for the 12-bit parallel-in, serial-in, serial-out shift register.

17.8.4 Serial-In, Parallel-Out Shift Register

The *serial-in, parallel-out* (SIPO) register is another typical synchronous iterative network containing *p* identical cells. For a right-shift operation, data enters the register from the left and shifts serially to the right through all *p* stages, one bit position per clock pulse. After *p* shifts, the register is fully loaded and the bits can be transferred in parallel to the destination.

The data input of each flip-flop is connected directly to the output of the preceding flip-flop with the exception of the leftmost flip-flop, which receives the external serial binary data. Shift registers of any type can be designed using *JK* flip-flops, *D* flip-flops, or *SR* latches — all are equally acceptable storage elements. Each stage of the machine is required to perform only one function: store the state of the storage element to its immediate left.

A typical application of a serial-in, parallel-out register is to deserialize binary data from a single-track peripheral subsystem. The resulting word of parallel bits is placed on the system data bus.

An 8-bit serial-in, parallel-out shift register is shown in Figure 17.98. The behavioral module is shown in Figure 17.99 and the test bench module is shown in Figure 17.100. The outputs and waveforms are shown in Figure 17.101 and Figure 17.102, respectively.

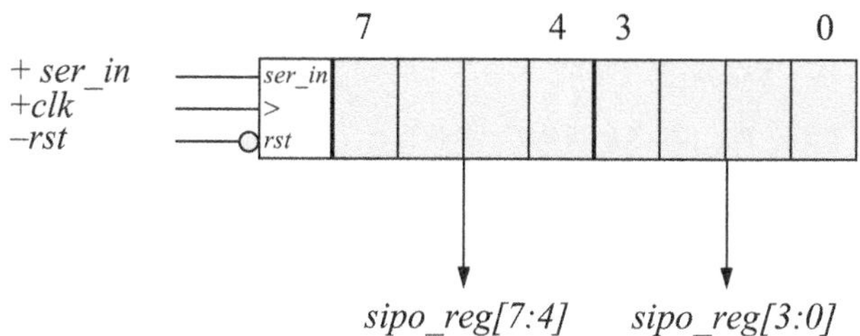

Figure 17.98 Organization of a serial-in, parallel-out shift register.

```
//behavioral serial-in, parallel-out shift register
module shift_reg_sipo (rst_n, clk, ser_in, sipo_reg);

//define inputs and outputs
input rst_n, clk, ser_in;
output [7:0] sipo_reg;

//variables used in an always block are declared as registers
reg [7:0] sipo_reg;

always @ (rst_n)
begin
   if (rst_n == 0)
      sipo_reg <= 8'b0000_0000;
end

always @ (posedge clk)
begin
   sipo_reg [7] <= ser_in;
   sipo_reg [6] <= sipo_reg [7];
   sipo_reg [5] <= sipo_reg [6];
   sipo_reg [4] <= sipo_reg [5];
   sipo_reg [3] <= sipo_reg [4];
   sipo_reg [2] <= sipo_reg [3];
   sipo_reg [1] <= sipo_reg [2];
   sipo_reg [0] <= sipo_reg [1];
end

endmodule
```

Figure 17.99 Behavioral module for the serial-in, parallel-out shift register.

```
//test bench for serial-in parallel-out shift register
module shift_reg_sipo_tb;

reg rst_n, clk, ser_in;
wire [7:0] sipo_reg;

initial     //define clock
begin
   clk = 1'b0;
   forever
      #10 clk = ~clk;
end

//display variables
initial
$monitor ("ser_in = %b, shift_reg = %b", ser_in, sipo_reg);

initial     //apply inputs
begin
   #0    rst_n = 1'b0;
         ser_in = 1'b0;
   #5    rst_n = 1'b1;
         ser_in = 1'b1;

   #10   ser_in = 1'b1;
   #10   ser_in = 1'b0;
   #10   ser_in = 1'b1;
   #10   ser_in = 1'b0;
   #10   ser_in = 1'b1;
   #10   ser_in = 1'b1;
   #10   ser_in = 1'b0;
   #10   ser_in = 1'b1;
   #10   ser_in = 1'b1;
   #10   ser_in = 1'b1;
   #20   ser_in = 1'b0;
   #10   ser_in = 1'b1;
   #30   $stop;
end

//instantiate the module into the test bench
shift_reg_sipo inst1 (
   .rst_n(rst_n),
   .clk(clk),
   .ser_in(ser_in),
   .sipo_reg(sipo_reg)
   );
endmodule
```

Figure 17.100 Test bench for the serial-in, parallel-out shift register.

```
ser_in = 0, shift_reg = 00000000
ser_in = 1, shift_reg = 00000000
ser_in = 1, shift_reg = 10000000
ser_in = 0, shift_reg = 10000000
ser_in = 0, shift_reg = 01000000
ser_in = 1, shift_reg = 01000000
ser_in = 0, shift_reg = 01000000
ser_in = 0, shift_reg = 00100000
ser_in = 1, shift_reg = 00100000
ser_in = 1, shift_reg = 10010000
ser_in = 0, shift_reg = 10010000
ser_in = 1, shift_reg = 10010000
ser_in = 1, shift_reg = 11001000
ser_in = 1, shift_reg = 11100100
ser_in = 0, shift_reg = 11100100
ser_in = 0, shift_reg = 01110010
ser_in = 1, shift_reg = 01110010
ser_in = 1, shift_reg = 10111001
```

Figure 17.101 Outputs for the serial-in, parallel-out shift register.

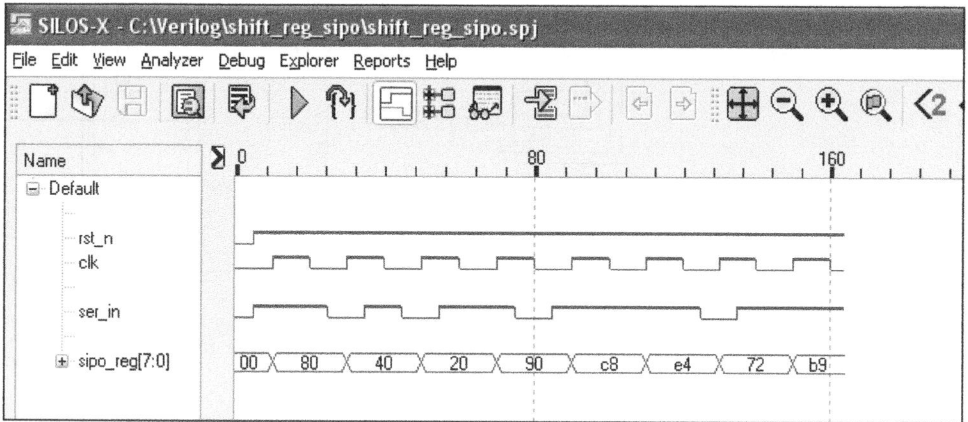

Figure 17.102 Waveforms for the serial-in, parallel-out shift register.

Another use for a serial-in, parallel-out shift register is to generate nonoverlapping pulses for certain applications. Additional logic is required to generate the serial input and to provide feedback to the serial input logic. This method will be used to create a design that will generate eight nonoverlapping pulses to provide eight unique states using the 8-bit SIPO register previously described. This provides a simple, yet effective state machine, where each pulse represents a different state. A small amount of additional logic is required as shown in the logical organization of Figure 17.103.

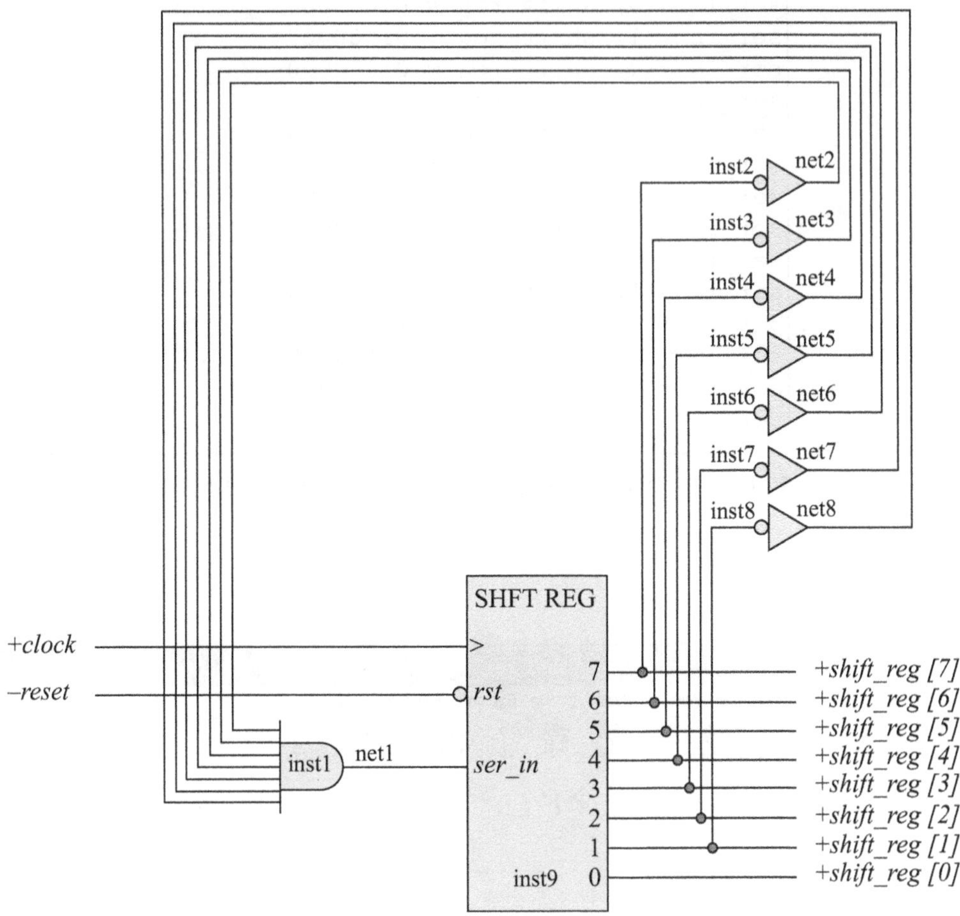

Figure 17.103 Logical organization for a SIPO shift register that generates eight nonoverlapping pulses.

The shift register is reset initially, providing low voltage levels for the eight outputs. The low voltage outputs are inverted and fed back to the 7-input AND gate that generates the serial input, thus asserting output *+shift_reg[7]* at a high voltage level. This results in a low voltage (0) for *net1*. As the 1 bit is shifted through the register, each successive output — *+shift_reg[7]* through *+shift_reg[1]* — produces a high voltage level, which generates a zero for the serial input. When the 1 bit reaches *+shift_reg[0]*, the other outputs are at a low voltage level, which generates a 1 bit for the serial input, and the process repeats.

The SIPO shift register previously designed will be instantiated into the structural module of Figure 17.104 to generate eight nonoverlapping pulses using the organization of Figure 17.103. The test bench is shown in Figure 17.105. The outputs and waveforms are shown in Figure 17.106 and Figure 17.107, respectively.

```
//structural shifter pulse generator
module shftr_pulse_gen (rst_n, clk, shift_reg);

input rst_n, clk;
output [7:0] shift_reg;

wire rst_n, clk;
wire [7:0] shift_reg;

//instantiate the logic for the shifter
and8_df inst1 (
   .x1(net2),
   .x2(net3),
   .x3(net4),
   .x4(net5),
   .x5(net6),
   .x6(net7),
   .x7(net8),
   .x8(1'b1),
   .z1(net1)
   );

not   inst2 (net2, shift_reg [7]),
      inst3 (net3, shift_reg [6]),
      inst4 (net4, shift_reg [5]),
      inst5 (net5, shift_reg [4]),
      inst6 (net6, shift_reg [3]),
      inst7 (net7, shift_reg [2]),
      inst8 (net8, shift_reg [1]);

//instantiate the shifter
shift_reg_sipo inst9 (
   .rst_n(rst_n),
   .clk(clk),
   .ser_in(net1),
   .sipo_reg(shift_reg)
   );

endmodule
```

Figure 17.104 Structural module for the pulse generator.

```
//test bench for the pulse generator
module shftr_pulse_gen_tb;

reg rst_n, clk;
wire [7:0] shift_reg;

//display outputs
initial
$monitor ("out = %b", shift_reg);

//generate reset
initial
begin
   #0    rst_n = 1'b0;
   #2    rst_n = 1'b1;
end

//generate clock
initial
begin
   clk = 1'b0;
   forever
      #10   clk = ~clk;
end

//determine length of simulation
initial
begin
   repeat (10) @ (posedge clk)
   #200  $stop;
end

//instantiate the module into the test bench
shftr_pulse_gen inst1 (
   .rst_n(rst_n),
   .clk(clk),
   .shift_reg(shift_reg)
   );
endmodule
```

Figure 17.105 Test bench for the pulse generator.

The test bench uses the **repeat** keyword to execute a loop a fixed number of times as specified by a constant contained within parentheses following the **repeat** keyword. The loop can be a single statement or a block of statements contained within **begin . . . end** keywords. The syntax is shown below.

repeat (loop count expression)
procedural statement or block of procedural statements

When the activity flow reaches the **repeat** construct, the expression in parentheses is evaluated to determine the number of times that the loop is to be executed. The *loop count expression* can be a constant, a variable, or a signal value. If the expression evaluates to **x** or **z**, then the value is treated as 0 and the loop is not executed. The **repeat** keyword can also be used as a *repeat event control* to delay the assignment of the right-hand expression to the left-hand target. This is a form of intrastatement delay. The syntax is shown below.

repeat (expression) @ (event_expression)

Execution of the **repeat** loop can be terminated by a **disable** statement before the loop has executed the specified number of times. The **disable** statement terminates a named block of procedural statements or a task, and transfers control to the statement immediately following the block or task.

The waveforms of Figure 17.107 clearly show the eight nonoverlapping disjoint pulses.

```
out = 00000000        out = 00010000        out = 00000001
out = 10000000        out = 00001000        out = 10000000
out = 01000000        out = 00000100        out = 01000000
out = 00100000        out = 00000010
```

Figure 17.106 Outputs for the pulse generator.

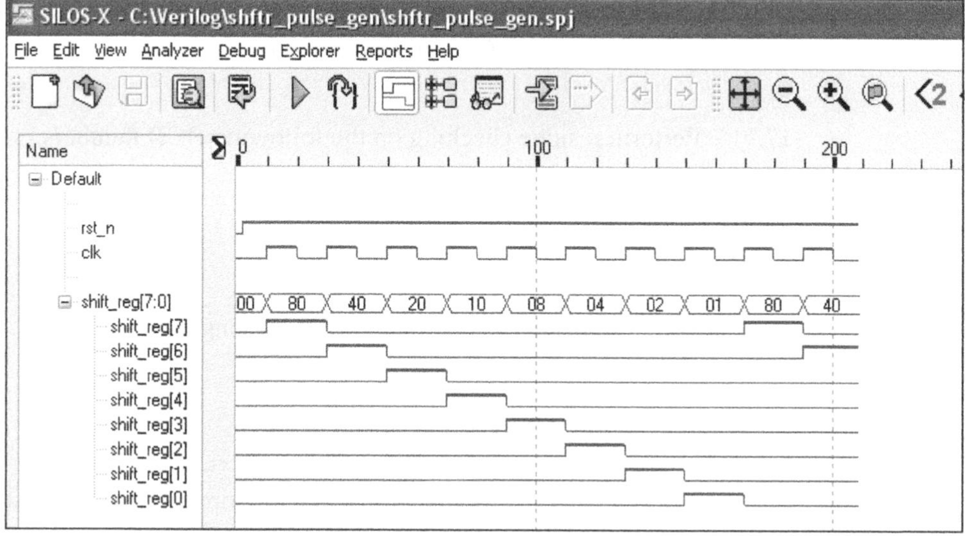

Figure 17.107 Waveforms for the pulse generator.

17.9 Problems

17.1 Perform residue checking on the following BCD numbers:

1000 0111 0101
0110 0100 0010

17.2 Perform residue checking on the following BCD numbers:

1001 1001 1001
1001 1001 1001

17.3 Perform residue checking on the following numbers:

0010 1010 Carry-in = 0
0101 0111 Carry-out = 0

17.4 Perform residue checking on the following numbers:

0010 1010 Carry-in = 1
0101 0111 Carry-out = 0

17.5 Perform residue checking on the following numbers:

0111 1010 Carry-in = 0
1001 0101 Carry-out = 1

17.6 Perform residue checking on the following numbers:

0111 1010 Carry-in = 1
1001 0100 Carry-out = 1

17.7 Perform residue checking on the following BCD numbers:

0110 0111 0100
0111 0011 1000

17.8 Perform residue checking on the following numbers:

0111 1010 Carry-in = 1
1001 0100 Carry-out = 1

Assume that an error occured in the low-order bit of the sum to indicate a residue error.

17.9 Perform parity prediction on the following operands with a carry-in = 1: 0111 and 0111.

17.10 Perform parity prediction on the following operands with a carry-in = 0: 1010 and 1010.

17.11 Perform parity prediction on the following operands with a carry-in = 0:

$$A = 0111\ \ 0001$$
$$B = 0010\ \ 0001$$

17.12 Perform parity prediction on the following operands with a carry-in = 1:

$$A = 1100\ \ 0110$$
$$B = 0101\ \ 1101$$

Then draw the parity prediction logic that checks for errors during addition.

17.13 Derive the truth table to determine if the sum of two 16-bit operands is positive. Then draw the logic that will generate a logic 1 if the sign is positive.

17.14 Design a Verilog HDL module that generates the following results when two 8-bit operands are added: sum, determine if overflow occurs, numeric part of sum (*sum[6:0]*) equals zero, the sign is positive, the sum is less than zero, the sum equals zero, and the sum is greater than zero.

Obtain the behavioral module, the test bench module, the outputs, and the waveforms.

17.15 Design a 16-bit adder using Verilog HDL. Obtain the dataflow design module and the test bench module that applies several input vectors for the augend and addend, including a carry-in. Have at least one set of inputs that produce a carry-out. Obtain the outputs and the waveforms.

17.16 Design an 8-bit shifter, using 4:1 multiplexers, that executes a shift left algebraic operation. Obtain the structural module and the test bench module that applies several different input vectors for the operand that is to be shifted. Obtain the outputs and the waveforms.

17.17 Design an 8-bit shifter using 4:1 multiplexers that executes a shift right logical operation. Obtain the structural module and the test bench module that applies several different input vectors for the operand that is to be shifted. Obtain the outputs and the waveforms.

17.18 Design an 8-bit shifter, using 8:1 multiplexers, that executes a shift right algebraic operation. Obtain the structural module and the test bench module that applies several different input vectors for the operand that is to be shifted. Obtain the outputs and the waveforms.

17.19 Use behavioral modeling with the **case** statement to design a 4-function arithmetic and logic unit to perform the following operations on 4-bit operands: add, subtract, multiply, and divide. Then obtain the test bench, the outputs, and the waveforms.

17.20 Draw the logic diagram for an arithmetic and logic unit (ALU) to perform addition, subtraction, AND, and OR on the two 4-bit operands *a[3:0]* and *b[3:0]*. Use an adder and multiplexers with additional logic gates. Use structural modeling to design the 4-function ALU. Then obtain the test bench, the outputs, and the waveforms.

Use two control inputs to determine the operation to be performed, according to the table shown below.

c[1]	*c[0]*	Operation
0	0	Add
0	1	Subtract
1	0	AND
1	1	OR

17.21 Using mixed-design modeling (behavioral and dataflow), design a 4-function arithmetic and logic unit to execute the following four operations: add, multiply, AND, and OR. The operands are eight bits defined as *a[7:0]* and *b[7:0]*. The result of all operations will be 16 bits, *rslt[15:0]*. Determine the state of the following flags:

The parity flag *pf* = 1 if the result has an even number of 1s, otherwise, *pf* = 0.

The sign flag *sf* represents the state of the leftmost result bit.

The zero flag *zf* = 1 if the result of an operation is zero, otherwise, *zf* = 0.

Obtain the design module, the test bench module using several values for each of the four operations, the outputs, and the waveforms.

17.22 Using mixed-design modeling (behavioral and dataflow), design a 20-function arithmetic and logic unit to execute the operations shown below. The operands are eight bits defined as *a[7:0]* and *b[7:0]*. The result of all operations will be 16 bits, *rslt[15:0]*.

Operation		Operation Code	Description
add_op	=	5'b00000	Add
add_bcd_op	=	5'b00001	Add BCD operands
and_op	=	5'b00010	AND
cmp_op	=	5'b00011	Compare
dec_op	=	5'b00100	Decrement
div_op	=	5'b00101	Divide
inc_op	=	5'b00110	Increment
mul_op	=	5'b00111	Multiply
neg_op	=	5'b01000	Negate (2s complement)
not_op	=	5'b01001	NOT (1s complement)
or_op	=	5'b01010	OR
rol_op	=	5'b01011	Rotate left
ror_op	=	5'b01100	Rotate right
sla_op	=	5'b01101	Shift left algebraic
sll_op	=	5'b01110	Shift left logical
sra_op	=	5'b01111	Shift right algebraic
srl_op	=	5'b10000	Shift right logical
sub_op	=	5'b10001	Subtract
xor_op	=	5'b10010	Exclusive-OR
xnor_op	=	5'b10011	Exclusive-NOR

Determine the state of the following flags:

The parity flag *pf* = 1 if the result has an even number of 1s, otherwise, *pf* = 0.

The sign flag *sf* represents the state of the leftmost result bit.

The zero flag *zf* = 1 if the result of an operation is zero, otherwise, *zf* = 0.

The carry-out flag *cf* = 1 if there is a carry out of the high-order bit position.

The auxiliary carry flag *af* = 1 if there is a carry out of the low-order BCD decade.

Obtain the design module, the test bench module, and the outputs.

17.23 A count-down counter was presented in Section 17.7. Another type of counter that is used for specific applications is a Johnson counter. This is a counter in which any two contiguous state codes (or code words) differ by only one variable. It is similar, in this respect, to a Gray code counter. A 4-bit Johnson counter counts in the sequence shown below, where y_4 is the low-order bit.

y_1	y_2	y_3	y_4
0	0	0	0
1	0	0	0
1	1	0	0
1	1	1	0
1	1	1	1
0	1	1	1
0	0	1	1
0	0	0	1
0	0	0	0

A Johnson counter is designed by connecting the complement of the output of the final storage element to the input of the first storage element. Using the design procedure presented in Section 2.4, design the logic for a 4-bit Johnson counter using *D* flip-flops.

Then design a *D* flip-flop using behavioral modeling that will be used in the structural module of the Johnson counter. For the structural module, obtain the test bench, the outputs, and the waveforms.

17.24 Design the Johnson counter of Problem 17.23 using behavioral modeling with the **case** statement. Obtain the design module, the test bench module, the outputs, and the waveforms.

17.25 Develop a reciprocal algorithm to convert BCD-to-binary, similar to the algorithm used to convert binary-to-BCD in Section 10.1. Then use the paper-and-pencil method to convert 999_{BCD} to binary.

17.26 Design a behavioral module that converts two BCD digits to binary using a memory. Obtain the test bench that applies several input BCD vectors, the outputs, and the waveforms.

Appendix A

Verilog HDL Designs for Select Logic Functions

Several logic functions will be designed in this Appendix to supplement the Verilog HDL designs that were implemented throughout the book. The first few designs consist primarily of AND and OR gates, including exclusive-OR and exclusive-NOR functions. Although the various gates are available as built-in primitives, they will be designed as separate modules. Other designs are considered as logic macro functions, such as multiplexers, decoders, encoders, code converters, and adders.

It is assumed that the reader has an adequate background in combinational and sequential logic design; therefore, only the Verilog HDL implementations will be presented, not the actual synthesis (design) procedure for the logic functions.

In most cases, each design will be implemented using dataflow modeling, behavioral modeling, and structural modeling (where applicable) in order to provide contrasting methods of designing the same logic function. Test benches, outputs, and waveforms are also given, if necessary.

A.1 AND Gate

The logic symbol for a 2-input AND gate is shown in Figure A.1 with inputs x_1 and x_2 and output z_1. When both inputs are at a logic 1 voltage level, output z_1 is at a logic 1 voltage level.

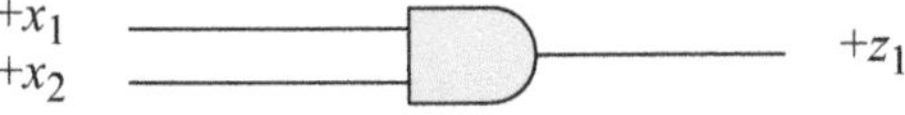

Figure A.1 Two-input AND gate.

Dataflow modeling The dataflow module is shown Figure A.2 using the continuous **assign** statement to design combinational logic. Continuous assignments can be applied only to nets. The left-hand side is declared as type **wire** not **reg**. When a variable on the right-hand side changes value, the right-hand side expression is evaluated and the value is assigned to the left-hand side net after any specified delay. The continuous assignment is used to place a value on a net. The test bench, outputs, and waveforms are shown in Figure A.3, Figure A.4, and Figure A.5, respectively.

```
//dataflow 2-input and gate
module and2_df (x1, x2, z1);

//list inputs and output
input x1, x2;
output z1;

//define signals as wire for dataflow
wire x1, x2;
wire z1;

//continuous assign for dataflow
assign z1 = x1 & x2;
endmodule
```

Figure A.2 Dataflow module for a 2-input AND gate.

```
//and2 test bench
module and2_df_tb;

reg x1, x2;
wire z1;

initial     //display inputs and outputs
$monitor ("inputs: x1x2 = %b, output = %b",
            {x1, x2}, z1);

initial     //apply input vectors
begin
   #0    x1 = 1'b0;  x2 = 1'b0;
   #10   x1 = 1'b0;  x2 = 1'b1;
   #10   x1 = 1'b1;  x2 = 1'b0;
   #10   x1 = 1'b1;  x2 = 1'b1;
   #10   x1 = 1'b0;  x2 = 1'b0;

   #10   $stop;
end

//instantiate the module into the test bench
and2_df inst1 (
   .x1(x1),
   .x2(x2),
   .z1(z1)
   );
endmodule
```

Figure A.3 Test bench for the 2-input AND gate.

```
inputs: x1x2 = 00, output = 0
inputs: x1x2 = 01, output = 0
inputs: x1x2 = 10, output = 0
inputs: x1x2 = 11, output = 1
inputs: x1x2 = 00, output = 0
```

Figure A.4 Outputs for the 2-input AND gate.

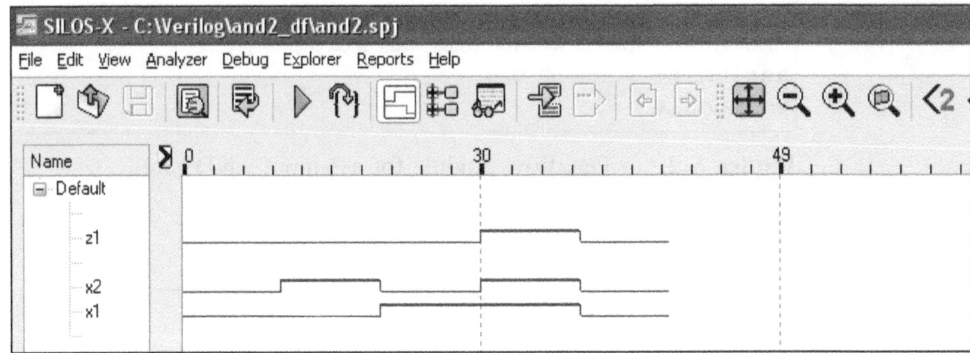

Figure A.5 Waveforms for the 2-input AND gate.

An alternative way to design a test bench is shown in Figure A.6. Since there are two inputs to the AND gate, all four combinations of two variables can be applied to the circuit. This is accomplished by a **for**-loop statement, which is similar in construction to the **for** loop in the C programming language. The outputs are in Figure A.7.

Following the keyword **begin** is the name of the block: *apply_stimulus*. In this block, a 3-bit **reg** variable is declared called *invect*. This guarantees that all combinations of the two inputs will be tested by the **for** loop, which applies input vectors of $x_1x_2 = 00, 01, 10, 11$ to the circuit. The **for** loop stops when the pattern 100 is detected by the test segment (*invect* < 4). If only a 2-bit vector were applied, then the expression (*invect* < 4) would always be true and the loop would never terminate. The increment segment of the **for** loop does not support an increment designated as *invect++*; therefore, the long notation must be used: *invect = invect + 1*.

```
//test bench for dataflow 2-input AND gate
module and2_dataflow_tb;

reg x1, x2;
wire z1;
//continued on next page
```

Figure A.6 Alternative way to design a test bench for the 2-input AND gate.

```
//generate stimulus and display variables
initial
begin: apply_stimulus
   reg [2:0] invect;
   for (invect=0; invect<4; invect=invect+1)
      begin
         {x1, x2} = invect [1:0];
         #10 $display ("x1x2 = %b%b, z1 = %b", x1, x2, z1);
      end
end

//instantiate the module into the test bench
and2_dataflow inst1 (
   .x1(x1),
   .x2(x2),
   .z1(z1)
   );

endmodule
```

Figure A.6 (Continued)

```
x1x2 = 00, z1 = 0
x1x2 = 01, z1 = 0
x1x2 = 10, z1 = 0
x1x2 = 11, z1 = 1
```

Figure A.7 Outputs for the 2-input AND gate obtained from the test bench of Figure A.6.

Behavioral modeling A 3-input AND gate is shown in Figure A.8. In this design, output z_1 is asserted when the inputs are $x_1x_2x_3 = 101$. The behavioral module for the 3-input AND gate is shown in Figure A.9 with an intrastatement delay of five time units. The test bench is shown in Figure A.10. The outputs and waveforms are shown in Figure A.11 and Figure A.12, respectively.

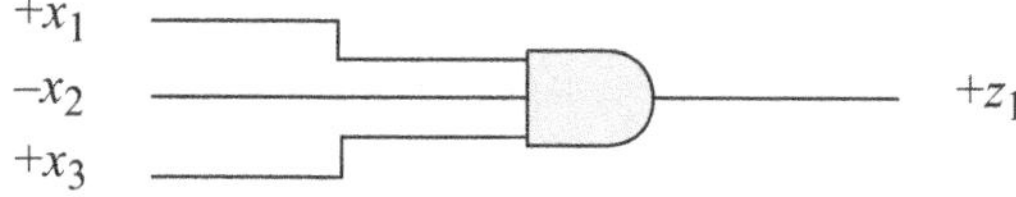

Figure A.8 Three-input AND gate.

```
//behavioral 3-input AND gate
module and3_behav (x1, x2, x3, z1);

input x1, x2, x3;
output z1;

wire x1, x2, x3;      //alternatively do not declare wires
                      //because inputs are wire by default

reg z1;               //variables used in an always block are
                      //declared as registers

always @ (x1 or x2 or x3)
   z1 = #5 (x1 & ~x2 & x3);

endmodule
```

Figure A.9 Behavioral module for a 3-input AND gate.

```
//test bench for bebavioral 3-input AND gate
module and3_behav_tb;

reg x1, x2, x3;
wire z1;

//generate stimulus and display variables
initial
begin: apply_stimulus
   reg [3:0] invect;
   for (invect=0; invect<8; invect=invect+1)
      begin
         {x1, x2, x3} = invect [2:0];
         #10 $display ("x1x2x3 = %b%b%b, z1 = %b",
                        x1, x2, x3, z1);
      end
end

//instantiate the module into the test bench
and3_behav inst1 (
   .x1(x1),
   .x2(x2),
   .x3(x3),
   .z1(z1)
   );
endmodule
```

Figure A.10 Test bench for the 3-input AND gate.

```
x1x2x3 = 000, z1 = 0
x1x2x3 = 001, z1 = 0
x1x2x3 = 010, z1 = 0
x1x2x3 = 011, z1 = 0
x1x2x3 = 100, z1 = 0
x1x2x3 = 101, z1 = 1
x1x2x3 = 110, z1 = 0
x1x2x3 = 111, z1 = 0
```

Figure A.11 Outputs for the 3-input AND gate.

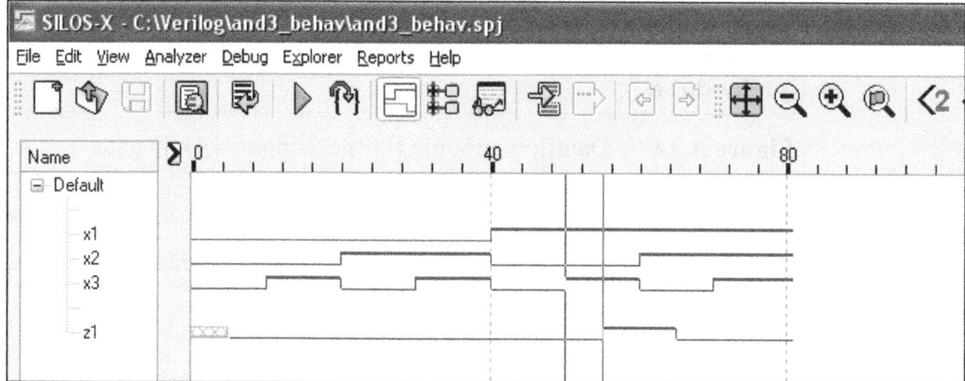

Figure A.12 Waveforms for the 3-input AND gate.

A.2 NAND Gate

A NAND gate is a universal logic gate, because it can be used to implement any of the following Boolean logic functions: AND, OR, and NOT, as shown in Figure A.13. Figure A.13(a) generates a low voltage output when x_1 and x_2 are both at a high voltage level; Figure A.13(b) generates a high voltage output when x_1 or x_2 is at a low voltage level; Figure A.13(c) performs the NOT function by inverting the input.

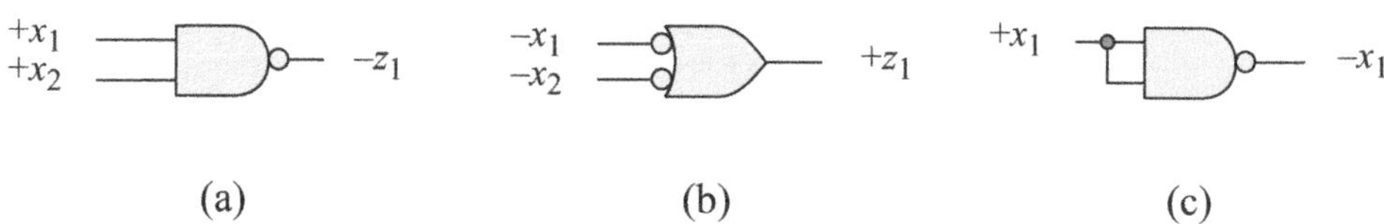

Figure A.13 NAND gate to implement logic functions: (a) AND; (b) OR, and (c) NOT.

The dataflow and behavioral modules for a 2-input NAND gate are shown in Figure A.14 and Figure A.15, respectively. The test bench, outputs, and waveforms for the dataflow module are shown in Figure A.16, Figure A.17, and Figure A.18, respectively.

```
//dataflow nand2 gate
module nand2_dataflow (x1, x2, z1);

input x1, x2;              //list inputs and output
output z1;

wire x1, x2;               //define signals as wire for dataflow
wire z1;

assign z1 = ~(x1 & x2); //continuous assign for dataflow
endmodule
```

Figure A.14 Dataflow module for the 2-input NAND gate.

```
//behavioral 2-input nand gate
module nand2_behav (x1, x2, z1);

input x1, x2;
output z1;

wire x1, x2;
reg z1;

always @ (x1 or x2)
begin
   z1 = ~(x1 & x2);
end
endmodule
```

Figure A.15 Behavioral module for the 2-input NAND gate.

```
//test bench for 2-input nand gate
module nand2_dataflow_tb;

reg x1, x2;
wire z1;
//continued on next page
```

Figure A.16 Test bench for the dataflow 2-input NAND gate.

```
initial
$monitor ("x1x2 = %b, z1 = %b", {x1, x2}, z1);

//apply input vectors
initial
begin
   #0    x1 = 1'b0;  x2 = 1'b0;
   #10   x1 = 1'b0;  x2 = 1'b1;
   #10   x1 = 1'b1;  x2 = 1'b0;
   #10   x1 = 1'b1;  x2 = 1'b1;

   #10   $stop;
end

//instantiate the module into the test bench
nand2_dataflow inst1 (
   .x1(x1),
   .x2(x2),
   .z1(z1)
   );
endmodule
```

Figure A.16 (Continued)

```
x1x2 = 00, z1 = 1
x1x2 = 01, z1 = 1
x1x2 = 10, z1 = 1
x1x2 = 11, z1 = 0
```

Figure A.17 Outputs for the dataflow 2-input NAND gate.

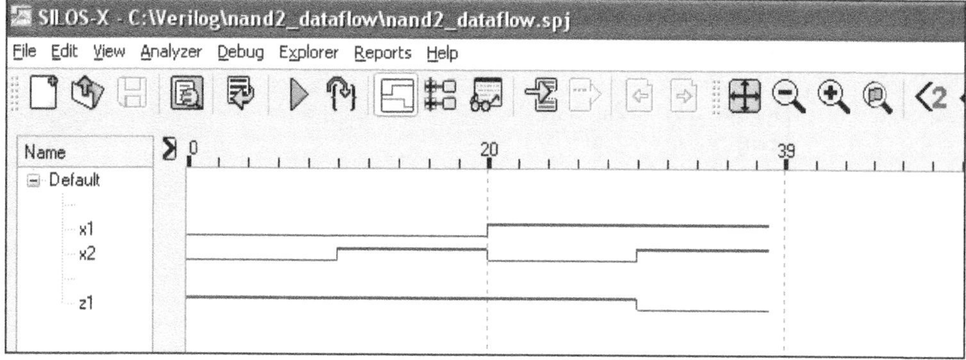

Figure A.18 Waveforms for the dataflow 2-input NAND gate.

A.3 OR Gate

A 3-input OR gate is shown in Figure A.19. Output z_1 is at its indicated polarity when one or more of the inputs is at a high voltage level. The dataflow and behavioral modules are shown in Figure A.20 and Figure A.21, respectively. The test bench, outputs, and waveforms for the behavioral module are shown in Figure A.22, Figure A.23, and Figure A.24, respectively.

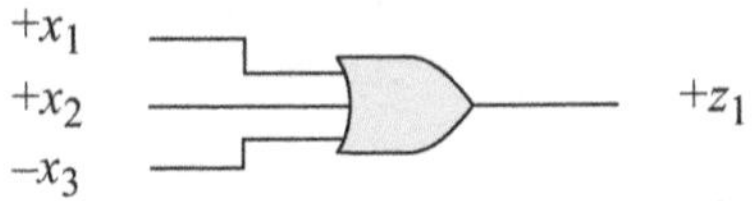

Figure A.19 Three-input OR gate.

```
//or3 dataflow
module or3_df (x1, x2, x3, z1);

input x1, x2, x3;
output z1;

wire x1, x2, x3;
wire z1;

assign z1 = x1 | x2 | ~x3;
endmodule
```

Figure A.20 Dataflow module for the 3-input OR gate.

```
//behavioral 3-input or gate
module or3_behav (x1, x2, x3, z1);

input x1, x2, x3;
output z1;

wire x1, x2, x3;
reg z1;

always @ (x1 or x2 or x3)
begin
   z1 = x1 | x2 | ~x3;
end
endmodule
```

Figure A.21 Behavioral module for the 3-input OR gate.

```
//or3 test bench
module or3_behav_tb;


reg x1, x2, x3;
wire z1;


//apply input vectors
initial
begin: apply_stimulus
   reg [3:0] invect;
   for (invect = 0; invect < 8; invect = invect + 1)
      begin
         {x1, x2, x3} = invect [3:0];
         #10 $display ("x1x2x3 = %b, z1 = %b",
                              {x1, x2, x3}, z1);
      end
end


//instantiate the module into the test bench
or3_behav inst1 (
   .x1(x1),
   .x2(x2),
   .x3(x3),
   .z1(z1)
   );

endmodule
```

Figure A.22 Test bench for the behavioral module of the 3-input OR gate.

```
x1x2x3 = 000, z1 = 1
x1x2x3 = 001, z1 = 0
x1x2x3 = 010, z1 = 1
x1x2x3 = 011, z1 = 1
x1x2x3 = 100, z1 = 1
x1x2x3 = 101, z1 = 1
x1x2x3 = 110, z1 = 1
x1x2x3 = 111, z1 = 1
```

Figure A.23 Outputs for the behavioral module of the 3-input OR gate.

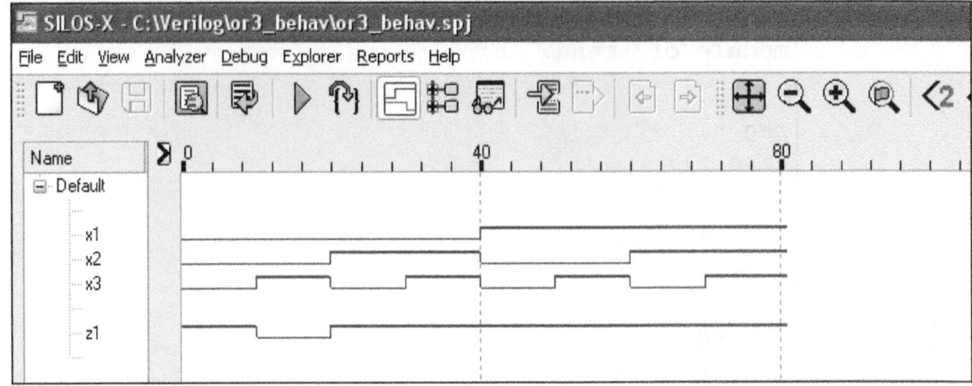

Figure A.24 Waveforms for the behavioral module of the 3-input OR gate.

A.4 NOR Gate

The NOR gate is also a universal gate, because it can be used to implement any of the following Boolean logic functions: OR, AND, and NOT, as shown in Figure A.25. Figure A.25(a) generates a low voltage output when either x_1 or x_2 or x_3 is at a high voltage level; Figure A.25(b) generates a high voltage output when x_1 and x_2 and x_3 are all at a low voltage level; Figure A.25(c) performs the NOT function by inverting the input.

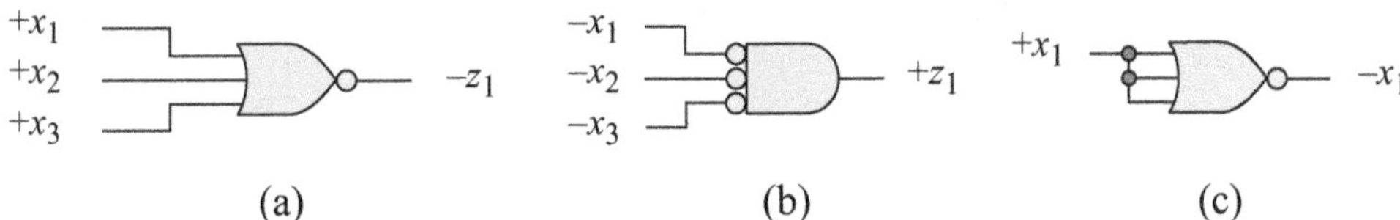

Figure A.25 NOR gate to implement logic functions: (a) OR; (b) AND, and (c) NOT.

The dataflow and behavioral modules for a 3-input NOR gate are shown in Figure A.26 and Figure A.27, respectively. The test bench, outputs, and waveforms for the dataflow module are shown in Figure A.28, Figure A.29, and Figure A.30, respectively.

```
//dataflow 3-input nor gate
module nor3_df (x1, x2, x3, z1);

input x1, x2, x3;
output z1;

wire x1, x2, x3;
wire z1;

assign z1 = ~(x1 | x2 | x3);
endmodule
```

Figure A.26 Dataflow module for the 3-input NOR gate.

```
//behavioral 3-input nor gate
module nor3_behav (x1, x2, x3, z1);

input x1, x2, x3;
output z1;

wire x1, x2;
reg z1;

always @ (x1 or x2 or x3)
begin
   z1 = ~(x1 | x2 | x3);
end
endmodule
```

Figure A.27 Behavioral module for the 3-input NOR gate.

```
//nor3_dataflow test bench
module nor3_df_tb;

reg x1, x2, x3;
wire z1;

initial      //display the variables
$monitor ("x1x2x3 = %b, z1 = %b",
         {x1, x2, x3}, z1);

//continued on next page
```

Figure A.28 Test bench for the dataflow module for the 3-input NOR gate.

```
initial      //apply input vectors
begin
   #0     x1 = 1'b0;  x2 = 1'b0;  x3 = 1'b0;
   #10    x1 = 1'b0;  x2 = 1'b0;  x3 = 1'b1;
   #10    x1 = 1'b0;  x2 = 1'b1;  x3 = 1'b0;
   #10    x1 = 1'b0;  x2 = 1'b1;  x3 = 1'b1;
   #10    x1 = 1'b1;  x2 = 1'b0;  x3 = 1'b0;
   #10    x1 = 1'b1;  x2 = 1'b0;  x3 = 1'b1;
   #10    x1 = 1'b1;  x2 = 1'b1;  x3 = 1'b0;
   #10    x1 = 1'b1;  x2 = 1'b1;  x3 = 1'b1;
   #10    $stop;
end

nor3_df inst1 (        //instantiate the module
   .x1(x1),
   .x2(x2),
   .x3(x3),
   .z1(z1)
   );
endmodule
```

Figure A.28 (Continued)

```
x1x2x3 = 000, z1 = 1      x1x2x3 = 100, z1 = 0
x1x2x3 = 001, z1 = 0      x1x2x3 = 101, z1 = 0
x1x2x3 = 010, z1 = 0      x1x2x3 = 110, z1 = 0
x1x2x3 = 011, z1 = 0      x1x2x3 = 111, z1 = 0
```

Figure A.29 Outputs for the dataflow module for the 3-input NOR gate.

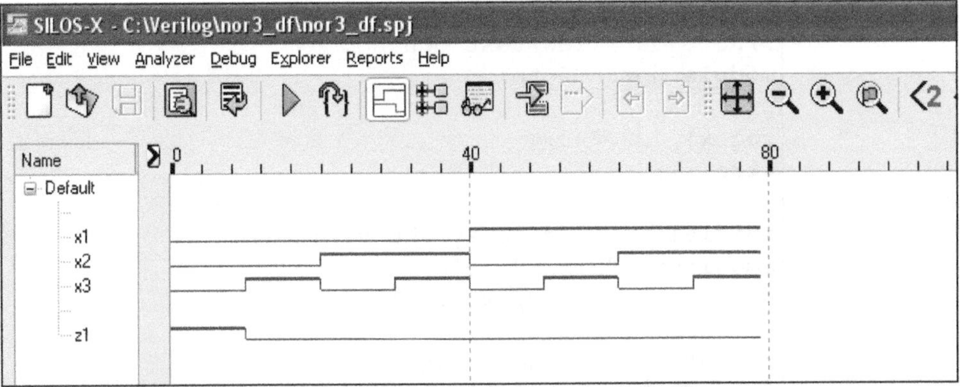

Figure A.30 Waveforms for the dataflow module for the 3-input NOR gate.

A.5 Exclusive-OR Function

Like the previous gates, this is a built-in primitive gate, but can also be designed using dataflow or behavioral modeling. The output of a 2-input exclusive-OR gate is a logical 1 whenever the two inputs are different. The exclusive-OR gate can also be used as an inverter by connecting one of the inputs to a logical 1. If the other input is a logical 0, then the inputs are different and the output is a logical 1, thus inverting the zero input. The converse is true when the second input is a logical 1 — the inputs are then identical providing a logical 0 output.

The symbol for an exclusive-OR function is shown in Figure A.31. The dataflow module and the test bench module are shown in Figure A.32 and Figure A.33, respectively. The outputs are shown in Figure A.34; the waveforms are shown in Figure A.35. The behavioral module is shown in Figure A.36.

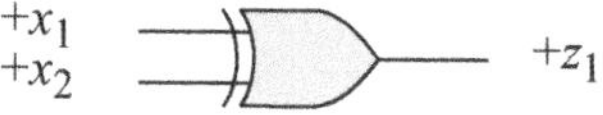

Figure A.31 Symbol for the exclusive-OR function.

```
//dataflow exclusive-or
module xor2_df (x1, x2, z1);

input x1, x2;           //list inputs and outputs
output z1;

wire x1, x2;
wire z1;                //define signals as wire for dataflow

assign z1 = x1 ^ x2; //continuous assignment for dataflow
endmodule
```

Figure A.32 Dataflow module for a 2-input exclusive-OR function.

```
//xor2_df test bench
module xor2_df_tb;

reg x1, x2;
wire z1;
//continued on next page
```

Figure A.33 Test bench for the 2-input exclusive-OR function.

```
initial
$monitor ("inputs: x1x2 = %b, output: = %b",
            {x1, x2}, z1);

initial
begin
   #0     x1 = 1'b0;   x2 = 1'b0;
   #10    x1 = 1'b0;   x2 = 1'b1;
   #10    x1 = 1'b1;   x2 = 1'b0;
   #10    x1 = 1'b1;   x2 = 1'b1;
   #10    $stop;
end

xor2_df inst1 (
   .x1(x1),
   .x2(x2),
   .z1(z1)
   );

endmodule
```

Figure A.33 (Continued)

```
inputs: x1x2 = 00, output: = 0
inputs: x1x2 = 01, output: = 1
inputs: x1x2 = 10, output: = 1
inputs: x1x2 = 11, output: = 0
```

Figure A.34 Outputs for the 2-input exclusive-OR function.

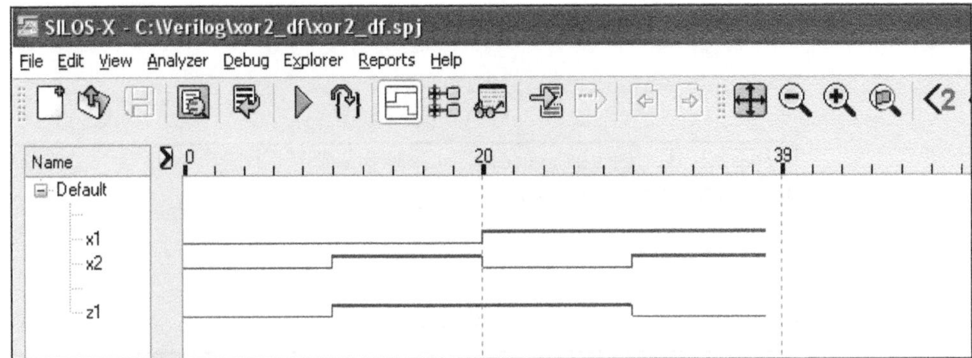

Figure A.35 Waveforms for the 2-input exclusive-OR function.

```
//behavioral 2-input exclusive-OR circuit
module xor2_bh (x1, x2, z1);

input x1, x2;
output z1;

reg z1;       //variables used in an always block are registers

always @ (x1 or x2)
   z1 = x1 ^ x2;
endmodule
```

Figure A.36 Behavioral module for a 2-input exclusive-OR function.

A multiple-input exclusive-OR function can be implemented in Verilog HDL. Figure A.37 illustrates a dataflow module for a 5-input exclusive-OR function. The test bench and outputs are shown in Figure A.38 and Figure A.39, respectively. Notice that output z_1 is asserted only when there are an odd number of logic 1s on the inputs.

```
//dataflow 5-input exclusive-OR
module xor5_df2 (x1, x2, x3, x4, x5, z1);

input x1, x2, x3, x4, x5;
output z1;

wire x1, x2, x3, x4, x5;
wire z1;

assign z1 = x1 ^ x2 ^ x3 ^ x4 ^ x5;

endmodule
```

Figure A.37 Dataflow module for a 5-input exclusive-OR function.

```
//test bench for 5-input exclusive-or function
module xor5_df2_tb;

reg x1, x2, x3, x4, x5;
wire z1;

//continued on next page
```

Figure A.38 Test bench for the 5-input exclusive-OR function.

```
//apply input vectors and display variables
initial
begin: apply_stimulus
   reg [5:0] invect;
   for (invect = 0; invect < 32; invect = invect + 1)
      begin
         {x1, x2, x3, x4, x5} = invect [5:0];
         #10 $display ("x1 x2 x3 x4 x5 = %b, z1 = %b",
                        {x1, x2, x3, x4, x5}, z1);
      end
end

//instantiate the module into the test bench
xor5_df2 inst1 (
   .x1(x1),
   .x2(x2),
   .x3(x3),
   .x4(x4),
   .x5(x5),
   .z1(z1)
   );

endmodule
```

Figure A.38 (Continued)

```
x1 x2 x3 x4 x5 = 00000, z1 = 0 | x1 x2 x3 x4 x5 = 10000, z1 = 1
x1 x2 x3 x4 x5 = 00001, z1 = 1 | x1 x2 x3 x4 x5 = 10001, z1 = 0
x1 x2 x3 x4 x5 = 00010, z1 = 1 | x1 x2 x3 x4 x5 = 10010, z1 = 0
x1 x2 x3 x4 x5 = 00011, z1 = 0 | x1 x2 x3 x4 x5 = 10011, z1 = 1
x1 x2 x3 x4 x5 = 00100, z1 = 1 | x1 x2 x3 x4 x5 = 10100, z1 = 0
x1 x2 x3 x4 x5 = 00101, z1 = 0 | x1 x2 x3 x4 x5 = 10101, z1 = 1
x1 x2 x3 x4 x5 = 00110, z1 = 0 | x1 x2 x3 x4 x5 = 10110, z1 = 1
x1 x2 x3 x4 x5 = 00111, z1 = 1 | x1 x2 x3 x4 x5 = 10111, z1 = 0
x1 x2 x3 x4 x5 = 01000, z1 = 1 | x1 x2 x3 x4 x5 = 11000, z1 = 0
x1 x2 x3 x4 x5 = 01001, z1 = 0 | x1 x2 x3 x4 x5 = 11001, z1 = 1
x1 x2 x3 x4 x5 = 01010, z1 = 0 | x1 x2 x3 x4 x5 = 11010, z1 = 1
x1 x2 x3 x4 x5 = 01011, z1 = 1 | x1 x2 x3 x4 x5 = 11011, z1 = 0
x1 x2 x3 x4 x5 = 01100, z1 = 0 | x1 x2 x3 x4 x5 = 11100, z1 = 1
x1 x2 x3 x4 x5 = 01101, z1 = 1 | x1 x2 x3 x4 x5 = 11101, z1 = 0
x1 x2 x3 x4 x5 = 01110, z1 = 1 | x1 x2 x3 x4 x5 = 11110, z1 = 0
x1 x2 x3 x4 x5 = 01111, z1 = 0 | x1 x2 x3 x4 x5 = 11111, z1 = 1
```

Figure A.39 Outputs for the 5-input exclusive-OR function.

A.6 Exclusive-NOR Function

This is also a built-in primitive gate, but can be designed using dataflow or behavioral modeling. The output of a 2-input exclusive-NOR gate is a logical 1 whenever the two inputs are the same. Therefore, the exclusive-NOR function is also referred to as an *equality* function.

The symbol for an exclusive-NOR function is shown in Figure A.40. The dataflow module and the test bench module are shown in Figure A.41 and Figure A.42, respectively. The outputs are shown in Figure A.43; the waveforms are shown in Figure A.44. The behavioral module is shown in Figure A.45.

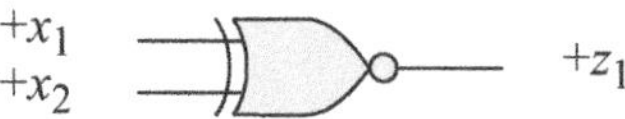

Figure A.40 Symbol for the exclusive-NOR function.

```
//dataflow 2-input exclusive-nor
module xnor2_df (x1, x2, z1);

input x1, x2;
output z1;

wire x1, x2;
wire z1;

assign z1 = ~(x1 ^ x2); //continuous assign used for dataflow
endmodule
```

Figure A.41 Dataflow module for a 2-input exclusive-NOR function.

```
//dataflow xnor2_df test bench
module xnor2_tb;

reg x1, x2;     //inputs are reg for test bench
wire z1;        //outputs are wire for test bench

//apply input vectors and display variables
initial
begin: apply_stimulus      //continued on next page
```

Figure A.42 Test bench for the 2-input exclusive-NOR function.

```
   reg [2:0] invect;
   for (invect = 0; invect < 4; invect = invect + 1)
      begin
         {x1, x2} = invect [2:0];
         #10 $display ("x1 x2 = %b, z1 = %b", {x1, x2}, z1);
      end
end

//instantiate the dataflow module into the test bench
xnor2_df inst1 (
   .x1(x1),
   .x2(x2),
   .z1(z1)
   );

endmodule
```

Figure A.42 (Continued)

```
x1 x2 = 00, z1 = 1
x1 x2 = 01, z1 = 0
x1 x2 = 10, z1 = 0
x1 x2 = 11, z1 = 1
```

Figure A.43 Outputs for the 2-input exclusive-NOR function.

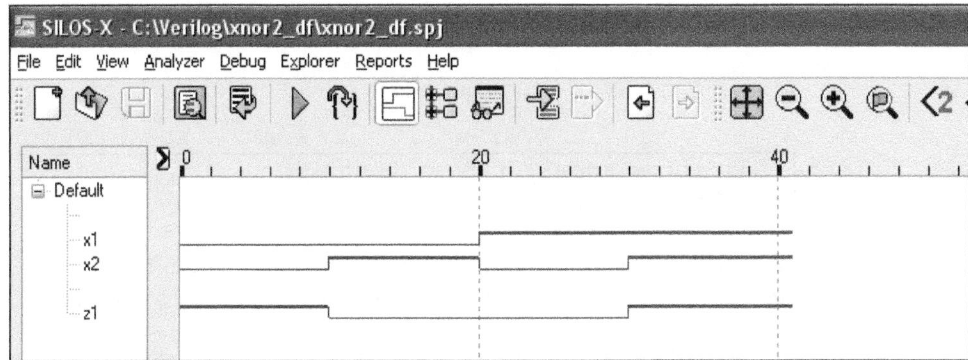

Figure A.44 Waveforms for the 2-input exclusive-NOR function.

```
//behavioral exclusive-nor circuit
module xnor2_bh (x1, x2, z1);

input x1, x2;
output z1;

wire x1, x2;        //inputs are wire by default for behavioral

//variables used in an always block are declared as reg
reg z1;

always @ (x1 or x2)
begin
   z1 = ~(x1 ^ x2);
end
endmodule
```

Figure A.45 Behavioral module for the 2-input exclusive-NOR function.

A multiple-input exclusive-NOR function can be implemented in Verilog HDL. Figure A.46 illustrates a dataflow module for a 5-input exclusive-NOR function. The test bench and outputs are shown in Figure A.47 and Figure A.48, respectively. Notice that output z_1 is asserted at a logic 1 voltage level only when there are an even number of logic 1s on the inputs.

```
//dataflow 5-input exclusive-nor function
module xnor5_df (x1, x2, x3, x4, x5, z1);


//list all inputs and outputs
input x1, x2, x3, x4, x5;
output z1;


//all signals are wire
wire x1, x2, x3, x4, x5;
wire z1;


//continuous assign used for dataflow
assign z1 = ~(x1 ^ x2 ^ x3 ^ x4 ^ x5);

endmodule
```

Figure A.46 Dataflow module for a 5-input exclusive-NOR function.

```
//test bench for 5-input exclusive-or function
module xnor5_df_tb;

reg x1, x2, x3, x4, x5;
wire z1;

initial      //apply input vectors and display variables
begin: apply_stimulus
   reg [5:0] invect;
   for (invect = 0; invect < 32; invect = invect + 1)
      begin
         {x1, x2, x3, x4, x5} = invect [5:0];
         #10 $display ("x1 x2 x3 x4 x5 = %b, z1 = %b",
                           {x1, x2, x3, x4, x5}, z1);
      end
end

xnor5_df inst1 (      //instantiate the module
   .x1(x1),
   .x2(x2),
   .x3(x3),
   .x4(x4),
   .x5(x5),
   .z1(z1)
   );
endmodule
```

Figure A.47 Test bench for the 5-input exclusive-NOR function.

```
x1 x2 x3 x4 x5 = 00000, z1 = 1   x1 x2 x3 x4 x5 = 10000, z1 = 0
x1 x2 x3 x4 x5 = 00001, z1 = 0   x1 x2 x3 x4 x5 = 10001, z1 = 1
x1 x2 x3 x4 x5 = 00010, z1 = 0   x1 x2 x3 x4 x5 = 10010, z1 = 1
x1 x2 x3 x4 x5 = 00011, z1 = 1   x1 x2 x3 x4 x5 = 10011, z1 = 0
x1 x2 x3 x4 x5 = 00100, z1 = 0   x1 x2 x3 x4 x5 = 10100, z1 = 1
x1 x2 x3 x4 x5 = 00101, z1 = 1   x1 x2 x3 x4 x5 = 10101, z1 = 0
x1 x2 x3 x4 x5 = 00110, z1 = 1   x1 x2 x3 x4 x5 = 10110, z1 = 0
x1 x2 x3 x4 x5 = 00111, z1 = 0   x1 x2 x3 x4 x5 = 10111, z1 = 1
x1 x2 x3 x4 x5 = 01000, z1 = 0   x1 x2 x3 x4 x5 = 11000, z1 = 1
x1 x2 x3 x4 x5 = 01001, z1 = 1   x1 x2 x3 x4 x5 = 11001, z1 = 0
x1 x2 x3 x4 x5 = 01010, z1 = 1   x1 x2 x3 x4 x5 = 11010, z1 = 0
x1 x2 x3 x4 x5 = 01011, z1 = 0   x1 x2 x3 x4 x5 = 11011, z1 = 1
x1 x2 x3 x4 x5 = 01100, z1 = 1   x1 x2 x3 x4 x5 = 11100, z1 = 0
x1 x2 x3 x4 x5 = 01101, z1 = 0   x1 x2 x3 x4 x5 = 11101, z1 = 1
x1 x2 x3 x4 x5 = 01110, z1 = 0   x1 x2 x3 x4 x5 = 11110, z1 = 1
x1 x2 x3 x4 x5 = 01111, z1 = 1   x1 x2 x3 x4 x5 = 11111, z1 = 0
```

Figure A.48 Outputs for the 5-input exclusive-NOR function.

A.7 Multiplexers

A *multiplexer* is a logic macro device that allows digital information from two or more data inputs to be directed to a single output. Data input selection is controlled by a set of select inputs that determine which data input is gated to the output. The select inputs are labeled *sel*[n–1], $\cdots$, *sel*[i], $\cdots$, *sel*[2], *sel*[1], *sel*[0], where *sel*[0] is the low-order select input with a binary weight of 2^0 and *sel*[n–1] is the high-order select input with a binary weight of 2^{n-1}. The data inputs are labeled as follows:

$$data[0],\ data[1],\ data[2],\ \cdots,\ data[j],\ \cdots,\ data[2^n-1]$$

Thus, if a multiplexer has n select inputs, then the number of data inputs will be 2^n and will be labeled *data*[0] through *data*[2^n–1]. For example, if $n = 2$, then the multiplexer has two select inputs, *sel*[1] and *sel*[0], and four data inputs, *data*[0], *data*[1], *data*[2], and *data*[3].

This section will design an 8:1 multiplexer using dataflow modeling. The design module is shown in Figure A.49. The test bench is shown in Figure A.50 and tests all inputs for the eight combinations of the select inputs. The outputs and waveforms are shown in Figure A.51 and Figure A.52, respectively.

```
//dataflow 8:1 multiplexer
module mux_8to1_df (sel, data, out);

input [2:0] sel;
input [7:0] data;
output out;

assign   out = data[0] & ~sel[2] & ~sel[1] & ~sel[0]) |

              (data[1] & ~sel[2] & ~sel[1] & sel[0]) |

              (data[2] & ~sel[2] & sel[1] & ~sel[0]) |

              (data[3] & ~sel[2] & sel[1] & sel[0]) |

              (data[4] & sel[2] & ~sel[1] & ~sel[0]) |

              (data[5] & sel[2] & ~sel[1] & sel[0]) |

              (data[6] & sel[2] & sel[1] & ~sel[0]) |

              (data[7] & sel[2] & sel[1] & sel[0]);

endmodule
```

Figure A.49 Dataflow module for an 8:1 multiplexer.

```
//test bench for dataflow 8:1 multiplexer
module mux_8to1_df_tb;

reg [2:0] sel;
reg [7:0] data;
wire out;

initial     //display variables
$monitor ("sel = %b, data = %b, out = %b", sel, data, out);

//apply input vectors
initial
begin
   #0    sel = 3'b000;  data = 8'b0000_1110;
   #10   sel = 3'b000;  data = 8'b1111_0001;

   #10   sel = 3'b001;  data = 8'b1111_0000;
   #10   sel = 3'b001;  data = 8'b1111_0010;

   #10   sel = 3'b010;  data = 8'b1111_0000;
   #10   sel = 3'b010;  data = 8'b1111_0110;

   #10   sel = 3'b011;  data = 8'b1111_0111;
   #10   sel = 3'b011;  data = 8'b1111_1010;

   #10   sel = 3'b100;  data = 8'b1110_0000;
   #10   sel = 3'b100;  data = 8'b1111_0010;

   #10   sel = 3'b101;  data = 8'b1101_0000;
   #10   sel = 3'b101;  data = 8'b1111_0010;

   #10   sel = 3'b110;  data = 8'b1011_0000;
   #10   sel = 3'b110;  data = 8'b1111_0010;

   #10   sel = 3'b111;  data = 8'b0011_0000;
   #10   sel = 3'b111;  data = 8'b1111_0010;

   #10   $stop;
end

//instantiate the module into the test bench
mux_8to1_df inst1 (
   .sel(sel),
   .data(data),
   .out(out)
   );
endmodule
```

Figure A.50 Test bench for the dataflow 8:1 multiplexer.

```
sel = 000, data = 00001110, out = 0
sel = 000, data = 11110001, out = 1

sel = 001, data = 11110000, out = 0
sel = 001, data = 11110010, out = 1

sel = 010, data = 11110000, out = 0
sel = 010, data = 11110110, out = 1

sel = 011, data = 11110111, out = 0
sel = 011, data = 11111010, out = 1

sel = 100, data = 11100000, out = 0
sel = 100, data = 11110010, out = 1

sel = 101, data = 11010000, out = 0
sel = 101, data = 11110010, out = 1

sel = 110, data = 10110000, out = 0
sel = 110, data = 11110010, out = 1

sel = 111, data = 00110000, out = 0
sel = 111, data = 11110010, out = 1
```

Figure A.51 Outputs for the dataflow 8:1 multiplexer.

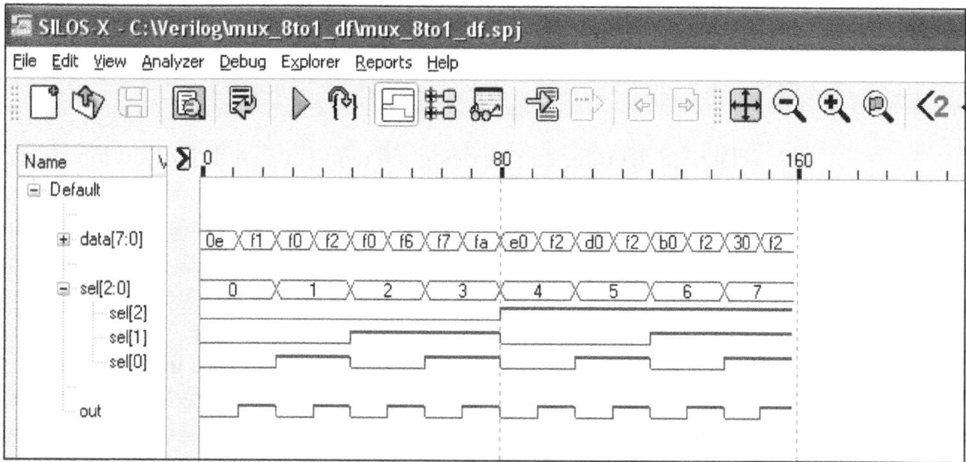

Figure A.52 Waveforms for the dataflow 8:1 multiplexer.

A.8 Decoders

A *decoder* is a combinational macro logic circuit that translates a binary input number to an equivalent output number for a specific radix. There is a one-to-one correspondence between the outputs and the combinations of the input signals. In general, there are n input lines and m output lines, where $m = 2^n$. For each combination of the 2^n input values, only one unique output signal is active — all other outputs are inactive. Thus, a fundamental characteristic of a decoder is the mutual exclusiveness of the outputs.

A decoder may also have an enable function which allows the selected output to be asserted. The enable function may be a single input or an AND gate with two or more inputs. If the enable function is deasserted, then the decoder outputs are deasserted. Table A.1 shows the truth table for a 3:8 decoder, which generates all eight minterms z_0 through z_7 of three binary variables x_1, x_2, and x_3. The truth table indicates the asserted output that represents the corresponding minterm.

Table A.1 Truth Table for a 3:8 Decoder

$x_1 x_2 x_3$	z_0	z_1	z_2	z_3	z_4	z_5	z_6	z_7
0 0 0	**1**	0	0	0	0	0	0	0
0 0 1	0	**1**	0	0	0	0	0	0
0 1 0	0	0	**1**	0	0	0	0	0
0 1 1	0	0	0	**1**	0	0	0	0
1 0 0	0	0	0	0	**1**	0	0	0
1 0 1	0	0	0	0	0	**1**	0	0
1 1 0	0	0	0	0	0	0	**1**	0
1 1 1	0	0	0	0	0	0	0	**1**

The 3:8 decoder can be designed using any of the following modeling techniques: built-in primitives, dataflow modeling, behavioral modeling, or structural modeling. In this section, the decoder will be implemented using built-in primitives that correspond to the logic diagram shown in Figure A.53 with optional instance names.

Verilog has a profuse set of built-in primitive gates that are used to model nets. The single output of each gate is declared as type **wire**. The inputs are declared as type **wire** or as type **reg** depending on whether they were generated by a structural or behavioral module. Built-in primitives are characterized by a low level of abstraction, where the logic hardware is described in terms of gates. Designing logic at this level is similar to designing logic by drawing gate symbols — there is a close correlation between the logic gate symbols and the Verilog built-in primitive gates. Each predefined primitive is declared by a keyword.

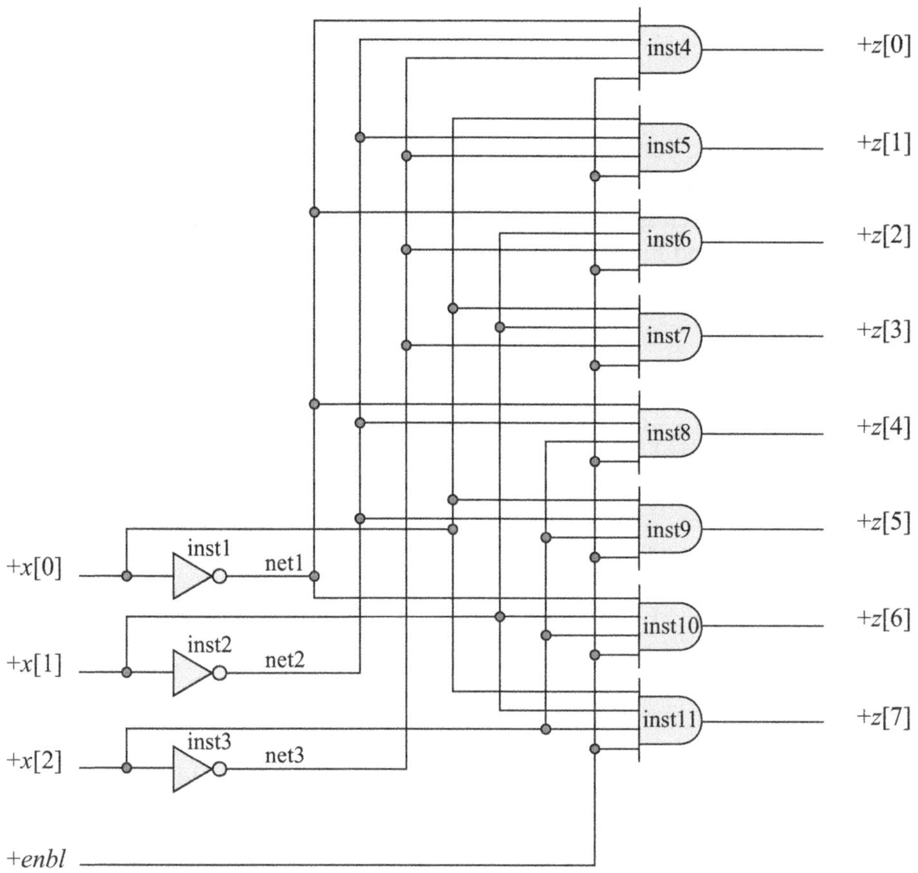

Figure A.53 Logic diagram for a 3:8 decoder.

The multiple-input gates are **and**, **nand**, **or**, **nor**, **xor**, and **xnor,** which are built-in primitive gates used to describe a net and have one or more scalar inputs, but only one scalar output. The output signal is listed first, followed by the inputs in any order. The outputs are declared as **wire**; the inputs can be declared as either **wire** or **reg**. The gates represent combinational logic functions and can be instantiated into a module, as follows, where the instance name is optional:

gate_type inst1 (output, input_1, input_2, . . . , input_*n*);

Two or more instances of the same type of gate can be specified in the same construct, as shown below. Note that only the last instantiation has a semicolon terminating the line. All previous lines are terminated by a comma.

gate_type inst1 (output_1, input_11, input_12, . . . , input_1*n*),
inst2 (output_2, input_21, input_22, . . . , input_2*n*),
. . .
inst*m* (output_*m*, input_*m*1, input_*m*2, . . . , input_*mn*);

The design module using built-in primitives is shown in Figure A.54 using the instantiation names and net names indicated in the logic diagram of Figure A.53. The test bench is shown in Figure A.55. The outputs and waveforms are shown in Figure A.56 and Figure A.57, respectively.

```
//3:8 decoder using built-in primitives
module decoder_3to8_bip2 (x, enbl, z);

input [2:0] x;
input enbl;
output [7:0] z;

//instantiate the inverters for the inputs
not    inst1     (net1, x[0]),
       inst2     (net2, x[1]),
       inst3     (net3, x[2]);

//instantiate the and gates for the decoder outputs
and    inst4     (z[0], net1, net2, net3, enbl),
       inst5     (z[1], net2, net3, x[0], enbl),
       inst6     (z[2], net1, x[1], net3, enbl),
       inst7     (z[3], net3, x[1], x[0], enbl),
       inst8     (z[4], x[2], net1, net2, enbl),
       inst9     (z[5], x[2], net2, x[0], enbl),
       inst10    (z[6], x[2], x[1], net1, enbl),
       inst11    (z[7], x[2], x[1], x[0], enbl);
endmodule
```

Figure A.54 Design module using built-in primitives for the 3:8 decoder.

```
//test bench for 3:8 decoder using built-in primitives
module decoder_3to8_bip2_tb;

reg [2:0] x;
reg enbl;
wire [7:0] z;
//continued on next page
```

Figure A.55 Test bench for the 3:8 decoder using built-in primitives.

```
//apply input vectors and display variables
initial
begin: apply_stimulus
   reg [4:0] invect;
   for (invect = 0; invect < 16; invect = invect + 1)
      begin
         {x[2], x[1], x[0], enbl} = invect [4:0];
         #10  $display ("x[2] x[1] x[0] enbl = %b, z = %b",
                        {x[2], x[1], x[0], enbl}, z);
      end
end

//instantiate the module into the test bench
decoder_3to8_bip2 inst1 (
   .x(x),
   .enbl(enbl),
   .z(z)
   );
endmodule
```

Figure A.55 (Continued)

```
x[2] x[1] x[0] enbl = 0000, z = 00000000
x[2] x[1] x[0] enbl = 0001, z = 00000001

x[2] x[1] x[0] enbl = 0010, z = 00000000
x[2] x[1] x[0] enbl = 0011, z = 00000010

x[2] x[1] x[0] enbl = 0100, z = 00000000
x[2] x[1] x[0] enbl = 0101, z = 00000100

x[2] x[1] x[0] enbl = 0110, z = 00000000
x[2] x[1] x[0] enbl = 0111, z = 00001000

x[2] x[1] x[0] enbl = 1000, z = 00000000
x[2] x[1] x[0] enbl = 1001, z = 00010000

x[2] x[1] x[0] enbl = 1010, z = 00000000
x[2] x[1] x[0] enbl = 1011, z = 00100000

x[2] x[1] x[0] enbl = 1100, z = 00000000
x[2] x[1] x[0] enbl = 1101, z = 01000000

x[2] x[1] x[0] enbl = 1110, z = 00000000
x[2] x[1] x[0] enbl = 1111, z = 10000000
```

Figure A.56 Outputs for the 3:8 decoder using built-in primitives.

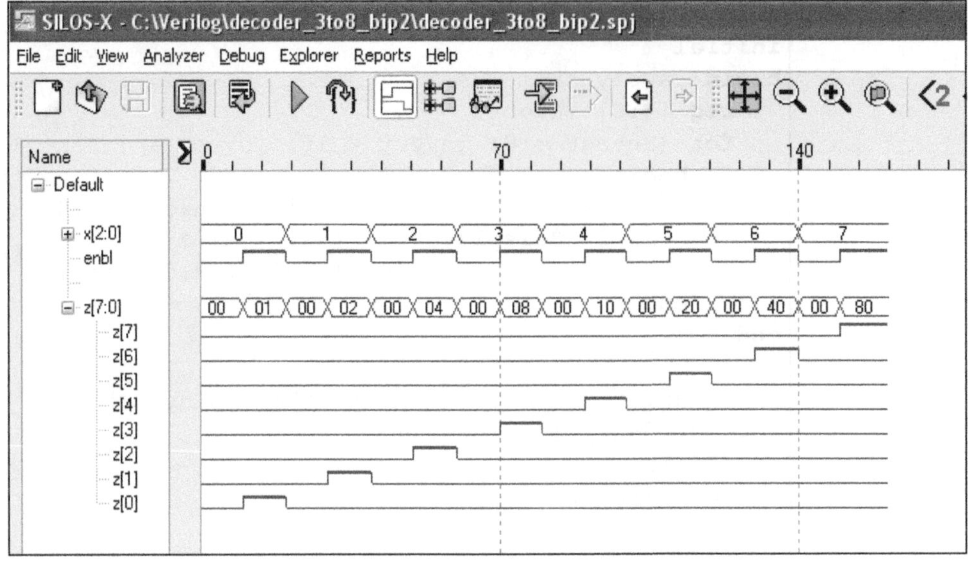

Figure A.57 Waveforms for the 3:8 decoder using built-in primitives.

A.9 Encoders

An *encoder* is a macro logic circuit with n mutually exclusive inputs and m binary outputs, where $n \leq 2^m$. The function of an encoder can be considered to be the inverse of a decoder; that is, the mutually exclusive inputs are encoded into a corresponding binary number. An encoder is also referred to as a *code converter*.

A general block diagram for an n:m encoder is shown in Figure A.58. The label X corresponds to the input code and Y corresponds to the output code. The general qualifying label X/Y is replaced by the input and output codes, respectively; thus, OCT/BIN refers to an octal-to-binary code converter. Only one input x_i is asserted at a time. The decimal value of x_i is encoded as a binary number which is specified by the m outputs.

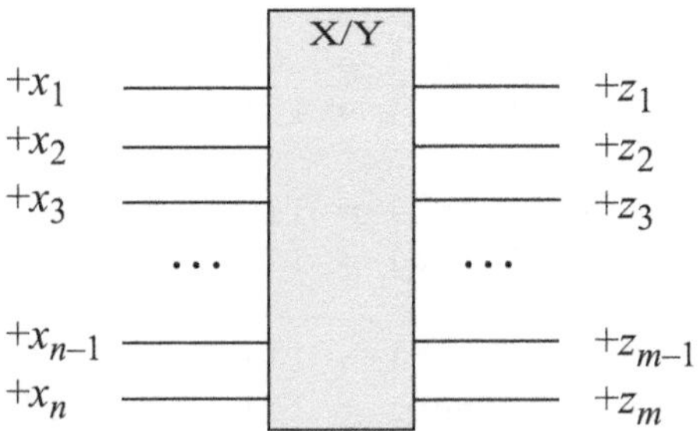

Figure A.58 An n:m encoder or code converter.

The truth table for an octal-to-binary encoder is shown in Table A.2, where the binary outputs represent the octal number of the active input. The equations for the encoder are obtained directly from the truth table and are shown in Equation A.1. The logic diagram is shown in Figure A.59 using three OR gates.

Table A.2 Truth Table for an Octal-to-Binary Encoder

Inputs								Outputs		
d[0]	*d[1]*	*d[2]*	*d[3]*	*d[4]*	*d[5]*	*d[6]*	*d[7]*	*b[2]*	*b[1]*	*b[0]*
1	0	0	0	0	0	0	0	0	0	0
0	**1**	0	0	0	0	0	0	0	0	1
0	0	**1**	0	0	0	0	0	0	1	0
0	0	0	**1**	0	0	0	0	0	1	1
0	0	0	0	**1**	0	0	0	1	0	0
0	0	0	0	0	**1**	0	0	1	0	1
0	0	0	0	0	0	**1**	0	1	1	0
0	0	0	0	0	0	0	**1**	1	1	1

$$\begin{aligned} b[0] &= d[1] + d[3] + d[5] + d[7] \\ b[1] &= d[2] + d[3] + d[6] + d[7] \\ b[2] &= d[4] + d[5] + d[6] + d[7] \end{aligned} \tag{A.1}$$

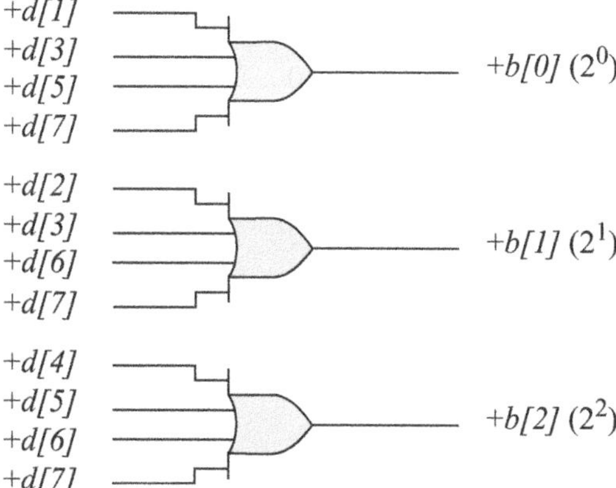

Figure A.59 Logic diagram for an 8:3 encoder.

The dataflow module for the octal-to-binary encoder is shown in Figure A.60 and the test bench is shown in Figure A.61. The outputs and waveforms are shown in Figure A.62 and Figure A.63, respectively.

```
//an octal-to-binary encoder
module encoder_8_to_3 (oct, bin);

input [0:7] oct;
output [2:0] bin;

wire [0:7] oct;
wire [2:0] bin;

assign   bin[0] = oct[1] | oct[3] | oct[5] | oct[7],
         bin[1] = oct[2] | oct[3] | oct[6] | oct[7],
         bin[2] = oct[4] | oct[5] | oct[6] | oct[7];

endmodule
```

Figure A.60 Dataflow module for an octal-to-binary encoder.

```
//test bench for the 8-to-3 encoder
module encoder_8_to_3_tb;

reg [0:7] oct;
wire [2:0] bin;

//display variables
initial
$monitor ("octal = %b, binary = %b", oct [0:7], bin [2:0]);

//apply input vectors
initial
begin
   #0    oct [0:7] = 8'b1000_0000;
   #10   oct [0:7] = 8'b0100_0000;
   #10   oct [0:7] = 8'b0010_0000;
   #10   oct [0:7] = 8'b0001_0000;
   #10   oct [0:7] = 8'b0000_1000;
   #10   oct [0:7] = 8'b0000_0100;
   #10   oct [0:7] = 8'b0000_0010;
   #10   oct [0:7] = 8'b0000_0001;
   #10   $stop;
end
//continued on next page
```

Figure A.61 Test bench for the octal-to-binary encoder.

```
//instantiate the module into the test bench
encoder_8_to_3 inst1 (
   .oct(oct),
   .bin(bin)
   );

endmodule
```

Figure A.61 (Continued)

```
octal = 10000000, binary = 000
octal = 01000000, binary = 001
octal = 00100000, binary = 010
octal = 00010000, binary = 011
octal = 00001000, binary = 100
octal = 00000100, binary = 101
octal = 00000010, binary = 110
octal = 00000001, binary = 111
```

Figure A.62 Outputs for the octal-to-binary encoder.

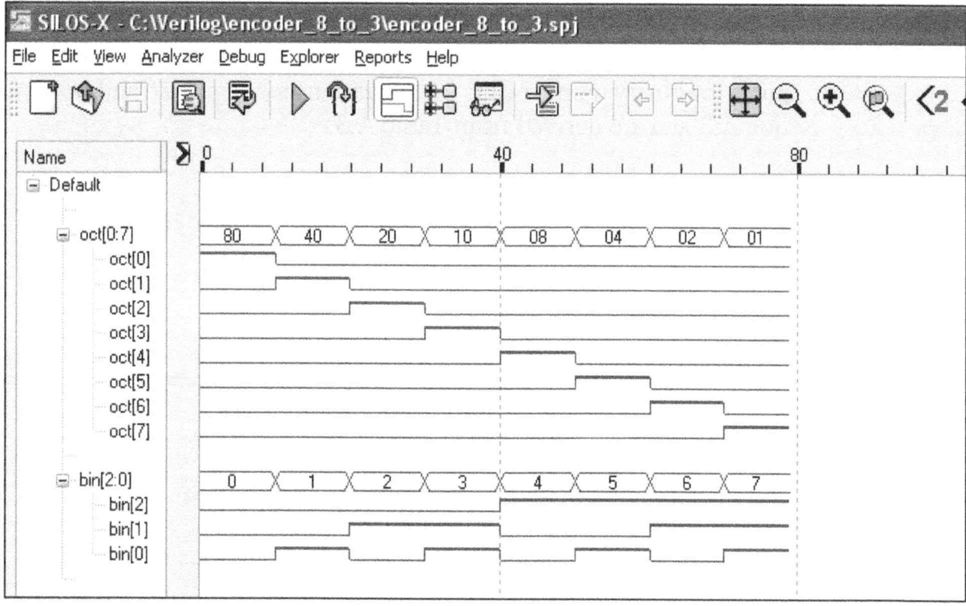

Figure A.63 Waveforms for the octal-to-binary encoder.

A.10 Priority Encoder

It was stated previously that encoder inputs are mutually exclusive. There may be situations, however, where more than one input can be active at a time. Then a priority must be established to select and encode a particular input. This is referred to as a *priority encoder*. Usually the input with the highest valued subscript is selected as highest priority for encoding. Thus, if x_i and x_j are active simultaneously and $i < j$, then x_j has priority over x_i. For example, assume that the octal-to-binary encoder of Figure A.64 is a priority encoder. If inputs *x[1]*, *x[5]*, and *x[7]* are asserted simultaneously, then the outputs will indicate the binary equivalent of decimal 7 such that, *z[2]z[1]z[0]* = 111.

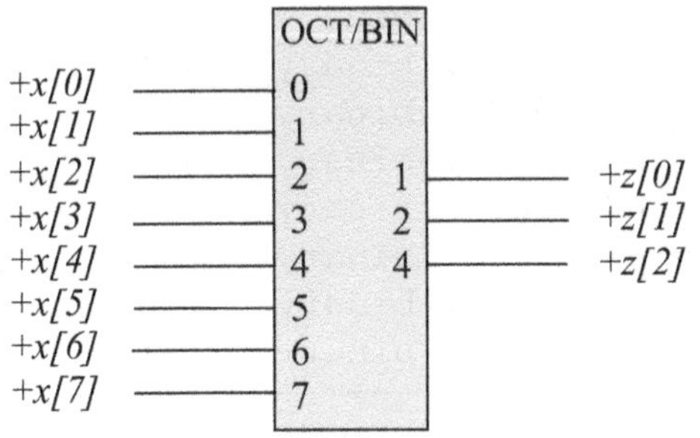

Figure A.64 Octal-to-binary encoder.

The truth table for an octal-to-binary priority encoder is shown in Table A.3. The outputs *z[2] z[1] z[0]* generate a binary number that is equivalent to the highest priority input. If *x[3]* = 1, the state of *x[0]*, *x[1]*, and *x[2]* is irrelevant ("don't care") and the output is the binary number 011. The equations for *z[2]*, *z[1]*, and *z[0]* are shown in Equation A.2 and are derived from Table A.3.

Table A.3 Octal-to-Binary Priority Encoder

Inputs								Outputs		
x[0]	*x[1]*	*x[2]*	*x[3]*	*x[4]*	*x[5]*	*x[6]*	*x[7]*	*z[2]*	*z[1]*	*z[0]*
1	0	0	0	0	0	0	0	0	0	0
–	**1**	0	0	0	0	0	0	0	0	1
–	–	**1**	0	0	0	0	0	0	1	0
–	–	–	**1**	0	0	0	0	0	1	1
–	–	–	–	**1**	0	0	0	1	0	0
–	–	–	–	–	**1**	0	0	1	0	1
–	–	–	–	–	–	**1**	0	1	1	0
–	–	–	–	–	–	–	**1**	1	1	1

$$z[2] = x[4] + x[5] + x[6] + x[7]$$

$$z[1] = x[2]\,x[4]'\,x[5]' + x[3]\,x[4]'\,x[5]' + x[6] + x[7]$$

$$z[0] = x[1]\,x[2]'\,x[4]'\,x[6]' + x[3]\,x[4]'\,x[6]' + x[5]\,x[6]' + x[7] \tag{A.2}$$

The design of the octal-to-binary priority encoder is shown in Figure A.65 using dataflow modeling. The **assign** statement directly implements Equation A.2. The test bench, shown in Figure A.66, applies several input vectors for *x[0:7]* to test the priority selection for the encoder. The outputs and waveforms are shown in Figure A.67 and Figure A.68, respectively.

```
//dataflow for a 8:3 priority encoder
module priority_encoder2 (x, z);

input [0:7] x;
output [2:0] z;

assign   z[2] = (x[4] | x[5] | x[6] | x[7]),

         z[1] = (x[2] & ~x[4] & ~x[5]) |
                (x[3] & ~x[4] & ~x[5]) | x[6] | x[7],

         z[0] = (x[1] & ~x[2] & ~x[4] & ~x[6]) |
                (x[3] & ~x[4] & ~x[6]) |
                (x[5] & ~x[6]) | x[7];
endmodule
```

Figure A.65 Dataflow module for the 8:3 priority encoder.

```
//test bench for 8:3 priority encoder
module priority_encoder2_tb;

reg [0:7] x;
wire [2:0] z;

//display variables
initial
$monitor ("x = %b, z = %b", x, z);

//continued on next page
```

Figure A.66 Test bench for the 8:3 priority encoder.

```
//apply input vectors
initial
begin
   #0     x = 8'b0000_0000;
   #10    x = 8'b1000_0000;
   #10    x = 8'b0100_0000;
   #10    x = 8'b0010_0000;
   #10    x = 8'b0001_0000;
   #10    x = 8'b0000_1000;
   #10    x = 8'b0000_0100;
   #10    x = 8'b0000_0010;
   #10    x = 8'b0000_0001;
//-------------------------
   #10    x = 8'b1001_0000;
   #10    x = 8'b0110_0000;
   #10    x = 8'b1110_0000;
   #10    x = 8'b0001_1100;
   #10    x = 8'b0011_1001;
   #10    x = 8'b0000_0111;
   #10    x = 8'b0011_0010;
   #10    x = 8'b0111_0001;
   #10    $stop;
end

//instantiate the module into the test bench
priority_encoder2 inst1 (
   .x(x),
   .z(z)
   );

endmodule
```

Figure A.66 (Continued)

```
x = 00000000, z = 000          x = 10010000, z = 011
x = 10000000, z = 000          x = 01100000, z = 010
x = 01000000, z = 001          x = 11100000, z = 010
x = 00100000, z = 010          x = 00011100, z = 101
x = 00010000, z = 011          x = 00111001, z = 111
x = 00001000, z = 100          x = 00000111, z = 111
x = 00000100, z = 101          x = 00110010, z = 110
x = 00000010, z = 110          x = 01110001, z = 111
x = 00000001, z = 111
```

Figure A.67 Outputs for the 8:3 priority encoder.

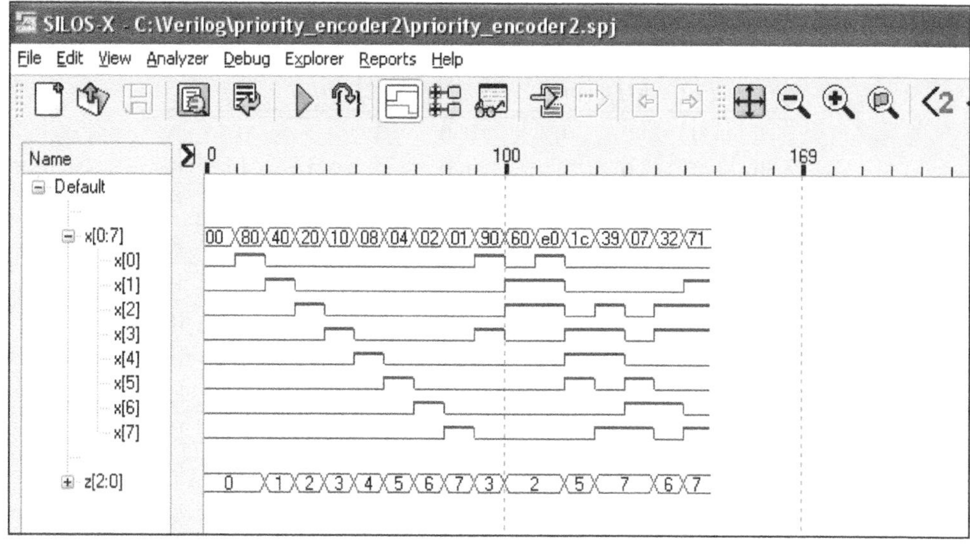

Figure A.68 Waveforms for the 8:3 priority encoder.

A.11 Binary-to-Gray Code Converters

The next-state table of Table A.4 shows the relationship between the binary 8421 code and the Gray code. The Gray code is an unweighted code and belongs to a class of cyclic codes called reflective codes. Notice in the first four rows, that *g[0]* reflects across the reflecting axis; that is, *g[0]* in rows 3 and 4 is the mirror image of *g[0]* in rows 1 and 2. In the same manner, *g[1]* and *g[0]* reflect across the reflecting axis under row 4. Thus, rows 5 through 8 reflect the state of rows 1 through 4 for *g[1]* and *g[0]*. The same is true for *g[2]*, *g[1]*, and *g[0]* relative to rows 9 through 16 and rows 1 through 8.

Table A.4 Binary-to-Gray Code Conversion

Binary				Gray				Row
b[3]	*b[2]*	*b[1]*	*b[0]*	*g[3]*	*g[2]*	*g[1]*	*g[0]*	
0	0	0	0	0	0	0	0	1
0	0	0	1	0	0	0	1	2
0	0	1	0	0	0	1	1	3
0	0	1	1	0	0	1	0	4

Continued on next page

Table A.4 Binary-to-Gray Code Conversion

Binary				Gray				Row
b[3]	*b[2]*	*b[1]*	*b[0]*	*g[3]*	*g[2]*	*g[1]*	*g[0]*	
0	1	0	0	0	1	1	0	5
0	1	0	1	0	1	1	1	6
0	1	1	0	0	1	0	1	7
0	1	1	1	0	1	0	0	8
1	0	0	0	1	1	0	0	9
1	0	0	1	1	1	0	1	10
1	0	1	0	1	1	1	1	11
1	0	1	1	1	1	1	0	12
1	1	0	0	1	0	1	0	13
1	1	0	1	1	0	1	1	14
1	1	1	0	1	0	0	1	15
1	1	1	1	1	0	0	0	16

Using *D* flip-flops A binary-to-Gray code converter will be designed using *D* flip-flops. From Table A.4, it is observed that the high-order bit of the Gray code word is the same as the high-order bit of the corresponding binary code word. Therefore, all the logic associated with flip-flop *g[3]* can be eliminated, and *g[3]* = *b[3]*. If all four flip-flops are an integral part of a macro logic function, then flip-flop *g[3]* is retained and acts as a 1-bit parallel-in, parallel-out register.

After examining Table A.4, an *n*-bit Gray code can be obtained from the corresponding *n*-bit binary code by the following algorithm:

$$g_{n-1} = b_{n-1}$$
$$g_i = b_i \oplus b_{i+1}$$

for $0 \le i \le n-2$, where the symbol $\oplus$ denotes modulo-2 addition defined as:

$$0 \oplus 0 = 0$$
$$0 \oplus 1 = 1$$
$$1 \oplus 0 = 1$$
$$1 \oplus 1 = 0$$

The *D* flip-flop that is used in the binary-to-Gray code converter is shown in Figure A.69. The structural module using four *D* flip-flops is shown in Figure A.70. The test bench is shown in Figure A.71 providing all combinations of 4-bit binary inputs. The outputs and waveforms are shown in Figure A.72 and Figure A.73, respectively.

```
//behavioral D flip-flop
module d_ff_bh (rst_n, clk, d, q, q_n);

input rst_n, clk, d;
output q, q_n;

wire rst_n, clk, d;

reg q;    //variables used in an always block are registers

assign q_n = ~q;

always @ (rst_n or posedge clk)
begin
   if (rst_n == 0)
      q <= 1'b0;
   else q <= d;
end
endmodule
```

Figure A.69 Behavioral module for a *D* flip-flop.

```
//structural binary-to-gray code converter
module bin_to_gray_struc (rst_n, clk, b, g);

input rst_n, clk;
input [3:0] b;
output [3:0] g;

//instantiate the logic for g[3]
d_ff_bh inst1 (
   .rst_n(rst_n),
   .clk(clk),
   .d(b[3]),
   .q(g[3])
   );

//instantiate the logic for g[2]
d_ff_bh inst2 (
   .rst_n(rst_n),
   .clk(clk),
   .d(b[3] ^ b[2]),
   .q(g[2])
   );                    //continued on next page
```

Figure A.70 Structural module for a binary-to-Gray code converter.

```
//instantiate the logic for g[1]
d_ff_bh inst3 (
   .rst_n(rst_n),
   .clk(clk),
   .d(b[2] ^ b[1]),
   .q(g[1])
   );

//instantiate the logic for g[0]
d_ff_bh inst4 (
   .rst_n(rst_n),
   .clk(clk),
   .d(b[1] ^ b[0]),
   .q(g[0])
   );
endmodule
```

Figure A.70 (Continued)

```
//test bench for binary-to-gray code converter
module bin_to_gray_struc_tb;

reg rst_n, clk;
reg [3:0] b;
wire [3:0] g;

//display variables
initial
$monitor ("bin = %b, gray = %b", b, g);

//define clock
initial
begin
   clk = 1'b0;
   forever
      #10   clk = ~clk;
end

//apply input vectors
initial
begin
   #0    rst_n = 1'b0;
   #5    rst_n = 1'b1;
         b = 4'b0000;       //continued on next page
```

Figure A.71 Test bench for the binary-to-Gray code converter.

```
    #20    b = 4'b0001;
    #20    b = 4'b0010;
    #20    b = 4'b0011;

    #20    b = 4'b0100;
    #20    b = 4'b0101;
    #20    b = 4'b0110;
    #20    b = 4'b0111;

    #20    b = 4'b1000;
    #20    b = 4'b1001;
    #20    b = 4'b1010;
    #20    b = 4'b1011;

    #20    b = 4'b1100;
    #20    b = 4'b1101;
    #20    b = 4'b1110;
    #20    b = 4'b1111;

    #20    $stop;
end


//instantiate the module into the test bench
bin_to_gray_struc inst1 (
    .rst_n(rst_n),
    .clk(clk),
    .b(b),
    .g(g)
    );
endmodule
```

Figure A.71 (Continued)

```
bin = 0000, gray = 0000          bin = 1000, gray = 1100
bin = 0001, gray = 0001          bin = 1001, gray = 1101
bin = 0010, gray = 0011          bin = 1010, gray = 1111
bin = 0011, gray = 0010          bin = 1011, gray = 1110

bin = 0100, gray = 0110          bin = 1100, gray = 1010
bin = 0101, gray = 0111          bin = 1101, gray = 1011
bin = 0110, gray = 0101          bin = 1110, gray = 1001
bin = 0111, gray = 0100          bin = 1111, gray = 1000
```

Figure A.72 Outputs for the binary-to-Gray code converter.

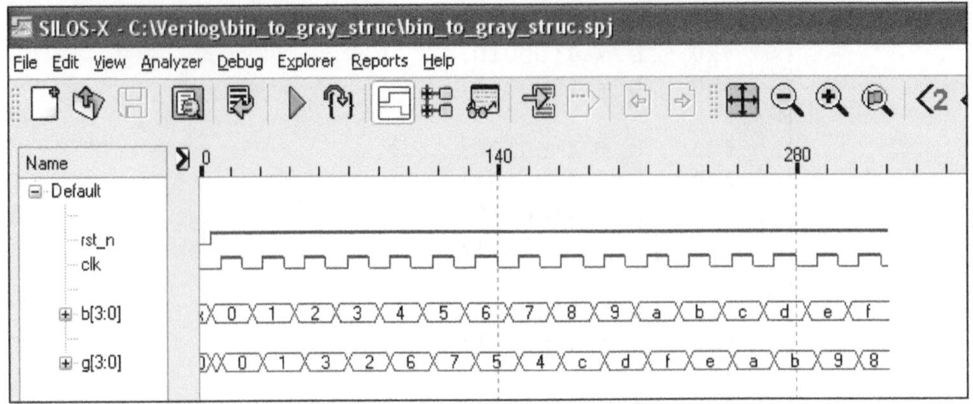

Figure A.73 Waveforms for the binary-to-Gray code converter.

Using built-in primitives A binary-to-Gray code converter will now be designed using built-in primitives. The binary-to-Gray code algorithm is repeated below.

$$g_{n-1} = b_{n-1}$$
$$g_i = b_i \oplus b_{i+1}$$

Therefore, for a 4-bit binary number, the Gray code bits are as follows:

$$g_3 = b_3$$
$$g_2 = b_2 \oplus b_3$$
$$g_1 = b_1 \oplus b_2$$
$$g_0 = b_0 \oplus b_1$$

The design module is shown in Figure A.74 using **buf** and **xor** built-in primitives. A **buf** gate is a noninverting primitive with one scalar input and one or more scalar outputs. The output terminals are listed first when instantiated; the input is listed last, as shown below. The instance name is optional.

```
buf inst1 (output, input);                    //one output
```

The test bench, shown in Figure A.75, applies all combinations of four binary bits. The outputs and waveforms are shown in Figure A.76 and Figure A.77, respectively.

```
//built-in primitives binary-to-gray converter
module bin_to_gray_bip (bin, gray);

input [3:0] bin;
output [3:0] gray;

buf   inst1 (gray[3], bin[3]);

xor   inst2 (gray[2], bin[3], bin[2]),
      inst3 (gray[1], bin[2], bin[1]),
      inst4 (gray[0], bin[1], bin[0]);

endmodule
```

Figure A.74 Design module for a binary-to-Gray code converter using built-in primitives.

```
//test bench for binary-to-gray converter
//using built-in primitives
module bin_to_gray_bip_tb;

reg [3:0] bin;
wire [3:0] gray;

//apply input vectors and display variables
initial
begin: apply_stimulus
   reg [4:0] invect;
   for (invect = 0; invect < 16; invect = invect + 1)
      begin
         bin = invect [4:0];
         #10 $display ("binary = %b, gray = %b", bin, gray);
      end
end

//instantiate the module into the test bench
bin_to_gray_bip inst1 (
   .bin(bin),
   .gray(gray)
   );

endmodule
```

Figure A.75 Test bench for the binary-to-Gray code converter using built-in primitives.

```
binary = 0000, gray = 0000
binary = 0001, gray = 0001
binary = 0010, gray = 0011
binary = 0011, gray = 0010
binary = 0100, gray = 0110
binary = 0101, gray = 0111
binary = 0110, gray = 0101
binary = 0111, gray = 0100
binary = 1000, gray = 1100
binary = 1001, gray = 1101
binary = 1010, gray = 1111
binary = 1011, gray = 1110
binary = 1100, gray = 1010
binary = 1101, gray = 1011
binary = 1110, gray = 1001
binary = 1111, gray = 1000
```

Figure A.76 Outputs for the binary-to-Gray code converter using built-in primitives.

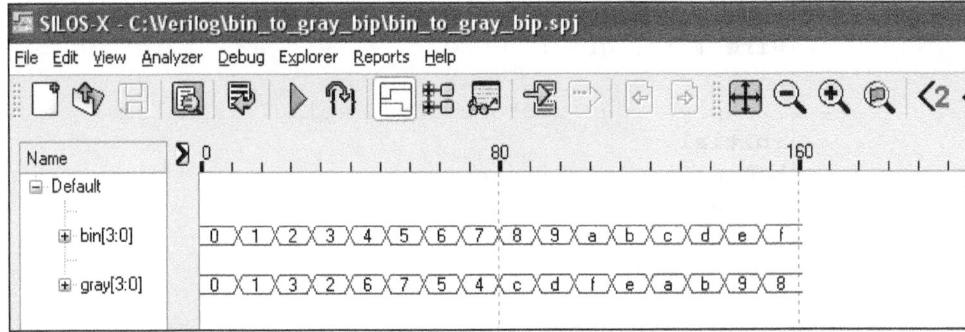

Figure A.77 Waveforms for the binary-to-Gray code converter using built-in primitives.

A.12 Adder/Subtractor

This section describes a 4-bit ripple adder/subtractor with the ability to detect overflow. A typical stage$_i$ of the adder/subtractor is shown in Figure A.78. The inputs are two 4-bit operands: the augend *a[3:0]* and the addend *b[3:0]*, and a mode control input m. If $m = 0$, then the operation is addition; if $m = 1$, then the operation is subtraction.

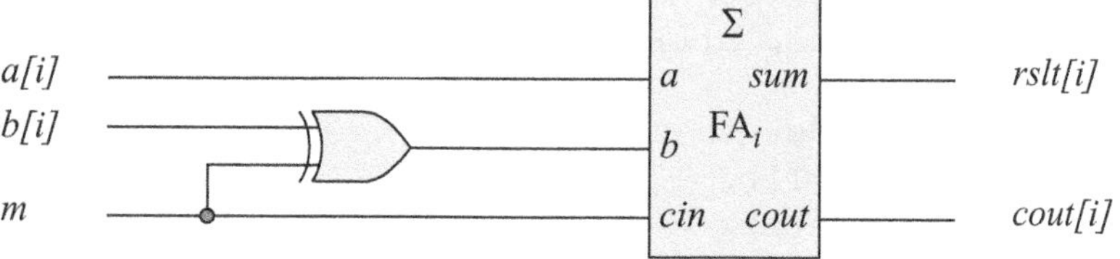

Figure A.78 Typical stage of a 4-bit adder/subtractor.

The structural module is shown in Figure A.79, which instantiates both the built-in primitive **xor** and the full adder four times. The test bench, shown in Figure A.80, applies input vectors for operands *a* and *b* for addition, subtraction, and overflow. The outputs and waveforms are shown in Figure A.81 and A.82, respectively.

```
//structural module for an adder-subtractor
module adder_subtr_struc (a, b, m, rslt, cout, ovfl);

input [3:0] a, b;
input m;
output [3:0] rslt, cout;
output ovfl;

//define internal nets
wire net0, net1, net2, net3;

//define overflow
xor (ovfl, cout[3], cout[2]);

//instantiate the xor and the full adder for FA0
xor (net0, b[0], m);

full_adder inst0 (
   .a(a[0]),
   .b(net0),
   .cin(m),
   .sum(rslt[0]),
   .cout(cout[0])
   );

//continued on next page
```

Figure A.79 Structural module for the 4-bit adder/subtractor.

```
//instantiate the xor and the full adder for FA1
xor (net1, b[1], m);

full_adder inst1 (
   .a(a[1]),
   .b(net1),
   .cin(cout[0]),
   .sum(rslt[1]),
   .cout(cout[1])
   );

//instantiate the xor and the full adder for FA2
xor (net2, b[2], m);

full_adder inst2 (
   .a(a[2]),
   .b(net2),
   .cin(cout[1]),
   .sum(rslt[2]),
   .cout(cout[2])
   );

//instantiate the xor and the full adder for FA3
xor (net3, b[3], m);

full_adder inst3 (
   .a(a[3]),
   .b(net3),
   .cin(cout[2]),
   .sum(rslt[3]),
   .cout(cout[3])
   );
endmodule
```

Figure A.79 (Continued)

```
//test bench for structural adder-subtractor
module adder_subtr_struc_tb;

reg [3:0] a, b;
reg m;
wire [3:0] rslt, cout;
wire  ovfl;
//continued on next page
```

Figure A.80 Test bench for the 4-bit adder/subtractor.

```
//display variables
initial
$monitor ("a=%b, b=%b, m=%b, rslt=%b, cout[3]=%b, cout[2]=%b,
            ovfl=%b",
         a, b, m, rslt, cout[3], cout[2], ovfl);


//apply input vectors
initial
begin
//addition
   #0    a = 4'b0000;   b = 4'b0001;   m = 1'b0;
   #10   a = 4'b0010;   b = 4'b0101;   m = 1'b0;
   #10   a = 4'b0110;   b = 4'b0001;   m = 1'b0;
   #10   a = 4'b0101;   b = 4'b0001;   m = 1'b0;


//subtraction
   #10   a = 4'b0111;   b = 4'b0101;   m = 1'b1;
   #10   a = 4'b0101;   b = 4'b0100;   m = 1'b1;
   #10   a = 4'b0110;   b = 4'b0011;   m = 1'b1;
   #10   a = 4'b0110;   b = 4'b0010;   m = 1'b1;


//overflow
   #10   a = 4'b0111;   b = 4'b0101;   m = 1'b0;
   #10   a = 4'b1000;   b = 4'b1011;   m = 1'b0;
   #10   a = 4'b0110;   b = 4'b1100;   m = 1'b1;
   #10   a = 4'b1000;   b = 4'b0010;   m = 1'b1;

   #10   $stop;
end


//instantiate the module into the test bench
adder_subtr_struc inst1 (
   .a(a),
   .b(b),
   .m(m),
   .rslt(rslt),
   .cout(cout),
   .ovfl(ovfl)
   );

endmodule
```

Figure A.80 (Continued)

```
m = 0 (ADD);  m = 1 (SUB)

ADD
a=0000, b=0001, m=0, rslt=0001, cout[3]=0, cout[2]=0, ovfl=0
a=0010, b=0101, m=0, rslt=0111, cout[3]=0, cout[2]=0, ovfl=0
a=0110, b=0001, m=0, rslt=0111, cout[3]=0, cout[2]=0, ovfl=0
a=0101, b=0001, m=0, rslt=0110, cout[3]=0, cout[2]=0, ovfl=0

SUBTRACT
a=0111, b=0101, m=1, rslt=0010, cout[3]=1, cout[2]=1, ovfl=0
a=0101, b=0100, m=1, rslt=0001, cout[3]=1, cout[2]=1, ovfl=0
a=0110, b=0011, m=1, rslt=0011, cout[3]=1, cout[2]=1, ovfl=0
a=0110, b=0010, m=1, rslt=0100, cout[3]=1, cout[2]=1, ovfl=0

OVERFLOW for addition
a=0111, b=0101, m=0, rslt=1100, cout[3]=0, cout[2]=1, ovfl=1
a=1000, b=1011, m=0, rslt=0011, cout[3]=1, cout[2]=0, ovfl=1

OVERFLOW for subtraction
a=0110, b=1100, m=1, rslt=1010, cout[3]=0, cout[2]=1, ovfl=1
a=1000, b=0010, m=1, rslt=0110, cout[3]=1, cout[2]=0, ovfl=1
```

Figure A.81 Outputs for the 4-bit adder/subtractor.

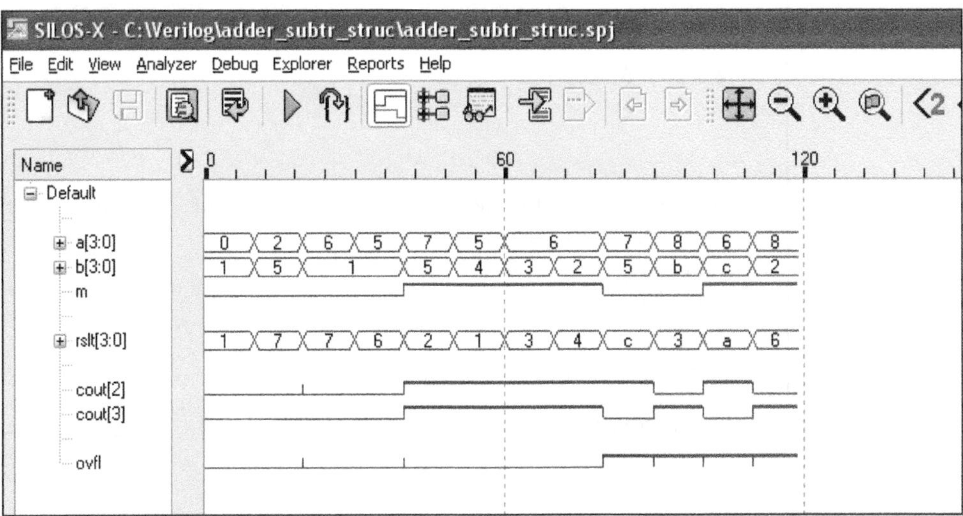

Figure A.82 Waveforms for the 4-bit adder/subtractor.

Appendix B

Event Queue

Event management in the Verilog hardware description language (HDL) is controlled by an event queue. Verilog modules generate events in the test bench, which provide stimulus to the module under test. These events can then produce new events by the modules under test. Since the Verilog HDL Language Reference Manual (LRM) does not specify a method of handling events, the simulator must provide a way to arrange and schedule these events in order to accurately model delays and obtain the correct order of execution. The manner of implementing the event queue is vendor-dependent.

Time in the event queue advances when every event that is scheduled in that time step is executed. Simulation is finished when all event queues are empty. An event at time t may schedule another event at time t or at time $t + n$.

B.1 Event Handling for Dataflow Assignments

Dataflow constructs consist of continuous assignments using the **assign** statement. The assignment occurs whenever simulation causes a change to the right-hand side expression. Unlike procedural assignments, continuous assignments are order independent — they can be placed anywhere in the module.

Consider the logic diagram shown in Figure B.1 which is represented by the two dataflow modules of Figure B.2 and Figure B.3. The test bench for both modules is shown in Figure B.4. The only difference between the two dataflow modules is the reversal of the two **assign** statements. The order in which the two statements execute is not defined by the Verilog HDL LRM; therefore, the order of execution is indeterminate.

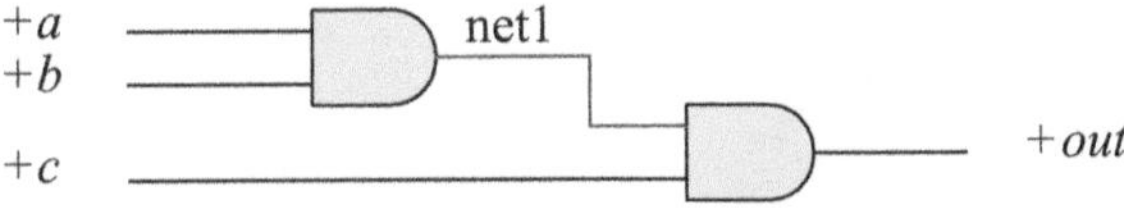

Figure B.1 Logic diagram to demonstrate event handling.

```
module dataflow (a, b, c, out);

input a, b, c;
output out;

wire a, b, c;
wire out;

//define internal net
wire net1;

assign net1 = a & b;
assign out = net1 & c;

endmodule
```

Figure B.2 Dataflow module 1.

```
module dataflow (a, b, c, out)

input a, b, c;
output out;

wire a, b, c;
wire out;

//define internal net
wire net1;

assign out = net1 & c;
assign net1 = a & b;

endmodule
```

Figure B.3 Dataflow module 2.

```
module dataflow_tb;

reg test_a, test_b, test_c;
wire test_out;

initial
begin
   test_a = 1'b1;
   test_b = 1'b0;
   test_c = 1'b0;

   #10   test_b = 1'b1;
         test_c = 1'b1;
   #10   $stop;
end
//instantiate the module
dataflow inst1
   .a(test_a),
   .b(test_b),
   .c(test_c),
   .out(test_out)
   );

endmodule
```

Figure B.4 Test bench for Figure B.2 and Figure B.3.

Assume that the simulator executes the assignment order shown in Figure B.2 first. When the simulator reaches time unit #10 in the test bench, it will evaluate the right-hand side of *test_b = 1'b1;* and place its value in the event queue for an immediate scheduled assignment. Since this is a blocking statement, the next statement will not execute until the assignment has been made. Figure B.5 represents the event queue after the evaluation. The input signal *b* will assume the value of *test_b* through instantiation.

Event queue					
Scheduled event 5	Scheduled event 4	Scheduled event 3	Scheduled event 2	Scheduled event 1	Time unit
				test_b ← 1'b1 b ← 1'b1	$t = \#10$
← Order of execution					

Figure B.5 Event queue after execution of *test_b = 1'b1;*.

After the assignment has been made, the simulator will execute the *test_c = 1'b1;* statement by evaluating the right-hand side, and then placing its value in the event queue for immediate assignment. The new event queue is shown in Figure B.6. The entry that is not shaded represents an executed assignment.

Event queue					
Scheduled event 5	Scheduled event 4	Scheduled event 3	Scheduled event 2	Scheduled event 1	Time unit
			test_c ← 1'b1 c ← 1'b1	test_b ← 1'b1 b ← 1'b1	$t = \#10$
← Order of execution					

Figure B.6 Event queue after execution of *test_c = 1'b1;*.

When the two assignments have been made, time unit #10 will have ended in the test bench, which is the top-level module in the hierarchy. The simulator will then enter the instantiated dataflow module during this same time unit and determine that events have occurred on input signals *b* and *c* and execute the two continuous assignments. At this point, inputs *a*, *b*, and *c* will be at a logic 1 level. However, *net1* will still contain a logic 0 level as a result of the first three assignments that executed at time #0 in the test bench. Thus, the statement *assign out = net1 & c;* will evaluate to a logic 0, which will be placed in the event queue and immediately assigned to *out*, as shown in Figure B.7.

Event queue					
Scheduled event 5	Scheduled event 4	Scheduled event 3	Scheduled event 2	Scheduled event 1	Time unit
		out ← 1'b0 test_out ← 1'b0	test_c ← 1'b1 c ← 1'b1	test_b ← 1'b1 b ← 1'b1	$t = \#10$
←				Order of execution	

Figure B.7 Event queue after execution of *assign out = net1 & c;*.

The simulator will then execute the *assign net1 = a & b;* statement in which the right-hand side evaluates to a logic 1 level. This will be placed on the queue and immediately assigned to *net1* as shown in Figure B.8.

Event queue					
Scheduled event 5	Scheduled event 4	Scheduled event 3	Scheduled event 2	Scheduled event 1	Time unit
	net1 ← 1'b1	out ← 1'b0 test_out ← 1'b0	test_c ← 1'b1 c ← 1'b1	test_b ← 1'b1 b ← 1'b1	$t = \#10$
←				Order of execution	

Figure B.8 Event queue after execution of *assign net1 = a & b;*.

When the assignment has been made to *net1*, the simulator will recognize this as an event on *net1*, which will cause all statements that use *net1* to be reevaluated. The only statement to be reevaluated is *assign out = net1 & c;*. Since both *net1* and *c* equal a logic 1 level, the right-hand side will evaluate to a logic 1, resulting in the event queue shown in Figure B.9.

Event queue					
Scheduled event 5	Scheduled event 4	Scheduled event 3	Scheduled event 2	Scheduled event 1	Time unit
out ← 1'b1 test_out ← 1'b1	net1 ← 1'b1	out ← 1'b0 test_out ← 1'b0	test_c ← 1'b1 c ← 1'b1	test_b ← 1'b1 b ← 1'b1	$t = \#10$
←			Order of execution		

Figure B.9 Event queue after execution of *assign out = net1 & c;*.

The test bench signal *test_out* must now be updated because it is connected to *out* through instantiation. Because the signal *out* is not associated with any other statements within the module, the output from the module will now reflect the correct output. Since all statements within the dataflow module have been processed, the simulator will exit the module and return to the test bench. All events have now been processed; therefore, time unit #10 is complete and the simulator will advance the simulation time.

Since the order of executing the **assign** statements is irrelevant, processing of the dataflow events will now begin with the *assign net1 = a & b;* statement to show that the result is the same. The event queue is shown in Figure B.10.

Event queue					
Scheduled event 5	Scheduled event 4	Scheduled event 3	Scheduled event 2	Scheduled event 1	Time unit
		net1 ← 1'b1	test_c ← 1'b1 c ← 1'b1	test_b ← 1'b1 b ← 1'b1	$t = \#10$
←			Order of execution		

Figure B.10 Event queue beginning with the statement *assign net1 = a & b;*.

Once the assignment to *net1* has been made, the simulator recognizes this as a new event on *net1*. The existing event on input *c* requires the evaluation of statement *assign out = net1 & c;*. The right-hand side of the statement will evaluate to a logic 1, and will be placed on the event queue for immediate assignment, as shown in Figure B.11.

Event queue					
Scheduled event 5	Scheduled event 4	Scheduled event 3	Scheduled event 2	Scheduled event 1	Time unit
	out ← 1'b1 test_out ← 1'b1	net1 ← 1'b1	test_c ← 1'b1 c ← 1'b1	test_b ← 1'b1 b ← 1'b1	$t = \#10$
←				Order of execution	

Figure B.11 Event queue after execution of *assign out = net1 & c;*.

B.2 Event Handling for Blocking Assignments

The blocking assignment operator is the equal (=) symbol. A blocking assignment evaluates the right-hand side arguments and completes the assignment to the left-hand side before executing the next statement; that is, the assignment *blocks* other assignments until the current assignment has been executed.

Example B.1 Consider the code segment shown in Figure B.12 using blocking assignments in conjunction with the event queue of Figure B.13. There are no interstatement delays and no intrastatement delays associated with this code segment. In the first blocking assignment, the right-hand side is evaluated and the assignment is scheduled in the event queue. Program flow is blocked until the assignment is executed. This is true for all blocking assignment statements in this code segment. The assignments all occur in the same simulation time step *t*.

```
always @ (x2 or x3 or x5 or x7)
begin
   x1 = x2 | x3;
   x4 = x5;
   x6 = x7;
end
```

Figure B.12 Code segment with blocking assignments.

Event queue					
Scheduled event 5	Scheduled event 4	Scheduled event 3	Scheduled event 2	Scheduled event 1	Time unit
		x6 ← x7 (t)	x4 ← x5 (t)	x1 ← x2 | x3 (t)	t
←				Order of execution	

Figure B.13 Event queue for Figure B.12.

Example B.2 The code segment shown in Figure B.14 contains an interstatement delay. Both the evaluation and the assignment are delayed by two time units. When the delay has taken place, the right-hand side is evaluated and the assignment is scheduled in the event queue as shown in Figure B.15. The program flow is blocked until the assignment is executed.

```
always @ (x2)
begin
   #2 x1 = x2;
end
```

Figure B.14 Blocking statement with interstatement delay.

Event queue					
Scheduled event 5	Scheduled event 4	Scheduled event 3	Scheduled event 2	Scheduled event 1	Time unit
					t
				x1 ← x2 ($t + 2$)	t + 2
←				Order of execution	

Figure B.15 Event queue for Figure B.14.

Example B.3 The code segment of Figure B.16 shows three statements with interstatement delays of $t + 2$ time units. The first statement does not execute until simulation time $t + 2$ as shown in Figure B.17. The right-hand side $(x_2 \mid x_3)$ is evaluated at the current simulation time which is $t+2$ time units, and then assigned to the left-hand side. At $t+2$, x_1 receives the output of $x_2 \mid x_3$.

```
always @ (x2 or x3 or x5 or x7)
begin
   #2 x1 = x2 | x3;
   #2 x4 = x5;
   #2 x6 = x7;
end
```

Figure B.16 Code segment for delayed blocking assignment with interstatement delays.

Event queue					
Scheduled event 5	Scheduled event 4	Scheduled event 3	Scheduled event 2	Scheduled event 1	Time unit
					t
				x1 ← x2 | x3 ($t+2$)	$t+2$
				x4 ← x5 ($t+4$)	$t+4$
				x6 ← x7 ($t+6$)	$t+6$
←				Order of execution	

Figure B.17 Event queue for Figure B.16.

Example B.4 The code segment in Figure B.18 shows three statements using blocking assignments with intrastatement delays. Evaluation of $x_3 = \#2\ x_4$ and $x_5 = \#2\ x_6$ is blocked until x_2 has been assigned to x_1, which occurs at $t+2$ time units. When the second statement is reached, it is scheduled in the event queue at time $t+2$, but the assignment to x_3 will not occur until $t+4$ time units. The evaluation in the third statement is blocked until the assignment is made to x_3. Figure B.19 shows the event queue.

```
always @ (x2 or x4 or x6)
begin
   x1 = #2 x2;      //first statement
   x3 = #2 x4;      //second statement
   x5 = #2 x6;      //third statement
end
```

Figure B.18 Code segment using blocking assignments with interstatement delays.

Event queue					
Scheduled event 5	Scheduled event 4	Scheduled event 3	Scheduled event 2	Scheduled event 1	Time unit
					t
				x1 ← x2 (t)	$t + 2$
				x3 ← x4 ($t + 2$)	$t + 4$
				x5 ← x6 ($t + 4$)	$t + 6$
←				Order of execution	

Figure B.19 Event queue for the code segment of Figure B.18.

B.3 Event Handling for Nonblocking Assignments

Whereas blocking assignments block the sequential execution of an **always** block until the simulator performs the assignment, nonblocking statements evaluate each statement in succession and place the result in the event queue. Assignment occurs when all of the **always** blocks in the module have been processed for the current time unit. The assignment may cause new events that require further processing by the simulator for the current time unit.

Example B.5 For nonblocking statements, the right-hand side is evaluated and the assignment is scheduled at the end of the queue. The program flow continues and the assignment occurs at the end of the time step. This is shown in the code segment of Figure B.20 and the event queue of Figure B.21.

```
always @ (posedge clk)
begin
   x1 <= x2;
end
```

Figure B.20 Code segment for a nonblocking assignment.

Event queue					
Scheduled event 5	Scheduled event 4	Scheduled event 3	Scheduled event 2	Scheduled event 1	Time unit
x1 ← x2 (t)					t
←				Order of execution	

Figure B.21 Event queue for Figure B.20.

Example B.6 The code segment of Figure B.22 shows a nonblocking statement with an interstatement delay. The evaluation is delayed by the timing control, and then the right-hand side expression is evaluated and assignment is scheduled at the end of the queue. Program flow continues and assignment is made at the end of the current time step as shown in the event queue of Figure B.23.

```
always @ (posedge clk)
begin
   #2 x1 <= x2;
end
```

Figure B.22 Nonblocking assignment with interstatement delay.

Event queue					
Scheduled event 5	Scheduled event 4	Scheduled event 3	Scheduled event 2	Scheduled event 1	Time unit
					t
x1 ← x2 ($t + 2$)					$t + 2$
←				Order of execution	

Figure B.23 Event queue for Figure B.22.

Example B.7 The code segment of Figure B.24 shows a nonblocking statement with an intrastatement delay. The right-hand side expression is evaluated, and assignment is delayed by the timing control and scheduled at the end of the queue. Program flow continues and assignment is made at the end of the current time step as shown in the event queue of Figure B.25.

```
always @ (posedge clk)
begin
   x1 <= #2 x2;
end
```

Figure B.24 Nonblocking assignment with intrastatement delay.

Event queue					
Scheduled event 5	Scheduled event 4	Scheduled event 3	Scheduled event 2	Scheduled event 1	Time unit
					t
x1 ← x2 (t)					$t+2$
←				Order of execution	

Figure B.25 Event queue for Figure B.24.

Example B.8 The code segment of Figure B.26 shows nonblocking statements with intrastatement delays. The right-hand side expressions are evaluated, and assignment is delayed by the timing control and scheduled at the end of the queue. Program flow continues and assignment is made at the end of the current time step as shown in the event queue of Figure B.27.

```
always @ (posedge clk)
begin
   x1 <= #2 x2;
   x3 <= #2 x4;
   x5 <= #2 x6;
end
```

Figure B.26 Nonblocking assignments with intrastatement delays.

Event queue					
Scheduled event 5	Scheduled event 4	Scheduled event 3	Scheduled event 2	Scheduled event 1	Time unit
					t
x5 ← x6 (t)	x3 ← x4 (t)	x1 ← x2 (t)			$t+2$
←				Order of execution	

Figure B.27 Event queue for Figure B.26.

Example B.9 Figure B.28 shows a code segment using nonblocking assignment with an intrastatement delay. The right-hand expression is evaluated at the current time. The assignment is scheduled, but delayed by the timing control #2. This method allows for propagation delay through a logic element; for example, a *D* flip-flop. The event queue is shown in Figure B.29.

```
always @ (posedge clk)
begin
   q <= #2 d;
end
```

d D q

Figure B.28 Code segment using intrastatement delay with blocking assignment.

Event queue					
Scheduled event 5	Scheduled event 4	Scheduled event 3	Scheduled event 2	Scheduled event 1	Time unit
					t
				q ← d (t)	t + 2
←				Order of execution	

Figure B.29 Event queue for the code segment of Figure B.28.

B.4 Event Handling for Mixed Blocking and Nonblocking Assignments

All nonblocking assignments are placed at the end of the queue, while all blocking assignments are placed at the beginning of the queue in their respective order of evaluation. Thus, for any given simulation time t, all blocking statements are evaluated and assigned first, then all nonblocking statements are evaluated.

This is the reason why combinational logic requires the use of blocking assignments, while sequential logic, such as flip-flops, requires the use of nonblocking assignments. In this way, Verilog events can model real hardware in which combinational logic at the input to a flip-flop can stabilize before the clock sets the flip-flop to the state of the input logic. Therefore, blocking assignments are placed at the top of the queue to allow the input data to be stable, whereas nonblocking assignments are placed at the bottom of the queue to be executed after the input data has stabilized.

The logic diagram of Figure B.30 illustrates this concept for two multiplexers connected to the D inputs of their respective flip-flops. The multiplexers represent combinational logic; the D flip-flops represent sequential logic. The behavioral module is shown in Figure B.31 and the event queue is shown in Figure B.32.

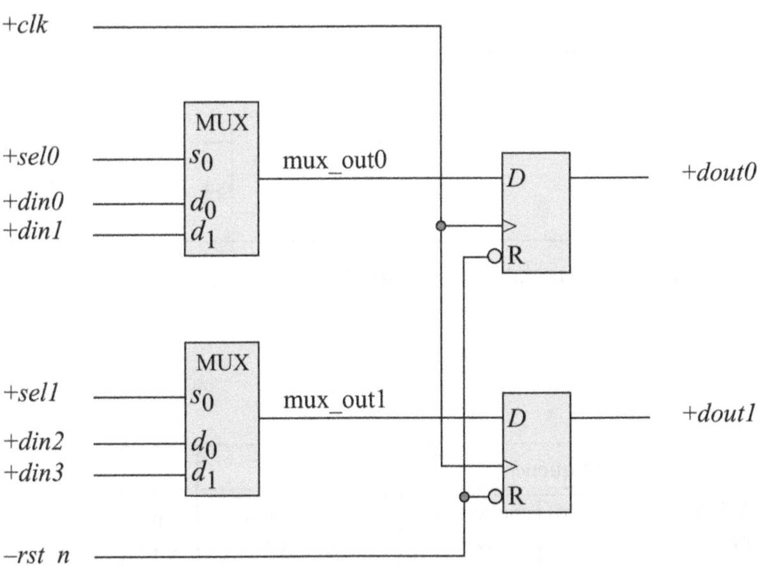

Figure B.30 Combinational logic connected to sequential logic to illustrate the use of blocking and nonblocking assignments.

Because multiplexers are combinational logic, the outputs *mux_out0* and *mux_out1* are placed at the beginning of the queue, as shown in Figure B.32. Nets *mux_out0* and *mux_out1* are in separate **always** blocks; therefore, the order in which they are placed in the queue is arbitrary and can differ with each simulator. The result, however, is the same. If *mux_out0* and *mux_out1* were placed in the same **always** block, then the order in which they are placed in the queue must be the same order as they appear in the **always** block.

Because *dout0* and *dout1* are sequential, they are placed at the end of the queue. Since they appear in separate **always** blocks, the order of their placement in the queue is irrelevant. Once the values of *mux_out0* and *mux_out1* are assigned in the queue, their values will then be used in the assignment of *dout0* and *dout1*; that is, the state of *mux_out0* and *mux_out1* will be set into the *D* flip-flops at the next positive clock transition and assigned to *dout0* and *dout1*.

```
//behavioral module with combinational and sequential logic
//to illustrate their placement in the event queue

module mux_plus_flop (clk, rst_n,
      din0, din1, sel0, dout0,
      din2, din3, sel1, dout1);

input clk, rst_n;
input din0, din1, sel0;
input din2, din3, sel1;
output dout0, dout1;

reg mux_out0, mux_out1;
reg dout0, dout1;

//combinational logic for multiplexers
always @ (din0 or din1 or sel0)
begin
   if (sel0)
      mux_out0 = din1;
   else
      mux_out0 = din0;
end

always @ (din2 or din3 or sel1)
begin
   if (sel1)
      mux_out1 = din3;
   else
      mux_out1 = din2;
end
//continued on next page
```

Figure B.31 Mixed blocking and nonblocking assignments that represent combinational and sequential logic.

```
//sequential logic for D flip-flops
always @ (posedge clk or negedge rst_n)
begin
   if (~rst_n)
      dout0 <= 1'b0;
   else
      dout0 <= mux_out0;
end

always @ (posedge clk or negedge rst_n)
begin
   if (~rst_n)
      dout1 <= 1'b0;
   else
      dout1 <= mux_out1;
end

endmodule
```

Figure B.31 (Continued)

Event queue					
Scheduled event 4	Scheduled event 3	N/A	Scheduled event 2	Scheduled event 1	Time unit
dout1 ← mux_out1 (t)	dout0 ← mux_out0 (t)		mux_out1 ← din3 (t)	mux_out0 ← din1 (t)	*t*
← Order of execution					

Figure B.32 Event queue for Figure B.31.

Appendix C

Verilog Project Procedure

- **Create a folder** (Do only once)

  Windows Explorer > C > New Folder <Verilog> > Enter > Exit Windows Explorer.

- **Create a project** (Do for each project)

  Bring up Silos Simulation Environment.
  File > Close Project. Minimize Silos.
  Windows Explorer > Verilog > File > New Folder <new folder name> Enter.
  Exit Windows Explorer. Maximize Silos.
  File > New Project.
  Create New Project. Save In: Verilog folder.
  Click new folder name. Open.
  Create New Project. Filename: Give project name — usually same name as the folder name. Save
  Project Properties > Cancel.

- **File > New**

  .
  . Design module code goes here
  .

- **File > Save As > File name: <filename.v> > Save**

- **Compile code**

  Edit > Project Properties > Add. Select one or more files to add.
  Click on the file > Open.
  Project Properties. The selected files are shown > OK.
  Load/Reload Input Files. This compiles the code.
  Check screen output for errors. "Simulation stopped at the end of time 0" indicates no compilation errors.

- **Test bench**

  File > New

  .

  . Test bench module code goes here

  .

- **File > Save As > File name: < filename.v> > Save.**

- **Compile test bench**

  Edit > Project Properties > Add. Select one or more files to add.
  Click on the file > Open
  Project Properties. The selected files are shown > OK.
  Load/Reload Input Files. This compiles the code.
  Check screen output for errors. "Simulation stopped at end of time 0" indicates no compilation errors.

- **Binary Output and Waveforms**

  For binary output: click on the GO icon.
  For waveforms: click on the Analyzer icon.
  Click on the Explorer icon. The signals are listed in Silos Explorer.
  Click on the desired signal names.
  Right click. Add Signals to Analyzer.
  Waveforms are displayed.
  Exit Silos Explorer.

- **Change Time Scale**

  With the waveforms displayed, click on Analyzer > Timeline > Timescale
  Enter Time / div > OK

- **Exit the project**.

  Close the waveforms, module, and test bench.
  File > Close Project.

Appendix D

Answers to Select Problems

Chapter 1 Number Systems and Number Representations

1.2 Convert the octal number 5476_8 to radix 10.

$$\begin{aligned} 5476_8 &= (5 \times 8^3) + (4 \times 8^2) + (7 \times 8^1) + (6 \times 8^0) \\ &= 2560 + 256 + 56 + 6 \\ &= 2878_{10} \end{aligned}$$

1.6 Convert the unsigned binary number 1100.110_2 to radix 10.

$$\begin{aligned} 1100.110_2 &= (1 \times 2^3) + (1 \times 2^2) + (0 \times 2^1) + (0 \times 2^0) + (1 \times 2^{-1}) + (1 \times 2^{-2}) \\ &= 8 + 4 + 0.5 + 0.25 \\ &= 12.75_{10} \end{aligned}$$

1.9 Convert the following decimal number to a 16-bit binary number:

$+127.5625_{10} = 0001111111.100100_2$

1.13 The numbers shown below are in 2s complement representation. Convert the numbers to sign-magnitude representation for radix 2 with the same numerical value using eight bits.

2s complement	Sign magnitude	
0111 1111	0111 1111	+117
1000 0001	1111 1111	–117
0000 1111	0000 1111	+15
1111 0001	1000 1111	–15
1111 0000	1001 0000	–16

1.21 Obtain the diminished-radix complement and the radix complement of the following numbers: 9834_{10}, 1000_{10}, and 0000_{10}.

9834_{10}

Diminished-radix complement: 0165_{10}

Radix complement: 0166_{10}

1000_{10}

Diminished-radix complement: 8999_{10}

Radix complement: 9000_{10}

0000_{10}

Diminished-radix complement: 9999_{10}

Radix complement: 0000_{10}

1.24 Convert the following unsigned radix 2 numbers to radix 10:

$100\ 0001.111_2 = 65.875_{10}$
$1111\ 1111.1111_2 = 255.9375_{10}$

Chapter 2 Logic Design Fundamentals

2.5 Minimize each of the following expressions:

(a) $(x_1'x_2 + x_3)(x_1'x_2 + x_3)'$
$= (x_1'x_2 + x_3)(x_1x_3' + x_2'x_3') = 0$

(b) $(x_1x_2' + x_3' + x_4x_5')(x_1x_2' + x_3')$
Let a $= x_1x_2' + x_3'$
Let b $= x_4x_5'$
Therefore, $(a + b)a = a + ab = a$
Substitute into original function: $x_1x_2' + x_3'$

(c) $(x_1'x_3 + x_2 + x_4 + x_5)(x_1'x_3 + x_4)'$
$= [(x_1'x_3 + x_4) + (x_2 + x_5)](x_1'x_3 + x_4)'$
$= (x_1'x_3 + x_4)'(x_2 + x_5)$

2.7 Prove algebraically that $x_1 + 1 = 1$ using only the axioms.

$x_1 + 1 = 1(x_1 + 1)$	Identity
$= (x_1 + x_1')(x_1 + 1)$	Complementation
$= x_1 + (x_1' \bullet 1)$	Distributive
$= x_1 + x_1'$	Identity
$= 1$	Complementation

2.10 Obtain the minimized product-of-sums expression for the function z_1 represented by the Karnaugh map shown below.

$x_5 = 0$

$x_1x_2 \backslash x_3x_4$	0 0	0 1	1 1	1 0
0 0	0 (0)	1 (2)	1 (6)	0 (4)
0 1	1 (8)	1 (10)	1 (14)	1 (12)
1 1	0 (24)	1 (26)	0 (30)	0 (28)
1 0	0 (16)	1 (18)	1 (22)	1 (20)

$x_5 = 1$

$x_1x_2 \backslash x_3x_4$	0 0	0 1	1 1	1 0
0 0	0 (1)	1 (3)	1 (7)	0 (5)
0 1	1 (9)	1 (11)	1 (15)	1 (13)
1 1	1 (25)	1 (27)	0 (31)	0 (29)
1 0	1 (17)	1 (19)	1 (23)	0 (21)

z_1

$$z_1 = (x_1' + x_2' + x_3')(x_1 + x_2 + x_4)(x_1' + x_3 + x_4 + x_5)(x_1' + x_3' + x_4 + x_5')$$

2.13 Write the equation for a logic circuit that generates a logic 1 output whenever a 4-bit unsigned binary number is greater than six. The equation is to be in a minimum sum-of-products notation and a minimum product-of-sums notation.

x_1x_2 \ x_3x_4	0 0	0 1	1 1	1 0
0 0	0 (0)	0 (1)	0 (3)	0 (2)
0 1	0 (4)	0 (5)	1 (7)	0 (6)
1 1	1 (12)	1 (13)	1 (15)	1 (14)
1 0	1 (8)	1 (9)	1 (11)	1 (10)

z_1

$$z_1 = x_1 + x_2x_3x_4$$
$$z_1 = (x_1 + x_2)(x_1 + x_3)(x_1 + x_4)$$

2.18 Obtain the disjunctive normal form for the function shown below using any method. Then use Boolean algebra to minimize the function.

$$z_1(x_1, x_2, x_3, x_4) = x_1x_2x_4 + x_1x_2'x_3$$

x_1x_2 \ x_3x_4	0 0	0 1	1 1	1 0
0 0	0 (0)	0 (1)	0 (3)	0 (2)
0 1	0 (4)	0 (5)	0 (7)	0 (6)
1 1	0 (12)	1 (13)	1 (15)	0 (14)
1 0	0 (8)	0 (9)	1 (11)	1 (10)

z_1

$$z_1 = x_1x_2x_3'x_4 + x_1x_2x_3x_4 + x_1x_2'x_3x_4 + x_1x_2'x_3x_4' \text{ Disjunctive form}$$
$$= x_1x_2x_4(x_3 + x_3') + x_1x_2'x_3(x_4 + x_4')$$
$$= x_1x_2x_4 + x_1x_2'x_3$$

2.24 Minimize the equation shown below then implement the equation: (a) using NAND gates and (b) NOR gates. Output z_1 is to be asserted high in both cases.

$$z_1 = x_1'x_2(x_3'x_4' + x_3'x_4) + x_1x_2(x_3'x_4' + x_3'x_4) + x_1x_2'x_3'x_4$$

$$\begin{aligned} z_1 &= x_1'x_2x_3'x_4' + x_1'x_2x_3'x_4 + x_1x_2x_3'x_4' + x_1x_2x_3'x_4 + x_1x_2'x_3'x_4 \\ &= x_2x_3'(x_1'x_4' + x_1'x_4 + x_1x_4' + x_1x_4) + x_1x_2'x_3'x_4 \\ &= x_2x_3' + x_1x_2'x_3'x_4 \\ &= x_3'[x_2 + x_2'(x_1x_3'x_4)] \\ &= x_3'[x_2 + x_1x_3'x_4] \\ &= x_2x_3' + x_1x_3'x_4 \end{aligned}$$

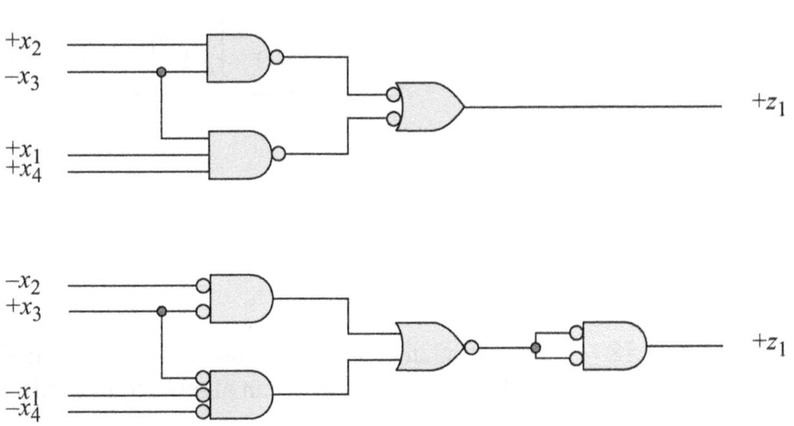

2.29 Implement the equation shown below using a decoder and a minimal amount of additional logic. The decoder has active-low outputs.

$$z_1(x_1, x_2, x_3) = x_1'x_2'x_3 = x_1x_3' + x_2x_3 + x_1x_2'x_3' + x_1x_2$$

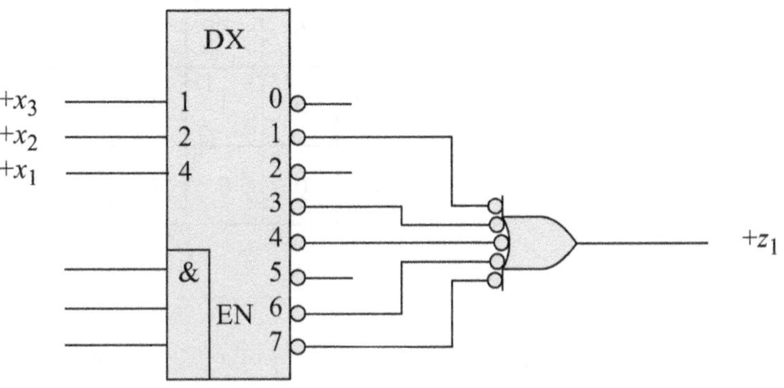

2.32 Given the input equations shown below for a counter that uses three D flip-flops $y_1y_2y_3$, obtain the counting sequence. The counter is reset initially such that $y_1y_2y_3 = 000$.

$$Dy_1 = y_1'y_2 + y_2'y_3$$
$$Dy_2 = y_2y_3' + y_1y_2 + y_1'y_2'y_3$$
$$Dy_3 = y_1'y_2' + y_1'y_3 + y_1y_2y_3'$$

Counting sequence = 0, 1, 7, 2, 6, 3, 5, 4, 0, . . .

2.35 Obtain the state diagram for the Moore synchronous sequential machine shown below. The flip-flops are reset initially; that is, $y_1y_2 = 00$.

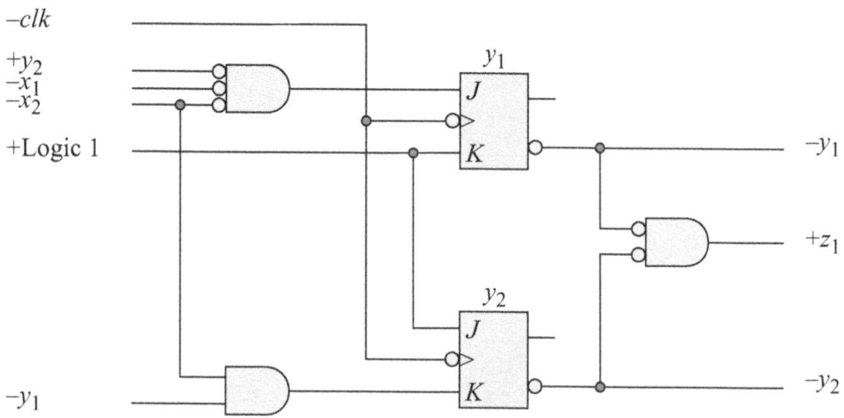

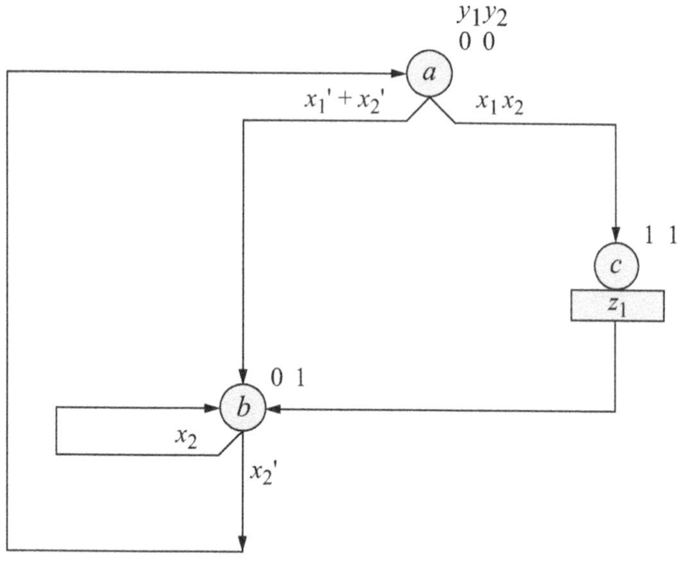

3.2 Design the circuit for the equation shown below using built-in primitives. Obtain the design module, the test bench module, and outputs.

$$z_1 = [x_1 x_2 + (x_1 \oplus x_2)]\, x_3$$

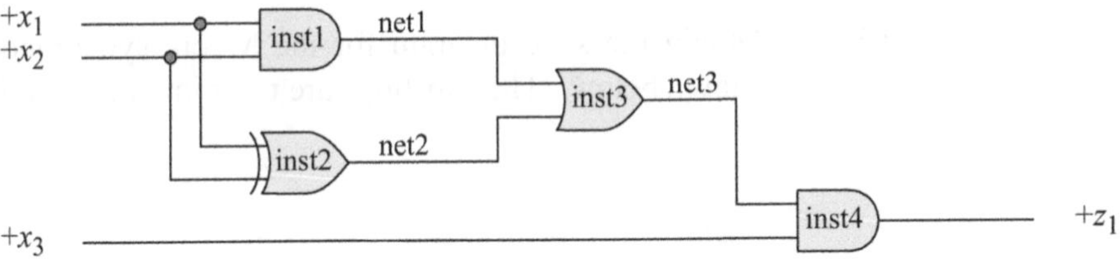


```
//xor product of sums
module xor_pos (x1, x2, x3, z1);

input x1, x2, x3;
output z1;

and inst1 (net1, x1, x2);
xor inst2 (net2, x1, x2);
or   inst3 (net3, net1, net2);
and inst4 (z1, net3, x3);
endmodule
```

```
//test bench for xor_pos
module xor_pos_tb;

reg x1, x2, x3;
wire z1;

initial
begin: apply_stimulus
reg [3:0] invect;
for (invect=0; invect<8; invect=invect+1)
   begin
      {x1, x2, x3} = invect [3:0];
      #10 $display ("{x1x2x3} = %b, z1 = %b",
                     {x1, x2, x3}, z1);
   end
end

//instantiate the module into the test bench
xor_pos inst1 (
   .x1(x1),
   .x2(x2),
   .x3(x3),
   .z1(z1)
   );

endmodule
```

```
{x1x2x3} = 000, z1 = 0
{x1x2x3} = 001, z1 = 0
{x1x2x3} = 010, z1 = 0
{x1x2x3} = 011, z1 = 1
{x1x2x3} = 100, z1 = 0
{x1x2x3} = 101, z1 = 1
{x1x2x3} = 110, z1 = 0
{x1x2x3} = 111, z1 = 1
```

3.5 Obtain the equation for a logic circuit that will generate a logic 1 whenever a 4-bit unsigned binary number N satisfies the following criteria:

$$N = x_1 x_2 x_3 x_4 \text{ (low order)}$$
$$2 < N \leq 6$$
$$11 \leq N < 14$$

Using NOR user-defined primitives, obtain the design module, test bench, and the outputs. Output z_1 will be active high if the above conditions are met.

x_1x_2 \ x_3x_4	0 0	0 1	1 1	1 0
0 0	0 (0)	0 (1)	1 (3)	0 (2)
0 1	1 (4)	1 (5)	0 (7)	1 (6)
1 1	1 (12)	1 (13)	0 (15)	0 (14)
1 0	0 (8)	0 (9)	1 (11)	0 (10)

z_1

$$z_1 = x_2 x_3' + x_2' x_3 x_4 + x_1' x_2 x_4'$$

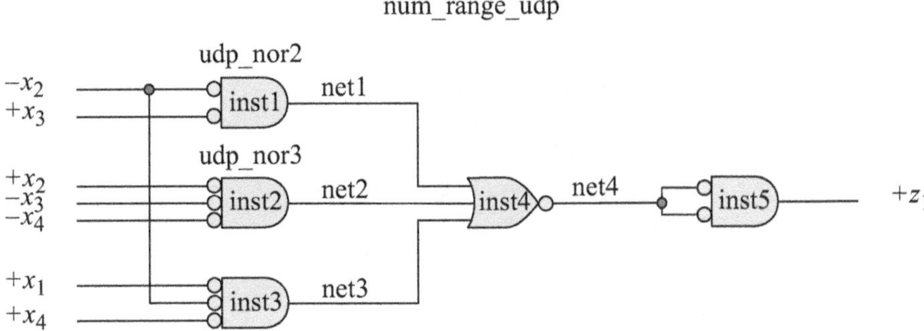

```
//2-input NOR gate as a user-defined primitive
primitive udp_nor2 (z1, x1, x2);

input x1, x2;
output z1;

//define state table
table
//inputs are in the same order as the input list
// x1 x2 :  z1;   comment is for readability
   0  0  :  1;
   0  1  :  0;
   1  0  :  0;
   1  1  :  0;
endtable

endprimitive
```

```
//3-input NOR gate as a user-defined primitive
primitive udp_nor3 (z1, x1, x2, x3);

input x1, x2, x3;
output z1;

//define state table
table
//inputs are in the same order as the input list
// x1 x2 x3 :  z1;   comment is for readability
   0  0  0  :  1;
   0  0  1  :  0;
   0  1  0  :  0;
   0  1  1  :  0;
   1  0  0  :  0;
   1  0  1  :  0;
   1  1  0  :  0;
   1  1  1  :  0;
endtable

endprimitive
```

```
//logic circuit to determine if a number
//is within a certain range
module num_range (x1, x2, x3, x4, z1);

input x1, x2, x3, x4;
output z1;

//instantiate the udps
udp_nor2 inst1 (net1, ~x2, x3);
udp_nor3 inst2 (net2, x2, ~x3, ~x4);
udp_nor3 inst3 (net3, x1, ~x2, x4);
udp_nor3 inst4 (net4, net1, net2, net3);
udp_nor2 inst5 (z1, net4, net4);

endmodule
```

```
//test bench for num_range module
module num_range_tb;

reg x1, x2, x3, x4;
wire z1;

//apply input vectors
initial
begin: apply_stimulus
   reg [4:0] invect;
   for (invect=0; invect<16; invect=invect+1)
      begin
         {x1, x2, x3, x4} = invect [4:0];
         #10 $display ("x1x2x3x4 = %b, z1 = %b",
                              {x1, x2, x3, x4}, z1);
      end
end

//instantiate the module into the test bench
num_range inst1 (
   .x1(x1),
   .x2(x2),
   .x3(x3),
   .x4(x4),
   .z1(z1)
   );

endmodule
```

```
x1x2x3x4 = 0000, z1 = 0
x1x2x3x4 = 0001, z1 = 0
x1x2x3x4 = 0010, z1 = 0
x1x2x3x4 = 0011, z1 = 1
x1x2x3x4 = 0100, z1 = 1
x1x2x3x4 = 0101, z1 = 1
x1x2x3x4 = 0110, z1 = 1
x1x2x3x4 = 0111, z1 = 0
x1x2x3x4 = 1000, z1 = 0
x1x2x3x4 = 1001, z1 = 0
x1x2x3x4 = 1010, z1 = 0
x1x2x3x4 = 1011, z1 = 1
x1x2x3x4 = 1100, z1 = 1
x1x2x3x4 = 1101, z1 = 1
x1x2x3x4 = 1110, z1 = 0
x1x2x3x4 = 1111, z1 = 0
```

3.8 Design a dataflow module that will generate a logic 1 when two 4-bit unsigned binary operands are unequal. The operands are: *A* = *[3:0]* and *B* = *[3:0]*; the output is z_1. Obtain the design module, the test bench module for 16 combinations of the two operands, and the outputs.

```
//a 4-bit comparator that checks for inequality
module comp_unequal (a, b, z1);

input[3:0]  a, b;
output z1;

assign z1 = (a[0]^b[0]) | (a[1]^b[1]) |
            (a[2]^b[2]) | (a[3]^b[3]);

endmodule
```

```
//test bench to compare for inequality
module compare_unequal2_tb;

reg [3:0] a, b;
wire z1;

//display variables
initial
$monitor ("a=%b, b=%b, z1 = %b", a, b, z1);

//apply input vectors
initial
begin
   #0    a=4'b0000;  b=4'b0000;
   #10   a=4'b0001;  b=4'b0010;
   #10   a=4'b0101;  b=4'b0101;
   #10   a=4'b1101;  b=4'b0010;
   #10   a=4'b1111;  b=4'b0010;
   #10   a=4'b1111;  b=4'b1111;
   #10   a=4'b0110;  b=4'b0110;
   #10   a=4'b1001;  b=4'b0111;

   #10   a=4'b0111;  b=4'b1000;
   #10   a=4'b0111;  b=4'b0111;
   #10   a=4'b1101;  b=4'b0101;
   #10   a=4'b1110;  b=4'b0010;
   #10   a=4'b1011;  b=4'b1011;
   #10   a=4'b0001;  b=4'b0001;
   #10   a=4'b0110;  b=4'b0111;
   #10   a=4'b1001;  b=4'b0111;

   #10   $stop;
end

//instantiate the module into the test bench
compare_unequal2 inst1 (
   .a(a),
   .b(b),
   .z1(z1)
   );

endmodule
```

```
a=0000, b=0000, z1 = 0
a=0001, b=0010, z1 = 1
a=0101, b=0101, z1 = 0
a=1101, b=0010, z1 = 1
a=1111, b=0010, z1 = 1
a=1111, b=1111, z1 = 0
a=0110, b=0110, z1 = 0
a=1001, b=0111, z1 = 1

a=0111, b=1000, z1 = 1
a=0111, b=0111, z1 = 0
a=1101, b=0101, z1 = 1
a=1110, b=0010, z1 = 1
a=1011, b=1011, z1 = 0
a=0001, b=0001, z1 = 0
a=0110, b=0111, z1 = 1
a=1001, b=0111, z1 = 1
```

3.12 Use behavioral modeling to design a Mealy synchronous sequential machine that operates according to the state diagram shown below. The outputs are asserted during the last half of the clock cycle. Obtain the behavioral module, the test bench module, the outputs, and the waveforms.

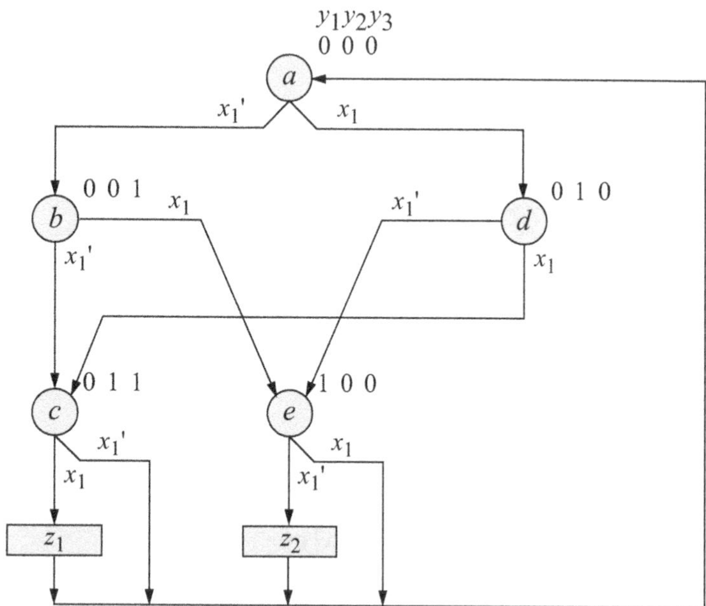

```
//behavioral mealy synchronous sequential machine
module mealy_ssm10 (rst_n, clk, x1, y, z1, z2);

input rst_n, clk, x1;
output [1:3] y;
output z1, z2;

reg [1:3] y, next_state;
wire z1, z2;

//assign state codes
parameter   state_a = 3'b000,
            state_b = 3'b001,
            state_c = 3'b011,
            state_d = 3'b010,
            state_e = 3'b100;

//set next state
always @ (posedge clk)
begin
   if (~rst_n)
      y <= state_a;
   else
      y <= next_state;
end

//define output
assign   z1 = (y[2] & y[3] & x1 & ~clk),
         z2 = (y[1] & ~x1 & ~clk);


//determine next state
always @ (y or x1)
begin
   case (y)
      state_a:
         if (x1==0)
            next_state = state_b;
         else
            next_state = state_d;

      state_b:
         if (x1==0)
            next_state = state_c;
         else
            next_state = state_e;

//continued on next page
```

```
      state_c: next_state = state_a;

      state_d:
         if (x1==0)
            next_state = state_e;
         else
            next_state = state_c;

      state_e: next_state = state_a;
      default: next_state = state_a;
   endcase
end
endmodule
```

```
x1=0, state= xxx, z1=0, z2=x
x1=0, state= 000, z1=0, z2=0
x1=1, state= 001, z1=0, z2=0
x1=0, state= 100, z1=0, z2=0
x1=0, state= 100, z1=0, z2=1
x1=1, state= 000, z1=0, z2=0
x1=0, state= 010, z1=0, z2=0
x1=1, state= 100, z1=0, z2=0
x1=0, state= 000, z1=0, z2=0
x1=1, state= 001, z1=0, z2=0
x1=1, state= 100, z1=0, z2=0
x1=1, state= 000, z1=0, z2=0
x1=1, state= 010, z1=0, z2=0
x1=1, state= 011, z1=0, z2=0
x1=1, state= 011, z1=1, z2=0
x1=1, state= 000, z1=0, z2=0
```

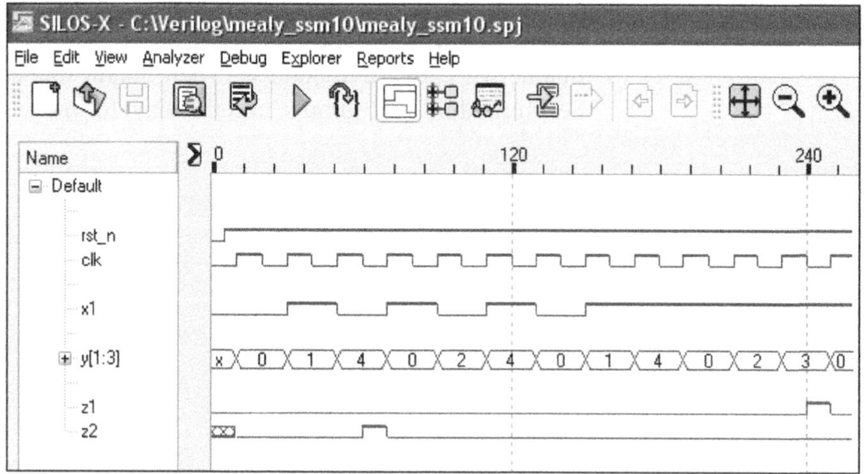

Chapter 4 Fixed-Point Addition

4.2 Perform the arithmetic operations shown below with fixed-point binary numbers in 2s complement representation. In each case, indicate if there is an overflow.

(a)

```
    0100 0000
+)  0100 0000
    ---------
    1000 0000        Overflow
```

(b)

```
    0011 0110
+)  1110 0011
    ---------
    0001 1001        No overflow
```

(c)

$+64_{10}$	0100 0000	
$+63_{10}$	+) 0011 1111	
	0111 1111	No overflow

4.6 Obtain the group generate and group propagate functions for stage $i + 4$ and stage $i + 5$ of a 6-bit carry lookahead adder.

$$GG = G_{i+5} + P_{i+5}\,G_{i+4} + P_{i+5}\,P_{i+4}\,G_{i+3} + P_{i+5}\,P_{i+4}\,P_{i+3}\,G_{i+2} + P_{i+5}\,P_{i+4}\,P_{i+3}\,P_{i+2}\,G_{i+1} + P_{i+5}\,P_{i+4}\,P_{i+3}\,P_{i+2}\,P_{i+1}\,G_i$$

$$GP = P_{i+5}\,P_{i+4}\,P_{i+3}\,P_{i+2}\,P_{i+1}\,P_i$$

4.8 Use only carry-save full adders in a Wallace tree configuration to design a circuit that will add the 2^ith bit of four different operands. Use the fewest number of carry-save adders. Then obtain the structural module, the test bench module, the outputs, and the waveforms.

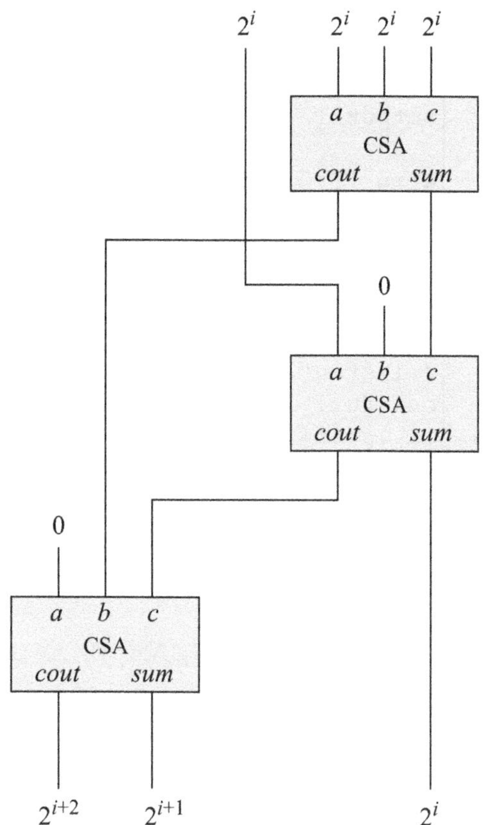

2^i
2^i
2^i
2^i
a b c
CSA
cout sum
0
a b c
CSA
cout sum
0
a b c
CSA
cout sum
2^{i+2}
2^{i+1}
2^i

```
//structural 4-input wallace tree
module wallace_tree_4_inputs (in, sum_w);

input [3:0] in;
output [2:0] sum_w;

wire [3:0] in;
wire [2:0] sum_w;

//define internal wires
wire net1, net2, net3;

//instantiate the carry-save full adders
csa_full_adder inst1 (
   .a(in[0]),
   .b(in[1]),
   .c(in[2]),
   .sum(net1),
   .cout(net2)
   );

csa_full_adder inst2 (
   .a(in[3]),
   .b(1'b0),
   .c(net1),
   .sum(sum_w[0]),
   .cout(net3)
   );

csa_full_adder inst3 (
   .a(1'b0),
   .b(net2),
   .c(net3),
   .sum(sum_w[1]),
   .cout(sum_w[2])
   );

endmodule
```

```
//test bench for 4-input wallace tree
module wallace_tree_4_inputs_tb;

//define inputs and outputs
reg [3:0] in;
wire [2:0] sum_w;

//apply input vectors and display variables
initial
begin: apply_stimulus
   reg [4:0] invect;
   for (invect = 0; invect < 16; invect = invect + 1)
      begin
         in = invect [4:0];
         #10 $display ("inputs = %b, sum = %b",
                        in, sum_w);
      end
end

//instantiate the module into the test bench
wallace_tree_4_inputs nst1 (
   .in(in),
   .sum_w(sum_w)
   );

endmodule
```

```
inputs = 0000, sum = 000     inputs = 1000, sum = 001
inputs = 0001, sum = 001     inputs = 1001, sum = 010
inputs = 0010, sum = 001     inputs = 1010, sum = 010
inputs = 0011, sum = 010     inputs = 1011, sum = 011
inputs = 0100, sum = 001     inputs = 1100, sum = 010
inputs = 0101, sum = 010     inputs = 1101, sum = 011
inputs = 0110, sum = 010     inputs = 1110, sum = 011
inputs = 0111, sum = 011     inputs = 1111, sum = 100
```

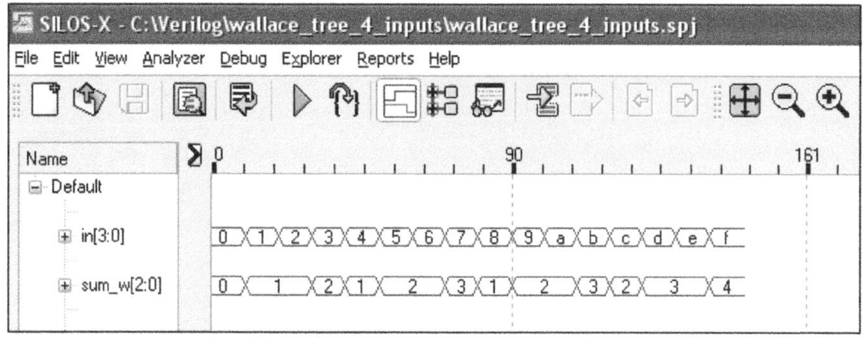

Chapter 5 Fixed-Point Subtraction

5.2 Perform the operation of subtraction on the operands shown below, which are in radix complementation for radix 3.

$$\begin{array}{rl} 0\,2\,0\,2\,1_3 & (+61) \\ -)\ \underline{2\,2\,1\,0\,0_3} & (-18) \end{array}$$

$2-2=\mathbf{0}$ $2-2=\mathbf{0}$ $2-1=\mathbf{1}$ $2-0=\mathbf{2}$ $2-0=\mathbf{2}+\underline{\mathbf{1}}=\mathbf{00200_3}$

$$\begin{array}{r} 0\,2\,0\,2\,1_3 \\ +)\ \underline{0\,0\,2\,0\,0_3} \\ 0\,2\,2\,2\,1_3 \end{array}$$

5.6 Perform the arithmetic operations shown below with fixed-point binary numbers in 2s complement representation. In each case, indicate if there is an overflow.

(a)

```
    1001 1000             1001 1000
-)  0010 0010   →    +)   1101 1110
                          0111 0110        Overflow
```

(b)

```
    0011 0110             0011 0110
-)  1110 0011   →    +)   0001 1101
                          0101 0011        No overflow
```

5.11 Perform the subtraction shown below using 8-bit operands in 2s complement representation and indicate if an overflow occurs.

(–62) – (+67)

```
     1100 0010
+)   1011 1101
     0111 1111        Overflow
```

5.12 Design a 2s complementer for an 8-bit operand *B[7:0]* as an iterative array. Obtain the dataflow design module, the test bench module for at least 20 different input vectors, and the outputs.

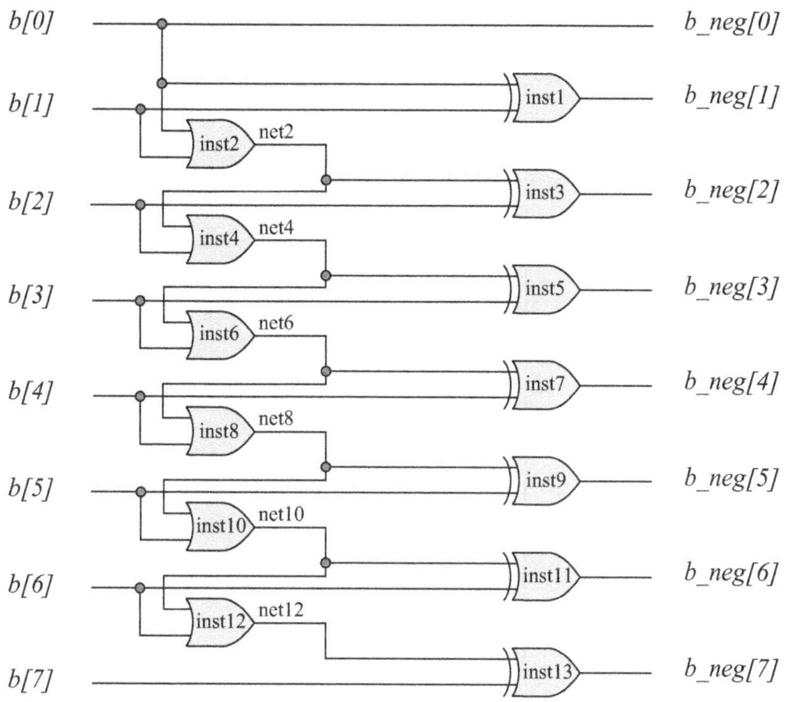


```
//dataflow and built-in primitive 8-bit 2s complementer
module twos_compl_8_bit (b, b_neg);

input [7:0] b;
output [7:0] b_neg;

assign b_neg[0] = b[0];

xor   inst1 (b_neg[1], b[1], b[0]);

or    inst2 (net2, b[1], b[0]);
xor   indt3 (b_neg[2], b[2], net2);

or    indt4 (net4, b[2], net2);
xor   inst5 (b_neg[3], b[3], net4);

or    inst6 (net6, b[3], net4);
xor   inst7 (b_neg[4], b[4], net6);

//continued on next page
```

```
or    inst8 (net8, b[4], net6);
xor   inst9 (b_neg[5], b[5], net8);
or    inst10(net10, b[5], net8);
xor   inst11(b_neg[6], b[6], net10);
or    inst12(net12, b[6], net10);
xor   inst13(b_neg[7], b[7], net12);

endmodule
```

```
//test bench for 8-bit 2s complementer
module twos_compl_8_bit_tb;

reg [7:0] b;
wire [7:0] b_neg;

//display variables
initial
$monitor ("b = %b, b_negated = %b", b, b_neg);

//apply input vectors
initial
begin
   #0    b = 8'b0000_0000;
   #10   b = 8'b0000_0001;
   #10   b = 8'b0000_0010;
   #10   b = 8'b0000_0011;

   #10   b = 8'b0000_0100;
   #10   b = 8'b0001_1010;
   #10   b = 8'b0001_1011;
   #10   b = 8'b0001_1100;

   #10   b = 8'b0001_1101;
   #10   b = 8'b0011_0100;
   #10   b = 8'b0011_0101;
   #10   b = 8'b0011_0110;

   #10   b = 8'b0011_0111;
   #10   b = 8'b0101_1000;
   #10   b = 8'b0101_1001;
   #10   b = 8'b0101_1010;

//continued on next page
```

```
   #10   b = 8'b0101_1011;
   #10   b = 8'b1100_1100;
   #10   b = 8'b1100_1101;
   #10   b = 8'b1100_1110;

   #10   b = 8'b1100_1111;
   #10   b = 8'b1000_0000;

   #10   $stop;
end

//instantiate the module into the test bench
twos_compl_8_bit inst1 (
   .b(b),
   .b_neg(b_neg)
   );

endmodule
```

```
b = 00000000, b_negated = 00000000
b = 00000001, b_negated = 11111111
b = 00000010, b_negated = 11111110
b = 00000011, b_negated = 11111101
b = 00000100, b_negated = 11111100
b = 00011010, b_negated = 11100110
b = 00011011, b_negated = 11100101
b = 00011100, b_negated = 11100100
b = 00011101, b_negated = 11100011
b = 00110100, b_negated = 11001100
b = 00110101, b_negated = 11001011
b = 00110110, b_negated = 11001010
b = 00110111, b_negated = 11001001
b = 01011000, b_negated = 10101000
b = 01011001, b_negated = 10100111
b = 01011010, b_negated = 10100110
b = 01011011, b_negated = 10100101
b = 11001100, b_negated = 00110100
b = 11001101, b_negated = 00110011
b = 11001110, b_negated = 00110010
b = 11001111, b_negated = 00110001
b = 10000000, b_negated = 10000000
```

6.2 Use the paper-and-pencil method to multiply the operands shown below which are in 2s complement representation.

Multiplicand = 0111 (+7)
Multiplier = 1110 (–2)

Twos complement the multiplicand to equal 0010 (+2), then 2s complement the product.

				0	1	1	1	(+7)
			×)	0	0	1	0	(+2)
0	0	0	0	0	0	0	0	
0	0	0	0	1	1	1		
0	0	0	0	0	0			
0	0	0	0	0				
0	0	0	0	1	1	1	0	(+14)
				↓				
1	1	1	1	0	0	1	0	(–14)

6.8 Multiply the two operands shown below using the sequential add-shift technique shown in Figure 6.2.

Multiplicand = 0101 (+5)
Multiplier = 0111 (+7)

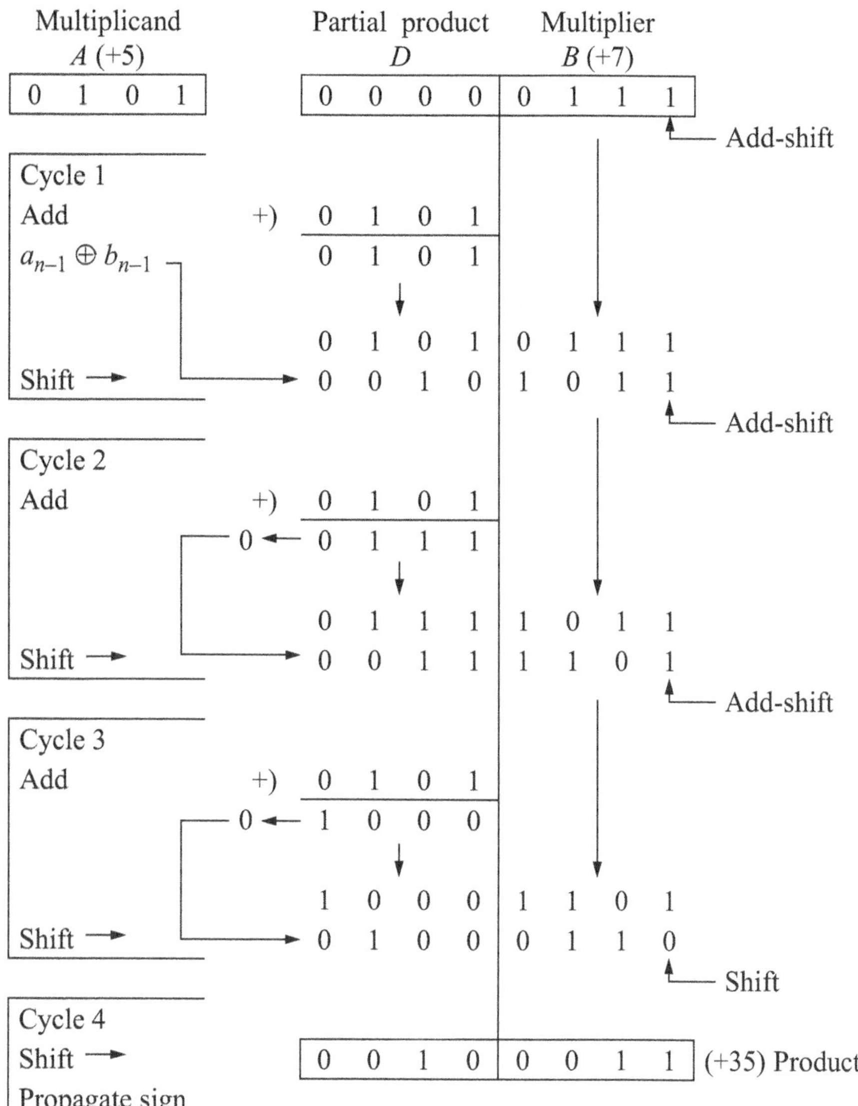

Multiplicand
A (+5)
0 1 0 1
Partial product
D
Multiplier
B (+7)
0 0 0 0 0 1 1 1
Add-shift
Cycle 1
Add
+) 0 1 0 1
0 1 0 1
$a_{n-1} \oplus b_{n-1}$
0 1 0 1 0 1 1 1
Shift →
0 0 1 0 1 0 1 1
Add-shift
Cycle 2
Add
+) 0 1 0 1
0 ← 0 1 1 1
0 1 1 1 1 0 1 1
Shift →
0 0 1 1 1 1 0 1
Add-shift
Cycle 3
Add
+) 0 1 0 1
0 ← 1 0 0 0
1 0 0 0 1 1 0 1
Shift →
0 1 0 0 0 1 1 0
Shift
Cycle 4
Shift →
Propagate sign
0 0 1 0 0 0 1 1
(+35) Product

6.12 Use the Booth algorithm to multiply the operands shown below, which are signed operands in 2s complement representation.

Multiplicand	0	0	1	1	1	(+7)
Multiplier	1	0	1	1	0	(−10)

Booth algorithm											
Multiplicand						0	1	0	1	1	
Recoded multiplier					×)	−1	+1	0	−1	0	
0	0	0	0	0	0	0	0	0	0		
1	1	1	1	1	1	0	0	1			
0	0	0	0	0	0	0	0				
0	0	0	0	1	1	1					
1	1	1	0	0	1						
1	1	1	0	1	1	1	0	1	0	(−70)	

6.17 Use bit-pair recoding to multiply the following operands which are in 2s complement representation:

Multiplicand	1	1	0	0	1	0	(−14)
Multiplier	0	1	1	0	0	1	(+25)

A =		1 1	0 0	1 0	(−14)
B =	×)	0 1	1 0	0 1 [0]	(+25)
			↓		
		1 1	0 0	1 0	
		[0 1]	[1 0]	[0 1 [0]]	
		+2	−2	+1	

1 1	1 1	1 1	1 1	0 0	1 0	
0 0	0 0	0 1	1 1	0 0		
1 1	1 0	0 1	0 0			
1 1	1 0	1 0	1 0	0 0	1 0	(−350)

6.23 Design an array multiplier that multiplies two 2-bit unsigned fixed-point operands: multiplicand $A = [1:0]$ and multiplier $B = [1:0]$. The product is $P[3:0]$. Obtain the structural design module using only AND gates and half adders. Obtain the test bench for all combinations of the multiplicand and multiplier, the outputs, and waveforms. The inputs and outputs are to be shown in binary notation.

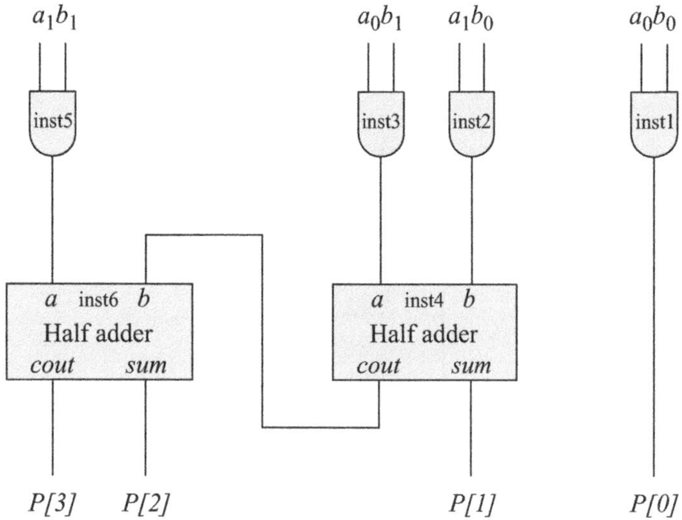

```
//structural module for an array multiplier
//using half adders and AND gates
module array_mul2_ha (a, b, p);

input [1:0] a, b;
output [3:0] p;

//instantiate the logic for p[0]
and2_df inst1 (
   .x1(a[0]),
   .x2(b[0]),
   .z1(p[0])
   );

//instantiate the logic for p[1]
and2_df inst2 (
   .x1(a[1]),
   .x2(b[0]),
   .z1(net2)
   );

and2_df inst3 (
   .x1(a[0]),
   .x2(b[1]),
   .z1(net3)
   );

half_adder_df inst4 (
   .a(net3),
   .b(net2),
   .sum(p[1]),
   .cout(net4)
   );

//instantiate the logic for p[2] and p[3]
and2_df inst5 (
   .x1(a[1]),
   .x2(b[1]),
   .z1(net5)
   );

half_adder_df inst6 (
   .a(net5),
   .b(net4),
   .sum(p[2]),
   .cout(p[3])
   );

endmodule
```

```
//test bench for the multiplier using half adders
module array_mul2_ha_tb;

reg [1:0] a, b;
wire [3:0] p;

//apply stimulus and display variables
initial
begin: apply_stimulus
   reg [4:0] invect;
      for (invect=0; invect<16; invect=invect+1)
      begin
         {a, b} = invect [4:0];
         #10 $display ("a = %b, b = %b, p = %b",
                           a, b, p);
      end
end

//instantiate the module into the test bench
array_mul2_ha inst1 (
   .a(a),
   .b(b),
   .p(p)
   );
endmodule
```

```
a = 00, b = 00, p = 0000     a = 10, b = 00, p = 0000
a = 00, b = 01, p = 0000     a = 10, b = 01, p = 0010
a = 00, b = 10, p = 0000     a = 10, b = 10, p = 0100
a = 00, b = 11, p = 0000     a = 10, b = 11, p = 0110
a = 01, b = 00, p = 0000     a = 11, b = 00, p = 0000
a = 01, b = 01, p = 0001     a = 11, b = 01, p = 0011
a = 01, b = 10, p = 0010     a = 11, b = 10, p = 0110
a = 01, b = 11, p = 0011     a = 11, b = 11, p = 1001
```

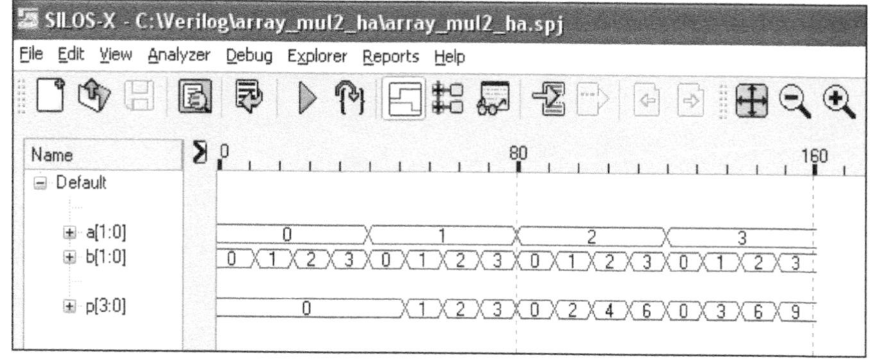

Chapter 7 Fixed-Point Division

7.2 Determine whether the following operands produce an overflow for fixed-point binary division.

(a) Dividend = 001 0110
Divisor = 0011 Overflow

(b) Dividend = 0110 1100
Divisor = 0101 Overflow

(c) Dividend = 0010 1010
Divisor = 0011 No overflow

7.5 Use sequential nonrestoring division to perform the following divide operation:

$A.Q$ (dividend) = 00100 1110 (+78)
B (divisor) = 0101 (+5)

Quotient = 15 Remainder = 3

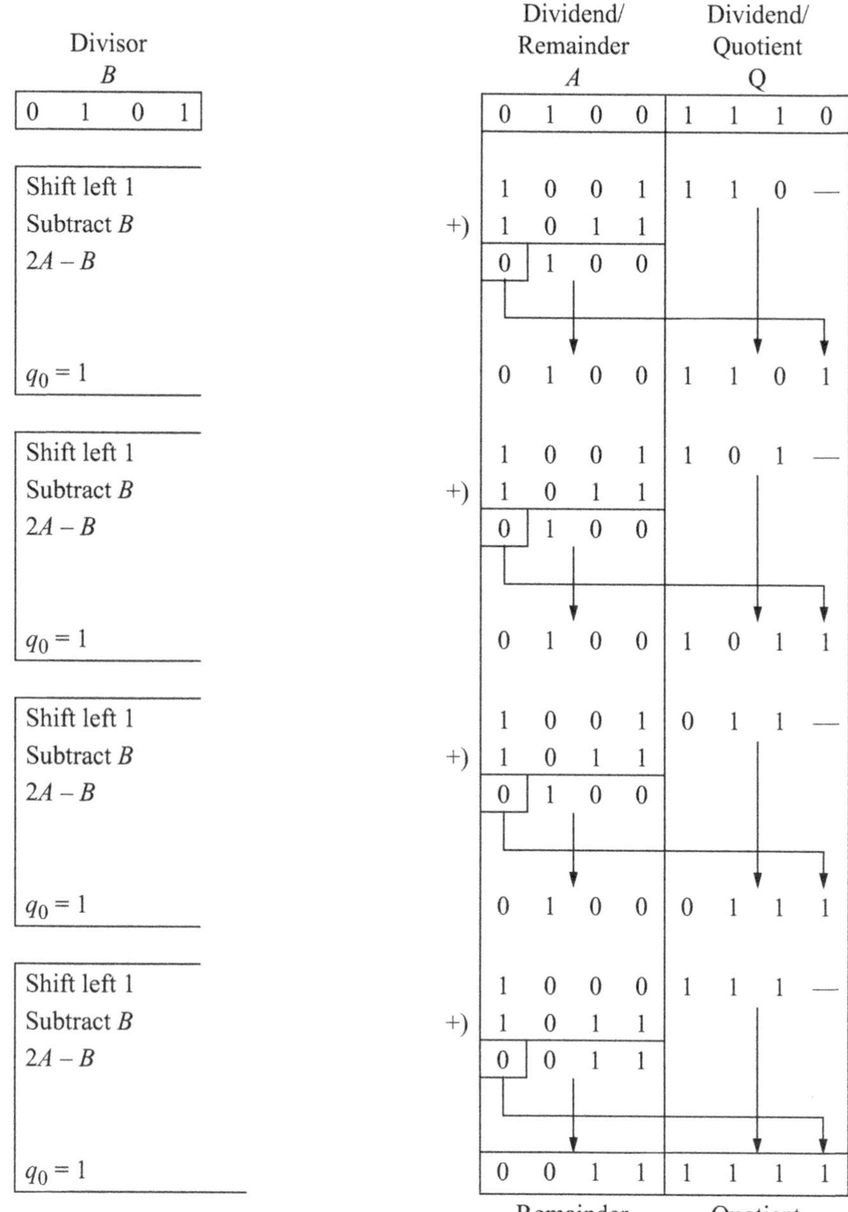
Divisor
B
0 1 0 1
Dividend/
Remainder
A
Dividend/
Quotient
Q
0 1 0 0
1 1 1 0
Shift left 1
Subtract B
2A – B
q_0 = 1
1 0 0 1
1 1 0 —
+)
1 0 1 1
0 1 0 0
0 1 0 0
1 1 0 1
Shift left 1
Subtract B
2A – B
q_0 = 1
1 0 0 1
1 0 1 —
+)
1 0 1 1
0 1 0 0
0 1 0 0
1 0 1 1
Shift left 1
Subtract B
2A – B
q_0 = 1
1 0 0 1
0 1 1 —
+)
1 0 1 1
0 1 0 0
0 1 0 0
0 1 1 1
Shift left 1
Subtract B
2A – B
q_0 = 1
1 0 0 0
1 1 1 —
+)
1 0 1 1
0 0 1 1
0 0 1 1
1 1 1 1
Remainder
Quotient

7.13 Use the paper-and-pencil method to perform SRT division on the following operands: dividend $A.Q = 0.011\ 101$ and divisor $B = 0.100$. Obtain the result as a 3-bit quotient and a 3-bit remainder.

$A.Q = 0.011101$

$B = 0.100$

Normalized $B = 0.100$

2s complement of normalized $B = 1.100$

		6		5	4	3	2	1	0
					A			Q	
		0	.	0	1	1	1	0	1
No adjustment		0	.	0	1	1	1	0	1
Shift over 0s ← 1		0	.	1	1	1	0	1	*
Subtract normalized B	+)	1	.	1	0	0			
		0	.	0	1	1			
Load A, shift ← 1, $q_0 = 1$		0	.	1	1	0	1	*	1
Subtract normalized B	+)	1	.	1	0	0			
		0	.	0	1	0			
Load A, shift ← 1, $q_0 = 1$		0	.	1	0	1	*	1	1
Subtract normalized B	+)	1	.	1	0	0			
		0	.	0	0	1			
Load A, shift ← 1, $q_0 = 1$		0	.	0	1	*	1	1	1
								Q	
Asterisk is at bit 3, ∴ shift A 1 →				0	0	1			
					R				

7.16 Use behavioral modeling to design an SRT divider using the table lookup method for a 4-bit dividend and a 2-bit divisor. Obtain the result as a 2-bit quotient and a 2-bit remainder. Obtain the behavioral module, the test bench module, the outputs, and the waveforms. Include two cases where overflow occurs.

```
//behavioral srt division using table lookup
module srt_div_tbl_lookup2 (opnds, quot_rem, ovfl);

input [5:0] opnds;    //dvdnd 4 bits; dvsr 2 bits
output [3:0] quot_rem;
output ovfl;

wire [5:0] opnds;
reg [3:0] quot_rem;
reg ovfl;

//check for overflow
always @ (opnds)
begin
   if (opnds[5:4] >= opnds[1:0])
      ovfl = 1'b1;
   else
      ovfl = 1'b0;
end

//define memory size
//mem_srt is an array of 64 four-bit registers
reg [3:0] mem_srt[0:63];

//define memory contents
//load mem_srt from file opnds.srt
initial
begin
   $readmemb ("opnds.srt", mem_srt);
end

//use the operands to access memory
always @ (opnds)
begin
   quot_rem = mem_srt[opnds];
end

endmodule
```

```
//test bench for srt division using table lookup
module srt_div_tbl_lookup2_tb;

reg [5:0] opnds;
wire [3:0] quot_rem;
wire ovfl;

//display variables
initial
$monitor ("opnds=%b, quot=%b, rem=%b, ovfl=%b",
            opnds, quot_rem[3:2], quot_rem[1:0], ovfl);

//apply stimulus
initial
begin
   #0    opnds = 6'b010110;
   #10   opnds = 6'b100011;
   #10   opnds = 6'b011111;
   #10   opnds = 6'b101011;
   #10   opnds = 6'b110010;
   #10   opnds = 6'b101110;

   #10   $stop;
end

//instantiate the module into the test bench
srt_div_tbl_lookup2 inst1 (
   .opnds(opnds),
   .quot_rem(quot_rem),
   .ovfl(ovfl)
   );

endmodule
```

```
Q R   Dvdnd_Dvsr                 Q R   Dvdnd_Dvsr
0000                             0000
 . . .                            . . .
0000                             0000
1001  0101_10                    1010  1000_11
0000                             0000
 . . .                            . . .
0000                             0000
1001  0111_11                    1101  1010_11
0000                             0000
 . . .                            . . .
```

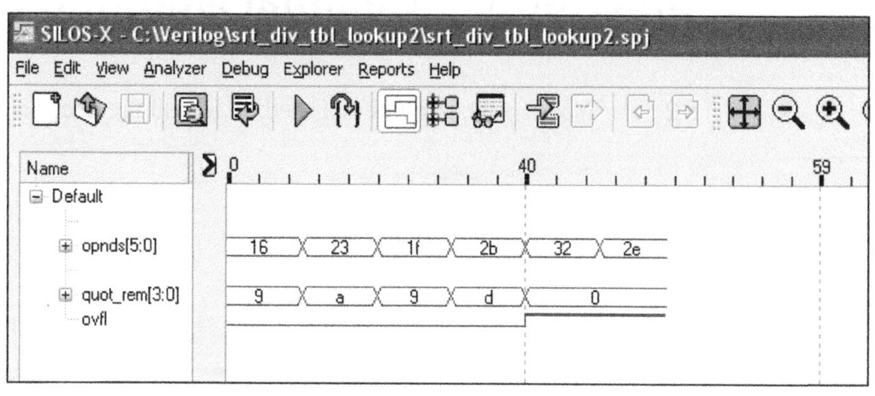

SILOS-X - C:\Verilog\srt_div_tbl_lookup2\srt_div_tbl_lookup2.spj
File Edit View Analyzer Debug Explorer Reports Help
Name
Default
opnds[5:0]
quot_rem[3:0]
ovfl
0
40
59
16
23
1f
2b
32
2e
9
a
9
d
0

Chapter 8 Decimal Addition

8.2 Use the paper-and-pencil method to add the decimal digits 4673 and 9245.

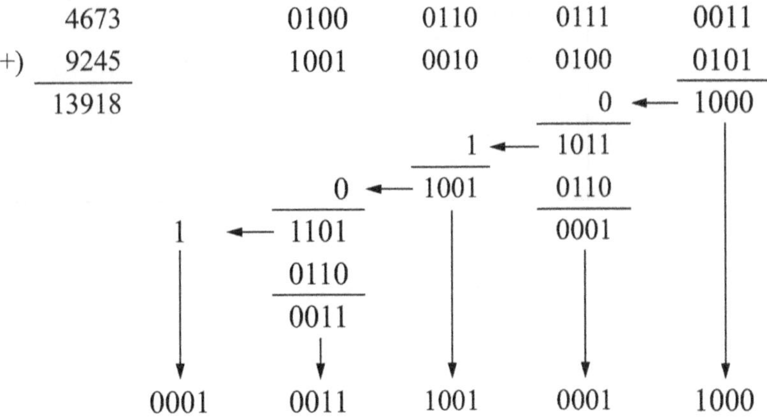

8.5 Add the decimal operands 786 and 956 using memories to correct the intermediate sum. Use the paper-and-pencil method to add the two 3-digit BCD numbers.

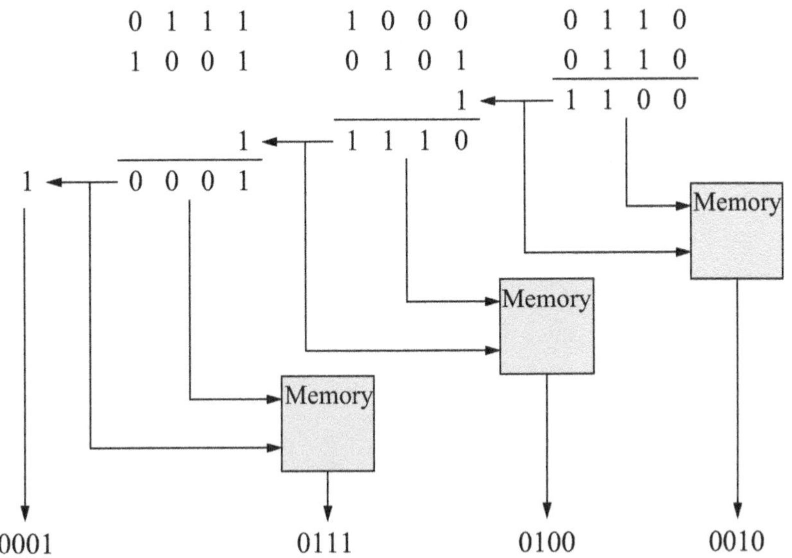

8.9 Use behavioral modeling to design a 1-digit BCD adder. Obtain the test bench for several input vectors, the outputs, and the waveforms.

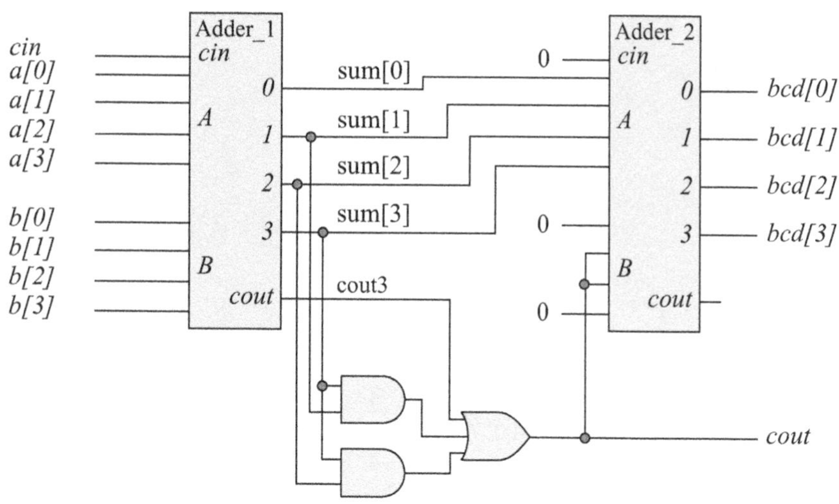


```
//behavioral for a bcd adder
module add_bcd_behav (a, b, cin, bcd, cout);

input [3:0] a, b;
input cin;
output [3:0] bcd;
output cout;

//variables are declared as reg if used in always
reg [3:0] sum;
reg [3:0] bcd;
reg cout3;
reg cout;

always @ (a or b)
begin
   {cout3, sum} = a + b + cin;
   cout = (cout3 || (sum[3] && sum[1]) ||
            (sum[3] && sum[2]));
   bcd = sum + {1'b0, cout, cout, 1'b0};
end
endmodule
```

```
//test bench for behavioral bcd addition
module add_bcd_behav_tb;

reg [3:0] a, b;
reg cin;
wire [3:0] bcd;
wire cout;

initial      //display variables
$monitor ("a=%b, b=%b, cin=%b, cout=%b, bcd=%b",
             a, b, cin, cout, bcd);

initial      //apply input vectors
begin
   #0    a = 4'b0110;   b = 4'b0011;   cin = 1'b0;
   #10   a = 4'b0101;   b = 4'b0010;   cin = 1'b0;
   #10   a = 4'b0111;   b = 4'b0110;   cin = 1'b0;
   #10   a = 4'b0011;   b = 4'b0100;   cin = 1'b1;
   #10   a = 4'b1001;   b = 4'b0110;   cin = 1'b1;
   #10   a = 4'b1000;   b = 4'b0010;   cin = 1'b0;
   #10   a = 4'b1001;   b = 4'b1001;   cin = 1'b1;
   #10   a = 4'b0101;   b = 4'b1000;   cin = 1'b0;
   #10   $stop;
end

//instantiate the module into the test bench
add_bcd_behav inst1 (
   .a(a),
   .b(b),
   .cin(cin),
   .bcd(bcd),
   .cout(cout)
   );

endmodule
```

```
a=0110, b=0011, cin=0, cout=0, bcd=1001
a=0101, b=0010, cin=0, cout=0, bcd=0111
a=0111, b=0110, cin=0, cout=1, bcd=0011
a=0011, b=0100, cin=1, cout=0, bcd=1000
a=1001, b=0110, cin=1, cout=1, bcd=0110
a=1000, b=0010, cin=0, cout=1, bcd=0000
a=1001, b=1001, cin=1, cout=1, bcd=1001
a=0101, b=1000, cin=0, cout=1, bcd=0011
```

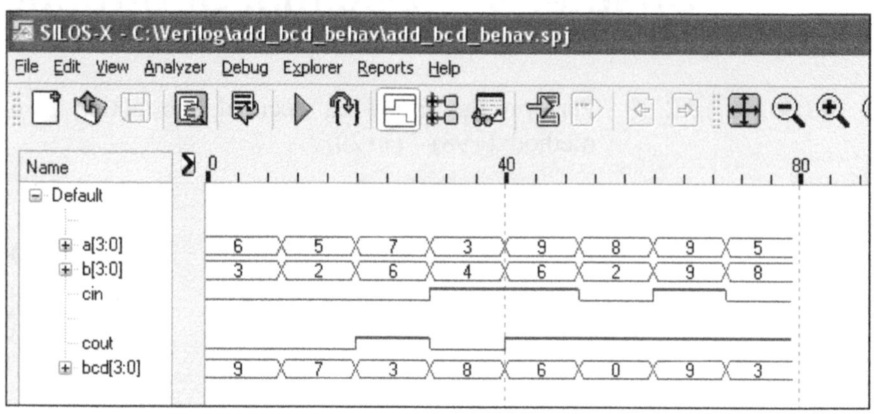
SILOS-X - C:\Verilog\add_bcd_behav\add_bcd_behav.spj
File Edit View Analyzer Debug Explorer Reports Help
Name
Default
a[3:0]
b[3:0]
cin
cout
bcd[3:0]
0
40
80
6 5 7 3 9 8 9 5
3 2 6 4 6 2 9 8
9 7 3 8 6 0 9 3

Chapter 9 Decimal Subtraction

9.2 Perform the following decimal subtraction using the paper-and-pencil method: (+98) – (+120).

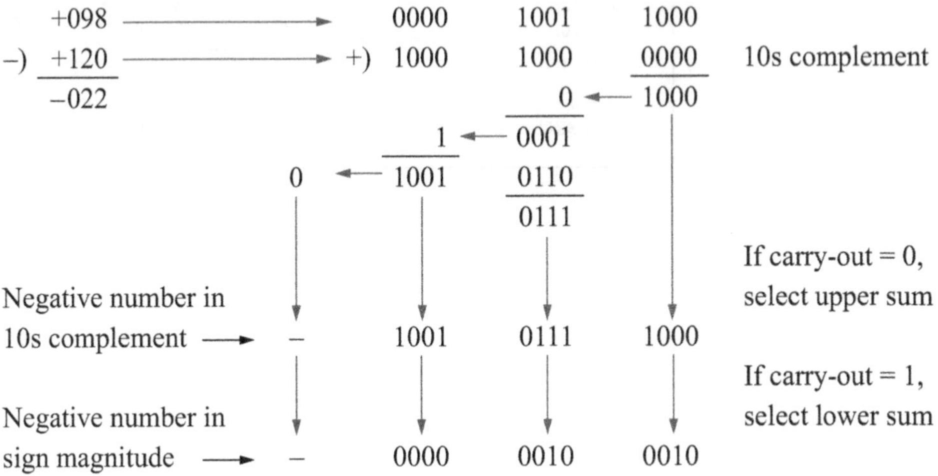

9.5 Perform the following decimal subtraction using the paper-and-pencil method: (+436) + (–825).

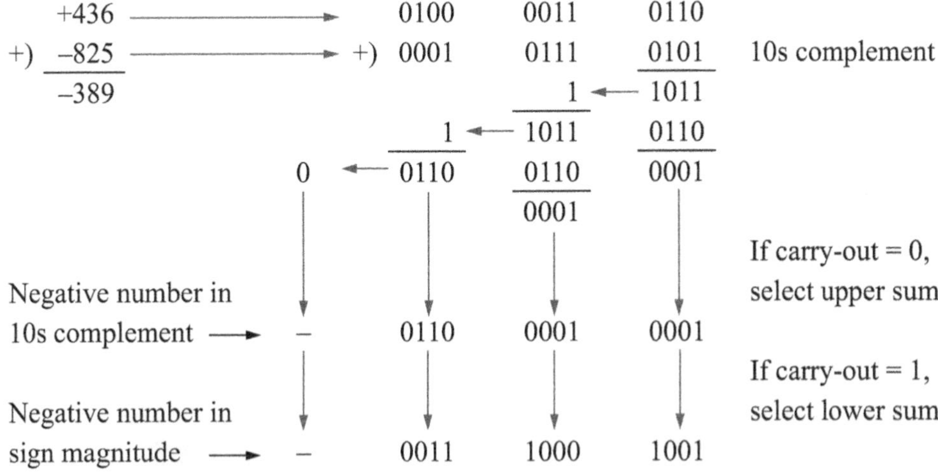

10.3 Using the paper-and-pencil method, multiply the decimal numbers 63 and 87 using the table lookup technique.

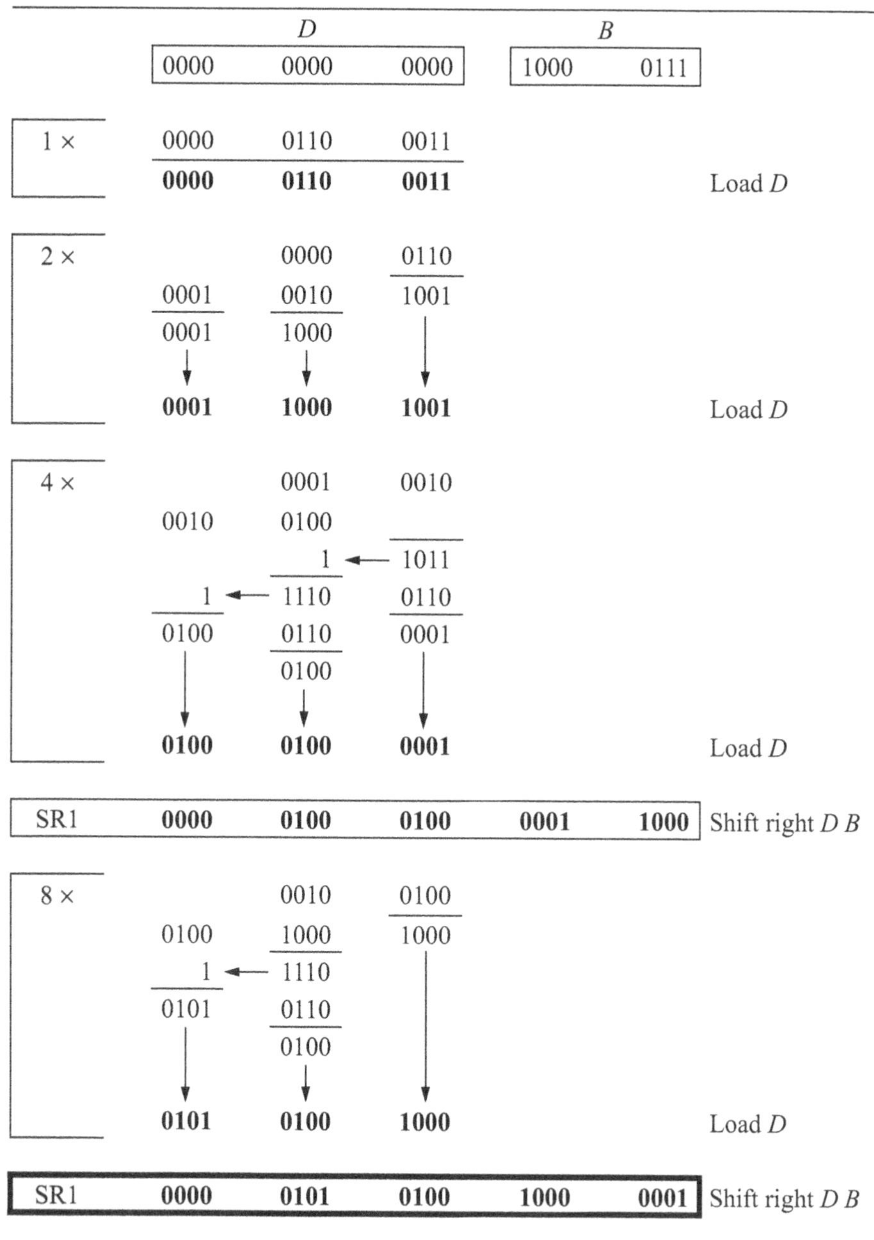

10.6 Design a 3-digit decimal adder using behavioral modeling. This adder can be used as part of decimal multiply algorithm using the table lookup technique. Obtain the design module, the test bench module, the outputs, and the waveforms.

```
//behavioral for a 3-digit bcd adder
module add_bcd_behav3 (a, b, cin, bcd);

input [11:0] a, b;
input cin;
output [12:0] bcd;

//define internal registers
reg [3:0] sum_units, sum_tens, sum_hunds;
reg cout_units, cout_tens, cout_hunds, cout;

always @ (a or b)
begin
   sum_units = a[3:0] + b[3:0] + cin;
   if ((a[3] && b[3]) || (sum_units[3] &&
      sum_units[2]) || (sum_units[3] && sum_units[1]))
      begin
         sum_units = sum_units + 4'b0110;
         cout_units = 1'b1;
      end
   else  cout_units = 1'b0;

   sum_tens = a[7:4] + b[7:4] + cout_units;
   if ((a[7] && b[7]) || (sum_tens[3] &&
         sum_tens[2]) || (sum_tens[3] && sum_tens[1]))
      begin
         sum_tens = sum_tens + 4'b0110;
         cout_tens = 1'b1;
      end
   else  cout_tens = 1'b0;

   {cout, sum_hunds} = a[11:8] + b[11:8] + cout_tens;
   if ((a[11] && b[11]) || (sum_hunds[3] &&
         sum_hunds[2]) || (sum_hunds[3] &&
         sum_hunds[1]) || cout)
      begin
         sum_hunds = sum_hunds + 4'b0110;
         cout_hunds = 1'b1;
      end
   else  cout_hunds = 1'b0;
end

//continued on next page
```

```
assign   bcd[12] = cout_hunds,
         bcd[11:8] = sum_hunds,
         bcd[7:4] = sum_tens,
         bcd[3:0] = sum_units;
endmodule
```

```
//test bench for add_bcd_behav2
module add_bcd_behav3_tb;

reg [11:0] a, b;
reg cin;
wire [12:0] bcd;

initial  //display variables
$monitor ("a=%b, b=%b, cin=%b, bcd_thous=%b,
         bcd_hunds=%b, bcd_tens=%b, bcd_units=%b",
            a, b, cin, {{3{1'b0}}, bcd[12]},
            bcd[11:8], bcd[7:4], bcd[3:0]);

initial  //apply input vectors
begin
         //735 + 967 = 1702
   #0    a = 12'b0111_0011_0101;
         b = 12'b1001_0110_0111; cin = 1'b0;

         //259 + 529 = 0788
   #10   a = 12'b0010_0101_1001;
         b = 12'b0101_0010_1001; cin = 1'b0;

         //999 + 999 = 1998
   #10   a = 12'b1001_1001_1001;
         b = 12'b1001_1001_1001; cin = 1'b0;

         //963 + 775 + 1 = 1739
   #10   a = 12'b1001_0110_0011;
         b = 12'b0111_0111_0101; cin = 1'b1;

   #10   $stop;
end

add_bcd_behav3 inst1 (      //instantiate the module
   .a(a),
   .b(b),
   .cin(cin),
   .bcd(bcd)
   );

endmodule
```

```
a=011100110101, b=100101100111, cin=0,
   bcd_thous=0001, bcd_hunds=0111,
   bcd_tens=0000, bcd_units=0010

a=001001011001, b=010100101001, cin=0,
   bcd_thous=0000, bcd_hunds=0111,
   bcd_tens=1000, bcd_units=1000

a=100110011001, b=100110011001, cin=0,
   bcd_thous=0001, bcd_hunds=1001,
   bcd_tens=1001, bcd_units=1000

a=100101100011, b=011101110101, cin=1,
   bcd_thous=0001, bcd_hunds=0111,
   bcd_tens=0011, bcd_units=1001
```

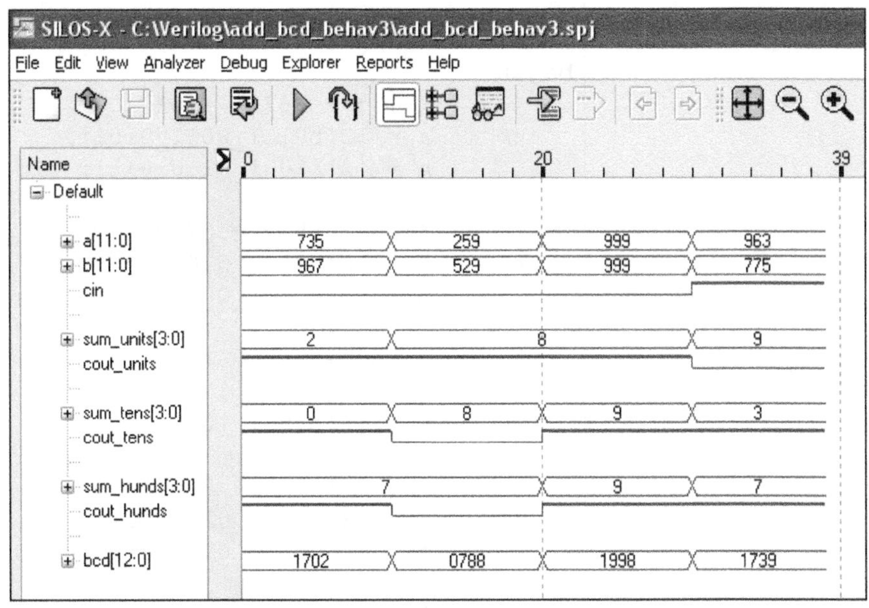

11.3 Use the table lookup method to perform the following divide operation: +77 ÷ +5.

	Divisor / Sign		Quotient: 1
	5	7	7
− Divisor × 8	−	4	0
	−	3	3
+ Divisor × 4	+	2	0
	−	1	3
+ Divisor × 2	+	1	0
	−		3
+ Divisor × [1]	+		5
	+		[2] ← Partial remainder

	Divisor / Sign		Quotient: 5
	5	2	7
− Divisor × 8	−	4	0
	−	1	3
+ Divisor × [4]	+	2	0
	+		7
− Divisor × 2	−	1	0
	−		3
+ Divisor × [1]	+		5
	+		[2] ← Remainder

Chapter 12 Floating-Point Addition

12.2 Add the two floating-point numbers shown below with no implied 1 bit.

$A = 0.1100\ 1000\ 0000 \ldots 0 \times 2^6$ (+50)
$B = 0.1000\ 1000\ 0000 \ldots 0 \times 2^3$ (+4.25)

$$
\begin{array}{rl}
A = & 0.1100\ 1000\ 0000 \ldots 0 \times 2^6 \\
+)\ \text{Align } B = & \underline{0.0001\ 0001\ 0000 \ldots 0 \times 2^6} \\
 & 0.1101\ 1001\ 0000 \ldots 0 \quad \times 2^6 \quad (+54.25)
\end{array}
$$

12.6 For a 23-bit fraction with an 8-bit exponent and sign bit, determine:

(a) The largest positive number with an unbiased exponent
0 0111 1111 1111 . . . 1111

(b) The largest positive number with a biased exponent
0 1111 1111 1111 . . . 1111

(c) The most negative number with the most negative unbiased exponent
1 1000 0000 1111 1111

12.10 Obtain the normalized floating-point representation for the decimal number shown below using the single-precision format. Use a biased exponent and an implied 1 bit.

$$
\begin{aligned}
+9.75 &= 1001.11 \\
&= 0.1001\ 1100 \times 2^4 \\
&= 0\quad 1000\ 0011\quad 1001\ 1100\ 0000 \ldots 0000 \\
&= 0\quad 1000\ 0011\quad 0011\ 1000\ 0000 \ldots 0000
\end{aligned}
$$

12.12 Add the decimal operands shown below using the single-precision floating-point format with biased exponents and an implied 1 bit.

$A = (-10),\ \ B = (-35)$

$A = 1.1010\ 0000\ \times 2^4$
$B = 1.1000\ 1100\ \times 2^6$

$A = 1.0010\ 1000\ \times 2^6$
$B = 1.\underline{1000\ 1100\ \times 2^6}$
$\text{Sum} = .1011\ 0100\ \times 2^6$

1 0000 0110 1011 0100 0000 0000 . . . 0000
Add bias = 1 1000 0110 0110 1000 0000 0000 . . . 0000

13.3 Perform the following operation on the two operands: (+127) – (–77).

Before alignment

$A =$ 0	.	1	1	1	1	1	1	1	0	$\times 2^7$	+127
$B =$ 1	.	1	0	0	1	1	0	1	0	$\times 2^7$	–77

After alignment (already aligned)

$A =$ 0	.	1	1	1	1	1	1	1	0	$\times 2^7$	+127
+) $B =$ 1	.	1	0	0	1	1	0	1	0	$\times 2^7$	–77
1 ←	.	1	0	0	1	1	0	0	0	$\times 2^7$	
0	.	1	1	0	0	1	1	0	0	$\times 2^8$	+204

13.7 Perform the following operation on the two operands: (+127) – (+255).

Before alignment

$A =$ 0	.	1	1	1	1	1	1	1	0	$\times 2^7$	+127
$B =$ 0	.	1	1	1	1	1	1	1	1	$\times 2^8$	+255

After alignment

$A =$ 0	.	0	1	1	1	1	1	1	1	$\times 2^8$	+127
$B =$ 0	.	1	1	1	1	1	1	1	1	$\times 2^8$	+255

Add fractions

$A' + 1 =$ 0	.	1	0	0	0	0	0	0	1	$\times 2^8$	
+) $B =$ 0	.	1	1	1	1	1	1	1	1	$\times 2^8$	
1 ←	.	1	0	0	0	0	0	0	0	$\times 2^8$	
1	.	1	0	0	0	0	0	0	0	$\times 2^8$	–128

13.11 Perform the following operation on the two operands: (–127) + (+150).

Before alignment													
	$A =$	1	.	1	1	1	1	1	1	1	0	$\times 2^7$	–127
	$B =$	0	.	1	0	0	1	0	1	1	0	$\times 2^8$	+150
After alignment													
	$A =$	1	.	0	1	1	1	1	1	1	1	$\times 2^8$	–127
	$B =$	0	.	1	0	0	1	0	1	1	0	$\times 2^8$	+150
Add fractions													
	$A' + 1 =$	1	.	1	0	0	0	0	0	0	1	$\times 2^8$	
	+) $B =$	0	.	1	0	0	1	0	1	1	0	$\times 2^8$	
	1 ←		.	0	0	0	1	0	1	1	1	$\times 2^8$	
Postnormalize		0	.	1	0	1	1	1	0	0	0	$\times 2^5$	+23

13.15 Perform the following operation on the two operands: (+85.5) + (–70.5).

Before alignment													
	$A =$	0	.	1	0	1	0	1	0	1	1	$\times 2^7$	+85.5
	$B =$	1	.	1	0	0	0	1	1	0	1	$\times 2^7$	–70.5
After alignment													
	$A =$	0	.	1	0	1	0	1	0	1	1	$\times 2^7$	+85.5
	$B =$	1	.	1	0	0	0	1	1	0	1	$\times 2^7$	–70.5
Add fractions													
	$A =$	0	.	1	0	1	0	1	0	1	1	$\times 2^7$	
	+) $B' + 1 =$	1	.	0	1	1	1	0	0	1	1	$\times 2^7$	
	1 ←		.	0	0	0	1	1	1	1	0	$\times 2^7$	
Postnormalize		0	.	1	1	1	1	0	0	0	0	$\times 2^4$	+15

13.20 Perform the following operation on the two operands: (+105) – (+5).

Before alignment

$A =$	0	.	1	1	0	1	0	0	1	0	$\times 2^7$	+105
$B =$	0	.	1	0	1	0	0	0	0	0	$\times 2^3$	+5

After alignment

$A =$	0	.	1	1	0	1	0	0	1	0	$\times 2^7$	+105
$B =$	0	.	0	0	0	0	1	0	1	0	$\times 2^7$	+5

Add fractions

$A =$	0	.	1	1	0	1	0	0	1	0	$\times 2^7$	
+) $B' + 1 =$	0	.	1	1	1	1	0	1	1	0	$\times 2^7$	
1 ←		.	1	1	0	0	1	0	0	0	$\times 2^7$	
	0	.	1	1	0	0	1	0	0	0	$\times 2^7$	+100

14.4 Represent the decimal number –1.75 in the single-precision format with an implied 1 bit and a biased exponent.

1 0111_1111 1100_0000_0000_0000_0000_000

14.8 Use the sequential add-shift method to multiply the two decimal numbers shown below as floating-point numbers using 8-bit fractions. Show multiple shifts as a single shift amount, not as separate shifts.

+34 × –68

+34 = 0 0000_0110 1000_1000
–68 = 1 0000_0111 1000_1000

fract_a (+34)		*prod*	*fract_b* (–68)	
1000 1000	*prod*	0000 0000	1000 1000	
Shift right 3		0000 0000	0001 0001	
Add	+)	1000 1000		Add-shift
	0←	1000 1000	0001 0001	
Shift		0100 0100	0000 1000	
				Shift
Shift right 3		0000 1000	1000 0001	
Add	+)	1000 1000		Add-shift
	0←	1001 0000	1000 0001	
Shift		0100 1000	0100 0000	
Postnormalize		1001 0000	1000 0000	$\times 2^{12} = -2312$

14.10 A floating-point format is shown below for 14 bits. Design a behavioral module for floating-point multiplication using the sequential add-shift method for two operands: *flp_a[13:0]* and *flp_b[13:0]*. Obtain the test bench for several different input vectors, including both positive and negative operands. Obtain the outputs and waveforms.

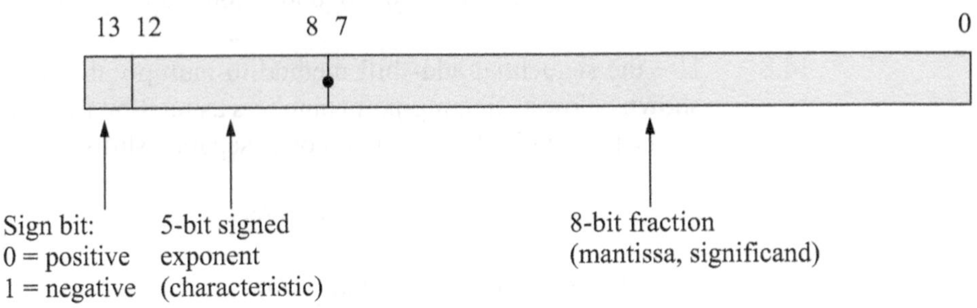


```
//behavioral floating-point multiplication
module mul_flp4 (flp_a , flp_b, start, sign,
                  exponent, exp_unbiased, cout, prod);


input [13:0] flp_a, flp_b;
input start;
output sign;
output [4:0] exponent, exp_unbiased;
output cout;
output [15:0] prod;

//variables used in an always block
//are declared as registers
reg sign_a, sign_b;
reg [4:0] exp_a, exp_b;
reg [4:0] exp_a_bias, exp_b_bias;
reg [4:0] exp_sum;
reg [7:0] fract_a, fract_b;
reg [7:0] fract_b_reg;
reg sign;
reg [4:0] exponent, exp_unbiased;
reg cout;
reg [15:0] prod;
reg [3:0] count;

//continued on next page
```

```
//define sign, exponent, and fraction
always @ (flp_a or flp_b)
begin
   sign_a = flp_a[13];
   sign_b = flp_b[13];

   exp_a = flp_a[12:8];
   exp_b = flp_b[12:8];

   fract_a = flp_a[7:0];
   fract_b = flp_b[7:0];

//bias exponents
   exp_a_bias = exp_a + 5'b01111;
   exp_b_bias = exp_b + 5'b01111;

//add exponents
   exp_sum = exp_a_bias + exp_b_bias;

//remove one bias
   exponent = exp_sum - 5'b01111;
   exp_unbiased = exponent - 5'b01111;
end

always @ (posedge start)   //multiply fractions
begin
   fract_b_reg = fract_b;
   prod = 0;
   count = 4'b1000;
      if ((fract_a != 0) && (fract_b != 0))
         while (count)
            begin
               {cout, prod[15:8]}=(({8{fract_b_reg[0]}}
                & fract_a) + prod[15:8]);
                prod = {cout, prod[15:8], prod[7:1]};
                fract_b_reg = fract_b_reg >> 1;
                count = count - 1;
            end

   while (prod[15] == 0)   //postnormalize result
      begin
         prod = prod << 1;
         exp_unbiased = exp_unbiased - 1;
      end

   sign = sign_a ^ sign_b;
end
endmodule
```

```
//test bench for floating-point multiplication
module mul_flp4_tb;

reg [13:0] flp_a, flp_b;
reg   start;
wire  sign;
wire  [4:0] exponent, exp_unbiased;
wire  [4:0] exp_sum;
wire  [15:0] prod;

//display variables
initial
$monitor ("sign = %b, exp_unbiased = %b, prod = %b",
         sign, exp_unbiased, prod);

//apply input vectors
initial
begin
   #0    start = 1'b0;
         //+5 x +3 = +15
         //          s   e     f
         flp_a = 14'b0_00011_1010_0000;
         flp_b = 14'b0_00010_1100_0000;
   #10   start = 1'b1;
   #10   start = 1'b0;


         //+7 x -5 = -35
         //          s   e     f
   #0    flp_a = 14'b0_00011_1110_0000;
         flp_b = 14'b1_00011_1010_0000;
   #10   start = 1'b1;
   #10   start = 1'b0;


         //+25 x +25 = +625
         //          s   e     f
   #0    flp_a = 14'b0_00101_1100_1000;
         flp_b = 14'b0_00101_1100_1000;
   #10   start = 1'b1;
   #10   start = 1'b0;


         //-7 x -15 = +105
         //          s   e     f
   #0    flp_a = 14'b1_00011_1110_0000;
         flp_b = 14'b1_00100_1111_0000;
   #10   start = 1'b1;
   #10   start = 1'b0;
         //continued on next page
```

```
        //-35 x +72 = -2520
        //          s   e       f
   #0   flp_a = 14'b1_00110_1000_1100;
        flp_b = 14'b0_00111_1001_0000;
   #10  start = 1'b1;
   #10  start = 1'b0;


        //+80 x +37 = +2960
        //          s   e       f
   #0   flp_a = 14'b0_00111_1010_0000;
        flp_b = 14'b0_00110_1001_0100;
   #10  start = 1'b1;
   #10  start = 1'b0;


        //+34 x -68 = -2312
        //          s   e       f
   #0   flp_a = 14'b0_00110_1000_1000;
        flp_b = 14'b1_00111_1000_1000;
   #10  start = 1'b1;
   #10  start = 1'b0;

   #10  $stop;

end

//instantiate the module into the test bench
mul_flp4 inst1 (
   .flp_a(flp_a),
   .flp_b(flp_b),
   .start(start),
   .sign(sign),
   .exponent(exponent),
   .exp_unbiased(exp_unbiased),
   .prod(prod)
   );

endmodule
```

```
flp_a (+5) x flp_b (+3) = +15
sign = 0, exp_unbiased = 00100,
   prod = 1111_0000_0000_0000

flp_a (+7) x flp_b (-5) = -35
sign = 1, exp_unbiased = 00110,
   prod = 1000_1100_0000_0000

flp_a (+25) x flp_b (+25) = +625
sign = 0, exp_unbiased = 01010,
   prod = 1001_1100_0100_0000

flp_a (-7) x flp_b (-15) = +105
sign = 0, exp_unbiased = 00111,
   prod = 1101_0010_0000_0000

flp_a (-35) x flp_b (+72) = -2520
sign = 1, exp_unbiased = 01100,
   prod = 1001_1101_1000_0000

flp_a (80) x flp_b (+37) = +2960
sign = 0, exp_unbiased = 01100,
   prod = 1011_1001_0000_0000

flp_a (+34) x flp_b (-68) = -2312
sign = 1, exp_unbiased = 01100,
   prod = 1001_0000_1000_0000
```

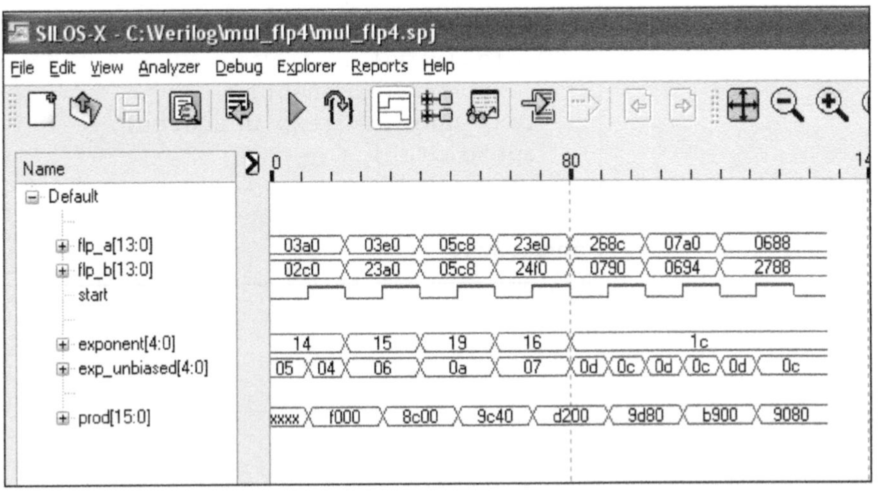

15.3 Comment on the biasing problem when the exponents are operated on during floating-point division.

Subtraction produces the difference of two biased exponents. This yields a difference without a bias. The problem is resolved by adding back the bias.

15.5 Let the dividend fraction $fract\_a = 0.1010\ 0010 \times 2^7$ (+81) be divided by a divisor fraction $fract\_b = 0.1001 \times 2^4$ (+9). Use the paper-and-pencil technique to obtain the quotient and remainder.

fract_b (+9)	*fract_a* (+81)	
1001	1010 0010	
Align	0101 0001	$\times 2^{(7+1)} = 2^8$

Shift left 1	1010 001–	
Subtract *B*	+) 0111	
	1 ← 0001	
No restore	0001 0011	

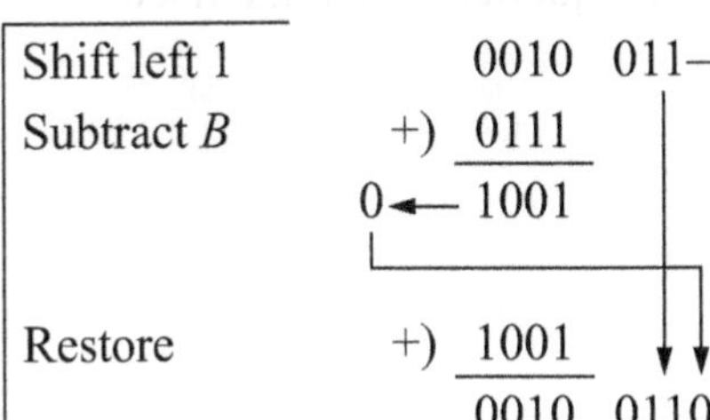

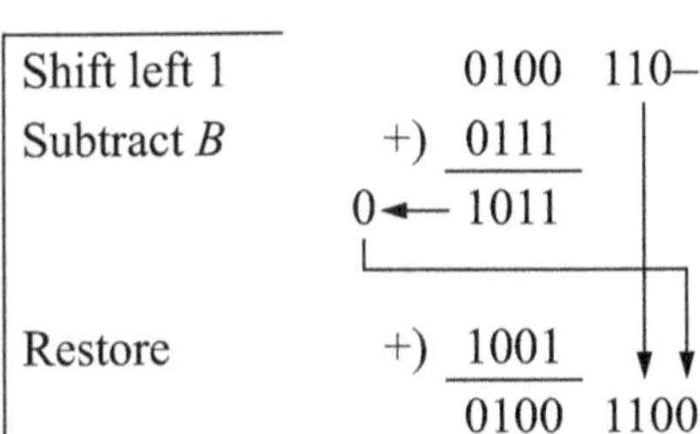

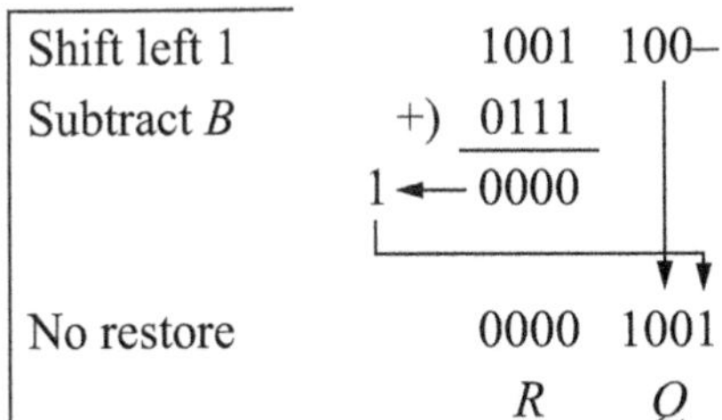

$\times 2^{(7-4)+1} = 2^4$

Chapter 16 Additional Floating-Point Topics

16.6 A memory address consists of the three low-order bits of the significand right-concatenated with the high-order bit of the three guard bits. The memory contents contain the rounded results of the adder-based rounding technique. Obtain the memory contents.

Memory address $b_2\ b_1\ b_0$	Guard bit g_2	Memory contents
000	0	000
000	1	001
001	0	001
001	1	010
010	0	010
010	1	011
011	0	011
011	1	100
100	0	100
100	1	101
101	0	101
101	1	110
110	0	110
110	1	111
111	0	111
111	1	111

16.8 Let the augend be $A = 0.1100\ 1101 \times 2^6$ (+51.2500).
Let the addend be $B = 0.1011\ 0001 \times 2^4$ (+11.0625).
Sum = +62.3125

Perform the addition operation and round the result using all three rounding methods.

	$A =$	0.11001101	$\times 2^6$	
Align	+) $B =$	0.00101100 010	$\times 2^6$	
		0.11111001	$\times 2^6$	(+62.25)
Truncation		0.11111001	$\times 2^6$	(+62.25)
Adder-based		0.11111001	$\times 2^6$	(+62.25)
von Neumann		0.11111001	$\times 2^6$	(+62.25)

16.11 Using the floating-point format shown below, design a Verilog HDL module to perform addition on two operands, then round the result using truncation. Let the fractions be *fract_a[10:0]* and *fract_b[10:0]*. Truncate the result to eight bits labeled *trunc_sum[7:0]*. Obtain the behavioral module, the test bench module containing integers and integers with fractions, and the outputs.

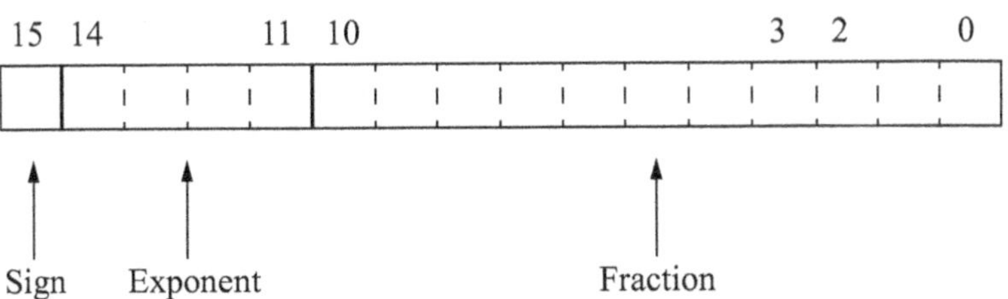

```
//behavioral truncation rounding
module truncation_rounding (flp_a, flp_b, sign,
            exponent, exp_unbiased, sum, trunc_sum);

input [15:0] flp_a, flp_b;
output [10:0] sum;
output [7:0] trunc_sum;
output sign;
output [3:0] exponent, exp_unbiased;

//variables in an always block are declared as registers
reg sign_a, sign_b;
reg [3:0] exp_a, exp_b;
reg [3:0] exp_a_bias, exp_b_bias;
reg [10:0] fract_a, fract_b;
reg [3:0] ctr_align;
reg [10:0] sum;
reg [7:0] trunc_sum;
reg sign;
reg [3:0] exponent, exp_unbiased;
reg cout;

always @ (flp_a or flp_b)
begin
   sign_a = flp_a[15];
   sign_b = flp_b[15];

   exp_a = flp_a[14:11];
   exp_b = flp_b[14:11];

   fract_a = flp_a[10:0];
   fract_b = flp_b[10:0];

//continued on next page
```

```
//bias exponents
   exp_a_bias = exp_a + 4'b0111;
   exp_b_bias = exp_b + 4'b0111;

//align fractions
   if (exp_a_bias < exp_b_bias)
      ctr_align = exp_b_bias - exp_a_bias;

      while (ctr_align)
         begin
            fract_a = fract_a >> 1;
            exp_a_bias = exp_a_bias + 1;
            ctr_align = ctr_align - 1;
         end

   if (exp_b_bias < exp_a_bias)
      ctr_align = exp_a_bias - exp_b_bias;

      while (ctr_align)
         begin
            fract_b = fract_b >> 1;
            exp_b_bias = exp_b_bias + 1;
            ctr_align = ctr_align - 1;
         end

//add fractions
   {cout, sum[10:3]} = fract_a[10:3] + fract_b[10:3];

//normalize result
   if (cout == 1)
      begin
         {cout, sum[10:3]} = {cout, sum[10:3]} >> 1;
         exp_b_bias = exp_b_bias + 1;
      end

//execute truncation rounding
   trunc_sum = sum[10:3];

   sign = sign_a;
   exponent = exp_b_bias;
   exp_unbiased = exp_b_bias - 4'b0111;

end

endmodule
```

```
//test bench for behavioral adder-based rounding
module truncation_rounding_tb;

reg [15:0] flp_a, flp_b;
wire [10:0] sum;
wire [7:0] trunc_sum;
wire sign;
wire [3:0] exponent, exp_unbiased;


//display variables
initial
$monitor ("sign=%b, exp_biased=%b, exp_unbiased=%b,
   sum=%b", sign, exponent, exp_unbiased, trunc_sum);

//apply input vectors
initial
begin
         //+12 + +35 = +47
         //           s --e- ------f------
   #0    flp_a = 16'b0_0100_1100_0000_000;
         flp_b = 16'b0_0110_1000_1100_000;

         //+26 + +20 = +46
         //           s --e- ------f------
   #10   flp_a = 16'b0_0101_1101_0000_000;
         flp_b = 16'b0_0101_1010_0000_000;

         //+26.5 + +4.375 = +30.875
         //           s --e- ------f------
   #10   flp_a = 16'b0_0101_1101_0100_000;
         flp_b = 16'b0_0011_1000_1100_000;

         //+11 + +34 = +45
         //           s --e- ------f------
   #10   flp_a = 16'b0_0100_1011_0000_000;
         flp_b = 16'b0_0110_1000_1000_000;

         //+58.2500 + +9.4375 = +67.6875
         //           s --e- ------f------
   #10   flp_a = 16'b0_0110_1110_1001_000;
         flp_b = 16'b0_0100_1001_0111_000;

         //+50.2500 + +9.4375 = +59.6875
         //           s --e- ------f------
   #10   flp_a = 16'b0_0110_1100_1001_000;
         flp_b = 16'b0_0100_1001_0111_000;

//continued on next page
```

```
        //+51.2500 + +11.0625 = +62.3125
        //          s --e- ------f------
   #10  flp_a = 16'b0_0110_1100_1101_000;
        flp_b = 16'b0_0100_1011_0001_000;

        //+12.75000 + +4.34375 = +17.09375
        //          s --e- ------f------
   #10  flp_a = 16'b0_0100_1100_1100_000;
        flp_b = 16'b0_0011_1000_1011_000;

   #10  $stop;
end

truncation_rounding inst1 (   //instantiate the module
   .flp_a(flp_a),
   .flp_b(flp_b),
   .sum(sum),
   .trunc_sum(trunc_sum),
   .sign(sign),
   .exponent(exponent),
   .exp_unbiased(exp_unbiased)
   );
endmodule
```

```
sign=0, exp_biased=1101,
   exp_unbiased=0110, sum=1011_1100

sign=0, exp_biased=1101,
   exp_unbiased=0110, sum=1011_1000

sign=0, exp_biased=1100,
   exp_unbiased=0101, sum=1111_0111

sign=0, exp_biased=1101,
   exp_unbiased=0110, sum=1011_0100

sign=0, exp_biased=1110,
   exp_unbiased=0111, sum=1000_0111

sign=0, exp_biased=1101,
   exp_unbiased=0110, sum=1110_1110

sign=0, exp_biased=1101,
   exp_unbiased=0110, sum=1111_1001

sign=0, exp_biased=1100,
   exp_unbiased=0101, sum=1000_1000
```

Chapter 17 Additional Topics in Computer Arithmetic

17.4 Perform residue checking on the following numbers:

```
0010 1010     Carry-in = 1
0101 0111     Carry-out = 0
```

```
                            Residue of
                             segments
0010        1010            2   1   → 0
                                            → 0 + carry-in → mod-3 = 1
0101        0111            2   1   → 0
----
0111           1
            ----
   1  ←     0010
----
1000                                    No error

 ↓           ↓

1000        0010            2   2   → 1   → mod-3 = 1
```

17.9 Perform parity prediction on the following operands with a carry-in = 1: 0111 and 0111.

```
        a =       0    1    1    1                 Pa   = 0 ┐
+)      b =       0    1    1    1                 Pb   = 0 │ ⊕ = 1
  Carries =       1 ↖  1 ↖  1 ↖  1 ←— (c_in)       Pcy  = 1 ┘   ↑
            --------------------
      Sum =       1    1    1    1                 Psum = 1 ←———┘
```

17.13 Derive the truth table to determine if the sum of two 16-bit operands is positive. Then draw the logic that will generate a logic 1 if the sign is positive.

a[15]	*b[15]*	*cy[14]*	*Sign*
0	0	0	+
0	1	1	+
1	0	1	+
1	1	0	+

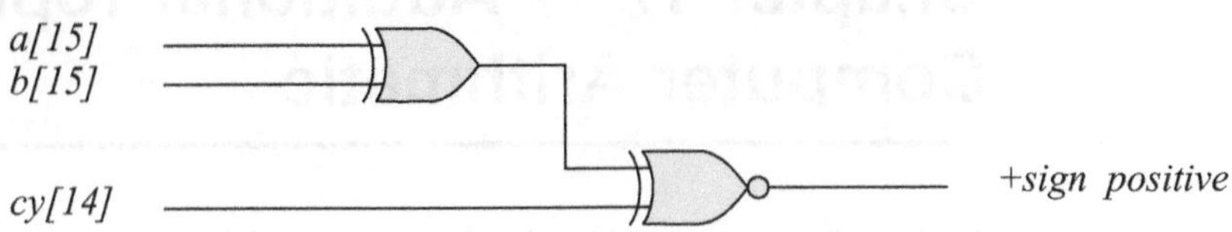

17.20 Draw the logic diagram for an arithmetic and logic unit (ALU) to perform addition, subtraction, AND, and OR on the two 4-bit operands *a[3:0]* and *b[3:0]*. Use an adder and multiplexers with additional logic gates. Use structural modeling to design the 4-function ALU. Then obtain the test bench, the outputs, and the waveforms.

Use two control inputs to determine the operation to be performed, according to the table shown below.

c[1]	*c[0]*	Operation
0	0	Add
0	1	Subtract
1	0	AND
1	1	OR

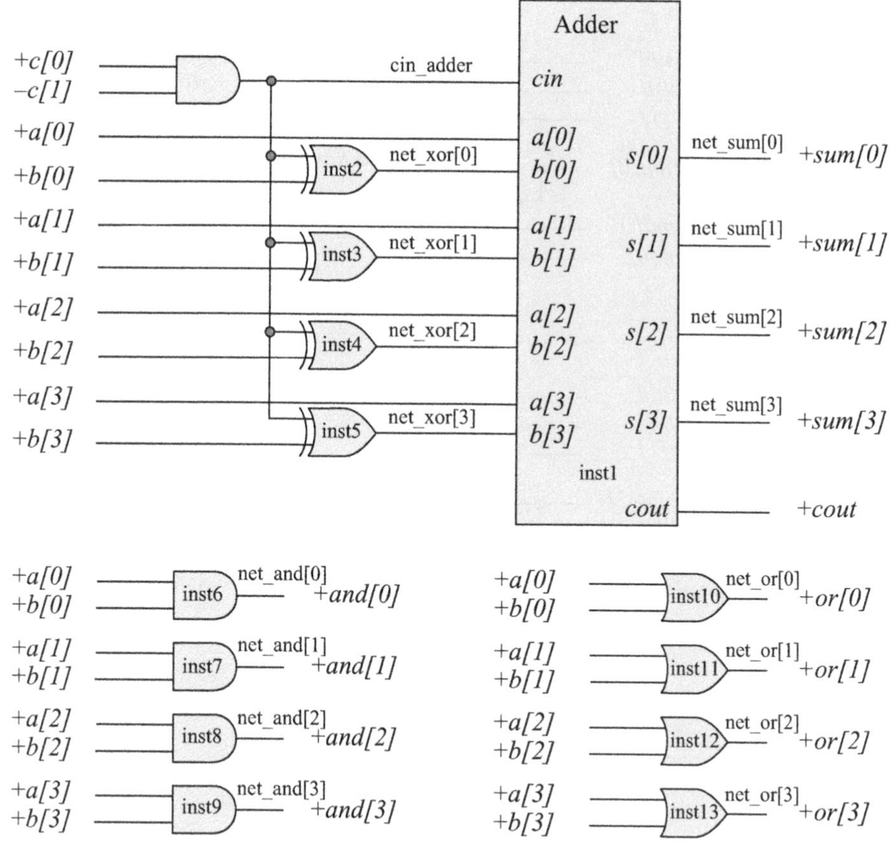

(continued on next page)

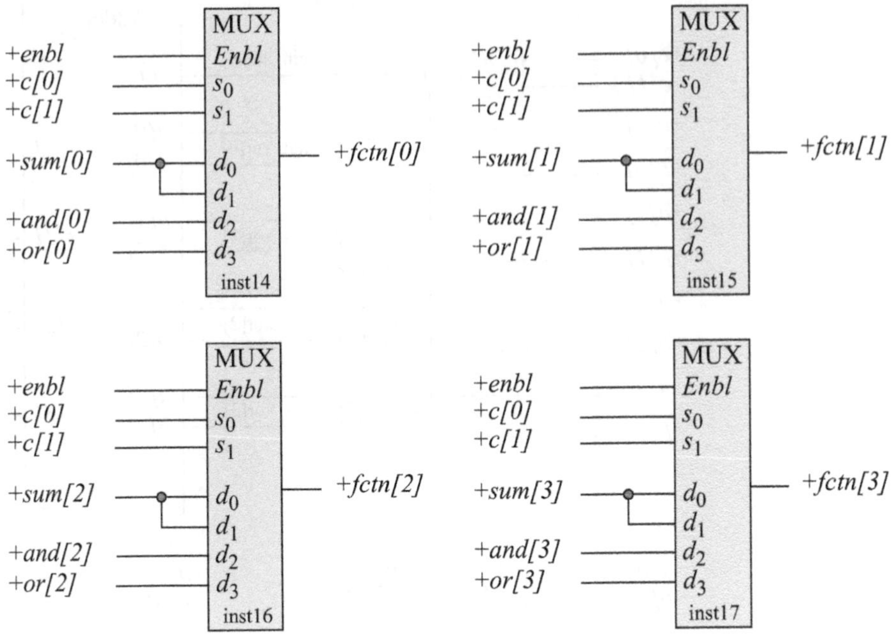

```
//structural 4-function alu
module alu_4function (a, b, ctrl, enbl, cout, fctn);

input [3:0] a, b;
input [1:0] ctrl;
input enbl;
output [3:0] fctn;
output cout;

wire [3:0] a, b;
wire [1:0] ctrl;
wire enbl;
wire [3:0] fctn;
wire cout;

//define internal nets
wire cin_adder;
wire [3:0] net_xor, net_sum, net_and, net_or;

assign cin_adder = (~ctrl [1] & ctrl [0]);

//continued on next page
```

```
adder4_df inst1 (
   .a(a),
   .b(net_xor),
   .cin(cin_adder),
   .sum(net_sum),
   .cout(cout)
   );

xor2_df inst2 (
   .x1(cin_adder),
   .x2(b[0]),
   .z1(net_xor[0])
   );

xor2_df inst3 (
   .x1(cin_adder),
   .x2(b[1]),
   .z1(net_xor[1])
   );

xor2_df inst4 (
   .x1(cin_adder),
   .x2(b[2]),
   .z1(net_xor[2])
   );

xor2_df inst5 (
   .x1(cin_adder),
   .x2(b[3]),
   .z1(net_xor[3])
   );

and2_df inst6 (
   .x1(a[0]),
   .x2(b[0]),
   .z1(net_and[0])
   );

and2_df inst7 (
   .x1(a[1]),
   .x2(b[1]),
   .z1(net_and[1])
   );

//continued on next page
```

```
and2_df inst8 (
   .x1(a[2]),
   .x2(b[2]),
   .z1(net_and[2])
   );

and2_df inst9 (
   .x1(a[3]),
   .x2(b[3]),
   .z1(net_and[3])
   );

or2_df inst10 (
   .x1(a[0]),
   .x2(b[0]),
   .z1(net_or[0])
   );

or2_df inst11 (
   .x1(a[1]),
   .x2(b[1]),
   .z1(net_or[1])
   );

or2_df inst12 (
   .x1(a[2]),
   .x2(b[2]),
   .z1(net_or[2])
   );

or2_df inst13 (
   .x1(a[3]),
   .x2(b[3]),
   .z1(net_or[3])
   );

mux4_df inst14 (
   .s(ctrl),
   .d({net_or[0], net_and[0], net_sum[0],
         net_sum[0]}),
   .enbl(enbl),
   .z1(fctn[0])
   );

//continued on next page
```

```
mux4_df inst15 (
   .s(ctrl),
   .d({net_or[1], net_and[1], net_sum[1],
         net_sum[1]}),
   .enbl(enbl),
   .z1(fctn[1])
   );

mux4_df inst16 (
   .s(ctrl),
   .d({net_or[2], net_and[2], net_sum[2],
         net_sum[2]}),
   .enbl(enbl),
   .z1(fctn[2])
   );

mux4_df inst17 (
   .s(ctrl),
   .d({net_or[3], net_and[3], net_sum[3],
         net_sum[3]}),
   .enbl(enbl),
   .z1(fctn[3])
   );
endmodule
```

```
//structural 4-function alu test bench
module alu_4function_tb;

reg [3:0] a, b;
reg [1:0] ctrl;
reg enbl;
wire [3:0] fctn;
wire cout;

initial
$monitor ("ctrl = %b, a = %b, b = %b, cout=%b,
            fctn = %b", ctrl, a, b, cout, fctn);

initial
begin
//add operation
   #0    ctrl=2'b00; a=4'b0000; b=4'b0000; enbl=1'b1;
   #10   ctrl=2'b00; a=4'b0001; b=4'b0011; enbl=1'b1;
   #10   ctrl=2'b00; a=4'b0111; b=4'b0011; enbl=1'b1;
   #10   ctrl=2'b00; a=4'b1101; b=4'b0110; enbl=1'b1;
   #10   ctrl=2'b00; a=4'b0011; b=4'b1111; enbl=1'b1;

//continued on next page
```

```
//subtract operation
   #10   ctrl=2'b01; a=4'b0111; b=4'b0011; enbl=1'b1;
   #10   ctrl=2'b01; a=4'b1101; b=4'b0011; enbl=1'b1;
   #10   ctrl=2'b01; a=4'b1111; b=4'b0011; enbl=1'b1;
   #10   ctrl=2'b01; a=4'b1111; b=4'b0001; enbl=1'b1;
   #10   ctrl=2'b01; a=4'b1100; b=4'b0111; enbl=1'b1;

//and operation
   #10   ctrl=2'b10; a=4'b1100; b=4'b0111; enbl=1'b1;
   #10   ctrl=2'b10; a=4'b0101; b=4'b1010; enbl=1'b1;
   #10   ctrl=2'b10; a=4'b1110; b=4'b0111; enbl=1'b1;
   #10   ctrl=2'b10; a=4'b1110; b=4'b1111; enbl=1'b1;
   #10   ctrl=2'b10; a=4'b1111; b=4'b0111; enbl=1'b1;

//or operation
   #10   ctrl=2'b11; a=4'b1100; b=4'b0111; enbl=1'b1;
   #10   ctrl=2'b11; a=4'b1100; b=4'b0100; enbl=1'b1;
   #10   ctrl=2'b11; a=4'b1000; b=4'b0001; enbl=1'b1;

   #10 $stop;
end

alu_4function inst1 (   //instantiate the module
   .a(a),
   .b(b),
   .ctrl(ctrl),
   .enbl(enbl),
   .fctn(fctn),
   .cout(cout)
   );
endmodule
```

```
ADD
ctrl = 00, a = 0000, b = 0000, cout=0, fctn = 0000
ctrl = 00, a = 0001, b = 0011, cout=0, fctn = 0100
ctrl = 00, a = 0111, b = 0011, cout=0, fctn = 1010
ctrl = 00, a = 1101, b = 0110, cout=1, fctn = 0011
ctrl = 00, a = 0011, b = 1111, cout=1, fctn = 0010

SUBTRACT
ctrl = 01, a = 0111, b = 0011, cout=1, fctn = 0100
ctrl = 01, a = 1101, b = 0011, cout=1, fctn = 1010
ctrl = 01, a = 1111, b = 0011, cout=1, fctn = 1100
ctrl = 01, a = 1111, b = 0001, cout=1, fctn = 1110
ctrl = 01, a = 1100, b = 0111, cout=1, fctn = 0101

//continued on next page
```

```
AND
ctrl = 10, a = 1100, b = 0111, cout=1, fctn = 0100
ctrl = 10, a = 0101, b = 1010, cout=0, fctn = 0000
ctrl = 10, a = 1110, b = 0111, cout=1, fctn = 0110
ctrl = 10, a = 1110, b = 1111, cout=1, fctn = 1110
ctrl = 10, a = 1111, b = 0111, cout=1, fctn = 0111

OR
ctrl = 11, a = 1100, b = 0111, cout=1, fctn = 1111
ctrl = 11, a = 1100, b = 0100, cout=1, fctn = 1100
ctrl = 11, a = 1000, b = 0001, cout=0, fctn = 1001
```

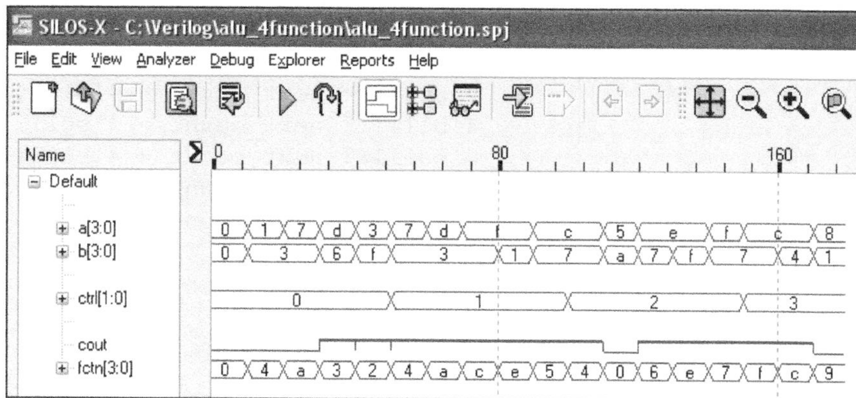

17.24 Design the Johnson counter of Problem 17.23 using behavioral modeling with the **case** statement. Obtain the design module, the test bench module, the outputs, and the waveforms.

```
//behavioral 4-bit johnson counter
module ctr_johnson4_behav (rst_n, clk, count);

//declare inputs and outputs
input rst_n, clk;
output [3:0] count;

//variables used in an always block
//are declared as registers
reg [3:0] count, next_count;

//continued on next page
```

```
//set next count
always @ (posedge clk or negedge rst_n)
begin
   if(~rst_n)
      count <= 4'b0000;
   else
      count <= next_count;
end

//determine next count
always @ (count)
begin
   case (count)
      4'b0000 : next_count = 4'b1000;
      4'b1000 : next_count = 4'b1100;
      4'b1100 : next_count = 4'b1110;
      4'b1110 : next_count = 4'b1111;
      4'b1111 : next_count = 4'b0111;
      4'b0111 : next_count = 4'b0011;
      4'b0011 : next_count = 4'b0001;
      4'b0001 : next_count = 4'b0000;

      default : next_count = 4'b0000;
   endcase
end
endmodule
```

```
//test bench for behavioral 4-bit johnson counter
module ctr_johnson4_behav_tb;

//inputs are reg and outputs are wire for test bench
reg rst_n, clk;
wire [3:0] count;

//display variables at simulation time
initial
$monitor ("count = %b", count);

//define reset
initial
begin
   #0    rst_n = 1'b0;
   #10   rst_n = 1'b1;
end

//continued on next page
```

```
//define clock
initial
begin
   clk = 1'b0;
   forever
      #10 clk = ~clk;
end

//define length of simulation
initial
begin
   #180 $finish;
end

//instantiate the module into the test bench
ctr_johnson4_behav inst1 (
   .rst_n(rst_n),
   .clk(clk),
   .count(count)
   );
endmodule
```

```
count = 0000        count = 0111
count = 1000        count = 0011
count = 1100        count = 0001
count = 1110        count = 0000
count = 1111        count = 1000
```

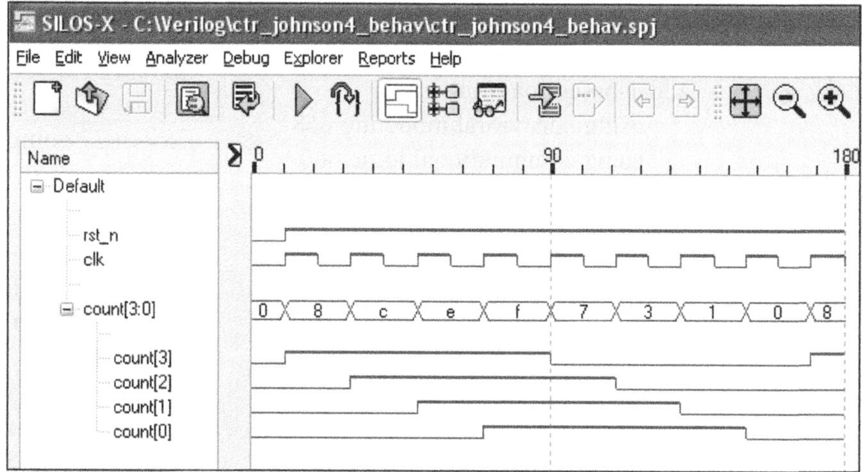

INDEX

Symbols

$finish 153, 154
$monitor 164
$random 145, 173
$readmemb 212, 214, 332, 340, 341, 448, 519
$readmemh 212
$stop 153, 165
$time 131
(r – 1) complement 15

Numerics

0 and 1 associated with a variable 28
0 and 1 complement 28
2s complementer 250
    truth table 251
8:3 encoder 57
9s complementer 467

A

absorption laws 1 29
absorption laws 2 29
addend 184
adder/subtractor 243, 843
adder-based rounding 651
    approach from above 651
    approach from below 651
adder-based rounding
    using behavioral modeling 668
    using combinational logic 660
    using memory 655
addition
    decimal 427
    fixed-point 183
    floating-point 551
addition with biased augend 454–460
addition with sum correction 427–436
    auxiliary carry 429
    intermediate sums 429
adjacency rules 75
algebraic manipulation 32
algebraic minimization 32
always 129, 618
    behavioral 345
    procedural 345
always block 129
always statement 129
AND gate 801
    behavioral modeling 804
    dataflow modeling 801
AND gates 46
AND operator 26
arithmetic and logic units 760–771
    4-function 760
    16-function 764
array division 408–423
array multiplication 318–328
assertion levels 45
assertion/deassertion times 72
assign statement 118, 119, 655
associative laws 27
augend 184
autonomous machine 71
auxiliary carry 429, 475
axiom 1: Boolean set definition 27
axiom 2: closure laws 27
axiom 3: identity laws 27
axiom 4: commutative laws 27
axiom 5: associative laws 27
axiom 6: distributive laws 27
axiom 7: complementation laws 28

B

base 1, 551
binary-coded decimal (BCD)
    adder/subtractor 473
BCD addition
    using multiplexers 437–444
    with memory-based correction 444–454
BCD division using table lookup 545–549

BCD multiplication
    using behavioral modeling 495–498
    using memory 510–523
    using structural modeling 498–510
    using table lookup 524–527
BCD restoring division
    version 1 529–538
    version 2 538–545
BCD subtraction examples 464–467
begin . . . end 506
behavior 129, 267, 668
behavioral addition/subtraction 267–271
behavioral modeling 129–154, 267
    always 129
    always statement 129
    begin . . . end 129
    blocking assignments 133
    case statement 141
    conditional statement 138
    event control list 130
    initial 129
    initial statement 129
    interstatement delay 133
    intrastatement delay 133
    loop statements 150
    nonblocking assignments 136
    parameter 145
    sensitivity list 130
bias constant 553, 554, 563, 564
biased exponents 554–557
binary counter 63
binary number system 4
    counting in binary 5
binary search 545
binary weights 5
binary weights for an 8-bit fraction 6
binary weights for an 8-bit integer 6
binary-coded decimal 9
binary-coded decimal (BCD)
        added/subtractor
    9s complementer 467
    BCD subtraction 463
binary-coded decimal digits 428
binary-coded decimal numbers 9
    packed 430
    unpacked 430
binary-coded hexadecimal 10
binary-coded hexadecimal numbers 10
binary-coded octal 8
binary-coded octal numbers 8
binary-to-BCD conversion 493–494
binary-to-Gray code converters 836–843
    using built-in primitives 841
    using *D* flip-flops 837
bit counter 525
bit-pair recoding 304
bit-pair recoding multiplication 304–318
    multiplicand versions 308
blocking procedural assignment 133
Boolean algebra 25–32
    0 and 1 associated with a variable 28
    0 and 1 complement 28
    absorption laws 1 29
    absorption laws 2 29
    associative laws 27
    axiom 26
    Boolean set definition 27
    closure laws 27
    identity laws 27
    commutative laws 27
    associative laws 27
    distributive laws 27
    complementation laws 28
    Boolean set definition 27
    closure laws 27
    commutative laws 27
    complementation laws 28
    DeMorgan's laws 29
    distributive laws 27
    dual 26
    idempotent laws 28
    involution law 28
    summary of axioms and theorems
        31
Boolean operators for two variables 26
Boolean product 27
Booth algorithm 289
    examples 291
Booth algorithm multiplication 289–304
    Booth multiplier recoding table 290
    recoded multiplier 291
    skipping over zeroes 289

built-in primitive gates 94, 825
built-in primitives 94, 94–108
burst mode 779

C

canonical product of sums 30
canonical sum of products 30
carry lookahead adder 191, 196
carry lookahead addition 191–201
  generate 192
  group generate 194
  group propagate 194
  propagate 192
  section generate 194
  section propagate 194
carry lookahead addition/subtraction 250–266
  2s complementer 250
carry lookahead generator 192
carry-save adders 201, 203
carry-save addition 201–211
  multiple-bit addition 201
  multiple-operand addition 206
carry-select addition 216–227
case statement 141, 312, 748
character sets for different radices 3
characteristic 552
check symbol 40
checksum 723
chopping 650
code converter 829
combinational logic 44–59
combinational shifter
  shift left algebraic 747, 748
  shift left logical 747, 748
  shift right algebraic 747, 748
  shift right logical 747
combined truncation, adder-based, and von Neumann rounding 674
commutative laws 27
comparators 58–59, 104
  equality function 59
complementation laws 28
condition codes 685
condition codes for addition 738–746
conditional statements 138
  if 138
  if ... else 138
  if ... else if 138
congruent 686
conjunctive normal form 31
continuous assignment 118, 655
count-down 4-bit binary counter 771–774
count-down modulo-8 counter 66
counter nonsequential 163
counters 62–70
  count-down 62
  count-up 62
  Gray code 63
  input maps 63
  Johnson 63
  modulo-10 63
  modulo-8 count-down 66
  nonsequential 68
  pulse generator 70
counting in binary 5
counting in hexadecimal 11
counting in octal 7
counting sequence for different radices 3

D

D flip-flop 61
  edge-triggered device 61
  excitation equation 61
dataflow modeling 118–129
  continuous assignment 118
dataflow module 690
decimal addition
  with sum correction 427
decimal division 529
decimal number system 8
  binary-coded decimal 9
decoders 53–56
  3:8 decoder 53, 825
  definition 825
  minterm 53
default 748
DeMorgan's laws 29
different radices 3
digit counter 525
diminished-radix complement 15

direct-memory access 779
disable 153
disjunctive normal form 30
distributive laws 27
divide-by-two operation 775
dividend 359
division
  combinational array 408
  decimal 529
  fixed-point 359
  floating-point 633
  nonrestoring 374
  overflow 361
  restoring 530
divisor 359
don't care 37, 43
double bias 613
double-precision format 552
duality 26

E

eight-bit carry lookahead adder 730
emitter-coupled logic (ECL) 46
encoder 56
encoders 56–58, 829
  8:3 encoder 57
  BCD-to-binary 57
  octal-to-binary 57, 830
  priority encoder 58, 833
end-around carry 239
endcase 748
endtable 109
equality function 59, 818
error detection and correction codes
  checksum 723
essential prime implicant 41
event control 214
event control list 130, 618
event queue 849
  blocking and nonblocking assignments 861
  blocking assignments 854
  dataflow assignments 849
  event handling for dataflow assignments 849
  nonblocking assignments 857
excitation maps 75
exclusive-NOR function 818
  behavioral modeling 820
  dataflow modeling 818
exclusive-OR function 814
  behavioral modeling 816
  dataflow modeling 814
exponent overflow 583, 612
exponent overflow/underflow 638–640
exponent underflow 583
exponents 551

F

first-in, first-out (FIFO) queue 779
fixed-point addition 183
  section generate 194
  section propagate 194
fixed-point division 633
fixed-point multiplication 275
  sequential add-shift 278
fixed-point subtraction 237
floating-point
  dividend alignment 633
floating-point addition 557–560, 564–568
  definition 558
floating-point addition algorithm 562
floating-point addition flowcharts 562
floating-point division
  definition 634
  flowcharts 641–642
floating-point division algorithm 641
floating-point division numerical examples 643–646
floating-point format 552–554
floating-point multiplication
  definition 611
  flowcharts 614–615
floating-point multiplication algorithm 614
floating-point multiplication
  version 1 618
  version 2 624
floating-point numerical examples 573–580, 616–617
floating-point subtraction 571

definition 571, 572
true addition 584
true subtraction version 1 589
true subtraction version 2 593
true subtraction version 3 598
true subtraction version 4 603
floating-point subtraction flowcharts 581–584
floating-point subtraction Verilog HDL implementations 584–607, 618–630
for loop 150
forever 150
forever loop 153, 506
four-bit adder behavioral module 710
fraction overflow 560, 583
fraction underflow 561, 583
full adder 184
full adder logic diagram 186
functionally complete gates 44

G

gate-level modeling
majority circuit 109
multiple-input gates 94, 826
general floating-point organization 561–564
generate 192, 725
George Boole 25
glitches 76
Gray code 33, 836
Gray code counter 63
group generate 194
group propagate 194
guard bits 654

H

half adder 184
half adder logic diagram 186
hexadecimal number system 10
binary-coded hexadecimal 10
hierarchical method 83
high-speed addition 738
high-speed shifter 754

I

idempotent laws 28
identity elements 27
identity laws 27
IEEE floating-point formats 552
if 138
if ... else 138
if ... else if 138
implied 1 553, 572
initial 129, 506
initial block 129
initial statement 129
input maps 75
instance 154
instantiation 93, 154
instantiation by name 154
intermediate sum for BCD addition 429
interstatement delay 133
intrastatement delay 133
involution law 28

J

jamming 653
JK flip-flop 61
edge-triggered device 61
excitation equation 62
excitation table 61
Johnson counter 63, 113

K

Karnaugh maps 33–39
alternative map for five variables 34
five variables 34
four variables 34
logically adjacent squares 33
map-entered variables 37
physically adjacent squares 33
three variables 34
two variables 34
unspecified entries 37
keywords
always 618
begin . . . end 506
case 748
default 748
disable 794
endcase 748
forever 506

initial 506
parameter 748
repeat 793

L

logic design fundamentals 25
logic macro functions
linear-select multiplexers 50
multiplexers 47, 822
nonlinear-select multiplexers 51
logic synthesis 118
logical and algebraic shifters 747–759
behavioral modeling 748
structural modeling 753
loop statements
$finish 154
for loop 150
forever loop 150, 153
repeat loop 150, 151
while loop 150, 151, 531

M

mantissa 552
map-entered variables 37
maxterm 30
maxterm expansion 30
Mealy machine 78
definition 78
general configuration 78
memory
$readmemb 212
$readmemh 212
instruction cache 213, 340
memory-based addition 212–216
memory-based multiplication 339–344
minimization techniques 32–44
algebraic minimization 32
Karnaugh maps 33
Quine-McCluskey algorithm 39
minterm 30, 41
minterm expansion 30
minuend 238
module 93
module instantiation 154
modulo-2 addition 837
modulo-3 residue 687
modulo-3 residue generator
structural module 701
modulo-8 count-down counter 66
modulo-10 counter 63
modulus 686
Moore machine 71
definition 71
general configuration 71
next state 72
Moore synchronous sequential machine 145, 171
multiple-bit addition 201
multiple-operand addition 206
multiple-operand multiplication 344–353
multiplexers 47–53, 822
data input selection 47, 822
linear-select 50
nonlinear-select 51
select inputs 47, 822
multiplicand 275
multiplicand versions for multiplier bit-pair recoding 308
multiplication
array multiplier 318
bit-pair recoding 304
decimal 493
fixed-point 275
floating-point 611
sequential add-shift 278
multiplier 275
Booth algorithm 289
multiply-by-two operation 775
multi-way branching
case 748
default 748
endcase 748
if ... else 138
if ... else if 138
multliplicative division 402–408

N

NAND gates 44, 46, 806
nonblocking procedural assignment 136, 771, 772
nonoverlapping pulses 790

nonrestoring division 374
nonrestoring division algorithm 693
nonsequential counter 68
NOR gates 44, 811
not a number (NaN) 556
number representations 1, 12–21
  diminished-radix complement 15
  radix complement 18
  sign magnitude 13
number representations for positive and negative integers in radix 2 19
number systems 1–12
  binary number system 4
  binary weights 5
  decimal number system 8
  hexadecimal number system 10
  octal number system 6

O

octal number system 6
  binary-coded octal 8
  counting in octal 7
octal-to-binary encoder 830
OR gate 809
output map 76
overflow 559
overflow and underflow 560–561

P

parallel-in, serial-in, serial-out shift register 782
parallel-in, serial-out register 775
parameter 145, 748, 764
parity prediction 685, 723–737
parity-checked shift register 717–723
partial product 279
Petrick algorithm 43
posedge clock 137
positional number systems 1
postnormalization 559, 563, 582
postulate 26
prime implicant 35, 40
prime implicant chart 41
primitives 94
priority encoder 58, 833
procedural constructs
  always statement 130
procedural flow control
  case 141
  forever 506
  repeat 793
product of maxterms 30
product of sums 31, 35, 36
product term 30
propagate 192, 725
pulse generator 70

Q

Quine-McCluskey algorithm 39
  check symbol 40
  don't care 43
  essential prime implicant 41
  Petrick algorithm 43
  prime implicant chart 41
  secondary essential prime implicant 41

R

race condition 136
radix 1, 14
radix 2 14
radix 10 14
radix complement 18
radix point 4, 560
radix *r* 2
registers 775
repeat loop 150, 151
residue 686
residue checker 708
residue checking 685, 686–717
  addition 687
  binary-coded decimal addition 689
  congruence 686
  dataflow modeling 690
  modulo-3 687
  modulus 686
  residue 686
  structural modeling 693
residue generator 698
restoring division 360, 529
  version 1 362
  version 2 368

ripple-carry addition 184–191
logical organization 187
ripple-carry subtraction 243–249
rounding methods 649–653
adder-based rounding 651
adder-based rounding using behavioral modeling 668
combined truncation, adder-based, and von Neumann rounding 674
truncation rounding 650
von Neumann rounding 653

S

scaling factor 551
secondary essential prime implicant 41
section generate 194
section propagate 194
serial addition 227–234
sensitivity list 130, 618
sequence counter 369
sequential add-shift multiplication 276–289
version 1 282
version 2 285
sequential add-shift multiplication hardware algorithm 278
sequential logic 60–83
sequential shift-add/subtract
nonrestoring division 374–382
restoring division 360–374
version 1 362
version 2 368
serial addition 227
serial-in, parallel-out shift register 787
serial-in, serial-out register 778
shift left algebraic 717, 747, 748, 749
shift left logical 717, 749
shift registers 775–794
nonoverlapping pulses 790
parallel-in, serial-in, serial-out shift register 782
parallel-in, serial-out shift register 775
serial-in, parallel-out shift register 787
serial-in, serial-out shift register 778
shift right algebraic 717, 747, 748, 749
shift right logical 717, 747, 748, 749
shifting factors 558, 572
shifting left 775
shifting right 775
sign magnitude 13
significand 552
significand digits 551
single-precision format 552, 584, 612
sixteen-function arithmetic and logic unit 764
skipping over zeroes 289
SR latch 60
complementary outputs 60
negative feedback 60
SRT division 382–402
using table lookup 393
using the case statement 397
standard product of sums 30
standard sum of products 30
state codes 73
adjacency 74
state diagram 73
sticky bit 654
structural modeling 154–179
combinational array 693
design examples 155
instances 154
instantiation 154
instantiation by name 154
module instantiation 154
stuck-at-1 fault 718
subtraction
decimal 463
fixed-point 237
floating-point 571
subtrahend 238
sum and carry equations 185
sum of minterms 30
sum of products 30, 35
sum term 30
summary of Boolean axioms and theorems 31
synchronous counters
modulo-10 63
synthesis of combinational logic 44–59

comparators 58
decoders 53
encoders 56
full adder 185
half adder 185
linear-select multiplexers 50
multiplexers 47
nonlinear-select multiplexers 51

T

table for user-defined primitive 108
table lookup multiplication 329–339
tabular method for Quine-McCluskey 39
test bench 93
theorem 1: 0 and 1 associated with a variable 28
theorem 2: 0 and 1 complement 28
theorem 3: idempotent laws 28
theorem 4: involution law 28
theorem 5: absorption law 1 29
theorem 6: absorption law 2 29
theorem 7: DeMorgans laws 29
three-input adder dataflow module 699
top-down approach 83
transistor-transistor logic (TTL) 46
true addition 464, 557, 560, 571, 573
rules 574
true subtraction 464, 573
rules 576
truncation rounding 650
truth table for AND, OR, NOT, exclusive-OR, and exclusive-NOR 26
two-decade BCD addition/subtraction unit 467–481, 481–490
twos complement subtraction 238–243

U

universal logic gate
NAND 806
NOR 811
unspecified entries 37
user-defined primitives 108–117
endprimitive 108
endtable 108
primitive 108
table 108

V

Verilog HDL rounding implementations 654–680
adder-based rounding
using combinational logic 660
using memory 655
Verilog HDL introduction 93
Verilog HDL project procedure 801, 865
von Neumann rounding 653

W

Wallace tree 201, 206, 209
add five bits 203
weight associated with 4-bit binary segments 6
while loop 151, 285, 495, 531

X

x value 151

Z

z value 151
zero bias 635–638